THE PRINCIPAL FUNCTIONAL GROUPS OF ORGANIC CHEMISTRY

	Example	*Acceptable Name(s) of Example*	*Characteristic Reaction Type*
Carboxylic acid derivatives			
Acyl halides	$\overset{\displaystyle O}{\overset{\|}{CH_3CCl}}$	Ethanoyl chloride or acetyl chloride	Nucleophilic acyl substitution
Acid anhydrides	$\overset{\displaystyle O\ \ O}{\overset{\|\ \ \ \|}{CH_3COCCH_3}}$	Ethanoic anhydride or acetic anhydride	Nucleophilic acyl substitution
Esters	$\overset{\displaystyle O}{\overset{\|}{CH_3COCH_2CH_3}}$	Ethyl ethanoate or ethyl acetate	Nucleophilic acyl substitution
Amides	$\overset{\displaystyle O}{\overset{\|}{CH_3CNHCH_3}}$	*N*-Methylethanamide or *N*-methylacetamide	Nucleophilic acyl substitution
Nitrogen-containing organic compounds			
Amines	$CH_3CH_2NH_2$	Ethanamine or Ethylamine	Nitrogen acts as a base or as a nucleophile
Nitriles	$CH_3C{\equiv}N$	Ethanenitrile or acetonitrile	Nucleophilic addition to carbon-nitrogen triple bond
Nitro compounds	$C_6H_5NO_2$	Nitrobenzene	Reduction of nitro group to amine

ORGANIC CHEMISTRY
A BRIEF COURSE

ORGANIC CHEMISTRY
A BRIEF COURSE

ROBERT C. ATKINS
Department of Chemistry
James Madison University

FRANCIS A. CAREY
Department of Chemistry
University of Virginia

McGRAW-HILL PUBLISHING COMPANY
New York   St. Louis   San Francisco   Auckland   Bogotá   Caracas
Hamburg   Lisbon   London   Madrid   Mexico   Milan
Montreal   New Delhi   Oklahoma City   Paris   San Juan   São Paulo
Singapore   Sydney   Tokyo   Toronto

ORGANIC CHEMISTRY
A BRIEF COURSE

2 3 4 5 6 7 8 9 0   VNH   VNH   9 4 3 2 1 0

ISBN 0-07-009919-7

This book was set in Times Roman by Progressive Typographers, Inc.
The editors were Randi B. Rossignol, Denise T. Schanck, and Jack Maisel;
the designer was Rafael Hernandez;
the production supervisor was Leroy A. Young.
Von Hoffmann Press, Inc., was printer and binder.

Library of Congress Cataloging-in-Publication Data

Atkins, Robert C. (Robert Charles), (date).
  Organic Chemistry: a brief course/Robert C. Atkins, Francis A. Carey.
    p.   cm.
  Bibliography: p.
  Includes index.
  ISBN 0-07-009919-7
  1. Chemistry, Organic.   I. Carey, Francis A., (date).
II. Title.
QD253.A815   1990
547—dc20   89-8083

ABOUT THE AUTHORS

Robert C. Atkins was born in Massachusetts and educated in Bergenfield, New Jersey. He received an S.B. in chemistry from the Massachusetts Institute of Technology in 1966 and a Ph.D. in organic chemistry from the University of Wisconsin, Madison, in 1970. Following a year of postdoctoral study at Columbia University, he was appointed to the faculty of James Madison University, rising to the rank of professor of chemistry. He has published two study guides including one coauthored with Professor Carey, and is the author of several research and science-education publications.

Francis A. Carey is a native of Pennsylvania, educated in the public schools of Philadelphia, at Drexel University (B.S. in chemistry, 1959), and at Penn State (Ph.D., 1963). Following postdoctoral work at Harvard and military service, he was appointed to the chemistry faculty of the University of Virginia. With his students, he has published over forty research papers in synthetic and mechanistic organic chemistry. He is author of *Organic Chemistry,* a text for the introductory two-semester course and coauthor (with Richard J. Sundberg) of *Advanced Organic Chemistry,* a two-volume treatment designed for graduate students and advanced undergraduates.

This book is dedicated to our families.

CONTENTS IN BRIEF

CONTENTS

CHAPTER *3*
INTRODUCTION TO ORGANIC CHEMICAL REACTIONS

CHAPTER *4*
ALKENES, ALKADIENES, AND ALKYNES
I. Structure and Preparation

PREFACE

There are challenges aplenty in the one-semester (or two-quarter) organic chemistry course. The student is challenged to switch gears from the intensive numerical problem solving which characterizes the introductory college chemistry course to the descriptive "chain-of-reasoning" approach encountered in organic chemistry. The instructor is challenged to be selective in what topics to cover and how to cover them. All teachers of organic chemistry agree that there is too much material to teach in the two-semester organic course. How can this material possibly be condensed to one semester? What topics are most important to this group of students? How much detail can be left out before the treatment becomes too superficial to be useful? It is to help both students and instructors that we undertook to write this book.

Who are the students for whom we wrote the book? The short organic course enrolls students with a wide range of backgrounds, majors, and career goals. Traditionally, these have included students in biology, nutrition, agricultural sciences, and the allied health sciences. This already diverse group is enriched by a growing presence of students in education, environmental science, engineering, and liberal arts programs who recognize a need to be conversant in the language of organic chemistry.

How can we help these students? Because the nature of problem solving in organic chemistry is different from that of the freshman course, students need to be shown the reasoning process involved and drilled in its application. The text includes numerous problems within the chapters accompanied by a step-by-step solution to part *a* of multipart problems. Answers to all in-chapter problems are included as an appendix at the back of the book and detailed solutions to these, as well as all end-of-chapter problems, are provided in a separate *Solutions Manual.*

Students seem to learn organic chemistry best when it is presented in a *functional-group* format, and we have retained the classical functional-group organization. We try to help the student organize the material by providing self-contained review and summary tables. Most chapters contain *annotated tables of reactions* that review reactions from previous chapters which lead to a particular compound type, and end-of-chapter summary tables of the new reactions appearing within the chapter.

Students retain their knowledge best when they understand the *mechanisms* of how organic reactions occur. Moreover, they are curious about how reactions take place. What we frequently hear from them is a request for more, rather than less, mechanistic discussion. Reaction mechanisms, including *electron counting* and *curved-arrow notation* are emphasized in this text. Most reaction mechanisms are portrayed as a sequence of steps in a figure along with an accompanying text discussion.

A list of *learning objectives* is included at the beginning of each chapter.

Supplementary essays, twenty-seven in all, are scattered throughout the text and offer a combination of historical perspective and additional insights into the contemporary relevance of organic chemistry and its impact on the economy and its potential for improving the quality of life.

How can we help those who teach the short organic course? A textbook sets the agenda for a course, and the problem is always too much material rather than not enough. We have tried to be selective in our coverage, thinking it most important to do a careful job of explaining thoroughly what we believe to be the core material of organic chemistry. Explanations within the text must be sufficiently developed so that the instructor can say to the class, "I want to spend most of today's lecture discussing the mechanism of electrophilic aromatic substitution, so will leave it to you to read the section on the nomenclature of aromatic compounds," and be confident that the students will not be shortchanged by the book's treatment. Thus if this text is somewhat longer than is customary for the short organic course, it is not because there is more material; it is because there are more and fuller explanations of the most important material.

We have tried to write a modern organic chemistry text, selective in its coverage and accurate in its presentation. Its purpose is to teach students who will not be chemists the language of organic chemistry, the structural features of organic compounds, the reactions they undergo, and the useful purposes to which they are applied.

Robert C. Atkins
Francis A. Carey

ACKNOWLEDGEMENTS

We both owe a great debt of gratitude to the reviewers, whose insightful comments at various stages of the preparation of this book have had a significant impact on its development. These reviewers are:

David L. Anderson, *Westbrook College*
Winfield M. Baldwin, Jr., *University of Georgia*
Joel M. Hawkins, *University of California — Berkeley*
Thomas R. Hays, *Texas A&I University*
John F. Helling, *University of Florida*
Victor Hoffman, *Northeast Missouri State University*
Elmer E. Jones, *Northeastern University*
Martin B. Jones, *University of North Dakota*
George A. Kraus, *Iowa State University of Science & Technology*
Kenneth L. Marsi, *California State University — Long Beach*
Roger K. Murray, *University of Delaware*
George V. Odell, *Oklahoma State University*
Philip D. Roskos, *Lakeland Community College*
K. Barbara Schowen, *University of Kansas*
Charles E. Sundin, *University of Wisconsin — Platteville*
Leroy G. Wade, Jr., *Colorado State University*
Frederick W. Wassmundt, *University of Connecticut — Storrs*
Max T. Wills, *California Polytechnic State University*
Gordon Wilson, *Western Kentucky University*
George K. Wittenberg, *Arizona State University*

Several figures are reproduced from other sources, and are acknowledged within the text. In addition, we wish to thank the Aldrich Chemical Company for permission to reproduce infrared and ^{1}H nuclear magnetic resonance spectra from the following sources:

"The Aldrich Library of NMR Spectra," Edition II, C. J. Pouchert, 1983.
"The Aldrich Library of FT-IR Spectra," Edition I, C. J. Pouchert, 1985.

ORGANIC CHEMISTRY
A BRIEF COURSE

CHAPTER 1

INTRODUCTION TO ORGANIC CHEMISTRY: CHEMICAL BONDING

LEARNING OBJECTIVES

This chapter reviews the principles of structure and bonding that will be useful as you learn about the chemistry of carbon compounds. Its emphasis will be on chemical bonds and electron "bookkeeping," and its main objective is to provide you with those skills that are essential in writing proper structural formulas. Upon completion of this chapter you should be able to:

■ Write the electron configuration corresponding to the neutral atom or to an ion derived from it when given the atomic number of any element between hydrogen and argon in the periodic table.

■ Identify elements as electronegative or electropositive on the basis of their location in the periodic table.

■ Describe the difference between ionic and covalent bonding.

■ State the octet rule and understand its significance.

■ Write Lewis structures for covalent molecules and for ionic species that contain covalent bonds, including those with multiple bonds.

■ Calculate formal charges on atoms in Lewis structures.

■ Determine the direction of polarization of a covalent bond on the basis of the difference in electronegativity of the atoms that it connects.

■ Convert an empirical formula to a molecular formula of a substance when given its molecular weight.

■ Given a molecular formula, write Lewis structures for constitutionally isomeric substances.

■ Use the resonance concept to describe electron delocalization in molecules and ions and understand the difference between resonance and isomerism.

■ Use the valence-shell electron pair repulsion model to predict the shapes of simple molecules.

Organic chemistry is the study of carbon compounds. It is a vast field, one that encompasses a great deal of information and one that is inherently useful. Research in organic chemistry continues to provide new materials that enrich our life, new drugs that extend it, and new knowledge that describes the chemical basis of life itself. The key to understanding organic chemistry must begin with an understanding of molecular *structure,* because the way a substance behaves is inextricably related to the way its atoms are bonded together. This chapter is designed to remind you of those aspects of chemical bonding that are basic to our discussions of molecular structure; it reviews the elements of covalent bonding as well as the procedures to be followed when writing *Lewis structures* for molecules.

1.1 ATOMS AND ELECTRONS

Before discussing bonding principles, let us first review some fundamental relationships between atoms and electrons. Each element is characterized by a unique **atomic number Z,** which is equal to the number of protons in its nucleus. A neutral atom has equal numbers of protons, which are positively charged, and electrons, which are negatively charged. The electrons spend 90 to 95 percent of their time near the nucleus in regions of space called **orbitals.**

Orbitals are described by specifying their size, shape, and spatial orientation. Those that are spherically symmetric are called *s* orbitals and are identified as $1s$, $2s$, $3s$, etc., according to the volume they enclose (Figure 1.1). The number that describes the **energy level** of the orbital (1, 2, 3, etc.) is termed the **principal quantum number.** An electron in a $1s$ orbital is likely to be found closer to the nucleus, is lower in energy, and is more strongly held by the atom than an electron in a $2s$ orbital.

A hydrogen atom ($Z = 1$) has one electron; a helium atom ($Z = 2$) has two. These electrons occupy a $1s$ orbital in each case. We say that the electron configurations of hydrogen and helium are

Hydrogen: $1s^1$ Helium: $1s^2$

In addition to being negatively charged, electrons possess the property of **spin.** The **spin quantum number** of an electron can have a value of either $+\frac{1}{2}$ or $-\frac{1}{2}$. According to the **Pauli exclusion principle,** only two electrons may occupy the same orbital, and then only when they have opposite, or "paired," spins. Since two electrons fill the $1s$ orbital, the third electron in lithium ($Z = 3$) must occupy a higher-energy orbital. The only orbital available at the 1 level is the $1s$ orbital. After $1s$, the next-higher-energy orbital is $2s$. Therefore, the third electron in lithium occupies the $2s$ orbital, and the electron configuration of lithium is

Lithium: $1s^2$, $2s^1$

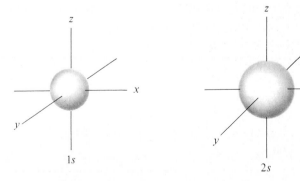

FIGURE 1.1
Representations of the boundary surfaces of a $1s$ and a $2s$ orbital. The boundary surfaces enclose the volume where there is a 90 to 95 percent probability of finding the electron.

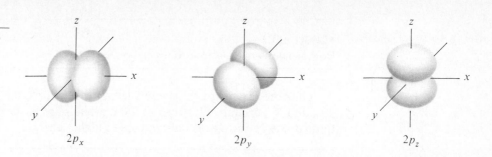

FIGURE 1.2
Representations of the boundary surfaces of the 2p orbitals.

The principal quantum number of the electron in the highest-energy orbital (1 in the case of hydrogen and helium, 2 in the case of lithium) corresponds to the **period** (or **row**) of the periodic table in which an element appears. Hydrogen and helium are first-row elements; lithium is a second-row element. A complete periodic table of the elements is presented on the inside back cover.

With beryllium ($Z = 4$), the $2s$ level becomes filled, and the next orbitals to be occupied in the remaining second-row elements are the $2p_x$, $2p_y$, and $2p_z$ orbitals. These orbitals, portrayed in Figure 1.2, are usually described as being "dumbbell-shaped." Each consists of two "lobes," i.e., slightly flattened spheres that touch each other along a surface passing through the nucleus. The $2p_x$, $2p_y$, and $2p_z$ orbitals are equal in energy and mutually perpendicular to each other.

The electron configurations of the first 12 elements, hydrogen through magnesium, are given in Table 1.1. In filling the $2p$ orbitals, notice that each orbital is singly occupied before any one orbital is doubly occupied. This is a general principle for orbitals of equal energy and is known as **Hund's rule.** *Of particular importance in Table 1.1 are hydrogen, carbon, nitrogen, and oxygen.* Countless organic compounds contain nitrogen, oxygen, or both in addition to carbon, the essential element of organic chemistry. Most of them also contain hydrogen.

It is often convenient to speak of the **valence electrons** of an atom. These are the outermost electrons, the ones most likely to be involved in chemical reactions. For second-row elements these are the $2s$ and $2p$ electrons. Since four orbitals ($2s$, $2p_x$, $2p_y$, $2p_z$) are involved, the maximum number of electrons in the **valence shell** of any second-row element is eight. Neon with all its $2s$ and $2p$ orbitals doubly occupied completes the second row of the periodic table.

TABLE 1.1

Electron Configurations of the First 12 Elements of the Periodic Table

Element	Atomic number Z	Number of electrons in indicated orbital					
		$1s$	$2s$	$2p_x$	$2p_y$	$2p_z$	$3s$
Hydrogen	1	1					
Helium	2	2					
Lithium	3	2	1				
Beryllium	4	2	2				
Boron	5	2	2	1			
Carbon	6	2	2	1	1		
Nitrogen	7	2	2	1	1	1	
Oxygen	8	2	2	2	1	1	
Fluorine	9	2	2	2	2	1	
Neon	10	2	2	2	2	2	
Sodium	11	2	2	2	2	2	1
Magnesium	12	2	2	2	2	2	2

PROBLEM 1.1
How many valence electrons does carbon have?

Once the $2s$ and $2p$ orbitals are filled, the next level is the $3s$, followed by the $3p_x$, $3p_y$, and $3p_z$ orbitals. Electrons in these orbitals are farther from the nucleus than those in the $2s$ and $2p$ orbitals and are of higher energy.

PROBLEM 1.2
Refer to the periodic table as needed and write electron configurations for all the third-row elements.

SAMPLE SOLUTION
The third row of the periodic table begins with sodium and ends with argon. The atomic number Z of sodium is 11, so a sodium atom has 11 electrons. The total number of electrons in the $1s$, $2s$, and $2p$ orbitals is 10, so sodium has one additional electron. Since levels 1 and 2 have their full complement of electrons, the eleventh electron occupies a $3s$ orbital. The electron configuration of sodium is $1s^2$, $2s^2$, $2p_x^2$, $2p_y^2$, $2p_z^2$, $3s^1$.

Neon in the second row and argon in the third possess eight electrons in their valence shell; they are said to have a complete **octet** of electrons. Helium, neon, and argon belong to the class of elements known as **noble gases** or **rare gases.** The noble gases are characterized by an extremely stable "closed-shell" electron configuration and are very unreactive.

1.2 IONIC BONDING

In general, elements with electron configurations close to that of a noble gas tend to lose or gain the requisite number of electrons necessary to achieve the electron configuration of that noble gas. Elements at the left of the periodic table tend to lose electrons to produce the electron configuration of the preceding noble gas. Elements at the right of the periodic table tend to gain electrons to give the electron configuration of the next-higher noble gas. The equations which follow illustrate these processes for a sodium atom and a chlorine atom. The species formed, Na^+ and Cl^-, have the same electron configuration as neon and argon, respectively.

$$
\begin{array}{lll}
Na\cdot & \longrightarrow Na^+ & + \ e^- \\
1s^2 & 1s^2 & \\
2s^2, 2p_x^2, 2p_y^2, 2p_z^2 & 2s^2, 2p_x^2, 2p_y^2, 2p_z^2 & \\
3s^1 & \text{(Ne electronic configuration)} &
\end{array}
$$

$$
\begin{array}{ll}
e^- + \ :\overset{\cdot\cdot}{\underset{\cdot\cdot}{Cl}}\cdot & \longrightarrow \ :\overset{\cdot\cdot}{\underset{\cdot\cdot}{Cl}}:^- \\
1s^2 & 1s^2 \\
2s^2, 2p_x^2, 2p_y^2, 2p_z^2 & 2s^2, 2p_x^2, 2p_y^2, 2p_z^2 \\
3s^2, 3p_x^2, 3p_y^2, 3p_z^1 & 3s^2, 3p_x^2, 3p_y^2, 3p_z^2 \\
& \text{(Ar electronic configuration)}
\end{array}
$$

Electrically charged species which are formed by the gain or loss of electrons from a neutral atom are called **ions;** positively charged ions are referred to as **cations** and negatively charged ions as **anions.** Transfer of an electron from a sodium atom to a

FIGURE 1.3 Lattice structure of crystalline sodium chloride. The smaller spheres represent sodium ions; the larger spheres are chloride ions.

chlorine atom yields a sodium cation and a chloride anion, both of which have a noble gas electron configuration.

$$Na \cdot (g) \quad + \quad \cdot \ddot{\underset{\cdot\cdot}{Cl}} : (g) \quad \longrightarrow \quad Na^+ [: \ddot{\underset{\cdot\cdot}{Cl}} :]^- (g)$$

Sodium atom   Chlorine atom   Sodium chloride

[(*g*) indicates that the species is present in the gas phase]

The electrostatic attraction between oppositely charged ions such as a sodium cation and a chloride anion is termed an **ionic bond** and can exist even in the gas phase. In the solid phase bonding forces are even stronger because, as shown in Figure 1.3, each sodium ion is surrounded by six chloride ions (and vice versa) in the crystal lattice.

Ionic bonds are very common in inorganic chemistry but very rare in organic chemistry. Instead, carbon compounds are characterized by a different kind of bonding called **covalent bonding** which involves the sharing of electrons.

1.3 COVALENT BONDING

An alternative to ionic bonding that is quite helpful in understanding the structure and properties of organic compounds is based on the **shared electron pair**, or **covalent**, bond. To illustrate the principles of covalent bonding consider first a simple example, the hydrogen molecule H_2. Here, where both atoms bonded together are the same, there is no real possibility of ionic bonding wherein one hydrogen atom transfers its electron to the other to give $H^+ H^-$. Instead, according to a suggestion by Gilbert N. Lewis of the University of California in 1916, the two hydrogens are joined because they share two electrons.

H·   ·H   H:H

Two hydrogen atoms,   Hydrogen molecule:
each with a single   covalent bonding by way of
electron   a shared electron pair

Likewise, the two fluorine atoms in F_2 share an electron pair.

$$: \ddot{\underset{\cdot\cdot}{F}} \cdot \qquad\qquad \cdot \ddot{\underset{\cdot\cdot}{F}} : \qquad\qquad\qquad : \ddot{\underset{\cdot\cdot}{F}} : \ddot{\underset{\cdot\cdot}{F}} :$$

Two fluorine atoms, each   Fluorine molecule:
with seven electrons in   covalent bonding by way of
their valence shells   a shared electron pair

G. N. LEWIS

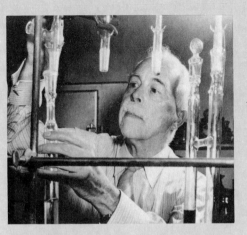

G. N. Lewis has been called the greatest and one of the most influential of American chemists. Born in Massachusetts on October 25, 1875, Gilbert Newton Lewis grew up in Nebraska and attended the University of Nebraska for two years before completing his education at Harvard (B.A. 1896; M.A. 1898; Ph.D. 1899). Except for a year of postdoctoral study in Germany, Lewis stayed at Harvard as an instructor until 1904, when he accepted a government position in weights and measures in the Philippines. One year later, Lewis was back in Cambridge, this time as a faculty member at the Massachusetts Institute of Technology, where he stayed until 1912 when he was appointed chairman of the department of chemistry and dean of the College of Chemistry at the University of California, Berkeley. Lewis' travels were over; he stayed at Berkeley for the remainder of his career.

Lewis' contributions to chemistry were many. He was primarily a physical chemist who was an early leader in the study of *thermodynamics,* the quantitative study of the role that energy changes play in chemical reactions. His theory of covalent bonding in molecules in 1916 introduced the *octet rule* and the *shared electron pair bond* and focused attention on the distribution of a molecule's electrons as being as important as the location of its atoms. Later he proposed a general theory of acids and bases that is permanently entrenched among the fundamental principles of chemistry. Lewis was an experimentalist as well as a theorist, not only in the area of thermodynamics but also in the study of isotopes.

As important as Lewis' scientific contributions are, he is equally regarded as a leading figure in chemical education. He is recognized as the person most responsible for building a chemistry department at Berkeley that ranks among the best in the world. Three of Lewis' graduate students (William F. Giauque, Glenn T. Seaborg, and Harold C. Urey) went on to win Nobel Prizes. Lewis continued in research after officially retiring from the Berkeley faculty and was working in his laboratory at the time of his death on March 23, 1946.

G. N. Lewis. *(Photograph provided by University Archives, The Bancroft Library, University of California, Berkeley.)*

Structural formulas of this type in which valence electrons are represented as dots are called *Lewis structures.*

The Lewis concept of electron sharing is based on the idea that a stable noble gas electron configuration can be achieved by a sharing of electrons. In H_2, the two electrons are associated with both hydrogens and together give each one the electron configuration of the noble gas helium. Similarly, each fluorine in F_2, with eight electrons in its valence shell, has an electron configuration equivalent to that of neon.

PROBLEM 1.3

Hydrogen is bonded to fluorine in hydrogen fluoride by a covalent bond. Write a Lewis structural formula for hydrogen fluoride.

An important feature of the Lewis approach is that second-row elements such as fluorine are limited to a total of eight electrons (shared plus unshared) in their valence shells. Hydrogen is limited to two.

> Most of the elements that we will encounter in this text obey the **octet rule;** in forming compounds they gain, lose, or share electrons to give a stable noble gas electron configuration characterized by eight valence electrons.

Lewis structures of organic substances are constructed on the basis of the two principles of shared electron pairs and attainment of noble gas electron configurations. Covalent bonding in methane and carbon tetrafluoride, for example, follows directly from the Lewis rules.

Combine ·C· and four H·   to write a Lewis structure for methane

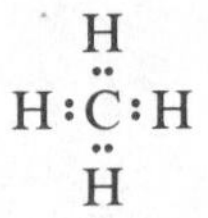

$$\begin{array}{c} H \\ H{:}C{:}H \\ H \end{array}$$

Combine ·C· and four ·F: to write a Lewis structure for carbon tetrafluoride

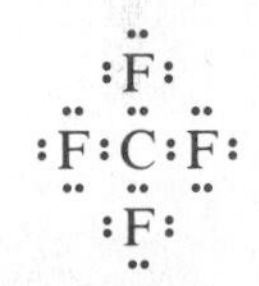

$$\begin{array}{c} {:}\ddot{F}{:} \\ {:}\ddot{F}{:}C{:}\ddot{F}{:} \\ {:}\ddot{F}{:} \end{array}$$

In each of these organic molecules, carbon has eight electrons in its valence shell. By forming covalent bonds to four other atoms, carbon has achieved a noble gas electron configuration.

PROBLEM 1.4

Write a satisfactory Lewis structure, showing all valence electrons, for each of the following:

(a) H_2O (b) NH_3 (c) NF_3 (d) PCl_3 (e) CH_3Cl (f) C_2H_6

SAMPLE SOLUTION

(a) Oxygen is in group VI of the periodic table and contributes six valence electrons. Each hydrogen contributes one.

Combine ·Ö· and two H·   to write a Lewis structure for water $H{:}\ddot{O}{:}H$

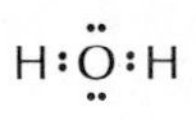

In the Lewis structure shown, each hydrogen is associated with two electrons and oxygen has eight. The structure represented by $H{:}H{:}\ddot{O}$ is incorrect because its central hydrogen has too many electrons in its valence shell (four).

It is customary to represent a two-electron covalent bond by a dash (—). Thus, the Lewis structures for hydrogen fluoride, fluorine, methane, and carbon tetrafluoride become

$$H—\overset{\cdot\cdot}{\underset{\cdot\cdot}{F}}: \qquad :\overset{\cdot\cdot}{\underset{\cdot\cdot}{F}}—\overset{\cdot\cdot}{\underset{\cdot\cdot}{F}}: \qquad H—\overset{\displaystyle H}{\underset{\displaystyle H}{\overset{|}{\underset{|}{C}}}}—H \qquad :\overset{\cdot\cdot}{\underset{\cdot\cdot}{F}}—\overset{\displaystyle :\overset{\cdot\cdot}{F}:}{\underset{\displaystyle :\overset{\cdot\cdot}{\underset{\cdot\cdot}{F}}:}{\overset{|}{\underset{|}{C}}}}—\overset{\cdot\cdot}{\underset{\cdot\cdot}{F}}:$$

Hydrogen   Fluorine   Methane   Carbon tetrafluoride
fluoride

1.4 POLAR COVALENT BONDS

Electrons in covalent bonds are not necessarily shared equally by the two atoms that they join together. If one atom has a greater tendency to attract electrons toward itself than the other, the electron distribution in the bond is said to be **polarized,** and the bond is referred to as a **polar** bond. Hydrogen fluoride, for example, has a polar covalent bond. Fluorine attracts electrons more strongly than does hydrogen. The center of negative charge in the molecule is closer to fluorine, while the center of positive charge is closer to hydrogen. This polarization of electron density in the hydrogen-fluorine bond is represented in various ways.

$$^{\delta+}H—F^{\delta-} \qquad\qquad H\overset{\longleftrightarrow}{—}F$$

(The symbols $\delta+$ and $\delta-$ (The symbol $\longleftrightarrow$ represents
indicate partial positive   the direction of polarization
and partial negative   of electrons in the H—F bond)
charge, respectively)

The separation of positive and negative charge in a polar bond produces a **dipole.** We refer to the tendency of an atom to draw the electrons in a covalent bond toward itself as its **electronegativity** and note that the more electronegative atom is the negatively polarized end of the dipole. A portion of the most widely used electronegativity scale, that devised by Linus Pauling, is presented in Table 1.2. Electronegativity increases across a row in the periodic table; fluorine is the most electronegative of the second-row elements. (Indeed, fluorine is the most electronegative of all the elements.) Electronegativity decreases when proceeding down a column; chlorine is less electronegative than fluorine but more electronegative than bromine. Metals, espe-

TABLE 1.2
Selected Values from the Pauling Electronegativity Scale

Period	Group number						
	I	II	III	IV	V	VI	VII
1	H 2.1						
2	Li 1.0	Be 1.5	B 2.0	C 2.5	N 3.0	O 3.5	F 4.0
3	Na 0.9	Mg 1.2	Al 1.5	Si 1.8	P 2.1	S 2.5	Cl 3.0
4	K 0.8	Ca 1.0					Br 2.8
5							I 2.5

cially those in groups I and II, tend to donate electrons rather than attract them and are said to be **electropositive.** Elements in the middle of the periodic table, including carbon, are neither very electronegative nor electropositive. Because the electronegativities of hydrogen (2.1) and carbon (2.5) are very similar, the carbon-hydrogen covalent bonds in organic molecules are not very polar. The bonds between carbon and highly electronegative elements such as oxygen and fluorine are relatively polar but remain, nonetheless, covalent bonds rather than ionic bonds.

1.5 FORMAL CHARGE

Lewis structures frequently contain atoms that bear a unit positive or negative charge. If the molecule as a whole is neutral, the sum of its positive charges must equal the sum of its negative charges. An example is ammonium chloride. Ammonium chloride, like sodium chloride, is an ionic compound. The positive ion (the *cation*) is NH_4^+. The negative ion (the *anion*) is Cl^-.

$$Na^+[:\ddot{C}l:]^- \qquad [NH_4]^+[:\ddot{C}l:]^-$$

Sodium chloride   Ammonium chloride

Bonding in an ammonium ion is represented by a Lewis structure in which nitrogen has an octet of electrons. Since nitrogen shares those eight electrons with four hydrogens, it can be said to have an *electron count* of four.

$$\left[\begin{array}{c} H \\ \ddot{} \\ H:N:H \\ \ddot{} \\ H \end{array} \right]^+$$

Ammonium ion

Electron count of nitrogen $= \frac{1}{2}(8) = 4$
Electron count of each hydrogen $= \frac{1}{2}(2) = 1$

A neutral nitrogen has five electrons in its valence shell. Nitrogen in an ammonium ion has an electron count equal to four electrons and so lacks one of the electrons of a neutral nitrogen. Nitrogen in an ammonium ion has a *formal charge* of +1.

Each hydrogen in an ammonium ion shares two electrons with nitrogen and so has an electron count of one. Since this is exactly equal to the number of electrons of a neutral hydrogen atom, none of the hydrogen substituents in an ammonium ion has a formal charge.

The total charge in an ammonium ion is equal to the sum of the formal charges of all its atoms. It is +1.

This charge is called a formal charge because it is based on Lewis structures in which electrons are considered to be shared equally between covalently bonded atoms. In actuality, polarization of the nitrogen-hydrogen bonds leads to some transfer of positive charge from nitrogen to each of its hydrogen substituents. Charge dispersal of this kind cannot be represented in a Lewis structure.

The formal charge of a chloride ion, which in this case is the same as the actual charge, is calculated in the same way. A chloride ion shares none of its electrons; it has an electron count of eight.

$$:\ddot{C}l: \qquad \text{Electron count} = 8$$

A neutral chlorine atom has seven electrons. Thus, a chloride ion has one electron in excess of that of a neutral chlorine. It has a charge of −1.

PROBLEM 1.5
Determine the formal charge at all the atoms in each of the following species and the net charge on the species as a whole.

(a) H—Ö—H   (c) H—C̈—H   (e) H—C̈—H
     | |
     H   H

(b) H—C̈—H   (d) H—C—H
     | |
     H   H

SAMPLE SOLUTION
(a) Each hydrogen has a formal charge of zero, as is always the case when hydrogen is covalently bonded to one substituent by a covalent bond. Oxygen has an electron count of five.

H—Ö—H   Electron count of oxygen $= 2 + \frac{1}{2}(6) = 5$
 |
 H

                               Unshared   Covalently
                                 pair   bonded electrons

A neutral oxygen atom has six valence electrons; therefore, oxygen in this species has a formal charge of +1. The species as a whole has a unit positive charge. It is the hydronium ion H_3O^+.

Counting electrons for the purpose of computing the formal charge of an atom differs from counting electrons to see if the octet rule is satisfied. A second-row element has a filled valence shell if the sum of all the electrons, shared and unshared, is eight. Electrons that connect two atoms by a covalent bond count toward filling the valence shell of both. When calculating the formal charge, however, only one-half the number of electrons in covalent bonds can be considered to be "owned" by an atom. Determination of formal charges on individual atoms of Lewis structures is an important element in good "electron bookkeeping." So much of organic chemistry can be made more understandable by keeping track of electrons that it is worth taking some time at the outset to develop a reasonable facility in the seemingly simple exercise of counting electrons.

1.6 MULTIPLE BONDING IN LEWIS STRUCTURES

An extension of the Lewis concept of shared electron pair bonds allows for four-electron **double bonds** and six-electron **triple bonds.** Carbon dioxide (CO_2) has two carbon-oxygen double bonds, and the octet rule is satisfied for both carbon and oxygen. Similarly, the most stable Lewis structure for hydrogen cyanide (HCN) has a carbon-nitrogen triple bond.

Carbon dioxide: :Ö::C::Ö: or   :Ö=C=Ö:
Hydrogen cyanide: H:C:::N: or   H—C≡N:

Multiple bonding is very common in organic chemistry. Ethylene (C_2H_4) contains a carbon-carbon double bond in its most stable Lewis structure, and each carbon has a completed octet. The most stable Lewis structure for acetylene (C_2H_2) contains a carbon-carbon triple bond.

Ethylene:

Acetylene: $H:C:::C:H$ or $H-C\equiv C-H$

Formal charges in Lewis structures containing double and triple bonds are calculated in exactly the same way as was described in the preceding section. A double bond contributes two electrons and a triple bond contributes three toward the electron count of an atom.

1.7 EMPIRICAL AND MOLECULAR FORMULAS

When we experimentally identify the elements that are present in a particular substance and measure the integral ratios of these elements, we determine the **empirical formula** of the substance. Formaldehyde and glucose, for example, are both found to have the empirical formula CH_2O. The ratio of hydrogen to carbon atoms in formaldehyde and in glucose is $2:1$; the ratio of oxygen to carbon is $1:1$. There is, however, only one sequence of atomic connections that allows a stable Lewis structure to be written for CH_2O. It is the one corresponding to formaldehyde.

Formaldehyde

If formaldehyde is the only compound that can have the formula CH_2O, how do we account for the fact that glucose has the same empirical formula? The answer is that empirical formulas tell us only the *relative* proportions of elements in a molecule, not their absolute values. To translate an empirical formula into a **molecular formula** of a substance, we need to know its molecular weight. Experimental determination of the molecular weights of formaldehyde and glucose reveals them to be 30 and 180, respectively. Thus, the molecular formula of formaldehyde is CH_2O — the same as its empirical formula — while the molecular formula of glucose is $(CH_2O)_6$ or $C_6H_{12}O_6$. [Molecular weights are calculated as the sum of the atomic weights of the constituent atoms. The atomic weight of carbon is 12, hydrogen is 1, and oxygen is 16. Thus the calculated molecular weight of CH_2O is $12 + (2 \times 1) + 16 = 30$.]

When writing empirical and molecular formulas for organic substances, carbon is cited first, hydrogen second, and the remaining elements are cited in alphabetical order.

PROBLEM 1.6
The empirical formula of caffeine (present in coffee, tea, and cola drinks) is $C_4H_5N_2O$. If the molecular weight of caffeine is determined to be 194, what is its molecular formula? (*Note:* The atomic weight of hydrogen = 1, carbon = 12, nitrogen = 14, and oxy-

gen = 16. The atomic weight of each element is included in the periodic table found on the inside back cover.)

1.8 ISOMERS AND ISOMERISM

As we just saw in the preceding section, two different substances such as formaldehyde and glucose can have the same empirical formula. We also saw there that these two compounds have quite different molecular formulas. Does it ever happen that two different compounds have the same molecular formula? The answer is an unqualified yes! There are, for example, scores of known substances that have the same molecular formula as glucose, $C_6H_{12}O_6$. Different compounds that have the same molecular formula are called **isomers** (from the Greek *isos + meros,* meaning "the same parts").

We can illustrate one aspect of isomerism by referring to two compounds, nitromethane and methyl nitrite, both of which have the molecular formula CH_3NO_2.

$$
\begin{array}{cc}
\underset{\text{Nitromethane}}{
\begin{array}{c}
\text{H} \quad\quad \ddot{\text{O}}: \\
| \quad\quad\; \parallel \\
\text{H}-\overset{|}{\underset{|}{\text{C}}}-\overset{+}{\text{N}} \\
\text{H} \quad\quad :\underset{..}{\text{O}}:^{-}
\end{array}
}
&
\underset{\text{Methyl nitrite}}{
\begin{array}{c}
\text{H} \\
| \\
\text{H}-\overset{|}{\underset{|}{\text{C}}}-\ddot{\underset{..}{\text{O}}}-\text{N} \\
\text{H} \quad\quad\quad \parallel \\
\quad\quad\quad\quad :\text{O}:
\end{array}
}
\end{array}
$$

Nitromethane and methyl nitrite are different compounds as evidenced by their properties. Nitromethane, used as a high-energy fuel for racing cars, is a liquid with a boiling point of 101°C. Methyl nitrite is a gas that causes dilation of the blood vessels when inhaled; its boiling point is −12°C. The two compounds have different properties because they have different structures. They are often referred to as **structural isomers** of one another; a more modern term is **constitutional isomer**. The order of atomic connections that defines a molecule is termed its *constitution,* and we say that two compounds are constitutionally isomeric when they have the same molecular formula but differ in the order of their atomic connections.

PROBLEM 1.7
Write a structural formula for the CH_3NO isomer characterized by the structural unit indicated:

    (a) C—N=O   (b) C=N—O   (c) O—C≡N   (d) O=C—N

SAMPLE SOLUTION
(a) The structural unit given defines the connections between carbon, nitrogen, and oxygen. All that is necessary is to add the three hydrogens required by the molecular formula to carbon so as to give it its full complement of four bonds.

$$
\begin{array}{c}
\text{H} \\
| \\
\text{H}-\overset{|}{\underset{|}{\text{C}}}-\ddot{\text{N}} \\
\text{H} \quad\quad \parallel \\
\quad\quad :\text{O}:
\end{array}
$$

The phenomenon of isomerism is closely linked to the development of organic chemistry as a science. At one time it was believed that organic compounds could arise only through the action of some "vital force." Organic chemistry, according to this view, was truly the chemistry of living systems. Inorganic chemistry, on the other hand, existed in a separate domain—the world of water, metal, minerals, and the like. The laboratory synthesis of organic compounds from inorganic ones was believed to be impossible unless some vital force was present to forge the link between living and nonliving matter.

The experiment that began the downfall of this doctrine of "vitalism" was the discovery by Friedrich Wöhler in 1828 that crystals of urea were formed when a solution of ammonium cyanate in water was evaporated.

Ammonium cyanate (CH_4N_2O)   Urea (CH_4N_2O)

The reaction that Wöhler observed was the conversion of a compound to its isomer; both ammonium cyanate and urea have the molecular formula (CH_4N_2O). What made the transformation noteworthy was that ammonium cyanate was classified as inorganic, while urea was accepted as an organic substance because it had been isolated earlier from urine. Without the aid of some vital force, an inorganic substance had been transformed into an organic one. Wöhler could not describe the nature of the transformation in structural terms because chemists had not yet begun to think of substances as having defined structures at that time. His work demonstrated a fundamental flaw in the doctrine of vitalism, however, and over the next 30 years organic chemistry outgrew vitalism.

Beginning approximately in 1858, what we now know as the **structural theory** of organic chemistry was independently proposed by August Kekule (Germany), Archibald Couper (Scotland), and Alexander Butlerov (Russia). Its fundamental tenets are that carbon has four bonds in its stable compounds and that carbon has the capacity to bond to other carbon atoms so as to form long chains. Constitutional isomers are possible because a particular elemental composition can accommodate more than one pattern of atoms and bonds. Later, in Chapter 7, we will discuss a second type of isomerism called **stereoisomerism** in which two compounds with the same constitution differ in the arrangement of their atoms in space.

1.9 RESONANCE

When we write a Lewis formula for a molecule, we restrict its electrons to certain well-defined locations, either between two nuclei linking them by a covalent bond or as unshared electrons localized on a single atom. Sometimes our reliance on Lewis formulas can be misleading when we use them in an attempt to infer structural properties. Consider, for example, nitromethane (a substance encountered in Section 1.8). The structural formula for nitromethane is fully consistent with the Lewis octet principle; carbon, nitrogen, and oxygen all have eight electrons in their respective valence shells.

Nitromethane:
$$\begin{array}{c} H \\ | \\ H-C-\overset{+}{N} \\ | \\ H \end{array} \begin{array}{c} \ddot{O}: \\ \diagup \diagdown \\ \diagdown \\ :\overset{..}{O}:^- \end{array}$$
also written as
$$CH_3-\overset{+}{N} \begin{array}{c} \ddot{O}: \\ \diagup \diagdown \\ \diagdown \\ :\overset{..}{O}:^- \end{array}$$

Either structure is an *inadequate* representation of the actual structure of nitromethane, however. It implies that the two oxygens are bonded to nitrogen differently, one being attached to nitrogen by a single bond, the other by a double bond. Since a double bond between two atoms is significantly shorter than a corresponding single bond, we would expect one nitrogen-oxygen bond in nitromethane to be shorter than the other. A typical N—O single bond length is 136 pm, while a typical N=O double bond length is 114 pm. What is actually observed is that both nitrogen-oxygen bonds are exactly the same length — 122 pm — intermediate in length between a nitrogen-oxygen single bond and a nitrogen-oxygen double bond.

> Bond distances in organic compounds are usually 1 to 2 Å (1 Å = 10^{-10} m). Since the angstrom unit (Å) is not an SI *(Système International)* unit, we will express bond distances in picometers (1 pm = 10^{-12} m). Thus, 122 pm = 1.22 Å.

To deal with circumstances such as this, the concept of **resonance** was developed. According to resonance theory, when more than one Lewis structure may be written for a molecule, then a single structure is not sufficient to describe the molecule. Rather, the molecule has properties of all the possible structures. In the case of nitromethane, two equivalent Lewis structures may be written. A double-headed arrow is used to represent the resonance depiction of the Lewis structures.

$$CH_3-\overset{+}{N} \begin{array}{c} \ddot{O}: \\ \diagup \diagdown \\ \diagdown \\ :\overset{..}{O}:^- \end{array} \longleftrightarrow CH_3-\overset{+}{N} \begin{array}{c} :\ddot{O}:^- \\ \diagup \diagdown \\ \diagdown \\ \overset{..}{O}: \end{array}$$

Resonance depiction of nitromethane

The oxygen that is singly bonded to nitrogen in one structure is doubly bonded in the other, and vice versa. The true structure of nitromethane is not accurately described by either Lewis structure but is said to be a **hybrid** of the two contributing structures. Thus, each oxygen is linked to nitrogen by a bond that is intermediate between a single bond and a double bond, and each oxygen bears one-half of a unit negative charge.

The rules to be applied when deciding whether resonance between Lewis structures is important are summarized in Table 1.3. In addition to these rules, it is important to remember that the double-headed resonance arrow does not indicate a process in which the two Lewis structures interconvert. Nitromethane, for example, has a single arrangement of its atoms and electrons; it does not oscillate back and forth between two Lewis structures. Nor are the two Lewis structures of nitromethane isomers of one another. They have the same constitution; they differ only in respect to their electron positions.

Resonance is a useful concept because of the nature of Lewis formulas. Lewis structures depict electrons as being localized, while, in fact, electrons distribute themselves in the way that leads to their most stable arrangement. This sometimes means that a pair of electrons is **delocalized** over several nuclei. Electrons are bound more strongly if they "feel" the attractive force of more than one nucleus. Electrons in a covalent bond are attracted by two nuclei and so are bound more strongly than

TABLE 1.3
Introduction to the Rules of Resonance*

Rule	Illustration
1. Atomic positions must be the same in all resonance structures; only the electron positions may vary among the various contributing structures.	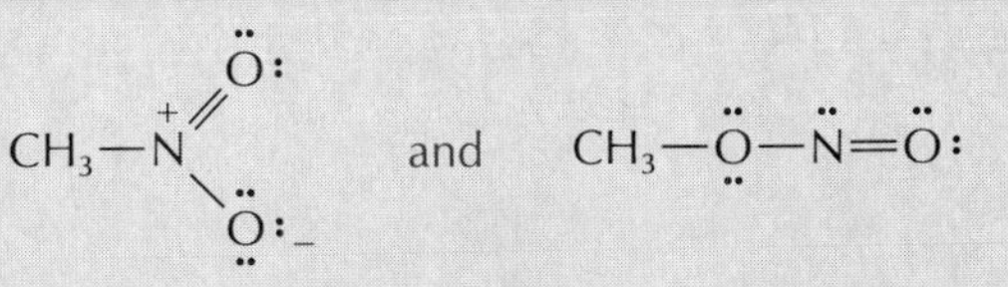 $CH_3—N^+$ (with O double bond and O^- single bond) and $CH_3—\ddot{O}—N=\ddot{O}$ are different compounds (constitutional isomers), *not* different resonance forms of the same compound.
2. Only valid Lewis structures are permitted.	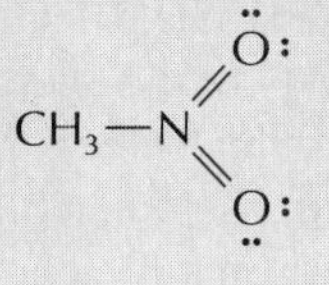 $CH_3—N$ (with two double-bonded O) ten electrons around nitrogen; *not* a permissible Lewis structure for nitromethane.
3. Each contributing Lewis structure must have the same number of electrons and the same *net* charge, although the formal charges of individual atoms may vary among the various Lewis structures.	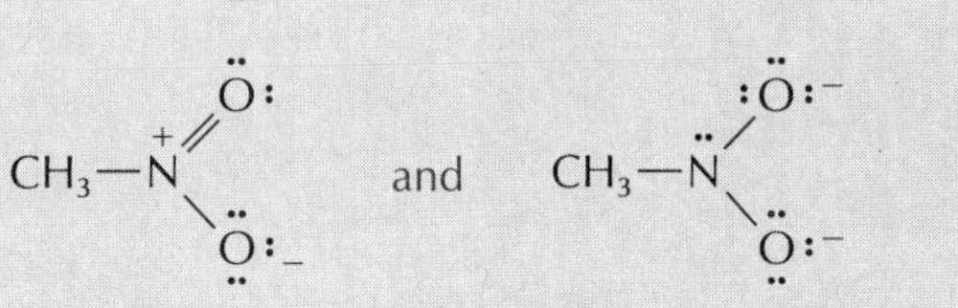 $CH_3—N^+$ (with O double bond and O^- single bond) and $CH_3—N$ (with two $\ddot{O}^-$) are *not* resonance forms of one another; the first has 24 valence electrons and a net charge of 0; the second has 26 valence electrons and a net charge of -2.
4. Each contributing Lewis structure must have the same number of unpaired electrons.	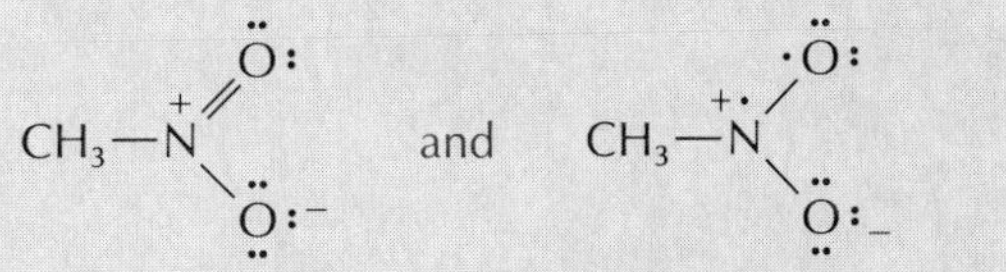 $CH_3—N^+$ (with O double bond and O^- single bond) and $CH_3—N^+·$ (with $·O$ and O^-) are *not* "in resonance" with one another. The structure on the right has two unpaired electrons. The structure on the left has all its electrons paired and is a more stable structure; it is a better representation of nitromethane.

* These are the most important rules to be concerned with at present. Additional aspects of electron delocalization via resonance, as well as additional rules, will be developed as needed in subsequent chapters.

electrons in separated atoms. Electrons are bound even more strongly if they can interact with more than two nuclei. What we try to show by the resonance formulation of nitromethane is the delocalization of the lone pair electrons of oxygen and the electrons in the double bond over the three atoms of the ONO unit. Organic chemists often use curved arrows to show this electron delocalization. Alternatively, a superposition of two Lewis structures is sometimes represented by using a dashed line to depict a "partial" bond

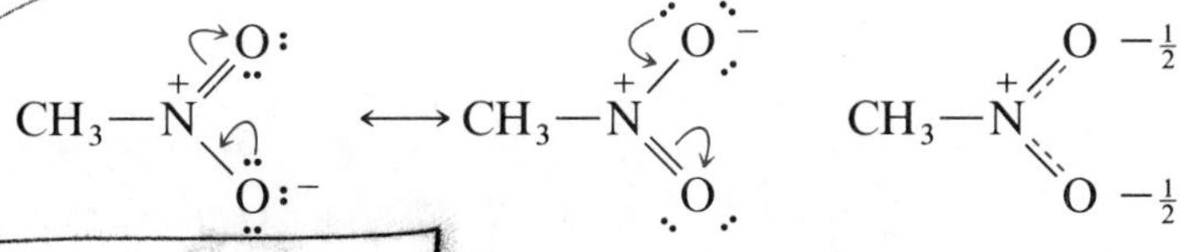

$CH_3—N^+$ ⟷ $CH_3—N^+$ $CH_3—N^+$
Curved-arrow notation   Dashed-line notation

Electron delocalization in nitromethane

(handwritten marginalia: "Brad", "is Imp.", "a⁻ mohan")

PROBLEM 1.8

Electron delocalization can be important in ions as well as in neutral molecules. Using curved arrows, show how an equivalent resonance structure can be generated for each of the following anions.

(a) $^{-}:\ddot{O}-\overset{+}{N}\overset{\nearrow O:}{\underset{\searrow \ddot{O}:^{-}}{}}$ (b) $^{-}:\ddot{O}-C\overset{\nearrow O:}{\underset{\searrow \ddot{O}-H}{}}$ (c) $^{-}:\ddot{O}-C\overset{\nearrow O:}{\underset{\searrow \ddot{O}:^{-}}{}}$

SAMPLE SOLUTION

(a) When using curved arrows to represent the reorganization of electrons, begin at a site of high electron density, preferably an atom that is negatively charged. Move electrons continuously until a proper Lewis structure results. For a nitrate ion, this can be accomplished in two ways.

Three equivalent Lewis structures are possible for a nitrate ion. The negative charge in nitrate is shared equally by all three oxygens.

We will find the use of curved arrows to be helpful on numerous occasions in organic chemistry, not only for representing electron delocalization in Lewis structures but also in subsequent chapters to provide a clearer understanding of how chemical reactions occur. It is customary to use a single-barbed "fishhook" arrow to show the movement of a single electron and a normal double-barbed arrow to show the movement of an electron pair.

A single-barbed arrow shows   A double-barbed arrow shows
movement of a single electron   movement of a pair of electrons

1.10 THE SHAPES OF SOME SIMPLE MOLECULES

Our concern up to this point has been with the electron distribution in bonds between atoms and has emphasized electron bookkeeping. We now shift our emphasis to a consideration of the directional properties of bonds, understanding that the shape of a molecule is governed by the spatial orientation of its bonds.

We use certain geometric terms to describe the shapes of simple molecules. These terms refer to atomic positions. For example, water is described as *bent*, ammonia as *trigonal pyramidal*, and methane as *tetrahedral* (Figure 1.4). These molecular shapes

FIGURE 1.4 The shapes of water, ammonia, and methane. (a) Shapes of the molecules as represented by ball-and-stick models. (b) Bonds can be represented by "wedges" and "dashes" to give the impression of three-dimensionality. A wedge projects from the plane of the paper toward you; a dash projects away from you. A bond represented by a line drawn in the customary way lies in the plane of the paper. The directional properties of the unshared electron pairs in water and ammonia are shown to indicate that four pairs of electrons are present in the valence shell of each molecule and that these electron pairs are directed toward the corners of a tetrahedron.

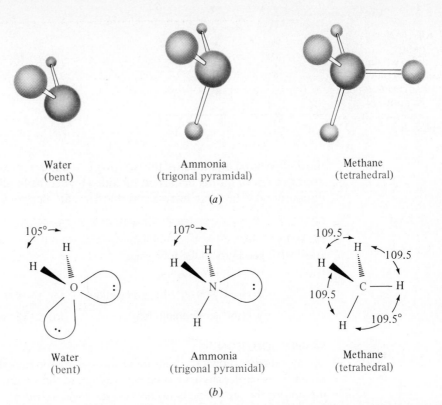

Water
(bent)

Ammonia
(trigonal pyramidal)

Methane
(tetrahedral)

(a)

Water
(bent)

Ammonia
(trigonal pyramidal)

Methane
(tetrahedral)

(b)

may be understood on the basis of the **valence-shell electron pair repulsion** (VSEPR) model. The VSEPR model is based on the idea that since electrons repel each other, an electron pair associated with a particular atom will be as far apart from the atom's other electron pairs as possible. Thus when four electron pairs surround a central atom, they are farthest apart when directed toward the corners of a tetrahedron. Water, ammonia, and methane share the common feature of having four electron pairs around a central atom, so they should all have an approximately tetrahedral arrangement of these electron pairs. Methane is a perfectly tetrahedral molecule. Each H—C—H bond angle of methane is 109.5°, a value referred to as the **tetrahedral angle**.

Boron trifluoride (BF_3) (Figure 1.5) is a *trigonal planar* molecule. There are six electrons, two for each B—F bond, associated with the valence shell of boron. These three bonded pairs are farthest apart when they are coplanar with F—B—F bond angles of 120°.

PROBLEM 1.9

The salt $NaBF_4$ has an ionic bond between Na^+ and the anion BF_4^-. What are the F—B—F angles in the anion?

Multiple bonds are treated as if they were single bonds in the VSEPR model. Formaldehyde (Figure 1.6a) is a trigonal planar molecule in which the electrons of

FIGURE 1.5 Molecular geometry of boron trifluoride. Boron and the three fluorines that are bonded to it lie in the same plane. Boron trifluoride is a trigonal planar molecule.

Boron and the three florines that are bonded to it lie in the same plane.
Boron trifluoride is a trigonal planar molecule.

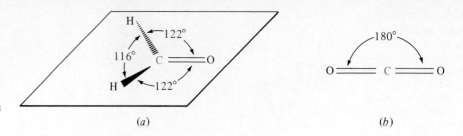

FIGURE 1.6 Molecular geometry of (a) formaldehyde, which is a trigonal planar molecule with bond angles that are close to 120°; and (b) carbon dioxide, which is a linear molecule.

(a)

(b)

the double bond and those of the two single bonds are maximally separated. A linear arrangement of atoms in carbon dioxide (Figure 1.6b) allows the electrons in one double bond to be as far away as possible from the electrons in the other double bond.

PROBLEM 1.10
Specify the geometry of the following:

(a) H—C≡N: (hydrogen cyanide)   (c) :N̄=N⁺=N̄: (azide ion)

(b) H_4N^+ (ammonium ion)   (d) CO_3^{2-} (carbonate ion)

SAMPLE SOLUTION
(a) The structure shown accounts for all the electrons in hydrogen cyanide. There are no unshared electron pairs associated with carbon, so the structure is determined by maximizing the separation between its single bond to hydrogen and the triple bond to nitrogen. Hydrogen cyanide is a *linear* molecule.

H—C≡N:
180°

1.11 THE MOLECULAR ORBITAL VIEW OF BONDING

We have described the Lewis view of a covalent bond as a mutual sharing of an electron pair by two atoms. Our picture of bonding has been expanded since Lewis' early suggestions to incorporate a more detailed description of how electrons are distributed among the orbitals of a molecule — its **molecular orbitals** — analogous to electron distribution in atomic orbitals. In the molecular orbital view of the hydrogen molecule (H_2) as illustrated in Figure 1.7, two hydrogen atoms approach each other so that their 1s atomic orbitals overlap. A molecular orbital encompassing both hydrogen atoms results. This orbital has rotational symmetry around the internuclear axis; a cross section of the orbital taken perpendicular to the internuclear axis is a circle. The orbital is described as a σ (sigma) **bonding molecular orbital.** When the

FIGURE 1.7 When the 1s atomic orbitals of two hydrogen atoms overlap, a bonding molecular orbital results.

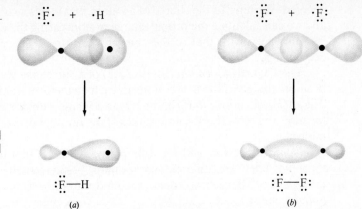

FIGURE 1.8 σ orbitals generated by overlap of a p orbital with (a) an s orbital as in HF (a σ orbital formed by s-p overlap and (b) another p orbital as in F_2 (a σ orbital formed by p-p overlap).

two electrons of H_2 occupy this orbital, they are more strongly held than when they occupy the $1s$ orbitals of two separate hydrogen atoms, resulting in the formation of a **σ bond** between the two atoms.

A second molecular orbital that also results from combining the two $1s$ atomic orbitals is an **antibonding σ^*** (sigma star) **molecular orbital.** It will not be necessary for us to consider explicitly the properties of antibonding orbitals in this text.

σ orbitals may also be formed by overlap of s orbitals with p orbitals and by p orbitals with other p orbitals (Figure 1.8).

While it is possible, in principle, to construct a molecular orbital description of any molecule, the complexity of a molecular orbital description increases enormously as the number of atoms increases. Consequently, chemists have adapted certain elements of molecular orbital theory and terminology to a modification of it known as the **orbital hybridization model.** The orbital hybridization model of bonding is particularly well-suited to providing insight into the structure and properties of carbon compounds and will be introduced in the following chapter.

1.12 SUMMARY

Chemical bonds are classified as ionic or covalent. An *ionic bond* (Section 1.2) is the electrostatic attraction between two oppositely charged ions and occurs in substances such as sodium chloride. An electronegative element, such as chlorine, gains an electron to form an *anion*. A metal, such as sodium, loses an electron to form a *cation*.

Carbon does not normally participate in ionic bonds; *covalent bonding* (Section 1.3) is observed instead. A covalent bond is the sharing of a pair of electrons between two atoms. Double bonds correspond to the sharing of four electrons, and triple bonds to the sharing of six.

Covalent bonds between atoms of different electronegativity are *polarized* in the sense that the electrons in the bond are drawn closer to the more electronegative atom (Section 1.4).

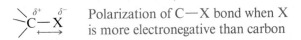

Polarization of C—X bond when X is more electronegative than carbon

Lewis structures for covalently bonded molecules are written on the basis of the *octet rule.* The most stable structures are those in which second-row elements are associated with eight electrons (shared plus unshared) in their valence shells. A particular atom in a Lewis structure may be neutral, positively charged, or negatively charged. We refer to the charge of an atom in a Lewis structure as its *formal charge,* and we can calculate the formal charge by comparing the electron count of an atom in a molecule with that of the neutral atom itself. The procedure is described in Section 1.5.

The *empirical formula* (Section 1.7) of a substance specifies its elemental composition and gives the proportions of each element present relative to one another. The *molecular formula* (Section 1.7) is a whole-number multiple of the empirical formula and gives the absolute values of the number of atoms of each element in the molecule.

Isomers (Section 1.8) are different compounds that have the same molecular formula. They are different compounds because their structures are different. Isomers that differ in the order of their atomic connections are described as *constitutional isomers.*

Resonance between Lewis structures is a device used to describe electron delocalization in molecules (Section 1.9). Many molecules are not adequately described on the basis of a single Lewis structure because the Lewis rules restrict electrons to the region between only two nuclei. In those cases, the true structure is better understood as a hybrid of all possible structures that can be written which have the same atomic positions but differ only in respect to their electron distribution. The most fundamental rules for resonance are summarized in Table 1.3.

The *valence-shell electron pair repulsion* method (Section 1.10) predicts molecular geometries on the basis of repulsive interactions between the pairs of electrons that surround a central atom. A tetrahedral arrangement provides for the maximum separation of four electron pairs, a trigonal planar geometry is optimal for three electron pairs, and a linear arrangement is best for two electron pairs.

| Tetrahedral arrangement of four electron pairs | Trigonal planar arrangement of three electron pairs | Linear arrangement of two electron pairs |

Molecular orbital theory (Section 1.11) is a modern approach to bonding in which electrons in molecules are assigned to molecular orbitals, just as electrons in atoms are assigned to atomic orbitals. When a σ molecular orbital encompassing two atoms is occupied by two electrons, the two atoms are connected by a σ bond. σ orbitals can arise by overlap of two s orbitals, two p orbitals, or by s-p orbital overlap.

ADDITIONAL PROBLEMS

Electronic Configuration and Lewis Structures

1.11 Write the electron configuration for each of the following ions. Which of these ions possesses a noble gas electron configuration?

(a) Li^+ (c) Mg^{2+} (e) H^- (g) O^{2-}
(b) Mg^+ (d) K^+ (f) O^- (h) S^{2-}

1.12 The atomic number Z of an element and the electron configuration corresponding to an ion derived from that element are given below. Identify the ion in each case.

(a) $(Z = 9)$: $1s^2 2s^2 2p^6$ (d) $(Z = 13)$: $1s^2 2s^2 2p^6 3s^1$
(b) $(Z = 12)$: $1s^2 2s^2 2p^6 3s^1$ (e) $(Z = 13)$: $1s^2 2s^2 2p^6$
(c) $(Z = 12)$: $1s^2 2s^2 2p^6$ (f) $(Z = 16)$: $1s^2 2s^2 2p^6 3s^2 3p^6$

1.13 Identify the compound in each of the following groups that is most likely to have an ionic bond.

(a) CO, NO, O_2, CaO   (c) CCl_4, $MgCl_2$, Cl_2O, Cl_2
(b) LiF, BF_3, CF_4, F_2 (d) PBr_3, $AlBr_3$, KBr, BrCl

1.14 Each of the following species will be encountered at some point in this text. They all have the same number of electrons binding the same number of atoms and the same arrangement of bonds; i.e., they are *isoelectronic*. Specify which atoms, if any, bear a formal charge in the Lewis structure given and the net charge for each species.

(a) :N≡N: (b) :C≡N: (c) :C≡C: (d) :N≡O: (e) :C≡O:

1.15 You will meet all the following isoelectronic species in this text. Repeat Problem 1.14 for these three structures.

(a) :Ö=C=Ö: (b) :N̈=N=N̈:   (c) :Ö=N=O:

1.16 Consider the structural formulas A, B, and C.

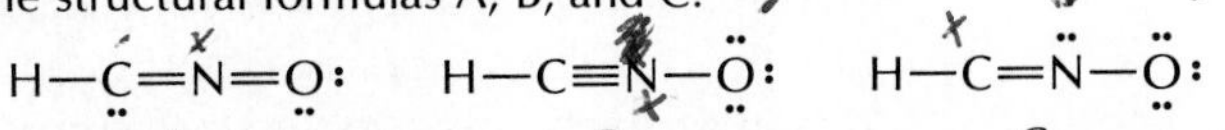

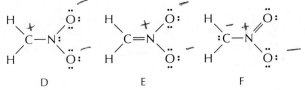

A   B   C

(a) Which structure or structures contain a positively charged carbon?
(b) Which structure or structures contain a positively charged nitrogen?
(c) Which structure or structures contain a positively charged oxygen?
(d) Which structure or structures contain a negatively charged carbon?
(e) Which structure or structures contain a negatively charged nitrogen?
(f) Which structure or structures contain a negatively charged oxygen?
(g) Which structure or structures are electrically neutral (contain equal numbers of positive and negative charges)?

1.17 Repeat Problem 1.16 for structures D, E, and F.

D   E   F

1.18 For each of the following molecules that contain polar covalent bonds, indicate the positive and negative ends of the dipole by using the symbol ↔. Refer to Table 1.2 as needed.

(a) HCl   (b) ICl   (c) HI   (d) H_2O (e) HOCl

1.19 Write Lewis formulas of the following molecules or ions, showing all electron pairs.

(a) PH_3
(b) AlH_4^-
(c) $COCl_2$ (all atoms bonded to carbon)
(d) HCO_3^- (hydrogen is bonded to oxygen)
(e) $^+NO_2$ (order of atoms is ONO)
(f) NO_2^- (order of atoms is ONO)

1.20 Using the VSEPR approach to molecular geometry, predict the shape of each of the species in Problem 1.19.

1.21 Write a Lewis structure for each of the following organic molecules.

(a) C_2H_5Cl (ethyl chloride: sprayed from aerosol cans onto skin to relieve pain of injury caused by bruises)
(b) C_2H_3Cl [vinyl chloride: starting material for the preparation of polyvinyl chloride (PVC) plastics]

(c) $C_2HBrClF_3$ (halothane: a nonflammable inhalation anesthetic; all three fluorines are bonded to the same carbon)

(d) $C_2Cl_2F_4$ (Freon 114; used as a refrigerant and aerosol propellant; each carbon bears one chlorine)

Molecular Formulas and Isomers

1.22 In each of the following exercises a pair of compounds with the same empirical formula is given, along with the molecular weight of each compound. Give the correct molecular formula for each one.

Exercise	Empirical formula	Compound	Molecular weight
(a)	C_2H_4O	Dioxane (an industrial solvent)	88
	C_2H_4O	Paraldehyde (used as a sedative)	132
(b)	C_5H_6O	Eugenol (present in oil of cloves)	164
	C_5H_6O	Ambrosin (obtained from ambrosia)	246
(c)	$C_{10}H_{14}O$	Thymol (present in oil of thyme)	150
	$C_{10}H_{14}O$	Retinoic acid (formed by oxidation of vitamin A)	300

1.23 Each of the following molecular formulas represents two constitutionally isomeric substances. Write Lewis structures for the two isomers in each case.

(a) C_4H_{10} (c) C_2H_6O (e) $C_2H_4Cl_2$

(b) C_3H_7Cl (d) C_2H_7N (f) $C_2H_2Cl_4$

1.24 Each of the following molecular formulas represents three constitutionally isomeric substances. Write Lewis structures for the three isomers in each case.

(a) C_5H_{12} (b) $C_2H_3Cl_2F$ (c) C_3H_8O (d) C_2H_4O

Resonance

1.25 In each of the following pairs, determine whether the two represent resonance forms of a single species or depict different substances. If two structures are not resonance forms, explain why.

(a) $:\ddot{N}-N{\equiv}N:$ and $:N{=}N{=}N:$

(b) $:\ddot{N}-N{\equiv}N:$ and $:\ddot{N}-N{=}\ddot{N}:$

(c) $:\ddot{N}-N{\equiv}N:$ and $:\ddot{N}-\ddot{N}-\ddot{N}:$

1.26 Write a resonance form that is more stable than the one given for each of the following:

(a) $:O{=}\overset{+}{C}-\ddot{O}:^-$

(b) $:O{=}C{=}\ddot{N}:^-$

(c) $:\overset{+}{O}-\ddot{O}-\ddot{O}:^-$

(d) $\underset{H}{\overset{H}{\diagdown}}\overset{+}{C}-\ddot{C}-\ddot{O}:^-$

(e) $\underset{H}{\overset{H}{\diagdown}}\overset{-}{C}-C\diagup\overset{\ddot{O}:}{}$

(f) $\underset{H}{\overset{H}{\diagdown}}\ddot{C}-C{\equiv}N:$

(g) $H-\overset{+}{C}{=}\ddot{O}:$

(h) $\underset{H}{\overset{H}{\diagdown}}\overset{+}{C}-\ddot{O}H$

(i) $\underset{H}{\overset{H}{\diagdown}}:\overset{-}{C}-\overset{+}{N}{=}NH_2$

CHAPTER 2

ALKANES AND CYCLOALKANES

This chapter uses the family of hydrocarbons known as alkanes as a vehicle to introduce preliminary concepts of structure, bonding, and nomenclature in organic chemistry. Upon completion of this chapter you should be able to:

- Understand the sp^3 orbital hybridization model of bonding in alkanes.
- Give the IUPAC names of the unbranched alkanes having up to 20 carbon atoms.
- Given an alkane or cycloalkane, write its IUPAC name.
- Given the IUPAC name for an alkane or cycloalkane, write its structural formula.
- Recognize by common name and structure the alkyl groups that contain up to four carbon atoms.
- Recognize and represent conformations of particular molecules by wedge-and-dash, Newman projection, and sawhorse formulas.
- Draw a chair conformation for a cyclohexane derivative, clearly showing substituent(s) in axial or equatorial orientations as appropriate.
- Know the meaning of the terms eclipsed conformation, staggered conformation, anti conformation, gauche conformation.
- Know the meaning of the terms angle strain, torsional strain, van der Waals strain.
- Given the chair conformation for a cyclohexane derivative, draw a structural formula for its ring-flipped form.
- Understand the difference between constitutional isomers and stereoisomers.
- Write a balanced chemical equation for the combustion of any alkane.

Now that we have reviewed the Lewis model of covalent bonding, we are ready to begin our discussion of organic compounds by focusing on the family of *hydrocarbons* known as *alkanes*. **Hydrocarbons** are compounds that contain only carbon and hydrogen, and **alkanes** are hydrocarbons in which all the bonds are single bonds. Alkanes have molecular formulas that conform to the general expression C_nH_{2n+2} (n is an integer). **Cycloalkanes** are alkanes in which the carbon chain closes upon itself to form a ring; cycloalkanes have the molecular formula C_nH_{2n}. The simplest alkane is methane (CH_4), and it is with methane that we begin the chapter.

2.1 METHANE (occurs naturally)

Methane occurs naturally in quite significant amounts. It is formed by the decomposition of the cellulose portion of plant material in the absence of air, a type of decomposition that frequently takes place in aqueous environments—marshes, bogs, and the ocean, for example. Marsh gas is almost entirely methane. Natural gas, which contains 75 to 85 percent methane, usually occurs in association with petroleum deposits, and these deposits are the legacy of marine plants that lived and died in inland seas millions of years ago. Far away from our own planet, methane constitutes a major portion of the atmospheres of Jupiter, Saturn, Uranus, and Neptune.

Methane is a gas with a boiling point of $-160°C$ at atmospheric pressure; its melting point is $-182.5°C$.

Structurally, as illustrated in Figure 2.1, methane is a tetrahedral molecule. Each H—C—H angle is 109.5°, and each C—H bond distance is 109 pm.

A detailed description of the bonding in methane presents certain difficulties when the electron configuration of carbon is considered. The stable electron configuration of carbon is $1s^2, 2s^2, 2p_x^1, 2p_y^1, 2p_z^0$. Thus, carbon has two unpaired electrons, and we would expect it to bond to two hydrogen atoms (each contributing one electron) to give the molecule CH_2. While this species is known, it is not the most stable hydride of carbon; CH_4 owns that distinction. Why does carbon form bonds to four hydrogen atoms, giving CH_4, rather than only to two, giving CH_2?

2.2 ORBITAL HYBRIDIZATION AND BONDING IN METHANE

In the 1930s Linus Pauling proposed a model for bonding in methane designed to account for the fact that carbon forms four bonds in its most stable compounds. First, in order to convert the stable electron configuration of carbon with its two half-filled $2p$ orbitals (Figure 2.2a) to one with four half-filled orbitals, he suggested that one of the $2s$ electrons was "promoted" to a $2p$ orbital (Figure 2.2b). Next, the $2s$ orbital and the three $2p$ orbitals were combined to give four equivalent sp^3 **hybrid orbitals** (Figure 2.2c). All four sp^3 hybrid orbitals are of equal energy and have 25 percent s character and 75 percent p character. Each of the four sp^3 hybrid orbitals is half-filled (contains one electron) and can overlap with the $1s$ orbital of a hydrogen atom to give the four equivalent σ bonds of methane, CH_4.

FIGURE 2.1 Various depictions of the structure of methane. The atoms and bonds are clearly shown in a ball-and-stick model (a), and it is easier to discern the identity of the molecule than in a space-filling model (b). Space-filling models show the approximate sizes of the atoms and are useful when one wants to know how closely nonbonded atoms come to one another. In (c), a wedge-and-dash description is a way of writing a structural formula so as to show the directional properties of the bonds that connect atoms. A wedge indicates a bond directed from the paper toward you, a solid line is a bond in the plane of the paper, and a dash is a bond directed away from you.

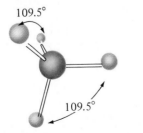

(a) Ball-and-stick model

(b) Space-filling model

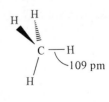

(c) Wedge-and-dash representation

FIGURE 2.2
Representation of orbital
hybridization of carbon.
Each box represents an
orbital. An orbital can
contain two electrons (■),
one electron (◨), or be
vacant (□). The stable
electron configuration of
carbon is shown in (a). In
(b) one of the 2s electrons
has been raised in energy
("promoted") and
occupies a previously
vacant 2p orbital. In (c)
the 2s and 2p orbitals are
mixed to give four equal
energy sp³ hybridized
orbitals, each of which
contains one electron.

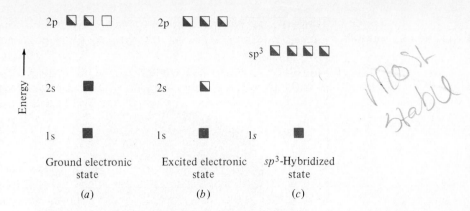

Ground electronic state (a)

Excited electronic state (b)

sp³-Hybridized state (c)

Most stable

Pauling did not intend that orbital hybridization in methane be thought of as a process with promotion, hybridization, and σ bond formation occurring sequentially as independent events. Rather, the electrons spontaneously adopt the configuration that provides the most stable structure. He analyzed hybridization in the manner described both as a convenience and because it allowed him to focus on the reasons why it seemed reasonable. First, the *sp³* hybrid state permits carbon to form four bonds rather than two. Second, each of the four bonds to *sp³* hybridized carbon is a *stronger* bond than it would be in the absence of orbital hybridization.

The shapes of the *sp³* hybrid orbitals are portrayed in Figure 2.3. They are not spherical as are *s* orbitals but have directional characteristics as do *p* orbitals. Unlike *p* orbitals, however, the two lobes of an *sp³* hybrid orbital are not of equal size; there is a higher probability of finding an electron on one side of the nucleus than on the other. This puts the electron in a region where, when the hybrid orbital overlaps with an atomic orbital of another atom, the combined attractive forces between the two bonded nuclei are at a maximum. In methane the four *sp³* orbitals of carbon are arranged in a tetrahedral geometry around carbon, where each overlaps with a 1*s* orbital of hydrogen to give a σ bond.

2.3 ETHANE AND PROPANE

The next two members of the alkane family are ethane (C_2H_6) and propane (C_3H_8). Ethane is, after methane, the second most abundant component of natural gas, where it occurs to the extent of 5 to 10 percent. It has been suggested on the basis of data collected during the Voyager 1 mission that Titan, the largest moon of Saturn, with a surface temperature of $-189\,°C$, is covered by an ocean that is predominantly liquid ethane. Small amounts of propane are present in natural gas.

Organic chemists have developed a number of shortcuts to speed the writing of structural formulas. Representing covalent bonds by dashes is one of them. Another simplification, shown below, deletes the bonds entirely and indicates the number of attached atoms or groups by a subscript.

Ethane:

$$H-\underset{\underset{H}{|}}{\overset{\overset{H}{|}}{C}}-\underset{\underset{H}{|}}{\overset{\overset{H}{|}}{C}}-H \qquad \text{or} \qquad CH_3CH_3$$

Propane:

$$H-\underset{\underset{H}{|}}{\overset{\overset{H}{|}}{C}}-\underset{\underset{H}{|}}{\overset{\overset{H}{|}}{C}}-\underset{\underset{H}{|}}{\overset{\overset{H}{|}}{C}}-H \qquad \text{or} \qquad CH_3CH_2CH_3$$

LINUS PAULING

The man whose name has been associated with important concepts in this and the preceding chapter — electronegativity and the orbital hybridization model of bonding in carbon compounds—is truly one of the most remarkable scientists of the twentieth century. Born in Oregon in 1901, Linus Pauling studied chemical engineering at Oregon Agricultural College (now Oregon State University), then pursued graduate work in chemistry at the California Institute of Technology, where he earned his Ph.D. in 1925. After postdoctoral work in Europe, Pauling returned to Caltech, serving on the faculty for 35 years.

During the 1930s Pauling made great strides in relating the principles of physics to the study of chemistry. His book *The Nature of the Chemical Bond*, first published in 1939 but anticipated by his earlier scientific articles of the same title, remains a classic. Pauling believed that molecular structure was the key to function and properties, and in 1954 he was honored with the award of the Nobel Prize in Chemistry for his work on the structure of proteins.

In the 1950s Pauling spoke out forcefully against the testing of nuclear weapons in the atmosphere. His efforts in this area were recognized when he received a second Nobel Prize, the Nobel Peace Prize, in 1962.

Since the middle 1960s Pauling has concentrated his efforts on the study of nutrition and disease. This has led to publication of his well-known, and among many scientists controversial, book *Vitamin C and the Common Cold* in 1970. Later books written by Pauling have included *Cancer and Vitamin C* in 1979 and *How to Live Longer and Feel Better* in 1986.

Linus Pauling. (*Photograph provided by Linus Pauling Institute of Science and Medicine, Palo Alto, Calif.*)

We call the resulting representations **condensed structural formulas.** Thus, the structural formula for ethane may be written in condensed form as CH_3CH_3, and propane may be written as $CH_3CH_2CH_3$. The group CH_3— is called a *methyl* group. The group —CH_2— is called a *methylene* group.

Ethane has a much higher boiling point than methane, and propane a higher boiling point than ethane.

$$CH_4 \qquad CH_3CH_3 \qquad CH_3CH_2CH_3$$

(bp $-160°C$)   (bp $-89°C$)   (bp $-42°C$)

[Boiling points cited in this text are at 1 atmosphere (760 mmHg) unless otherwise stated.]

The remaining sp^3 hybrid orbitals on each of two methyl groups

overlap to form the two-electron σ bond between the two carbon atoms of ethane

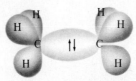

FIGURE 2.4 Description of σ bond formation between the two carbons of ethane by overlap of sp^3 hybridized orbitals. Each carbon contributes one electron (represented by an arrow) to the σ bond. There are two electrons (not shown) in each of the C—H σ bonds.

This will be generally true as we proceed to look at other alkanes: as the number of carbon atoms increases, so does the boiling point. All the alkanes with four carbons or less are gases at room temperature.

The orbital hybridization model of covalent bonding is readily extended to compounds such as ethane and propane that contain carbon-carbon bonds. As Figure 2.4 illustrates, ethane can be described in terms of a carbon-carbon bond joining two CH_3 (methyl) groups. Each methyl group consists of an sp^3 hybridized carbon attached to three hydrogens by sp^3-$1s$ σ bonds. Overlap of the remaining half-filled orbital of one carbon with that of the other generates a σ bond between them. Here is yet another kind of σ bond, one that has as its basis the overlap of two sp^3 hybridized orbitals. In general, you can expect that carbon will be sp^3 hybridized when it is bonded to four atoms or groups. The bond angles at carbon in ethane are approximately tetrahedral. The C—C bond distance is 153 pm, significantly longer than the C—H bond distance which, at 111 pm, is very much like that of methane.

PROBLEM 2.1

How many carbons are sp^3 hybridized in propane? How many σ bonds are there in this molecule? Identify the orbital overlaps that give rise to each σ bond.

2.4 CONFORMATIONS OF ETHANE AND PROPANE

In addition to its structure as described by its atomic connections, there is another aspect of the structure of ethane that commands our attention: its **conformation.** *Conformations are the nonidentical arrangements of atomic positions in a molecule that are generated by rotation about single bonds.* Two of the many conformations of ethane, the **eclipsed conformation** and the **staggered conformation,** are depicted in Figure 2.5. In the eclipsed conformation each C—H bond of one carbon is lined up with a C—H bond of the other carbon. In the staggered conformation each C—H bond of one carbon lies in a plane that bisects an H—C—H angle of the other carbon. Alternative representations of the eclipsed and staggered conformations of ethane are presented in Figures 2.6 to 2.8. You will find it very helpful at this point to construct a molecular model of ethane and view the two conformations from a variety of perspectives.

The eclipsed and staggered conformations of ethane are interconverted by rotation of one carbon with respect to the other around the bond that connects them. They are not considered to be isomers because they are not different compounds; they represent only different forms of the same compound.

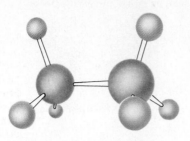

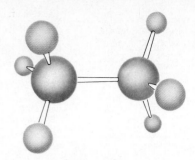

(a) Ball-and-stick model of eclipsed
   conformation
   (least stable conformation)

(b) Ball-and-stick model of staggered
   conformation
   (most stable conformation)

FIGURE 2.5 Depictions
of the conformations of
ethane. Ball-and-stick
models of (a) the eclipsed
and (b) the staggered
conformations. Space-
filling models of (c) the
eclipsed and (d) the
staggered conformations.

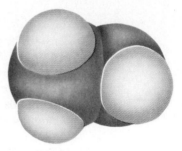

(c) Space-filling model of eclipsed
   conformation

(d) Space-filling model of staggered
   conformation

You may have noticed that there are, in fact, an infinite number of conformations of ethane differing by only tiny increments of rotation about the carbon-carbon
bond. Are all conformations possible? How fast is the process of rotation about the
carbon-carbon bond? Which conformation is the most stable? Which one is the least
stable? Questions of this type arise with almost all chemical substances (not just
organic compounds and not just alkanes), and their study is called **conformational
analysis.** In the case of ethane, *the staggered conformation is the most stable,* and the
eclipsed form is the least stable of all conformations. Rotation about the carbon-carbon bond is extremely fast (several million times per second at room temperature),
and conformations interconvert rapidly. At any instant, most of the ethane molecules exist in the staggered conformation; those that are not in the staggered conformation are in conformations closer to staggered than to eclipsed. It is believed that the
greater stability of the staggered conformation lies in the fact that it permits electron
pairs on adjacent atoms to be maximally separated. Recall from Section 1.10 that the
VSEPR model predicts molecular shapes on the basis of maximum separation of
electron pairs on a single atom. In ethane the electron pairs of the C—H bonds of one
carbon are farthest away from the electron pairs of the C—H bonds of the adjacent
carbon when the bonds are staggered.

Because it is less stable than the staggered conformation, we say the eclipsed
conformation of ethane is *strained* and identify that strain as being due to the
eclipsing of bonds on adjacent atoms. This type of strain is called **torsional strain.** In
the following section we will see a second type of strain that, in combination with

FIGURE 2.6   Wedge-and-dash
representations of (a) the eclipsed
conformation (least stable conformation) and
(b) the staggered conformation (most stable
conformation) of ethane.

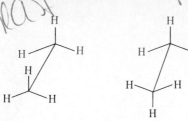

least *most*

eclipsed *staggered*

FIGURE 2.7 Sawhorse drawings of (a) the
eclipsed conformation (least stable
conformation) and (b) the staggered
conformation (most stable conformation) of
ethane. Sawhorse drawings permit the
conformation to be illustrated without resorting
to different styles of bonds.

(a)   (b)

torsional strain, is an important consideration in the conformational analysis of
higher alkanes.

2.5 ISOMERIC ALKANES. THE BUTANES

Methane is the only alkane of molecular formula CH_4, ethane is the only one that is
C_2H_6, and propane is the only one that is C_3H_8. Beginning with C_4H_{10}, however, the
possibility of constitutional isomerism (Section 1.8) arises, and there are two alkanes
that have this molecular formula. One has its four carbons joined in a continuous
chain and is called *n*-butane. The *n* is a notational device that stands for "normal,"
meaning that the carbon chain is unbranched. The second isomer has a branched
carbon chain and is called *isobutane*.

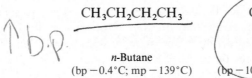

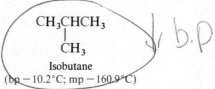

↑ b.p.   ↓ b.p.

n-Butane   Isobutane
(bp −0.4°C; mp −139°C)   (bp −10.2°C; mp −160.9°C)

The two compounds have the same molecular formula but differ in respect to their
atomic connections. They also have different properties. While both are gases at
room temperature, *n*-butane has a boiling point which is almost 10° higher than that
of isobutane and a melting point which is over 20° higher. "Butane" lighters contain
a mixture of *n*-butane (about 5 percent) and isobutane (about 95 percent) in a sealed
container. The pressure produced by the two compounds is on the order of 3 to 4 atm
and is sufficient to maintain them in the liquid state until a small valve is opened to
emit a fine stream of the vaporized mixture across a spark which ignites it.

    The bonding in *n*-butane and isobutane is similar to that of ethane and propane.
The bond angles are close to tetrahedral, each carbon atom is sp^3 hybridized, and all
the bonds are σ bonds.

    The most stable conformation of *n*-butane is shown in Figure 2.9a. It has a zigzag
arrangement of its carbon chain, and all the bonds are staggered. A Newman projection formula of this conformation, sighting down the C-2—C-3 bond, is shown in
Figure 2.9b. As can be seen in the Newman projection, not only are the bonds to C-2

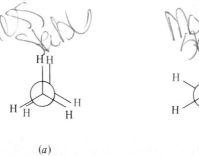

least stable

most stable

FIGURE 2.8 Newman projection formulas
of (a) the eclipsed conformation (least
stable conformation) and (b) the staggered
conformation (most stable conformation)
of ethane. In a Newman projection we
sight down the carbon-carbon bond,
represent the front carbon atom by a point
and the back carbon by a circle. The
bonds to each carbon are arranged
symmetrically around each carbon.

(a)   (b)

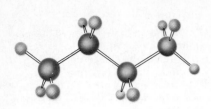

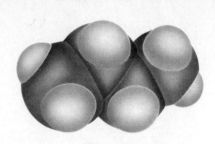

Most Stable

CH₃
H ⟍ ⟋ H
H ⟋ ⟍ H
CH₃

(a) Ball-and-stick model

(b) Newman projection formula

(c) Space-filling model

FIGURE 2.9 (a) A ball-and-stick model and (b) a Newman projection formula of the most stable (anti) conformation of *n*-butane. The Newman projection is developed by sighting down the bond between C-2 and C-3. A space-filling model is shown in (c).

and C-3 staggered with respect to each other, but the angle between the methyl groups appears to be 180°. We call this the *anti* conformation.

Figure 2.10 depicts a second staggered conformation of *n*-butane called the *gauche* conformation. The methyl groups are much closer together in the gauche conformation, where the angle between them is only 60°, than they are in the anti, and the gauche conformation is slightly less stable than the anti. The molecule resists having its methyl groups too close together and relieves the repulsive force between them by rotation about the C-2—C-3 bond. Repulsions between atoms or groups that are too close together in space are called **van der Waals repulsions.** The anti conformation of *n*-butane is more stable than the gauche conformation because it has less **van der Waals strain.** Since the anti and gauche are both staggered conformations, however, they are free of torsional strain. They are rapidly interconverted at room temperature, and at any instant approximately 65 percent of the molecules of *n*-butane exist in the anti conformation and 35 percent in the gauche. As in the case of ethane, the percentage of molecules in eclipsed conformations is vanishingly small.

PROBLEM 2.2
Draw a Newman projection formula of the most stable conformation of isobutane.

2.6 HIGHER ALKANES

n-Alkanes are characterized by the presence of methyl groups at both ends of a continuous series of methylene groups. *n*-Pentane and *n*-hexane are *n*-alkanes possessing five and six carbon atoms, respectively.

$$CH_3CH_2CH_2CH_2CH_3 \qquad CH_3CH_2CH_2CH_2CH_2CH_3$$
n-Pentane   *n*-Hexane

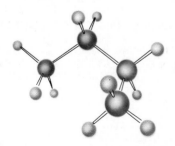

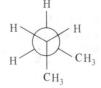

H
H ⟍ ⟋ H
H ⟋ ⟍ CH₃
CH₃

(a) Ball-and-stick model

(b) Newman projection formula

(c) Space-filling model

FIGURE 2.10 (a) A ball-and-stick model and (b) a Newman projection formula of the gauche conformation of *n*-butane. A space-filling model is shown in (c).

Their condensed structural formulas can be further abbreviated by indicating within parentheses the number of methylene groups in the chain. Thus, *n*-pentane may be written as $CH_3(CH_2)_3CH_3$ and *n*-hexane as $CH_3(CH_2)_4CH_3$. This shortcut is especially convenient with alkanes possessing longer carbon chains such as *n*-heptane $CH_3(CH_2)_5CH_3$, a compound present in the heartwood of several species of pine trees. The laboratory synthesis of the "ultralong" alkane $CH_3(CH_2)_{388}CH_3$ was achieved in 1985 — imagine trying to write a structural formula for this compound in anything other than an abbreviated way!

PROBLEM 2.3

An *n*-alkane of molecular formula $C_{28}H_{58}$ has been isolated from a certain fossil plant. Write a condensed structural formula for this alkane.

n-Alkanes have the general formula $CH_3(CH_2)_xCH_3$ and are said to belong to a **homologous series** of compounds. A homologous series of organic compounds is one in which successive members differ by a $—CH_2—$ group.

n-Alkanes are also called *unbranched* alkanes and will be referred to by that term in this text. They are also referred to as "straight-chain alkanes," but, as we saw in the case of *n*-butane in Section 2.5, their chains are not straight but instead tend to adopt a zigzag shape. The zigzag arrangement of a carbon chain provides for another notational device that is useful in representing hydrocarbon structures. According to the **carbon-skeleton, bond-line,** or **line-segment** method, we represent a chain of carbon atoms by a zigzag line, understanding that there is a carbon at each end of the chain and at every bend in the chain. The structures that result can be simplified even further by omitting the hydrogens. Thus, the structural formulas for *n*-pentane and *n*-hexane may be written as shown.

n-Pentane: $CH_3CH_2CH_2CH_2CH_3$ or

n-Hexane: $CH_3(CH_2)_4CH_3$ or

PROBLEM 2.4

Much of the communication between insects is achieved through the use of chemical messengers, or *pheromones*. A species of cockroach secretes an aggregation pheromone from its mandibular glands that alerts other cockroaches to its presence and causes them to congregate. One of the principal components of this *aggregation pheromone* is the alkane shown below. Give the molecular formula of this substance and represent it by its condensed structural formula.

There are three isomeric alkanes of molecular formula C_5H_{12}. The unbranched isomer is, as we have seen, *n-pentane.* The isomer with a single methyl branch is called *isopentane.*

$CH_3CHCH_2CH_3$ or $(CH_3)_2CHCH_2CH_3$ or

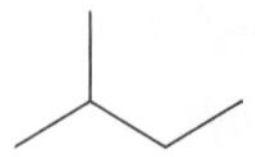

    |
   CH_3

The third isomer has a 3-carbon chain with two methyl branches. It is called *neopentane.*

neopentene

$$CH_3-\overset{\displaystyle CH_3}{\underset{\displaystyle CH_3}{C}}-CH_3 \quad \text{or} \quad (CH_3)_4C \quad \text{or} \quad \times$$

Are there any additional isomers of C_5H_{12}? Table 2.1 presents the number of isomers possible for several examples. According to the table, there are only three isomers of C_5H_{12}: the three indicated by the names *n*-pentane, isopentane, and neopentane. As the data in the table indicate, the number of isomers increases enormously with the number of carbon atoms in the alkane and raises two important questions:

1. How can we tell when we have written all the possible isomers corresponding to a particular molecular formula?
2. How can we name alkanes so that each one has a name that is uniquely its own?

The answer to the first question is that you cannot calculate the number of isomers. The data in Table 2.1 were computed by a mathematician who summarized the results of his efforts by concluding that there was no simple expression from which the number of isomers could be calculated. The best way to ensure that you have written all the isomers of a particular molecular formula is to work systematically, beginning with the unbranched chain, then shortening it while adding branches one by one. It is important that you be able to recognize when two different-looking structural formulas are actually the same molecule written in different ways. The key point is the *connectivity* of the carbon chain. For example, the structural formulas

$$CH_3CH_2CH_2\underset{\displaystyle CH_3}{CH_2} \quad \text{and} \quad \underset{\displaystyle CH_3}{CH_2}CH_2CH_2CH_3$$

are nothing more than alternative representations of *n*-pentane. Because we have written them in a "bent" fashion does not alter the fact that they, like *n*-pentane, have a *continuous* chain of five carbon atoms. Likewise, the four structural formulas that

TABLE 2.1

The Number of Constitutionally Isomeric Alkanes of Particular Molecular Formulas

Molecular formula	Number of constitutional isomers
CH_4	1
C_2H_6	1
C_3H_8	1
C_4H_{10}	2
C_5H_{12}	3
C_6H_{14}	5
C_7H_{16}	9
C_8H_{18}	18
C_9H_{20}	35
$C_{10}H_{22}$	75
$C_{15}H_{32}$	4,347
$C_{20}H_{42}$	366,319
$C_{40}H_{82}$	62,491,178,805,831

follow do not represent different compounds; they are the same compound (isopentane) drawn in four different ways.

$$CH_3CHCH_2CH_3 \quad \underset{CH_3}{\overset{CH_3}{CH_3CHCH_2CH_3}} \quad CH_3CH_2CHCH_3 \quad \underset{CH_3}{\overset{CH_3}{CH_3CH_2CHCH_3}}$$

Each structural formula is characterized by a methyl branch located one carbon from the end of a continuous chain of four carbons.

PROBLEM 2.5

Write condensed and carbon-skeleton structural formulas for the five isomeric C_6H_{14} alkanes.

$$CH_3(CH_2)_4CH_3$$

SAMPLE SOLUTION

When writing isomeric alkanes, it is best to begin with the unbranched isomer.

$$CH_3CH_2CH_2CH_2CH_2CH_3 \quad \text{or} \quad \text{/\\/\\/}$$

Next, remove a carbon from the chain and use it as a one-carbon (methyl) branch at the carbon atom next to the end of the chain.

$$\underset{CH_3}{\overset{}{CH_3CHCH_2CH_2CH_3}} \quad \text{or} \quad \text{/\\/\\}$$

Now, write structural formulas for the remaining three. Be sure that each one is a unique compound and not simply a different representation of one written previously.

The answer to the second question — how to provide a name that is unique to a particular structure — is presented in the following section. It is worth noting here, however, that being able to name compounds in a *systematic* way is a great help in deciding whether two structural formulas represent isomeric substances or are the same compound represented in two different ways. By following a precise set of rules, one will always get the same systematic name for a compound, regardless of how it is written. Likewise, two different compounds will always have different names.

2.7 SYSTEMATIC IUPAC NOMENCLATURE OF UNBRANCHED ALKANES

Nomenclature in organic chemistry is of two types: **common** (or "trivial") and **systematic.** Common names existed long before organic chemistry became an organized branch of chemical science. Methane, ethane, propane, *n*-butane, isobutane, *n*-pentane, isopentane, and neopentane are common names. One simply memorizes the name that goes with a particular structure in just the same way that we match names with faces. As long as there are only a few names and a few structures, the task is manageable. But there are literally millions of organic compounds already known, and the list continues to grow! A system built on common names is not adequate to the task of communicating structural information. In 1892 a set of rules was adopted for naming organic compounds. These rules have been the subject of periodic revision, and their most recent version was published in 1979. The rules are called the

TABLE 2.2

IUPAC Names of Some Unbranched Alkanes

Number of carbon atoms	Name	Number of carbon atoms	Name
1	Methane	11	Undecane
2	Ethane	12	Dodecane
3	Propane	13	Tridecane
4	Butane	14	Tetradecane
5	Pentane	15	Pentadecane
6	Hexane	16	Hexadecane
7	Heptane	17	Heptadecane
8	Octane	18	Octadecane
9	Nonane	19	Nonadecane
10	Decane	20	Icosane*

(handwritten annotations: CH_4, CH_3CH_3, $CH_3CH_2CH_3$; "memorize all 12")

* Spelled "eicosane" prior to 1979 version of IUPAC rules.

IUPAC rules, where IUPAC stands for the International Union of Pure and Applied Chemistry.

The IUPAC rules assign names to unbranched alkanes, as shown in Table 2.2. Methane, ethane, propane and butane are retained for CH_4, CH_3CH_3, $CH_3CH_2CH_3$, and $CH_3CH_2CH_2CH_3$, respectively. Thereafter, the number of carbon atoms in the chain is specified by a Latin or Greek prefix preceding the suffix *-ane* which identifies the compound as a member of the alkane family. Notice that the prefix *n-* is not part of the IUPAC system. The IUPAC name for $CH_3CH_2CH_2CH_3$ is butane, not *n*-butane.

PROBLEM 2.6

What is the IUPAC name of the alkane described in Problem 2.4 as a component of the cockroach aggregation pheromone?

In Problem 2.5, you were asked to write structural formulas for the five isomeric alkanes of molecular formula C_6H_{14}. In the next section you will see how to apply the IUPAC rules to provide a unique name for each isomer.

2.8 APPLYING THE IUPAC RULES. THE NAMES OF THE C_6H_{14} ISOMERS

We can present and illustrate the most important of the IUPAC rules for alkane nomenclature by naming the five C_6H_{14} isomers. The IUPAC rules name branched alkanes as substituted derivatives of the unbranched alkanes listed in Table 2.2. Thus, by definition the unbranched C_6H_{14} isomer is hexane.

$$CH_3CH_2CH_2CH_2CH_2CH_3$$

IUPAC name: hexane

(Common name: *n*-hexane)

Consider next the isomer represented by the structure

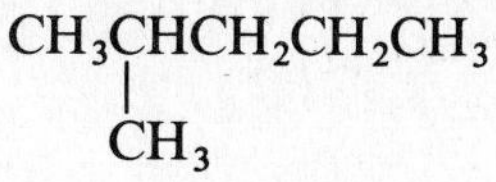

$$CH_3CHCH_2CH_2CH_3$$
$$| \atop CH_3$$

STEP 1

Identify the longest continuous carbon chain and find the IUPAC name in Table 2.2 that corresponds to the unbranched alkane having that number of carbons. This is the parent alkane from which the IUPAC name is to be derived.

In this case, the longest continuous chain has *five* carbon atoms; the compound is named as a derivative of pentane. The key word here is *continuous*. It does not matter whether the carbon skeleton is drawn in an extended straight-chain form or in one with many bends and turns. All that counts is the number of carbons linked together in an uninterrupted sequence.

STEP 2

Identify the substituent groups attached to the parent chain.

The parent pentane chain bears a methyl (CH_3) group as a substituent. Alkyl groups are named by dropping the *-ane* ending of the corresponding alkane and replacing it by *-yl*.

STEP 3

Locate the position of the substituent groups by number. Number the longest continuous chain in the direction that gives the lowest number to the substituent groups at the first point of branching.

The numbering scheme

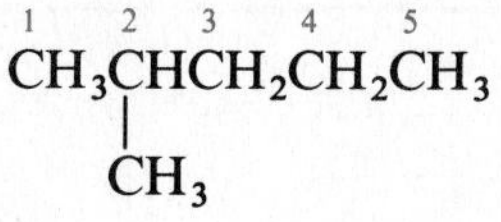

$$\overset{1}{C}H_3\overset{2}{C}H\overset{3}{C}H_2\overset{4}{C}H_2\overset{5}{C}H_3 \quad \text{is equivalent to} \quad \overset{2}{C}H_3\overset{3}{C}H\overset{4}{C}H_2\overset{5}{C}H_2\overset{}{C}H_3$$
$$\underset{CH_3}{|} \qquad\qquad\qquad\qquad\qquad \underset{\underset{1}{C}H_3}{|}$$

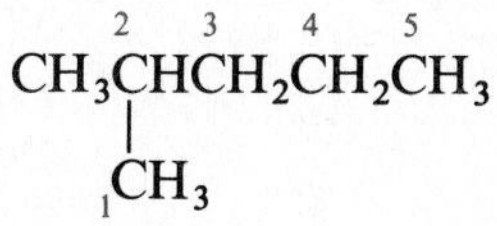

Both schemes count five carbon atoms in their longest continuous chain and bear a methyl group as a substituent at the second carbon. An alternative numbering sequence, one that begins at the other end of the chain, is incorrect.

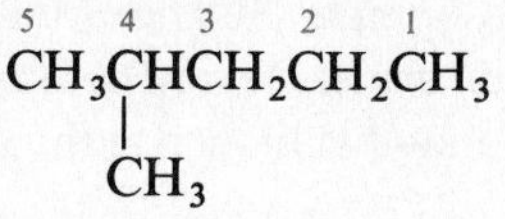

$$\overset{5}{C}H_3\overset{4}{C}H\overset{3}{C}H_2\overset{2}{C}H_2\overset{1}{C}H_3 \qquad \text{(methyl group attached to C-4)}$$
$$\underset{CH_3}{|}$$

STEP 4

Write the name of the compound. The parent alkane is the last part of the name and is preceded by the names of the substituent groups and their numerical locations (locants). No punctuation is used between names, but hyphens separate the locants from the names.

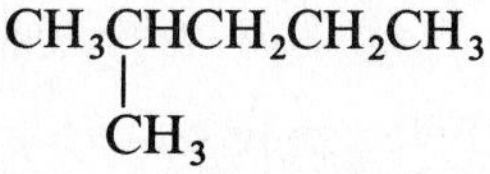

$$CH_3CHCH_2CH_2CH_3$$
$$| \atop CH_3$$

IUPAC name: 2-methylpentane
(Common name: isohexane)

Applying the same sequence of four steps leads to the IUPAC name for the isomer that has its methyl group attached to the middle carbon of the five-carbon chain.

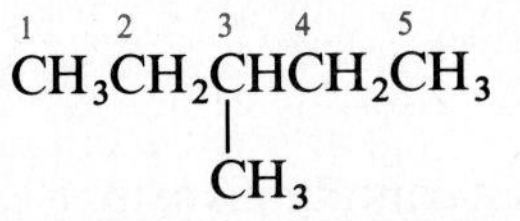

$$\overset{1}{C}H_3\overset{2}{C}H_2\overset{3}{C}H\overset{4}{C}H_2\overset{5}{C}H_3 \qquad \text{IUPAC name: 3-methylpentane}$$
$$|$$
$$CH_3$$

Both remaining C_6H_{14} isomers have two methyl groups as substituents on a four-carbon chain. Thus the parent chain is butane. When the same group appears more than once as a substituent, the multiplying prefixes *di-, tri-, tetra-,* etc., are appended. A separate locant is used for each substituent, and the locants are separated in the IUPAC name by commas.

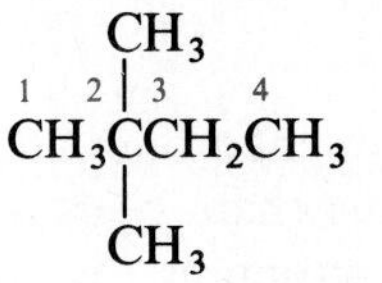

$$CH_3$$
$$|$$
$$\overset{1}{C}H_3\overset{2}{C}\overset{3}{C}H_2\overset{4}{C}H_3$$
$$|$$
$$CH_3$$

IUPAC name: 2,2-dimethylbutane
(Common name: neohexane)

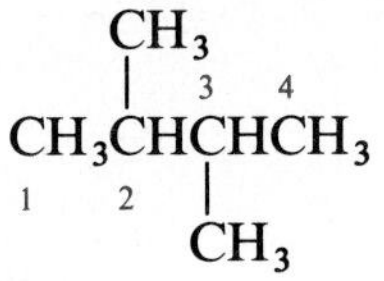

$$CH_3$$
$$|$$
$$\overset{}{C}H_3\overset{3}{C}H\overset{4}{C}HCH_3$$
$$\overset{1}{}\ \overset{2}{|}$$
$$CH_3$$

IUPAC name: 2,3-dimethylbutane

PROBLEM 2.7

Phytane is a naturally occurring alkane produced by the alga *Spirogyra* and is a constituent of petroleum. The IUPAC name for phytane is 2,6,10,14-tetramethylhexadecane. Write a structural formula for phytane.

PROBLEM 2.8

Derive the IUPAC names for

(a) The isomers of C_4H_{10} (c) $(CH_3)_3CCH_2CH(CH_3)_2$
(b) The isomers of C_5H_{12} (d) $(CH_3)_3CC(CH_3)_3$

SAMPLE SOLUTION

(a) There are two C_4H_{10} isomers. Butane (Table 2.2) is the IUPAC name for the isomer that has an unbranched carbon chain. The other isomer has three carbons in its longest continuous chain with a methyl branch at the central carbon; thus its IUPAC name is 2-methylpropane.

$$CH_3CH_2CH_2CH_3 \qquad\qquad \overset{1}{C}H_3\overset{2}{C}H\overset{3}{C}H_3$$
$$|$$
$$CH_3$$

IUPAC name: butane   IUPAC name: 2-methylpropane
(Common name: *n*-butane)   (Common name: isobutane)

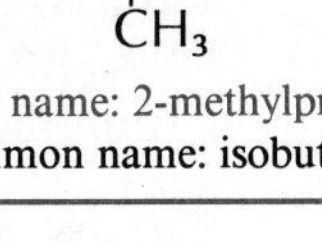

Before proceeding further with a discussion of the IUPAC rules for alkanes, let us shift our focus from the examination of entire molecules to the consideration of substituent groups other than methyl that may be attached to the main carbon chain.

2.9 ALKYL GROUPS

Alkyl groups are structural units that lack one of the hydrogen substituents of an alkane. Unbranched alkyl groups in which the point of attachment is at the end of the chain are named in systematic nomenclature by replacing the *-ane* endings of Table 2.2 by *-yl*. Their common names also include the prefix *n-* when the alkyl group contains three or more carbons.

$$
\underset{\substack{\text{Ethyl group}}}{\overset{\displaystyle H-\overset{\displaystyle H}{\underset{\displaystyle H}{C}}-\overset{\displaystyle H}{\underset{\displaystyle H}{C}}-}{}} \quad \text{or} \quad CH_3CH_2- \qquad \underset{\substack{\text{Pentyl group}\\ \text{(Common name: } n\text{-pentyl)}}}{CH_3CH_2CH_2CH_2CH_2-}
$$

The dash at the end of the chain represents a potential point of attachment for some other atom or group.

Branched alkyl groups are named by using the longest continuous chain that begins at the point of attachment as the base name. Thus, the systematic names of the two C_3H_7 alkyl groups are propyl and 1-methylethyl. Both are better known by their common names, *n*-propyl and isopropyl, respectively.

$$
\underset{\substack{\text{Propyl group}\\ \text{(Common name: } n\text{-propyl)}}}{CH_3CH_2CH_2-} \qquad \underset{\substack{\text{1-Methylethyl group}\\ \text{(Common name: isopropyl)}}}{\overset{\displaystyle CH_3}{\underset{}{CH_3CH-}}} \quad \text{or} \quad (CH_3)_2CH-
$$

The C_4H_9 alkyl groups may be derived either from the unbranched carbon skeleton of butane or from the branched carbon skeleton of isobutane. Those derived from butane are the butyl (*n*-butyl) group and the 1-methylpropyl (*sec*-butyl) group.

$$
\underset{\substack{\text{Butyl group}\\ \text{(Common name: } n\text{-butyl)}}}{CH_3CH_2CH_2CH_2-} \qquad \underset{\substack{\text{1-Methylpropyl group}\\ \text{(Common name: } sec\text{-butyl)}}}{\overset{\displaystyle CH_3}{\underset{}{CH_3CH_2CH-}}}
$$

Those derived from isobutane are the 2-methylpropyl (isobutyl) group and the 1,1-dimethylethyl (*tert*-butyl) group.

$$
\underset{\substack{\text{2-Methylpropyl group}\\ \text{(Common name: isobutyl)}}}{\overset{\displaystyle CH_3}{\underset{}{CH_3CHCH_2-}}} \quad \text{or} \quad (CH_3)_2CHCH_2- \qquad \underset{\substack{\text{1,1-Dimethylethyl group}\\ \text{(Common name: } tert\text{-butyl)}}}{\overset{\displaystyle CH_3}{\underset{\displaystyle CH_3}{CH_3C-}}} \quad \text{or} \quad (CH_3)_3C-
$$

In addition to methyl and ethyl groups, we will encounter *n*-propyl, isopropyl, *n*-butyl, *sec*-butyl, isobutyl, and *tert*-butyl groups many times throughout this text. While these are common names, they have been integrated into the IUPAC system and are an acceptable adjunct to systematic nomenclature. You should be able to recognize these groups on sight and to produce their structures when needed.

2.10 IUPAC NAMES OF HIGHLY BRANCHED ALKANES

By combining the fundamental principles of IUPAC notation with the names of the various alkyl groups, we can develop systematic names for highly branched alkanes. Take, for example, the alkane

$$\underset{1}{CH_3}\underset{2}{CH_2}\underset{3}{CH_2}\underset{4}{CH}\underset{5}{CH_2}\underset{6}{CH_2}\underset{7}{CH_2}\underset{8}{CH_3}$$

with CH_2CH_3 at C-4

As numbered on the structural formula, the longest continuous chain contains eight carbons, so the compound is named as a derivative of octane. Since it bears an ethyl group at C-4, its IUPAC name is **4-ethyloctane.**

What happens to the IUPAC name when another substituent, for example, a methyl group at C-3, is added to the structure?

$$\underset{1}{CH_3}\underset{2}{CH_2}\underset{3}{CH}\underset{4}{CH}\underset{5}{CH_2}\underset{6}{CH_2}\underset{7}{CH_2}\underset{8}{CH_3}$$

with CH_2CH_3 at C-3 and CH_3 at C-4

The compound is named as an octane derivative that bears a C-3 methyl group and a C-4 ethyl group. When two or more different substituents are present, they are listed in alphabetical order in the name. The IUPAC name for this compound is **4-ethyl-3-methyloctane.**

Replicating prefixes such as *di-*, *tri-*, and *tetra-* are ignored when substituents are arranged alphabetically. Adding a second methyl group to the original structure, at C-5, for example, converts it to **4-ethyl-3,5-dimethyloctane.**

$$\underset{1}{CH_3}\underset{2}{CH_2}\underset{3}{CH}\underset{4}{CH}\underset{5}{CH}\underset{6}{CH_2}\underset{7}{CH_2}\underset{8}{CH_3}$$

with CH_2CH_3 at C-4 and CH_3 at C-3 and C-5

Italicized prefixes such as *sec-* and *tert-* are ignored except when compared with each other. *tert*-Butyl precedes isobutyl, and *sec*-butyl precedes *tert*-butyl.

PROBLEM 2.9
Give an acceptable IUPAC name for each of the following alkanes:

(a) $CH_3CH_2CHCHCHCH_2CHCH_3$ with CH_2CH_3 and three CH_3 groups

(b) $(CH_3CH_2)_2CHCH_2CH(CH_3)_2$

(c) $CH_3CH_2CHCH_2CHCH_2CHCH(CH_3)_2$ with CH_3, CH_2CH_3, and $CH_2CH(CH_3)_2$ groups

SAMPLE SOLUTION
(a) This problem extends the preceding discussion by adding a third methyl group to 4-ethyl-3,5-dimethyloctane, the compound just described. It is, therefore, an ethyltrimethyloctane derivative. Notice, however, that the numbering sequence needs to be changed to adhere to the rule that the order must give the lowest number to the

substituent at the first point of difference. The first-appearing substituent in the former numbering scheme was located at C-3. When numbered from the other end, this compound has a methyl group at C-2 as its first-appearing substituent.

CH₂CH₃
8 7 6 4 3 2 1
CH₃CH₂CHCHCHCH₂CHCH₃
5
CH₃ CH₃ CH₃

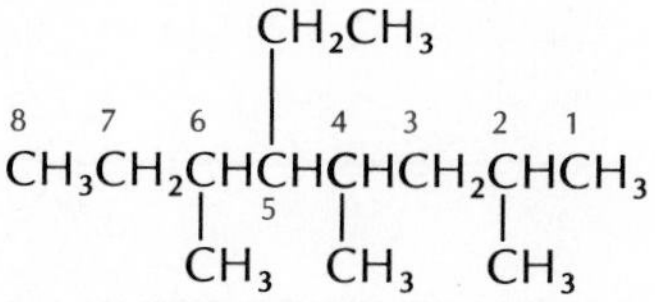

5-Ethyl-2,4,6-trimethyloctane

The IUPAC nomenclature system is inherently logical and incorporates healthy elements of common sense into its rules. Granted, some long, funny-looking, hard-to-pronounce names are generated. Once one knows the code (rules of grammar) though, it becomes a simple matter to convert that long name to a unique structural formula.

2.11 CYCLOALKANE NOMENCLATURE

Cyclic alkanes and their derivatives are frequently encountered in organic chemistry. These **cycloalkanes** are compounds in which the carbon skeleton forms a ring and have the molecular formula C_nH_{2n}. Some examples of cycloalkanes include

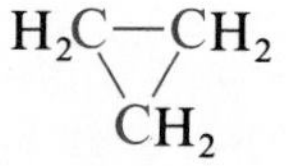

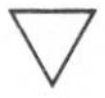

H₂C—CH₂
CH₂

usually represented as

Cyclopropane

H₂
C
H₂C
CH₂
H₂C
CH₂
C
H₂

usually represented as

Cyclohexane

As you can see, cycloalkanes are named, under the IUPAC system, by adding the prefix *cyclo-* to the name of the unbranched alkane with the same number of carbons as the ring. Substituent groups are identified in the usual way. Their positions are specified by numbering the carbon atoms of the ring in the direction that gives the lowest number to the substituent groups at the first point of difference between the two directions.

CH₂CH₃

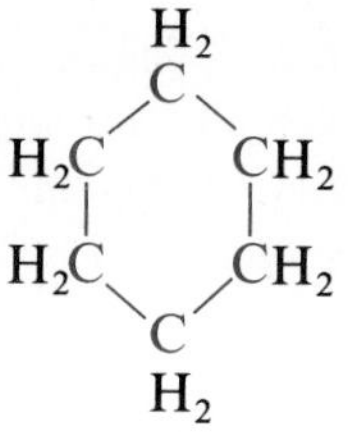

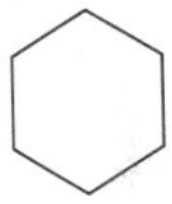

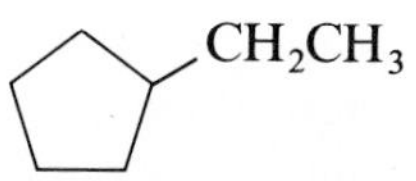

Ethylcyclopentane

H₃C
CH₃
1
2
3
4
5
6
CH₃

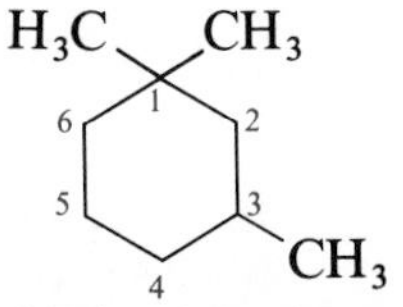

1,1,3-Trimethylcyclohexane

When the ring contains fewer carbon atoms than an alkyl group attached to it, the compound is named as an alkane and the ring is treated as a cycloalkyl substituent.

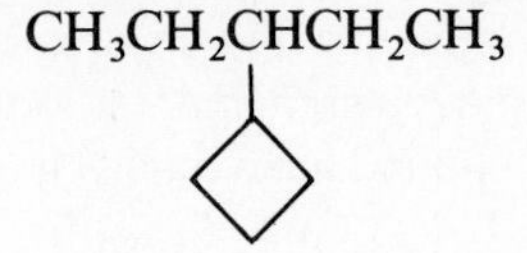

$$CH_3CH_2CHCH_2CH_3$$

3-Cyclobutylpentane

PROBLEM 2.10

Name each of the following compounds:

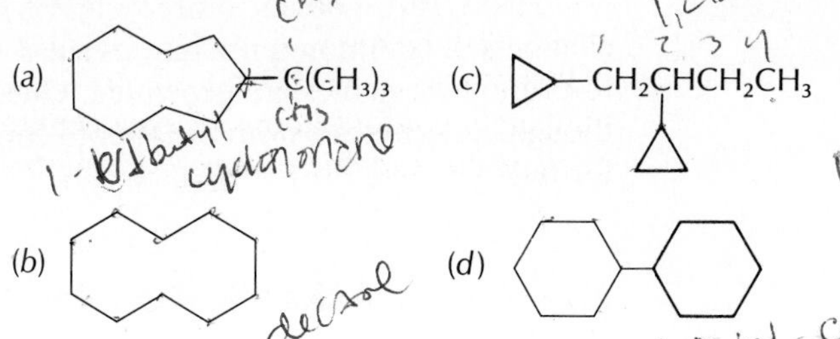

(a) *[handwritten: 1-tert-butyl cyclononene, CH_3, C(CH_3)_3, CH_3]*

(b) *[handwritten: cyclodecane]*

(c) —CH_2CHCH_2CH_3 *[handwritten: dicyclopropyl, 1,2 dimethyl butane]*

(d) *[handwritten: 1,2 dicyclo propylbutane, cyclohexyl-cyclohexane]*

SAMPLE SOLUTION

(a) The molecule has a *tert*-butyl group bonded to a nine-membered cycloalkane. It is *tert*-butylcyclononane.

2.12 CONFORMATIONS OF CYCLOPROPANE, CYCLOBUTANE, AND CYCLOPENTANE

During the nineteenth century it was widely believed (erroneously, as we will see) that the carbon skeletons of cycloalkanes were planar. Cyclopropane must, of course, have all its carbon atoms in a single plane because, as we know from geometry, three points determine a plane.

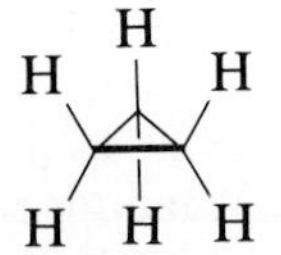

Cyclopropane
(A gas sometimes used as a general anesthetic)

As can be seen from its structural formula, but which is more clearly evident from a molecular model, the three C—H bonds on the upper face of the ring are mutually eclipsed as are the three on the bottom face. Thus, cyclopropane incorporates an element of *torsional strain* into its structure. There is, however, a more serious source of strain in cyclopropane. An equilateral triangle must have angles of 60°, yet we have seen that the bond angles at carbon when attached to four atoms or groups are ideally 109.5°. This distortion of the bond angles at carbon from the tetrahedral value is referred to as **angle strain** and makes cyclopropane less stable than other members of the alkane and cycloalkane family.

Cyclobutane is also subject to angle strain. It is, however, able to reduce some of its torsional strain by adopting a nonplanar "puckered" shape.

Planar conformation of
cyclobutane (less stable):
all bonds are eclipsed

Nonplanar conformation of
cyclobutane (more stable):
some staggering of bonds

Since the angles of a regular pentagon are 108° and therefore very close to the tetrahedral value, the planar conformation of cyclopentane is not subject to much angle strain. There is, however, a fair amount of torsional strain associated with sets of five eclipsed C—H bonds on the top and bottom faces of the ring. There are two nonplanar conformations of cyclopentane that are of comparable stability; each is more stable than the planar conformation because it has less torsional strain.

Envelope conformation
of cyclopentane

Half-chair conformation
of cyclopentane

In the envelope conformation four of the carbon atoms are coplanar. The fifth carbon is out of the plane defined by the other four. There are three coplanar carbons in the half-chair conformation, with one carbon atom displaced above that plane and another below it. These conformations are best seen by constructing molecular models.

2.13 CONFORMATIONS OF CYCLOHEXANE

Six-membered rings occur more often than rings of any other size in organic compounds. Consequently, six-membered rings have been studied more intently and their conformations are better understood than those of any other ring size (except for cyclopropane, for which only one conformation is possible). As you might expect from our discussion of cyclopropane, cyclobutane, and cyclopentane, the planar conformation of cyclohexane is not expected to be its most stable conformation. Since a regular hexagon has 120° angles, planar cyclohexane will suffer from angle strain. It will also have a significant amount of torsional strain arising from the two sets of six eclipsed C—H bonds that encircle its top and bottom faces. If you make a molecular model of cyclohexane using tetrahedral carbon atoms, you will find that it spontaneously adopts a nonplanar shape with no apparent distortion of the bond angles. The most stable conformation of cyclohexane is the **chair conformation;** it is represented in Figure 2.11. A second, less stable, conformation is the **boat,** shown in Figure 2.12.

While both are free of angle strain, the chair has less torsional strain than the boat. The chair has a staggered arrangement of its bonds, most readily seen in a Newman-style presentation. The corresponding bonds in the boat are eclipsed.

FIGURE 2.11 A ball-and-stick model (*a*) and a space-filling model (*b*) of the chair conformation of cyclohexane.

(*a*) Ball-and-stick model of chair conformation of cyclohexane

(*b*) Space-filling model of the chair conformation of cyclohexane

Staggered arrangement of bonds in chair conformation of cyclohexane

Eclipsed bonds in boat conformation give it torsional strain

b/c they're all staggered
∴ no torsional
strain
95% of time
it's in
chair position

The difference in stability between the chair and the boat forms is sufficiently pronounced that virtually all the molecules in a sample of cyclohexane exist in the chair conformation.

Considering the chair conformation in detail reveals some surprising features. The 12 hydrogen atoms are *not* all identical but are divided into two groups, as shown in Figure 2.13. Six of the hydrogens, called **axial** hydrogens, have their bonds parallel and alternately directed up and down on adjacent carbons. The second set of six hydrogens, called **equatorial** hydrogens, are located approximately along the "equator" of the molecule. These positions arise as a natural result of the tetrahedral geometry of sp^3 hybridized carbon and the staggering of bonds on adjacent atoms.

Because six-membered rings are so common in organic chemistry, it is important to be familiar with their conformational features and to have a clear understanding of the directional properties of axial and equatorial bonds. Figure 2.14 offers some guidance on the drawing of chair conformations of cyclohexane.

PROBLEM 2.11

Given the chair conformations of cyclohexane shown below, draw the indicated carbon-hydrogen bonds.

A

B

(*a*) Axial C—H on C-1 of A

(*b*) Equatorial C—H on C-3 of A

(*c*) Equatorial C—H on C-1 of B

(*d*) Axial C—H on C-5 of B

SAMPLE SOLUTION

(*a*) Carbon in position 1 lies below its nearest neighbors; it is "down." Axial bonds point alternately straight up and straight down and take their direction from the carbon atom to which they are attached. Draw the axial bond to C-1 straight down.

H

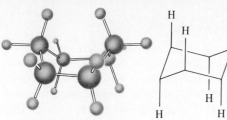

(a) Ball-and-stick model of boat conformation of cyclohexane

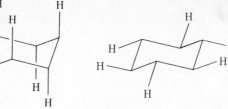 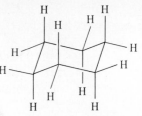

Axial C-H bonds Equatorial C-H bonds Axial and equatorial bonds together

FIGURE 2.13 Axial and equatorial bonds in cyclohexane.

These hydrogens are close enough to touch one another

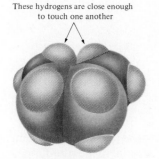

(b) Space-filling model of boat conformation of cyclohexane

FIGURE 2.12 A ball-and-stick model (*a*) and a space-filling model (*b*) of the boat conformation of cyclohexane. The close approach of the two uppermost hydrogen substituents is clearly evident in the space-filling model.

(1) Begin with the chair conformation of cyclohexane.

(2) Draw the axial bonds before the equatorial ones, alternating their direction on adjacent carbon atoms. Always start by placing an axial bond "up" on the uppermost carbon or "down" on the lowest carbon.

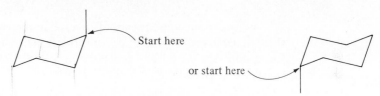

Start here

or start here

then alternate to give

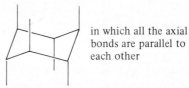

in which all the axial bonds are parallel to each other

(3) Place the equatorial bonds so as to approximate a tetrahedral arrangement of the bonds to each carbon. The equatorial bond of each carbon should be parallel to the ring bonds of its two nearest-neighbor carbons.

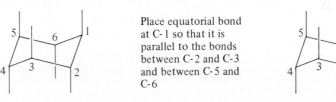

Place equatorial bond at C-1 so that it is parallel to the bonds between C-2 and C-3 and between C-5 and C-6

Following this pattern gives the complete set of equatorial bonds.

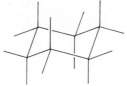

(4) Practice drawing cyclohexane chairs oriented in either direction.

and

FIGURE 2.14 A guide to representing the orientations of the bonds in the chair conformation of cyclohexane.

2.14 CONFORMATIONAL INVERSION (RING FLIPPING) in CYCLOHEXANE

Recall from our discussion of ethane that rotation around the carbon-carbon bond is extremely rapid, occurring millions of times each second. Will a more complex molecule such as cyclohexane show similar mobility, or will it be a static structure?

Cyclohexane is conformationally mobile. Through a process known as **ring inversion, chair-chair interconversion,** or, more simply, **ring flipping,** one chair conformation is converted to another chair. Ring flipping in cyclohexane is an extremely rapid process, although somewhat slower than rotation about single bonds in ethane and *n*-butane.

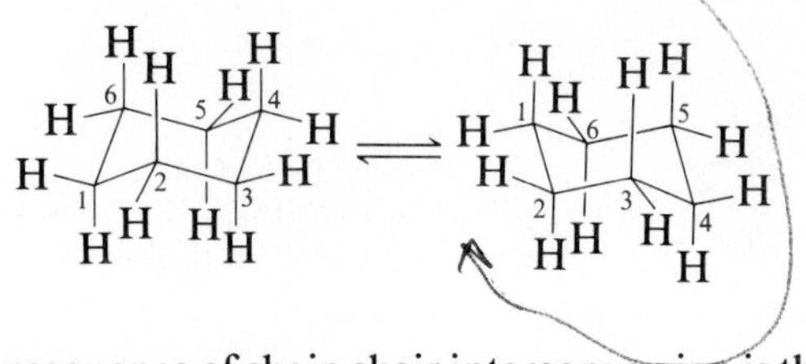

An important consequence of chair-chair interconversion is that any substituent that is axial in the original chair conformation becomes equatorial in the ring-flipped form, and vice versa. The process does *not* require any bond breaking; only ring inversion is involved.

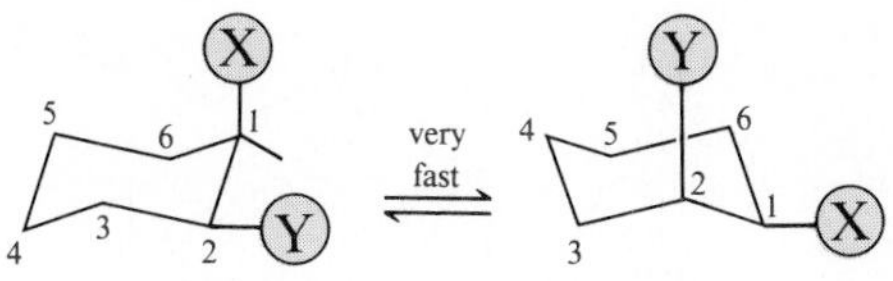

X axial; Y equatorial   X equatorial; Y axial

Ring inversion in a substituted derivative of cyclohexane, such as methylcyclohexane, differs from that of cyclohexane itself in that the two chair conformations are not equivalent. In one, the methyl group occupies an axial position; in the other it is equatorial. Structural studies have established that approximately 95 percent of the molecules of methylcyclohexane are in the chair conformation that has an equatorial methyl group, while only 5 percent of the molecules have an axial methyl group at room temperature.

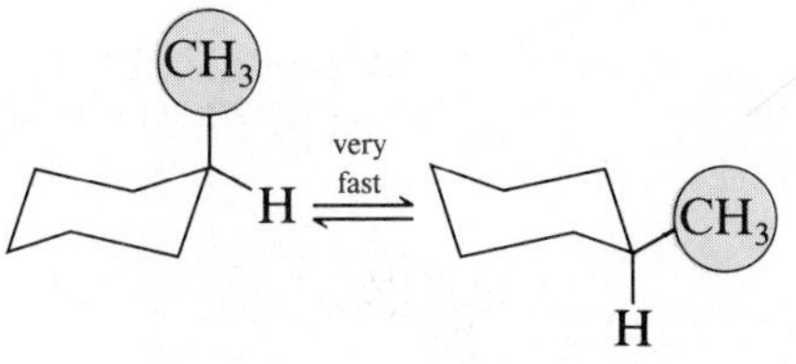

5%   95%
(Less stable chair   (More stable chair
conformation)   conformation)

In any equilibrium process, the species present in greatest amount is the most stable one. Therefore, we conclude that equatorial methylcyclohexane is more stable than axial methylcyclohexane. Is there a structural reason for this experimentally determined fact?

It appears that a methyl group is more stable when it occupies an equatorial site because it is less crowded there than as an axial substituent. An axial methyl group at

C-1 is relatively close to the axial hydrogens at C-3 and C-5. The van der Waals repulsion between the methyl group and these hydrogens is relieved by the ring-flipping process which places the methyl group in an equatorial orientation.

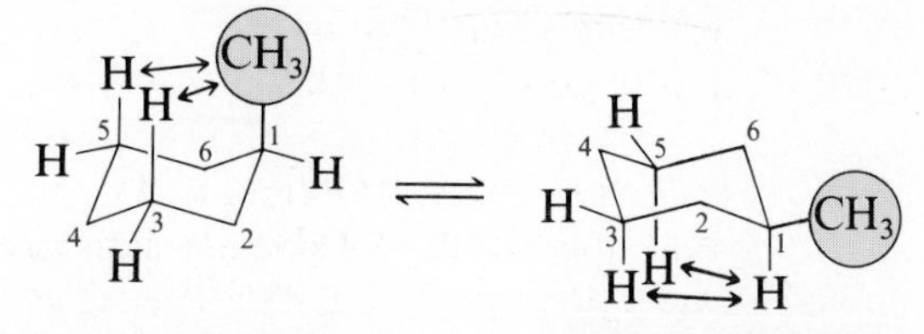

Van der Waals repulsions
between axial CH_3 and
axial hydrogens at C-3
and C-5

Smaller van der Waals
repulsions between hydrogen
at C-1 and axial hydrogens
at C-3 and C-5

The same reasoning can explain the observed conformations of other substituted cyclohexanes. The larger group tends to be equatorial, and this tendency increases as the group becomes progressively "bulkier." While the ratio of equatorial to axial methylcyclohexane conformations is 95:5, that ratio increases to greater than 9999:1 for *tert*-butylcyclohexane.

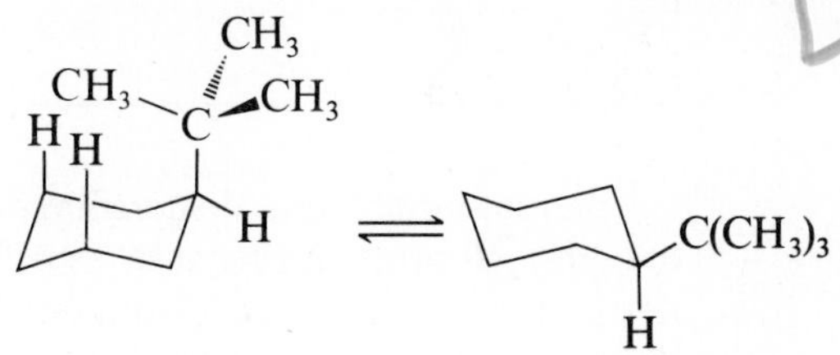

Less than 0.01%

Greater than 99.99%

PROBLEM 2.12

Draw the most stable conformation of 1-*tert*-butyl-1-methylcyclohexane.

In general, a group is "bulky" in organic chemistry if it is highly branched. A very long carbon chain is no bulkier than a shorter one, but a branched alkyl group is bulkier than an unbranched one.

2.15 DISUBSTITUTED CYCLOALKANES AND STEREOISOMERISM

It is possible to have constitutional isomerism in hydrocarbons that contain rings just as it is in alkanes. Thus, 1,1-dimethylcyclopropane, methylcyclobutane, and cyclopentane all have the molecular formula C_5H_{10} and are constitutionally isomeric.

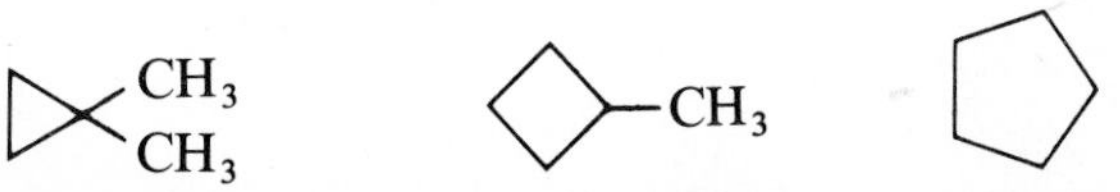

1,1-Dimethylcyclopropane   Methylcyclobutane   Cyclopentane

These three are not the only isomers of C_5H_{10}, however. There are two others, both of which are 1,2-dimethyl derivatives of cyclopropane. In one of these, called *cis*-1,2-dimethylcyclopropane, the methyl groups are on the same face of the ring. (It does not matter whether they are both on the top face or the bottom face.) The other isomer is called *trans*-1,2-dimethylcyclopropane and has its methyl groups on opposite faces of the ring. Both terms come from the Latin, in which *cis* means "on this side" and *trans* means "across."

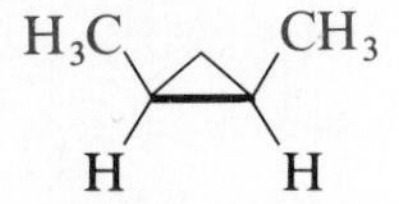

cis-1,2-Dimethylcyclopropane
(bp 37°C)

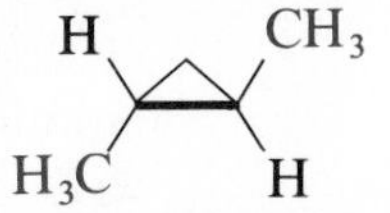

trans-1,2-Dimethylcyclopropane
(bp 29°C)

The cis and trans forms of 1,2-dimethylcyclopropane are **stereoisomers** of one another. They have the same constitution but differ in the arrangement of their atoms in space. They are different compounds, not different conformations of the same compound. To interconvert the two, bonds must be broken and reformed; the two stereoisomers are not interconvertible by a simple rotation about a carbon-carbon bond in the absence of a bond-breaking process.

PROBLEM 2.13

trans-1,2-Dimethylcyclopropane is more stable than *cis*-1,2-dimethylcyclopropane because the cis stereoisomer suffers from van der Waals strain. Can you identify the source of the van der Waals strain in *cis*-1,2-dimethylcyclopropane?

Stereoisomerism in disubstituted cyclohexanes is somewhat more complicated than in cyclopropanes because, as we have seen, the cyclohexane ring is not planar. Let us first examine *cis*-1,2-dimethylcyclohexane. Both methyl groups in the cis stereoisomer are on the same face of the molecule. As shown below, both are "up," i.e., above the hydrogen substitutent on the same carbon. Recalling that ring flipping interconverts axial and equatorial positions, we can see that the molecule can adopt either of two equivalent chair conformations. In each one, one methyl group is axial and the other equatorial.

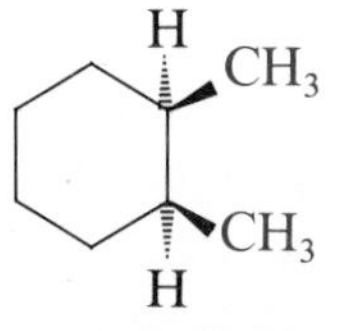

cis-1,2-Dimethylcyclohexane

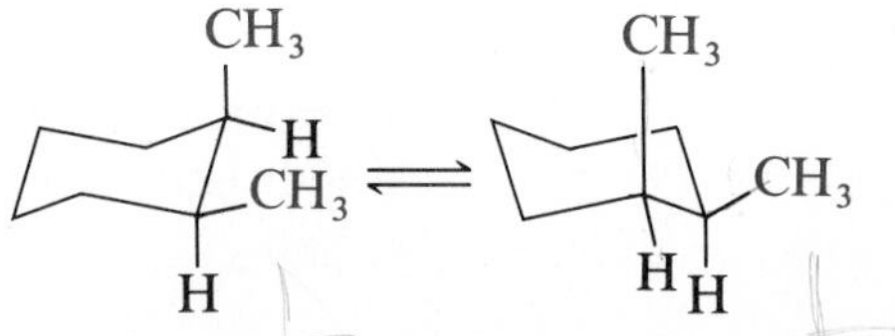

cis-1,2-Dimethylcyclohexane

The situation is different for *trans*-1,2-dimethylcyclohexane, however. The two chair conformations are not equivalent. In one, both methyl groups are axial; in the other, both are equatorial.

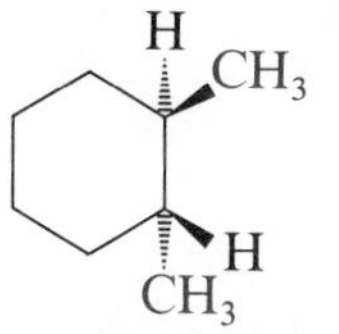

trans-1,2-Dimethylcyclohexane

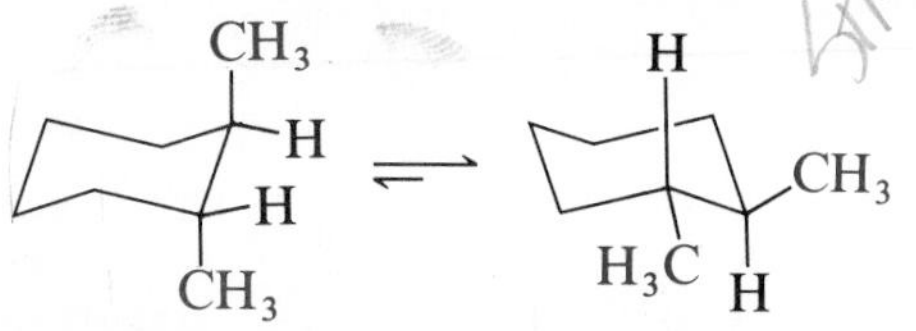

(Both methyl groups are axial; less stable chair conformation)

(Both methyl groups are equatorial; more stable chair conformation)

trans-1,2-Dimethylcyclohexane

The chair conformation in which both methyl groups are equatorial is much more stable than the one in which both are axial, and is the predominant one at equilibrium. We can understand why by recalling that substituent groups are more stable, because they are less crowded, in the equatorial position.

If two substituent groups are different, the preferred conformation will be the one in which the larger group is equatorial. Thus the most stable conformation of *cis*-1-*tert*-butyl-2-methylcyclohexane has an equatorial *tert*-butyl group and an axial methyl group.

PROBLEM 2.14
Draw the most stable conformation of:

(a) *cis*-1-*tert*-Butyl-2-methylcyclohexane.
(b) *trans*-1-*tert*-Butyl-2-methylcyclohexane.

All the properties of a molecule ultimately depend on its structure. Constitution or connectivity is an important element of molecular structure, but it is not the only one. The three-dimensional shape of a molecule—the arrangement of its atoms in space, or its **stereochemistry**—is also important. Many organic reactions and biochemical processes are known in which one stereoisomeric form of a substance reacts readily, while the other form is essentially inert under the same conditions.

2.16 POLYCYCLIC ALKANES

Organic compounds are not limited to only a single ring, and many compounds are known that contain many rings. Substances that contain two rings are referred to as *bicyclic,* those with three rings are called *tricyclic,* those with four are *tetracyclic,* and so forth. Collectively these compounds are referred to as *polycyclic.* We will not go into their systematic nomenclature because in most cases polycyclic alkanes are known by their common names, as the examples below illustrate.

Prismane

Cubane

Adamantane

Chemists synthesize compounds such as prismane and cubane not only to probe the effects of strain on molecular stability and reactivity, but also because their structures

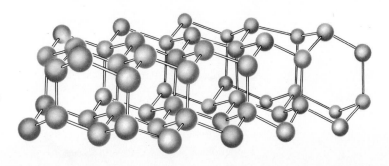

FIGURE 2.15 A ball-and-stick model illustrating the assembly of six-membered rings in their chair forms that comprise the diamond lattice.

are intrinsically interesting and esthetically pleasing and their construction presents a challenge. Adamantane occurs naturally in petroleum and is an assembly of cyclohexane rings in their chair conformation.

PROBLEM 2.15
Give the molecular formulas of prismane, cubane, and adamantane.

Diamonds are a form of elemental carbon in which each carbon is bonded to four other carbons in an extended three-dimensional network of adamantane units. The *diamond lattice* is portrayed in Figure 2.15.

2.17 PHYSICAL PROPERTIES OF ALKANES

As we have seen earlier in this chapter, methane, ethane, propane, and butane are gases at room temperature. The unbranched alkanes pentane (C_5H_{12}) through heptadecane ($C_{17}H_{36}$) are liquids, while those of higher molecular weight are solids. As shown in Figure 2.16, boiling points of unbranched alkanes increase as the number of carbon atoms in the chain increases. Figure 2.16 also shows that the boiling points for 2-methyl branched alkanes are lower than those of the unbranched isomer. The effect of chain branching in lowering the boiling points of alkanes is clearly seen by comparing the three C_5H_{12} isomers.

$$CH_3CH_2CH_2CH_2CH_3 \qquad CH_3\overset{\displaystyle |}{\underset{\displaystyle CH_3}{C}}HCH_2CH_3 \qquad CH_3\overset{\displaystyle CH_3}{\underset{\displaystyle CH_3}{\overset{|}{\underset{|}{C}}}}CH_3$$

Pentane	Isopentane	Neopentane
(bp 36°C)	(bp 28°C)	(bp 9°C)

The most instructive way to consider the relation between boiling point and molecular structure is to ask yourself why any substance, pentane, for example, is a

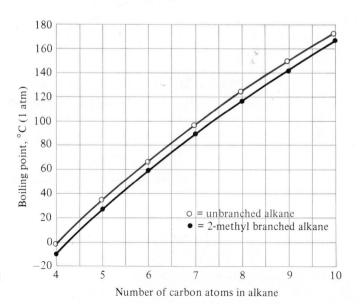

FIGURE 2.16 Boiling points of unbranched alkanes and their 2-methyl branched isomers.

liquid rather than a gas. Pentane is a liquid at room temperature and atmospheric pressure because there are cohesive forces between molecules that are greater in the liquid state than in the vapor. These attractive **intermolecular forces** must be overcome to vaporize pentane, or any other substance.

The strength of the intermolecular forces is directly related to the surface area of the molecule. Branched isomers have lower boiling points than their unbranched counterparts because they are more compact and have a smaller surface area. Neopentane, shown above, is the most compact of the pentane isomers and therefore has the fewest points of contact with neighboring molecules. This is reflected in neopentane having the lowest boiling point of the pentane isomers.

PROBLEM 2.16
Match the boiling points with the appropriate alkanes.

Alkanes: octane, 2-methylheptane, 2,2,3,3-tetramethylbutane, nonane
Boiling points (°C, 1 atm): 106, 116, 126, 151

The forces of attraction between alkane molecules are van der Waals forces. While the van der Waals forces we have seen prior to this point are repulsive, they are not always so. At modest distances van der Waals forces between molecules are weakly attractive but become repulsive when the molecules approach each other too closely and act to resist closer approach. Van der Waals forces are not the only means by which two molecules may attract each other, but they are the only ones available to hydrocarbons. We will encounter additional attractive forces when we discuss other classes of organic compounds in subsequent chapters.

Alkanes are insoluble in water and less dense than water. Thus, when a liquid alkane is added to water, it floats on top of the water as a separate phase.

2.18 CHEMICAL PROPERTIES OF ALKANES. COMBUSTION

As a group, alkanes are relatively unreactive. An old name for alkanes was **paraffin hydrocarbons.** The name *paraffin* is derived from the Latin *parum affinis* and means "with little affinity." Alkanes do, however, burn readily in air. Their combination with oxygen is called **combustion.** On combustion in air, alkanes are converted to carbon dioxide and water.

$$CH_3CH_2CH_2CH_3 + \tfrac{13}{2}O_2 \longrightarrow 4CO_2 + 5H_2O$$

| Butane | Oxygen | Carbon dioxide | Water |

PROBLEM 2.17
Write a balanced chemical equation for the combustion of pentane.

The combustion of alkanes is **exothermic,** meaning that it gives off heat, and is a principal source of energy in our society. Natural gas is, as we have noted, predominantly methane accompanied by smaller amounts of ethane, propane, and butane. Petroleum, from the Latin words *petra* ("rock") and *oleum* ("oil"), is the source of many of the fuels we use every day. The complex mixture of materials present in petroleum (also called *crude oil*) can be separated into simpler mixtures by distillation. The fraction boiling in the range 30 to 150°C is called *straight-run gasoline* and contains, among other substances, alkanes with 5 to 10 carbon atoms. Straight-run gasoline is not a satisfactory fuel for automobile engines because its "octane rating" is too low. Premature ignition of the fuel gives rise to engine "knock" and robs high-compression engines of their power. Unbranched alkanes such as heptane have quite poor performance characteristics and are assigned an octane rating of 0. Branched alkanes are much better fuels for automobile engines; a value of 100 is assigned to 2,2,4-trimethylpentane (isooctane) as the standard. The octane rating of straight-run gasoline is raised by adding substances which serve as octane "boosters."

Kerosene is the petroleum fraction boiling between 175 to 325°C; it is principally C_8–C_{14} hydrocarbons and is used as diesel fuel. Higher boiling fractions are used as lubricating oils, greases, and asphalt.

It is important to recognize that petroleum is much more than a source of gasoline and that refineries do much more than make automobile fuel. Petroleum is far more valuable as a source of *petrochemicals* than as a source of gasoline. Petroleum fractions can be "cracked" to give ethylene, and from ethylene are derived a host of products that we come into contact with every day. We will describe some materials derived from ethylene and other petrochemicals in later chapters.

2.19 SUMMARY

Alkanes are hydrocarbons having the general molecular formula C_nH_{2n+2}; cycloalkanes have the general formula C_nH_{2n}. According to the orbital hybridization model, carbon is sp^3 hybridized when it bonds to four atoms or groups. The sp^3 hybridization state is derived by mixing the $2s$ and the three $2p$ orbitals of carbon to give a set of four equivalent orbitals. The bonds to sp^3 hybridized carbon are σ (sigma) bonds and are directed toward the corners of a tetrahedron (Sections 2.2 and 2.3).

A single alkane may have different names; a name may be a *common* name or it may be a *systematic* name developed by a well-defined set of rules. The system that is the most widely used in chemistry is *IUPAC nomenclature* (Sections 2.7 to 2.11). According to IUPAC nomenclature, alkanes are named as derivatives of unbranched parents. Substituents on the longest continuous chain are identified and their positions specified by number.

Conformations are different spatial arrangements of a molecule that are generated by rotation about single bonds. The most stable and least stable conformations of ethane are the *staggered* and the *eclipsed,* respectively (Section 2.4).

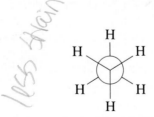

Staggered conformation	Eclipsed conformation
of ethane (most stable conformation)	of ethane (least stable conformation)

Torsional strain is the destabilization that results from the eclipsing of bonds. Staggered conformations are more stable than eclipsed because they have less torsional strain.

The two staggered conformations of butane are not equivalent. The *anti* conformation is more stable than the *gauche* (Section 2.5).

Anti conformation
of butane (more stable)

Gauche conformation
of butane (less stable)

Neither conformation incorporates any torsional strain because each is a staggered form. The gauche conformation is less stable because of van der Waals repulsions between the methyl groups.

The chair is by far the most stable conformation for cyclohexane and its derivatives (Sections 2.13 to 2.15). The chair conformation is free of angle strain, torsional strain, and van der Waals strain. The C—H bonds in cyclohexane are not all equivalent but are divided into two sets of six each, called *axial* and *equatorial*. Cyclohexane undergoes a rapid conformational change referred to as *ring inversion* or *ring flipping*. The process of ring inversion causes all axial bonds to become equatorial, and vice versa.

Blue
= equat

Substituents on a cyclohexane ring are more stable when they occupy equatorial sites than when they are axial. Branched substituents, especially *tert*-butyl, have a pronounced preference for the equatorial position. The relative stabilities of stereoisomeric disubstituted (and more highly substituted) cyclohexanes can be assessed by analyzing chair conformations of van der Waals repulsion involving axial substitutents.

Cyclopropane is planar and strained (angle strain and torsional strain). Cyclobutane is nonplanar and less strained than cyclopropane. Cyclopentane has two nonplanar conformations which are of similar stability, the envelope and the half-chair (Section 2.12).

Alkanes and cycloalkanes are essentially nonpolar. The forces of attraction between molecules are relatively weak van der Waals forces. Because of their smaller surface area (Section 2.17), branched alkanes have lower boiling points than their unbranched isomers.

Alkanes and cycloalkanes burn in air to give carbon dioxide, water, and heat. This process is called *combustion* (Section 2.18).

ADDITIONAL PROBLEMS

Structure and Nomenclature

2.18 Write the structures and give the IUPAC names for all the alkanes of molecular formula C_7H_{16} that:

(a) Are named as methyl-substituted derivatives of hexane.

(b) Are named as dimethyl derivatives of pentane.

(c) Are named as ethyl-substituted derivatives of pentane

2.19 Give the molecular formula and the IUPAC name for each of the following compounds:

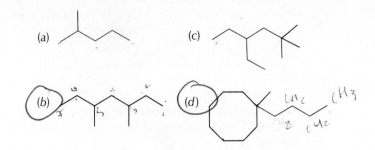

(a) (c)

(b) (d)

2.20 Rewrite the following condensed structural formulas as carbon-skeleton (bond-line) formulas and give the IUPAC name for each one.

(a) $(CH_3)_3CCH_2CH_2CH_3$

(c) $CH_3CH_2CHCH_2CHCH_2CH_2CH_3$
$\qquad\qquad\qquad |\qquad\quad |$
$\qquad\qquad\quad CH_3\quad CH_2CH_3$

(b) $CH_3CH_2CH_2CH(CH_2CH_3)_2$

(d) $CH_3CH_2CHCH_2C(CH_3)_3$
$\qquad\qquad |$
$\qquad\quad CH_2CH(CH_3)_2$

2.21 Write a structural formula for each of the following compounds:
(a) 3-Ethylhexane
(b) 6-Isopropyl-2,3-dimethylnonane
(c) 4-*tert*-Butyl-3-methylheptane
(d) 4-Isobutyl-1,1-dimethylcyclohexane
(e) *sec*-Butylcycloheptane
(f) Dicyclopropylmethane
(g) Cyclobutylcyclopentane

2.22 Which of the compounds in each of the following groups are isomers?
(a) Butane, cyclobutane, isobutane, 2-methylbutane
(b) Cyclopentane, neopentane, 2,2-dimethylpentane, 2,2,3-trimethylbutane
(c) Cyclohexane, hexane, methylcyclopentane, 1,1,2-trimethylcyclopropane
(d) Ethylcyclopropane, 1,1-dimethylcyclopropane, 1-cyclopropylpropane, cyclopentane
(e) 4-Methyltetradecane, 2,3,4,5-tetramethyldecane, pentadecane, 4-cyclobutyldecane

2.23 A certain alkane isolated from a species of blue-green alga has a molecular weight of 240 and an unbranched carbon chain. Identify this alkane.

2.24 Female tiger moths signify their presence to male moths by giving off a sex attractant. The sex attractant has been isolated and found to be a 2-methyl-branched alkane having a molecular weight of 254. What is this material?

2.25 Pristane is an alkane that is present to the extent of about 14 percent in shark liver oil. Its systematic IUPAC name is 2,6,10,14-tetramethylpentadecane.
(a) Write a structural formula for pristane.
(b) What is the molecular formula of pristane?

2.26 Beeswax contains 8 to 9 percent of an alkane with the IUPAC name *hentriacontane*. Hentriacontane has 31 carbon atoms.
(a) What is the molecular formula of hentriacontane?
(b) Write a condensed structural formula for hentriacontane.

2.27 How many σ bonds are there in pentane? In cyclopentane?

2.28 Hectane is the IUPAC name for the unbranched alkane which contains 100 carbon atoms.

(a) What is the molecular formula of hectane?

(b) Write the condensed molecular formula for hectane in the form $CH_3(CH_2)_nCH_3$.

(c) How many σ bonds are there in hectane?

(d) How many alkanes have names of the type X-methylhectane? (Examples include 2-methylhectane, 3-methylhectane, etc.)

(e) How many alkanes have names of the type 2,X-dimethylhectane?

Conformations of Alkanes and Cycloalkanes

2.29 Which has more torsional strain, cyclopropane or the planar conformation of cyclopentane? Which has more angle strain?

2.30 Draw a Newman projection formula (looking down the C-1—C-2 bond) for the most stable conformation of 2,2-dimethylpropane.

2.31 Write Newman projection formulas for two different staggered conformations of 2,3-dimethylbutane (as viewed down the C-2—C-3 bond).

2.32 Draw Newman projection formulas for the three most stable conformations of 2-methylbutane (as viewed down the C-2—C-3 bond). One of these conformations is less stable than the other two. Which one? Why?

2.33 Determine whether the two structures in each of the following pairs represent constitutional isomers, stereoisomers, or different conformations of the same compound.

(a)

and

(b) and

(c)

CH_3CH_2— and CH_3CH_2— CH_3

2.34 Write structural formulas for:

(a) All the constitutional isomers of cis-1,2-dimethylcyclobutane that contain four-membered rings.

(b) All the constitutional isomers of cis-1,2-dimethylcyclobutane that contain rings larger than four-membered.

(c) A stereoisomer of cis-1,2-dimethylcyclobutane

Cyclohexane Conformations and Stereochemistry

2.35 Draw a clear conformational depiction of the most stable conformation of:

(a) 1,1,3-Trimethylcyclohexane   (b) 1,1,4-Trimethylcyclohexane

2.36 Draw chair conformations of 1,1,3-trimethylcyclohexane and 1,1,4-trimethylcyclohexane that are less stable than those of Problem 2.35.

2.37 Draw both possible chair conformations for each of the following compounds, clearly showing the orientation of the substituent (axial or equatorial). Indicate which conformation is the more stable one.

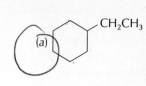

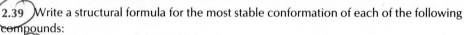

(a) CH₂CH₃

(c) CH₃ ... CH₃

(b) H₃C ⟨⟩ CH₃   (d) H₃C ⟨⟩ C(CH₃)₃

2.38 (a) Which stereoisomer of 1,3-dimethylcyclohexane exists in two equivalent chair conformations?

(b) Draw a clear conformational depiction of this stereoisomer.

(c) Draw clear conformational depictions of the 1,3-dimethylcyclohexane stereoisomer that has two nonequivalent chair conformations and specify which conformation is more stable.

2.39 Write a structural formula for the most stable conformation of each of the following compounds:

(a) cis-1-Isopropyl-3-methylcyclohexane

(b) trans-1-Isopropyl-3-methylcyclohexane

(c) cis-1-tert-Butyl-4-ethylcyclohexane

2.40 Identify the more stable stereoisomer in each of the following pairs and give the reason for your choice:

(a) cis- or trans-1-Isopropyl-2-methylcyclohexane

(b) cis- or trans-1-Isopropyl-3-methylcyclohexane

(c) cis- or trans-1-Isopropyl-4-methylcyclohexane

(d)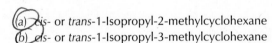

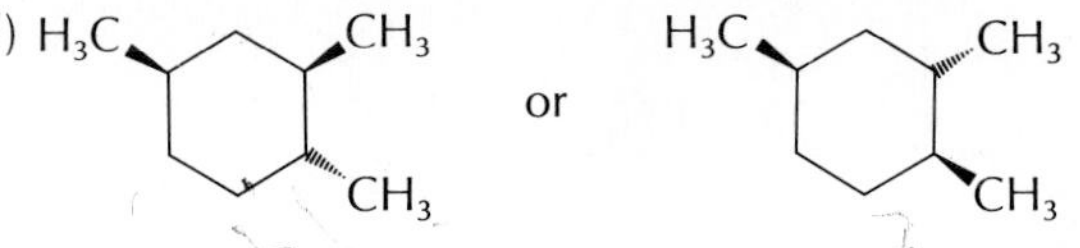

H₃C CH₃ CH₃   or   H₃C CH₃ CH₃

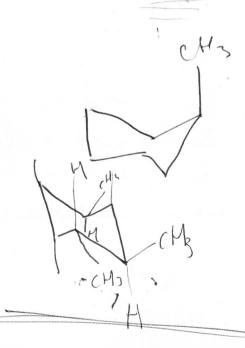

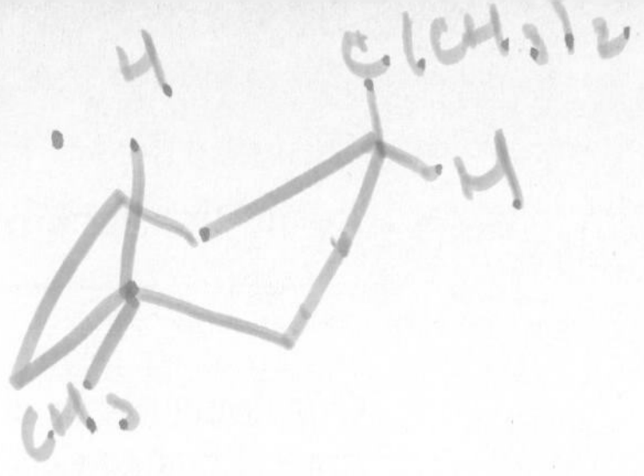

CHAPTER 3

INTRODUCTION TO ORGANIC CHEMICAL REACTIONS

This chapter introduces concepts of organic chemical reactivity by describing two functional group transformations, the formation of alkyl halides by the reaction of alcohols with hydrogen halides and by halogenation of alkanes. Upon completion of this chapter you should be able to:

■ Recognize the hydrocarbon families alkanes, alkenes, alkynes, and arenes on the basis of the structural units they contain.

■ Write systematic IUPAC names for alcohols and alkyl halides on the basis of a given structural formula.

■ Write structural formulas for alcohols and alkyl halides given their IUPAC names.

■ Explain what is meant by hydrogen bonding and cite examples of the kinds of compounds in which it is important.

■ Give the Brönsted-Lowry definitions for acids and bases.

■ Describe the relationship between acidity, K_a, and pK_a.

■ Write a balanced chemical equation for the reaction of an alcohol with a hydrogen halide.

■ Write chemical equations describing the mechanism of the reaction of an alcohol with a hydrogen halide.

■ Explain the meaning of the terms alkyloxonium ion, carbocation, nucleophile, and electrophile.

■ Describe how the rate of reaction of alcohols with hydrogen halides depends on the structure of the alcohol.

■ Write a balanced chemical equation for the reaction of an alkane with chlorine.

■ Write chemical equations describing the mechanism of the reaction of an alkane with chlorine.

■ Describe what is meant by the terms free radical, initiation step, propagation step, and termination step.

- Describe the sp^2 hybridization model of bonding in carbocations and free radicals.
- Relate the stability of carbocations and free radicals to their structure.

Our concerns in studying organic chemistry are *structure, reactions,* and *applications.* The first two chapters established some fundamental principles concerning the structure of organic molecules. In the present chapter we begin our discussion of organic chemical **reactions.**

Estimates place the number of known organic compounds at more than 6 million. Were we to list the reactions available to each one separately, it would tax the capacity of the most powerful computers. Yet someone who is trained in organic chemistry can look at the structure of a substance and make reasonably confident predictions about how it will behave under a particular set of reaction conditions. The basis for this predictive power rests in the fact that the more than 6 million known organic compounds belong to a relatively few structural types and that there are even fewer reaction types than structural types.

The structural types of organic chemistry are divided into families of compounds characterized by the presence of particular **functional groups.** A functional group is a structural unit in a molecule which is responsible for its chemical reactivity under a given set of conditions. In this chapter we develop the concept of the functional group, describe several useful reactions by which one functional group is transformed into another, and examine how these reactions occur. Time and experience have shown that by organizing the reactions of organic compounds according to functional groups, the task of associating structural type with reaction type is considerably simplified.

3.1 FAMILIES OF ORGANIC COMPOUNDS. HYDROCARBONS

We saw in Chapter 2 that hydrocarbons are compounds which contain only carbon and hydrogen and that alkanes constitute an important class of hydrocarbons. Al-

TABLE 3.1

Class	Representative example	Name of example	Generalized abbreviation*
Alkane	$CH_3CH_2CH_3$	Propane	RH
Alkene	$CH_2{=}CHCH_3$	Propene	$\underset{R}{\overset{R}{\diagdown}}C{=}C\underset{R}{\overset{R}{\diagup}}$
Alkyne	$HC{\equiv}CCH_3$	Propyne	$R{-}C{\equiv}C{-}R$
Arene	benzene ring structure	Benzene	ArH

* Primes may be used when desired to indicate that R groups in alkenes and alkynes are different from one another as in R, R′, R″, R‴, etc.

kanes are not the only members of the hydrocarbon family, however. Other hydrocarbons are **alkenes, alkynes,** and **arenes.** Table 3.1 gives examples of each of these hydrocarbon types.

All the bonds in alkanes are single bonds. *Alkanes* have the molecular formula C_nH_{2n+2}. *Alkenes* are characterized by the presence of a carbon-carbon double bond and have the molecular formula C_nH_{2n}. *Alkynes* contain a carbon-carbon triple bond and have the molecular formula C_nH_{2n-2}. *Arenes* are hydrocarbons based on the benzene ring (Table 3.1) as the simplest structural unit. Each of these classes of hydrocarbons will be discussed separately: alkenes and alkynes in Chapters 4 and 5 and arenes in Chapter 6.

3.2 FUNCTIONAL GROUPS IN HYDROCARBONS

A functional group may be as small as a single hydrogen atom, or it can encompass several atoms. The functional group of an alkane is any one of its hydrogen substituents. A reaction that we will discuss later in this chapter, illustrated below for the case of ethane, is one in which an alkane reacts with chlorine.

$$CH_3CH_3 + Cl_2 \longrightarrow CH_3CH_2Cl + HCl$$

| Ethane | Chlorine | Chloroethane | Hydrogen chloride |

In this reaction one of the hydrogen substituents of the alkane is replaced by chlorine. We call this a **substitution** reaction. Substitution of hydrogen by chlorine is a characteristic reaction of all alkanes and can be represented for the general case by the equation

$$R-H + Cl_2 \longrightarrow R-Cl + HCl$$

| Alkane | Chlorine | Alkyl chloride | Hydrogen chloride |

In the general equation the functional group ($-H$) is shown explicitly, while the remainder of the alkane molecule is abbreviated as **R**. This is a commonly used notation which allows us to focus attention on the functional group transformation without being distracted by the parts of the molecule that remain unaffected under the reaction conditions. A hydrogen atom in one alkane is very much like the hydrogen atom of any other alkane in respect to its reactivity toward chlorine. Our ability to write general equations such as the one shown illustrates why the functional group approach is so useful in organic chemistry.

PROBLEM 3.1
On the basis of the general equation shown above, write a balanced equation for the reaction of each of the following alkanes with chlorine:

    (a) 2,2-dimethylpropane   (b) 2,2,3,3-tetramethylbutane   (c) cyclopentane

SAMPLE SOLUTION
(a) First write the structure of the starting material 2,2-dimethylpropane.

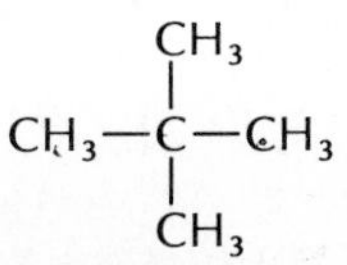

$$CH_3-\underset{\underset{CH_3}{|}}{\overset{\overset{CH_3}{|}}{C}}-CH_3 \qquad \text{may also be written as} \qquad (CH_3)_4C$$

All the hydrogen substituents of 2,2-dimethylpropane are equivalent to one another, so it does not matter which one is replaced by chlorine.

$$CH_3-\overset{\underset{\displaystyle CH_3}{|}}{\underset{\underset{\displaystyle CH_3}{|}}{C}}-CH_3 + Cl_2 \longrightarrow CH_3-\overset{\underset{\displaystyle CH_3}{|}}{\underset{\underset{\displaystyle CH_3}{|}}{C}}-CH_2Cl + HCl$$

[*Note:* The product may also be written as $(CH_3)_3CCH_2Cl$.]

A hydrogen substituent is a functional group in alkenes and alkynes as well as in alkanes. However, these hydrocarbons contain a second functional group as well. The carbon-carbon double bond is a functional group in alkenes, and the carbon-carbon triple bond is a functional group in alkynes. The chemical reactions of alkenes and alkynes are different from those of alkanes and will be discussed in detail in Chapter 5.

A hydrogen substituent is a functional group in arenes, and we represent arenes as **ArH** to reflect this. What will become apparent when we discuss arenes in Chapter 6, however, is that the chemistry of arenes is much richer than that of alkanes, and it is more appropriate to consider the ring in its entirety as the functional group.

It is very common for a particular molecule to possess structural units characteristic of two or more classes. Styrene, for example, is a hydrocarbon that contains both an arene and an alkene structural unit.

Styrene

Styrene is an industrial chemical prepared on a massive scale and used as a starting material for the preparation of *polystyrene* plastics and films. Disposable coffee cups and the "clamshell" sandwich containers so common in fast-food restaurants are made from polystyrene.

3.3 FUNCTIONALLY SUBSTITUTED DERIVATIVES OF ALKANES

As a class, alkanes are not particularly reactive compounds, and a hydrogen substituent of an alkene is not a particularly reactive functional group. *Indeed when a group other than hydrogen is a substituent on an alkane framework, that group is almost always the functional group.* Table 3.2 lists examples of some compounds of this type.

The first two classes of compounds in the table, **alcohols** and **alkyl halides,** are especially useful and often serve as starting materials for the preparation of other types of organic compounds. The present chapter describes the preparation of alkyl halides from alkanes and from alcohols. The preparation of chloroethane by the reaction of ethane with chlorine described in the preceding section is an example of the conversion of an alkane to an alkyl halide and is discussed in more detail in Sections 3.15 to 3.18.

TABLE 3.2
Functionally Substituted Alkanes

Class	Representative example	Name of example*	Generalized abbreviation
Alcohol	CH_3CH_2OH	Ethanol (ethyl alcohol)	ROH
Alkyl halide	CH_3CH_2Cl	Chloroethane (ethyl chloride)	RCl
Amine†	$CH_3CH_2NH_2$	Ethanamine (ethylamine)	RNH_2
Epoxide	$CH_2—CH_2$ $\quad$ O	Oxirane (ethylene oxide)	
Ether	$CH_3CH_2OCH_2CH_3$	Ethoxyethane (diethyl ether)	ROR
Nitroalkane	$CH_3CH_2NO_2$	Nitroethane	RNO_2
Thiol	CH_3CH_2SH	Ethanethiol (ethyl mercaptan)	RSH

* The preferred IUPAC name is given along with the acceptable synonym in parentheses.
† The example given is a *primary* amine (RNH_2). *Secondary* amines have the general structure R_2NH; *tertiary* amines are R_3N.

Like the formation of alkyl chlorides from alkanes, the formation of alkyl chlorides from alcohols is a substitution reaction.

$$R—OH + \quad HCl \quad \longrightarrow \quad R—Cl \quad +H—OH$$

Alcohol $\quad$ Hydrogen chloride $\quad$ Alkyl chloride $\quad$ Water

The **hydroxyl group (—OH)** is the functional group of an alcohol and is replaced by a chlorine substituent. This reaction will be described in detail in Sections 3.10 to 3.12.

As we progress through this chapter you will see that the conversions RH → RCl and ROH → RCl, while both are functional group transformations of the substitution type, proceed by **mechanisms** that are distinctly different from one another. A **reaction mechanism** is a description, in as much detail as experimental data permit, of the sequence of steps that occur during the conversion of reactants to products. The development of the fundamental principles of reaction mechanism is a principal objective of this chapter. Organizing organic chemical reactions according to changes in functional group, coupled with an understanding of *how* organic reactions take place, provides the basis for making confident predictions concerning the products that are most likely to be formed from a particular set of reactants and guides the choice of the best conditions for carrying out the preparation of a desired substance.

Before we begin to discuss the functional group transformations whereby alkanes and alcohols are converted to alkyl halides, we need to introduce the nomenclature of alcohols and alkyl halides, describe aspects of their structure and bonding, and review some fundamental principles of acid-base chemistry.

3.4 NOMENCLATURE OF ALCOHOLS AND ALKYL HALIDES

Several alcohols are commonplace substances, well known by familiar names that reflect their origin (wood alcohol, grain alcohol) or use (rubbing alcohol). The common name of wood alcohol is *methyl alcohol,* grain alcohol is *ethyl alcohol,* and rubbing alcohol is *isopropyl alcohol.*

THE COMMON ALCOHOLS:
METHANOL, ETHANOL, AND ISOPROPYL ALCOHOL

Until the 1920s, the major source of methanol (CH_3OH) was its isolation as a by-product in the production of charcoal from wood — hence the name, *wood alcohol*. Now most of the 7 billion pounds of methanol used annually in the United States is synthetic, prepared directly from carbon monoxide and hydrogen.

$$CO \quad + \quad 2H_2 \quad \xrightarrow[\text{400°C, pressure}]{\text{ZnO—Cr}_2\text{O}_3} \quad CH_3OH$$

Carbon monoxide   Hydrogen   Methanol

Almost one-half of this methanol is converted to formaldehyde for incorporation into various resins and plastics. Methanol is also used as a solvent, as an antifreeze, and as a convenient clean-burning liquid fuel. Methanol is a colorless liquid, boiling at 65°C, and is miscible with water in all proportions. It is poisonous; drinking as little as 30 mL has been fatal. Ingestion of sublethal amounts can lead to blindness.

When vegetable matter ferments, its carbohydrates are converted to ethanol and carbon dioxide by enzymes present in yeast. Using glucose as a representative carbohydrate, the reaction may be written as

$$C_6H_{12}O_6 \xrightarrow{\text{enzymes}} 2CH_3CH_2OH + \quad 2CO_2$$

Glucose   Ethanol   Carbon dioxide

Fermentation of barley produces beer; grapes give wine. The maximum ethanol content is on the order of 15% because higher concentrations inactivate the enzymes, halting fermentation. Since ethanol boils at 78°C and water at 100°C, distillation of the fermentation broth can be used to give "distilled spirits" of increased ethanol content. Whiskey is the aged distillate of fermented grain and contains slightly less than 50% ethanol. Brandy and cognac are made by aging the distilled spirits from fermented grapes and other fruits and are about 70% ethanol. The characteristic flavors, odors, and colors of the various alcoholic beverages depend both on their origin and the way they are aged.

Synthetic ethanol is derived from petroleum via ethylene (CH_2=CH_2) by a process to be described in Chapter 5. In the United States, some 1.2 billion pounds of synthetic ethanol is produced annually. It is relatively inexpensive and useful for industrial applications. To render it unfit for drinking, ethanol can be *denatured* by deliberately adding any of a number of noxious materials, thereby exempting it from the high taxes most governments impose upon ethanol used in beverages.

Our bodies are reasonably well equipped to metabolize ethanol, making it less dangerous than methanol. Alcohol abuse and alcoholism have been and remain, however, persistent problems in human societies.

Isopropyl alcohol [$(CH_3)_2CHOH$] is prepared commercially from petroleum via propene (CH_3CH=CH_2) by a process similar to that of ethanol (Chapter 5). With a boiling point of 82°C, isopropyl alcohol evaporates quickly from the skin, producing a cooling effect. Often containing dissolved oils and fragrances, it is the major component of rubbing alcohol. Isopropyl alcohol possesses weak antibacterial properties and is used to maintain medical instruments in a sterile condition and to clean the skin before minor surgery.

Alcohols are among the most readily available organic compounds, both through isolation from natural sources and by synthesis. Because of this availability, alcohols are valuable starting materials for the preparation of a variety of other classes of compounds.

$$CH_3OH \qquad CH_3CH_2OH \qquad \begin{matrix} CH_3CHCH_3 \\ | \\ OH \end{matrix}$$

Methyl alcohol   Ethyl alcohol   Isopropyl alcohol

The common names of alcohols are derived by naming the alkyl group to which the hydroxyl group is attached, then adding the separate word *alcohol*.

Alkyl halides are named in a similar way. After specifying the alkyl group, the halide is identified in a separate word as *fluoride, chloride, bromide,* or *iodide,* as appropriate.

$$CH_3CH_2CH_2F \qquad (CH_3)_3CCl$$

n-Propyl   *tert*-Butyl   Cyclopropyl   Cyclohexyl
fluoride   chloride   bromide   iodide

Alcohols are given systematic IUPAC names by replacing the *-e* ending of the corresponding alkane name by *-ol*. The position of the hydroxyl group is indicated by number, choosing the sequence that assigns the lower locant to the carbon that bears the hydroxyl group.

$$\overset{3}{CH_3}\overset{2}{CH_2}\overset{1}{CH_2}OH \qquad \overset{1}{CH_3}\overset{2}{\underset{\underset{OH}{|}}{CH}}\overset{3}{CH_2}\overset{4}{CH_2}\overset{5}{CH_3} \qquad \overset{1}{CH_3}\overset{2}{CH_2}\overset{3}{\underset{\underset{OH}{|}}{CH}}\overset{4}{CH_2}\overset{5}{CH_3}$$

1-Propanol   2-Pentanol   3-Pentanol

Hydroxyl groups take precedence over alkyl groups in determining the direction in which a carbon chain is numbered.

$$\overset{7}{CH_3}\overset{6}{\underset{\underset{CH_3}{|}}{CH}}\overset{5}{CH_2}\overset{4}{CH_2}\overset{3}{\underset{\underset{OH}{|}}{CH}}\overset{2}{CH_2}\overset{1}{CH_3}$$

6-Methyl-3-heptanol
(not 2-methyl-5-heptanol)   *trans*-2-Methylcyclopentanol

PROBLEM 3.2
Give systematic IUPAC names to all the isomeric $C_4H_{10}O$ alcohols.

Systematic nomenclature of alkyl halides treats the halogen as a substituent on an alkane chain. The carbon chain is numbered in the direction that gives the carbon bearing the halogen substituent the lower locant.

$$\overset{5}{C}H_3\overset{4}{C}H_2\overset{3}{C}H_2\overset{2}{C}H_2\overset{1}{C}H_2F$$

1-Fluoropentane

$$\overset{1}{C}H_3\overset{2}{C}HCH_2\overset{4}{C}H_2\overset{5}{C}H_3$$
$$\underset{Br}{|}$$

2-Bromopentane

$$\overset{1}{C}H_3\overset{2}{C}H_2\overset{3}{C}HCH_2\overset{5}{C}H_3$$
$$\underset{I}{|}$$

3-Iodopentane

When the carbon chain bears both a halogen and an alkyl substituent, the two substituents are considered of equal rank and the chain is numbered so as to give the lower number to the substituent nearer the end of the chain.

$$\overset{1}{C}H_3\overset{2}{C}HCH_2\overset{4}{C}H_2\overset{5}{C}HCH_2\overset{7}{C}H_3$$
$$\underset{CH_3}{|}\qquad\underset{Cl}{|}$$

5-Chloro-2-methylheptane

$$\overset{1}{C}H_3\overset{2}{C}HCH_2\overset{4}{C}H_2\overset{5}{C}HCH_2\overset{7}{C}H_3$$
$$\underset{Cl}{|}\qquad\underset{CH_3}{|}$$

2-Chloro-5-methylheptane

PROBLEM 3.3
Give the systematic names for all the isomeric alkyl chlorides having the molecular formula C_4H_9Cl.

Hydroxyl groups have precedence over halogen substituents in determining the direction of numbering. $FCH_2CH_2CH_2OH$, for example, is 3-fluoro-1-propanol, not 1-fluoro-3-propanol.

3.5 CLASSES OF ALCOHOLS AND ALKYL HALIDES

The reactivity of a substance is often affected by the degree of substitution at the carbon atom which bears the functional group. Alkyl groups are specified as **primary, secondary,** or **tertiary** according to the number of carbons *directly bonded* to the carbon that bears the functional group.

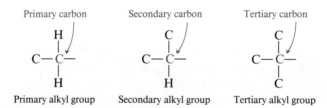

| Primary carbon | Secondary carbon | Tertiary carbon |

Primary alkyl group — Secondary alkyl group — Tertiary alkyl group

The point of attachment of a primary alkyl group originates at a primary carbon, that is, a carbon which is directly bonded to only one other carbon. The point of attachment of a secondary alkyl group originates at a secondary carbon, a carbon which is directly bonded to two other carbons. A tertiary carbon is directly bonded to three other carbons, and the point of attachment of a tertiary alkyl group originates at a tertiary carbon. The terms primary, secondary, and tertiary may be represented by the symbols 1°, 2°, and 3°, respectively.

PROBLEM 3.4
A *quaternary* carbon is one which is directly bonded to four other carbons. Which one of the C_5H_{12} alkanes contains four primary carbons and one quaternary carbon?

Among the $C_4H_{10}O$ alcohols, 1-butanol is a primary alcohol, 2-butanol is a secondary alcohol, and 2-methyl-2-propanol is a tertiary alcohol.

Primary carbon

$$CH_3CH_2CH_2-\overset{\overset{\displaystyle H}{|}}{\underset{\underset{\displaystyle H}{|}}{C}}-OH$$

1-Butanol
(*n*-Butyl alcohol)

Secondary carbon

$$CH_3-\overset{\overset{\displaystyle CH_2CH_3}{|}}{\underset{\underset{\displaystyle H}{|}}{C}}-OH$$

2-Butanol
(*sec*-Butyl alcohol)

Tertiary carbon

$$CH_3-\overset{\overset{\displaystyle CH_3}{|}}{\underset{\underset{\displaystyle CH_3}{|}}{C}}-OH$$

2-Methyl-2-propanol
(*tert*-Butyl alcohol)

Notice that the common names of the *sec*-butyl and *tert*-butyl groups reflect the degree of substitution at the carbon that bears the hydroxyl group.

PROBLEM 3.5
The fourth isomeric $C_4H_{10}O$ alcohol is 2-methyl-1-propanol (isobutyl alcohol). Is 2-methyl-1-propanol a primary, secondary, or tertiary alcohol?

PROBLEM 3.6
The European bark beetle is the insect most responsible for the spread of Dutch elm disease. The beetle bores into the bark of an elm tree and emits an aggregation phero- mone which attracts other beetles to the site. The beetles carry with them a fungus which, if it becomes established and grows uncontrollably, can kill the tree. One component of the aggregation pheromone is 4-methyl-3-heptanol. Write a structural formula for this alcohol. Is it a primary, secondary, or tertiary alcohol?

Alkyl halides are classified in the same way. A tertiary alkyl chloride, for exam- ple, is one in which the carbon atom which bears the chlorine substituent is directly bonded to three other carbons.

3.6 BONDING AND PHYSICAL PROPERTIES OF ALCOHOLS AND ALKYL HALIDES

The carbon that bears the functional group is sp^3 hybridized in alcohols and alkyl halides. The hydroxyl group of an alcohol is attached to carbon by a σ bond generated by overlap of an sp^3 hybrid orbital of carbon with an sp^3 hybrid orbital of oxygen. Figure 3.1 illustrates this for the specific case of bonding in methanol. The bonds to carbon are arranged in a tetrahedral geometry. Bonding in alkyl halides is similar to that of alcohols. The halogen substituent is connected to sp^3 hybridized carbon by a σ bond.

Oxygen and halogen substituents are more electronegative than carbon, so elec- trons in carbon-oxygen and carbon-halogen bonds are drawn away from carbon toward the more electronegative atom. Carbon-oxygen and carbon-halogen bonds are polar bonds (Section 1.4), and alcohols and alkyl halides are polar molecules.

Water   Methanol   Chloromethane

Alkyl halides and alcohols have higher boiling points than alkanes of similar molecular weight, as the following examples demonstrate.

FIGURE 3.1 Orbital hybridization model of bonding in methanol. (*a*) The orbitals used in bonding are the 1*s* orbitals of hydrogen, *sp³* hybridized orbitals of carbon, and *sp³* hybridized orbitals of oxygen. (*b*) The bond angles at carbon and oxygen are close to tetrahedral, and the carbon-oxygen σ bond distance is slightly shorter than that of a carbon-carbon single bond.

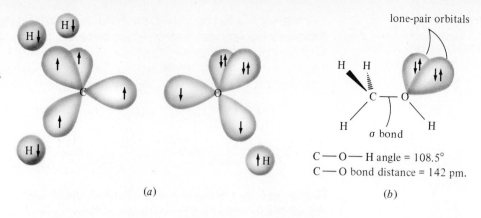

lone-pair orbitals

C — O — H angle = 108.5°
C — O bond distance = 142 pm.

(*a*)

(*b*)

	CH₃CH₂CH₃ Propane	CH₃CH₂F Ethyl fluoride	CH₃CH₂OH Ethyl alcohol
Molecular weight	44	48	46
Boiling point	−42°C	−32°C	78°C

The boiling point of a substance depends on the strength of the forces of attraction between molecules. As described in Section 2.17, intermolecular attractions in alkanes are limited to van der Waals forces and are relatively weak. Van der Waals forces result from temporary dipole-dipole attractions mutually induced between two molecules. Attractive forces between molecules with permanent dipoles are stronger than van der Waals forces, and polar molecules such as alkyl halides have higher boiling points than alkanes. In ethyl fluoride, for example, the positively polarized carbon of one ethyl fluoride molecule is attracted to the negatively polarized fluorine of another. An extended network of these dipole-dipole attractive forces is present in liquid ethyl fluoride, and in order for individual ethyl fluoride molecules to escape from the liquid phase and enter the gas phase, energy (heat) must be added to disrupt this network.

$$\overset{\delta+}{\text{---CH}_2}\text{---}\overset{\delta-}{\overset{..}{\underset{..}{\text{F}}}}\text{:---}\overset{\delta+}{\text{CH}_2}\text{---}\overset{\delta-}{\overset{..}{\underset{..}{\text{F}}}}\text{:---}\overset{\delta+}{\text{CH}_2}\text{---}\overset{\delta-}{\overset{..}{\underset{..}{\text{F}}}}\text{:---}\overset{\delta+}{\text{CH}_2}\text{---}\overset{\delta-}{\overset{..}{\underset{..}{\text{F}}}}\text{:---etc.}$$

$$\underset{\text{CH}_3}{|} \quad \underset{\text{CH}_3}{|} \quad \underset{\text{CH}_3}{|} \quad \underset{\text{CH}_3}{|}$$

Dipole-dipole attractive forces in ethyl fluoride

The positively polarized hydrogen of a hydroxyl (—OH) group is an especially effective participant in interactions of this type. The term **hydrogen bonding** is used to describe dipole-dipole attractive forces involving protons bonded to electronegative elements. The considerably higher boiling point of ethyl alcohol, as compared with propane and ethyl fluoride, for example, is primarily a result of hydrogen bonding. A network of hydrogen bonds between the positively polarized hydrogen and the negatively polarized oxygen of ethyl alcohol molecules must be broken in order to vaporize the liquid.

$$\overset{\delta+}{\text{---H}}\text{---}\overset{\delta-}{\overset{..}{\underset{..}{\text{O}}}}\text{:---}\overset{\delta+}{\text{H}}\text{---}\overset{\delta-}{\overset{..}{\underset{..}{\text{O}}}}\text{:---}\overset{\delta+}{\text{H}}\text{---}\overset{\delta-}{\overset{..}{\underset{..}{\text{O}}}}\text{:---}\overset{\delta+}{\text{H}}\text{---}\overset{\delta-}{\overset{..}{\underset{..}{\text{O}}}}\text{:---etc.}$$

$$\underset{\text{CH}_3\text{CH}_2}{|} \quad \underset{\text{CH}_3\text{CH}_2}{|} \quad \underset{\text{CH}_3\text{CH}_2}{|} \quad \underset{\text{CH}_3\text{CH}_2}{|}$$

Hydrogen bonding in ethyl alcohol

Hydrogen bonding can be expected in molecules that have polar covalent bonds between hydrogen and electronegative elements with unshared electron pairs, especially compounds with O—H and N—H groups. As will be seen in Chapters 16 and 18, the three-dimensional structures adopted by proteins and nucleic acids, the organic molecules of life, are dictated by patterns of hydrogen bonds.

3.7 ACID-BASE PROPERTIES OF ORGANIC MOLECULES

Understanding how theories of acidity and basicity apply to organic substances is an important element in understanding chemical reactivity. Let us therefore review some principles and properties of acids and bases.

According to the theory proposed by Syante Arrhenius, a Swedish chemist and winner of the 1903 Nobel Prize in Chemistry, an acid is any substance that liberates protons (hydrogen ions) in aqueous solution, while a base is any substance that liberates hydroxide ions in aqueous solution.

$$H\overset{\frown}{-}A \rightleftharpoons H^+ + :A^-$$
Arrhenius acid

$$M\overset{\frown}{-}\ddot{O}H \rightleftharpoons M^+ + {}^-:\ddot{O}H$$

Arrhenius base

Note the use of curved arrows here to show the nature of the ionization. A covalent bond in acid HA is broken, with both electrons in that bond becoming an unshared pair of the ion :A$^-$. Similarly, the metal-oxygen bond in the base MOH cleaves, with both electrons becoming an unshared pair of the hydroxide ion.

A more general theory of acids and bases was devised independently by Johannes Brönsted (Denmark) and Thomas M. Lowry (England) in 1923. In the Brönsted-Lowry approach, an **acid** is a **proton donor** and a **base** is a **proton acceptor.**

$$B\overset{\frown}{:} + H\overset{\frown}{-}A \rightleftharpoons \overset{+}{B}-H + :A^-$$
Base   Acid   Conjugate   Conjugate
            acid   base

The curved arrows in this equation show the electron pair of the base abstracting a proton from the acid. The pair of electrons in the H—A covalent bond becomes an unshared electron pair in the conjugate base :A$^-$.

The Brönsted-Lowry definitions are more general than the Arrhenius definition of acids and bases and are more widely used in organic chemistry. We will refer to proton donors and proton acceptors as Brönsted acids and Brönsted bases, respectively. As noted in the above equation, the **conjugate acid** of a substance is formed when it accepts a proton from a suitable donor. Conversely, the proton donor is converted to its **conjugate base.**

PROBLEM 3.7
Write an equation for the reaction of ammonia (:NH$_3$) with hydrogen chloride (HCl). Use curved arrows to track electron movement, and identify the acid, base, conjugate acid, and conjugate base.

In aqueous solution, an acid transfers a proton to water. Water acts as a Brönsted base.

$$\text{HO:} + \text{H—A} \rightleftharpoons \text{HOH} + \text{:A}^-$$

Water	Acid	Conjugate	Conjugate
(base)		acid of water	base

The conjugate acid of water (H_3O^+) is systematically named as the **oxonium ion.** It is more commonly known as the **hydronium ion.**

The strength of an acid in dilute aqueous solution is given by the **equilibrium constant** K for the reaction

$$HA + H_2O \rightleftharpoons H_3O^+ + A^-$$

where

$$K = \frac{[H_3O^+][A^-]}{[HA][H_2O]}$$

But, since water is the solvent and the solution is dilute, the large water concentration (55 M) does not change much. The concentration of water can thus be treated as a constant and incorporated into a new constant K_a called the **acid dissociation (or acidity) constant** or the **acid ionization constant.**

$$K_a = K[H_2O] = \frac{[H_3O^+][A^-]}{[HA]}$$

TABLE 3.3
Acid Dissociation Constants K_a and pK_a Values for Some Brönsted Acids*

Acids	Formula†	Dissociation constant, K_a	pK_a	Conjugate base
Hydrogen iodide	HI	$\sim 10^{10}$	~ -10	I^-
Hydrogen bromide	HBr	$\sim 10^9$	~ -9	Br^-
Hydrogen chloride	HCl	$\sim 10^7$	~ -7	Cl^-
Sulfuric acid	$HOSO_2OH$	1.6×10^5	-4.8	$HOSO_2O^-$
Nitric acid	$HONO_2$	2.5×10^1	-0.6	$^-ONO_2$
Phosphoric acid	$(HO)_2P(O)(OH)$	6×10^{-3}	2.2	$(HO)_2PO_2^-$
Hydrogen fluoride	HF	3.5×10^{-4}	3.5	F^-
Acetic acid	$CH_3\overset{O}{\overset{\|}{C}}OH$	1.8×10^{-5}	4.7	$CH_3\overset{O}{\overset{\|}{C}}O^-$
Water	HOH	1.8×10^{-16}	15.7	HO^-
Methanol	CH_3OH	$\sim 10^{-16}$	~ 16	CH_3O^-
Ethanol	CH_3CH_2OH	$\sim 10^{-16}$	~ 16	$CH_3CH_2O^-$
Isopropyl alcohol	$(CH_3)_2CHOH$	$\sim 10^{-17}$	~ 17	$(CH_3)_2CHO^-$
tert-Butyl alcohol	$(CH_3)_3COH$	$\sim 10^{-18}$	~ 18	$(CH_3)_3CO^-$
Ammonia	H_2NH	$\sim 10^{-36}$	~ 36	H_2N^-

* Acid strength decreases from top to bottom of the table. Strength of conjugate base increases from top to bottom of the table.

† The most acidic proton—the one that is lost on ionization—is indicated in color.

Strong acids are characterized by large values of K_a; essentially every molecule of a strong acid transfers a proton to water in dilute aqueous solution. Weak acids have small K_a values. Table 3.3 lists a number of Bönsted acids and their acid dissociation constants. Most alcohols have K_a's in the range 10^{-16} to 10^{-18}; they are extremely weak acids.

A convenient way to express acid strength is through the use of pK_a, defined as

$$pK_a = -\log_{10} K_a$$

This permits acidity to be expressed in numbers that are not exponentials. Thus, an alcohol with $K_a = 10^{-16}$ has $pK_a = 16$. The stronger the acid, the smaller the value of pK_a. The weaker the acid, the larger the value of pK_a. Table 3.3 includes pK_a as well as K_a values for acids. Both K_a and pK_a values are used extensively as measures of acid strength. Our custom will be to cite K_a but to include pK_a in parentheses.

There is much to be said for being conversant with both.

PROBLEM 3.8

Calculate K_a for each of the following acids, given its pK_a. Rank the compounds in order of decreasing acidity.

    (a) Aspirin: $pK_a = 3.48$
    (b) Vitamin C (ascorbic acid): $pK_a = 4.17$
    (c) Formic acid (present in sting of ants): $pK_a = 3.75$
    (d) Oxalic acid (poisonous substance found in certain berries): $pK_a = 1.19$

SAMPLE SOLUTION

(a) This problem reviews the relationship between logarithms and exponential numbers. We need to determine K_a, given pK_a. The equation that relates the two is

$$pK_a = -\log_{10} K_a$$

Therefore
$$K_a = 10^{-pK_a}$$
$$= 10^{-3.48} = 10^{-4} \times 10^{0.52}$$
$$= 3.3 \times 10^{-4}$$

An important corollary of the Brönsted-Lowry view of acids and bases involves the relative strengths of an acid and its conjugate base. *The stronger the acid, the weaker the conjugate base; the weaker the acid, the stronger the conjugate base.* Referring to Table 3.3, we see that ammonia is a very weak acid, its K_a being only 10^{-36} ($pK_a = 36$). Therefore, an amide anion (NH_2^-) is a very strong base. An ethoxide ion ($CH_3CH_2O^-$) and a methoxide ion (CH_3O^-) are comparable with a hydroxide ion in basicity. Halide ions are very weak bases: a fluoride ion is the strongest base of the halide ions but is 10^{12} times less basic than a hydroxide ion based on a comparison of their K_a's.

In any proton-transfer process the position of equilibrium favors formation of the weaker acid and the weaker base. Thus, when *tert*-butoxide ion is introduced into aqueous solution, the predominant base that is present is the hydroxide ion.

$$(CH_3)_3CO^- \ + \quad H_2O \quad \rightleftharpoons \quad (CH_3)_3COH \ + \quad HO^-$$

tert-Butoxide ion	Water	*tert*-Butyl alcohol	Hydroxide ion
(stronger base)	(stronger acid: $K_a \cong 10^{-16}$)	(weaker acid: $K_a \cong 10^{-18}$)	(weaker base)

PROBLEM 3.9

Write an equation for the reaction between an amide ion and methanol. What species are present in greatest concentration at equilibrium? Identify the various species as acids or bases and specify which are stronger and which are weaker.

3.8 ALCOHOLS AS BRÖNSTED BASES

Several important reactions of alcohols, including their reaction with hydrogen halides, involve treatment of the alcohol with a strong acid. Alcohols are structurally similar to water and can act as Brönsted bases. Just as proton transfer to a water molecule gives an oxonium ion (called the hydronium ion), proton transfer to an alcohol gives an **alkyloxonium ion.**

$$
\overset{H}{\underset{\cdot\cdot}{R\overset{\cdot\cdot}{O}:}} + H{-}A \rightleftharpoons \overset{H}{\underset{+}{R\overset{|}{O}H}} + :A^-
$$

| Alcohol | Acid | Alkyloxonium ion | Conjugate base |

Strong acids such as sulfuric acid and hydrogen halides transfer a proton to the oxygen atom of alcohols in an acid-base reaction.

$$
R\overset{\cdot\cdot}{O}H + H{-}\overset{\cdot\cdot}{\underset{\cdot\cdot}{I}}: \rightleftharpoons R\overset{+}{O}H_2 + :\overset{\cdot\cdot}{\underset{\cdot\cdot}{I}}:^-
$$

| Alcohol | Hydrogen iodide | Alkyloxonium ion | Iodide ion |
| (base) | (acid) | (conjugate acid of ROH) | (conjugate base of HI) |

PROBLEM 3.10

Write an equation for the transfer of one of the protons of sulfuric acid to methyl alcohol. Identify the acid, the base, the conjugate acid, and the conjugate base in your equation. Using curved arrows, show the movement of the electrons.

The conversion to an alkyloxonium ion is usually the first step in the mechanism of acid-promoted reactions of alcohols. The alkyloxonium ion is not isolated but reacts further to be eventually transformed into the observed product.

3.9 PREPARATION OF ALKYL HALIDES

Much of what organic chemists do is directed toward practical goals. Chemists in the pharmaceutical industry, for example, synthesize new compounds as potential drugs for the treatment of diseases. Agricultural chemicals designed to increase crop yields include organic compounds used as herbicides, insecticides, and fungicides. Among the "building block" molecules used as starting materials for the preparation of new substances, alcohols and alkyl halides are especially valuable. How alkyl halides are prepared is described in the following sections, where chemical equations will be used extensively to illustrate the reactions that take place. In those examples, the **percent yield** of isolated product will often be cited. The efficiency of a synthetic transformation is normally expressed as a percentage of the **theoretical yield,** which is the amount of product that could be formed if the reaction proceeded to completion and did not lead to the formation of any products other than those given in the equation. The yield data are taken from experiments reported in the chemical literature and are frequently well below 100 percent. The difference between the theoretical yield and that actually obtained results from competing side reactions that divert the reactants to other products, as well as from mechanical losses of product during its isolation and purification.

The procedures to be described use either an alkane or an alcohol as the starting material for the preparation of an alkyl halide. By knowing how alkyl halides are prepared we can better appreciate the material in succeeding chapters, where alkyl halides are featured prominently in key chemical transformations. The preparation of alkyl halides serves also as a focal point to develop some principles of reaction mechanisms.

3.10 ALKYL HALIDES FROM ALCOHOLS AND HYDROGEN HALIDES

Alcohols react with hydrogen halides according to the general equation shown:

$$\underset{\text{Alcohol}}{ROH} + \underset{\substack{\text{Hydrogen} \\ \text{halide}}}{HX} \longrightarrow \underset{\substack{\text{Alkyl} \\ \text{halide}}}{R} + \underset{\text{Water}}{H_2O}$$

As the examples given below illustrate, the alcohol may be primary, secondary, or tertiary and the hydrogen halide may be hydrogen chloride, hydrogen bromide, or hydrogen iodide.

$$\underset{\substack{\text{2-Methyl-1-propanol} \\ \text{(isobutyl alcohol),} \\ \text{a primary alcohol}}}{(CH_3)_2CHCH_2OH} + \underset{\text{Hydrogen iodide}}{HI} \longrightarrow \underset{\substack{\text{1-Iodo-2-methylpropane} \\ \text{(isobutyl iodide) (64\%)}}}{(CH_3)_2CHCH_2I} + \underset{\text{Water}}{H_2O}$$

$$\underset{\substack{\text{2-Butanol} \\ \text{(sec-butyl alcohol),} \\ \text{a secondary alcohol}}}{\underset{\underset{OH}{|}}{CH_3CHCH_2CH_3}} + \underset{\text{Hydrogen bromide}}{HBr} \longrightarrow \underset{\substack{\text{2-Bromobutane} \\ \text{(sec-butyl bromide) (73\%)}}}{\underset{\underset{Br}{|}}{CH_3CHCH_2CH_3}} + \underset{\text{Water}}{H_2O}$$

$$\underset{\substack{\text{2-Methyl-2-propanol} \\ \text{(tert-butyl alcohol),} \\ \text{a tertiary alcohol}}}{(CH_3)_3COH} + \underset{\text{Hydrogen chloride}}{HCl} \longrightarrow \underset{\substack{\text{2-Chloro-2-methylpropane} \\ \text{(tert-butyl chloride) (78–88\%)}}}{(CH_3)_3CCl} + \underset{\text{Water}}{H_2O}$$

PROBLEM 3.11
Write chemical equations expressing the reaction that takes place between each of the following pairs of reactants:

(a) Cyclohexanol and hydrogen iodide
(b) 3-Ethyl-3-pentanol and hydrogen chloride
(c) 1-Tetradecanol and hydrogen bromide

SAMPLE SOLUTION

(a) An alcohol and a hydrogen halide react to form an alkyl halide and water. In this case, iodocyclohexane was isolated in 79% yield.

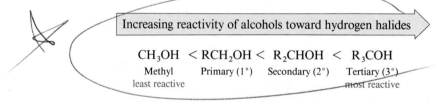

| Cyclohexanol | Hydrogen iodide | Iodocyclohexane (cyclohexyl iodide) | Water |

As noted in Section 3.3, the reaction of an alcohol with a hydrogen halide is a substitution reaction. A halogen—chloride, bromide, or iodide—replaces a hydroxyl group as a substituent on carbon. The reaction is not effective as a method for the preparation of alkyl fluorides.

3.11 MECHANISM OF THE REACTION OF ALCOHOLS WITH HYDROGEN HALIDES

Identifying a reaction as one of substitution simply tells us the relationship between the organic reactant and its product but does not reveal how the reaction takes place. The "how" of a chemical reaction is what we try to explain when we propose a **mechanism** for it. A reaction mechanism is a structural description of the individual reaction steps during the conversion of reactants to products.

A mechanism is developed by combining experimental observations with some basic principles of chemical reactivity. An important observation in the reaction of alcohols with hydrogen halides is that tertiary alcohols are the most reactive, while methyl alcohol is the least reactive.

Increasing reactivity of alcohols toward hydrogen halides

$$CH_3OH \ < \ RCH_2OH < \ R_2CHOH \ < \ R_3COH$$

| Methyl | Primary (1°) | Secondary (2°) | Tertiary (3°) |
| least reactive | | | most reactive |

Consider a specific reaction, the reaction of *tert*-butyl alcohol with hydrogen chloride.

$$(CH_3)_3COH \ + \ \ HCl \ \longrightarrow \ (CH_3)_3CCl \ + H_2O$$

| *tert*-Butyl alcohol | Hydrogen chloride | *tert*-Butyl chloride | Water |

The generally accepted mechanism for this reaction is presented as a series of three steps in Figure 3.2.

We already know something about step 1 of this mechanism; it is a Brönsted acid-base reaction. As the curved arrows indicate, an unshared electron pair of the alcohol oxygen abstracts a proton from hydrogen chloride to give *tert*-butyloxonium chloride. In step 2, the curved arrow depicts cleavage of the carbon-oxygen bond as the oxonium ion dissociates to a molecule of water and a **carbocation,** a species that contains a positively charged carbon. (Carbocations are also sometimes referred to as **carbonium ions.**) In this dissociation, the shared electron pair of the carbon-oxygen bond becomes an unshared electron pair on oxygen, leaving carbon with only six

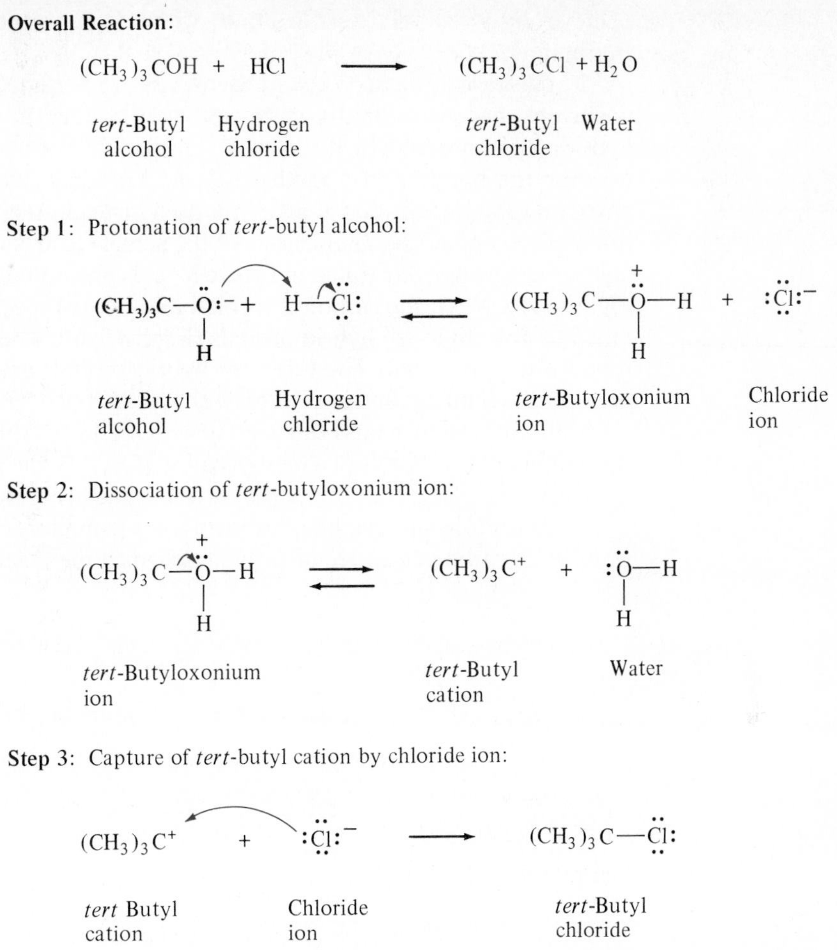

Overall Reaction:

$$(CH_3)_3COH + HCl \longrightarrow (CH_3)_3CCl + H_2O$$

tert-Butyl   Hydrogen   *tert*-Butyl   Water
alcohol   chloride   chloride

Step 1: Protonation of *tert*-butyl alcohol:

$$(CH_3)_3C{-}\overset{..}{\underset{|}{\overset{..}{O}}}{:}^- + \quad H{-}\overset{..}{\underset{..}{Cl}}{:} \quad \rightleftharpoons \quad (CH_3)_3C{-}\overset{+}{\underset{|}{\overset{..}{O}}}{-}H \quad + \quad :\overset{..}{\underset{..}{Cl}}{:}^-$$

$$\text{H} \qquad\qquad\qquad\qquad\qquad\qquad\qquad\qquad\qquad \text{H}$$

tert-Butyl   Hydrogen   *tert*-Butyloxonium   Chloride
alcohol   chloride   ion   ion

Step 2: Dissociation of *tert*-butyloxonium ion:

$$(CH_3)_3C{-}\overset{+}{\underset{|}{\overset{..}{O}}}{-}H \quad \rightleftharpoons \quad (CH_3)_3C^+ \quad + \quad :\overset{..}{O}{-}H$$

$$\text{H} \qquad\qquad\qquad\qquad\qquad\qquad\qquad\qquad \text{H}$$

tert-Butyloxonium   *tert*-Butyl   Water
ion   cation

Step 3: Capture of *tert*-butyl cation by chloride ion:

$$(CH_3)_3C^+ \quad + \quad :\overset{..}{\underset{..}{Cl}}{:}^- \quad \longrightarrow \quad (CH_3)_3C{-}\overset{..}{\underset{..}{Cl}}{:}$$

tert Butyl   Chloride   *tert*-Butyl
cation   ion   chloride

FIGURE 3.2 Sequence of steps that describes the mechanism of formation of *tert*-butyl chloride from *tert*-butyl alcohol and hydrogen chloride.

electrons and a positive charge. The formation of the carbon-chlorine bond of *tert*-butyl chloride by combination of the positively charged carbocation with the negatively charged chloride ion in step 3 completes the process. The curved arrow in step 3 illustrates how an unshared electron pair of chloride ions combine with the positively charged carbon to form a covalent bond.

Both the oxonium ion and the carbocation are said to be **intermediates** in the reaction. They are not isolated but are formed in one step and consumed in another during the passage of reactants to products. Since the formation of a carbocation in step 2 and its reaction in step 3 are critical to the mechanism of Figure 3.2 and because we have not encountered species of this type prior to this point, it is necessary that we now look at some features of their structure, bonding, and stability.

3.12 STRUCTURE AND STABILITY OF CARBOCATIONS

A carbocation is a species that incorporates a positively charged carbon. The structural features of carbocations can be described by considering the simplest example, a methyl cation (CH_3^+). Carbon is bonded to each of its three hydrogen substituents by a two-electron covalent bond. The valence shell of carbon contains only six electrons instead of the full complement of eight. According to the valence-shell electron pair repulsion rationale (Section 1.10), maximum separation of three pairs of electrons is

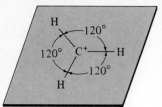

FIGURE 3.3 The electron pairs of the three C—H bonds of a methyl cation are farthest apart in a trigonal planar geometry.

achieved by a trigonal planar geometry with bond angles of 120° at carbon, as shown in Figure 3.3.

In contrast to alkanes where each carbon is bonded to four other atoms, the positively charged carbon in carbocations is bonded to only three atoms. The sp^3 hybridization model which serves well in the case of alkanes is not appropriate to describe the bonding in a methyl cation. An alternative, but related, description based on **sp^2 hybridization** at carbon is used instead. According to the sp^2 hybridization model, one of the $2s$ electrons of the stable electron configuration of C^+ (three valence electrons), illustrated in Figure 3.4a, is promoted to a $2p$ orbital, as shown in Figure 3.4b. Next, the carbon $2s$ orbital and the two occupied $2p$ orbitals are combined to give three sp^2 hybrid orbitals (Figure 3.4c). The third $2p$ orbital of carbon remains unhybridized. The three sp^2 hybrid orbitals are of equal energy and have one-third s character and two-thirds p character. Each sp^2 hybrid orbital is half-filled (contains one electron) and can overlap with the $1s$ orbital of a hydrogen atom to give three equivalent σ bonds of a methyl cation (CH_3^+). The vacant $2p$ orbital is perpendicular to the plane containing the three σ bonds (Figure 3.5).

Carbocations are classified as primary, secondary, or tertiary according to the number of carbons that are directly attached to the positively charged carbon.

$$C—\overset{H}{\underset{H}{C^+}} \qquad C—\overset{H}{\underset{C}{C^+}} \qquad C—\overset{C}{\underset{C}{C^+}}$$

Primary carbocation   Secondary carbocation   Tertiary carbocation

Thus, the *tert*-butyl cation (Figure 3.2) is a tertiary carbocation because the positively charged carbon is directly bonded to three other carbons.

PROBLEM 3.12
Write structural formulas for all the isomeric carbocations of formula $C_4H_9^+$. Which ones are primary carbocations? Which are secondary? Which are tertiary?

FIGURE 3.4
Representation of orbital hybridization of carbon in methyl cation. Each box represents an orbital. An orbital can contain two electrons (■), one electron (◣), or be vacant (□). The stable electron configuration of C^+ is shown in (a). In (b) one of the $2s$ electrons has been raised in energy ("promoted") and occupies a previously vacant $2p$ orbital. In (c) the $2s$ and two of the $2p$ orbitals are mixed to give three equal energy sp^2 hybridized orbitals, each of which contains one electron. One of the $2p$ orbitals remain unhybridized.

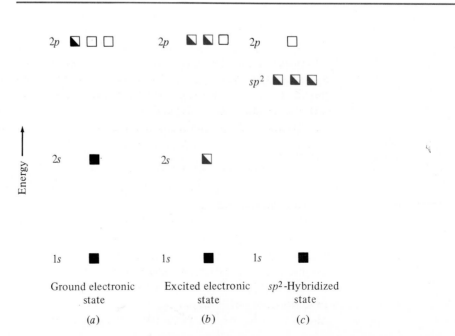

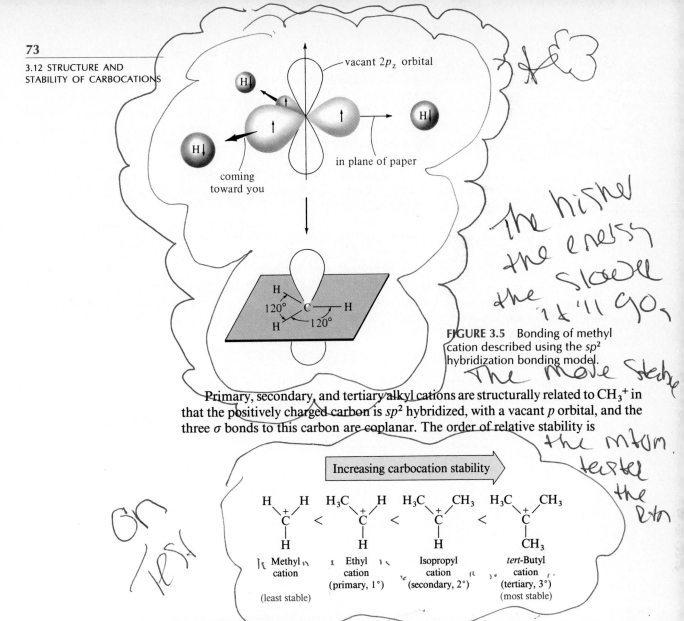

FIGURE 3.5 Bonding of methyl cation described using the sp^2 hybridization bonding model.

The higher the energy the slower it'll go.

Primary, secondary, and tertiary alkyl cations are structurally related to CH_3^+ in that the positively charged carbon is sp^2 hybridized, with a vacant p orbital, and the three σ bonds to this carbon are coplanar. The order of relative stability is

The more stable

the more tester the rxn

on test

Increasing carbocation stability

Methyl cation
(least stable)

< Ethyl cation
(primary, 1°)

< Isopropyl cation
(secondary, 2°)

< *tert*-Butyl cation
(tertiary, 3°)
(most stable)

Carbocations are stabilized by substituents that release or donate electrons to the positively charged carbon. Alkyl groups release electrons better than do hydrogen substituents, so the more alkyl groups that are attached to the positively charged carbon, the more stable the carbocation.

PROBLEM 3.13
Of the isomeric $C_5H_{11}^+$ carbocations, which one is the most stable?

When we say that tertiary carbocations are more stable than their secondary and primary counterparts, understand that we are speaking here in a relative sense. Even tertiary carbocations are highly reactive species, normally not capable of being isolated under conditions of their formation. The effect of carbocation stability is observed in the relative rates at which various alcohols react with hydrogen halides. Tertiary alcohols react faster than secondary alcohols because tertiary carbocations, being more stable than secondary carbocations, are formed faster and the rate of reaction is governed by the rate of carbocation formation.

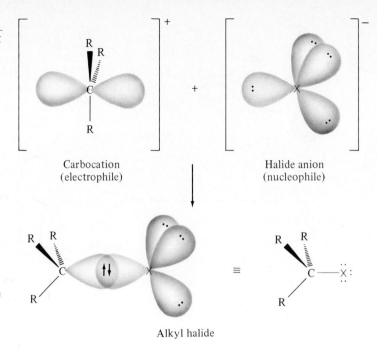

FIGURE 3.6 Combination of a carbocation and a halide anion to give an alkyl halide.

PROBLEM 3.14
Of the isomeric $C_5H_{12}O$ alcohols, which one reacts at the fastest rate with hydrogen chloride?

3.13 ELECTROPHILES AND NUCLEOPHILES

The structural features that dominate the chemistry of carbocations are the positive charge on carbon and the vacant *p* orbital. These features combine to make carbocations strongly **electrophilic** ("electron-loving" or "electron-seeking"). **Nucleophiles** have precisely complementary characteristics. A nucleophile is "nucleus-seeking"; it has an unshared pair of electrons that it can use to form a covalent bond. Step 3 of the mechanism of the reaction of *tert*-butyl alcohol with hydrogen chloride is an example of a reaction between an electrophile and a nucleophile and is depicted from a structural perspective in Figure 3.6. The crucial electronic interaction is between an unshared electron pair of the nucleophilic chloride anion and the vacant $2p$ orbital of the electrophilic carbocation.

A number of years ago G. N. Lewis (Chapter 1) extended our understanding of acid-base behavior to include reactions other than proton transfers. According to Lewis, **an acid is an electron–pair acceptor** and **a base is an electron–pair donor.** Thus, carbocations are electron-pair acceptors and are **Lewis acids.** Halide ions are electron-pair donors and are **Lewis bases.** It is generally true that electrophiles are Lewis acids and nucleophiles are Lewis bases.

3.14 OTHER METHODS FOR CONVERTING ALCOHOLS TO ALKYL HALIDES

Because alkyl halides are useful starting materials for the preparation of a large number of other functional group types, chemists have developed several different methods for the conversion of alcohols to alkyl halides. Two methods, based on the inorganic reagents **thionyl chloride** and **phosphorus tribromide,** bear special mention.

Thionyl chloride reacts with alcohols to give alkyl chlorides. The inorganic by-products in the reaction, sulfur dioxide and hydrogen chloride, are both gases at room temperature and are easily removed, thus facilitating purification of the alkyl halide product.

$$\underset{\text{Alcohol}}{\text{ROH}} + \underset{\substack{\text{Thionyl} \\ \text{chloride}}}{\text{SOCl}_2} \longrightarrow \underset{\substack{\text{Alkyl} \\ \text{chloride}}}{\text{RCl}} + \underset{\substack{\text{Sulfur} \\ \text{dioxide}}}{\text{SO}_2(g)} + \underset{\substack{\text{Hydrogen} \\ \text{chloride}}}{\text{HCl}(g)}$$

$$\underset{\substack{| \\ \text{OH} \\ \text{2-Octanol}}}{\text{CH}_3\text{CH(CH}_2)_5\text{CH}_3} \xrightarrow{\text{SOCl}_2} \underset{\substack{| \\ \text{Cl} \\ \text{2-Chlorooctane (81\%)}}}{\text{CH}_3\text{CH(CH}_2)_5\text{CH}_3} + \text{SO}_2(g) + \text{HCl}(g)$$

Phosphorus tribromide reacts with alcohols to give alkyl bromides and phosphorous acid. Phosphorous acid is water-soluble and may be removed by washing the alkyl halide with water or with a dilute aqueous base.

$$\underset{\text{Alcohol}}{3\text{ROH}} + \underset{\substack{\text{Phosphorus} \\ \text{tribromide}}}{\text{PBr}_3} \longrightarrow \underset{\substack{\text{Alkyl} \\ \text{bromide}}}{3\text{RBr}} + \underset{\substack{\text{Phosphorous} \\ \text{acid}}}{\text{P(OH)}_3}$$

$$\underset{\text{Isobutyl alcohol}}{(\text{CH}_3)_2\text{CHCH}_2\text{OH}} \xrightarrow{\text{PBr}_3} \underset{\text{Isobutyl bromide (55–60\%)}}{(\text{CH}_3)_2\text{CHCH}_2\text{Br}} + \text{P(OH)}_3$$

Thionyl chloride and phosphorus tribromide are specialized reagents used to bring about particular functional group transformations. For this reason, we will not discuss the mechanisms by which they convert alcohols to alkyl halides. Rather, we will limit our coverage to those mechanisms which have broad applicability and thus enhance our knowledge of fundamental principles. In those instances you will find that a mechanistic understanding is of immeasurable help in organizing the reaction types of organic chemistry.

3.15 CHLORINATION OF METHANE

Most alkyl halides are prepared in the laboratory from alcohols as the starting material. There are, however, a number of industrially important alkyl chlorides that are prepared by direct chlorination of alkanes. In the chlorination of alkanes a chlorine replaces hydrogen as a substituent on carbon.

$$\underset{\text{Alkane}}{\text{RH}} + \underset{\text{Chlorine}}{\text{Cl}_2} \xrightarrow[\text{heat}]{\text{light or}} \underset{\text{Alkyl chloride}}{\text{RCl}} + \underset{\text{Hydrogen chloride}}{\text{HCl}}$$

The various *chloro-* derivatives of methane, for example, are prepared in this way.

CH_3Cl	CH_2Cl_2	CHCl_3	CCl_4
Chloromethane (methyl chloride)	Dichloromethane (methylene dichloride)	Trichloromethane (chloroform)	Tetrachloromethane (carbon tetrachloride)

The last three compounds are used as solvents and in paint removers and degreasing agents. Dichloromethane has been used to extract caffeine from coffee and as a propellant in aerosol cans, but recent concerns about its toxicity are causing it to be replaced by other compounds for these purposes.

Methane is the ultimate starting material in the preparation of each compound in the series of chloromethane derivatives. All four hydrogen atoms of methane may be replaced sequentially by reaction with chlorine at elevated temperatures.

$$CH_4 + Cl_2 \xrightarrow{400-440°C} CH_3Cl + HCl$$

Methane · Chlorine · Chloromethane (bp −24°C) · Hydrogen chloride

$$CH_3Cl + Cl_2 \xrightarrow{400-440°C} CH_2Cl_2 + HCl$$

Chloromethane · Chlorine · Dichloromethane (bp 40°C) · Hydrogen chloride

$$CH_2Cl_2 + Cl_2 \xrightarrow{400-440°C} CHCl_3 + HCl$$

Dichloromethane · Chlorine · Trichloromethane (bp 61°C) · Hydrogen chloride

$$CHCl_3 + Cl_2 \xrightarrow{400-440°C} CCl_4 + HCl$$

Trichloromethane · Chlorine · Tetrachloromethane (bp 77°C) · Hydrogen chloride

Indeed, it is difficult to control the chlorination of methane so as to introduce exactly a prescribed number of chlorine atoms. A mixture of chlorinated methane derivatives is obtained, and the various components are then separated by distillation.

3.16 FREE RADICALS

The mechanism of chlorination of methane is fundamentally different from the mechanism by which alcohols react with hydrogen halides. Alcohols are converted to alkyl halides in reactions involving ionic (or "polar") intermediates—oxonium ions and carbocations. The intermediates in the chlorination of methane and other alkanes are quite different; they are neutral ("nonpolar") species called **free radicals.** Before we describe the mechanism of chlorination of methane, let us examine the structure and bonding of free radicals.

Free radicals are characterized by the presence of one or more unpaired electrons. We will discuss only those which have one unpaired electron, examples of which are shown below.

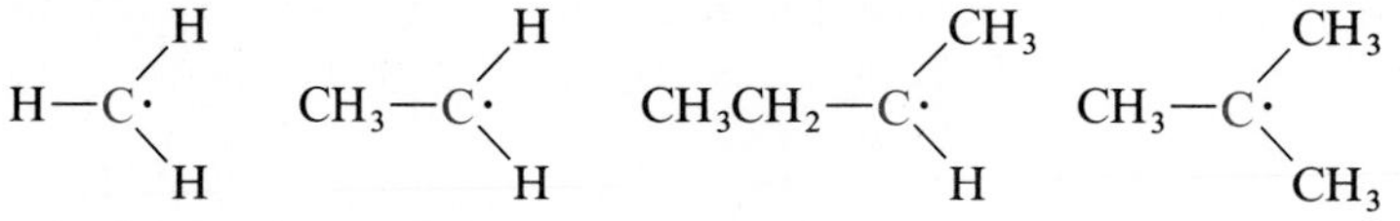

$$H-\overset{H}{\underset{H}{C}}\cdot \qquad CH_3-\overset{H}{\underset{H}{C}}\cdot \qquad CH_3CH_2-\overset{CH_3}{\underset{H}{C}}\cdot \qquad CH_3-\overset{CH_3}{\underset{CH_3}{C}}\cdot$$

Methyl radical · Ethyl radical · sec-Butyl radical · tert-Butyl radical

Alkyl radicals are classified as primary, secondary, or tertiary according to the number of carbon atoms directly attached to the carbon that bears the unpaired electron. Thus, an ethyl radical is a primary alkyl radical, a *sec*-butyl radical is secondary, and a *tert*-butyl radical is tertiary.

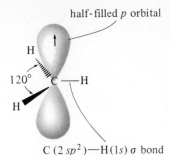

half-filled *p* orbital

H

120° C —H

H

C (2 *sp*²)—H(1*s*) σ bond

FIGURE 3.7 Orbital hybridization model of bonding in a methyl radical. Carbon is *sp*² hybridized with an unpaired electron in the 2*p* orbital.

Bonding in a methyl radical may be approached by comparing it to a methyl cation, bearing in mind that a methyl radical has one more electron than a methyl cation and is electrically neutral. As described in Section 3.12, a methyl cation is planar and its six valence electrons are used to form σ bonds between the *sp*² hybridized carbon and the three hydrogen substituents. This leaves an empty 2*p* orbital on carbon. It is this available 2*p* orbital that is occupied by the single additional electron in a methyl radical. Figure 3.7 depicts this bonding model. The bonding in other alkyl radicals is analogous to a methyl radical; the carbon that bears the unpaired electron is *sp*² hybridized, and the three bonds to this carbon are coplanar (or nearly coplanar).

Like carbocations, free radicals are stabilized by substituents such as alkyl groups that release electrons to carbon. Consequently the order of free radical stability parallels that of carbocations.

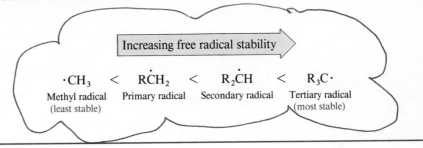

Increasing free radical stability

$\cdot CH_3$ < $R\dot{C}H_2$ < $R_2\dot{C}H$ < $R_3C\cdot$

Methyl radical (least stable)   Primary radical   Secondary radical   Tertiary radical (most stable)

PROBLEM 3.15
Write a structural formula for the most stable of the free radicals having the formula C_5H_{11}.

Since one carbon lacks an octet of electrons, an alkyl free radical is electrophilic. A free radical is not nearly as electrophilic as a carbocation, however, because it has one more electron. Like carbocations, most free radicals are exceedingly reactive species, too reactive to be isolated but capable of being formed as transient intermediates in chemical reactions.

3.17 MECHANISM OF CHLORINATION OF METHANE

The generally accepted mechanism for the chlorination of methane is presented in Figure 3.8. As indicated in the equations in Section 3.15, the reaction is normally carried out at high temperature. The reaction itself is strongly exothermic (heat is evolved), but energy must be put into the system to get the reaction going. This energy goes into breaking the covalent bond that links two chlorine atoms, and the step in which this occurs is called the **initiation step.**

Each chlorine atom formed in the initiation step has seven valence electrons and is a very reactive species. Once formed, a chlorine atom abstracts a hydrogen atom from methane as shown in step 2. Hydrogen chloride, one of the isolated products from the overall reaction, is formed in this step. A methyl radical is also formed which then attacks a molecule of Cl_2 in step 3. Attack of methyl radical on Cl_2 gives chloromethane, the other product of the overall reaction, along with a chlorine atom which then cycles back to step 2, repeating the process. Steps 2 and 3 are called the **propagation steps** of the reaction. Since one initiation step can result in a great many propagation cycles, the overall process is called a **free radical chain reaction.**

In practice, side reactions intervene to reduce the efficiency of the propagation steps. The chain sequence is interrupted whenever two odd-electron species combine to give an even-electron product. Reactions of this type are called **chain-terminating**

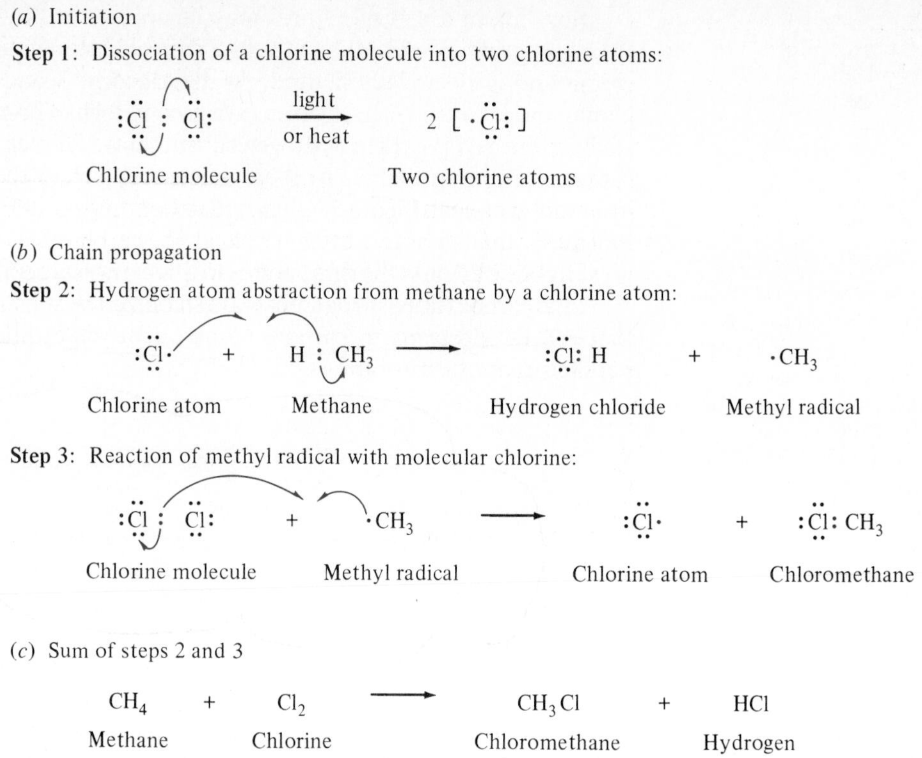

(*a*) Initiation

Step 1: Dissociation of a chlorine molecule into two chlorine atoms:

$$:\!\overset{..}{\underset{..}{Cl}}\!:\overset{..}{\underset{..}{Cl}}\!: \quad \xrightarrow[\text{or heat}]{\text{light}} \quad 2\,[\,\cdot\overset{..}{\underset{..}{Cl}}\!:\,]$$

Chlorine molecule   Two chlorine atoms

(*b*) Chain propagation

Step 2: Hydrogen atom abstraction from methane by a chlorine atom:

$$:\!\overset{..}{\underset{..}{Cl}}\!\cdot \quad + \quad H:CH_3 \quad \longrightarrow \quad :\!\overset{..}{\underset{..}{Cl}}\!:H \quad + \quad \cdot CH_3$$

Chlorine atom   Methane   Hydrogen chloride   Methyl radical

Step 3: Reaction of methyl radical with molecular chlorine:

$$:\!\overset{..}{\underset{..}{Cl}}\!:\overset{..}{\underset{..}{Cl}}\!: \quad + \quad \cdot CH_3 \quad \longrightarrow \quad :\!\overset{..}{\underset{..}{Cl}}\!\cdot \quad + \quad :\!\overset{..}{\underset{..}{Cl}}\!:CH_3$$

Chlorine molecule   Methyl radical   Chlorine atom   Chloromethane

FIGURE 3.8 Equations describing the (*a*) initiation and (*b*) propagation steps in the free radical chain mechanism for the chlorination of methane. Together the two propagation steps give the balanced equation for (c) the overall reaction.

(*c*) Sum of steps 2 and 3

$$CH_4 \quad + \quad Cl_2 \quad \longrightarrow \quad CH_3Cl \quad + \quad HCl$$

Methane   Chlorine   Chloromethane   Hydrogen

steps. The chain-terminating steps that occur during the chlorination of methane are shown in the following equations.

Combination of a methyl radical with a chlorine atom:

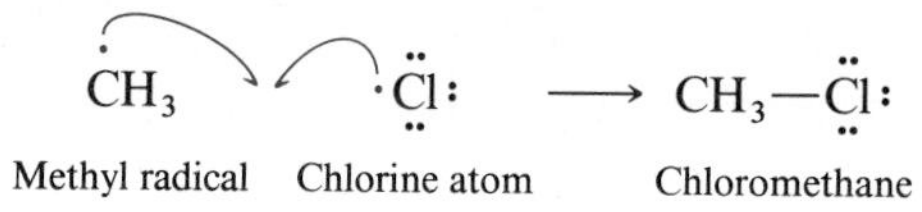

$$\dot{C}H_3 \quad \cdot\overset{..}{\underset{..}{Cl}}\!: \quad \longrightarrow \quad CH_3\!-\!\overset{..}{\underset{..}{Cl}}\!:$$

Methyl radical   Chlorine atom   Chloromethane

Combination of two methyl radicals:

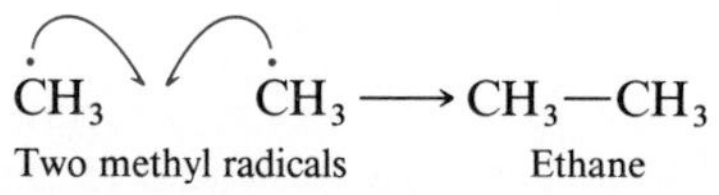

$$\dot{C}H_3 \quad \dot{C}H_3 \longrightarrow CH_3\!-\!CH_3$$

Two methyl radicals   Ethane

Combination of two chlorine atoms:

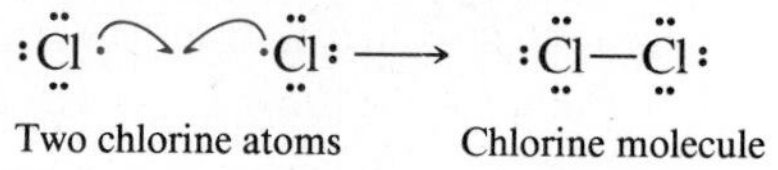

$$:\!\overset{..}{\underset{..}{Cl}} \quad \cdot\overset{..}{\underset{..}{Cl}}\!: \longrightarrow \quad :\!\overset{..}{\underset{..}{Cl}}\!-\!\overset{..}{\underset{..}{Cl}}\!:$$

Two chlorine atoms   Chlorine molecule

3.18 CHLORINATION OF ALKANES

The chlorination of methane, and ethane as well, is carried out on an industrial scale in a high-temperature reaction.

$$CH_3CH_3 + \quad Cl_2 \quad \xrightarrow{420°C} \quad CH_3CH_2Cl \quad + \quad HCl$$

Ethane   Chlorine   Chloroethane (78%)   Hydrogen chloride
(ethyl chloride)

HALOGENATED HYDROCARBONS AND THE ENVIRONMENT

A number of halogenated hydrocarbons have been used for a variety of commercial purposes. Some of these substances have been found to be detrimental to the environment, however, and their use has been restricted and, in some cases, banned. While *chlorofluorocarbons* are widely used as refrigerant gases, they have been found to accumulate in the upper atmosphere, where they cause damage to the earth's ozone layer. Solar irradiation cleaves the carbon-chlorine bond of dichlorodifluoromethane (a typical chlorofluorocarbon), and the resulting chlorine atom attacks ozone.

Dichlorodifluoromethane   Chlorine atom   Chlorodifluoromethyl
                                                           radical

It has been estimated that 100,000 ozone molecules are destroyed by conversion to O_2 in a free radical chain reaction for every chlorine atom formed. The ozone layer absorbs much of the sun's ultraviolet radiation, and the National Academy of Sciences estimates that a 1 percent decrease in ozone concentration could lead to as many as 10,000 additional cases of skin cancer in the United States each year. The use of chlorofluorocarbons as aerosol propellants has been banned in the United States and many other countries. An international agreement signed in 1987 is expected to reduce worldwide production of chlorofluorocarbons by 35 percent by the end of this century.

Until the mid-1970s *polychlorinated biphenyls* (PCBs) were widely used as insulating fluids in electrical equipment. The structure of a typical PCB is shown.

A polychlorinated biphenyl

Waste fluid and discarded equipment caused large amounts of PCBs to enter the environment and wash into rivers, where they were deposited on the river bottom. Because PCBs are very stable, they survive a long time in this environment and are injested by microorganisms. They enter the food chain when small fish eat the microorganisms, large fish eat the small fish, and mammals eat the fish. PCBs are health hazards, having been shown to cause mutations in the offspring of affected individuals. The use of PCBs has been banned in many countries, and the disposal of electrical equipment containing PCBs regulated.

Many pesticides are halogenated hydrocarbons, and some, like DDT, were used on a massive scale.

Dichlorodiphenyltrichloromethane (DDT)

DDT was introduced as a pesticide for mosquitoes in order to control malaria and was exceedingly successful in this role for a period of about 30 years. Eventually, however, resistance to DDT emerged in the pests it was developed to control. When it became apparent in the 1960s that species of birds such as the peregrine falcon and the bald eagle were threatened because DDT that had accumulated in their fatty tissue caused them to lay eggs with thin shells that broke prematurely, the use of DDT became subject to strict regulation.

In the laboratory it is more convenient to use light, either visible or ultraviolet, as the source of energy to break the chlorine-chlorine bond in the initiation step. Reactions that occur when light energy is absorbed by a molecule are called **photochemical reactions.** Photochemical techniques permit the reaction of alkanes with chlorine to be performed at room temperature.

$$\square \quad + \quad Cl_2 \quad \xrightarrow{\text{light}} \quad \square\!-\!Cl \quad + \quad HCl$$

| Cyclobutane | Chlorine | Chlorocyclobutane (73%) (cyclobutyl chloride) | Hydrogen chloride |

Chlorination of alkanes such as butane is more complicated than that of methane, ethane, and cyclobutane in that it yields a mixture of isomeric chlorides.

$$CH_3CH_2CH_2CH_3 \xrightarrow[\text{light, 35°C}]{Cl_2} CH_3CH_2CH_2CH_2Cl + \underset{\underset{Cl}{|}}{CH_3CHCH_2CH_3}$$

| Butane | 1-Chlorobutane (28%) | 2-Chlorobutane (72%) |

Butane has hydrogen substituents in two different environments. Four hydrogens are part of two equivalent methylene (CH_2) groups, and six are part of two equivalent methyl (CH_3) groups. Substitution of a methylene hydrogen gives 2-chlorobutane, while substitution of a methyl hydrogen gives 1-chlorobutane. Chlorination of alkanes is not a very selective process. It is generally observed that a mixture of every possible monochloride is formed from alkanes with nonequivalent hydrogens.

PROBLEM 3.16

Give the structure of the radical that leads to 1-chlorobutane in the photochemical chlorination of butane. Give the structure of the radical that leads to 2-chlorobutane. Classify these radicals as primary, secondary, or tertiary.

PROBLEM 3.17

How many isomeric *monochloro-* derivatives would you expect to obtain in the photochemical reaction of each of the following alkanes with chlorine?

(a) 2-Methylbutane   (b) Pentane   (c) 2-Methylpentane   (d) 2,4-Dimethylpentane

SAMPLE SOLUTION

(a) First examine the structure of 2-methylbutane to determine the number of nonequivalent sets of hydrogen substituents. There are four different hydrogen environments in 2-methylbutane, indicated as (a), (b), (c), and (d) in the structural formula

$$\underset{(a)}{CH_3} - \overset{\overset{\displaystyle (c)}{\overset{\displaystyle H}{|}}}{\underset{\underset{\displaystyle CH_3}{|}}{C}} - \underset{(d)}{CH_2} - \underset{(b)}{CH_3}$$

(a)

Two of the three methyl groups are equivalent, and substitution of any of the six hydrogens indicated as (a) gives 1-chloro-2-methylbutane. Substitution of any of the three hydrogens in the methyl group designated as (b) gives 1-chloro-3-methylbutane. Substitution of the lone tertiary hydrogen (c) gives 2-chloro-2-methylbutane, and substitution at (d) gives 2-chloro-3-methylbutane.

$\underset{\underset{\displaystyle CH_3}{|}}{ClCH_2CHCH_2CH_3}$	$\underset{\underset{\displaystyle CH_3}{|}}{CH_3CHCH_2CH_2Cl}$	$\underset{\underset{\displaystyle CH_3}{|}}{\overset{\overset{\displaystyle Cl}{|}}{CH_3CCH_2CH_3}}$	$\underset{\underset{\displaystyle CH_3}{|}}{\overset{\overset{\displaystyle Cl}{|}}{CH_3CHCHCH_3}}$
1-Chloro-2-methylbutane	1-Chloro-3-methylbutane	2-Chloro-2-methylbutane	2-Chloro-3-methylbutane

3.19 SUMMARY

Chemical reactivity is the unifying theme of this chapter. Many chemical reactions involve the conversion of one *functional group* to another. A functional group is a structural unit of a molecule that undergoes chemical change under a given set of reaction conditions. The reactions discussed in this chapter convert the *hydroxyl* (—OH) functional group of an alcohol (ROH) to the *halogen* (—X = —Cl, —Br, —I) functional group of an alkyl halide (R—X) (Sections 3.10 to 3.14) and the *hydrogen* (—H) functional group of an alkane (R—H) to the chlorine (—Cl) functional group of an alkyl chloride (R—Cl) (Sections 3.15 to 3.18). Examples of these reactions are presented in Table 3.4.

Alcohols and alkyl halides are named systematically as functional derivatives of alkanes, but in slightly different ways (Section 3.4). Alkyl halides are named as halo-substituted alkanes. Alcohols are named by replacing the -*e* ending of an alkane with -*ol*. In both types of compounds the hydrocarbon chain is numbered in the direction that gives the lower number to the carbon that bears the functional group.

$$\underset{\underset{\displaystyle OH}{|}}{CH_3CH_2CH_2CH_2CHCH_3} \qquad \underset{\underset{\displaystyle Br}{|}}{CH_3CH_2CH_2CH_2CHCH_3}$$

2-Hexanol   2-Bromohexane

Alcohols and alkyl halides may also be named on the basis of the common names of their alkyl groups. The alkyl group is named followed by the words *alcohol* or *fluoride, chloride, bromide,* or *iodide* as appropriate.

$$\underset{\underset{\displaystyle OH}{|}}{CH_3CHCH_3} \qquad \underset{\underset{\displaystyle Br}{|}}{CH_3CHCH_3}$$

Isopropyl alcohol   Isopropyl bromide

The reaction of an alcohol with a hydrogen halide proceeds by way of two charged intermediates, an *oxonium ion* and a *carbocation.* The oxonium ion is

TABLE 3.4

Conversions of Alcohols and Alkanes to Alkyl Halides

Reaction (section) and comments	General equation and specific example(s)
Reactions of alcohols with hydrogen halides (Section 3.10) Alcohols react with hydrogen halides to yield alkyl halides. The reaction is useful as a synthesis of alkyl halides. Alcohol reactivity decreases in the order tertiary > secondary > primary > methyl.	$ROH + HX \longrightarrow R + H_2O$ Alcohol Hydrogen halide Alkyl halide Water 1-Methylcyclopentanol 1-Chloro-1-methylcyclopentane (96%)
Reaction of alcohols with thionyl chloride (Section 3.14) Thionyl chloride is a synthetic reagent used to convert alcohols to alkyl chlorides.	$ROH + SOCl_2 \longrightarrow RCl + SO_2 + HCl$ Alcohol Thionyl chloride Alkyl chloride Sulfur dioxide Hydrogen chloride $CH_3CH_2CH_2CH_2CH_2OH \xrightarrow{SOCl_2} CH_3CH_2CH_2CH_2CH_2Cl$ 1-Pentanol 1-Chloropentane (80%)
Reaction of alcohols with phosphorus tribromide (Section 3.14) As an alternative to converting alcohols to alkyl bromides with hydrogen bromide, the inorganic reagent phosphorus tribromide is sometimes used.	$3ROH + PBr_3 \longrightarrow 3RBr + P(OH)_3$ Alcohol Phosphorus tribromide Alkyl bromide Phosphorous acid $\underset{\underset{OH}{\mid}}{CH_3CHCH_2CH_2CH_3} \xrightarrow{PBr_3} \underset{\underset{Br}{\mid}}{CH_3CHCH_2CH_2CH_3}$ 2-Pentanol 2-Bromopentane (67%)
Free radical chlorination of alkanes (Sections 3.15 to 3.18) Alkanes react with halogens by substitution of a halogen for a hydrogen on the alkane. Chlorination is not very selective and so is only used when all the hydrogens of the alkane are equivalent.	$RH + X_2 \longrightarrow RX + HX$ Alkane Halogen Alkyl halide Hydrogen halide Cyclodecane Cyclodecyl chloride (64%)

formed by transfer of a proton from the hydrogen halide to the alcohol. The hydrogen halide acts as a *Brönsted acid* (Section 3.7), while the alcohol acts as a *Brönsted base* (Section 3.8). The carbocation is formed by dissociation of an oxonium ion and is then captured by a halide ion to give an alkyl halide. Steps 1 to 3 describe the *mechanism* for the reaction of an alcohol with a hydrogen halide (Section 3.11).

1. $ROH + HX \longrightarrow R\overset{+}{O}H_2 + X^-$
   Alcohol   Hydrogen halide   Oxonium cation   Halide anion

2. $R\overset{+}{O}H_2 \longrightarrow R^+ + H_2O$
   Oxonium cation   Carbocation   Water

3. $R^+ + X^- \longrightarrow RX$
   Carbocation   Halide anion   Alkyl halide

A different kind of intermediate, a *free radical,* is involved in the chlorination of alkanes. Steps 1 to 3 describe a *free radical chain* mechanism for the reaction of an alkane with chlorine (Section 3.17).

1. Initiation step:

$$Cl_2 \longrightarrow 2[\cdot \ddot{Cl}\colon]$$

Chlorine   Two chlorine
molecule   atoms

2. Propagation step:

$$RH + \cdot \ddot{Cl}\colon \longrightarrow R\cdot + HCl$$

Alkane   Chlorine   Alkyl   Hydrogen
      atom   radical   chloride

3. Propagation step:

$$R\cdot + Cl_2 \longrightarrow RCl + \cdot \ddot{Cl}\colon$$

Alkyl   Chlorine   Alkyl   Chlorine
radical   molecule   chloride   atom

The combination of any two radical species is a *termination step* that interrupts the chain process.

Termination step $P\cdot + Q\cdot \longrightarrow P—Q$

Both carbocations and carbon free radicals are characterized by the presence of a carbon atom that has only three atoms or groups bonded to it. Such carbon atoms are sp^2 hybridized, and the three σ orbitals are coplanar. Carbocations are positively charged and have a vacant $2p$ orbital whose axis is perpendicular to the plane of the three σ orbitals. Carbocations are strongly *electrophilic* and thus act as *Lewis acids* (Section 3.13). Carbocations react with *nucleophiles,* substances with an unshared electron pair available for covalent bond formation. Nucleophiles are therefore *Lewis bases.* Free radicals are neutral and have one electron in their $2p$ orbital.

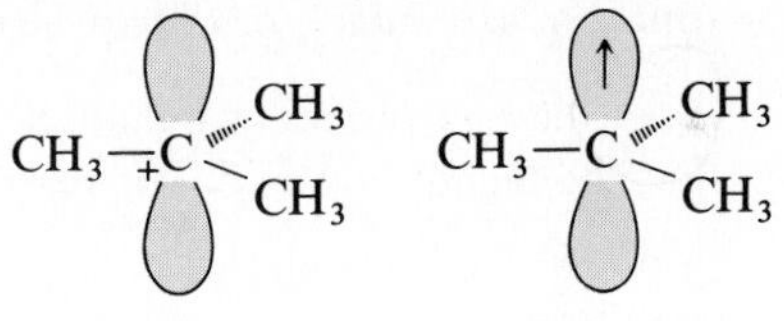

tert-Butyl cation   *tert*-Butyl radical

ADDITIONAL PROBLEMS

Structure and Nomenclature

3.18 Write structural formulas for each of the following compounds:

(a) Cyclobutanol
(b) 3-Heptanol
(c) *trans*-2-Chlorocyclopentanol
(d) 4-Methyl-2-hexanol
(e) 1-Bromo-3-iodobutane
(f) 2,6-Dichloro-4-methyl-4-octanol

3.19 Classify each halide and alcohol group in Problem 3.18 as primary, secondary, or tertiary.

3.20 Give a systematic IUPAC name for each of the following:

(a) $(CH_3)_2CHCH_2CH_2CH_2OH$
(b) $Cl_2CHCHBr$
         |
        Cl

(c) CF$_3$CH$_2$OH

(d)

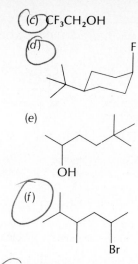

(e)

(f)

3.21 Although fluorinated hydrocarbons (fluorocarbons) are no longer used as aerosol propellants in the United States, other uses are still permitted. Assign a systematic name to each of the compounds listed below.

    (a) F$_3$CCHBrCl (halothane, an anesthetic)

    (b) ClCF$_2$CF$_2$Cl (Freon 114, a refrigerant)

    (c)

$$\begin{array}{c} \text{F} \quad \text{F} \\ \text{F} - \!\!\!+\!\!\!-\!\!\!+\!\!\! - \text{F} \\ \text{F} - \!\!\!+\!\!\!-\!\!\!+\!\!\! - \text{F} \\ \text{F} \quad \text{F} \end{array}$$ A refrigerant

3.22 Write structural formulas for all the constitutionally isomeric alcohols of molecular formula C$_5$H$_{12}$O. Assign a systematic IUPAC name to each one and specify whether it is a primary, secondary, or tertiary alcohol.

3.23 Both 1-propanol and ethyl chloride have similar molecular weights. Which of these has the higher boiling point? Why?

3.24 The molecular weight of hydrogen fluoride is the same as the atomic weight of neon (20). One of these substances has a boiling point of $-246°$C, the other a boiling point of $19°$C. Match the substance with its boiling point and explain the reason for your choice.

Acid–Base

3.25 Write the structure of the conjugate acid and the conjugate base of:

    (a) Water   (b) Methanol   (c) Ammonia

3.26 For each of the following pairs of ions, identify which is the stronger base:

    (a) HO$^-$ or NH$_2^-$ (c) CH$_3$CH$_2$O$^-$ or CH$_3\overset{\displaystyle O}{\overset{\|}{C}}O^-$

    (b) F$^-$ or Cl$^-$ (d) HSO$_4^-$ or NO$_3^-$

3.27 Each of the following pairs of compounds undergoes a Brönsted acid-base reaction whose equilibrium constant is greater than unity. Give the products of each reaction and identify the acid, the base, the conjugate acid, and the conjugate base. Use the acid dissociation constants in Table 3.3 as a guide.

    (a) HI + HO$^-$ $\rightleftharpoons$

    (b) CH$_3$CH$_2$O$^-$ + CH$_3\overset{\displaystyle O}{\overset{\|}{C}}$OH $\rightleftharpoons$

    (c) HF + H$_2$N$^-$ $\rightleftharpoons$

(d) $CH_3\overset{\displaystyle O}{\overset{\|}{C}}O^- + HCl \rightleftharpoons$

3.28 For each of the following acid-base reactions, indicate whether the reaction favors products ($K > 1$) or reactants ($K < 1$). Refer to Table 3.3 for the relevant acid dissociation constants.

(a) $(CH_3)_3CO^- + H_2O \rightleftharpoons$

(b) $NH_3 + (CH_3)_2CHO^- \rightleftharpoons$

(c) $HF + HO^- \rightleftharpoons$

3.29 (a) The acid dissociation constant of benzoic acid (a carboxylic acid) is 6.3×10^{-5}. What is the pK_a?

(b) The pK_a of chloroacetic acid is 2.9. Which acid is stronger, chloroacetic or benzoic?

(c) The conjugate base of chloroacetic acid is the chloroacetate ion; that of benzoic acid is the benzoate ion. Which conjugate base is stronger, chloroacetate or benzoate?

3.30 The approximate values for the pK_a of CH_4 and CH_3OH are 60 and 16, respectively.

(a) Label the stronger and weaker acids and bases in the following equation:

$$CH_3O^- + CH_4 \rightleftharpoons CH_3^- + CH_3OH$$

(b) Is the value of the equilibrium constant K greater or less than 1 for this system?

Alcohol Reactions; Carbocations

3.31 Write a chemical equation for the reaction of 1-butanol with each of the following:

(a) Hydrogen bromide, heat

(b) Thionyl chloride

(c) Phosphorus tribromide

3.32 Repeat Problem 3.31 for the corresponding reactions of cyclohexanol.

3.33 Select the compound in each of the following pairs that will exhibit the greater reactivity on being treated with hydrogen bromide. Explain the reason for your choice.

(a) 1-Butanol or 2-butanol.

(b) 2-Methyl-2-butanol or 2-butanol

(c) 2-Methylbutane or 2-butanol

3.34 Show the structure of the carbocation formed from the reaction of each of the following alcohols with hydrogen chloride. Rank the cations in order of decreasing stability.

$$(CH_3)_2CHCH_2CH_2OH \qquad (CH_3)_2\underset{\underset{\displaystyle OH}{|}}{C}CH_2CH_3 \qquad (CH_3)_2CH\underset{\underset{\displaystyle OH}{|}}{C}HCH_3$$

3.35 Which among the following carbocations is the most stable? Which is the least stable?

3.36 Experimental evidence indicates that cyclopropyl cation is much less stable than isopropyl cation. Why?

Alkane Chlorination; Free Radicals

3.37 Write structural formulas for all the isomeric alkyl free radicals of formula C_4H_9. Which one is the most stable?

3.38 Write the structures of all the possible monochlorides formed during the photochemical chlorination of $(CH_3)_3CCH_2CH_3$.

3.39 Among the isomeric alkanes of molecular formula C_5H_{12}, identify the one that on reaction with chlorine in the presence of light yields only:

(a) A single monochloride

(b) Three isomeric monochlorides

(c) Four isomeric monochlorides

(d) Two isomeric dichlorides

3.40 Write out the steps describing the mechanism of the free radical chlorination of ethane. Clearly identify the initiation, propagation, and termination steps of the process.

3.41 Chlorination of ethane yields, in addition to the major monochlorination product, a mixture of two isomeric dichlorides. What are the structures of these two dichlorides?

CHAPTER 4

ALKENES, ALKADIENES, AND ALKYNES
I. STRUCTURE AND PREPARATION

LEARNING OBJECTIVES

The topic of alkene, alkadiene, and alkyne chemistry is sufficiently broad to warrant two chapters. The reactions of hydrocarbons that contain carbon-carbon double and triple bonds are presented in Chapter 5, while the present chapter describes the nomenclature, formation, and structural features of these compounds. Upon completion of Chapter 4 you should be able to:

- Write a structural formula for an alkene, alkadiene, or alkyne on the basis of its systematic IUPAC name.

- Write a correct IUPAC name for an alkene, alkadiene, or alkyne on the basis of a given structural formula.

- Write a chemical equation expressing the formation of an alkene by dehydration of an alcohol.

- Write a chemical equation expressing the formation of an alkene by dehydrohalogenation of an alkyl halide.

- State and give an example of Zaitsev's rule.

- Describe the E2 mechanism of dehydrohalogenation, clearly identifying the role of the base, and show the bonding changes which occur at the transition state.

- Describe the E1 mechanism for dehydrohalogenation of alkyl halides.

- Recognize alkenes capable of existing in stereoisomeric forms and identify these forms as cis or trans as appropriate.

- Compare isomeric alkenes with respect to their relative stability according to the degree of substitution and stereochemistry at the double bond.

- Describe the sp^2 hybridization model for bonding in alkenes.

- Explain the structural differences between conjugated, cumulated, and isolated double bonds.

- Describe the sp hybridization model for bonding in alkynes.

■ Write a chemical equation for the formation of an alkyne by the dehydrohalogenation of a geminal or vicinal dihalide.

Alkenes were described in Section 3.1 as hydrocarbons that contain a carbon-carbon double bond and **alkynes** as hydrocarbons that contain a carbon-carbon triple bond. Alkenes are characterized by the molecular formula C_nH_{2n} and alkynes by the molecular formula C_nH_{2n-2}. The two-carbon hydrocarbons ethylene (C_2H_4) and acetylene (C_2H_2) are the simplest alkene and alkyne, respectively.

Ethylene   Acetylene

Carbon-carbon double bonds and triple bonds are important structural units and functional groups in organic chemistry. A double bond or triple bond affects the shape of a molecule and is the site of some interesting and useful reactions. This chapter and the next introduce the principal features of alkene and alkyne chemistry. The present chapter describes the structure, bonding, and synthesis of alkenes and alkynes, while the reactions of these two hydrocarbon classes are presented in Chapter 5.

Aspects of the chemistry of **alkadienes,** alkenes that contain two double bonds, are also included in these two chapters, where it will be seen that the presence of two alkene functions appropriately positioned in a molecule can lead to properties which are qualitatively different from those of individual double bonds.

4.1 ALKENES

4.1.1 Alkene Nomenclature

Alkenes are named in the IUPAC system by replacing the **-ane** ending in the name of the corresponding alkane with **-ene.** The longest continuous chain that includes the double bond forms the base name of the alkene. Thus, the two simplest alkenes are named **ethene** and **propene.** The IUPAC rules permit the common name **ethylene** to be used as a synonym for ethene. Propylene, butylene, and other common *-ylene* names are not acceptable IUPAC names. The chain is numbered in the direction that gives the doubly bonded carbons their lower numbers. The locant (or numerical position) of only one of the doubly bonded carbons is specified in the name; it is understood that the other doubly bonded carbon must follow in sequence. No locants are required for ethene and propene.

$$CH_2{=}CH_2 \qquad CH_2{=}CHCH_3 \qquad \overset{1}{C}H_2{=}\overset{}{C}H\overset{3}{C}H_2\overset{4}{C}H_3$$

Ethene   Propene   1-Butene
(Ethylene)   (not 1,2-butene)

Alkyl substituents are identified in the usual way and the chain numbered in the direction that gives the lower number to the doubly bonded carbons.

2-Methyl-2-butene

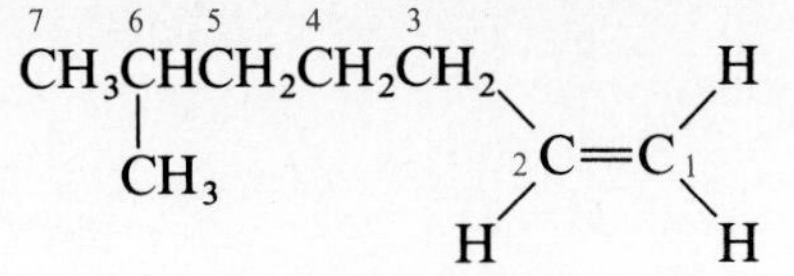

6-Methyl-1-heptene
(not 2-methyl-6-heptene)

PROBLEM 4.1
Name the following alkenes using systematic IUPAC nomenclature:

(a) $(CH_3)_2C{=}C(CH_3)_2$

(b) $(CH_3)_3CCH{=}CH_2$

(c) $(CH_3)_2C{=}CHCH_2CH_2CH_3$

(d) $(CH_3)_2C{=}CHC(CH_3)_3$

(e) $(CH_3)_2CH$
$\quad\quad\quad\quad\ C{=}CHCH_2CH(CH_3)_2$
$(CH_3)_2CH$

SAMPLE SOLUTION
(a) The longest continuous chain in this alkene contains four carbon atoms. The double bond is between C-2 and C-3, so it is named as a derivative of 2-butene.

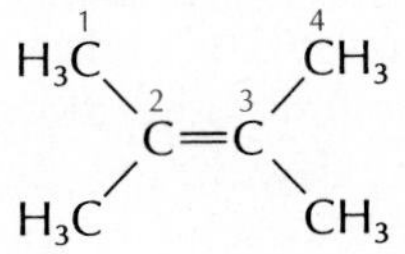

2,3-Dimethyl-2-butene

Identifying the alkene as a derivative of 2-butene leaves two methyl groups to be accounted for as substituents attached to the main chain. This alkene is 2,3-dimethyl-2-butene. (It is sometimes called tetramethylethylene, but that is a common name, not a systematic IUPAC name.)

The groups $CH_2{=}CH{-}$ (**vinyl**) and $CH_2{=}CHCH_2{-}$ (**allyl**) are two frequently encountered **alkenyl groups.** You should familiarize yourself with their structures so that you can readily associate their names with their structures.

$CH_2{=}CHCl$ $\quad$ $CH_2{=}CHCH_2OH$
Vinyl chloride $\quad$ Allyl alcohol

Cycloalkenes and their derivatives are named by adapting cycloalkane terminology to the principles of alkene nomenclature.

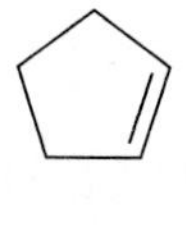

Cyclopentene 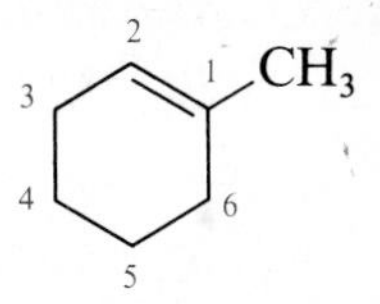 $\quad$ 1-Methylcyclohexene 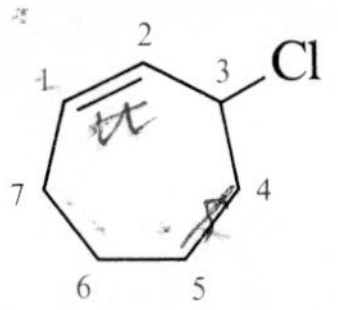 $\quad$ 3-Chlorocycloheptene
(not 1-chloro-2-cycloheptene)

No locants are needed in the absence of substituents; it is understood that the double bond connects C-1 and C-2. Substituted cycloalkenes are numbered beginning with

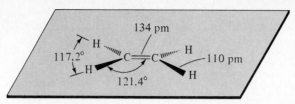

FIGURE 4.1 All the atoms of ethylene lie in the same plane. Bond angles are close to 120°, and the carbon-carbon bond distance is significantly shorter than that of ethane.

the double bond, proceeding through it, and continuing in sequence around the ring. The direction of numbering is chosen so as to give the lower of two possible locants to the substituent group.

PROBLEM 4.2

Write structural formulas and give the IUPAC names of all the monochloro-substituted derivatives of cyclopentene.

4.1.2 Structure and Bonding in Ethylene

As shown in Figure 4.1, ethylene is a planar molecule with a carbon-carbon bond distance (134 pm) that is significantly shorter than that of ethane (153 pm). The bond angles are close to 120°.

While both carbons of ethylene have four bonds, each carbon is directly attached to only three atoms. As we have seen in our discussion of bonding in the methyl cation (CH_3^+, Section 3.12) and the methyl radical ($\cdot CH_3$, Section 3.16), a carbon which is bonded to three other atoms or groups is sp^2 hybridized. Thus, as depicted in Figure 4.2, we can view the two carbon atoms of ethylene as being connected by a σ bond generated by overlap of an sp^2 hybridized orbital of one carbon with the corresponding sp^2 hybridized orbital of the other. The unhybridized $2p$ orbital on one carbon overlaps with its counterpart on the other carbon to give an orbital containing two electrons corresponding to the second component of the carbon-carbon double bond. The orbital generated by this "side-by-side" $p - p$ overlap is called a π (pi) orbital, and the bond formed by a sharing of two electrons in a π orbital is called a **π bond.** According to the sp^2 hybridization model, the carbon-carbon double bond in ethylene has two components, a σ bond and a π bond.

4.1.3 Bonding in Propene and Higher Alkenes

The C—H bonds in ethylene are σ bonds formed by $C(sp^2)$—$H(1s)$ overlap. Higher alkenes are related to ethylene by replacement of one or more of its hydrogen substituents by alkyl groups. Thus, there are two different types of carbon-carbon bonds in propene, $CH_3CH{=}CH_2$; the double bond is of the $\sigma + \pi$ type, while the methyl group is attached by a σ bond formed by sp^3-sp^2 overlap.

sp^3 hybridized carbon

Propene

sp^2 hybridized carbon

PROBLEM 4.3

We can use carbon-skeleton formulas to represent alkenes in much the same way that we use them to represent alkanes. Consider the following alkene:

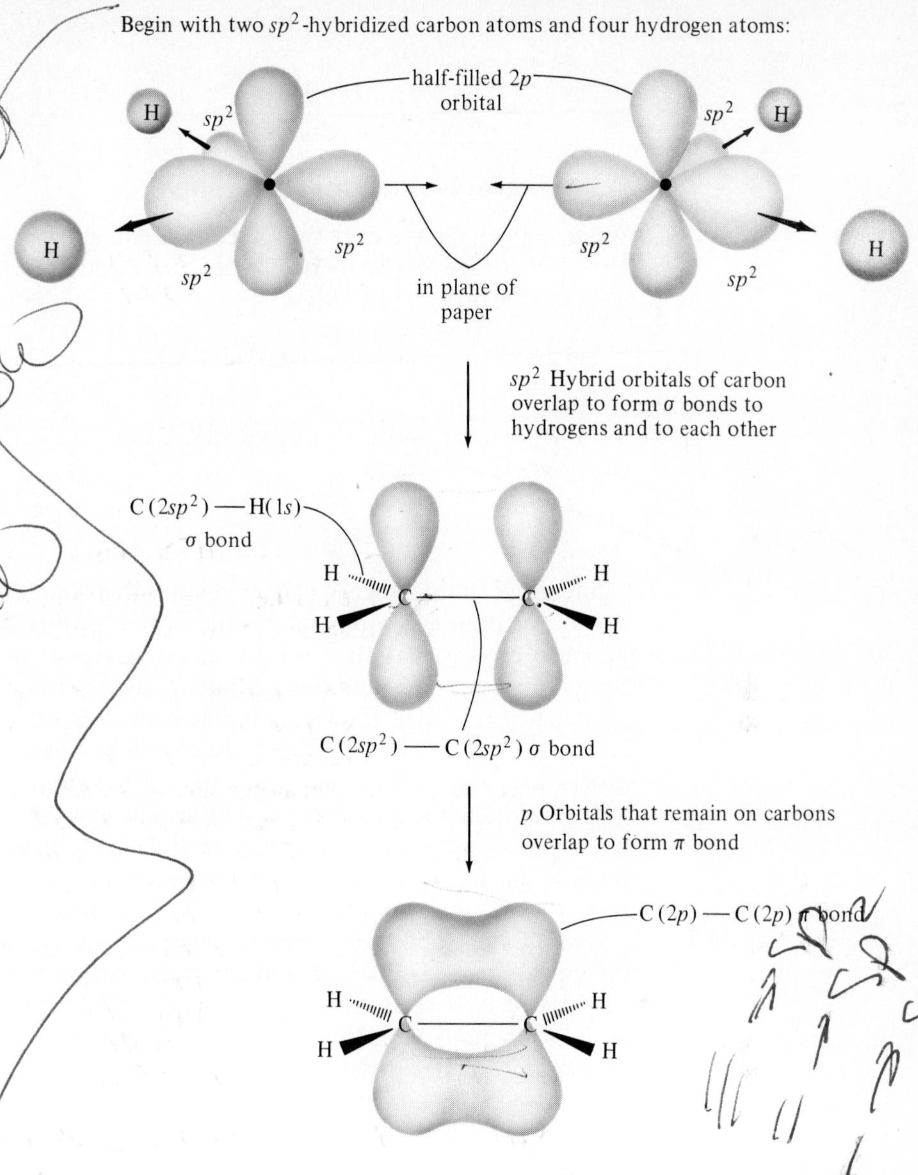

Begin with two sp^2-hybridized carbon atoms and four hydrogen atoms:

sp^2 Hybrid orbitals of carbon overlap to form σ bonds to hydrogens and to each other

p Orbitals that remain on carbons overlap to form π bond

FIGURE 4.2 Orbital hybridization model of bonding in ethylene.

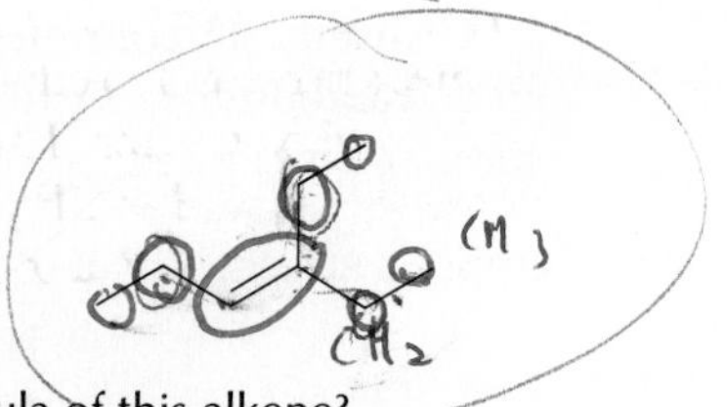

(a) What is the molecular formula of this alkene?
(b) What is its IUPAC name?
(c) How many carbon atoms are sp^2 hybridized in this alkene? How many are sp^3 hybridized?
(d) How many of the σ bonds in this substance are of the sp^2-sp type? How many are of the sp^3-sp^3 type?

SAMPLE SOLUTION

(a) Recall when writing carbon-skeleton formulas for hydrocarbons that there is a carbon at each end and at each bend in a carbon chain. The appropriate number of hydrogen substituents is attached so that each carbon has four bonds. Thus the compound shown may also be written as

$$CH_3CH_2CH{=}C(CH_2CH_3)_2$$

and its molecular formula is C_8H_{16}.

4.1.4 Stereoisomerism in Alkenes

While ethylene is the only two-carbon alkene and propene the only three-carbon alkene, there are *four* isomeric alkenes of molecular formula C_4H_8.

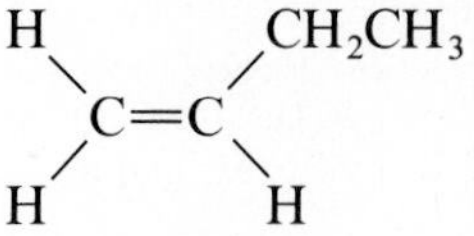

1-Butene
bp (1 atm) −6°C

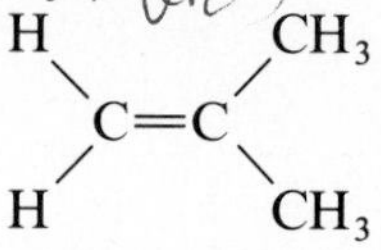

2-Methylpropene
bp (1 atm) −7°C

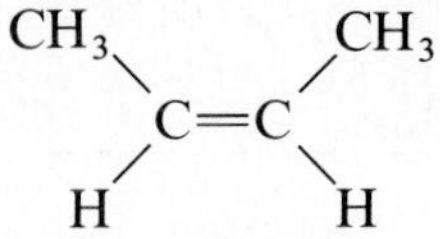

cis-2-Butene
bp (1 atm) 4°C

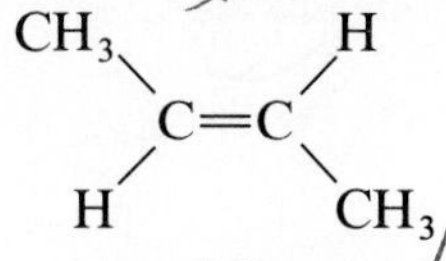

trans-2-Butene
bp (1 atm) 1°C

1-Butene has an unbranched carbon chain with a double bond between C-1 and C-2. It is a constitutional isomer of the other three. Similarly, 2-methylpropene with a branched carbon chain is a constitutional isomer of the other three. The pair of isomers designated *cis*- and *trans*-2-butene have the same constitution; both have an unbranched carbon chain with a double bond connecting C-2 and C-3. They differ from one another, however, in that the cis isomer has both its methyl groups on the same side of the double bond, while the methyl groups in the trans isomer are on opposite sides of the double bond. The arrangement of their atoms in space is what makes *cis*-2-butene and *trans*-2-butene different from each other. Recall from Section 2.15 that isomers which have the same constitution but differ in the arrangement of their atoms in space are classified as *stereoisomers.*

 cis-2-Butene and *trans*-2-butene are related by rotation about their carbon-carbon double bond. Unlike rotation about the carbon-carbon single bond in butane, which as we have seen in Section 2.5 is quite fast, interconversion of the stereoisomeric 2-butenes is exceedingly slow—so slow as to not be observable under normal circumstances. We say that rotation about a carbon-carbon double bond is *restricted.* As shown in Figure 4.3, rotation about a double bond would require twisting the *p* orbitals from their stable parallel alignment, in effect breaking the π component of the double bond.

 Cis-trans stereoisomerism in alkenes is not possible when one of the doubly bonded carbons bears two identical substituents. Thus, neither 1-butene nor 2-methylpropene is capable of existing in stereoisomeric forms.

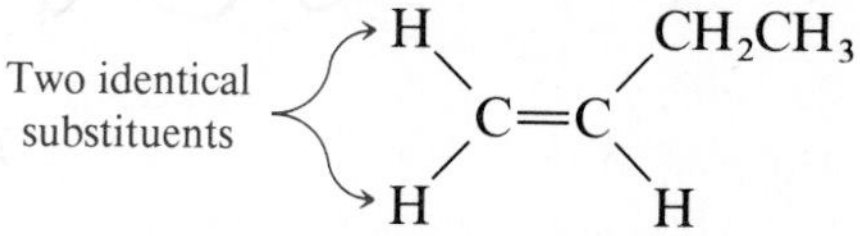

Two identical
substituents

1-Butene
(no stereoisomers possible)

Two identical
substituents

Two identical
substituents

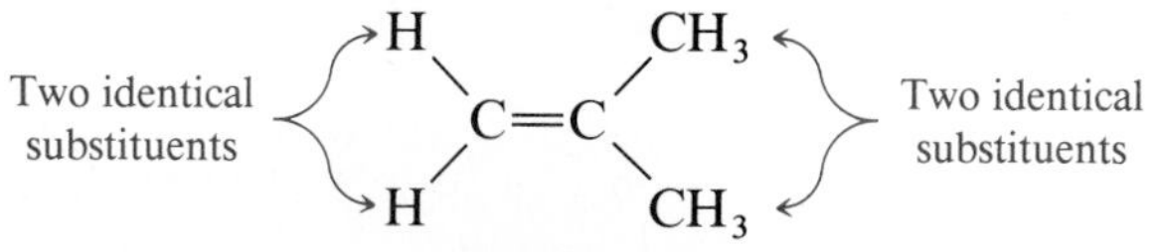

2-Methylpropene
(no stereoisomers possible)

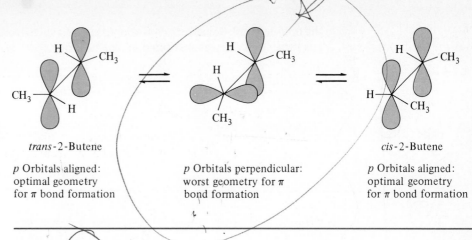

FIGURE 4.3
Interconversion of *cis*- and
trans-2-butene does not
occur; it would require
cleavage of the π
component of the double
bond.

trans-2-Butene

p Orbitals aligned:
optimal geometry
for π bond formation

p Orbitals perpendicular:
worst geometry for π
bond formation

cis-2-Butene

p Orbitals aligned:
optimal geometry
for π bond formation

PROBLEM 4.4
For which of the following alkenes are stereoisomeric forms possible?

(*a*) 1-Chloropropene   (*b*) 2-Chloropropene   (*c*) 1,2-Dichloropropene

SAMPLE SOLUTION
(*a*) Two stereoisomers are possible. In one, the methyl group and the chlorine substituent
are cis to each other; in the second isomer, methyl and chlorine are trans.

cis-1-Chloropropene   *trans*-1-Chloropropene

Comparing stereoisomeric alkenes, trans alkenes are more stable than cis because alkyl substituents crowd one another when they are cis. In the case of *cis*- and
trans-2-butene, the difference in stability is on the order of 1 kcal/mol. As the
substituents become larger, the van der Waals repulsion between them becomes
greater and the stability difference is magnified. Thus when *tert*-butyl groups are
present as substituents on a double bond, the trans stereoisomer is more than 10
kcal/mol more stable than the cis.

Less stable   More stable

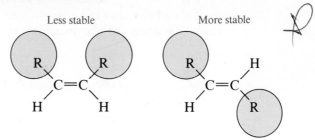

van der Waals repulsions are
larger when R = $(CH_3)_3C$—
than when R = CH_3

4.1.5 Classification of Alkenes

It is often useful to describe alkenes as having a certain degree of substitution at the
site of the double bond. A **monosubstituted** alkene has a single alkyl group attached to

the double bond, a **disubstituted** alkene has two such groups, a **trisubstituted** alkene has three, and a **tetrasubstituted** alkene has four.

Monosubstituted alkene: $RCH{=}CH_2$

Disubstituted alkenes:

$$\underset{R'}{\overset{R}{>}}C{=}C\underset{H}{\overset{H}{<}} \qquad \underset{H}{\overset{R}{>}}C{=}C\underset{H}{\overset{R'}{<}} \qquad \underset{H}{\overset{R}{>}}C{=}C\underset{R'}{\overset{H}{<}}$$

Trisubstituted alkene:

$$\underset{R''}{\overset{R}{>}}C{=}C\underset{H}{\overset{R'}{<}}$$

Tetrasubstituted alkene:

$$\underset{R''}{\overset{R}{>}}C{=}C\underset{R'''}{\overset{R'}{<}}$$

In general, alkyl substituents stabilize a double bond. Thus, among isomeric alkenes, one with a tetrasubstituted double bond is more stable than one with a trisubstituted double bond, and so on. This behavior parallels that seen earlier with carbocations (Section 3.12) and free radicals (Section 3.16). An alkyl group bonded to an sp^2 hybridized carbon donates electrons to it and stabilizes the species.

PROBLEM 4.5

Arrange the following alkenes in order of decreasing stability: 1-pentene; *cis*-2-pentene; *trans*-2-pentene; 2-methyl-2-butene.

4.1.6 Preparation of Alkenes. Elimination Reactions

The reactions used to prepare ethylene and propene on an industrial scale (see boxed essay) are not applicable to the laboratory preparation of alkenes, nor is the scope of those reactions such as to serve as a general preparation of alkenes. Alkenes are most commonly prepared in the laboratory from alcohols and alkyl halides by **elimination** reactions, such as

$$X{-}\overset{|}{\underset{|}{C}}{-}\overset{|}{\underset{|}{C}}{-}Y \longrightarrow ^{\backslash}_{/}C{=}C^{/}_{\backslash} + X{-}Y$$

Alkenes are prepared from alcohols by **dehydration.** The elements of water (X = H and Y = OH) are lost or "eliminated" from *adjacent* carbons with formation of a double bond between the two carbons. Alkenes are prepared from alkyl halides by **dehydrohalogenation.** The elements of a hydrogen halide (H and Cl, Br, or I) are eliminated from *adjacent* carbons.

Elimination reactions are not only useful for the synthesis of alkenes but are one of the fundamental reaction types of organic chemistry. Therefore, we will look at the dehydration of alcohols and the dehydrohalogenation of alkyl halides in some detail.

4.1.7 Dehydration of Alcohols

Alkenes are prepared from alcohols by heating in the presence of an acid such as sulfuric acid (H_2SO_4) or phosphoric acid (H_3PO_4). Before its preparation from

ETHYLENE

Ethylene was known to chemists in the eighteenth century and was isolated in pure form in 1795. An early name for ethylene was *gaz olefiant* (French for "oil-forming gas"), a term suggested to describe the fact that an oily liquid product is formed when two gases—ethylene and chlorine—react with each other.

$$CH_2{=}CH_2 + \quad Cl_2 \quad \longrightarrow \quad ClCH_2CH_2Cl$$

Ethylene	Chlorine	1,2-Dichloroethane
(bp $-104°C$)	(bp $-34°C$)	(bp $83°C$)

The term *gaz olefiant* survives in the general term **olefin,** formerly used as the name of the class of compounds we now call **alkenes.**

Ethylene occurs naturally in small amounts as a plant hormone. Hormones are substances that act as messengers and play regulatory roles in biological processes. Ethylene is involved in the ripening of many fruits, where it is formed in a complex series of steps from a compound containing a cyclopropane ring.

$$\underset{\substack{\text{1-Aminocyclopropane-}\\\text{carboxylic acid}}}{\triangleright\!\!<\genfrac{}{}{0pt}{}{NH_3^+}{CO_2^-}} \xrightarrow[\text{steps}]{\text{several}} \underset{\text{Ethylene}}{CH_2{=}CH_2} + \text{other products}$$

Even minute amounts of ethylene can stimulate ripening, and the rate of ripening increases with the concentration of ethylene. This property is used to advantage, for example, in the marketing of bananas. Bananas are picked green in the tropics, kept green by being stored with adequate ventilation to limit the amount of ethylene present, then induced to ripen at their destination by passing ethylene over the fruit.

Ethylene is the cornerstone substance of the world's mammoth petrochemical industry. Ethylene, and propene as well, is produced on an industrial scale in vast quantities. In a typical year the amount of ethylene produced in the United States is approximately the same as the combined weight of all its people (3×10^{10} lb). In one process, ethane from natural gas is heated to bring about its dissociation into ethylene and hydrogen.

$$\underset{\text{Ethane}}{CH_3CH_3} \xrightarrow{750°C} \underset{\text{Ethylene}}{CH_2{=}CH_2} + \underset{\text{Hydrogen}}{H_2}$$

This reaction is known as **dehydrogenation** and is simultaneously both a source of ethylene and the principal method by which hydrogen is prepared on an industrial scale. Most of the hydrogen so generated is subsequently used to reduce nitrogen to ammonia for the preparation of fertilizer.

Similarly, dehydrogenation of propane gives propene.

$$\underset{\text{Propane}}{CH_3CH_2CH_3} \xrightarrow{750°C} \underset{\text{Propene}}{CH_3CH{=}CH_2} + \underset{\text{Hydrogen}}{H_2}$$

Propene is the second most important petrochemical and is produced in amounts that are about one-half that of ethylene.

Almost any hydrocarbon can serve as a starting material for ethylene and propene production. By a process known as **thermal cracking** in which hydrocarbons are heated (not burned) in the absence of air, ethylene and propene are formed by the cleavage of carbon-carbon bonds of higher-molecular-weight hydrocarbons.

The major uses of ethylene and propene are as starting materials for the preparation of polyethylene and polypropylene plastics, fibers, and films. The chemical bases for these and other applications will be described in Chapter 5.

petroleum became the dominant route, ethylene was prepared by heating ethanol (obtained from fermentation of grain) with sulfuric acid.

$$CH_3CH_2OH \xrightarrow[160°C]{H_2SO_4} CH_2{=}CH_2 + H_2O$$

Ethyl alcohol   Ethylene   Water

Other alcohols behave similarly.

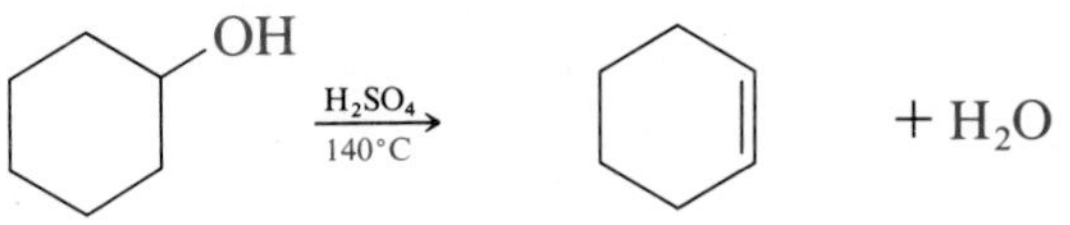

Cyclohexanol   Cyclohexene (79–87%)   Water

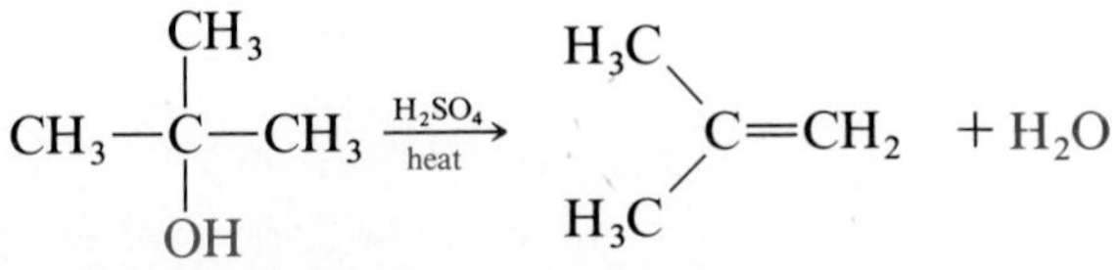

2-Methyl-2-propanol   2-Methylpropene (82%)   Water
(*tert*-butyl alcohol)

The acid acts as a **catalyst** for the dehydration of alcohols. A catalyst increases the rate of a reaction but does not undergo any net chemical change as a result of the reaction. Notice that the conversion of an alcohol to an alkene and water gives a balanced equation. The sulfuric acid catalyst in the examples given above is not consumed in the reaction and is indicated above the arrow. It is essential that it be shown, however, because the dehydration of alcohols normally occurs too slowly to be useful in the absence of acid catalysis.

PROBLEM 4.6

Identify the alkene obtained on dehydration of each of the following alcohols:

(a) 3-Ethyl-3-pentanol   (c) 2-Propanol
(b) 1-Propanol   (d) 2,3,3-Trimethyl-2-butanol

SAMPLE SOLUTION

(a) The hydrogen and the hydroxyl are lost from adjacent carbons in the dehydration of 3-ethyl-3-pentanol.

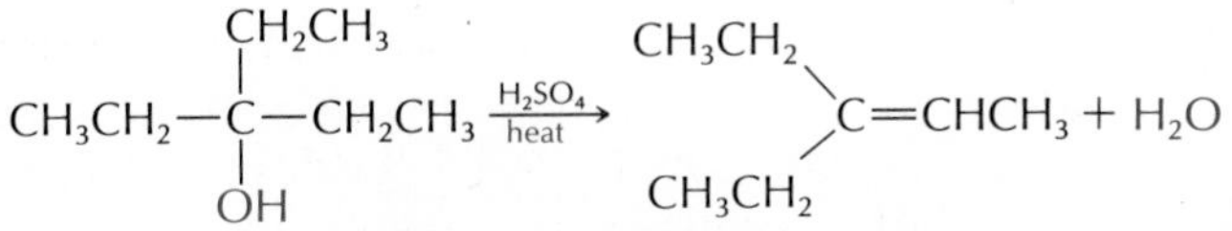

3-Ethyl-3-pentanol   3-Ethyl-2-pentene   Water

Notice that the hydroxyl group is attached to a carbon which bears three equivalent CH_2 substituents. It does not matter from which one of these the hydrogen is lost; the same alkene 3-ethyl-2-pentene is formed in each case.

When the alcohol is one which is capable of yielding two or more different alkenes by loss of water from adjacent carbons, a mixture is formed in which the alkene with the most highly substituted double bond predominates.

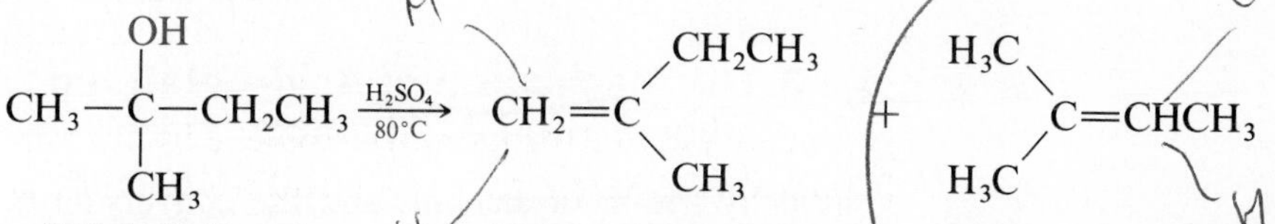

2-Methyl-2-butanol → 2-Methyl-1-butene (10%) (minor alkene; disubstituted) + 2-Methyl-2-butene (90%) (major alkene; trisubstituted)

We say that *elimination reactions that proceed in the direction to give the alkene with the most highly substituted double bond obey the* **Zaitsev rule.** This rule is named after Alexander Zaitsev, a nineteenth-century Russian chemist credited with making the first general statement to this effect. Since the alkene with the most highly substituted double bond is also the most stable one, Zaitsev's rule is sometimes expressed as a preference for elimination in the direction that gives the most stable alkene.

PROBLEM 4.7
Each of the following alcohols has been subjected to acid-catalyzed dehydration and yields a mixture of two isomeric alkenes. Identify the two alkenes in each case and predict which one is the major product on the basis of the Zaitsev rule.

(a) $(CH_3)_2CCH(CH_3)_2$ with OH (b) 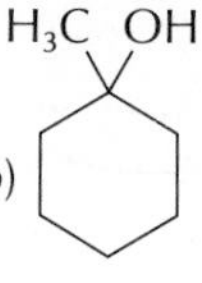   (c)

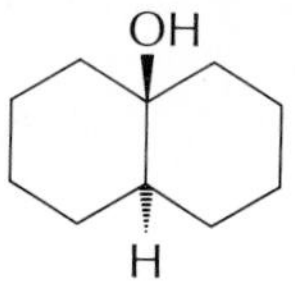

SAMPLE SOLUTION
(a) Dehydration of 2,3-dimethyl-2-butanol can lead to either 2,3-dimethyl-1-butene by removal of a C-1 hydrogen or to 2,3-dimethyl-2-butene by removal of the C-3 hydrogen.

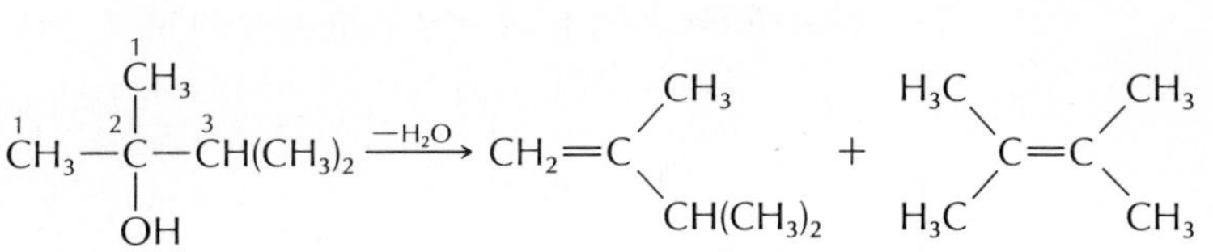

2,3-Dimethyl-2-butanol → 2,3-Dimethyl-1-butene (minor product) + 2,3-Dimethyl-2-butene (major product)

The major product is 2,3-dimethyl-2-butene. It has a tetrasubstituted double bond and is more stable than 2,3-dimethyl-1-butene, which has a disubstituted double bond. The major alkene product arises by loss of a hydrogen from the carbon that has fewer hydrogen substituents (C-3) rather than from the carbons that have the greater number of hydrogen substituents (either C-1).

When dehydration can lead to a mixture of stereoisomers, it is the more stable trans alkene that predominates.

$$CH_3CH_2CHCH_2CH_3 \xrightarrow[\text{heat}]{H_2SO_4}$$

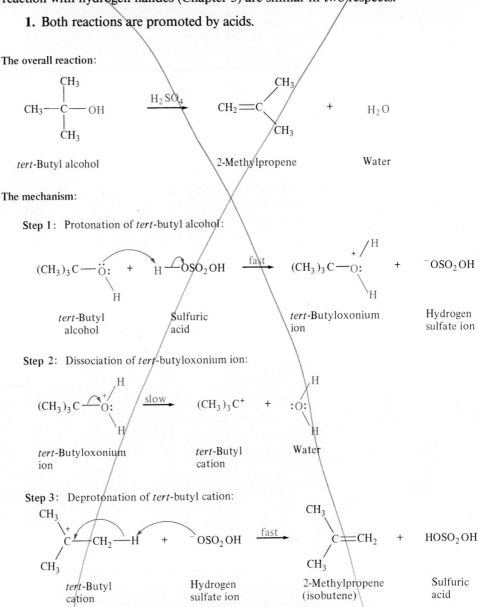

3-Pentanol

cis-2-Pentene (25%)
(minor product)

trans-2-Pentene (75%)
(major product)

4.1.8 The Mechanism of Acid-Catalyzed Dehydration of Alcohols

The dehydration of alcohols and the conversion of alcohols to alkyl halides by reaction with hydrogen halides (Chapter 3) are similar in two respects:

1. Both reactions are promoted by acids.

The overall reaction:

tert-Butyl alcohol

2-Methylpropene

Water

The mechanism:

Step 1: Protonation of tert-butyl alcohol:

tert-Butyl
alcohol

Sulfuric
acid

tert-Butyloxonium
ion

Hydrogen
sulfate ion

Step 2: Dissociation of tert-butyloxonium ion:

tert-Butyloxonium
ion

tert-Butyl
cation

Water

Step 3: Deprotonation of tert-butyl cation:

tert-Butyl
cation

Hydrogen
sulfate ion

2-Methylpropene
(isobutene)

Sulfuric
acid

FIGURE 4.4 Sequence of steps that describes the mechanism of the acid-catalyzed dehydration of tert-butyl alcohol.

2. The relative reactivity of alcohols decreases in the order tertiary > secondary > primary.

These common features suggest that carbocations are key intermediates in alcohol dehydration, just as they are in the conversion of alcohols to alkyl halides. Figure 4.4 portrays a three-step mechanism for the sulfuric acid–catalyzed dehydration of *tert*-butyl alcohol. Steps 1 and 2 describe the generation of the *tert*-butyl cation. This process is similar to the formation of the same cation in the reaction of *tert*-butyl alcohol with hydrogen chloride. Step 3 in Figure 4.4, however, is new to us and is the step in which the alkene product is formed from the carbocation intermediate.

Step 3 is an acid-base reaction. In this step the carbocation acts as a Brönsted acid, transferring a proton to a Brönsted base (hydrogen sulfate anion). This acidic behavior of carbocations is of the most significance to elimination reactions. Carbocations readily lose a proton to form alkenes.

PROBLEM 4.8
Write a structural formula for the carbocation intermediate formed in the dehydration of each of the alcohols in Problem 4.7. Using curved-arrow notation, show how each carbocation is deprotonated by the hydrogen sulfate anion ($^-OSO_2OH$) to give a mixture of alkenes.

SAMPLE SOLUTION
(a) The carbon that bears the hydroxyl group in the starting alcohol is the one that becomes positively charged in the carbocation.

$$(CH_3)_2CCH(CH_3)_2 \xrightarrow[-H_2O]{H^+} (CH_3)_2\overset{+}{C}CH(CH_3)_2$$
$$|$$
$$OH$$

The hydrogen sulfate ion may remove a proton from either C-1 or C-3 of this carbocation. Loss of a proton from C-1 yields the minor product 2,3-dimethyl-1-butene. (This alkene has a disubstituted double bond.)

$$HOSO_2O^- \quad H-\overset{1}{C}H_2-\overset{2}{\underset{+}{C}}\overset{CH_3}{\underset{\overset{3}{C}H(CH_3)_2}{}} \longrightarrow HOSO_2OH + CH_2=C\overset{CH_3}{\underset{CH(CH_3)_2}{}}$$

2,3-Dimethyl-1-butene

Loss of the proton from C-3 yields the major product 2,3-dimethyl-2-butene. (This alkene has a tetrasubstituted double bond.)

$$H_3\overset{1}{C}\underset{H_3C}{\overset{}{\diagdown}}\overset{+}{\underset{2}{C}}\overset{3}{-}\overset{}{\underset{\overset{|}{H}}{C}}(CH_3)_2 \longrightarrow \overset{H_3C}{\underset{H_3C}{\diagup}}C=C\overset{CH_3}{\underset{CH_3}{\diagdown}} + HOSO_2OH$$
$$\overset{\frown}{^-OSO_2OH}$$

2,3-Dimethyl-2-butene

Secondary alcohols, like tertiary alcohols, normally undergo dehydration by way of carbocation intermediates. Primary carbocations, however, are too unstable to be intermediates in most chemical reactions, and primary alcohols dehydrate by a mechanism in which a proton is removed from carbon in the same step in which the carbon-oxygen bond of the oxonium ion is cleaved. For example, this step in the sulfuric acid–catalyzed dehydration of ethanol may be represented as

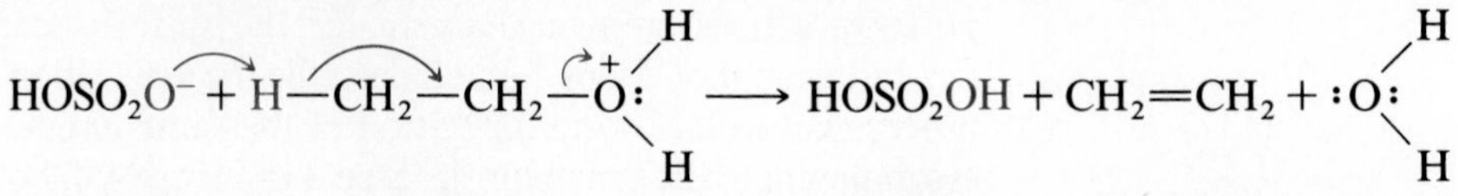

$$HOSO_2O^- + H-CH_2-CH_2-\overset{+}{O}: \longrightarrow HOSO_2OH + CH_2=CH_2 + :O:$$

| Hydrogen sulfate ion | Ethyloxonium ion | Sulfuric acid | Ethylene | Water |

4.1.9 Dehydrohalogenation of Alkyl Halides

Unlike the dehydration of alcohols which is carried out in the presence of a strong acid, dehydrohalogenation of alkyl halides takes place on heating in the presence of a strong base. Sodium ethoxide is a frequently used strong base, and reactions that employ it are commonly carried out in ethanol as the solvent.

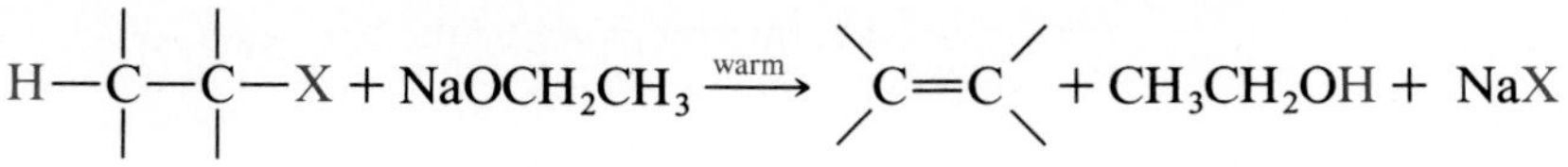

$$H-\overset{|}{\underset{|}{C}}-\overset{|}{\underset{|}{C}}-X + NaOCH_2CH_3 \xrightarrow{\text{warm}} \overset{}{C}=\overset{}{C} + CH_3CH_2OH + NaX$$

| Alkyl halide | Sodium ethoxide | Alkene | Ethyl alcohol | Sodium halide |

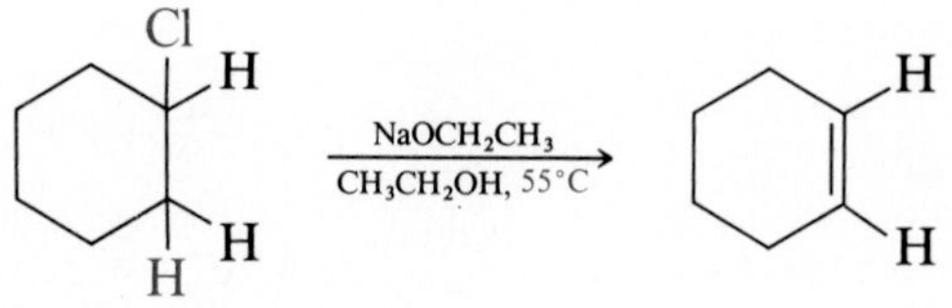

Cyclohexyl chloride   Cyclohexene (100%)

The dehydrohalogenation of alkyl halides follows the Zaitsev rule: *elimination proceeds in the direction that gives the most highly substituted alkene as the major product.*

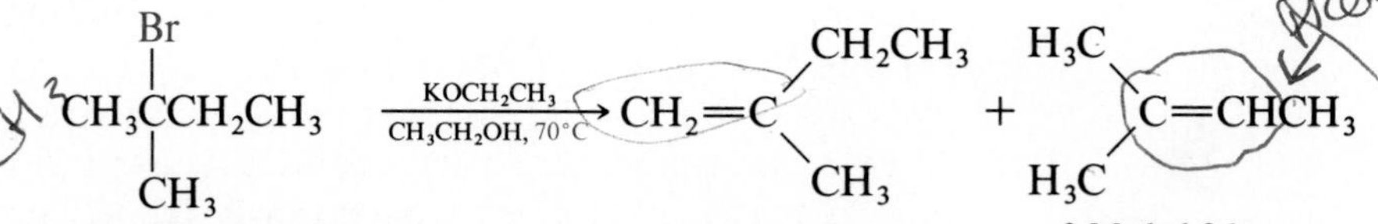

2-Bromo-2-methylbutane   2-Methyl-1-butene (29%)   2-Methyl-2-butene (71%)

PROBLEM 4.9

Write structural formulas for all the alkenes capable of being produced by dehydrohalogenation of each of the following alkyl halides with sodium ethoxide.

(a) 2-Bromobutane   (c) 2-Iodo-4-methylpentane
(b) 2-Bromohexane   (d) 3-Iodo-2,2-dimethylpentane

SAMPLE SOLUTION

(a) Elimination in 2-bromobutane can take place between C-1 and C-2 or between C-2 and C-3. There are three alkenes capable of being formed, namely, 1-butene, cis-2-butene, and trans-2-butene.

$$CH_3CHCH_2CH_3 \longrightarrow CH_2{=}CHCH_2CH_3 +$$

(with Br below the first carbon)

$$\underset{\text{2-Bromobutane}}{} \qquad \underset{\text{1-Butene}}{} \qquad \underset{cis\text{-2-Butene}}{\overset{H_3C}{\underset{H}{}}C{=}C\overset{CH_3}{\underset{H}{}}} + \underset{trans\text{-2-Butene}}{\overset{H_3C}{\underset{H}{}}C{=}C\overset{H}{\underset{CH_3}{}}}$$

The Zaitsev rule tells us that the mixture of *cis*- and *trans*-2-butene will be present in greater amounts than 1-butene. In the actual experiment the major product was *trans*-2-butene (60 percent); *cis*-2-butene and 1-butene each constituted 20 percent of the product.

4.1.10 The E2 Mechanism of Dehydrohalogenation

There are two principal mechanisms, identified by the symbols E1 **(elimination unimolecular)** and E2 **(elimination bimolecular),** by which alkyl halides undergo dehydrohalogenation. The molecularity of a particular step in a chemical reaction is equal to the number of species that undergoes covalency change in that step. Thus, a one-step transformation in which a single molecule of one species is transformed to another is a unimolecular reaction.

$$A \longrightarrow X \qquad \text{(unimolecular if one step)}$$

So also is the single-step dissociation of one molecule to two or more other species unimolecular.

$$A \longrightarrow X + Y \qquad \text{(unimolecular if one step)}$$

Two species are involved in a single-step bimolecular process.

$$A + B \longrightarrow X \qquad \text{(bimolecular if one step)}$$

The molecularity of a chemical reaction which proceeds by a mechanism involving more than one step is given by the molecularity of its slowest, or *rate-determining*, step. Thus, the unimolecular (E1) and bimolecular (E2) mechanisms for dehydrohalogenation of alkyl halides differ in that only a single molecule — the alkyl halide — is involved in the rate-determining step of the E1 mechanism, while two chemical species — the alkyl halide and the base — participate in the rate-determining step of the E2 mechanism. Reactions designed to convert alkyl halides to alkenes are almost always carried out in a strong base, and the E2 pathway is normally followed. Therefore, we will examine that mechanism first and defer discussion of the E1 mechanism to Section 4.1.11.

The E2 mechanism is a single-step, or **concerted,** process in which three structural changes represented by the curved arrows occur simultaneously.

$$\text{Base:}^- \quad H \quad \overset{|}{\underset{|}{C}}{-}\overset{|}{\underset{|}{C}} \quad \longrightarrow \quad \text{base}{-}H + \quad C{=}C \quad + \,{:}X^-$$
(with X below the second carbon)

- Abstraction of a proton from carbon by the base
- π bond formation between two adjacent carbons
- Expulsion of a halide ion from carbon

The high-energy species present at the transition state, called the **activated complex,** is often shown by a dashed-line notation, where the dashed lines represent "partial bonds," that is, bonds in the process of being formed or broken.

$$Base: ---- H$$
$$\overset{\delta^-}{}$$

The carbon-hydrogen and carbon-halogen bonds are partially broken at the transition state, the π component of the double bond is partially developed, and the hybridization of carbon is changing from sp^3 to sp^2. Figure 4.5 depicts the electronic changes which occur in an E2 reaction in more detail by showing the orbitals involved in each of the reacting species, at the transition state, and in the products.

Among the experimental facts which led to the proposal of the E2 pathway, the dependence of the *rate* of elimination on several factors was significant. One factor is the concentration of the reactants. The rate of dehydrohalogenation in the presence of base is directly proportional to the concentration of both reactants.

$$\text{Rate} = k \, [\text{alkyl halide}][\text{base}]$$

where k is a constant of proportionality called the *rate constant.* Doubling the concentration of either the alkyl halide or the base causes the rate of reaction to increase by a factor of 2. The **kinetic** (rate) behavior is first-order in alkyl halide and first-order in base. The exponent of each concentration term in the rate expression is 1. The overall **kinetic order** of the reaction is given by the sum of the exponents of all the concentration terms in the rate expression; we say that the dehydrohalogenation of an alkyl halide by a base exhibits **second-order kinetics** (first-order in alkyl halide + first-order in base). Kinetic studies provide an experimental determination of how many molecules, and which molecules, participate in the transition state of the slow step. The transition state of the E2 mechanism is bimolecular; the rate of elimination should be first-order in alkyl halide and first-order in base, and, indeed, it is observed to be just that.

The rate of elimination of an alkyl halide depends not only on its concentration but also on the type of halogen. Alkyl iodides are the most reactive, alkyl fluorides the least reactive.

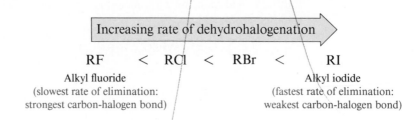

Increasing rate of dehydrohalogenation

| RF | < | RCl | < | RBr | < | RI |

Alkyl fluoride
(slowest rate of elimination:
strongest carbon-halogen bond)

Alkyl iodide
(fastest rate of elimination:
weakest carbon-halogen bond)

Among the halogens, iodide is the best **leaving group** in an elimination reaction, fluoride the poorest. The E2 mechanism is consistent with this observation because it incorporates partial cleavage of the carbon-halogen bond at the transition state. The weakest of the carbon-halogen bonds is the C-I bond, so the transition state in which it undergoes cleavage is lower in energy and formed faster than that for an alkyl bromide, chloride, or fluoride.

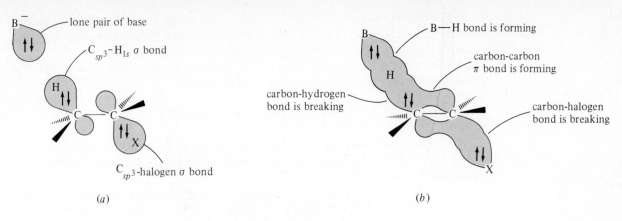

C_{sp^3}-H_{1s} σ bond

lone pair of base

C_{sp^3}-halogen σ bond

(a)

B—H bond is forming

carbon-carbon
π bond is forming

carbon-hydrogen
bond is breaking

carbon-halogen
bond is breaking

(b)

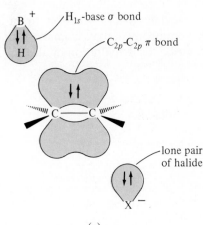

H_{1s}-base σ bond

C_{2p}-C_{2p} π bond

lone pair
of halide

(c)

FIGURE 4.5 Orbitals involved in bimolecular elimination (E2). (a) The relevant orbitals of the reactants are those of the σ bonds between carbon and hydrogen and carbon and halogen in the alkyl halide, along with the orbital that contains an unshared electron pair of the base B. (b) At the transition state, the σ bond between carbon and hydrogen becomes weaker as a bond between the base and hydrogen begins to form. A π bond begins to form as the two adjacent carbons undergo a hybridization change from sp^3 to sp^2 and their developing p orbitals begin to overlap with each other. The σ bond to the halogen substituent progressively becomes longer and weaker, with the electrons in that bond being attracted closer to the halogen. (c) At the completion of the reaction, the electron pair that was initially an unshared electron pair of the base has become a bonded pair in the covalent bond to the hydrogen which was removed from carbon. The electrons in the original C—H σ bond are now the electrons in the π component of the double bond of the alkene. The electrons in the carbon-halogen bond have become an unshared electron pair on a halide anion.

4.1.11 The E1 Mechanism of Dehydrohalogenation

Alkyl halides can undergo dehydrohalogenation even when a strong base is not present.

$$\underset{\text{Alkyl halide}}{H-\overset{|}{\underset{|}{C}}-\overset{|}{\underset{|}{C}}-X} \longrightarrow \underset{\text{Alkene}}{\diagdown C=C \diagup} + \underset{\text{Hydrogen halide}}{H-X}$$

These reactions occur much more slowly than those in which a strong base such as sodium ethoxide is present and are never employed as a synthetic method for the preparation of alkenes.

The mechanism by which elimination occurs in the absence of a strong base is presented in Figure 4.6. Step 1 is rate-determining and is the unimolecular dissociation of the alkyl halide to a carbocation and a halide anion. The rate of this step depends only on the concentration of the alkyl halide, and the overall reaction is therefore characterized by a first-order rate expression.

$$\text{Rate} = k \, [\text{alkyl halide}]$$

Once formed, the carbocation loses a proton to form an alkene. When carried out in a solvent such as ethanol, it is a molecule of ethanol which acts as a weak Brönsted base to remove a proton from the carbocation (step 2).

There is a strong similarity between the mechanism shown in Figure 4.6 and the one shown for alcohol dehydration in Figure 4.4. The principal difference between the two is the source of the carbocation. In the E1 reaction of an alkyl halide, it is the alkyl halide which dissociates to a carbocation. In the acid-catalyzed dehydration of an alcohol, it is the oxonium ion derived from the alcohol which dissociates. In fact, it is appropriate to describe the dehydration of alcohols as an E1 reaction of an oxonium ion.

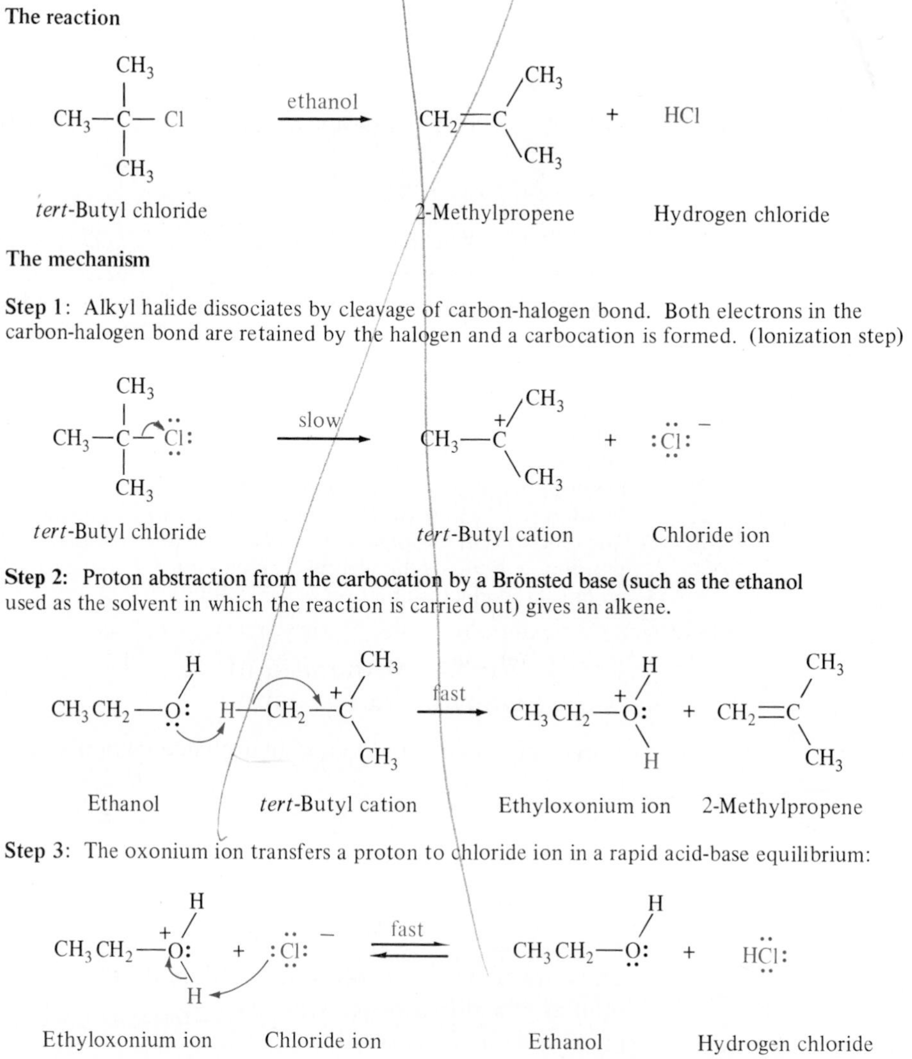

The reaction

tert-Butyl chloride   2-Methylpropene   Hydrogen chloride

The mechanism

Step 1: Alkyl halide dissociates by cleavage of carbon-halogen bond. Both electrons in the carbon-halogen bond are retained by the halogen and a carbocation is formed. (Ionization step)

tert-Butyl chloride   *tert*-Butyl cation   Chloride ion

Step 2: Proton abstraction from the carbocation by a Brönsted base (such as the ethanol used as the solvent in which the reaction is carried out) gives an alkene.

Ethanol   *tert*-Butyl cation   Ethyloxonium ion   2-Methylpropene

Step 3: The oxonium ion transfers a proton to chloride ion in a rapid acid-base equilibrium:

Ethyloxonium ion   Chloride ion   Ethanol   Hydrogen chloride

FIGURE 4.6 Sequence of steps that describes the E1 mechanism of dehydrohalogenation of *tert*-butyl chloride.

4.2 ALKADIENES

4.2.1 Classes of Dienes

Hydrocarbons that contain two carbon-carbon double bonds are called **alkadienes** (formal), or **dienes** (informal), and the relationship between the double bonds may be described as *cumulated, conjugated,* or *isolated.* In **cumulated dienes,** a single carbon is doubly bonded to two others. In **conjugated dienes,** two carbon-carbon double bond units are joined by a single bond. In **isolated dienes,** two carbon-carbon double bond units are separated from each other by one or more sp^3 hybridized carbon atoms.

$$CH_2=C=CHCH_2CH_3 \qquad CH_2=CH-CH=CHCH_3$$

1,2-Pentadiene (a cumulated diene)   1,3-Pentadiene (a conjugated diene)

$$CH_2=CH-CH_2-CH=CH_2$$

1,4-Pentadiene (an isolated diene)

Cumulated dienes are a relatively rare class of organic compounds and will not be discussed in this text.

Compounds with two or more double bonds are commonplace natural products. Figure 4.7 depicts some of these substances. Hydrocarbons with three double bonds are called **alkatrienes,** those with four are called **alkatetraenes,** and so on. **Polyenes** is a general term for compounds with several double bonds.

4.2.2 Diene Nomenclature

As may be surmised from the names of the pentadienes given in the preceding section, alkadienes are named in the same manner as alkenes except that the ending **-adiene** is used in place of **-ene** and separate locants are applied for each double bond.

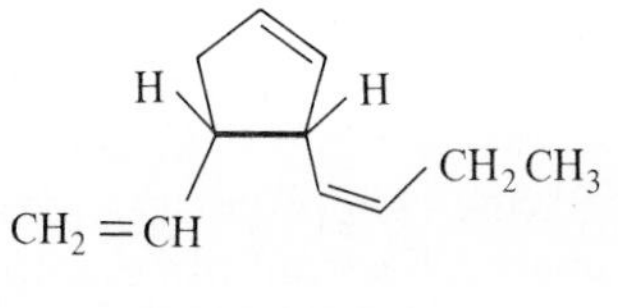

Multifidene
(atttracts male sperm cells of a species of brown alga to female)

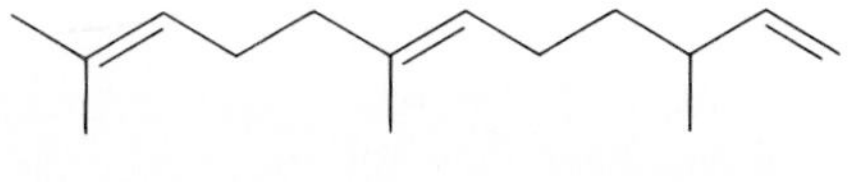

Farnesene
(present in the waxy coating found on apple skins)

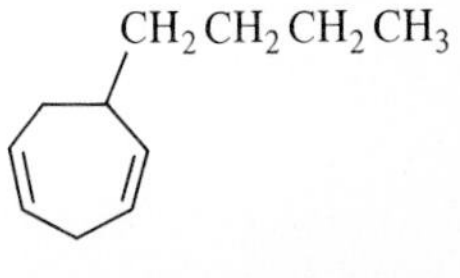

Dictyopterene C′
(obtained from oil of a species of marine alga that grows on Hawaiian reefs)

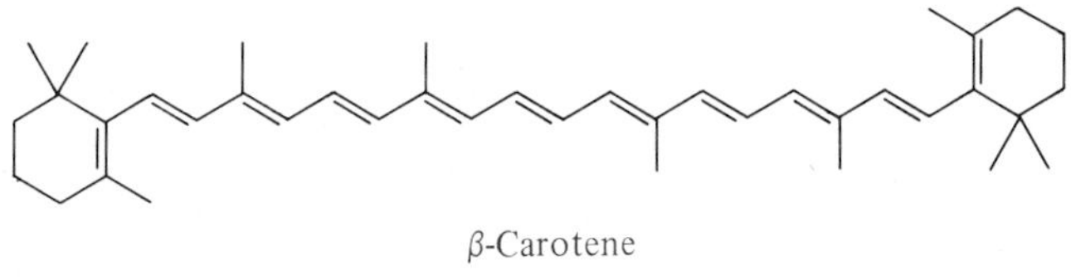

β-Carotene

(yellow pigment present in carrots and other vegetables)

FIGURE 4.7 Some naturally occurring hydrocarbons that contain two or more double bonds.

PROBLEM 4.10

Give IUPAC names to the following and specify whether they are isolated or conjugated dienes.

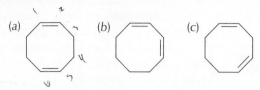

(a)   (b)   (c)

SAMPLE SOLUTION

(a) The compound is a diene derived from cyclooctane; it is a *cyclooctadiene*. Beginning with a doubly bonded carbon, number through the double bond and proceed around the ring in the direction that gives the lower number to the first carbon of the next-appearing double bond.

The IUPAC name for this compound is 1,5-cyclooctadiene. Its two double bonds are separated from each other by sp^3 hybridized carbons; 1,5-cyclooctadiene is an isolated diene.

4.2.3 Bonding in Conjugated Dienes

With few exceptions, the properties of isolated dienes are like those of alkenes, and the double bonds in an isolated diene are best viewed as independent structural units. Conjugated dienes, on the other hand, exhibit properties that indicate a significant interaction between the two double bonds. A conjugated diene, for example, is slightly more stable than an isomeric isolated diene. We will also see in Chapter 5 that conjugated dienes undergo certain reactions that isolated dienes do not.

The factor most responsible for the increased stability of conjugated double bonds is the greater delocalization of their π electrons. As shown in Figure 4.8a, an isolated diene system is characterized by two separate π bonds. An sp^3 hybridized carbon insulates the two π bonds from each other. In a conjugated diene, however, mutual overlap of the two π systems, as shown in Figure 4.8b, generates an extended π system encompassing four adjacent carbons. The four π electrons are said to be **delocalized** over the four carbons. Whenever electrons are delocalized over several carbon atoms, the system is more stable than when these electrons feel the attractive force of fewer nuclei.

4.2.4 Preparation of Alkadienes

The conjugated diene 1,3-butadiene is the starting material in the manufacture of synthetic rubber and is prepared on an industrial scale in vast quantities. Production in the United States is presently 2.5×10^9 lb/year. One industrial process is similar to

FIGURE 4.8 (a) Isolated double bonds are separated from one another by one or more sp^3 hybridized carbons and cannot overlap to give an extended π orbital. (b) In a conjugated diene, overlap of two π orbitals gives an extended π system encompassing four carbon atoms.

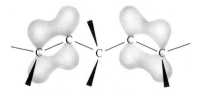

(a) Isolated double bonds

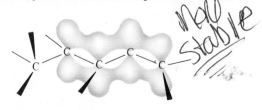

(b) Conjugated double bonds

that used for the preparation of ethylene; in the presence of a suitable catalyst, butane undergoes thermal dehydrogenation to yield 1,3-butadiene.

$$CH_3CH_2CH_2CH_3 \xrightarrow[\text{chromia-alumina}]{590-675°C} CH_2\!=\!CHCH\!=\!CH_2 + 2H_2$$

Laboratory syntheses of dienes are accomplished by adapting elimination reactions to molecules which already contain one double bond. When elimination could occur in two different directions, the more stable conjugated diene is normally formed in preference to a diene in which the double bonds are isolated.

$$\underset{\text{3-Methyl-5-hexen-3-ol}}{CH_2\!=\!CHCH_2\overset{\overset{\displaystyle CH_3}{|}}{\underset{\underset{\displaystyle OH}{|}}{C}}CH_2CH_3} \xrightarrow{\text{KHSO}_4,\text{ heat}} \underset{\text{4-Methyl-1,3-hexadiene (88\%)}}{CH_2\!=\!CHCH\!=\!\overset{\overset{\displaystyle CH_3}{|}}{C}CH_2CH_3}$$

$$\underset{\text{4-Bromo-4-methyl-1-hexene}}{CH_2\!=\!CHCH_2\overset{\overset{\displaystyle CH_3}{|}}{\underset{\underset{\displaystyle Br}{|}}{C}}CH_2CH_3} \xrightarrow{\text{KOH, heat}} \underset{\text{4-Methyl-1,3-hexadiene (78\%)}}{CH_2\!=\!CHCH\!=\!\overset{\overset{\displaystyle CH_3}{|}}{C}CH_2CH_3}$$

4.3 ALKYNES

4.3.1 Alkyne Nomenclature

In naming alkynes the usual IUPAC rules for hydrocarbons are followed and the **-ane** suffix is replaced by **-yne**. Both acetylene and ethyne are acceptable IUPAC names for HC≡CH. The position of the triple bond in the chain is specified by number as with alkenes.

$$\underset{\text{Propyne}}{HC\!\equiv\!CCH_3} \qquad \underset{\text{1-Butyne}}{HC\!\equiv\!CCH_2CH_3} \qquad \underset{\text{2-Butyne}}{CH_3C\!\equiv\!CCH_3} \qquad \underset{\text{4,4-Dimethyl-2-pentyne}}{(CH_3)_3CC\!\equiv\!CCH_3}$$

Propyne and 1-butyne are examples of **monosubstituted,** or **terminal, alkynes.** The triple bond in these compounds is at the end of the carbon chain. 2-Butyne and 4,4-dimethyl-2-pentyne have **disubstituted,** or **internal,** triple bonds.

PROBLEM 4.11
Write structural formulas and give the IUPAC names for all the alkynes of molecular formula C_5H_8.

4.3.2 Structure and Bonding in Alkynes

Acetylene is a linear molecule with a carbon-carbon bond distance of 120 pm.

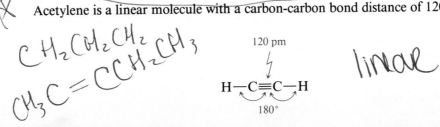

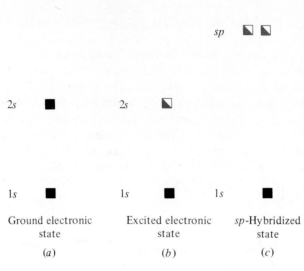

FIGURE 4.9 Representation of orbital hybridization of carbon in acetylene. Each box represents an orbital. An orbital can contain two electrons (■), one electron (◤), or be vacant (□). The stable electron configuration of carbon is shown in *a*. In *b* one of the 2*s* electrons has been raised in energy ("promoted") and occupies a previously vacant 2*p* orbital. In *c* the 2*s* and one of the 2*p* orbitals are mixed to give two equal energy *sp* hybridized orbitals, each of which contains one electron. Two of the 2*p* orbitals remain unhybridized.

There is a progressive shortening of the carbon-carbon bond distance in the series ethane (153 pm), ethylene (134 pm), and acetylene (120 pm).

We have earlier described bonding in alkanes and alkenes on the basis of orbital hybridization models that picture carbon as sp^3 hybridized when it is bonded to four atoms or groups, as in ethane, and sp^2 hybridized when it is bonded to three, as in

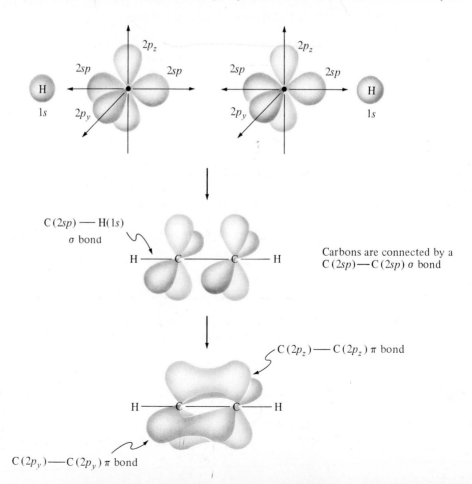

FIGURE 4.10 A description of bonding in acetylene based on *sp* hybridization of carbon.

ethylene. Each carbon in acetylene is bonded to two atoms, and extension of the orbital hybridization model to systems of this type is based on *sp* hybridization of the orbitals of carbon.

According to the *sp* hybridization model, one of the 2*s* electrons of the stable electron configuration of carbon (Figure 4.9*a*) is promoted to a 2*p* orbital, as shown in Figure 4.9*b*. Next, the carbon 2*s* orbital and *one* of the three 2*p* orbitals are combined to give two *sp* hybrid orbitals (Figure 4.9*c*). Two of the 2*p* orbitals of carbon remain unhybridized, and their axes are perpendicular to each other and to the axis of the two *sp* hybrid orbitals, as shown in Figure 4.10. The two *sp* hybrid orbitals are of equal energy, and each has one-half *s* character and one-half *p* character. Each *sp* hybrid orbital is half-filled (contains one electron) and, as depicted in Figure 4.10, can overlap with the 1*s* orbital of a hydrogen atom and the *sp* hybridized orbital of a second carbon to give the collection of σ bonds in an H—C—C—H framework. The unhybridized 2*p* orbitals on one carbon overlap with their counterparts on the other carbon to form two π bonds. The carbon-carbon triple bond in acetylene, and in higher alkynes as well, is viewed as a multiple bond of the $\sigma + \pi + \pi$ type.

4.3.3 Preparation of Alkynes by Elimination Reactions

Just as it is possible to prepare alkenes by dehydrohalogenation of alkyl halides, so may alkynes be prepared by a *double* dehydrohalogenation of dihaloalkanes. The dihalide may have both halogen substituents on the same carbon (a **geminal** dihalide) or on adjacent carbons (a **vicinal** dihalide). A very strong base such as sodium amide ($NaNH_2$) is required. Sodium amide is usually used in liquid ammonia as the solvent; it is a far stronger base than sodium hydroxide or sodium ethoxide.

Double dehydrohalogenation of a geminal dihalide:

$$(CH_3)_3C—\overset{\overset{\displaystyle H}{|}}{\underset{\underset{\displaystyle H}{|}}{C}}—\overset{\overset{\displaystyle Cl}{|}}{\underset{\underset{\displaystyle Cl}{|}}{C}}—H \quad \xrightarrow[\text{2. H}_2\text{O}]{\text{1. NaNH}_2\text{, NH}_3} \quad (CH_3)_3C—C\equiv C—H$$

1,1-Dichloro-3,3-dimethylbutane   3,3-Dimethyl-1-butyne (56–60%)

Double dehydrohalogenation of a vicinal dihalide:

$$CH_3(CH_2)_7—\overset{\overset{\displaystyle H}{|}}{\underset{\underset{\displaystyle Br}{|}}{C}}—\overset{\overset{\displaystyle H}{|}}{\underset{\underset{\displaystyle Br}{|}}{C}}—H \xrightarrow[\text{2. H}_2\text{O}]{\text{1. NaNH}_2\text{, NH}_3} CH_3(CH_2)_7—C\equiv C—H$$

1,2-Dibromodecane   1-Decyne (54%)

PROBLEM 4.12

Give the structures of three isomeric dibromides that could be used as starting materials for the preparation of 3,3-dimethyl-1-butyne.

Double dehydrohalogenation of geminal or vicinal dihalides is most often used for the preparation of terminal alkynes. The procedures used to prepare alkynes with internal triple bonds involve attaching alkyl groups to acetylene or a terminal alkyne and will be described in Chapter 5 after some fundamental chemical properties of alkynes have been discussed.

4.4 SUMMARY

Multiple bonds are commonly encountered structural units in organic compounds. Hydrocarbons that contain carbon-carbon double bonds are classified as *alkenes*. Alkenes with more than one double bond are referred to as *alkadienes, alkatrienes, etc. Alkynes* are hydrocarbons that contain carbon-carbon triple bonds. According to IUPAC nomenclature, the base name is derived from the alkane having the same number of carbon atoms as the longest continuous chain that includes the multiple bond; the *-ane* ending of the corresponding alkane is replaced by *-ene* (for alkenes, Section 4.1.1), *-adiene* (for alkadienes, Section 4.2.2), or *-yne* (for alkynes, Section 4.3.1). The chain is numbered in the direction which gives the lowest number to the first-appearing carbon of the multiple bond.

2,3-Dimethyl-2-pentene   3,4-Dimethyl-1,3-pentadiene   2-Pentyne

The double bonds of alkenes unite two sp^2 hybridized carbon atoms and are made up of a σ component and a π component (Section 4.1.3). The σ bond results from overlap of an sp^2 hybrid orbital of each carbon, the π bond results from a side-by-side overlap of p orbitals.

also
represented
as

Representations of bonding in ethylene

Rotation about the double bond of alkenes does not normally occur as a spontaneous process, and *stereoisomerism* is possible in alkenes (Section 4.1.4). Stereoisomeric alkenes have the same constitution (the same order of atomic connections) but differ in the arrangement of their atoms in space. When analogous substituents are on the same side of the double bond, they are said to be *cis* to one another; when they are on opposite sides of the double bond, they are said to be *trans*. The terms cis and trans are used as prefixes to differentiate the names of stereoisomeric alkenes.

cis-2-Pentene   *trans*-2-Pentene

The double bonds in dienes are classified as *cumulated, conjugated,* or *isolated* (Section 4.2.1). Only conjugated and isolated dienes are covered in this text. Conjugated dienes are characterized by a C=C—C=C structural unit; two C=C units are joined together by a single bond. In isolated dienes two C=C units are separated from each other by one or more sp^3 hybridized carbons. The double bonds in isolated dienes are similar to double bonds in alkenes. A conjugated diene is stabilized by delocalization of the π electrons over the four carbons of the C=C—C=C unit (Section 4.2.3).

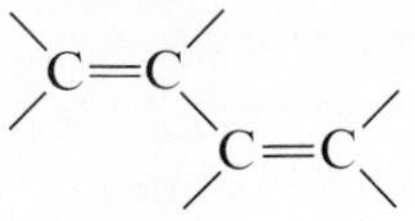

Conjugated diene

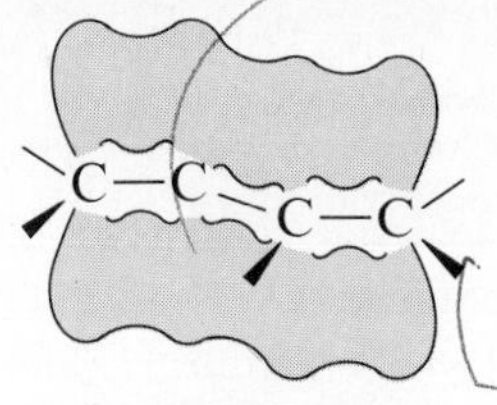

Extended π orbital
of conjugated diene

The carbon-carbon triple bond in alkynes is composed of a σ and two π components (Section 4.3.2). The σ component contains two electrons in an orbital generated by the overlap of sp hybridized orbitals on adjacent atoms. Each of these carbons also has two $2p$ orbitals which overlap in pairs so as to give two π orbitals. Alkynes have a *linear* arrangement of their bonds in the $-C\equiv C-$ unit.

Multiple bonds are introduced by *elimination* reactions. The elimination reactions used most frequently to prepare alkenes in the laboratory are the *dehydration of*

TABLE 4.1
Preparation of Alkenes by Elimination Reactions of Alcohols and Alkyl Halides

Reaction (section) and comments	General equation and specific example
Dehydration of alcohols (Sections 4.1.7 through 4.1.8) Dehydration requires an acid catalyst; the order of reactivity of alcohols is tertiary > secondary > primary. Elimination proceeds in the direction that produces the most highly substituted double bond. When stereoisomeric alkenes are possible, the more stable one is formed in greater amounts. A carbocation intermediate is involved.	$R_2CHCR'_2 \xrightarrow{H^+} R_2C{=}CR'_2 + H_2O$ $\;\;\;\;\overset{\textstyle |}{OH}$ Alcohol　　　　Alkene　　　Water HO 2-Methyl-2-hexanol $\downarrow$ $H_2SO_4, 80°C$ 2-Methyl-1-hexene (19%)　　2-Methyl-2-hexene (81%)
Dehydrohalogenation of alkyl halides (Sections 4.1.9 through 4.1.11) Strong bases cause a proton and a halide to be lost from adjacent carbons of an alkyl halide to yield an alkene. Orientation is in accord with the Zaitsev rule. The order of halide reactivity is I > Br > Cl > F and tertiary > secondary > primary. A concerted E2 reaction pathway is followed, and carbocations are not involved.	$R_2CHCR'_2 + :B^- \longrightarrow R_2C{=}CR'_2 + H{-}B + X^-$ $\;\;\;\;\overset{\textstyle |}{X}$ Alkyl　　Base　　　Alkene　　Conjugate　Halide halide　　　　　　　　　　acid of 　　　　　　　　　　　　base CH$_3$ Cl 1-Chloro-1-methylcyclohexane $\downarrow$ KOCH$_2$CH$_3$, CH$_3$CH$_2$OH, 100°C CH$_2$　　　　　　　CH$_3$ Methylenecyclohexane (6%)　1-Methylcyclohexene (94%)

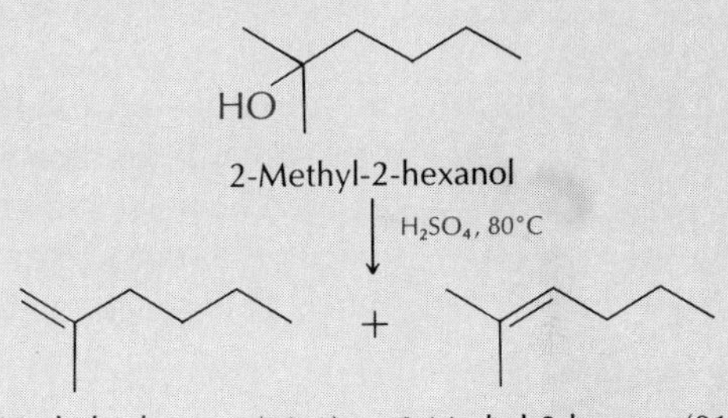

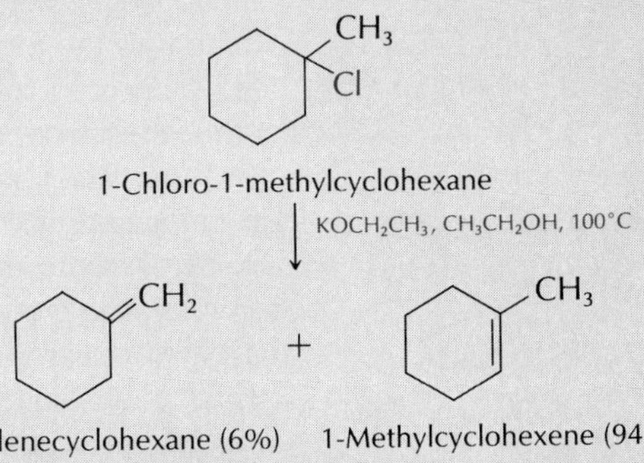

TABLE 4.2
Preparation of Alkynes

Reaction (section) and comments	General equation and specific example
Double dehydrohalogenation of geminal dihalides (Section 4.3.3) An E2 elimination reaction of a geminal dihalide yields a compound of the type $RCH=CR'$. If a strong enough base is used, sodium amide, for example, a second elimination step follows the first and the alkenyl halide is converted to an alkyne.	$$\begin{array}{c} H \quad X \\ \mid \quad \mid \\ RC-CR' + 2NaNH_2 \longrightarrow RC\equiv CR' + 2NaX \\ \mid \quad \mid \\ H \quad X \end{array}$$ Geminal dihalide Sodium amide Alkyne Sodium halide $(CH_3)_3CCH_2CHCl_2 \xrightarrow[\text{2. H}_2\text{O}]{\text{1. NaNH}_2,\ \text{NH}_3} (CH_3)_3CC\equiv CH$ 1,1-Dichloro-3,3-dimethylbutane 3,3-Dimethyl-1-butyne (56–60%)
Double dehydrohalogenation of vicinal dihalides (Section 4.3.3) Dihalides in which the halogens on adjacent carbons undergo two elimination processes analogous to those of geminal dihalides.	$$\begin{array}{c} H \quad H \\ \mid \quad \mid \\ RC-CR' + 2NaNH_2 \longrightarrow RC\equiv CR' + 2NaX \\ \mid \quad \mid \\ X \quad X \end{array}$$ Vicinal dihalide Sodium amide Alkyne Sodium halide $CH_3CH_2CHCH_2Br \xrightarrow[\text{2. H}_2\text{O}]{\text{1. NaNH}_2,\ \text{NH}_3} CH_3CH_2C\equiv CH$ $\qquad\qquad$ Br 1,2-Dibromobutane 1-Butyne (78–85%)

alcohols (Sections 4.1.7 and 4.1.8) and the *dehydrohalogenation of alkyl halides* (Sections 4.1.9 to 4.1.11). These reactions are summarized in Table 4.1. When mixtures of alkenes are possible, the one formed in greatest proportion is usually the one which bears the greatest number of alkyl groups as substituents on the double bond. The most substituted alkene is generally the most stable one and corresponds to loss of a proton from the adjacent carbon which has the fewest hydrogen substituents. This generalization is known as the *Zaitsev rule* (Section 4.1.7).

$$R_2CH-\overset{\displaystyle X}{\underset{\displaystyle CH_3}{C}}-CH_2R \longrightarrow R_2C=C\overset{\displaystyle CH_2R}{\underset{\displaystyle CH_3}{\diagdown}} \qquad X=OH,\ Cl,\ Br,\ or\ I$$

Hydrogen is lost from the adjacent carbon having the fewest hydrogen substituents

Alkene present in greatest amount in product

Carbocations are intermediates in the acid-catalyzed dehydration of alcohols, where they are formed by dissociation of an oxonium ion of the type ROH_2^+ generated by proton transfer from the acid catalyst to the alcohol. Once generated, the carbocation loses a proton to form an alkene (Section 4.1.8).

When alkyl halides are treated with strong bases such as sodium ethoxide, dehydrohalogenation occurs by the *E2 (elimination bimolecular) mechanism* (Section 4.1.10). According to the E2 mechanism the three elements of elimination, (1) removal of a proton from carbon by the base, (2) π bond formation, and (3) loss of halide from carbon, all occur in the same transition state. A second mechanism for dehydrohalogenation of alkyl halides, the E1 mechanism (Section 4.1.11), is a stepwise process involving a carbocation intermediate. It is observed only in the absence of a strong base.

Alkadienes are available by reactions similar to those used for the preparation of alkenes. Alkynes are prepared from *geminal* dihalides and from *vicinal* dihalides by double dehydrohalogenation reactions (Section 4.3.3), as summarized in Table 4.2.

ADDITIONAL PROBLEMS

Alkene Nomenclature

4.13 Write structural formulas for each of the following, clearly showing the stereochemistry around the double bond where indicated:

(a) 1-Heptene
(b) 3-Ethyl-1-pentene
(c) 3-Isopropyl-2-methyl-2-heptene
(d) *cis*-3-Octene
(e) *trans*-2-Hexene
(f) *trans*-3-Methyl-2-pentene
(g) 1-Bromocyclohexene
(h) Vinylcycloheptane
(j) 1,1-Diallylcyclopropane

4.14 Give an acceptable IUPAC name for each of the following compounds:

(a) $(CH_3CH_2)_2C=CHCH_3$
(b) $(CH_3CH_2)_2C=C(CH_2CH_3)_2$

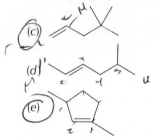

(c)

(d)

(e)

4.15 Hydroxyl groups take precedence over carbon-carbon double bonds when numbering the longest continuous chain that incorporates both functional groups. Double bonds, however, take precedence over halogen substituents. On the basis of this information write structural formulas for:

(a) 3-Buten-2-ol
(b) 3-Chloro-1-butene
(c) 4-Chloro-2-cyclohepten-1-ol

Alkene Structure and Bonding

4.16 Specify the hybridization of each carbon in the molecules shown. How many carbon-carbon σ bonds and carbon-carbon π bonds are present in each molecule?

(a) $(CH_3)_2C=C(CH_3)_2$
(b) $CH_2=CHCH_2CH_2CH=CH_2$
(c)

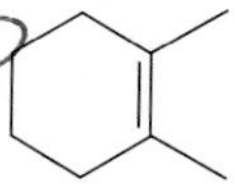

4.17 Specify each alkene in Problem 4.14 as being mono-, di-, tri-, or tetrasubstituted.

4.18 How many alkenes are there of molecular formula C_5H_{10}? Write their structures and give their IUPAC names. Which of these exist as pairs of stereoisomers? Specify the stereoisomers as cis or trans, as appropriate.

4.19 Arrange the alkenes in Problem 4.18 in order of decreasing stability.

4.20 Write the formula of the most stable alkene isomer having the formula C_6H_{12}.

Preparation of Alkenes

4.21 What three alkenes may be formed from the acid-catalyzed dehydration of 2-pentanol? Which one will predominate?

4.22 (a) Write the structures of all the isomeric alcohols having the molecular formula $C_5H_{12}O$.
   (b) Which one will undergo acid-catalyzed dehydration most readily?
   (c) Write the structure of the most stable C_5H_{11} carbocation.
   (d) Which alkenes may be derived from the carbocation in (c)? Which of these will predominate?
   (e) Which alcohol in (a) will dehydrate to give only a pair of stereoisomers?

4.23 Write a sequence of steps describing the mechanism of the dehydration of 2-butanol in phosphoric acid, H_3PO_4. Use the curved-arrow notation to show electron movement in each step.

4.24 How many alkenes would you expect to be formed by dehydrohalogenation of each of the following alkyl bromides? Identify the alkenes in each case.
   (a) 1-Bromohexane
   (b) 2-Bromohexane
   (c) 3-Bromo-3-methylpentane
   (d) 2-Bromo-3,3-dimethylbutane

4.25 Write the structures of all the alkenes capable of being produced when each of the following alkyl halides undergoes dehydrohalogenation. Apply the Zaitsev rule to predict the alkene formed in greatest amount in each case.
   (a) 2-Bromo-2,3-dimethylbutane
   (b) 3-Bromo-3-ethylpentane
   (c) 1-Bromo-3-methylbutane
   (d) 1-Iodo-1-methylcyclohexane

4.26 Choose the compound of molecular formula $C_7H_{13}Br$ that gives each alkene shown as the exclusive product of E2 elimination.

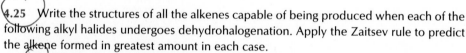

(a)

(b) =CH₂

(c) —CH₃

4.27 Give the structures of two different alkyl bromides, both of which yield the indicated alkene as the exclusive product of E2 elimination.
   (a) $CH_3CH{=}CH_2$
   (b) $(CH_3)_2C{=}CH_2$
   (c)

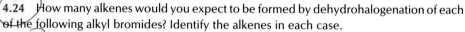

Alkadienes

4.28 Many naturally occurring substances contain several carbon-carbon double bonds: some isolated, some conjugated, and some cumulated. Identify the types of carbon-carbon double bonds found in each of the following substances:

(a) β-Springene (a scent substance obtained from the dorsal gland of springboks)

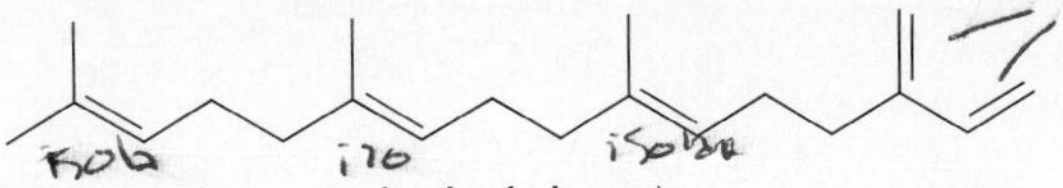

cont. dbl.

cob i?o iso√

(b) Humulene (found in hops and oil of cloves)

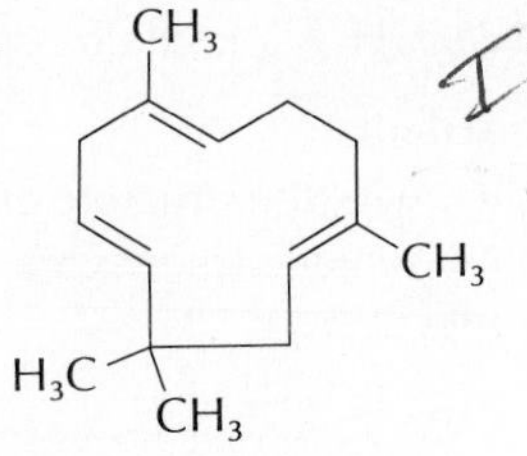

Isolated

(c) Cembrene (occurs in pine resin)

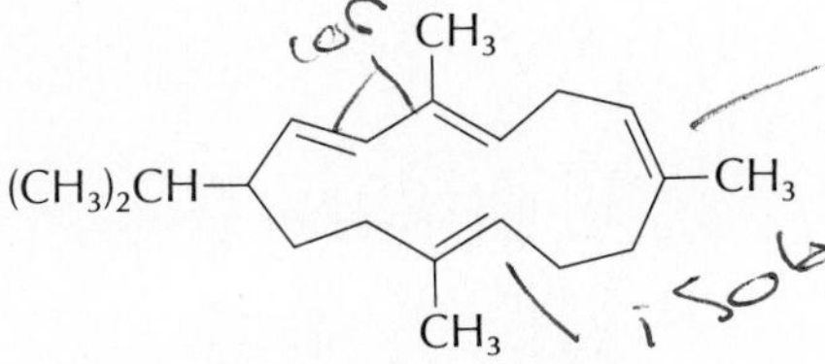

iso' isolated

(d) The sex attractant of the male dried-bean beetle

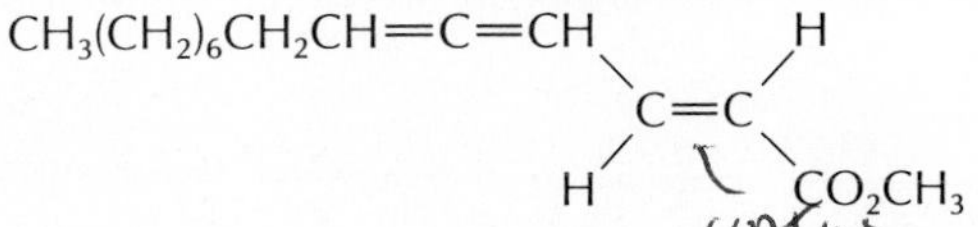

cumulated

$$CH_3(CH_2)_6CH_2CH=C=CH$$

untus.

4.29 Give the IUPAC names for each of the following compounds:

(a) $CH_2=CH(CH_2)_5CH=CH_2$

(b)
$$\begin{array}{c} CH_3 \\ | \\ (CH_3)_2C=CC=C(CH_3)_2 \\ | \\ CH_3 \end{array}$$

(c) $(CH_2=CH)_3CH$

4.30 Draw the structure and provide an IUPAC name for the diene formed by the following reaction:

$$(CH_3)_2C-C(CH_3)_2 \xrightarrow{\text{HBr, heat}}$$
$$\quad\;\; | \quad\;\; |$$
$$\quad\; HO \;\; OH$$

2,3-Dimethyl-2,3-butanediol

Alkynes

4.31 Write structural formulas and give the IUPAC names for all the alkynes of molecular formula C_6H_{10}.

4.32 Provide a systematic IUPAC name for each of the following alkynes:

(a) $CH_3CH_2CH_2C\equiv CH$

(b) $CH_3CH_2C\equiv CCH_3$

(c)
$$\begin{array}{c} CH_3C\equiv CCHCH(CH_3)_2 \\ | \\ CH_3 \end{array}$$

4.33 Write a structural formula corresponding to each of the following:

(a) 1-Octyne

(b) 2-Octyne

(c) 2,5-Dimethyl-3-hexyne

(d) 4-Ethyl-1-hexyne

(e) 3-Ethyl-3-methyl-1-pentyne

4.34 Identify the compounds in Problem 4.33 as either internal or terminal alkynes.

4.35 By writing chemical equations, show how two isomeric substances having the formula $C_6H_{12}Cl_2$ could be converted exclusively to 1-hexyne.

Reactions

4.36 Give the structure of the organic product of each of the following reactions. If two or more isomeric alkenes are formed, show all of them and indicate which one is the major product.

(a) 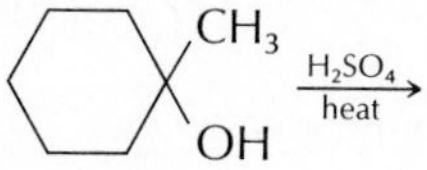 $\xrightarrow[\text{heat}]{\text{H}_2\text{SO}_4}$

(b) 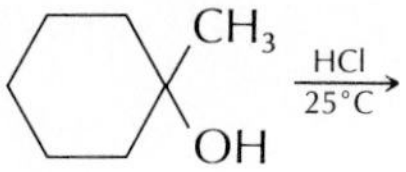 $\xrightarrow[\text{25°C}]{\text{HCl}}$

(c) Product of (b) + $NaOCH_2CH_3$ $\xrightarrow{\text{warm}}$

(d) 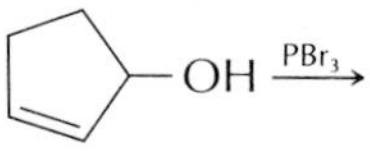 —OH $\xrightarrow{\text{PBr}_3}$

(e) Product of (d) + $NaOCH_2CH_3$ $\xrightarrow{\text{warm}}$

(f) 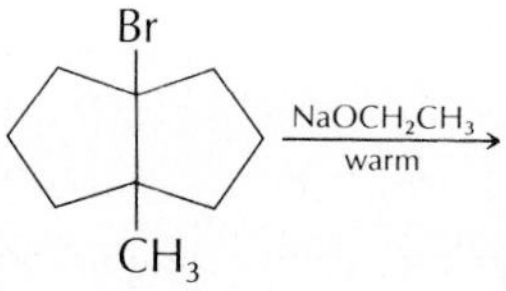 $\xrightarrow[\text{warm}]{\text{NaOCH}_2\text{CH}_3}$

(g) $(CH_3)_3CCH_2CHCl$ (with Br substituent) $\xrightarrow[\text{2. H}_2\text{O}]{\text{1. NaNH}_2,\ \text{NH}_3}$

CHAPTER 5

ALKENES, ALKADIENES, AND ALKYNES II. REACTIONS

There is a rich chemistry associated with the reactions of carbon-carbon double and triple bonds. This chapter includes numerous reactions which are learned most readily if one recognizes that they belong to a few fundamental types. A few of the reactions involve *cleavage* of carbon-carbon multiple bonds. Most of the reactions, however, are *addition* reactions, and most of these addition reactions are *electrophilic* additions. With the aid of some simple empirical rules and mechanistic arguments you will be able to predict the products of a great many useful transformations. More specifically, upon completion of this chapter you should be able to:

■ Write a chemical equation expressing the addition of each of the following to a representative alkene, alkadiene, or alkyne:

> Hydrogen in the presence of a suitable metal catalyst
> Hydrogen chloride, hydrogen bromide, hydrogen iodide
> Chlorine and bromine

■ State and give an example of Markovnikov's rule.

■ Contrast the regioselectivity of addition of hydrogen bromide to an alkene under conditions of electrophilic addition and free radical addition.

■ Describe, by writing appropriate chemical equations, how to convert alkenes to alcohols with control of the regioselectivity of hydration.

■ Write chemical equations expressing the cleavage of:

> An alkene by ozonolysis
> An alkyne by ozonolysis
> An alkene by potassium permanganate
> An alkyne by potassium permanganate

■ Contrast the acidity of acetylene and terminal alkynes with other hydrocarbons and describe reaction conditions suitable for converting acetylene and terminal alkynes to their derived carbanions.

- Write a chemical equation expressing the alkylation of acetylene or a terminal alkyne via the corresponding carbanion.
- Describe, by writing appropriate chemical equations, how to convert an alkyne to a cis alkene or to a trans alkene without formation of a mixture of the two.
- Contrast 1,2 addition and 1,4 addition to a conjugated diene.
- Write a chemical equation expressing a Diels-Alder reaction.
- Explain what is meant by the following terms:

  Halonium ion
  Dienophile
  Syn addition
  Anti addition
  Monomer, dimer, polymer
  Enol
  Regioselective reaction; regiospecific reaction
  Lindlar palladium

Now that we know something of the structural nature of alkenes, alkadienes, and alkynes and have seen how they can be prepared by elimination reactions, we are ready to look at their chemical reactions. The characteristic reaction of each of these classes is **addition** to the double or triple bond. The general form of an addition reaction of an alkene may be represented as

$$A-B + ^{\backslash}C=C^{\diagup} \longrightarrow A-\overset{|}{C}-\overset{|}{C}-B$$

The range of compounds represented as A—B in the above equation is quite large, and addition reactions offer a wealth of opportunity for the conversion of carbon-carbon double and triple bonds to a variety of other functional group types.

Addition reactions and elimination reactions are the reverse of one another. Alkenes, alkynes, and alkadienes are commonly described as **unsaturated** hydrocarbons because they have the capacity to react with substances which add to them. Alkanes, on the other hand, are said to be **saturated** hydrocarbons and are incapable of undergoing addition reactions.

5.1 ADDITION REACTIONS OF ALKENES

5.1.1 Hydrogenation

In terms of the relationship between reactants and products, addition reactions are best exemplified by the **hydrogenation** of alkenes to yield alkanes. Hydrogenation is the addition of H_2 to a multiple bond. The overall reaction is exothermic; two C—H σ bonds of an alkane are formed at the expense of the H—H σ bond and the π component of the carbon-carbon double bond.

$$^{\backslash}C\!\!=\!\!C^{\diagup}_{\pi} + H-H \longrightarrow H-\overset{|}{C}-\overset{|}{C}-H$$

Alkene   Hydrogen   Alkane

The uncatalyzed addition of hydrogen to an alkene, although exothermic, is a very slow process; too slow to be a useful synthetic reaction. The rate of hydrogena-

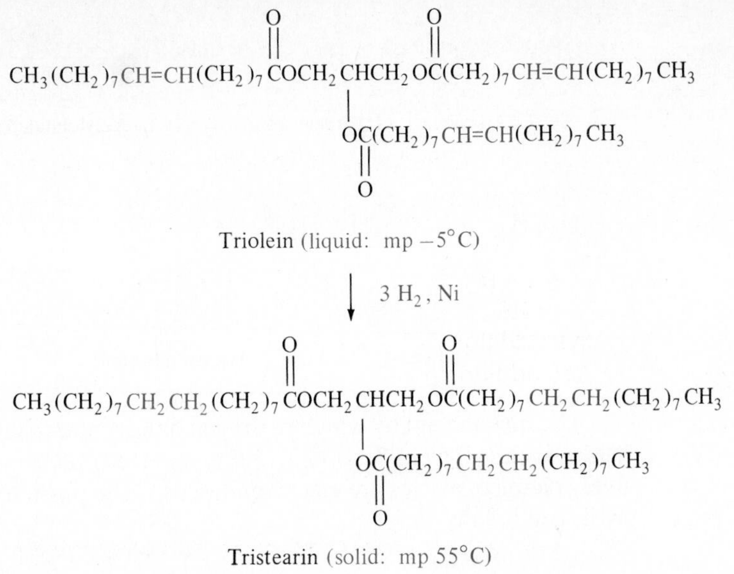

$$CH_3(CH_2)_7CH{=}CH(CH_2)_7\overset{\overset{\displaystyle O}{\|}}{C}OCH_2\,CHCH_2\,O\overset{\overset{\displaystyle O}{\|}}{C}(CH_2)_7CH{=}CH(CH_2)_7\,CH_3$$

$$\underset{\underset{\displaystyle O}{\|}}{O}C(CH_2)_7CH{=}CH(CH_2)_7\,CH_3$$

Triolein (liquid: mp $-5°C$)

$\downarrow$ 3 H$_2$, Ni

$$CH_3(CH_2)_7\,CH_2\,CH_2\,(CH_2)_7\overset{\overset{\displaystyle O}{\|}}{C}OCH_2\,CHCH_2\,O\overset{\overset{\displaystyle O}{\|}}{C}(CH_2)_7\,CH_2\,CH_2\,(CH_2)_7\,CH_3$$

$$\underset{\underset{\displaystyle O}{\|}}{O}C(CH_2)_7\,CH_2\,CH_2\,(CH_2)_7\,CH_3$$

Tristearin (solid: mp $55°C$)

FIGURE 5.1 Addition of hydrogen to the three carbon-carbon double bonds of triolein converts it to tristearin.

tion is dramatically increased, however, by carrying out the reaction in the presence of certain finely divided metal catalysts. Platinum, palladium, and nickel are all effective as hydrogenation catalysts.

$$(CH_3)_2C{=}CHCH_3 + H_2 \xrightarrow{Pt} (CH_3)_2CHCH_2CH_3$$

2-Methyl-2-butene   Hydrogen   2-Methylbutane (100%)

PROBLEM 5.1
What three alkenes yield 2-methylbutane on catalytic hydrogenation?

The hydrogenation of carbon-carbon double bonds finds commercial application in the conversion of vegetable oils to margarines and solid shortenings. A representative vegetable oil, *triolein,* is shown in Figure 5.1. Triolein is a liquid substance obtained from olive oil and contains three C=C units. On hydrogenation of these three double bonds, the solid substance *tristearin* is formed. Natural vegetable oils are mixtures of substances analogous to triolein, and their hydrogenation can be controlled to various degrees of "unsaturation" in the solid shortening which is formed. Margarine was introduced over 100 years ago as an inexpensive substitute for butter and is now produced in amounts two to three times that of butter by hydrogenation of soybean, corn, and other vegetable oils. While part of the reason for the increasing use of margarine is related to its lower cost, other factors such as a longer shelf life than butter and the realization that unsaturated fats make for a healthier diet have contributed as well.

5.1.2 Electrophilic Addition of Hydrogen Halides to Alkenes

In a large number of addition reactions the attacking reagent, unlike H$_2$, is a polar molecule or one which is easily polarizable. Hydrogen halides are among the simplest examples of polar substances which add to alkenes. The general equation describing the process and a specific example are shown:

$$\ce{\underset{\text{Alkene}}{>C=C<} + \underset{\substack{\text{Hydrogen halide} \\ (X = Cl,\ Br,\ or\ I)}}{H-X} \longrightarrow \underset{\text{Alkyl halide}}{H-\overset{|}{\underset{|}{C}}-\overset{|}{\underset{|}{C}}-X}}$$

$$\ce{\underset{\textit{cis}\text{-3-Hexene}}{\overset{CH_3CH_2}{\underset{H}{>}}C=C\overset{CH_2CH_3}{\underset{H}{<}}} + \underset{\substack{\text{Hydrogen bromide}}}{HBr} \longrightarrow \underset{\substack{\text{3-Bromohexane (76\%)}}}{CH_3CH_2CH_2\overset{|}{\underset{Br}{C}}HCH_2CH_3}}$$

The mechanism by which hydrogen halides add to alkenes is called **electrophilic addition.** Recall from Section 3.13 that an electrophile is an "electron seeker." Positively charged species are electrophiles, as is the positively polarized hydrogen of a hydrogen halide.

$$^{\delta+}H-X^{\delta-}$$

The electrons in the hydrogen halide bond are attracted away from hydrogen
toward the electronegative halogen so that hydrogen bears a partial
positive charge and is electrophilic.

Electrophiles react with substances that can act as a source of electrons. The π electrons of an alkene are more weakly held than the σ electrons and are capable of being attacked by electrophiles such as hydrogen halides. As indicated by the curved arrows, an alkene and a hydrogen halide interact by using a pair of π electrons of the alkene to form a σ bond to the proton of the hydrogen halide. A carbocation and a halide ion result. The alkene is acting as a Lewis base, donating an electron pair (the π electrons) to a Lewis acid, the hydrogen halide.

$$\ce{\underset{\substack{\text{Alkene} \\ \text{(Lewis base)}}}{R_2C=CR_2} + \underset{\substack{\text{Hydrogen halide} \\ \text{(Lewis acid)}}}{H-\overset{..}{\underset{..}{X}}:} \rightleftharpoons \underset{\substack{\text{Carbocation} \\ \text{(conjugate acid)}}}{R_2\overset{+}{C}-\overset{\overset{\displaystyle H}{|}}{C}R_2} + \underset{\substack{\text{Anion} \\ \text{(conjugate base)}}}{:\overset{..}{\underset{..}{X}}:^-}}$$

As we saw in Sections 3.11 to 3.13, carbocations generated in the presence of nucleophiles such as negatively charged halide ions react with them to give alkyl halides.

$$\ce{\underset{\substack{\text{Carbocation (electrophile)}}}{R_2\overset{+}{C}-\overset{\overset{\displaystyle H}{|}}{C}R_2} + \underset{\substack{\text{Halide ion (nucleophile)}}}{:\overset{..}{\underset{..}{X}}:^-} \longrightarrow \underset{\substack{\text{Alkyl halide}}}{R_2C-\overset{\overset{\displaystyle H}{|}}{\underset{\underset{\displaystyle :\overset{..}{X}:}{|}}{C}}R_2}}$$

In the two-step mechanism just shown, the rate-determining step is the first one, carbocation formation. The slow step is characterized by the addition of an electrophilic species, a proton, to a carbon-carbon double bond. The overall process is referred to as an electrophilic addition because the addition is triggered by electrophilic attack of the hydrogen halide on the alkene. As we proceed through the chapter, you will see that other electrophilic species react with alkenes by variations on this mechanism.

5.1.3 Orientation of Hydrogen Halide Addition to Alkenes. Markovnikov's Rule

In principle, a hydrogen halide can add to an unsymmetrical alkene (an alkene in which the two carbons of the double bond are not equivalently substituted) in either of two ways. In practice, however, only one mode of addition is observed for an individual alkene.

$$RCH{=}CH_2 + HX \longrightarrow \underset{\underset{X}{|}\quad\underset{H}{|}}{RCH{-}CH_2} \quad \text{rather than} \quad \underset{\underset{H}{|}\quad\underset{X}{|}}{RCH{-}CH_2}$$

$$R_2C{=}CH_2 + HX \longrightarrow \underset{\underset{X}{|}\quad\underset{H}{|}}{R_2C{-}CH_2} \quad \text{rather than} \quad \underset{\underset{H}{|}\quad\underset{X}{|}}{R_2C{-}CH_2}$$

$$R_2C{=}CHR + HX \longrightarrow \underset{\underset{X}{|}\quad\underset{H}{|}}{R_2C{-}CHR} \quad \text{rather than} \quad \underset{\underset{H}{|}\quad\underset{X}{|}}{R_2C{-}CHR}$$

In 1870 Vladimir Markovnikov, a colleague of Alexander Zaitsev, noted the pattern of hydrogen halide addition to alkenes and assembled his observations into a simple statement. **Markovnikov's rule** states that when an unsymmetrically substituted alkene reacts with a hydrogen halide, the *hydrogen adds to the carbon that has the greater number of hydrogen substituents*. A necessary corollary is that the halogen adds to the carbon having the fewer hydrogen substituents (the **more-substituted carbon**). Some examples illustrate additions in accordance with Markovnikov's rule.

$$CH_3CH_2CH{=}CH_2 + \quad HBr \quad \longrightarrow \quad \underset{\underset{Br}{|}}{CH_3CH_2CHCH_3}$$

1-Butene	Hydrogen bromide	2-Bromobutane (80%)

$$\underset{H_3C}{\overset{H_3C}{>}}C{=}CH_2 + \quad HBr \quad \longrightarrow \quad \underset{\underset{CH_3}{|}}{\overset{\overset{CH_3}{|}}{CH_3{-}C{-}Br}}$$

2-Methylpropene	Hydrogen bromide	2-Bromo-2-methylpropane (90%)

[structure of 1-methylcyclopentene] —CH₃ + HCl ⟶ [structure of 1-chloro-1-methylcyclopentane with CH₃ and Cl]

1-Methylcyclopentene	Hydrogen chloride	1-Chloro-1-methylcyclopentane (100%)

PROBLEM 5.2

Write the structure of the principal organic product formed in the addition of hydrogen chloride to each of the following:

(a) 2-Methyl-2-butene   (d) $CH_3CH{=}$ [cyclohexane ring]
(b) 2-Methyl-1-butene
(c) *cis*-2-Butene

SAMPLE SOLUTION

(a) Hydrogen chloride adds to the double bond of 2-methyl-2-butene in accordance with Markovnikov's rule. Hydrogen adds to the carbon that has one hydrogen substituent, chlorine to the carbon that has none.

$$H_3C \diagdown \diagup H$$
$$C{=}C$$
$$H_3C \diagup \diagdown CH_3$$

2-Methyl-2-butene

Chlorine becomes attached   Hydrogen becomes attached
to this carbon   to this carbon

$$CH_3$$
$$|$$
$$CH_3{-}\underset{|}{C}{-}CH_2CH_3$$
$$Cl$$

2-Chloro-2-methylbutane

(Major product from Markovnikov addition
of hydrogen chloride to 2-methyl-2-butene)

A useful term in organic chemistry which refers to reactions such as the addition of hydrogen halides to unsymmetrical alkenes is **regioselectivity**. A reaction is regioselective if two or more constitutionally isomeric products are possible but one is formed in greater amounts than the other(s). The possible products are sometimes called **regioisomers**. A reaction which gives only one regioisomer to the exclusion of all others is said to be **regiospecific**. Thus, Markovnikov's rule describes the regioselectivity of addition of hydrogen halides to alkenes, and the overall addition is regiospecific.

Markovnikov's rule, like Zaitsev's rule, is an empirical one. It collects experimental observations into a general statement having predictive value but does not offer an explanation for why things happen the way they do. To understand why hydrogen halides add to alkenes in accordance with Markovnikov's rule, we need to return to the mechanism for electrophilic addition.

5.1.4 Mechanistic Basis for Markovnikov's Rule

In the reaction of a hydrogen halide HX with an unsymmetrically substituted alkene $RCH{=}CH_2$, let us compare the carbocation intermediates for addition of HX according to Markovnikov's rule and contrary to Markovnikov's rule.

Addition according to Markovnikov's rule

$$RCH{=}CH_2 \longrightarrow R\overset{+}{CH}{-}\underset{H}{CH_2} + X^- \longrightarrow \underset{X}{RCHCH_3}$$
$$H{-}X$$

Secondary   Halide   Observed product
carbocation   ion

Addition contrary to Markovnikov's rule

$$RCH{=}CH_2 \longrightarrow RCH{-}\overset{+}{CH_2} + X^- \longrightarrow RCH_2CH_2X$$
$$X{-}H \qquad\qquad H$$

Primary   Halide   Not formed
carbocation   ion

Recall from Section 3.12 that secondary carbocations are more stable than primary ones. The observed product of hydrogen halide addition is formed from the **more stable of the two possible carbocation intermediates.** The secondary carbocation is more stable and is formed faster than the less stable primary carbocation, and the observed product is the secondary alkyl halide. Thus, the regioselectivity expressed in Markovnikov's rule results because a proton adds to the carbon that has the greater number of hydrogen substituents to give the more stable of two possible carbocations, and the product is formed via this intermediate.

PROBLEM 5.3

Give a structural formula for the carbocation intermediate leading to the principal product in each of the reactions of Problem 5.2.

SAMPLE SOLUTION

(a) Proton transfer to the carbon-carbon double bond can lead either to a tertiary carbocation or to a secondary one.

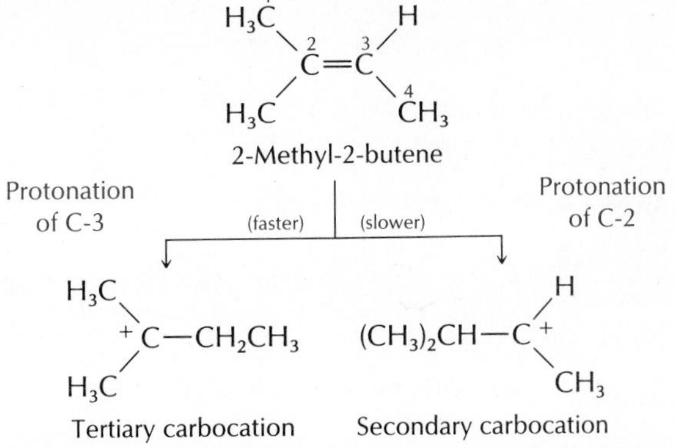

2-Methyl-2-butene

| Protonation of C-3 | (faster) | (slower) | Protonation of C-2 |

Tertiary carbocation   Secondary carbocation

The product of the reaction is derived from the more stable carbocation; in this case, it is a tertiary carbocation that is formed in preference to a secondary one.

5.1.5 Acid-Catalyzed Hydration of Alkenes

Alkenes may be converted to alcohols through the addition of a molecule of water across the carbon-carbon double bond in the presence of an acid catalyst.

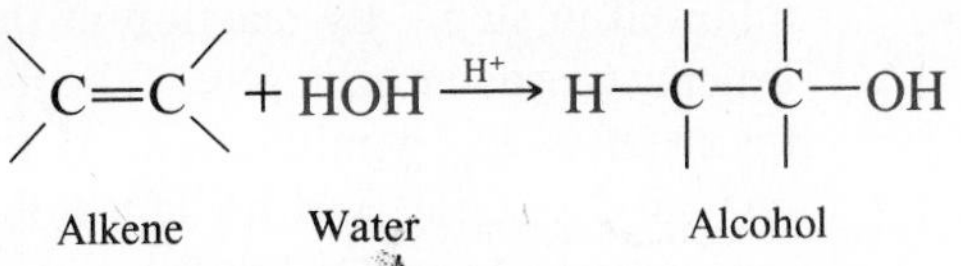

Alkene   Water   Alcohol

A proton adds to the carbon of the double bond that has the greater number of hydrogens attached to it, and a hydroxyl group adds to the other carbon in accordance with Markovnikov's rule.

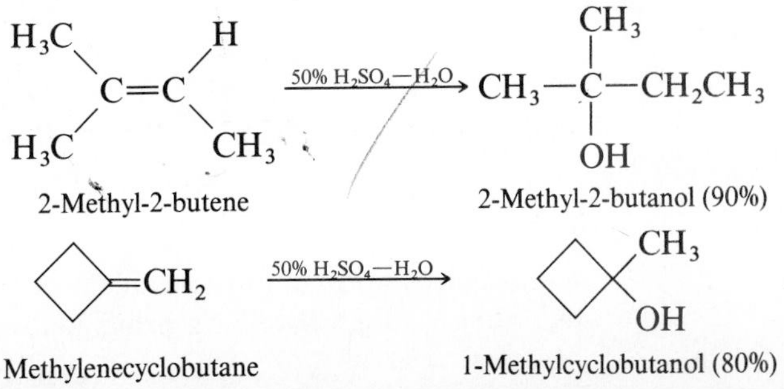

2-Methyl-2-butene   2-Methyl-2-butanol (90%)

Methylenecyclobutane   1-Methylcyclobutanol (80%)

The overall reaction

$$(CH_3)_2C=CH_2 \quad + \quad H_2O \quad \xrightarrow{H_3O^+} \quad (CH_3)_3COH$$

2-Methylpropene   Water   2-Methyl-2-propanol

The mechanism

Step 1: Protonation of the carbon-carbon double bond in the direction that leads to the more stable carbocation:

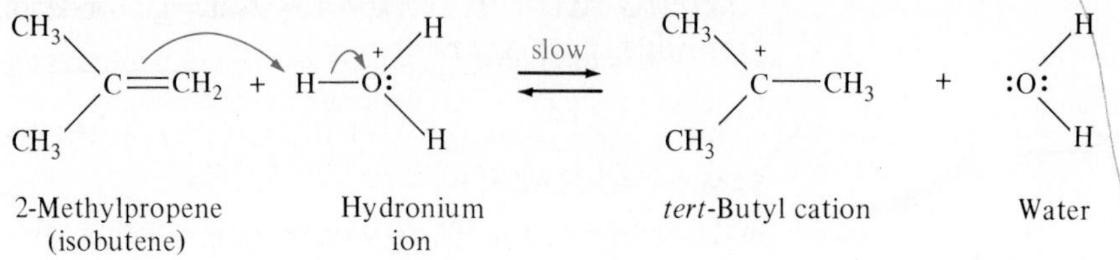

2-Methylpropene   Hydronium   *tert*-Butyl cation   Water
(isobutene)   ion

Step 2: Water acts as a nucleophile to capture *tert*-butyl cation:

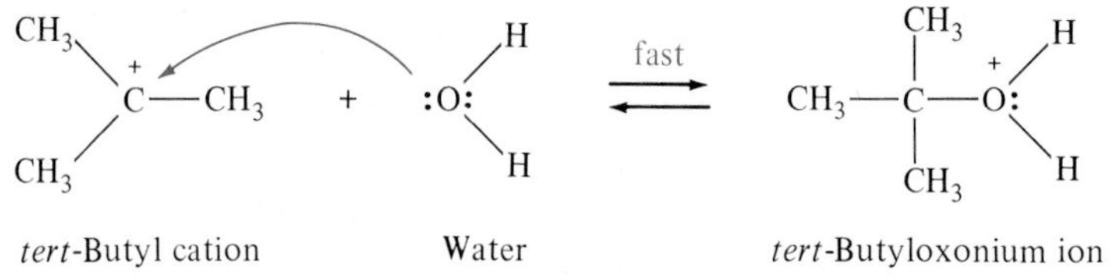

tert-Butyl cation   Water   *tert*-Butyloxonium ion

Step 3: Deprotonation of *tert*-butyloxonium ion. Water acts as a Bronsted base:

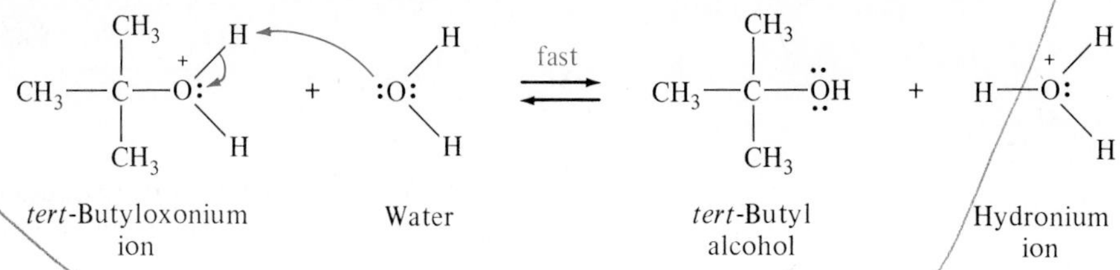

tert-Butyloxonium   Water   *tert*-Butyl   Hydronium
ion   alcohol   ion

FIGURE 5.2 Sequence of steps describing the mechanism of acid-catalyzed hydration of 2-methylpropene.

We may extend the general principles of electrophilic addition to alkenes to their acid-catalyzed hydration, as shown in Figure 5.2. In the specific example cited, proton transfer to 2-methylpropene forms a *tert*-butyl cation in the first step. This is followed in step 2 by reaction of the carbocation intermediate with a molecule of water acting as a nucleophile. The product of nucleophilic capture of the carbocation by water is an oxonium ion. The oxonium ion is simply the conjugate acid of *tert*-butyl alcohol, and its deprotonation in step 3 yields the alcohol product and regenerates the acid catalyst.

You may have noticed that the acid-catalyzed hydration of an alkene and the acid-catalyzed dehydration of an alcohol (Section 4.1.7) are the reverse of one another.

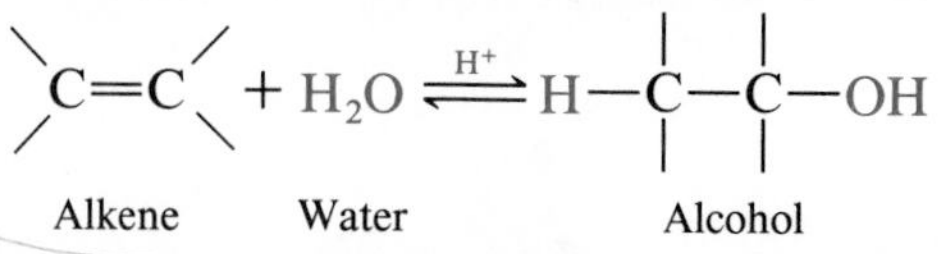

Alkene   Water   Alcohol

How does one decide whether the equilibrium expressed in the equation favors alcohol dehydration to form an alkene or hydration of an alkene to form an alcohol? According to **Le Chatelier's principle,** *a system at equilibrium adjusts so as to* *minimize any stress applied to it.* If the concentration of water is increased, the system responds by consuming water. The position of equilibrium shifts to the right, and more alkene is converted to alcohol. Thus, when we wish to prepare an alcohol from an alkene, we employ a reaction medium in which the molar concentration of water is high. Notice that the reactions used as examples of alkene hydration in this section were carried out in a 50% sulfuric acid solution—a solution which contains equal volumes of sulfuric acid and water—as the reaction medium. On the other hand, when we wish to convert an alcohol to an alkene, we employ a water-poor reaction medium such as concentrated (>98%) sulfuric acid. The system responds to the lowered water concentration by causing more alcohol molecules to dehydrate, thus forming more alkene.

5.1.6 Synthesis of Alcohols from Alkenes by Hydroboration-Oxidation

Acid-catalyzed hydration adds the components of water (H— and —OH), to alkenes according to Markovnikov's rule. Suppose you had 1-decene as a starting material for the preparation of pure samples of the alcohols 1-decanol and 2-decanol.

$$CH_3(CH_2)_7CH{=}CH_2$$
1-Decene

$\xrightarrow{H_2O,\ H_2SO_4}$

$$CH_3(CH_2)_7CHCH_3$$
$$|$$
$$OH$$
2-Decanol

$\xrightarrow{\ ?\ }$

$$CH_3(CH_2)_7CH_2CH_2OH$$
1-Decanol

Acid-catalyzed hydration would be feasible for the preparation of 2-decanol because this conversion requires hydration of the double bond according to Markovnikov's rule. How, though, could we prepare 1-decanol from 1-decene? Here the elements of water must be added to the double bond with a regiochemistry opposite to that of Markovnikov's rule. (Addition opposite to that of Markovnikov's rule is referred to as **anti-Markovnikov addition**.)

The synthetic procedure which effects anti-Markovnikov hydration of alkenes is an indirect one, known as **hydroboration-oxidation,** and does not, in fact, involve water as a reactant at all. It was developed by Prof. Herbert C. Brown and his coworkers at Purdue University in the 1950s as part of a broad program designed to demonstrate how boron-containing reagents could be applied to organic chemical synthesis. The number of applications is so large (hydroboration-oxidation is just one of them) and the work so fundamental and novel that Brown was a corecipient of the 1979 Nobel Prize in Chemistry.

In the first stage of the hydroboration-oxidation procedure an alkene is treated with diborane (B_2H_6). Diborane is a dimer [from the Greek *di* ("two") and *meros* ("parts")] of borane (BH_3).

$$2BH_3 \rightleftharpoons B_2H_6$$

Borane	Diborane
(monomer)	(dimer)

Diborane reacts with alkenes by a process in which a carbon-hydrogen and a carbon-boron bond are formed from each boron-hydrogen bond. This reaction is known as **hydroboration**.

$$3CH_2{=}CH_2 + \tfrac{1}{2}B_2H_6 \longrightarrow CH_3CH_2{-}B\overset{\displaystyle CH_2CH_3}{\underset{\displaystyle CH_2CH_3}{\Big\backslash}}$$

Ethylene   Diborane   Triethylborane

In the hydroboration of unsymmetrical alkenes, boron becomes attached to the less-substituted carbon of the double bond, i.e., the carbon which bears the greater number of hydrogens.

$$3\ CH_3(CH_2)_7CH{=}CH_2 + \tfrac{1}{2}B_2H_6 \longrightarrow [CH_3(CH_2)_7CH_2CH_2]_3B$$

1-Decene   Diborane   Tridecylborane

PROBLEM 5.4
Write a structural formula for the trialkylborane formed by addition of diborane to 2-methylpropene [$(CH_3)_2C{=}CH_2$].

After hydroboration a basic solution of hydrogen peroxide is added. Under these conditions the trialkylborane is converted to an alcohol. This is the **oxidation** stage of the sequence; hydrogen peroxide is the oxidizing agent.

$$[CH_3(CH_2)_7CH_2CH_2]_3B \xrightarrow[\text{NaOH}]{H_2O_2} CH_3(CH_2)_7CH_2CH_2OH$$

Tridecylborane   1-Decanol

(When sodium hydroxide is used as the base, boron of the trialkylborane is converted to the water-soluble and easily removed sodium salt of boric acid.)

PROBLEM 5.5
Write a structural formula for the alcohol isolated on oxidation of the trialkylborane formed in Problem 5.4.

The overall result is hydration of the double bond with a regioselectivity opposite to that obtained by acid-catalyzed hydration. It is customary to combine the two reactions, hydroboration and oxidation, in a single equation, indicating the reagents sequentially above and below the arrow.

$$CH_3(CH_2)_7CH{=}CH_2 \xrightarrow[\text{2. } H_2O_2,\ HO^-]{\text{1. } B_2H_6} CH_3(CH_2)_7CH_2CH_2OH$$

1-Decene   1-Decanol (93%)

(Notice that it is necessary to separate the two reactions by number when indicating reagents above and below the reaction arrow. Were each reaction not numbered, it could be erroneously inferred that all the reagents were added at the same time.)

PROBLEM 5.6
Write an equation showing the hydroboration-oxidation of each of the following alkenes:

    (a) 2-Methyl-2-butene   (b) 3,3-Dimethyl-1-butene   (c) cis-2-Butene

SAMPLE SOLUTION
(a) First write a structural formula for the alkene.

handwritten: hydration of alkene → alcohol (sub)

$$\overset{1}{C}H_3 \diagdown \underset{\underset{CH_3}{|}}{\overset{2}{C}} = \underset{\underset{\underset{4}{CH_3}}{|}}{\overset{3}{C}} \diagup \overset{H}{} \qquad \text{2-Methyl-2-butene}$$

Keep in mind that hydroboration-oxidation accomplishes the addition of the components of water (H and OH) to the double bond so that the hydroxyl group is bonded to the less-substituted carbon—the one that bears the greater number of hydrogen substituents. Thus, the hydroxyl group becomes bonded to C-3 of 2-methyl-2-butene and the hydrogen to C-2.

$$(CH_3)_2C{=}CHCH_3 \xrightarrow[\text{2. } H_2O_2,\ HO^-]{\text{1. } B_2H_6} (CH_3)_2CHCHCH_3 \atop \qquad\qquad\qquad\qquad\qquad \underset{OH}{|}$$

2-Methyl-2-butene   3-Methyl-2-butanol (98%)

An aspect of hydroboration-oxidation that bears mention has to do with its stereochemistry. As illustrated for the case of 1-methylcyclopentene, the H and OH groups are added to the same face of the double bond.

1-Methylcyclopentene $\xrightarrow[\text{2. } H_2O_2,\ HO^-]{\text{1. } B_2H_6}$ trans-2-Methylcyclopentanol
(only product, 86% yield)

When two atoms or groups add to the same face of a double bond, we say that **syn addition** has occurred. The converse situation, where they add to opposite faces of the double bond, is called **anti addition.** An example of anti addition is given in the next section.

5.1.7 Addition of Halogens to Alkenes

We have seen earlier (Sections 3.15 to 3.18) that chlorine reacts with alkanes to give alkyl chlorides by a free radical substitution reaction.

$$R{-}H + Cl_2 \xrightarrow[\text{light}]{\text{heat or}} R{-}Cl + HCl$$

Alkane   Chlorine   Alkyl chloride   Hydrogen chloride

In contrast to its behavior toward alkanes, chlorine, and bromine as well, react with alkenes by *addition* rather than substitution. The products of these reactions are vicinal dihalides (Section 4.3.3).

$$\underset{\text{Alkene}}{\diagup\diagdown}{C}{=}{C}\diagdown\diagup + \underset{\text{Halogen}}{X_2} \longrightarrow \underset{\text{Vicinal dihalide}}{X{-}\overset{|}{\underset{|}{C}}{-}\overset{|}{\underset{|}{C}}{-}X} \qquad (X = Cl \text{ or } Br)$$

For example,

$$(CH_3)_3CCH{=}CH_2 + Cl_2 \longrightarrow (CH_3)_3CCH{-}CH_2Cl \atop \qquad\qquad\qquad\qquad\qquad \underset{Cl}{|}$$

3,3-Dimethyl-l-butene   Chlorine   1,2-Dichloro-3,3-
dimethylbutane (53%)

Addition of bromine to ethylene yields 1,2-dibromoethane.

$$CH_2\!=\!CH_2 + Br_2 \longrightarrow BrCH_2CH_2Br$$

Ethylene   Bromine   1,2-Dibromoethane
(ethylene dibromide)

The common name for this substance is ethylene dibromide (EDB). Until it was banned in 1984, about 12×10^6 lb/year of EDB was produced in the United States for use as an agricultural pesticide and soil fumigant.

PROBLEM 5.7

Write a series of equations showing how you could prepare 1,2-dibromo-2-methylpropane from 2-bromo-2-methylpropane.

When addition of chlorine and bromine to cycloalkenes is examined, an important stereochemical feature of these reactions becomes apparent. The halogen substituents in the product are trans to one another.

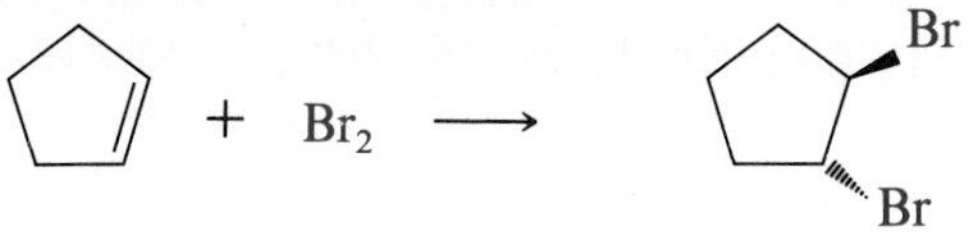

Cyclopentene   Bromine   *trans*-1,2-Dibromocyclopentane
(none of the cis
isomer is formed)

The addition is exclusively anti; the two halogens add to opposite faces of the double bond.

The mechanism depicted in Figure 5.3 for the addition of bromine to cyclopentene is generally accepted as correct for electrophilic addition of halogens to alkenes.

Step 1: The two reactants approach each other and mutually polarize the electron distribution in the π component of the alkene double bond and the bromine-bromine bond. On passing the transition state, a bridged bromonium ion is formed.

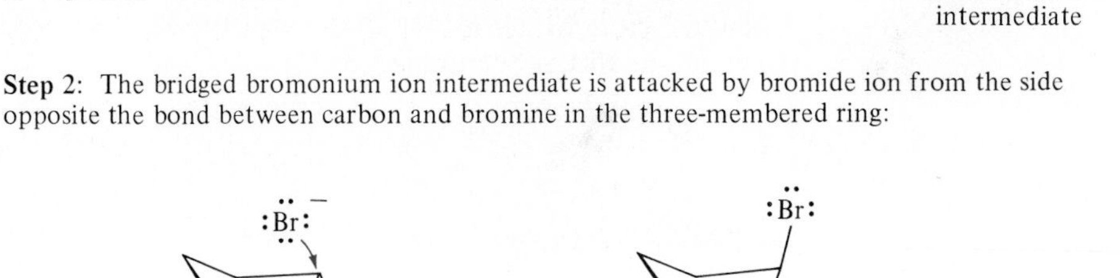

Cyclopentene + bromine   Polarized transition state   Bromonium ion
intermediate

Step 2: The bridged bromonium ion intermediate is attacked by bromide ion from the side opposite the bond between carbon and bromine in the three-membered ring:

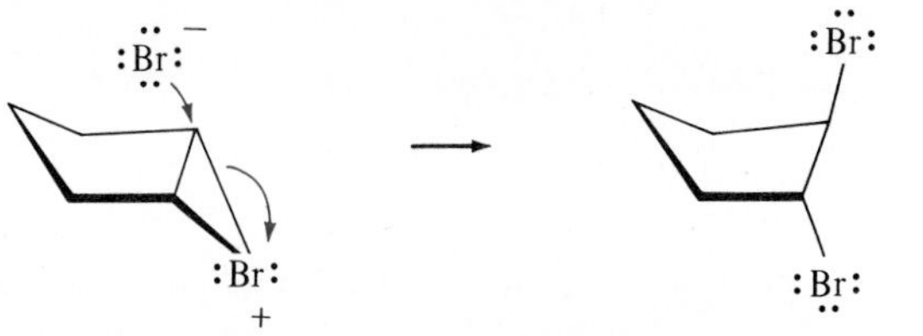

Bromonium ion
intermediate   *trans*-1,2-Dibromocyclopentane

FIGURE 5.3 Mechanistic depiction of electrophilic addition of bromine to cyclopentene showing the role of the bridged bromonium ion.

The key intermediate formed in step 1 of this mechanism is a bridged **halonium ion** (in this case, a bridged **bromonium ion**). The bromine and both carbon atoms of the three-membered ring in the bromonium ion have octets of electrons. Because of this the bromonium ion is more stable than any carbocation intermediate capable of being formed under the reaction conditions. Step 2 of the mechanism proposes that a negatively charged bromide ion attacks the bromonium ion at carbon from the side opposite the carbon-bromine bond. Notice in Figure 5.3 that the two bromines in the vicinal dibromide product formed in this step are trans to one another as the experimentally observed anti addition requires.

5.1.8 Free Radical Addition of Hydrogen Bromide to Alkenes

With the exception of catalytic hydrogenation, all the reactions of alkenes described to this point have involved attack by some electrophilic reagent on the π electrons of the double bond. Certain other kinds of reactive species, such as free radicals, are also capable of attacking double bonds. An example of a free radical addition to an alkene is the peroxide-induced addition of hydrogen bromide. Peroxides are compounds that contain O—O bonds whose cleavage can serve to initiate a free radical chain reaction. In the presence of peroxides, hydrogen bromide (but none of the other hydrogen halides) adds to alkenes with a regioselectivity *opposite* to that of Markovnikov's rule.

Addition of HBr in the absence of peroxides:

$$CH_2{=}CHCH_2CH_3 \;+\; \quad HBr \quad \xrightarrow[\text{peroxides}]{\text{no}} \quad CH_3CHCH_2CH_3$$
$$\overset{|}{Br}$$

| 1-Butene | Hydrogen bromide | 2-Bromobutane
(only product; 90% yield) |

Addition of HBr in the presence of peroxides: *Gives Hee radicals*

$$CH_2{=}CHCH_2CH_3 \;+\; \quad HBr \quad \xrightarrow{\text{peroxides}} BrCH_2CH_2CH_2CH_3$$

| 1-Butene | Hydrogen bromide | 1-Bromobutane
(only product; 95% yield) |

PROBLEM 5.8
Write a series of equations showing how you could prepare 1-bromo-2-methylpropane from 2-bromo-2-methylpropane.

The mechanism of the free radical addition of hydrogen bromide to 1-butene is presented in Figure 5.4. The regioselectivity of addition is determined in step 3 of the mechanism where a bromine atom adds to the alkene. The bromine adds to the double bond in the direction that leads to the more stable of two possible free radical intermediates. In the case of 1-butene, the bromine atom adds to C-1 because this generates a secondary radical. Had bromine added to C-2, the free radical produced would have been a less stable primary radical. Once the carbon-bromine bond is formed in step 3, hydrogen abstraction from HBr in step 4 completes the process.

The regioselectivity of addition of hydrogen bromide to alkenes in the presence of peroxides is different from that observed in the absence of peroxides because the mechanism of addition is different for the two cases. In the presence of peroxides, the

$$CH_3CH_2CH{=}CH_2 \quad + \quad HBr \quad \xrightarrow{\text{ROOR}} \quad CH_3CH_2CH_2CH_2\,Br$$

1-Butene   Hydrogen bromide   1-Bromobutane

The mechanism

(a) Initiation

  Step 1: Dissociation of a peroxide into two alkoxy radicals:

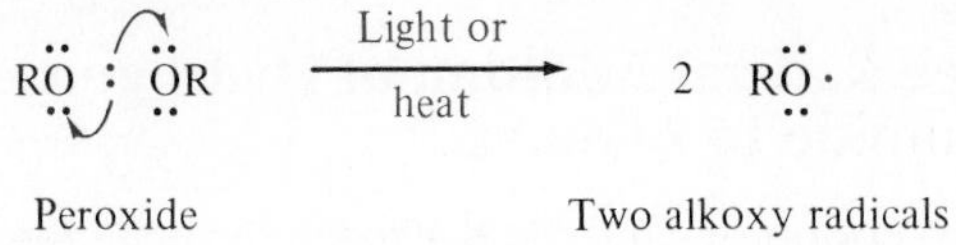

$$R\ddot{O} : \ddot{O}R \quad \xrightarrow[\text{heat}]{\text{Light or}} \quad 2\ \ R\ddot{O}\cdot$$

  Peroxide   Two alkoxy radicals

  Step 2: Hydrogen atom abstraction from hydrogen bromide by an alkoxy radical:

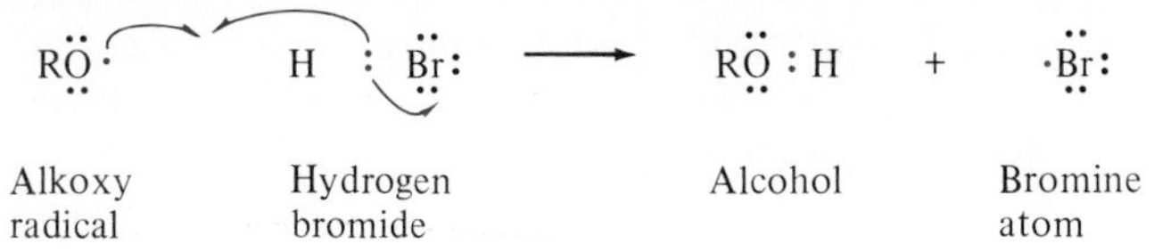

$$R\ddot{O}\cdot \quad H : \ddot{B}r: \quad \longrightarrow \quad R\ddot{O}:H \quad + \quad \cdot\ddot{B}r:$$

  Alkoxy   Hydrogen   Alcohol   Bromine
  radical   bromide   atom

(b) Chain propagation

  Step 3: Addition of a bromine atom to the alkene:

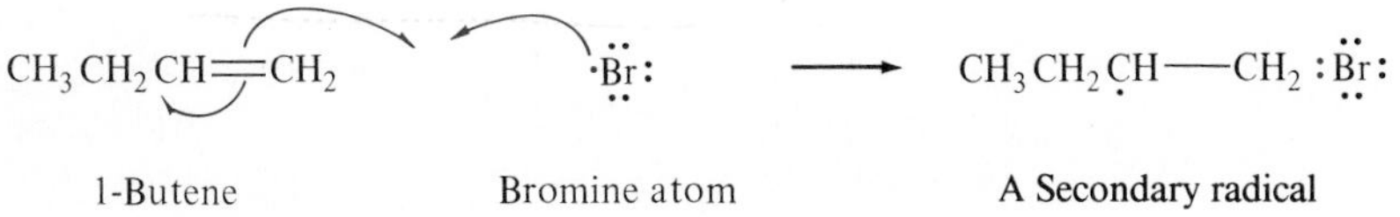

$$CH_3CH_2CH{=}CH_2 \quad \cdot\ddot{B}r: \quad \longrightarrow \quad CH_3CH_2\overset{\cdot}{C}H{-\!\!-}CH_2:\ddot{B}r:$$

  1-Butene   Bromine atom   **A Secondary radical**

  Step 4: Abstraction of a hydrogen atom from hydrogen bromide by the free radical formed in step 3:

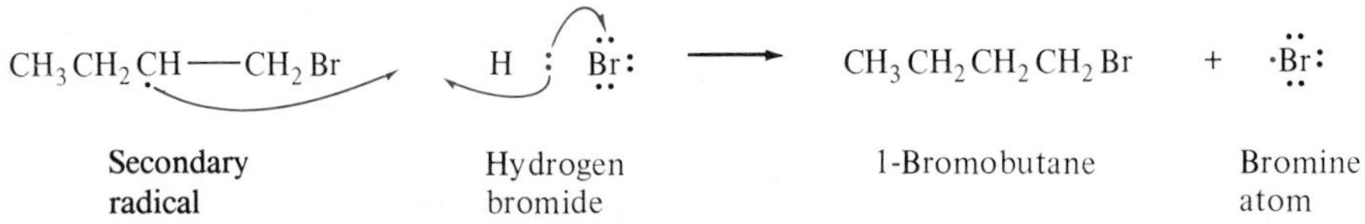

$$CH_3CH_2\overset{\cdot}{C}H{-\!\!-}CH_2Br \quad H : \ddot{B}r: \quad \longrightarrow \quad CH_3CH_2CH_2CH_2Br \quad + \quad \cdot\ddot{B}r:$$

  Secondary   Hydrogen   1-Bromobutane   Bromine
  radical   bromide   atom

FIGURE 5.4 The initiation and propagation steps in the free-radical addition of hydrogen bromide to 1-butene.

first species to add to the double bond is a bromine atom in a step which yields a free radical; in the absence of peroxides, the first species to add is a proton in a step which yields a carbocation.

5.1.9 Polymerization of Alkenes

Peroxides can initiate other free radical addition reactions of alkenes. One that is exceptionally important is free radical polymerization [from the Greek *poly* ("many") and *meros* ("parts")], a reaction in which alkene molecules add to one another repeatedly to form long chains. We have noted earlier (Section 4.1.6) the enormous volume of ethylene production in the petrochemical industry. Most of this ethylene is used to prepare **polyethylene,** a high-molecular-weight polymer formed by heating ethylene at high pressure in the presence of oxygen or a peroxide.

$$n\mathrm{CH_2{=}CH_2} \xrightarrow[\substack{O_2\ \text{or} \\ \text{peroxides}}]{\substack{200^\circ C \\ 2000\ \text{atm}}} {-}\mathrm{CH_2{-}CH_2{-}(CH_2{-}CH_2)}_{n-2}{-}\mathrm{CH_2{-}CH_2{-}}$$

   Ethylene   Polyethylene

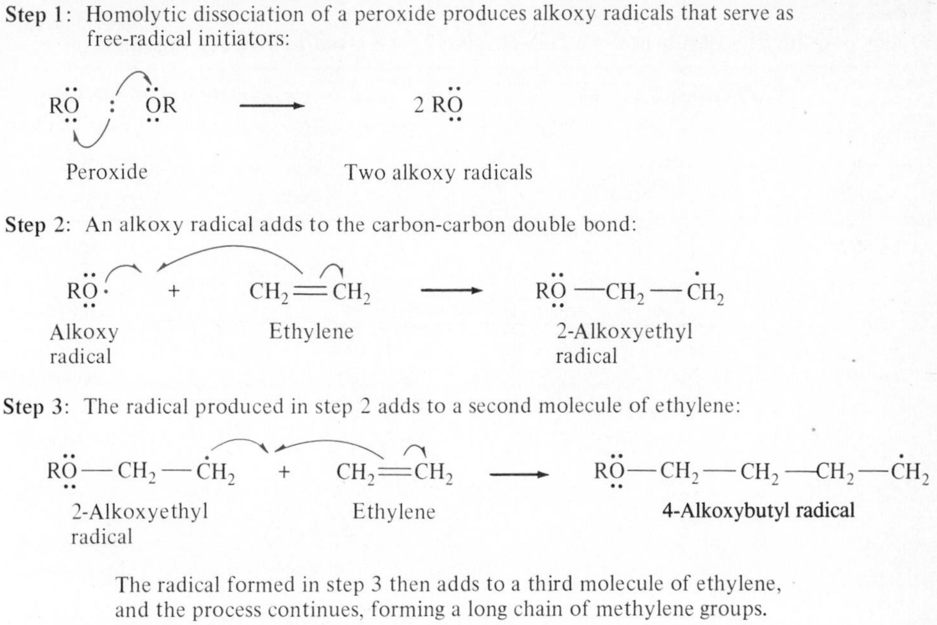

Step 1: Homolytic dissociation of a peroxide produces alkoxy radicals that serve as free-radical initiators:

$$\ddot{R\ddot{O}} : \ddot{O}R \longrightarrow 2\ R\ddot{O}$$

Peroxide   Two alkoxy radicals

Step 2: An alkoxy radical adds to the carbon-carbon double bond:

$$R\ddot{\ddot{O}}\cdot \ + \ CH_2 = CH_2 \longrightarrow R\ddot{\ddot{O}} - CH_2 - \dot{C}H_2$$

Alkoxy   Ethylene   2-Alkoxyethyl
radical   radical

Step 3: The radical produced in step 2 adds to a second molecule of ethylene:

$$R\ddot{\ddot{O}} - CH_2 - \dot{C}H_2 \ + \ CH_2 = CH_2 \longrightarrow R\ddot{\ddot{O}} - CH_2 - CH_2 - CH_2 - \dot{C}H_2$$

2-Alkoxyethyl   Ethylene   4-Alkoxybutyl radical
radical

The radical formed in step 3 then adds to a third molecule of ethylene, and the process continues, forming a long chain of methylene groups.

FIGURE 5.5 Mechanistic description of peroxide-induced free-radical polymerization of ethylene.

In this reaction *n* can have a value of thousands.

An outline of the mechanism of free radical polymerization of ethylene is shown in Figure 5.5. Dissociation of a peroxide initiates the process in step 1. The resulting peroxy radical adds to the carbon-carbon double bond in step 2, giving a new radical, which then adds to a second molecule of ethylene in step 3. The carbon-carbon bond-forming process in step 3 can be repeated thousands of times to give long carbon chains.

In spite of the *-ene* ending to its name, polyethylene is much more closely related to alkanes than to alkenes. It is simply a long chain of CH_2 groups bearing at its ends an alkoxy group (from the initiator) or a carbon-carbon double bond.

A large number of compounds with carbon-carbon double bonds have been polymerized to yield materials having useful properties. Some of the more important or familiar of these are listed in Table 5.1. Not all of them are effectively polymerized under free radical conditions, and much research has been carried out to develop alternative polymerization techniques. One of these, **coordination polymerization,** utilizes a mixture of titanium tetrachloride ($TiCl_4$) and triethylaluminum $[(CH_3CH_2)_3Al]$ as a catalyst. Polyethylene produced by coordination polymerization has a higher density than that produced by free radical polymerization and somewhat different — in many applications, more desirable — properties. The catalyst system used in coordination polymerization was developed independently by Karl Ziegler in Germany and Giulio Natta in Italy in the early 1950s. They shared the Nobel Prize in Chemistry in 1963 for this work. The Ziegler-Natta catalyst system also permits polymerization of propene to be achieved in a way that gives a form of *polypropylene* suitable for plastics and fibers. When propene is polymerized under free radical conditions, the polypropylene has physical properties (such as a low melting point) that preclude its use in plastics and fibers.

5.2 CLEAVAGE OF ALKENES

A number of electrophilic reagents attack the carbon-carbon double bond of alkenes but are sufficiently aggressive so that reaction continues beyond the addition stage,

TABLE 5.1

Some Compounds with Carbon-Carbon Double Bonds Used to Prepare Polymers

A. Alkenes of the type $CH_2{=}CH{-}X$ used to form polymers of the type $(-CH_2-\underset{\underset{\textstyle X}{|}}{CH}-)_n$

Compound	Structure	$-X$ in polymer	Application
Ethylene	$CH_2{=}CH_2$	$-H$	Polyethylene films as packaging material; "plastic" squeeze bottles are molded from high-density polyethylene
Propene	$CH_2{=}CH{-}CH_3$	$-CH_3$	Polypropylene fibers for use in carpets and automobile tires; consumer items (luggage, appliances, etc.); packaging material
Styrene	$CH_2{=}CH{-}$⬡ 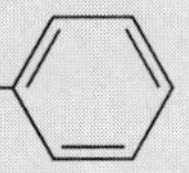	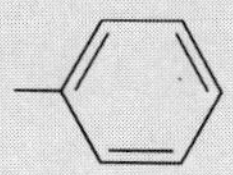 ⬡	Polystyrene packaging, housewares, luggage, radio and television cabinets
Vinyl chloride	$CH_2{=}CH{-}Cl$	$-Cl$	Polyvinyl chloride (PVC) has replaced leather in many of its applications; PVC tubes and pipes often used in place of copper
Acrylonitrile	$CH_2{=}CH{-}C{\equiv}N$	$-C{\equiv}N$	Wool substitute in sweaters, blankets, etc.

B. Alkenes of the type $CH_2{=}CX_2$ used to form polymers of the type $(-CH_2-CX_2-)n$

Compound	Structure	X in polymer	Application
1,1-Dichloroethene (Vinylidene chloride)	$CH_2{=}CCl_2$	Cl	Saran used as air- and water-tight packaging film
2-Methylpropene	$CH_2{=}C(CH_3)_2$	CH_3	Polyisobutene is component of "butyl rubber," one of earliest synthetic rubber substitutes

C. Others

Compound	Structure	Polymer	Application
Tetrafluoroethene	$CF_2{=}CF_2$	$(-CF_2-CF_2-)_n$ (Teflon)	Nonstick coating for cooking utensils; bearings, gaskets, and fittings
Methyl methacrylate	$CH_2{=}\underset{\underset{\textstyle CH_3}{|}}{C}CO_2CH_3$	$(-CH_2-\overset{\overset{\textstyle CO_2CH_3}{|}}{\underset{\underset{\textstyle CH_3}{|}}{C}}-)_n$	When cast in sheets is transparent; used as glass substitute (Lucite, Plexiglas)
2-Methyl-1,3-butadiene	$CH_2{=}\underset{\underset{\textstyle CH_3}{|}}{C}CH{=}CH_2$	$(-CH_2\underset{\underset{\textstyle CH_3}{|}}{C}{=}CH{-}CH_2-)_n$ (Polyisoprene)	Synthetic rubber

ETHYLENE AND PROPENE: THE MOST IMPORTANT INDUSTRIAL ORGANIC CHEMICALS

Two major components of the petrochemical industry are (1) the cracking of hydrocarbons to ethylene and propene and (2) the use of ethylene and propene as starting materials for the preparation of other products. We discussed ethylene and propene production in an earlier boxed essay (Sec. 4.1.6) and now turn our attention to some of the uses to which these two alkenes are put.

ETHYLENE Approximately 90 percent of the 3×10^{10} lb of ethylene produced annually in the United States is used for the preparation of four compounds (polyethylene, ethylene oxide, vinyl chloride, and styrene), with polymerization to polyethylene accounting for one-half the total.

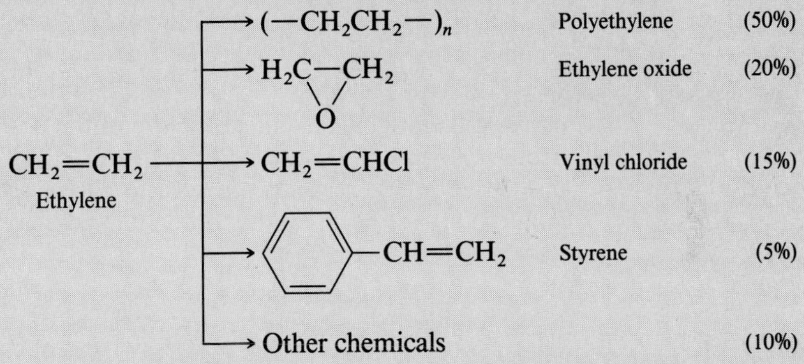

$(-CH_2CH_2-)_n$	Polyethylene	(50%)
H_2C-CH_2 (O)	Ethylene oxide	(20%)
$CH_2=CHCl$	Vinyl chloride	(15%)
⬡$-CH=CH_2$	Styrene	(5%)
Other chemicals		(10%)

$CH_2=CH_2$ (Ethylene)

Both vinyl chloride and styrene are polymerized to give polyvinyl chloride and polystyrene, respectively (Table 5.1). Ethylene glycol is prepared from ethylene oxide for use as an antifreeze in automobile radiators and in the production of polyester fibers.

Among the "other chemicals" prepared from ethylene are the following:

$ClCH_2CH_2Cl$	CH_3CH_2OH	CH_3CH (with $\overset{O}{\parallel}$)
1,2-Dichloroethane (additive in leaded gasoline)	Ethanol (industrial solvent; used in preparation of ethyl acetate; unleaded gasoline additive)	Acetaldehyde (used in preparation of acetic acid)

PROPENE The major use of propene is in the production of polypropylene.

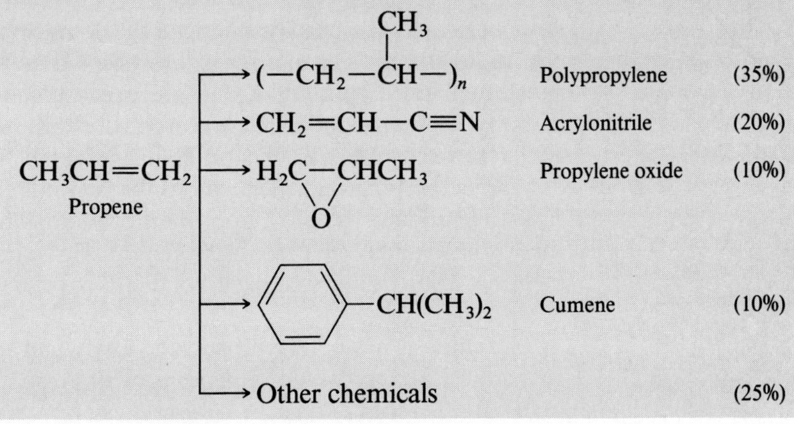

$(-CH_2-\overset{\displaystyle CH_3}{\overset{\displaystyle \vert}{CH}}-)_n$	Polypropylene	(35%)
$CH_2=CH-C\equiv N$	Acrylonitrile	(20%)
$H_2C-CHCH_3$ (O)	Propylene oxide	(10%)
⬡$-CH(CH_3)_2$	Cumene	(10%)
Other chemicals		(25%)

$CH_3CH=CH_2$ (Propene)

Two of the propene-derived organic chemicals shown, acrylonitrile and propylene oxide, are also starting materials for polymer synthesis. Acrylonitrile is used to make acrylic fibers (Table 5.1), while propylene oxide is one component in the preparation of *polyurethane* polymers. Cumene itself has no direct uses but rather serves as the starting material in a process which yields two valuable industrial chemicals, acetone and phenol.

We have not indicated the reagents employed in the reactions by which ethylene and propene are converted to the compounds shown. Because of patent requirements, different companies often use different processes. While the processes may be different, they share the common characteristic of being extremely efficient. The industrial chemist faces the challenge of producing valuable materials and at low cost as well. Thus, success in the industrial environment requires both an understanding of chemistry and an appreciation for the economics associated with alternative procedures.

cleaving the original alkene at the site of the double bond. Two such reagents are ozone (O_3) and potassium permanganate ($KMnO_4$). Procedures have been developed where these reagents are used to cleave alkenes so that each of the doubly bonded carbons becomes the carbon of a **carbonyl group** ($\diagdown C{=}O$).

Cleavage occurs here

$$\diagdown C{=}C \diagup \longrightarrow \diagdown C{=}O \ + \ O{=}C \diagup$$

Alkene   Two carbonyl-group-containing compounds

Ozone-induced cleavage is called **ozonolysis** and is carried out in two stages. In the first stage, the alkene is treated with ozone whereupon a rapid reaction ensues to give a species called an **ozonide.** This ozonide is then subjected to hydrolysis in the presence of zinc to give the carbonyl-containing reaction products.

$$\underset{H}{\overset{R}{\diagdown}} C{=}C \underset{R''}{\overset{R'}{\diagup}} \xrightarrow[\text{2. H}_2\text{O, Zn}]{\text{1. O}_3} \underset{H}{\overset{R}{\diagdown}} C{=}O + O{=}C \underset{R''}{\overset{R'}{\diagup}}$$

Alkene   Aldehyde   Ketone

Notice that a hydrogen substituent on the double bond becomes a hydrogen substituent on a carbonyl group after ozonolysis. Compounds which contain a hydrogen substituent on a carbonyl group are called **aldehydes.** Compounds in which the carbonyl group is bonded to two carbon substituents, alkyl groups, for example, are called **ketones.** We will encounter numerous classes of organic compounds that contain carbonyl groups throughout the remainder of the text, so it is useful at this time to introduce the most important of them. These are listed in Table 5.2. In all these compounds, the carbon-oxygen double bond is viewed in the same way as a carbon-carbon double bond; carbon is sp^2 hybridized, and the double bond is of the $\sigma + \pi$ type.

Ozonolysis has both synthetic and analytical applications in organic chemistry. In synthesis, ozonolysis of alkenes provides a method for the preparation of aldehydes and ketones.

TABLE 5.2
Carbonyl-Containing Functional Groups

Class	Representative example	Name of example*	Generalized abbreviation
Aldehyde *(easily oxidized)*	$\overset{O}{\overset{\|}{CH_3CH}}$	Ethanal (Acetaldehyde)	$\overset{O}{\overset{\|}{RCH}}$
Ketone *(stable)*	$\overset{O}{\overset{\|}{CH_3CCH_3}}$	2-Propanone (Acetone)	$\overset{O}{\overset{\|}{RCR}}$
Carboxylic acid	$\overset{O}{\overset{\|}{CH_3COH}}$	Ethanoic acid (Acetic acid)	$\overset{O}{\overset{\|}{RCOH}}$
Carboxylic acid derivatives:			
Acyl halide	$\overset{O}{\overset{\|}{CH_3CCl}}$	Ethanoyl chloride (Acetyl chloride)	$\overset{O}{\overset{\|}{RCX}}$
Acid anhydride	$\overset{O\ \ \ O}{\overset{\|\ \ \ \|}{CH_3COCCH_3}}$	Ethanoic anhydride (Acetic anhydride)	$\overset{O\ \ \ O}{\overset{\|\ \ \ \|}{RCOCR}}$
Ester	$\overset{O}{\overset{\|}{CH_3COCH_2CH_3}}$	Ethyl ethanoate (Ethyl acetate)	$\overset{O}{\overset{\|}{RCOR}}$
Amide	$\overset{O}{\overset{\|}{CH_3CNH_2}}$	Ethanamide (Acetamide)	$\overset{O}{\overset{\|}{RCNR_2}}$

* The preferred IUPAC name is given along with an acceptable synonym in parentheses.

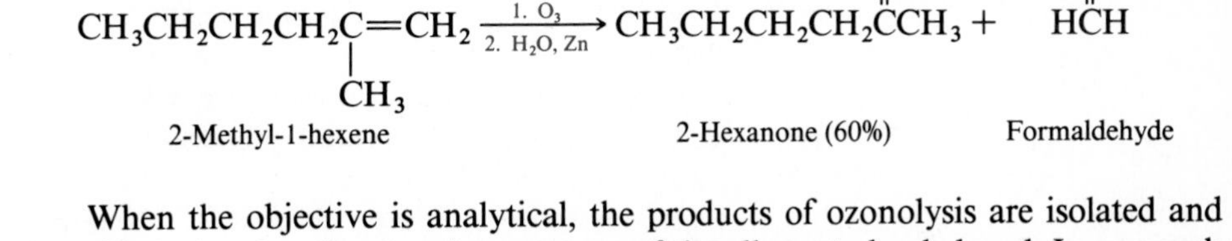

[handwritten annotations: "It's an aldehyde" "B–C—H / Formaldehyde" "Carbonyl"]

$$(CH_3)_2CHCH_2CH_2CH_2CH{=}CH_2 \xrightarrow[\text{2. H}_2\text{O, Zn}]{\text{1. O}_3}$$

6-Methyl-1-heptene

$$(CH_3)_2CHCH_2CH_2CH_2\overset{O}{\overset{\|}{CH}} + \overset{O}{\overset{\|}{HCH}}$$

5-Methylhexanal (62%) Formaldehyde

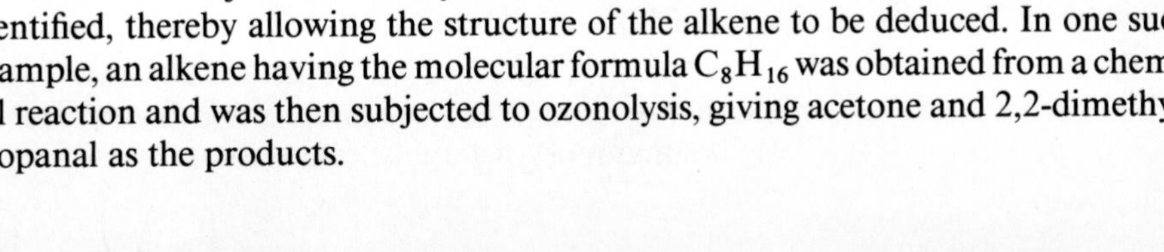

$$CH_3CH_2CH_2CH_2\underset{\underset{CH_3}{|}}{C}{=}CH_2 \xrightarrow[\text{2. H}_2\text{O, Zn}]{\text{1. O}_3} CH_3CH_2CH_2CH_2\overset{O}{\overset{\|}{C}}CH_3 + \overset{O}{\overset{\|}{HCH}}$$

2-Methyl-1-hexene   2-Hexanone (60%) Formaldehyde

When the objective is analytical, the products of ozonolysis are isolated and identified, thereby allowing the structure of the alkene to be deduced. In one such example, an alkene having the molecular formula C_8H_{16} was obtained from a chemical reaction and was then subjected to ozonolysis, giving acetone and 2,2-dimethylpropanal as the products.

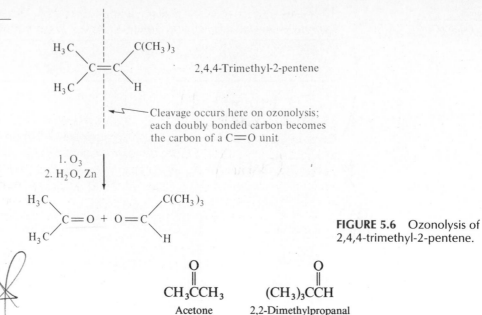

FIGURE 5.6 Ozonolysis of 2,4,4-trimethyl-2-pentene.

Together, these two products contain all eight carbons of the starting alkene. The two carbonyl carbons correspond to those that were doubly bonded in the original alkene. Therefore, one of the doubly bonded carbons bears two methyl substituents, the other bears a hydrogen and a *tert*-butyl group. The alkene is identified as 2,4,4-trimethyl-2-pentene, as shown in Figure 5.6.

PROBLEM 5.9

The same reaction that gave 2,4,4-trimethyl-2-pentene also yielded an isomeric alkene. This second alkene produced formaldehyde and 4,4-dimethyl-2-pentanone on ozonolysis. Identify this alkene.

$$\underset{\text{4,4-Dimethyl-2-pentanone}}{CH_3\overset{\overset{\displaystyle O}{\|}}{C}CH_2C(CH_3)_3}$$

Alkenes are also cleaved by heating with potassium permanganate.

$$\underset{\text{Alkene}}{\overset{R}{\underset{R'}{>}}C=C\overset{R''}{\underset{H}{<}}} \xrightarrow[\text{2. } H^+]{\text{1. } KMnO_4} \underset{\text{Ketone}}{\overset{R}{\underset{R'}{>}}C=O} + \underset{\text{Carboxylic acid}}{O=C\overset{R''}{\underset{OH}{<}}}$$

Notice that a hydrogen substituent on the double bond is not retained in the carbonyl compound under these conditions. Cleavage by permanganate causes a hydrogen substituent on the double bond to be replaced by a hydroxyl group, and the corresponding cleavage product is a carboxylic acid.

5.3 ACIDITY OF ALKYNES

Before describing addition reactions of alkynes, let us first discuss the property of alkynes that most distinguishes these compounds from other hydrocarbons—their acidity.

[handwritten notes in margins: "form C≡C bonds more easily", "Good b/c can build more things"]

As measured by their acid-dissociation constants K_a (Section 3.7), hydrocarbons are exceedingly weak acids. A carbon-hydrogen bond has little tendency to ionize according to the equation

$$R-H \rightleftharpoons R:^- + H^+$$

Hydrocarbon   Carbanion   Proton

Loss of a proton from a hydrocarbon leaves an anion in which the negative charge is borne by carbon; such a species is called a **carbanion.** Carbon is not very electronegative, and the formation of carbanions by ionization of C—H bonds is characterized by very small values of K_a. It is noteworthy, however, that the *range* of K_a values exhibited by hydrocarbons varies enormously when one compares ethane, ethylene, and acetylene as a representative alkane, alkene, and alkyne, respectively. As measured by K_a, acetylene is 10^{19} times more acidic than ethylene and 10^{36} times more acidic than ethane.

HC≡CH	$> CH_2=CH_2 >$	CH_3CH_3
Acetylene	Ethylene	Ethane
$K_a = 10^{-26}$	$\sim 10^{-45}$	$\sim 10^{-62}$
$pK_a = 26$	~ 45	~ 62
(Strongest acid)		(Weakest acid)

The greater acidity of acetylene is attributed to a hybridization effect. When carbon is sp hybridized, it is more electronegative than when it is sp^2 or sp^3 hybridized. Thus, the carbanion which results from ionization of acetylene binds its electron pair more strongly than do the carbanions from ethylene and ethane, is more stable than them, and its formation is associated with a higher equilibrium constant. **Terminal alkynes,** compounds of the type RC≡CH, have values of K_a which are similar to that of acetylene.

$$(CH_3)_3C-C≡C-H \rightleftharpoons (CH_3)_3C-C≡C:^- + H^+$$

3,3-Dimethyl-1-butyne
$K_a = 3 \times 10^{-26}$
$pK_a = 25.5$

Hydrocarbons with **internal triple bonds,** compounds of the type RC≡CR′, lack protons bonded to sp hybridized carbon and are more like alkanes in respect to their acid strength.

While acetylene and terminal alkynes are far stronger acids than other hydrocarbons, it must be remembered that they are, nevertheless, very weak acids—much weaker than water and alcohols, for example. Hydroxide ion is too weak a base to convert acetylene to its anion in meaningful amounts. The position of the equilibrium described by the following equation lies overwhelmingly to the left.

$$H-C≡C-H + \ :\overset{..}{O}H \rightleftharpoons H-C≡C:^- + \ H\overset{..}{O}H$$

Acetylene	Hydroxide ion	Acetylide ion	Water
(Weaker acid)	(Weaker base)	(Stronger base)	(Stronger acid)
$K_a = 10^{-26}$			$K_a = 1.8 \times 10^{-16}$
$pK_a = 26$			$pK_a = 15.7$

Because acetylene is a far weaker acid than water and alcohols, these substances are not suitable solvents for reactions involving acetylide ions. Acetylide is instantly converted to acetylene by proton transfer from compounds that contain hydroxyl groups.

An amide ion is a much stronger base than an acetylide ion and converts acetylene to its conjugate base quantitatively.

$$H-C\equiv C-H + \quad ^-:\ddot{N}H_2 \quad \rightleftharpoons H-C\equiv C:^- + \quad \ddot{N}H_3$$

Acetylene	Amide ion	Acetylide ion	Ammonia
(Stronger acid)	(Stronger base)	(Weaker base)	(Weaker acid)
$K_a = 10^{-26}$			$K_a = 10^{-36}$
$pK_a = 26$			$pK_a = 36$

Solutions of sodium acetylide ($HC\equiv CNa$) may be prepared by adding sodium amide to acetylene in liquid ammonia as the solvent. Terminal alkynes react similarly to give species of the type $RC\equiv CNa$.

5.4 PREPARATION OF ALKYNES BY ALKYLATION REACTIONS

The reactions described so far for the preparation of alkynes involved introducing a triple bond into an existing carbon chain by double dehydrohalogenation of a geminal dihalide or a vicinal dihalide. In this section we will see how alkynes are prepared by substitution of a terminal hydrogen atom by an alkyl group.

$H-C\equiv C-H$	$R-C\equiv C-H$	$R-C\equiv C-R'$
Acetylene	Monosubstituted or terminal alkyne	Disubstituted derivative of acetylene

Reactions that lead to attachment of alkyl groups to a molecule are called **alkylation** reactions.

Alkylation of acetylene is a synthetic process comprising two separate reactions carried out in sequence. In the first stage, acetylene is converted to its conjugate base by treatment with sodium amide.

$$HC\equiv CH + \quad NaNH_2 \quad \longrightarrow \quad HC\equiv CNa \quad + \quad NH_3$$

Acetylene	Sodium amide	Sodium acetylide	Ammonia

Next, an alkyl halide such as 1-bromobutane is added to the solution of sodium acetylide. A reaction of a type to be described in Chapter 8 ensues in which the alkyl group of the alkyl halide becomes bonded to the negatively charged carbon of sodium acetylide.

$$HC\equiv C^-Na^+ + CH_3CH_2CH_2CH_2Br \xrightarrow{NH_3} CH_3CH_2CH_2CH_2C\equiv CH + NaBr$$

Sodium acetylide	1-Bromobutane	1-Hexyne (70–77%)	Sodium bromide

For reasons to be explained in Chapter 8, *the alkyl halide must be a methyl halide or a primary alkyl halide.*

An analogous sequence using terminal alkynes as starting materials yields alkynes of the type $RC\equiv CR'$.

$$(CH_3)_2CHCH_2C\equiv CH \xrightarrow[NH_3]{NaNH_2}$$

4-Methyl-1-pentyne

$$(CH_3)_2CHCH_2C\equiv CNa \xrightarrow{CH_3Br} (CH_3)_2CHCH_2C\equiv CCH_3$$

5-Methyl-2-hexyne (81%)

Dialkylation of acetylene can be achieved by carrying out the sequence twice.

$$HC\equiv CH \xrightarrow[2.\ CH_3CH_2Br]{1.\ NaNH_2,\ NH_3} HC\equiv CCH_2CH_3 \xrightarrow[2.\ CH_3Br]{1.\ NaNH_2,\ NH_3} CH_3C\equiv CCH_2CH_3$$

Acetylene	1-Butyne	2-Pentyne (81%)

PROBLEM 5.10

Outline efficient syntheses of each of the following alkynes from acetylene and any necessary organic or inorganic reagents:

(a) 1-Heptyne   (b) 2-Heptyne   (c) 3-Heptyne

SAMPLE SOLUTION

(a) An examination of the structural formula of 1-heptyne reveals it to have a pentyl group attached to an acetylene unit. Alkylation of acetylene, by way of its anion, with a pentyl halide is a suitable synthetic route to 1-heptyne.

$$HC\equiv CH \xrightarrow[NH_3]{NaNH_2} HC\equiv CNa \xrightarrow{CH_3CH_2CH_2CH_2CH_2Br} HC\equiv CCH_2CH_2CH_2CH_2CH_3$$

Acetylene   Sodium acetylide   1-Heptyne

5.5 ADDITION REACTIONS OF ALKYNES

Alkynes react with many of the same reagents that react with alkenes. Thus, in the sections that follow you will see the triple bond act as a functional group toward addition of hydrogen, hydrogen halides, and halogens in much the same way that the double bond of an alkene does. With respect to hydration, however, you will see that addition of a molecule of water to a triple bond sets in motion a process which gives an unexpected product.

5.5.1 Hydrogenation

The conditions for hydrogenation of alkynes are similar to those employed for alkenes. In the presence of finely divided nickel or platinum catalysts, two molecules of H_2 add to the triple bond of an alkyne to yield an alkane.

$$RC\equiv CR' + 2H_2 \xrightarrow[\text{catalyst}]{\text{metal}} RCH_2CH_2R'$$

Alkyne   Hydrogen   Alkane

$$CH_3CH_2CHCH_2C\equiv CH + 2H_2 \xrightarrow{Ni} CH_3CH_2CHCH_2CH_2CH_3$$
$$\qquad\quad | \qquad\qquad\qquad\qquad\qquad\qquad\qquad | $$
$$\qquad\quad CH_3 \qquad\qquad\qquad\qquad\qquad\qquad\quad CH_3$$

4-Methyl-1-hexyne   Hydrogen   3-Methylhexane (77%)

Alkenes are intermediates in the hydrogenation of alkynes to alkanes, a fact which has led to the development of methods for preparation of alkenes from alkynes as starting materials. Using a specially prepared palladium catalyst known as **Lindlar palladium,** it is possible to halt the hydrogenation of an alkyne cleanly after 1 mol of H_2 has been consumed per mole of alkyne. Syn addition of hydrogen to the triple bond takes place at the surface of the metal and the product is a cis alkene.

$$CH_3(CH_2)_3C\equiv C(CH_2)_3CH_3 \xrightarrow[\text{Lindlar Pd}]{H_2} \begin{array}{c} CH_3(CH_2)_3 \qquad (CH_2)_3CH_3 \\ \diagdown \qquad\qquad \diagup \\ C=C \\ \diagup \qquad\qquad \diagdown \\ H \qquad\qquad\quad H \end{array}$$

5-Decyne   cis-5-Decene (87%)

PROBLEM 5.11

Write a series of equations showing how you could prepare cis-2-pentene from 1-butyne.

Lindlar palladium is said to be a "poisoned" catalyst. Treatment of palladium with a combination of lead acetate, barium sulfate, and a substance called *quinoline* degrades its catalytic activity so that, while it is still an effective catalyst for the hydrogenation of alkynes to alkenes, it is insufficiently reactive to catalyze the hydrogenation of alkenes to alkanes.

5.5.2 Metal-Ammonia Reduction of Alkynes

A second method for converting alkynes to alkenes does not involve H_2 as a reagent, so it is not proper to refer to it as a hydrogenation reaction. In this reaction, a metal such as lithium, sodium, or potassium in liquid ammonia is used as a reducing agent to transfer electrons to the alkyne which then abstracts a proton from solvent. The net result is similar; an alkyne is converted to an alkene, but the mechanism is different and the stereochemistry is different. Anti addition of two hydrogen atoms to the triple bond is observed and the product is a trans alkene.

$$CH_3CH_2C{\equiv}CCH_2CH_3 \xrightarrow[NH_3]{Na} \begin{array}{c} CH_3CH_2 \\ \diagup \\ H \end{array} C{=}C \begin{array}{c} H \\ \diagdown \\ CH_2CH_3 \end{array}$$

3-Hexyne   *trans*-3-Hexene (82%)

PROBLEM 5.12
Write a series of equations showing how you could prepare *trans*-2-pentene from 1-butyne.

5.5.3 Addition of Hydrogen Halides to Alkynes

Alkynes react with hydrogen halides in a manner analogous to that of alkenes. Because an alkyne has two π bonds, however, it has the capacity to react with one molecule of a hydrogen halide or with two molecules. The product from addition of one molecule of an alkyl halide to an alkyne is an **alkenyl halide.**

$$RC{\equiv}CR' + \quad HX \quad \longrightarrow \quad RCH{=}\underset{\underset{X}{|}}{C}R'$$

Alkene   Hydrogen halide   Alkenyl halide

The regioselectivity of addition follows Markovnikov's rule. A proton adds to the carbon that has the greater number of hydrogen substituents, while halide adds to the carbon with the fewer hydrogen substituents.

$$CH_3CH_2CH_2CH_2C{\equiv}CH + \quad HI \quad \longrightarrow \quad CH_3CH_2CH_2CH_2\underset{\underset{I}{|}}{C}{=}CH_2$$

1-Hexyne   Hydrogen iodide   2-Iodo-1-hexene (73%)

In the presence of excess hydrogen halide, geminal dihalides are formed by sequential addition of two molecules of hydrogen halide to the carbon-carbon triple bond.

$$RC{\equiv}CR' \xrightarrow{HX} RCH{=}\underset{\underset{X}{|}}{C}R' \xrightarrow{HX} RCH_2\underset{\underset{X}{|}}{\overset{\overset{X}{|}}{C}}R'$$

Alkene   Alkenyl halide   Geminal dihalide

The hydrogen halide adds to the initially formed alkenyl halide in accordance with Markovnikov's rule. Overall, both protons become bonded to the same carbon and both halogens to the adjacent carbon.

$$CH_3CH_2C{\equiv}CCH_2CH_3 + \quad 2HF \quad \longrightarrow CH_3CH_2CH_2\overset{\overset{\displaystyle F}{|}}{\underset{\underset{\displaystyle F}{|}}{C}}CH_2CH_3$$

3-Hexyne   Hydrogen fluoride   3,3-Difluorohexane (76%)

PROBLEM 5.13
Write a series of equations showing how you could prepare 1,1-dichloroethane from:

(a) Ethylene   (c) Vinyl bromide ($CH_2{=}CHBr$)
(b) Vinyl chloride ($CH_2{=}CHCl$)   (d) 1,1-Dibromoethane

SAMPLE SOLUTION
(a) Reasoning backward, we recognize 1,1-dichloroethane as the product of addition of two molecules of hydrogen chloride to acetylene. Thus, the synthesis requires converting ethylene to acetylene as a key feature. As described in Section 4.3.3, this may be accomplished by conversion of ethylene to a vicinal dihalide, followed by double dehydrohalogenation. A suitable synthesis based on this analysis is as shown:

$$CH_2{=}CH_2 \overset{Br_2}{\longrightarrow} \quad BrCH_2CH_2Br \quad \overset{1.\ NaNH_2}{\underset{2.\ H_2O}{\longrightarrow}} HC{\equiv}CH \overset{2HCl}{\longrightarrow} \quad CH_3CHCl_2$$

Ethylene   1,2-Dibromoethane   Acetylene   1,1-Dichloroethane

---

5.5.4 Addition of Halogens to Alkynes

Alkynes react with chlorine and bromine to yield tetrahaloalkanes. Two molecules of the halogen add to the triple bond.

$$RC{\equiv}CR' + \quad 2X_2 \quad \longrightarrow \quad \overset{\overset{\displaystyle X \quad X}{\displaystyle |\quad\ |}}{\underset{\underset{\displaystyle X \quad X}{\displaystyle |\quad\ |}}{RC{-}CR'}}$$

Alkyne   Halogen   Tetrahaloalkane
         (chlorine or
         bromine)

$$CH_3C{\equiv}CH + \ 2Cl_2 \quad \longrightarrow \quad CH_3\overset{\overset{\displaystyle Cl}{|}}{\underset{\underset{\displaystyle Cl}{|}}{C}}CHCl_2$$

Propyne   Chlorine   1,1,2,2-Tetrachloropropane (63%)

5.5.5 Hydration of Alkynes

By analogy to the hydration of alkenes, addition of the elements of water to the triple bond is expected to yield an alcohol. The kind of alcohol produced by hydration of an alkyne, however, is of a special kind, one in which the hydroxyl group is a substituent on a carbon-carbon double bond. This type of alcohol is called an *enol* (the double bond suffix *-ene* plus the alcohol suffix *-ol*). An important property of enols is that they are rapidly converted to aldehydes or ketones under the conditions of their formation.

$$RC{\equiv}CR' + H_2O \xrightarrow{\text{slow}} \underset{\substack{\text{Enol} \\ \text{(not isolated)}}}{RCH{=}\overset{\displaystyle OH}{\overset{|}{C}}R'} \xrightarrow{\text{fast}} \underset{\substack{R' = H,\ \text{aldehyde} \\ R' = \text{alkyl, ketone}}}{RCH_2\overset{\displaystyle O}{\overset{\|}{C}}R'}$$

Alkyne   Water

The process by which enols are converted to aldehydes or ketones is called **keto-enol isomerism** (or **keto-enol tautomerism**). We will defer discussion of this process until we treat aldehydes and ketones in more detail in Chapter 11.

PROBLEM 5.14
Give the structure of the ketone formed by hydration of 2-butyne. What is the structure of the enol intermediate?

Typically, a combination of sulfuric acid and mercury (II) sulfate is used as the hydration catalyst. Hydration occurs in accordance with Markovnikov's rule, and terminal alkynes yield ketones in which oxygen is bonded to C-2.

$$HC{\equiv}C(CH_2)_5CH_3 + H_2O \xrightarrow[\text{HgSO}_4]{\text{H}_2\text{SO}_4} \underset{\text{2-Octanone (91\%)}}{CH_3\overset{\displaystyle O}{\overset{\|}{C}}(CH_2)_5CH_3} \left(\text{via } \underset{\text{1-Octen-2-ol}}{CH_2{=}\overset{\displaystyle OH}{\overset{|}{C}}(CH_2)_5CH_3} \right)$$

1-Octyne   Water

Because of the regioselectivity of alkyne hydration, acetylene is the only alkyne structurally capable of yielding an aldehyde under these conditions.

$$HC{\equiv}CH + H_2O \xrightarrow{\text{slow}} \underset{\substack{\text{Vinyl alcohol} \\ \text{(not isolated)}}}{CH_2{=}CHOH} \xrightarrow{\text{fast}} \underset{\text{Acetaldehyde}}{CH_3\overset{\displaystyle O}{\overset{\|}{C}}H}$$

Acetylene   Water

At one time acetaldehyde was prepared on an industrial scale by this method. More modern methods involve direct oxidation of ethylene and are more economical.

5.6 CLEAVAGE OF ALKYNES

Reagents and reaction conditions that cleave alkenes also lead to cleavage of carbon-carbon triple bonds. Carboxylic acids are produced when alkynes are subjected to ozonolysis or to oxidation by potassium permanganate.

$$RC{\equiv}CR' \xrightarrow[\text{cleavage}]{\text{oxidative}} R\overset{\displaystyle O}{\overset{\|}{C}}OH + HO\overset{\displaystyle O}{\overset{\|}{C}}R'$$

$$CH_3CH_2CH_2CH_2C{\equiv}CH \xrightarrow[\text{2. H}_2\text{O}]{\text{1. O}_3} \underset{\text{Pentanoic acid (51\%)}}{CH_3CH_2CH_2CH_2CO_2H} + \underset{\text{Formic acid}}{HCO_2H}$$

1-Hexyne

$$\underset{\text{Stearolic acid}}{CH_3(CH_2)_7C{\equiv}C(CH_2)_7CO_2H} \xrightarrow[\text{2. H}^+]{\text{1. KMnO}_4,\ \text{HO}^-}$$

$$\underset{\text{Nonanoic acid}}{CH_3(CH_2)_7CO_2H} + \underset{\text{Azelaic acid}}{HO_2C(CH_2)_7CO_2H}$$

Oxidative cleavage reactions are used primarily as a tool in structure determination. By identifying the carboxylic acids produced, it becomes a simple matter to deduce the structure of the alkyne.

PROBLEM 5.15

A certain hydrocarbon had the molecular formula $C_{16}H_{26}$ and contained two triple bonds. Ozonation followed by hydrolysis gave $CH_3(CH_2)_4CO_2H$ and $HO_2CCH_2CH_2CO_2H$ as the only products. Suggest a reasonable structure for this hydrocarbon.

5.7 ELECTROPHILIC ADDITION REACTIONS OF DIENES

Our discussion of chemical reactions of alkadienes will be limited to those of conjugated dienes. Electrophilic addition is the characteristic chemical reaction of alkenes, and conjugated dienes undergo addition reactions with electrophiles such as hydrogen halides and halogens. As the reaction of 1,3-butadiene with hydrogen chloride illustrates, however, conjugated dienes exhibit a richer spectrum of reactivity than do simple alkenes.

$$CH_2{=}CHCH{=}CH_2 \xrightarrow{HCl} CH_3\underset{\underset{Cl}{|}}{C}HCH{=}CH_2 + CH_3CH{=}CHCH_2Cl$$

1,3-Butadiene	3-Chloro-1-butene	1-Chloro-2-butene

Two products are formed. One of them, 3-chloro-1-butene, is said to be the product of **direct addition,** or **1,2 addition.** The other, 1-chloro-2-butene, is said to be the product of **conjugate addition,** or **1,4 addition.** Both correspond to proton addition to the end of the conjugated diene unit but differ in respect to the position of attachment of the halogen and the location of the double bond. The proton and the chloride add to adjacent atoms in the formation of the 1,2-addition product; they add to the ends of the diene unit in forming the 1,4-addition product.

PROBLEM 5.16

Write structural formulas corresponding to the 1,2- and 1,4-addition products formed on reaction of each of the following alkadienes with hydrogen bromide.

(a) 2,4-Hexadiene   (b) 2,3-Dimethyl-1,3-butadiene   (c) 1,3-Cyclopentadiene

SAMPLE SOLUTION

(a) The numbers 1 and 2 in 1,2 addition and 1 and 4 in 1,4 addition do not refer to the locants in the IUPAC name of the compound but to the relative positions of carbons within the conjugated diene structural unit.

1,2 Addition:

$$CH_3{-}CH{=}CH{-}CH{=}CH{-}CH_3 \xrightarrow{HBr} CH_3{-}CH_2\underset{\underset{Br}{|}}{C}H{-}CH{=}CH{-}CH_3$$

4-Bromo-2-hexene

1,4 Addition:

$$CH_3-CH=CH-CH=CH-CH_3 \xrightarrow{HBr} CH_3-CH_2CH=CH-\underset{\underset{Br}{|}}{CH}-CH_3$$

2-Bromo-3-hexene

We can account for the formation of both kinds of products on the basis of the customary mechanism for hydrogen halide addition to alkenes once we recognize a unique feature of the carbocation intermediate that is formed by protonation of 1,3-butadiene.

$$H^+ + CH_2=CHCH=CH_2 \longrightarrow CH_3\overset{+}{C}HCH=CH_2$$

The carbocation is of a type called an **allylic carbocation** that is stabilized by resonance (Section 1.9). Its positive charge is shared by two carbons.

$$CH_3-\overset{+}{C}H-CH=CH_2 \longleftrightarrow CH_3-CH=CH-\overset{+}{C}H_2$$

A negatively charged chloride ion can react with the resonance-stabilized carbocation at either of the two carbons that share the positive charge. When chloride attacks the carbocation at its secondary carbon, 3-chloro-1-butene is formed; when chloride attacks the primary carbon, 1-chloro-2-butene is the product.

$$\boxed{\begin{array}{c} CH_3-\overset{+}{C}H-CH=CH_2 \\ \updownarrow \\ CH_3-CH=CH-\overset{+}{C}H_2 \end{array}} \xrightarrow{Cl^-} \begin{array}{l} CH_3-\underset{\underset{Cl}{|}}{CH}-CH=CH_2 \quad \text{3-Chloro-1-butene} \\ + \\ CH_3-CH=CH-CH_2Cl \quad \text{1-Chloro-2-butene} \end{array}$$

The sharing of positive charge between two carbons in an allylic carbocation is caused by delocalization of the π electrons of the double bond. Allylic carbocations are more stable than simple alkyl carbocations and are formed more readily. Thus, reactions which proceed through carbocation intermediates occur relatively rapidly when the carbocation is allylic.

A mixture of 1,2 and 1,4 addition is observed as well in the addition of chlorine or bromine to conjugated dienes.

$$\underset{\text{1,3-Butadiene}}{CH_2=CHCH=CH_2} + \underset{\text{Bromine}}{Br_2} \longrightarrow$$

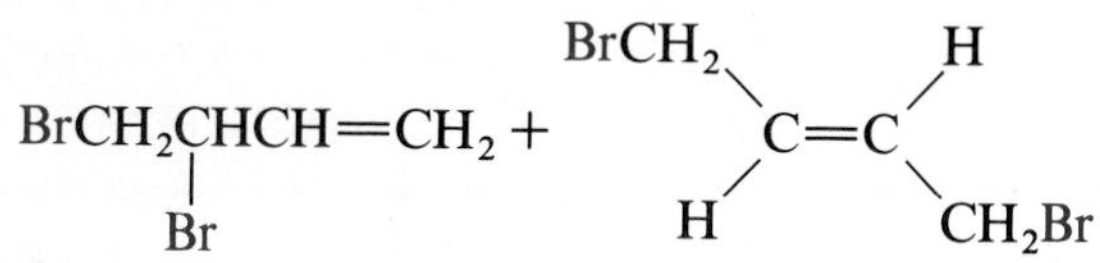

$$\underset{\underset{\underset{\text{3,4-Dibromo-1-butene}}{Br}}{|}}{BrCH_2CHCH=CH_2} + \underset{\textit{trans-1,4-Dibromo-2-butene}}{\overset{BrCH_2}{\underset{H}{}}C=C\overset{H}{\underset{CH_2Br}{}}}$$

PROBLEM 5.17

Exclusive of stereoisomers, how many products are possible in the electrophilic addition of 1 mol of bromine to 2-methyl-1,3-butadiene?

5.8 THE DIELS-ALDER REACTION

A particular kind of conjugate addition reaction earned the Nobel Prize in Chemistry for Otto Diels and Kurt Alder of the University of Kiel in 1950. The Diels-Alder reaction is the *conjugate addition of an alkene to a diene.* Using 1,3-butadiene as a typical diene, the Diels-Alder reaction may be represented by the general equation

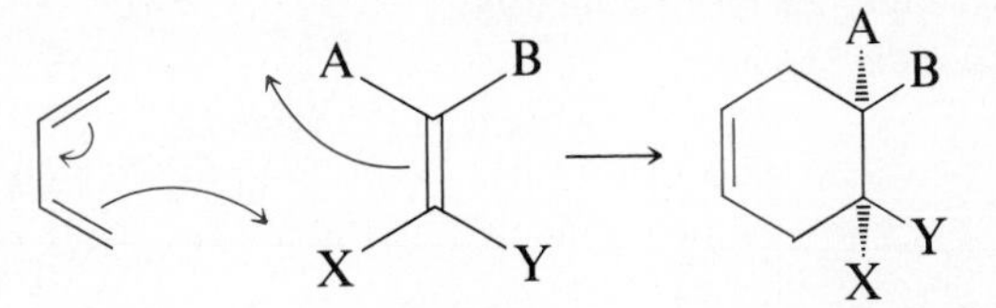

1,3-Butadiene   Dienophile   Diels-Alder adduct

The alkene that adds to the diene is called the *dienophile.* Because the Diels-Alder reaction leads to the formation of a ring, it is termed a *cycloaddition reaction.*

    The simplest of all Diels-Alder reactions, the addition of ethylene to 1,3-butadiene, does not proceed readily. It has a high activation energy and a low reaction rate. Certain substituents on the double bond of the dienophile, however, enhance its reactivity. Relatively reactive dienophiles are alkenes that bear one or more $\mathrm{C{=}O}$ or $-\mathrm{C{\equiv}N}$ substituents on the double bond. Compounds of this type, illustrated in the following examples, react readily with dienes.

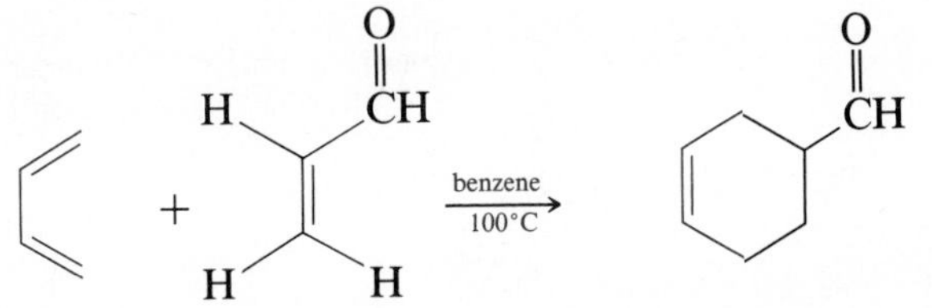

1,3-Butadiene   Acrolein   Cyclohexene-4-carboxaldehyde
(100%)

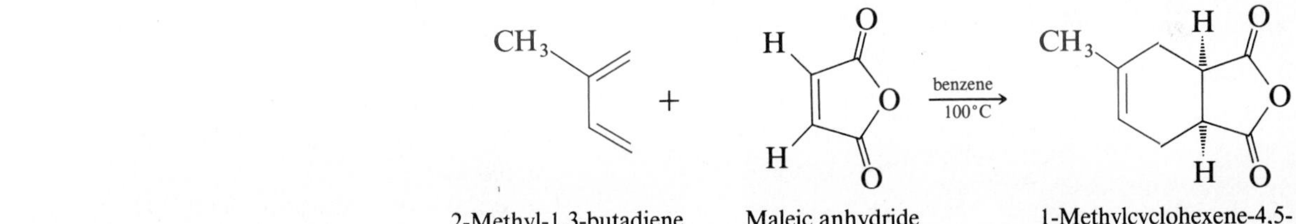

2-Methyl-1,3-butadiene   Maleic anhydride   1-Methylcyclohexene-4,5-dicarboxylic anhydride (100%)

    The Diels-Alder reaction is exceptionally useful in synthetic organic chemistry because two new carbon-carbon σ bonds are formed in a six-membered ring in a single operation without using expensive reagents or strong acids or bases that might react with a functional group elsewhere in the molecule. All that is required is to heat the diene and the dienophile together.

5.9 SUMMARY

Alkenes, alkadienes, and alkynes are *unsaturated hydrocarbons* and undergo reactions with substances which add to their multiple bonds. Representative addition reactions of alkenes are summarized in the first five entries of Table 5.3. Except for catalytic hydrogenation, these reactions proceed by *electrophilic* attack of the reagent

TABLE 5.3
Reactions of Alkenes

Reaction (section) and comments	General equation and specific example
Catalytic hydrogenation (Section 5.1.1) Alkenes react with hydrogen in the presence of a platinum, palladium, or nickel catalyst to form the corresponding alkane.	$R_2C=CR_2 + H_2 \longrightarrow R_2CHCHR_2$ Alkene Hydrogen Alkane 5,5-Dimethyl(methylene)cyclononane Hydrogen $\xrightarrow{Pt}$ 1,1,5-Trimethylcyclononane (73%)

Addition of hydrogen halides (Sections 5.1.2 through 5.1.4) A proton and a halide ion add to the double bond of an alkene to yield an alkyl halide. Addition proceeds in accordance with Markovnikov's rule; hydrogen adds to the carbon that has the greater number of hydrogen substituents, halide to the carbon that has the fewer hydrogen substituents.	$RCH=CR'_2 + HX \longrightarrow RCH_2-\underset{\underset{X}{\mid}}{CR'_2}$
	 Alkene Hydrogen Alkyl
	       halide halide
	
	 Methylenecyclo-hexane $=CH_2$ + HCl $\longrightarrow$ 1-Chloro-1-methylcyclohexane (75–80%)
	    Hydrogen chloride

Acid-catalyzed hydration (Section 5.1.5) Addition of water to the double bond of an alkene takes place in aqueous acidic solution. Addition occurs according to Markovnikov's rule. A carbocation is an intermediate and is captured by a molecule of water acting as a nucleophile.	$RCH=CR'_2 + H_2O \xrightarrow{H^+} RCH_2\underset{\underset{OH}{\mid}}{CR'_2}$ Alkene Water Alcohol $CH_2=C(CH_3)_2 \xrightarrow{50\% \ H_2SO_4-H_2O} (CH_3)_3COH$ 2-Methylpropene *tert*-Butyl alcohol (55–58%)
Hydroboration-oxidation (Section 5.1.6) This two-reaction sequence achieves hydration of alkenes with a regioselectivity opposite to that of Markovnikov's rule. An organoborane is formed by electrophilic addition of diborane to an alkene. Oxidation of the organoborane intermediate with hydrogen peroxide completes the process. The overall sequence is a syn addition of H— and —OH to the alkene.	$RCH=CR'_2 \xrightarrow[2. \ H_2O_2, \ HO^-]{1. \ B_2H_6} RCH\underset{\underset{OH}{\mid}}{CHR'_2}$ Alkene Alcohol $(CH_3)_2CHCH_2CH=CH_2 \xrightarrow[2. \ H_2O_2, \ HO^-]{1. \ B_2H_6} (CH_3)_2CHCH_2CH_2CH_2OH$ 4-Methyl-1-pentene 4-Methyl-1-pentanol (80%)

TABLE 5.3 *(continued)*

Reaction (section) and comments	General equation and specific example
Addition of halogens (Section 5.1.7) Bromine and chlorine add across the carbon-carbon double bond of alkenes to form vicinal dihalides. A cyclic halonium ion is an intermediate. The reaction is stereospecific; anti addition is observed.	$R_2C{=}CR_2 + X_2 \longrightarrow$ X—C—C—X (with R groups) Alkene　Halogen　Vicinal dihalide $CH_2{=}CHCH_2CH_2CH_2CH_3 + Br_2 \longrightarrow$ 1-Hexene　　　　　　　　Bromine $BrCH_2{-}CHCH_2CH_2CH_2CH_3$ with Br 1,2-Dibromohexane (100%)
Ozonolysis (Section 5.2) Ozone reacts with alkenes to form compounds known as ozonides. Hydrolysis converts ozonides to two carbonyl-containing compounds. Each carbon atom of the original double bond becomes the carbon atom of a carbonyl group.	$RCH{=}CR'_2 \xrightarrow[\text{2. }H_2O,\ Zn]{\text{1. }O_3} RCH{=}O + R'CR'{=}O$ Alkene　　　　　Aldehyde　Ketone $CH_3CH{=}C(CH_2CH_3)_2 \xrightarrow[\text{2. }Zn,\ H_2O]{\text{1. }O_3} CH_3CHO + CH_3CH_2CCH_2CH_3$ 3-Ethyl-2-pentene　　　　Acetaldehyde　　3-Pentanone 　　　　　　　　　　　　　(38%)　　　　　(57%)
Permanganate oxidation (Section 5.2) Potassium permanganate cleaves alkenes to carbonyl compounds. Unlike ozonolysis, aldehydes cannot be isolated because they are oxidized to carboxylic acids under the reaction conditions.	$RCH{=}CR'_2 \xrightarrow[\text{2. }H^+]{\text{1. }KMnO_4} RCOOH + R'CR'{=}O$ Alkene　　　　Carboxylic　Ketone 　　　　　　　acid *trans*-4,5-Dimethylcyclohexene $\xrightarrow[\text{2. }H^+]{\text{1. }KMnO_4}$ $HOCCH_2CHCHCH_2COH$ (with CH$_3$ groups) 3,4-Dimethyladipic acid (57%)

on the π electrons of the double bond. As described in the table the *regioselectivity* of addition of hydrogen halides to alkenes and the hydration of alkenes can be predicted by applying *Markovnikov's rule*. The last two entries of Table 5.3 illustrate reactions in which the double bond is cleaved by reaction with the strong oxidizing agents ozone and potassium permanganate, respectively.

In the presence of peroxides, hydrogen bromide adds to alkenes with a regioselectivity opposite to that predicted by Markovnikov's rule (Section 5.1.8).

$$RCH{=}CH_2 + HBr \xrightarrow{\text{peroxides}} RCH_2CH_2Br$$
Alkene　　　　Hydrogen bromide　　Alkyl bromide

This reaction proceeds by a free radical mechanism.

Table 5.4 summarizes the reactions of alkynes. While many of the reactions of alkynes are analogous to those of alkenes, an especially noteworthy property of alkynes is the relative acidity of compounds of the type $RC{\equiv}CH$. These compounds

TABLE 5.4
Reactions of Alkynes

Reaction (section) and comments	General equation and specific example
Hydrogenation of alkynes to alkanes (Section 5.5.1) Alkynes are completely hydrogenated, yielding alkanes, in the presence of the customary metal hydrogenation catalysts.	$RC \equiv CR' + 2H_2 \xrightarrow{\text{metal catalyst}} RCH_2CH_2R'$ Alkyne Hydrogen Alkane 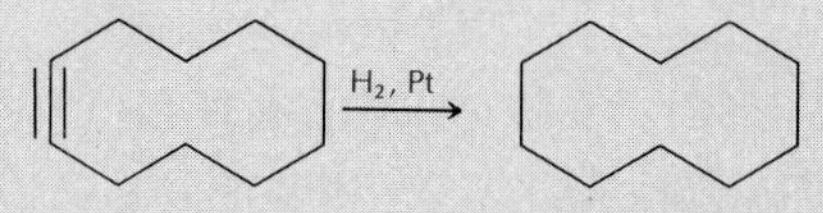 Cyclodecyne $\xrightarrow{H_2,\ Pt}$ Cyclodecane (71%)
Hydrogenation of alkynes to alkenes (Section 5.5.1) Hydrogenation of alkynes may be halted at the alkene stage by using special catalysts. Lindlar palladium is the metal catalyst employed most often. Hydrogenation occurs with syn selectivity and yields a cis alkene.	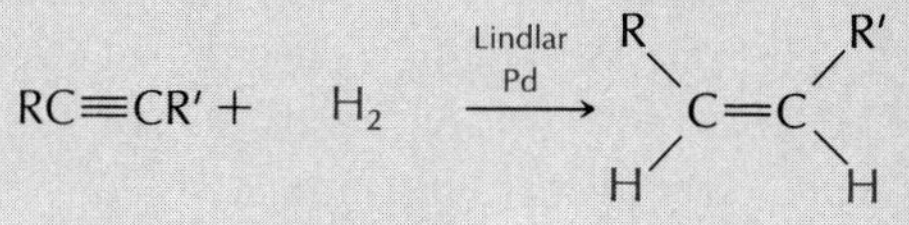 $RC \equiv CR' + H_2 \xrightarrow{\text{Lindlar Pd}}$ Cis alkene $\left(\begin{array}{c} R \\ \diagdown C = C \diagup R' \\ H \diagup \qquad \diagdown H \end{array}\right)$ Alkyne Hydrogen Cis alkene 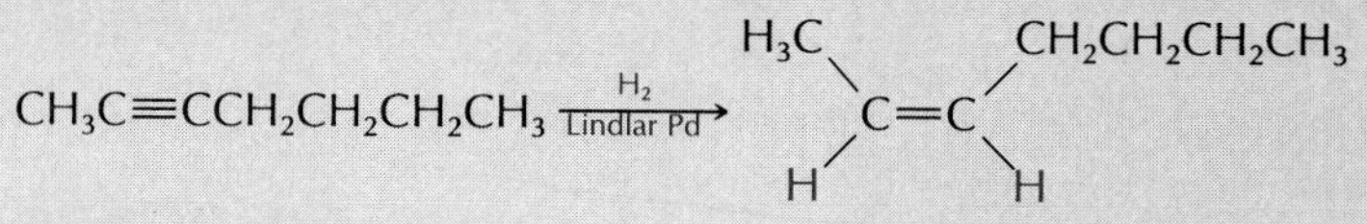 $CH_3C \equiv CCH_2CH_2CH_2CH_3 \xrightarrow[\text{Lindlar Pd}]{H_2}$ 2-Heptyne *cis*-2-Heptene (59%)
Metal-ammonia reduction (Section 5.5.2) Group I metals—sodium is the one usually employed—in liquid ammonia as the solvent convert alkynes to trans alkenes.	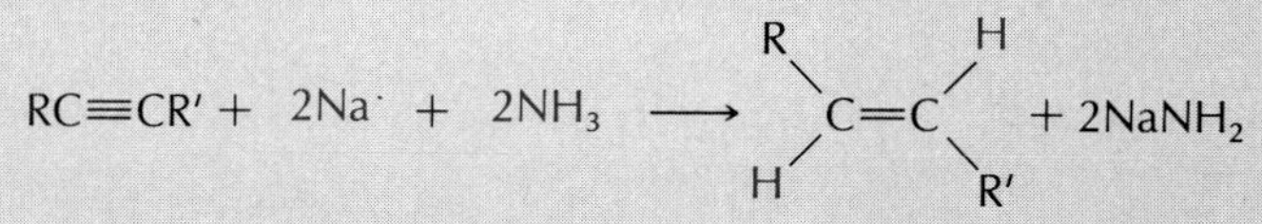 $RC \equiv CR' + 2Na\cdot + 2NH_3 \longrightarrow$ Trans alkene $+ 2NaNH_2$ Alkyne Sodium Ammonia Trans alkene Sodium amide 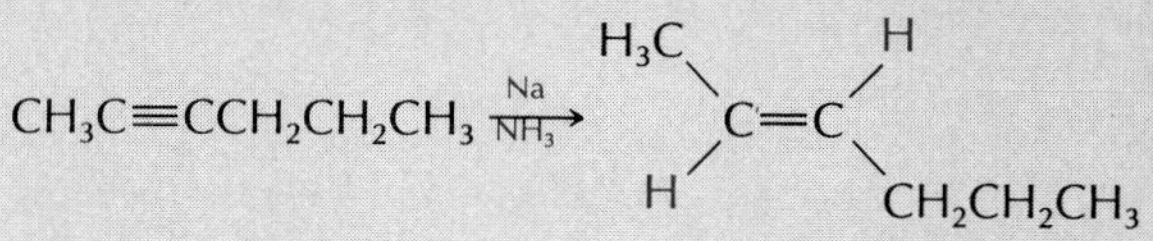 $CH_3C \equiv CCH_2CH_2CH_3 \xrightarrow[NH_3]{Na}$ 2-Hexyne *trans*-2-Hexene (69%)
Addition of hydrogen halides (Section 5.5.3) Hydrogen halides add to alkynes in accordance with Markovnikov's rule to give alkenyl halides. In the presence of 2 mol of hydrogen halide a second addition occurs to give a geminal dihalide.	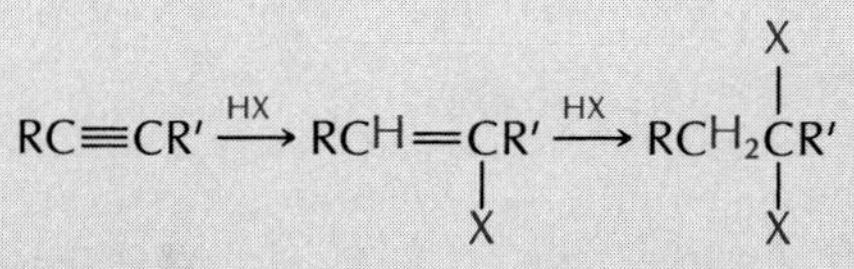 $RC \equiv CR' \xrightarrow{HX} RCH = CR'(X) \xrightarrow{HX} RCH_2CR'(X)(X)$ Alkene Alkenyl halide Geminal dihalide  $CH_3C \equiv CH + 2HBr \longrightarrow CH_3CCH_3$ (with Br, Br on central C) Propyne Hydrogen bromide 2,2-Dibromo-propane (100%)

TABLE 5.4 *(continued)*

Reaction (section) and comments	General equation and specific example
Halogenation (Section 5.5.4) Addition of two equivalents of chlorine or bromine to an alkyne yields a tetrahalide.	$RC\equiv CR \xrightarrow{2X_2}$ tetrahaloalkane (structure with X substituents) $RC-CR'$ with X, X on both carbons Alkyne — Tetrahaloalkane $CH_3C\equiv CH + 2Cl_2 \longrightarrow CH_3CCHCl_2$ (with Cl, Cl) Propyne — Chlorine — 1,1,2,2-Tetrachloro-propane (63%)
Acid-catalyzed hydration (Section 5.5.5) Water adds to the triple bond of alkynes to yield ketones by way of an unstable enol intermediate. The enol arises by Markovnikov hydration of the alkyne, followed by rapid isomerization of the enol to a ketone.	$RC\equiv CR' + H_2O \xrightarrow[Hg^{2+}]{H_2SO_4} RCH_2CR'$ ($C=O$) Alkyne — Water — Ketone $HC\equiv CCH_2CH_2CH_2CH_3 + H_2O \xrightarrow[HgSO_4]{H_2SO_4} CH_3CCH_2CH_2CH_2CH_3$ ($C=O$) 1-Hexyne — Water — 2-Hexanone (80%)
Alkylation of acetylene and terminal alkynes (Section 5.4) The acidity of acetylene and terminal alkynes permits them to be converted to their conjugate bases on treatment with sodium amide. These anions are good nucleophiles and react with primary or methyl alkyl halides to form carbon-carbon bonds.	$RC\equiv CH + NaNH_2 \longrightarrow RC\equiv CNa + NH_3$ Alkyne — Sodium amide — Sodium alkynide — Ammonia $RC\equiv CNa + R'CH_2X \longrightarrow RC\equiv CCH_2R' + NaX$ Sodium alkynide — Primary alkyl halide — Alkyne — Sodium halide $(CH_3)_3CC\equiv CH \xrightarrow[2.\ CH_3I]{1.\ NaNH_2,\ NH_3} (CH_3)_3CC\equiv CCH_3$ 3,3-Dimethyl-1-butyne — 4,4-Dimethyl-2-pentyne (96%)

are far more acidic than alkanes and alkenes and are characterized by K_a of approximately 10^{-26} for the equilibrium shown.

$$RC\equiv CH \rightleftharpoons H^+ + RC\equiv C:^-$$
Alkyne — Proton — Alkynide ion

Acetylene and terminal alkynes may be converted to the corresponding alkynide ion by treatment with sodium amide (usually in liquid ammonia as the solvent).

$$RC\equiv CH + NaNH_2 \xrightarrow{NH_3} RC\equiv C:^-Na^+ + NH_3$$
Alkyne — Sodium amide — Sodium alkynide — Ammonia

The alkynide ion is then *alkylated* by reaction with a methyl halide or primary alkyl halide. The final entry in Table 5.4 gives an example of this procedure.

Conjugated alkadienes react with many of the same species which react with alkenes. What is unusual about conjugated dienes, however, is their capacity to undergo both *1,2 addition* and *1,4 addition* (Section 5.7).

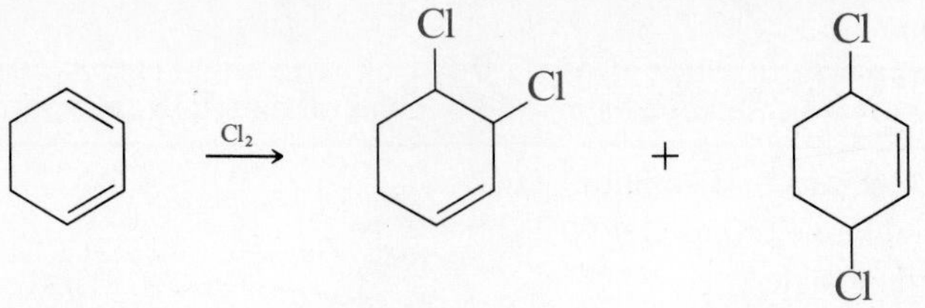

1,3-Cyclohexadiene   3,4-Dichlorocyclohexene   3,6-Dichlorocyclohexene
                          (mixture of cis and trans)   (mixture of cis and trans)

The Diels-Alder reaction (Section 5.8) is a novel kind of *cycloaddition* in which an alkene adds to a diene in a 1,4 fashion to give a cyclohexene derivative.

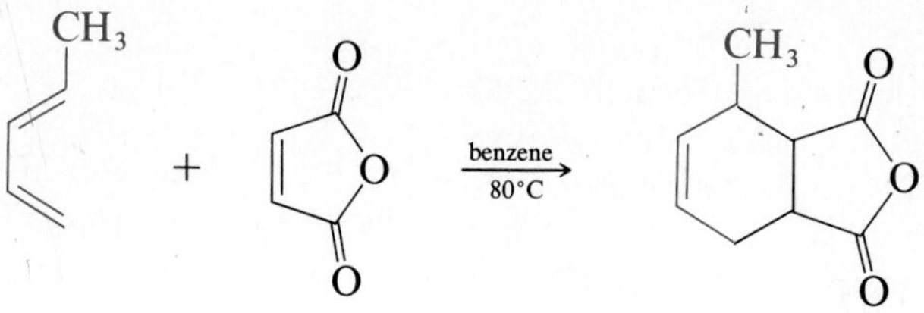

1,3-Pentadiene   Maleic anhydride   3-Methylcyclohexene-4,5-dicar-
                                            boxylic anhydride (81%)

ADDITIONAL PROBLEMS

Alkene Reactions

5.18 Write the structure of the principal organic product formed in the following addition reactions of 1-pentene:

(a) Hydrogen chloride
(b) Hydrogen bromide
(c) Hydrogen bromide in the presence of peroxides
(d) Hydrogen iodide
(e) Dilute sulfuric acid
(f) Diborane, followed by basic hydrogen peroxide
(g) Bromine

5.19 Repeat Problem 5.18 for 2-methyl-2-butene.

5.20 Repeat Problem 5.18 for 1-methylcyclohexene.

5.21 Write the structure of the principal organic product or products formed in each of the following cleavage reactions of 1-pentene:

(a) Ozone, followed by treatment with zinc and water
(b) Potassium permanganate ($KMnO_4$), followed by acidification.

5.22 Repeat Problem 5.21 for 2-methyl-2-butene.

5.23 Repeat Problem 5.21 for 1-methylcyclohexene.

5.24 (a) How many alkenes yield 2,3-dimethylbutane on catalytic hydrogenation? Write their structures.

(b) How many yield methylcyclobutane? Draw them.
(c) Several alkenes undergo hydrogenation to yield a mixture of cis- and trans-1,4-dimethylcyclohexane. Only one substance, however, gives only cis-1,4-dimethylcyclohexane. What compound is this?

5.25 Specify the reagents necessary for converting 3-ethyl-2-pentene to each of the following:

(a) Br, Br

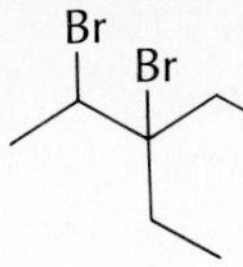

(b) OH

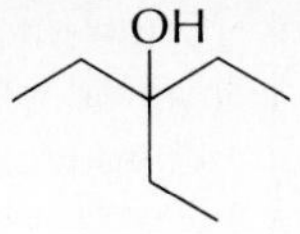

(c) Cl

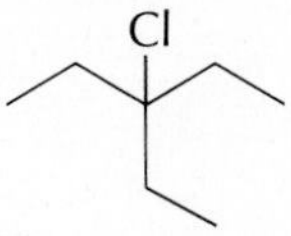

(d) OH

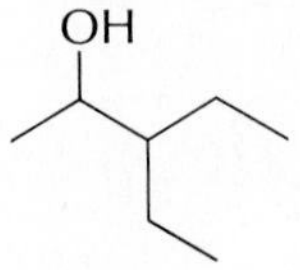

(e)

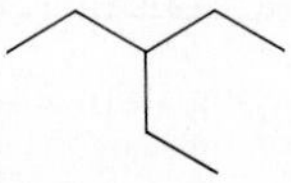

(f) Br

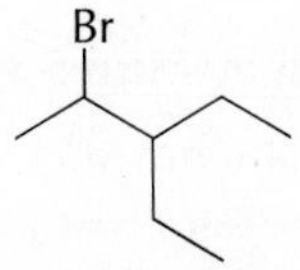

5.26 The addition of HBr to an alkene, both in the presence and in the absence of peroxides, proceeds through formation of the most stable intermediate species. These two reactions yield different regioisomers with an unsymmetrical alkene, however. In what fundamental way do these two reactions differ? Explain by writing a series of equations outlining the mechanisms of:

(a) Addition of HBr to 2-methyl-2-butene
(b) Addition of HBr in the presence of peroxides to 2-methyl-2-butene

Alkyne Reactions

5.27 Write the structure of the major product isolated from the reaction of 1-hexyne with:

(a) Hydrogen (2 mol), platinum
(b) Hydrogen (1 mol), Lindlar palladium
(c) Sodium in liquid ammonia
(d) Sodium amide in liquid ammonia
(e) Product of (c) with 1-bromobutane
(f) Hydrogen chloride (1 mol)
(g) Hydrogen chloride (2 mol)
(h) Chlorine (2 mol)
(i) Aqueous sulfuric acid, mercuric sulfate
(j) Ozone, followed by hydrolysis

5.28 Write the structure of the major product isolated from the reaction of 3-hexyne with:

(a) Hydrogen (2 mol), platinum
(b) Hydrogen (1 mol), Lindlar palladium
(c) Sodium in liquid ammonia
(d) Hydrogen chloride (1 mol)
(e) Hydrogen chloride (2 mol)
(f) Chlorine (2 mol)
(g) Aqueous sulfuric acid, mercuric sulfate
(h) Ozone, followed by hydrolysis

5.29 When 2-heptyne was treated with aqueous sulfuric acid containing mercuric sulfate, two products, each having the molecular formula $C_7H_{14}O$, were obtained in approximately equal amounts. What are these two compounds?

5.30 Write the structures of the enol intermediates which form in the reaction described in Problem 5.29.

Conjugated Diene Reactions

5.31 Give the structure, exclusive of stereochemistry, of the principal organic product formed on reaction of 2,3-dimethyl-1,3-butadiene with each of the following:

(a) H_2 (2 mol), platinum catalyst
(b) 1 mol HCl (product of 1,2 addition)

(c) 1 mol HCl (product of 1,4 addition)

(d) 1 mol Br_2 (product of 1,2 addition)

(e) 1 mol Br_2 (product of 1,4 addition)

(f) 2 mol Br_2

(g)

5.32 Repeat Problem 5.31 for the reactions of 1,3-cyclohexadiene.

5.33 Write the structure, clearly showing stereochemistry, from the reaction of 1,3-butadiene, $CH_2=CHCH=CH_2$, with:

(a) *cis*-Cinnamic acid:

$$\underset{H}{\overset{C_6H_5}{}} C=C \underset{H}{\overset{CO_2H}{}}$$

(b) *trans*-Cinnamic acid:

$$\underset{H}{\overset{C_6H_5}{}} C=C \underset{CO_2H}{\overset{H}{}}$$

Synthesis Using Alkenes and Alkynes

5.34 Suggest a sequence of reactions suitable for preparing each of the following compounds from the indicated starting material. You may use any necessary organic or inorganic reagents.

(a) 2-Methylhexane from 2-methyl-2-hexanol

(b) 1-Propanol from 2-propanol

(c) 2-Propanol from 1-propanol

(d) 1-Bromopropane from 2-bromopropane

(e) 1,2-Dibromopropane from 2-bromopropane

(f)
$$\underset{}{\overset{O}{\underset{\|}{}}} \quad \underset{}{\overset{O}{\underset{\|}{}}}$$
$CH_3CCH_2CH_2CH_2CH$ from 1-bromo-1-methylcyclopentane

(g) 2,2-Dibromopropane from 1,2-dibromopropane

5.35 The ketone 2-heptanone has been identified as contributing to the odor of a number of dairy products, including condensed milk and cheddar cheese. Describe a synthesis of 2-heptanone from acetylene and any necessary organic or inorganic reagents.

$$\overset{O}{\underset{\|}{}}$$
$$CH_3CCH_2CH_2CH_2CH_2CH_3$$
2-Heptanone

5.36 *cis*-9-Tricosene [*cis*-$CH_3(CH_2)_7CH=CH(CH_2)_{12}CH_3$] is the sex pheromone of the female housefly. Synthetic *cis*-9-tricosene is used as bait to lure male flies to traps that contain insecticide. Using acetylene and alcohols of your choice as starting materials, along with any necessary inorganic reagents, show how you could prepare *cis*-9-tricosene.

Structure Elucidation

For many years prior to the advent of spectroscopic techniques (Chapter 9), the structure of an unknown substance was often determined by chemical methods. The compound of interest was converted by known chemical reactions to other substances which could be identified. Ozonolysis and permanganate cleavage were frequently used to determine the structure of an unsaturated substance, as illustrated in the following problems.

5.37 Consider two isomeric alkenes A and B. Ozonolysis of the lower-boiling alkene A gives formaldehyde (CH_2=O) and 2,2,4,4-tetramethyl-3-pentanone. Ozonolysis of B gives formaldehyde and 3,3,4,4-tetramethyl-2-pentanone. Identify the structures of A and B.

$$(CH_3)_3C\overset{\displaystyle O}{\overset{\|}{C}}C(CH_3)_3 \qquad CH_3\overset{\displaystyle OCH_3}{\overset{\|\,|}{C}}C(CH_3)_3$$
$$\overset{|}{CH_3}$$

2,2,4,4-Tetramethyl-3-pentanone   3,3,4,4-Tetramethyl-2-pentanone

5.38 Compound C ($C_7H_{13}Br$) is a tertiary bromide. On treatment with sodium ethoxide in ethanol, C is converted into D (C_7H_{12}). Ozonolysis of D gives E as the only product. Deduce the structures of C and D. What is the symbol for the reaction mechanism by which C is converted to D under the reaction conditions?

$$CH_3\overset{\displaystyle O}{\overset{\|}{C}}CH_2CH_2CH_2CH_2\overset{\displaystyle O}{\overset{\|}{C}}H$$

Compound E

5.39 East Indian sandalwood oil contains a hydrocarbon given the name *santene* (C_9H_{14}). Ozonation of santene followed by hydrolysis gives compound F. What is the structure of santene?

$$CH_3\overset{\displaystyle O}{\overset{\|}{C}}\quad\overset{\displaystyle O}{\overset{\|}{C}}CH_3$$

Compound F

5.40 *Sabinene*, Δ^3-*carene*, and *α-pinene* are isomeric natural products with the molecular formula $C_{10}H_{16}$.

(a) Ozonolysis of sabinene followed by hydrolysis in the presence of zinc gives compound G. What is the structure of sabinene? What other compound is formed on ozonolysis

(b) Ozonolysis of Δ^3-carene followed by hydrolysis in the presence of zinc gives compound H. What is the structure of Δ^3-carene?

(c) Treatment of α-pinene with potassium permanganate, followed by acidification, gives compound I. What is the structure of α-pinene?

Compound G   Compound H   Compound I

5.41 The sex attractant by which the female housefly attracts the male has the molecular formula $C_{23}H_{46}$. Catalytic hydrogenation yields an alkane of molecular formula $C_{23}H_{48}$.

Ozonation followed by hydrolysis in the presence of zinc yields $CH_3(CH_2)_7\overset{\displaystyle O}{\overset{\|}{C}}H$ and $CH_3(CH_2)_{12}\overset{\displaystyle O}{\overset{\|}{C}}H$. What is the structure of the housefly sex attractant?

5.42 A certain compound of molecular formula $C_{19}H_{38}$ was isolated from fish oil and from plankton. On hydrogenation it gave 2,6,10,14-tetramethylpentadecane. Ozonation followed by hydrolysis in the presence of zinc gave $(CH_3)_2C$=O and a 16-carbon aldehyde. What is the structure of the natural product? What is the structure of the aldehyde?

5.43 The sex attractant of the female arctiid moth contains, among other components, a compound of molecular formula $C_{21}H_{40}$ that yields $CH_3(CH_2)_{10}\overset{\displaystyle O}{\overset{\displaystyle \|}{C}}H$, $CH_3(CH_2)_4\overset{\displaystyle O}{\overset{\displaystyle \|}{C}}H$, and $H\overset{\displaystyle O}{\overset{\displaystyle \|}{C}}CH_2\overset{\displaystyle O}{\overset{\displaystyle \|}{C}}H$ on ozonolysis. What is the constitution of this material?

5.44 Cedrene is a pleasant-smelling constituent of cedarwood oil and has the molecular formula $C_{15}H_{24}$. Treatment of cedrene with potassium permanganate followed by acidification of the reaction mixture gives compound J. Deduce the structure of cedrene.

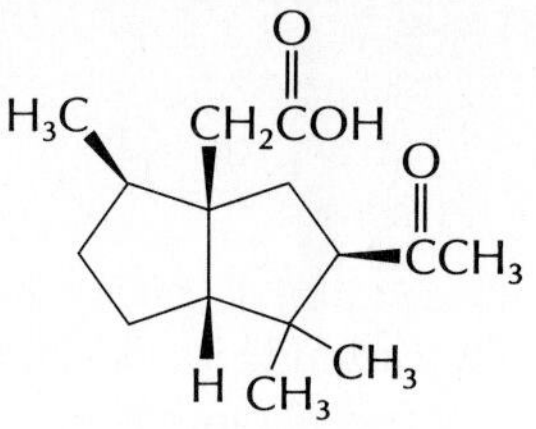

Compound J

5.45 Compound K has the molecular formula $C_{14}H_{25}Br$ and was obtained by reaction of sodium acetylide with 1,12-dibromododecane. On treatment of compound K with sodium amide, it was converted to compound L ($C_{14}H_{24}$). Ozonolysis of compound L gave the diacid $HO_2C(CH_2)_{12}CO_2H$. Catalytic hydrogenation of compound L over Lindlar palladium gave compound M ($C_{14}H_{26}$), while hydrogenation over platinum gave compound N ($C_{14}H_{28}$). Sodium-ammonia reduction of compound L gave compound O ($C_{14}H_{26}$). Both M and O yielded $HO_2C(CH_2)_{12}CO_2H$ on oxidation with potassium permanganate. Compound N was inert to potassium permanganate. Assign structures to compounds K through O so as to be consistent with the observed transformations.

CHAPTER 6

ARENES
AND
AROMATICITY

LEARNING OBJECTIVES

Aromaticity is one of the fundamental concepts of organic chemistry. You will see in this chapter how aromaticity leads to the unique chemical reactivity of a family of hydrocarbons called **arenes.** Upon completion of this chapter you should be able to:

■ Describe the structure of benzene using resonance.

■ Explain the bonding in benzene using the orbital hybridization model.

■ Write a structural formula for an aromatic compound on the basis of its systematic IUPAC name.

■ Write a correct IUPAC name for an aromatic compound on the basis of a given structural formula.

■ Write chemical equations describing the electrophilic aromatic substitution reactions of benzene: halogenation, nitration, sulfonation, Friedel-Crafts alkylation, and Friedel-Crafts acylation.

■ Explain the mechanistic basis for the action of activating ortho, para-directing groups.

■ Explain the mechanistic basis for the action of deactivating meta-directing groups.

■ Explain why the halogens are deactivating ortho, para-directing groups.

■ Write chemical equations describing the synthesis of disubstituted aromatic compounds.

■ Write chemical equations for electrophilic addition and oxidation reactions of aromatic side chains.

■ Predict whether a substance is aromatic based on its structure and Hückel's rule.

This chapter presents the structure and chemistry of *arenes,* also referred to as *aromatic hydrocarbons.* Although historically the word *aromatic* referred to the origin of many of these compounds in pleasant-smelling plant materials (see boxed material in Section 6.1), today the term has a very different meaning, as you will see when the structure and stability of benzene are discussed.

Arenes exhibit chemical properties substantially different from the alkenes we studied in the previous chapter. The predominant reaction of arenes is one of *substitution,* not *addition* as with alkenes. We will focus in this chapter on the class of reactions known as *electrophilic aromatic substitution.*

6.1 AROMATIC COMPOUNDS

Benzene (C_6H_6), first isolated in the early 1800s, is the simplest member of the class of compounds known as **aromatic hydrocarbons,** or **arenes.** Although most aromatic compounds are derivatives of benzene, you will see in Section 6.15 that other compounds that meet certain criteria are also aromatic.

Benzene and its close structural relative toluene are very important industrial chemicals. Annual U.S. production of benzene is approximately 10×10^9 lb and of

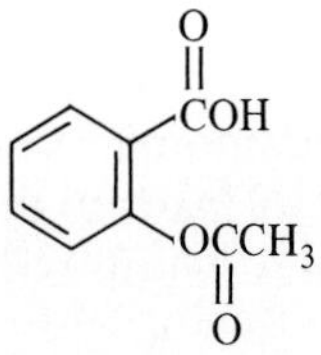

Acetylsalicylic acid
(Aspirin)

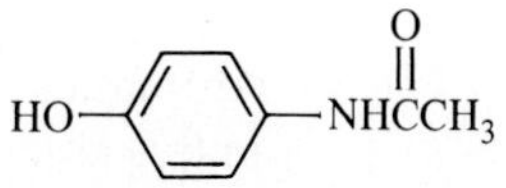

Acetaminophen
(Analgesic in Tylenol)

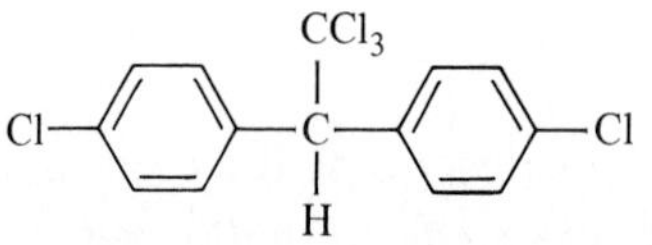

DDT
(DichloroDiphenylTrichloro-
ethane)

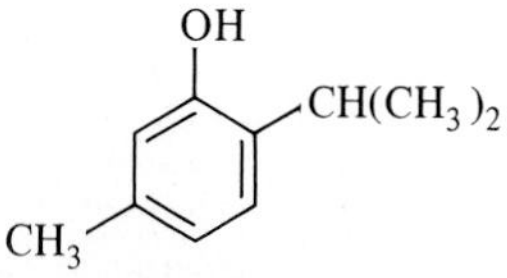

Thymol
(Found in the herb thyme)

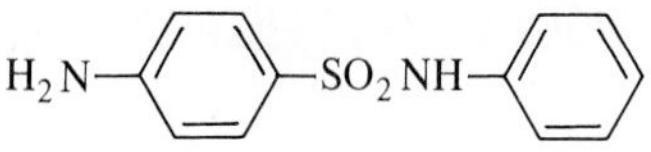

Sulfabenz
(Sulfa drug antibiotic)

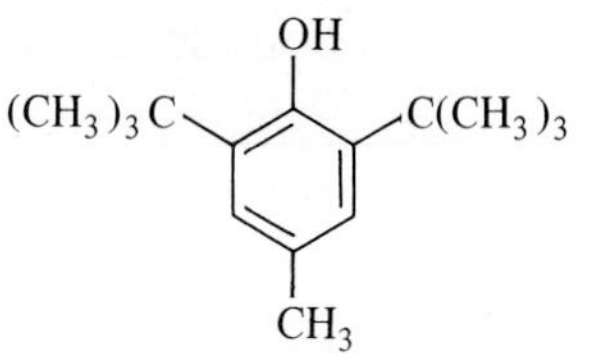

BHT
(Antioxidant used as a
food preservative)

FIGURE 6.1 Some important aromatic compounds.

HISTORY OF AROMATIC HYDROCARBONS

In 1825 Michael Faraday isolated a new hydrocarbon from the gas used to light street lamps, which he called "bicarburet of hydrogen." Nine years later this same hydrocarbon was prepared by Eilhardt Mitscherlich of the University of Berlin, who obtained it by heating benzoic acid with lime and determined its molecular formula to be C_6H_6.

$$C_6H_5CO_2H + \quad CaO \quad \xrightarrow{\text{heat}} C_6H_6 + \quad CaCO_3$$

Benzoic acid   Calcium oxide   Benzene   Calcium carbonate

Eventually, because of its relationship to benzoic acid, this hydrocarbon came to be named *benzin,* then later *benzene,* the name by which it is known today in the IUPAC nomenclature system.

Benzoic acid had been known for several hundred years by the time of Mitscherlich's experiment. Many trees exude resinous materials called balsams when cuts are made in their bark. Some of these balsams are very fragrant, a property which once made them highly prized articles of commerce, especially when the trees which produced them could be found only in exotic, faraway lands. *Gum benzoin* is such a substance and is obtained from a tree indigenous to the area around Java and Sumatra. *Benzoin* is a word derived from the French equivalent *benjoin,* which in turn comes from the Arabic *luban jawi,* meaning "incense from Java." Benzoic acid is itself odorless but can be easily isolated from the mixture that makes up the material known as *gum benzoin.*

Compounds related to benzene were obtained from similar plant extracts. For example, a pleasant-smelling resin known as *tolu balsam* was obtained from a South American tree. In the 1840s it was discovered that distillation of tolu balsam gave a methyl derivative of benzene, which, not surprisingly, came to be named *toluene.*

Although benzene and toluene are not particularly fragrant compounds themselves, their origins in aromatic plant extracts led them and compounds related to them to be classified as *aromatic hydrocarbons.* Alkanes, alkenes, and alkynes belong to another class, the *aliphatic hydrocarbons.* The word *aliphatic* comes from the Greek word *aleiphar* (meaning "oil" or "unguent") and arose from the observation that hydrocarbons of this class could be obtained by the chemical degradation of fats.

Benzene was prepared from coal tar by August W. von Hofmann in 1845. This remained the primary source for the industrial production of benzene for many years, until petroleum-based technologies became competitive about 1950. Although free benzene does not occur naturally, it is efficiently prepared from coal and petroleum and is thus a relatively inexpensive chemical.

Toluene is also an important organic chemical. Like benzene, its early production was from coal tar, but more recently petroleum sources have come to supply most of the toluene used for industrial purposes.

toluene is approximately 5×10^9 lb. Both are used as additives in unleaded gasoline. Benzene is the starting material for a diverse variety of consumer products including styrofoam plastics, nylon fibers, and aspirin. Toluene is used as a solvent for paints and lacquers, and is the starting material for polyurethane foams used in furniture cushions and mattresses.

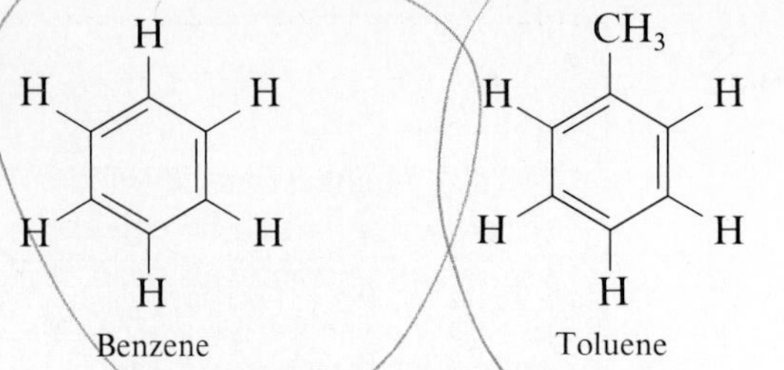

Benzene   Toluene

The structures of several additional aromatic compounds are shown in Figure 6.1. Acetylsalicylic acid (aspirin) and acetaminophen are the most widely used *analgesics,* or medications for the relief of pain. Aspirin has the additional advantage of reducing inflammation. DDT was once the most widely used insecticide. Harmful effects to the environment (see boxed material in Section 3.18) have led to a ban on DDT's use in most of the world, although it is still used to control the spread of malaria-carrying mosquitos in southeast Asia.

6.2 STRUCTURE AND BONDING OF BENZENE

One of the earliest structural formulas for benzene was proposed by August Kekulé in 1866. Kekulé suggested a structure having alternating single and double bonds in a ring of six carbons, each carbon having one hydrogen attached.

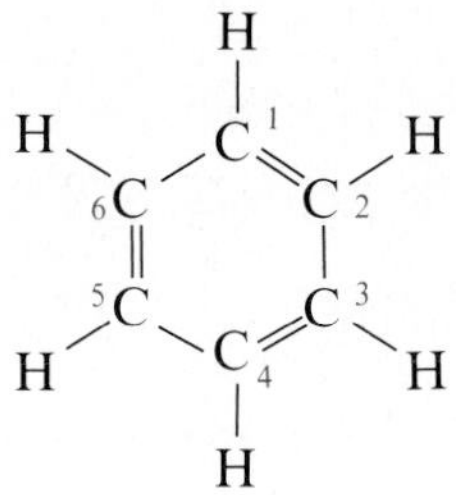

We know today that Kekulé's view does not explain all the observations regarding benzene's structure and reactivity. All six carbon-carbon bonds of benzene are the *same* length (140 pm), intermediate between sp^2-sp^2 single bonds (146 pm) and double bonds (134 pm). The fact that all the bond lengths are equal is not explained by the alternating double and single bonds of Kekulé's structure. The structure of benzene is better represented as a hybrid of the two equivalent *resonance forms* shown.

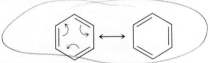

Remember from Section 1.9 that the double-headed arrow does *not* indicate an equilibrium process. It is a device used to show *electron-delocalized* structures as composites of Lewis structures written with localized electrons. In this case the Lewis structures are the two possible Kekulé representations of benzene. A convenient way to represent the two Kekulé structures as resonance forms of benzene is by inscribing a circle inside a hexagon.

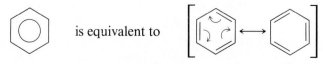

is equivalent to

PROBLEM 6.1
Draw the two Kekulé forms and the circle-in-a-ring representation of toluene.

In terms of reactivity, benzene is far less reactive than expected on the basis of the structural formulas written for it. Were benzene literally a "1,3,5-cyclohexatriene" we would expect it to react with the same kinds of substances that react with alkenes. In fact, benzene is inert or only sluggishly reactive toward many of the substances which react readily with alkenes. Low reactivity suggests high stability. Estimates suggest that benzene is 30 to 36 kcal/mol more stable than it would be if it were simply a polyene. The "special" stability of benzene and its derivatives is the modern definition of the term **aromaticity.**

6.3 AN ORBITAL HYBRIDIZATION MODEL OF BONDING IN BENZENE

Each carbon of benzene is attached to three other atoms, and all the atoms lie in the same plane. With these observations and since all the bond angles are 120°, the framework of carbon-carbon σ bonds is reasonably described as arising from the overlap of sp^2 hybrid orbitals. Figure 6.2 illustrates the σ framework of benzene.

An unhybridized $2p$ orbital containing one electron remains on each carbon. As shown in Figure 6.3, overlap of these parallel $2p$ orbitals generates a continuous π system encompassing all the carbon atoms of the ring. The six π electrons are delocalized over all six carbons. That is, benzene does not have three double bonds isolated from one another, but rather the six π electrons are shared by all six carbon atoms.

As you have seen previously (Sections 1.9 and 4.2.3), when π electrons are delocalized over several carbon atoms, the molecule is more stable than if these electrons were localized and felt the attractive force of fewer nuclei. The delocalization of the π electrons in a benzene ring results in a cyclic π cloud, almost like a "doughnut" of electrons, as shown in Figure 6.3b. Aromaticity is attributed to this cyclic delocalization of the conjugated system of 6π electrons.

6.4 POLYCYCLIC AROMATIC HYDROCARBONS

Naphthalene is the simplest member of the class of arenes known as **polycyclic aromatic hydrocarbons.** It contains two fused benzene rings sharing a common side.

FIGURE 6.2 (a) The assembly of σ bonds in benzene. Each carbon is sp^2 hybridized and forms σ bonds to two other carbons and to one hydrogen. (b) The six carbon atoms form a planar regular hexagon, with internal angles of 120°.

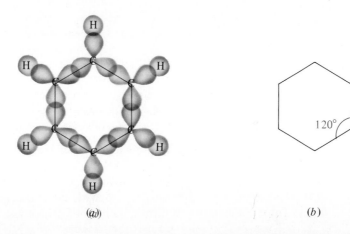

(a)   (b)

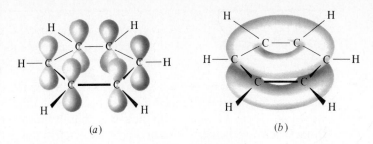

FIGURE 6.3 (a) The 2p orbitals of benzene's carbon atoms are suitably aligned for effective π overlap. (b) Overlap of the 2p orbitals generates a π system encompassing the entire ring. There are regions of high-π electron density above and below the plane of the ring.

Naphthalene

Among other uses naphthalene is employed as a moth repellant.

Polycyclic aromatic hydrocarbons may contain more than two rings. Two examples are anthracene and phenanthrene. Both compounds are tricyclic aromatic hydrocarbons used in the preparation of synthetic dyes.

Anthracene   Phenanthrene

One particular polycyclic aromatic hydrocarbon that has received widespread interest is benz[a]pyrene because of its role as a chemical carcinogen (see boxed material).

6.5 NOMENCLATURE OF SUBSTITUTED DERIVATIVES OF BENZENE

A benzene ring that has one of its hydrogens replaced by some other group is named as a substituted derivative of benzene. For example,

Bromobenzene   *tert*-Butylbenzene   Nitrobenzene

Many simple monosubstituted derivatives of benzene have common names of long standing that have been retained in the IUPAC system. The common names benzaldehyde and benzoic acid, for example, are used far more frequently than their systematic counterparts benzenecarbaldehyde and benzenecarboxylic acid, respectively. Table 6.1 gives the common names of some others.

CHEMICAL CARCINOGENS

A **carcinogen** is an agent capable of causing cancer. A number of chemical carcinogens have received considerable study in recent years, both in an effort to understand the causes of cancer and in an attempt to find ways to prevent its occurrence.

In 1775 Percivall Pott, an English surgeon, wrote of the incidence of scrotal cancer among chimney sweeps, blaming their disease on exposure to soot. Nearly 150 years later experiments showed that mice painted with soot extract developed cancer. These experiments and others in the 1930s led to the identification of numerous polycyclic aromatic hydrocarbons and the isolation of the first chemical identified as a carcinogen, benz[a]pyrene.

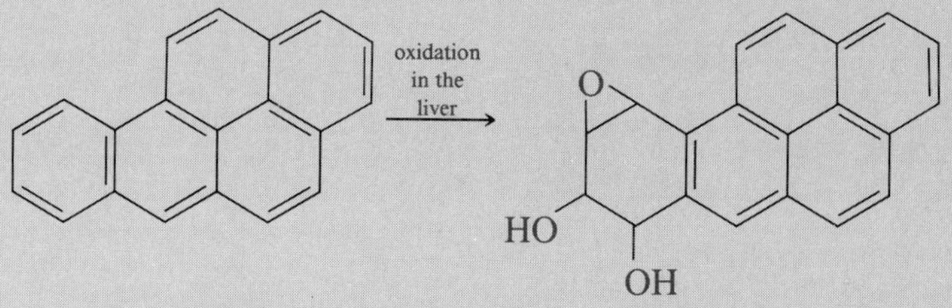

Benz[a]pyrene
A diol-epoxide

Current theories of carcinogenesis suggest that chemicals such as benz[a]pyrene are oxidized in the body to form other compounds called *metabolites,* and it is these metabolites that are the actual carcinogens. In the case of benz[a]pyrene, enzymes in the liver catalyze its conversion to a diol-epoxide which is able to react with DNA in cells. If these reactions cause a nonlethal change, or *mutation,* in the structure of the cell, uncontrolled cell reproduction may result, giving rise to tumor formation.

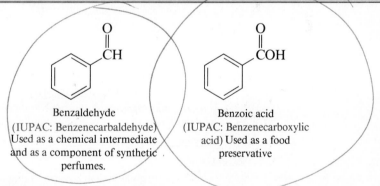

Benzaldehyde
(IUPAC: Benzenecarbaldehyde)
Used as a chemical intermediate
and as a component of synthetic
perfumes.

Benzoic acid
(IUPAC: Benzenecarboxylic
acid) Used as a food
preservative

Dimethyl derivatives of benzene are called **xylenes.** There are three xylene isomers, the ortho- (*o*-), meta- (*m*-), and para- (*p*-) substituted derivatives.

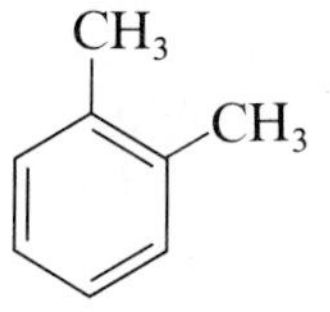

o-Xylene
(1,2-dimethylbenzene)

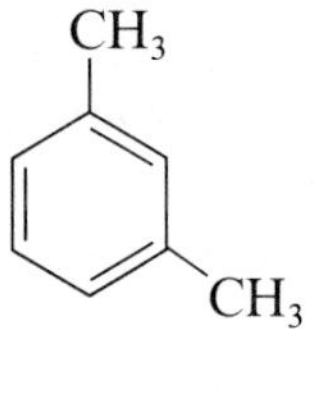

m-Xylene
(1,3-dimethylbenzene)

p-Xylene
(1,4-dimethylbenzene)

TABLE 6.1

Names and Uses of Some Common Benzene Derivatives

Structure	Name	Commercial use
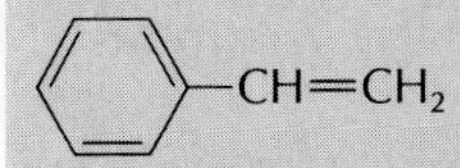 C₆H₅—CH=CH₂	Styrene	The basic component of polystyrene and styrofoam plastics.
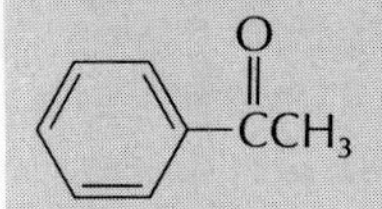 C₆H₅—C(=O)CH₃	Acetophenone	Used in perfumes to impart an orange-blossom-like odor.
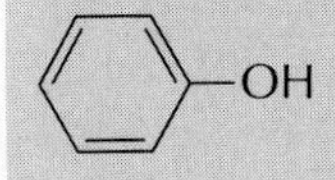 C₆H₅—OH	Phenol	Used as a germicidal agent and general disinfectant; also a component of epoxy resins.
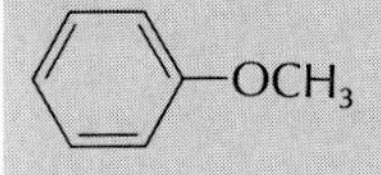 C₆H₅—OCH₃	Anisole	Intermediate in the synthesis of perfumes and flavorings.
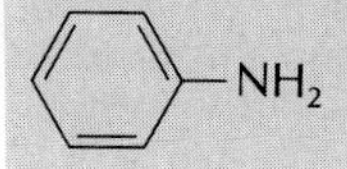 C₆H₅—NH₂	Aniline	Intermediate in preparation of dyes, urethane foams, and photographic chemicals.

Xylenes are used as additives in aviation gasoline, as well as being starting materials for the preparation of plasticizers and polyester fibers.

The prefix **ortho (o)** is used to identify a 1,2-disubstituted benzene ring, **meta (m)** signifies 1,3-disubstitution, and **para (p)** signifies 1,4-disubstitution. The o, m, and p prefixes can be used for all disubstituted benzenes, whether the substance is named as a benzene derivative or with a common name (such as acetophenone). For example,

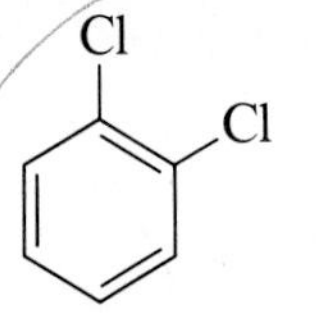

o-Dichlorobenzene
(1,2-dichlorobenzene)

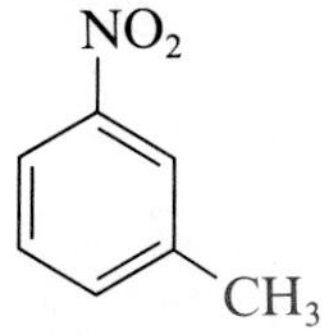

m-Nitrotoluene
(3-nitrotoluene)

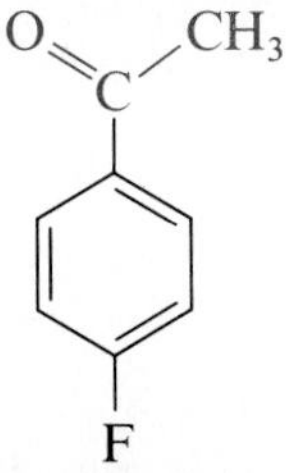

p-Fluoroacetophenone
(4-fluoroacetophenone)

PROBLEM 6.2

Write a structural formula for each of the following compounds:

(a) *o*-Ethylanisole   (b) *m*-Chlorostyrene   (c) *p*-Nitroaniline

SAMPLE SOLUTION

(a) The parent compound in *o*-ethylanisole is anisole. Anisole, as shown in Table 6.1, has a methoxy (CH_3O—) substituent on the benzene ring. The ethyl group in *o*-ethylanisole is attached to the carbon adjacent to the one that bears the methoxy substituent.

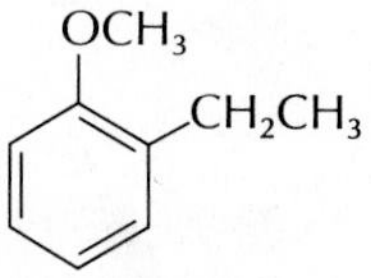

OCH₃
CH₂CH₃

o-Ethylanisole

The *o*, *m*, and *p* prefixes are not used when three or more substituents are present on benzene. Multiple substitution is described by identifying substituent positions on the ring by number.

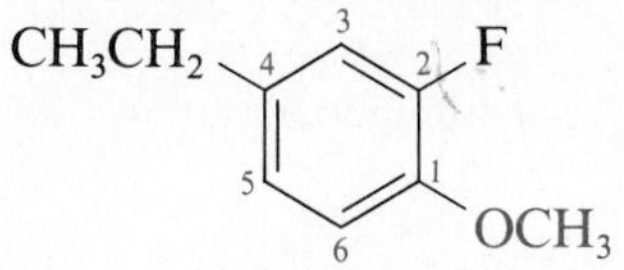

4-Ethyl-2-fluoro-
anisole

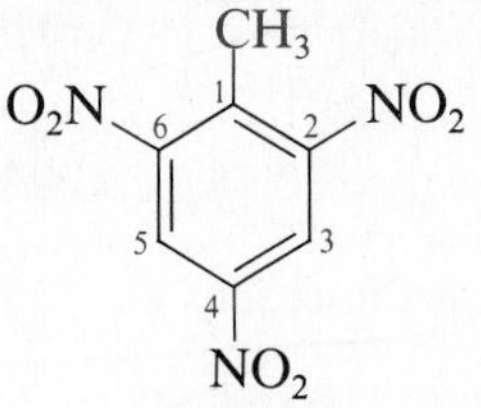

2,4,6-Trinitro-
toluene

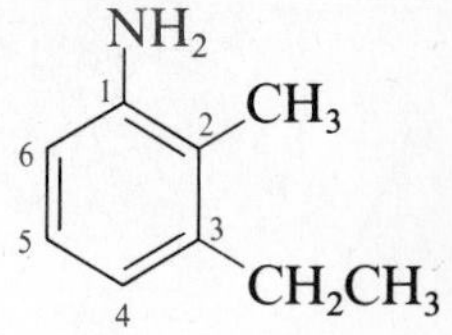

3-Ethyl-2-methyl-
aniline

As the preceding examples illustrate, the substituents are numbered on the ring beginning with the group which forms the basis for the compound name. Thus the substituted anisole derivative has its methoxy group at C-1, toluene its methyl group at C-1, and aniline its amino group at C-1. The direction of numbering is chosen to give the next substituted position the lowest number irrespective of what substituent it bears. The substituents appear in alphabetical order in the name.

In some compounds the benzene ring is named as a substituent and is called a **phenyl group.** For example, the compound $C_6H_5CH_2CH_2OH$ is called 2-phenyletha-nol, where C_6H_5— stands for the phenyl group.

6.6 REACTIONS OF ARENES: ELECTROPHILIC AROMATIC SUBSTITUTION

Many experiments have demonstrated that benzene behaves differently from other unsaturated compounds. Conditions under which an alkene undergoes *addition* with a reagent such as bromine, for example, give no reaction with benzene. In the presence of a catalyst a reaction takes place, but it is one of *substitution.*

Characteristically, the reagents that react with the aromatic ring of benzene and its derivatives are electron-deficient species called *electrophiles.* We have already discussed in Chapter 5 how electrophiles react with alkenes; electrophilic reagents *add* to alkenes.

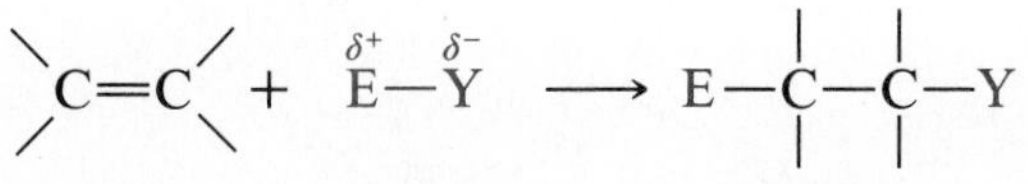

$$\text{Alkene} \quad + \quad \overset{\delta^+}{E}-\overset{\delta^-}{Y} \quad \longrightarrow \quad E-\overset{|}{\underset{|}{C}}-\overset{|}{\underset{|}{C}}-Y$$

Alkene　　　Electrophilic　　　Product of
　　　　　　reagent　　　electrophilic addition

A different reaction takes place when electrophiles react with arenes; arenes undergo *substitution instead of addition.*

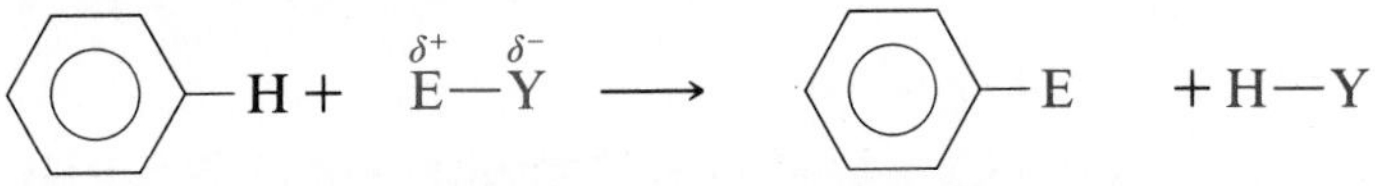

Arene　　　Electrophilic　　　Product of
　　　　　　reagent　　　electrophilic substitution

This reaction is called **electrophilic aromatic substitution** and is one of the fundamental processes of organic chemistry. The stability of the aromatic ring is such that it tends to be retained in reactions of benzene and its derivatives.

Some of the more frequently encountered electrophilic aromatic substitution reactions are shown below. In each, a group substitutes for one of the hydrogens of benzene. Each of these reactions will be discussed further in this chapter.

Nitration:

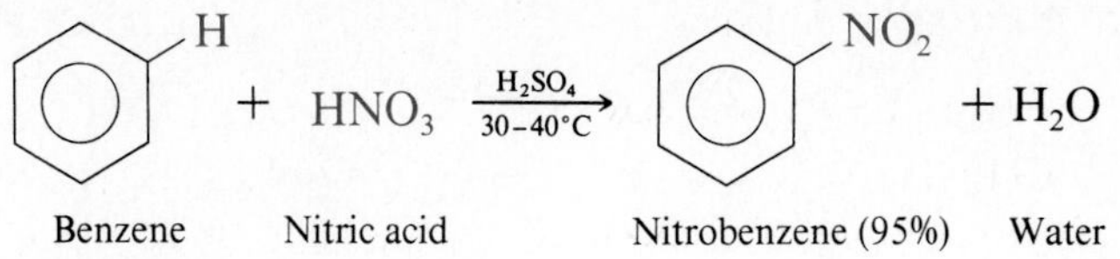

Benzene + HNO$_3$ $\xrightarrow[\text{30–40°C}]{\text{H}_2\text{SO}_4}$ Nitrobenzene (95%) + H$_2$O

Benzene   Nitric acid   Nitrobenzene (95%)   Water

Sulfonation:

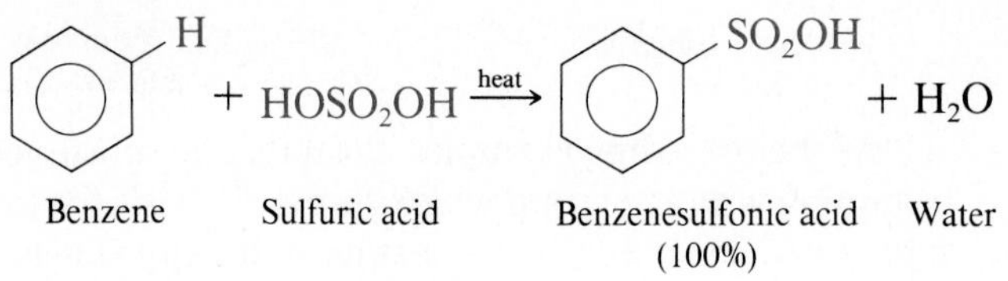

Benzene + HOSO$_2$OH $\xrightarrow{\text{heat}}$ Benzenesulfonic acid + H$_2$O

Benzene   Sulfuric acid   Benzenesulfonic acid   Water
                                (100%)

Bromination (halogenation):

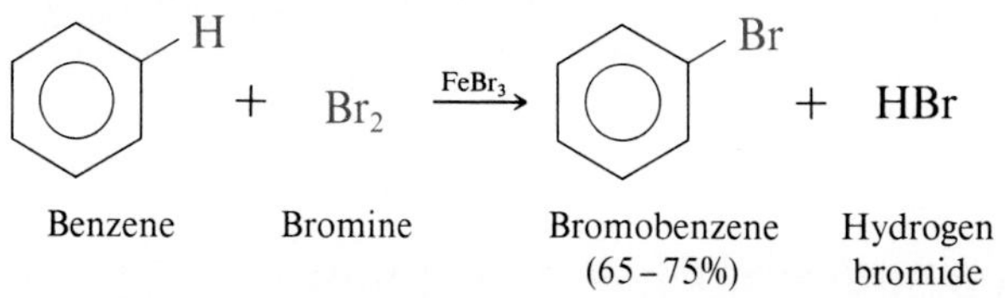

Benzene + Br$_2$ $\xrightarrow{\text{FeBr}_3}$ Bromobenzene + HBr

Benzene   Bromine   Bromobenzene   Hydrogen
                       (65–75%)   bromide

Friedel-Crafts alkylation:

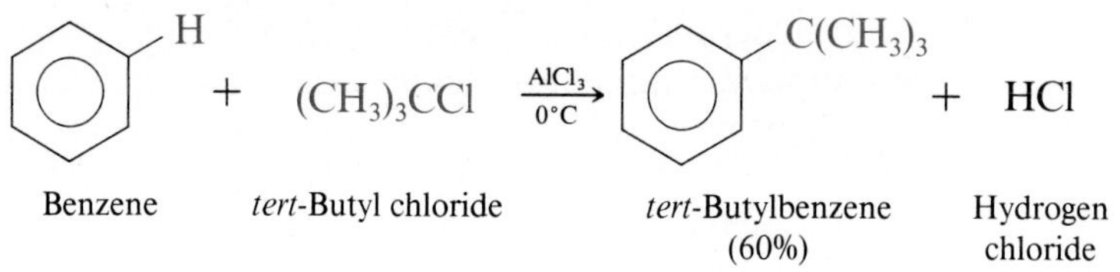

Benzene + (CH$_3$)$_3$CCl $\xrightarrow[\text{0°C}]{\text{AlCl}_3}$ tert-Butylbenzene + HCl

Benzene   *tert*-Butyl chloride   *tert*-Butylbenzene   Hydrogen
                                    (60%)   chloride

Friedel-Crafts acylation:

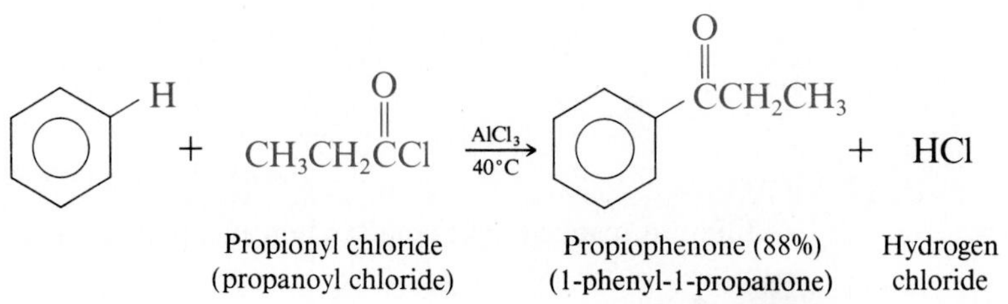

Benzene + CH$_3$CH$_2$CCl $\xrightarrow[\text{40°C}]{\text{AlCl}_3}$ Propiophenone + HCl

            Propionyl chloride   Propiophenone (88%)   Hydrogen
            (propanoyl chloride)   (1-phenyl-1-propanone)   chloride

6.7 MECHANISM OF ELECTROPHILIC AROMATIC SUBSTITUTION

Electrophilic aromatic substitution reactions are two-step processes. Recall from Chapter 5 that the first step in the electrophilic addition of an alkene is formation of a carbocation by addition of the electrophile to the π bond of the alkene. Similarly, in the first step of the reaction of an electrophilic reagent with benzene the electrophile accepts an electron pair from the π system of benzene to form a carbocation.

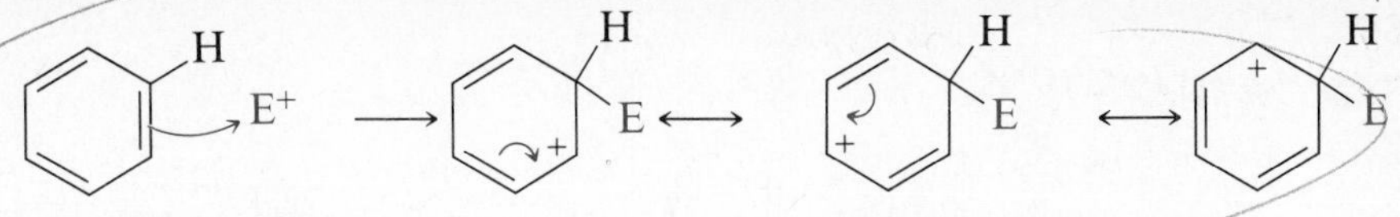

Benzene and electrophile   Cyclohexadienyl cation

This carbocation intermediate is stabilized by electron delocalization and is called a **cyclohexadienyl cation,** or σ **(sigma) complex.** The intermediate in each of the electrophilic aromatic substitution reactions mentioned in the preceding section is a cyclohexadienyl cation.

In the second step of the reaction, a proton is lost from the carbon to which the electrophile has added, giving rise to the substitution product and restoring the aromatic π-electron system to the ring.

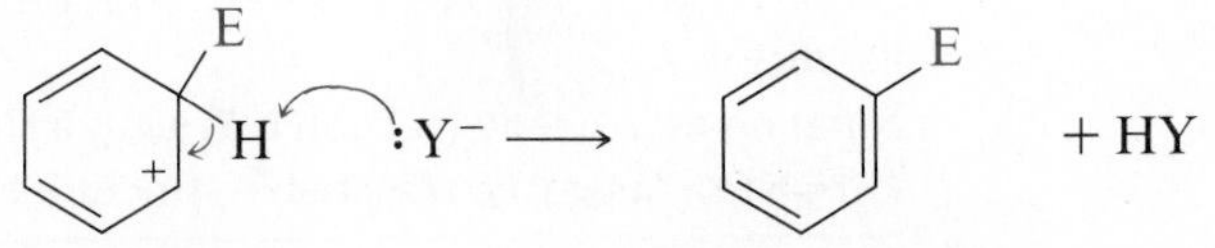

Cyclohexadienyl   Product of electrophilic
cation   aromatic substitution

Although stabilized by resonance, the cyclohexadienyl cation intermediate is *not* aromatic. The continuous cyclic array of *p* orbitals necessary for the aromatic system (Section 6.3) has been broken; the carbon to which the electrophile has added is sp^3 hybridized. Thus the stability associated with aromaticity has been lost in formation of the intermediate. Returning to a resonance-stabilized aromatic structure is the major reason why benzene undergoes substitution and not addition.

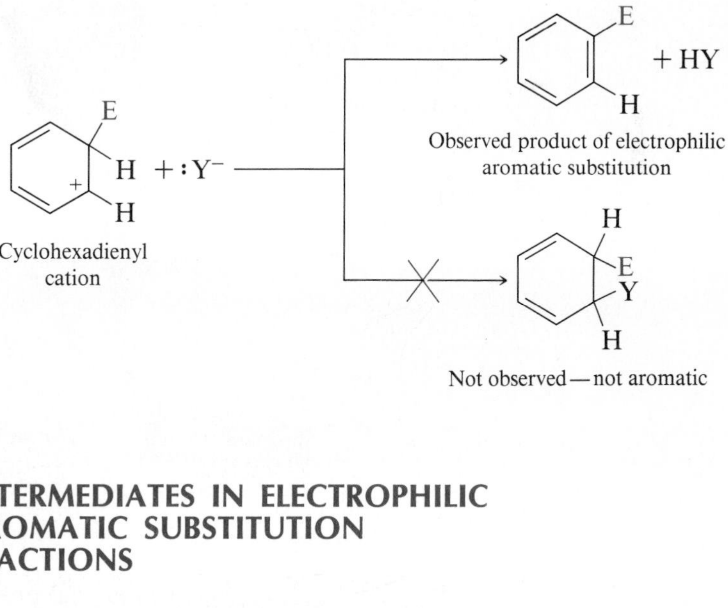

Observed product of electrophilic
aromatic substitution

Cyclohexadienyl
cation

Not observed — not aromatic

6.8 INTERMEDIATES IN ELECTROPHILIC AROMATIC SUBSTITUTION REACTIONS

We can now describe more fully the electrophilic aromatic substitution reactions shown in Section 6.6 by looking more closely at the electrophilic reagent which attacks benzene in each of these reactions.

6.8.1 Nitration

The electrophilic species in the nitration reaction is NO_2^+, called the **nitronium ion.**

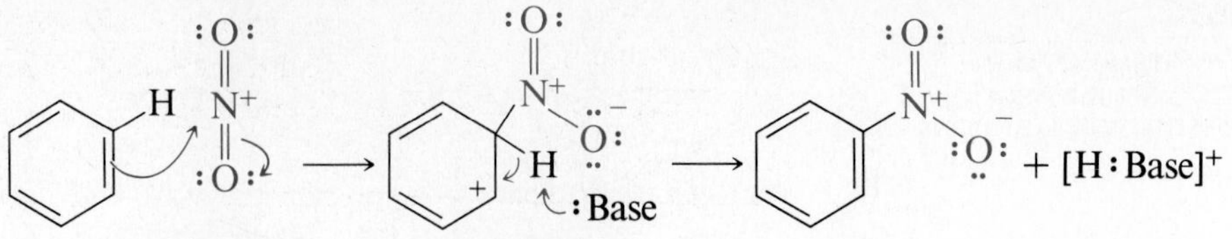

Benzene and nitronium cation — Cyclohexadienyl cation intermediate — Nitrobenzene — Protonated base

Nitric acid alone does not provide a high-enough concentration of nitronium ion for nitration of benzene to occur at an appreciable rate. However, addition of sulfuric acid causes the nitronium ion to form as shown in the following equation:

$$HNO_3 + 2H_2SO_4 \longrightarrow NO_2^+ + H_3O^+ + 2HSO_4^-$$

PROBLEM 6.3

Nitration of p-xylene (Section 6.5) gives a single product having the molecular formula $C_8H_9NO_2$. Suggest a reasonable structure for this product.

6.8.2 Sulfonation

When sulfuric acid is heated, sulfur trioxide (SO_3) is formed, and it is believed to be the electrophilic species which attacks the aromatic ring.

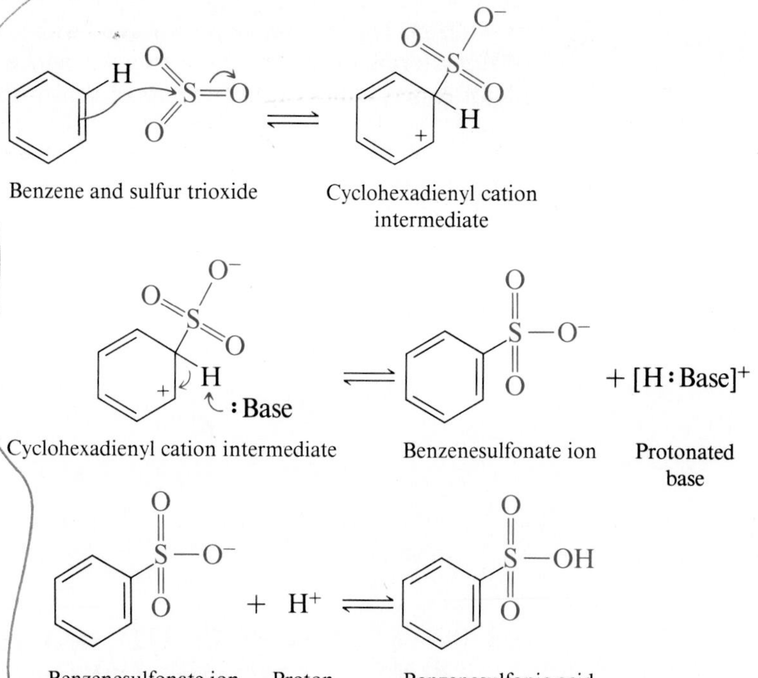

Benzene and sulfur trioxide — Cyclohexadienyl cation intermediate

Cyclohexadienyl cation intermediate — Benzenesulfonate ion — Protonated base

Benzenesulfonate ion — Proton — Benzenesulfonic acid

Sulfonations are important industrial processes in the manufacture of detergents, and in these reactions a solution of sulfur trioxide in sulfuric acid, called *oleum,* is often used.

6.8.3 Halogenation

Halogenation is brought about through the reaction of benzene with bromine or chlorine in the presence of a **Lewis acid catalyst.** The catalyst is usually ferric bromide ($FeBr_3$) for brominations and ferric chloride ($FeCl_3$) for chlorinations.

A catalyst is necessary to increase the electrophilic character of the halogen molecule. This is brought about by formation of a Lewis acid–Lewis base complex, as illustrated for the bromination reaction.

$$:\ddot{Br}-\ddot{Br}: + \quad FeBr_3 \rightleftharpoons :\ddot{Br}-\overset{+}{\ddot{Br}}-\overset{-}{FeBr_3}$$

Lewis base   Lewis acid   Lewis acid – Lewis
base complex

The bromine-bromine bond in this complex is polarized, making bromine a better electrophile.

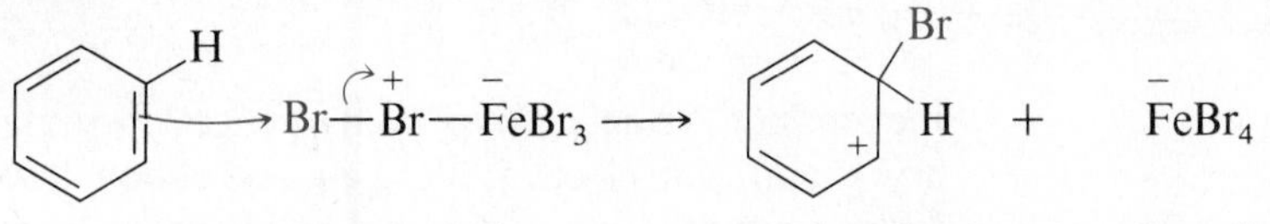

Benzene and bromine – ferric   Cyclohexadienyl   Tetrabromoferrate
bromide complex   cation intermediate   ion

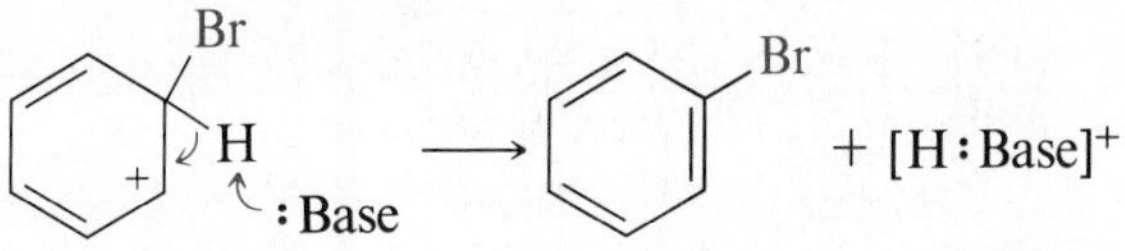

Cyclohexadienyl   Bromobenzene   Protonated
cation intermediate   base

6.8.4 Friedel-Crafts Alkylation

This reaction was discovered through the collaborative efforts of Charles Friedel (in France) and James Crafts (an American chemist working in Friedel's laboratory) in 1877. The electrophile is a carbocation formed by reaction of an alkyl halide with aluminum chloride which acts as a Lewis acid catalyst.

$$(CH_3)_3C-\ddot{Cl}: + \quad AlCl_3 \longrightarrow (CH_3)_3C-\overset{+}{\ddot{Cl}}-\overset{-}{AlCl_3}$$

tert-Butyl chloride   Aluminum   Lewis acid – Lewis base
chloride   complex

$$(CH_3)_3C-\overset{+}{\ddot{Cl}}-\overset{-}{AlCl_3} \longrightarrow (CH_3)_3C^+ + \quad \overset{-}{AlCl_4}$$

tert-Butyl chloride–   *tert*-Butyl   Tetrachloroaluminate
aluminum chloride complex   cation   anion

Once formed, the carbocation reacts with benzene in the same way as other electrophiles do.

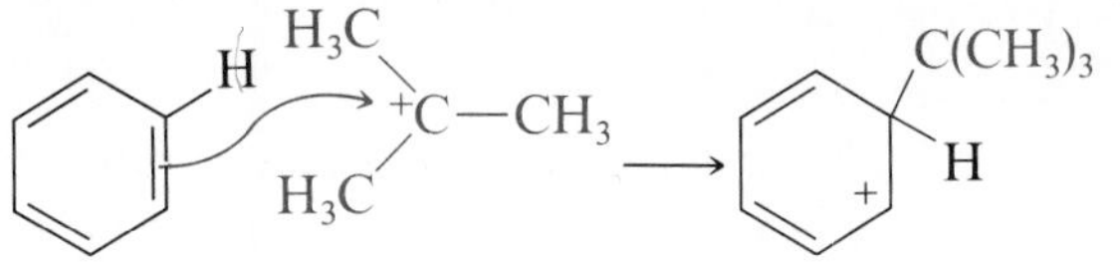

Benzene and *tert*-butyl cation   Cyclohexadienyl
cation intermediate

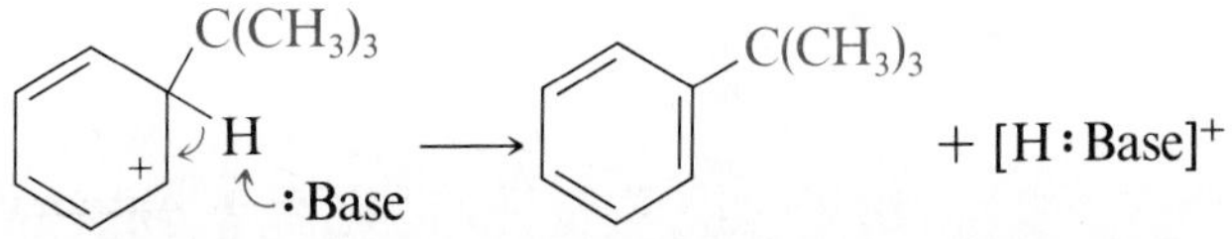

Cyclohexadienyl cation   *tert*-Butylbenzene   Protonated
intermediate   base

PROBLEM 6.4

About one-half the benzene produced in the United States each year is converted to ethylbenzene to be used in the preparation of styrene (Table 6.1). Describe with a chemical equation how ethylbenzene could be prepared from benzene and any other necessary reagents.

6.8.5 Friedel-Crafts Acylation

Another version of the Friedel-Crafts reaction uses **acyl halides** and yields a ketone as the product. An acyl group has the general formula $R\overset{\overset{\displaystyle O}{\|}}{C}-$, where R can be an alkyl or aryl group. The electrophile is an **acyl cation,** also known as an **acylium ion,** which forms from reaction of the acyl halide with aluminum chloride.

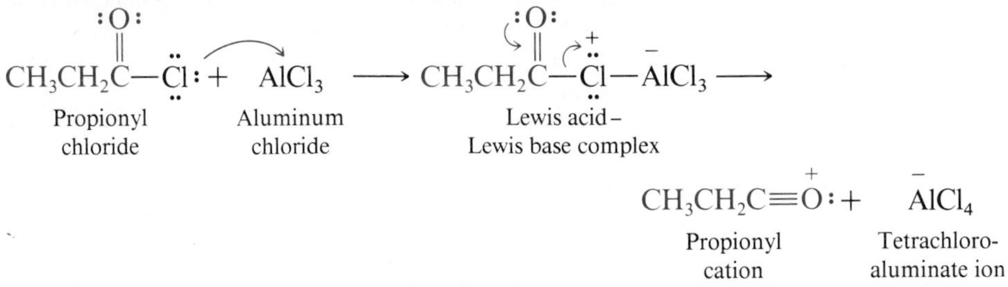

$$CH_3CH_2\overset{\overset{\displaystyle :O:}{\|}}{C}-\ddot{\underset{..}{C}}l: + \quad AlCl_3 \longrightarrow CH_3CH_2\overset{\overset{\displaystyle :O:}{\|}}{C}-\overset{+}{\underset{..}{C}}l-AlCl_3 \longrightarrow$$

| Propionyl chloride | Aluminum chloride | Lewis acid– Lewis base complex |

$$CH_3CH_2C\equiv\overset{+}{O}: + \quad \overset{-}{AlCl_4}$$

| Propionyl cation | Tetrachloro- aluminate ion |

The electrophilic site of an acyl cation is its acyl carbon; it reacts with benzene in a manner analogous to that of other electrophilic reagents.

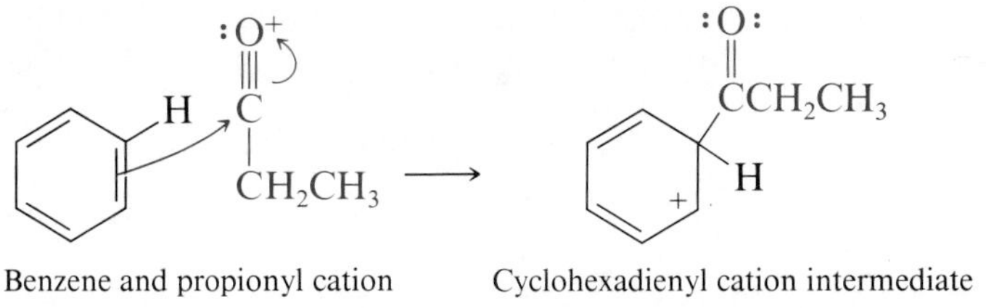

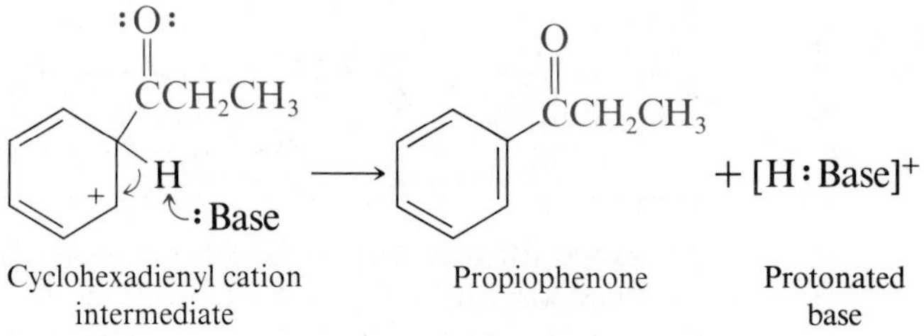

Benzene and propionyl cation   Cyclohexadienyl cation intermediate

Cyclohexadienyl cation intermediate   Propiophenone   Protonated base

Carboxylic acid anhydrides, compounds of the type $R\overset{\overset{\displaystyle O}{\|}}{C}O\overset{\overset{\displaystyle O}{\|}}{C}R$, can also serve as sources of acyl cations and, in the presence of aluminum chloride, can acylate benzene. One acyl unit of an acid anhydride becomes attached to the benzene ring, while the other becomes part of a carboxylic acid.

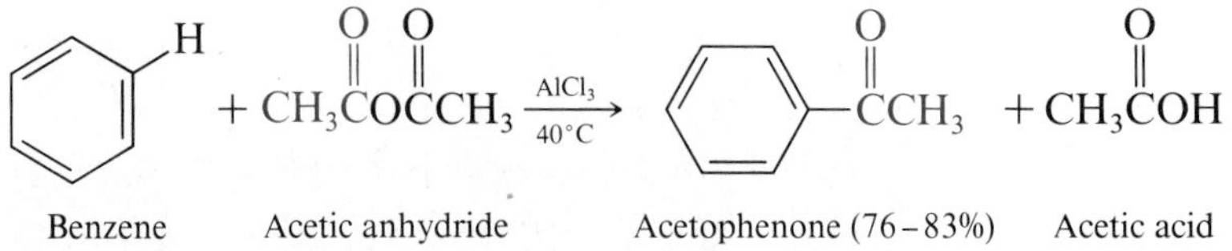

Benzene   Acetic anhydride   Acetophenone (76–83%)   Acetic acid

PROBLEM 6.5

The preceding equation shows one method for preparing acetophenone. Write an equation describing the preparation of acetophenone from benzene using an acyl halide.

6.9 RATE AND ORIENTATION IN ELECTROPHILIC AROMATIC SUBSTITUTION REACTIONS

So far we have been concerned only with electrophilic substitution on benzene. What if the reactant is an arene that already bears at least one substituent? Two questions arise with regard to electrophilic aromatic substitution of a substituted benzene derivative:

1. What is the effect of a substituent on the **rate;** will the reaction proceed faster or slower than the comparable reaction of benzene?
2. What is the effect of a substituent on the **orientation;** will the incoming electrophile attack the ring at a position ortho, meta, or para to the existing substituent?

6.9.1 Rate Effects of Substituents

Consider the nitration of benzene, toluene, and (trifluoromethyl)benzene.

Toluene	Benzene	(Trifluoromethyl)benzene
(most reactive)		(least reactive)

Toluene is *more* reactive than benzene. It undergoes nitration some 20 to 25 times faster than benzene. Because toluene is more reactive than benzene, we say that a methyl group **activates** the ring toward electrophilic substitution. (Trifluoromethyl)benzene is much *less* reactive than benzene; it undergoes nitration thousands of times more slowly than benzene. We say that a trifluoromethyl group **deactivates** the ring toward electrophilic substitution.

In fact, all electrophilic substitution reactions of toluene occur faster than the same reactions of benzene. Likewise, all electrophilic substitution reactions of (trifluoromethyl)benzene occur slower than the same reaction of benzene.

Table 6.2 lists several atoms and groups frequently encountered as substituents on a benzene ring. In the first column the substituents are classified as being either activating or deactivating. Benzene derivatives bearing **activating substituents** react *faster* in an electrophilic substitution reaction than benzene itself. Benzene derivatives bearing **deactivating substituents** undergo electrophilic aromatic substitution *slower* than benzene.

In the discussion of orientation which follows we will see in more detail how substituents exert their activating and deactivating effects.

PROBLEM 6.6

For each of the following pairs of aromatic compounds, specify which one will react faster in a nitration reaction.

TABLE 6.2
Classification of Substituents in Electrophilic Aromatic Substitution Reactions

Effect on rate	Substituent		Effect on orientation
Very strongly activating	—$\overset{..}{\underset{..}{N}}H_2$	(amino)	Ortho,para-directing
	—$\overset{..}{N}HR$	(alkylamino)	
	—$\overset{..}{N}R_2$	(dialkylamino)	
	—$\overset{..}{\underset{..}{O}}H$	(hydroxyl)	
Strongly activating	—$\overset{\displaystyle O}{\overset{\|}{\underset{..}{N}HC}}R$	(acylamino)	Ortho,para-directing
	—$\overset{..}{\underset{..}{O}}R$	(alkoxy)	
	—$\overset{\displaystyle O}{\overset{\|}{\underset{..}{\overset{..}{O}}C}}R$	(acyloxy)	
Activating	—R	(alkyl)	Ortho,para-directing
Standard of comparison	—H	(hydrogen)	
Deactivating	—X (X = F, Cl, Br, I)	(halogen)	Ortho,para-directing
Strongly deactivating	—$\overset{\displaystyle O}{\overset{\|}{C}}H$	(formyl)	Meta-directing
	—$\overset{\displaystyle O}{\overset{\|}{C}}R$	(acyl)	
	—$\overset{\displaystyle O}{\overset{\|}{C}}OH$	(carboxylic acid)	
	—$\overset{\displaystyle O}{\overset{\|}{C}}OR$	(ester)	
	—$\overset{\displaystyle O}{\overset{\|}{C}}Cl$	(acyl chloride)	
	—C≡N	(cyano)	
	—SO_3H	(sulfonic acid)	
Very strongly deactivating	—CF_3	(trifluoromethyl)	Meta-directing
	—NO_2	(nitro)	

(a) Toluene or bromobenzene
(b) Benzene or phenol (C_6H_5OH)
(c) Nitrobenzene or aniline ($C_6H_5NH_2$)

SAMPLE SOLUTION

(a) In toluene the substituent is an alkyl group (methyl), which is activating when compared to hydrogen. Therefore toluene undergoes substitution faster than benzene. Bromobenzene reacts slower than benzene since the halogen is a deactivating group. Toluene reacts faster than bromobenzene in an electrophilic substitution reaction.

6.9.2 Orientation Effects of Substituents

Let us consider again the nitration of toluene. Unlike benzene, the ring positions of toluene are no longer equivalent. Substitution of toluene may lead to products in which the new substituent is ortho, meta, or para to the methyl group. When toluene is nitrated and the products are examined, the ortho and para isomers are found to predominate, with only a trace of the meta being formed.

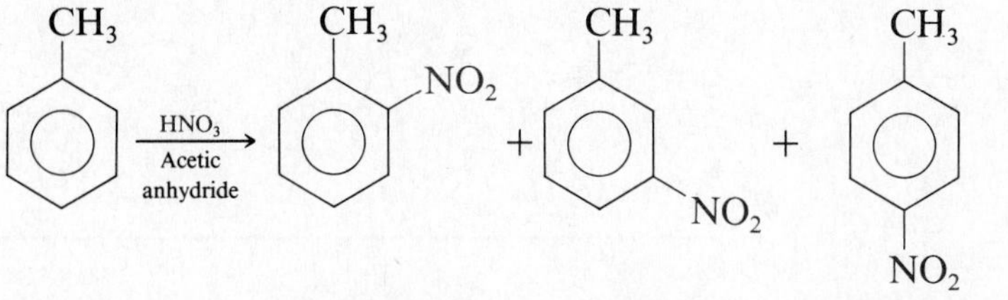

Toluene   o-Nitrotoluene (63%)   m-Nitrotoluene (3%)   p-Nitrotoluene (34%)

We say that a methyl substituent is an **ortho,para director.** By this we mean that the methyl group of toluene orients, or directs, an incoming substituent ortho and para to itself. As Table 6.2 illustrates, all activating substituents are ortho,para–directing groups.

Nitration of (trifluoromethyl)benzene gives a very different result. The meta isomer is formed almost exclusively; only small amounts of the ortho and para isomers are found. The trifluoromethyl group is said to be a **meta director.** That is, a trifluoromethyl group orients an incoming substituent meta to itself in an electrophilic aromatic substitution reaction.

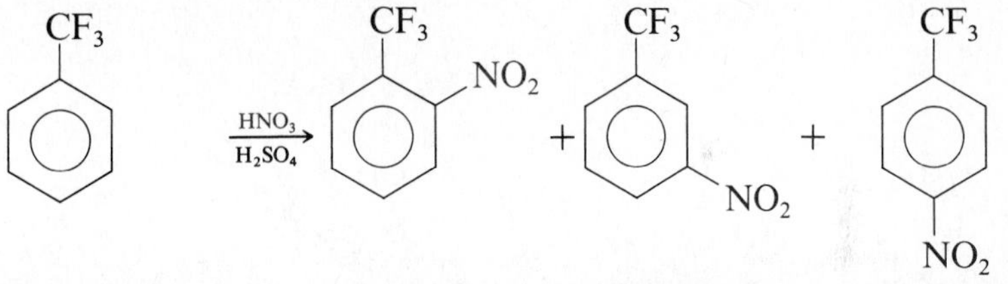

(Trifluoromethyl)benzene   o-Nitro(trifluoro-methyl)benzene (6%)   m-Nitro(trifluoro-methyl)benzene (91%)   p-Nitro(trifluoro-methyl)benzene (3%)

Table 6.2 lists a number of substituents which are meta directors. All substituents which are more deactivating than halogen are meta-directing groups.

PROBLEM 6.7

Specify whether the major product or products of each of the following reactions is the meta isomer or a mixture of ortho and para isomers.

     (a) Nitration of anisole ($C_6H_5OCH_3$)

     (b) Nitration of benzoic acid ($C_6H_5\overset{\displaystyle O}{\overset{\displaystyle \|}{C}}OH$)

     (c) Bromination of ethylbenzene

SAMPLE SOLUTION

(a) From Table 6.2 you can see that an alkoxy group such as methoxy ($-OCH_3$) is an ortho,para director and therefore directs incoming substituents ortho and para to itself.

The principal products of nitration of anisole will be a mixture of *ortho*-nitroanisole and *para*-nitroanisole.

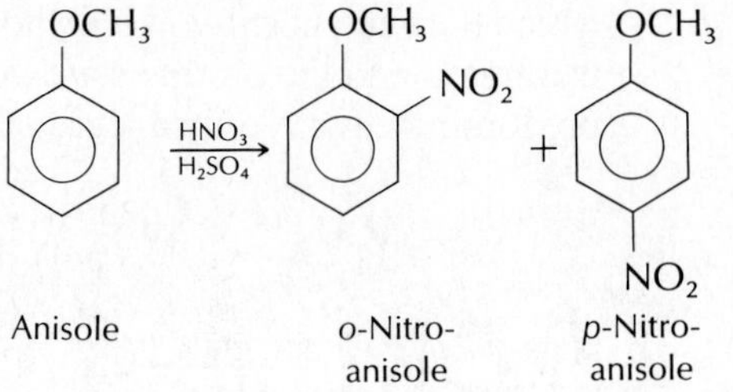

Anisole   *o*-Nitro-anisole   *p*-Nitro-anisole

6.10 MECHANISTIC EXPLANATION OF RATE AND ORIENTATION EFFECTS: ACTIVATING GROUPS

As you saw in Section 6.7 electrophilic aromatic substitution reactions proceed through formation of a cyclohexadienyl cation intermediate. The key to understanding the influence of a substituent on both the rate and orientation of a substitution reaction is to further examine the structure and stability of the cation intermediate.

Consider, for example, the intermediate that forms in the nitration of toluene. Three resonance structures may be written to describe electron delocalization in the intermediates resulting from ortho, meta, or para attack. The contributing resonance structures for ortho and para attack include one which is a tertiary carbocation.

Ortho attack:

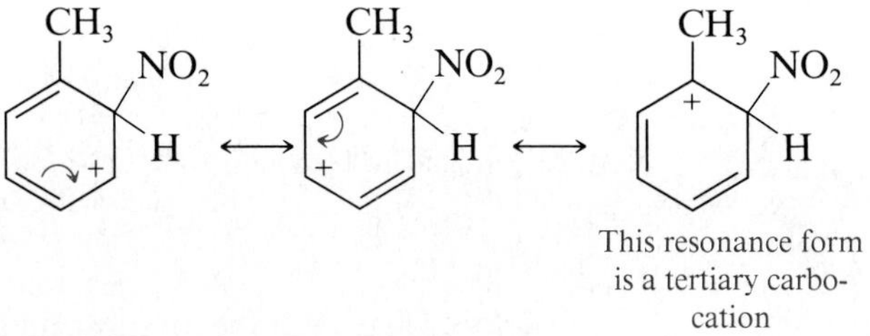

This resonance form is a tertiary carbo-cation

Para attack:

This resonance form is a tertiary carbocation

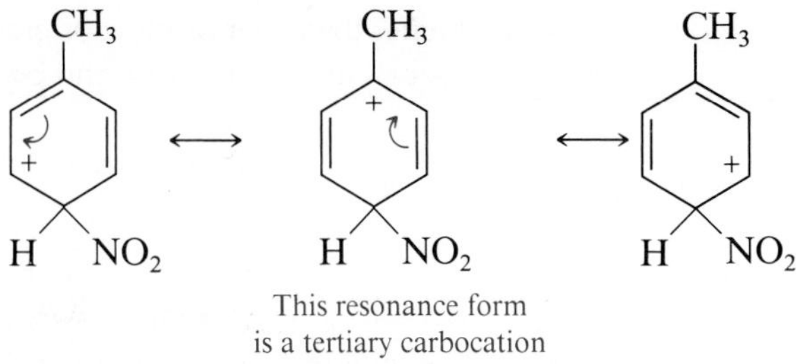

The three resonance structures of the cyclohexadienyl cation intermediate leading to meta substitution are all secondary carbocations.

Meta attack:

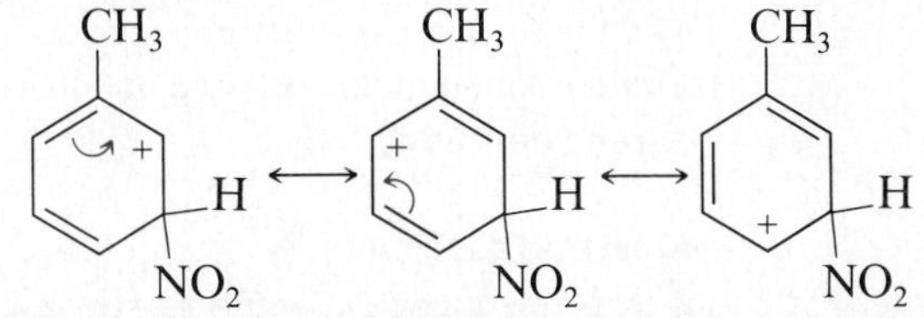

Recall from Chapter 3 (Section 3.12) that alkyl groups stabilize carbocations by releasing electrons to the positively charged carbon, causing tertiary carbocations to be more stable than secondary ones. Thus the cyclohexadienyl intermediates from either ortho or para attack are more stable than the intermediate formed from meta attack because they have some of the character of a tertiary carbocation.

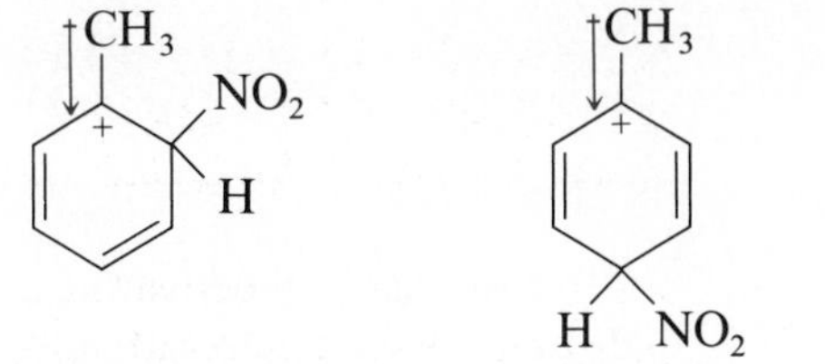

More stable tertiary
carbocation resonance
structure from ortho
attack

More stable tertiary
carbocation resonance
structure from para
attack

The carbocations formed as intermediates during ortho and para substitution of toluene are also more stable than the cyclohexadienyl cation intermediate in the corresponding reaction of benzene and are formed faster. This increased stability of the carbocation intermediate is responsible for the rate-enhancing effect of a methyl substituent.

You can see in Table 6.2 that several of the activating ortho,para–directing substituents contain an oxygen or nitrogen atom attached to the benzene ring. The cyclohexadienyl cations formed by ortho and para attack on these compounds are stabilized by donation of an unshared electron pair to the ring. We can illustrate this effect by looking at the intermediate formed by ortho attack on a ring containing an alkoxy (—$\ddot{\text{O}}$R) group.

Ortho attack:

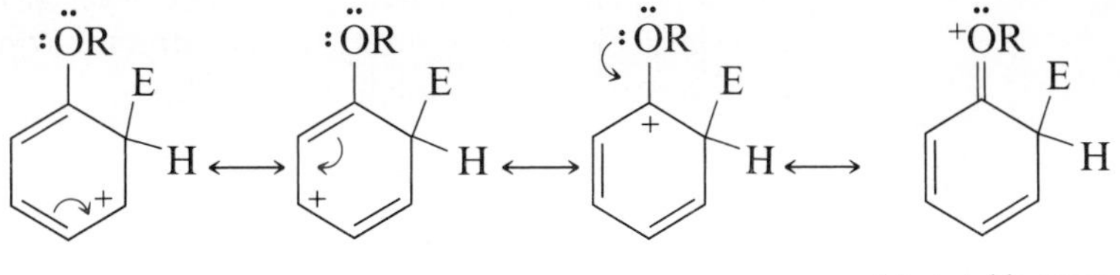

Most stable resonance
form; oxygen and all
carbons have octets of
electrons

While the unshared electron pair on oxygen is able to stabilize the cation from para attack in a similar manner, the cation formed from meta attack does not receive comparable stabilization by the electron pair. As this example illustrates, substituents with unshared electron pairs on the atom attached to the ring are ortho,para–directing.

PROBLEM 6.8

Draw the resonance structures corresponding to the cyclohexadienyl cation intermediate formed from (a) meta and (b) para attack of an electrophile on an alkoxy-substituted benzene derivative.

SAMPLE SOLUTION

(a) Three resonance structures may be drawn that contribute to the cyclohexadienyl cation intermediate formed by meta attack. The unshared electron pair of the alkoxy group is not able to stabilize the intermediate.

Meta attack

In summary, all activating substituents stabilize the cyclohexadienyl intermediate by donating electrons to the ring. Also, **all activating substituents are ortho,para directors.**

6.11 MECHANISTIC EXPLANATION OF RATE AND ORIENTATION EFFECTS: DEACTIVATING GROUPS

Turning now to meta-directing substituents, let us consider electrophilic aromatic substitution in (trifluoromethyl)benzene. First consider the electronic properties of a trifluoromethyl group: because of their high electronegativity the three fluorine atoms polarize the electron distribution in their σ bonds to carbon so that carbon bears a partial positive charge.

Unlike a methyl group, which is slightly electron-releasing, a trifluoromethyl group is a powerful electron-withdrawing substituent. Consequently, a CF_3 group destabilizes a carbocation site to which it is attached.

$$CH_3 \!-\! \overset{+}{C} \qquad \begin{array}{c} \text{more} \\ \text{stable} \\ \text{than} \end{array} \qquad H \!-\! \overset{+}{C} \qquad \begin{array}{c} \text{more} \\ \text{stable} \\ \text{than} \end{array} \qquad F_3C \!-\! \overset{+}{C}$$

Methyl group releases electrons, stabilizes carbocation

Trifluoromethyl group withdraws electrons, destabilizes carbocation

When we examine the cyclohexadienyl cation intermediates formed during the nitration of (trifluoromethyl)benzene, we find that those leading to ortho and para substitution are strongly *destabilized* (made less stable). This is due to the presence of a positive charge on the ring carbon bearing the strongly electron-withdrawing trifluoromethyl group.

Ortho attack:

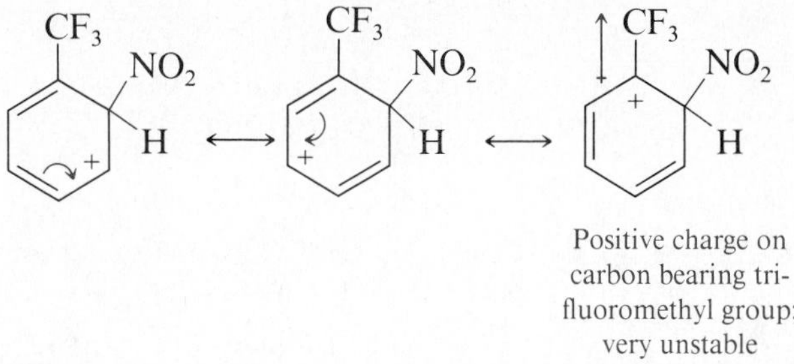

Positive charge on
carbon bearing tri-
fluoromethyl group;
very unstable

Para attack:

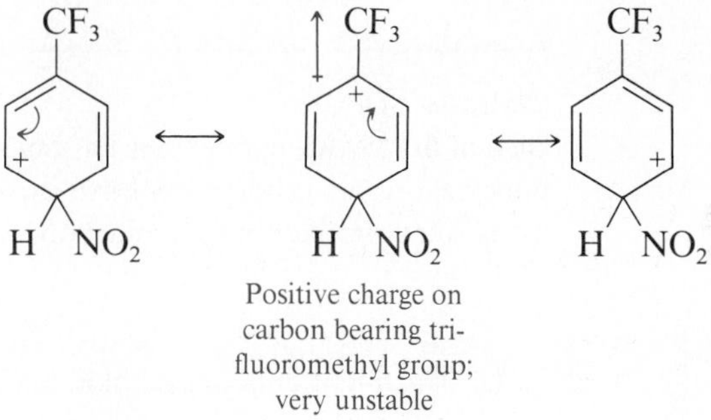

Positive charge on
carbon bearing tri-
fluoromethyl group;
very unstable

None of the three major resonance forms of the intermediate formed during attack at the meta position has a positive charge on the carbon bearing the trifluoromethyl substituent.

Meta attack:

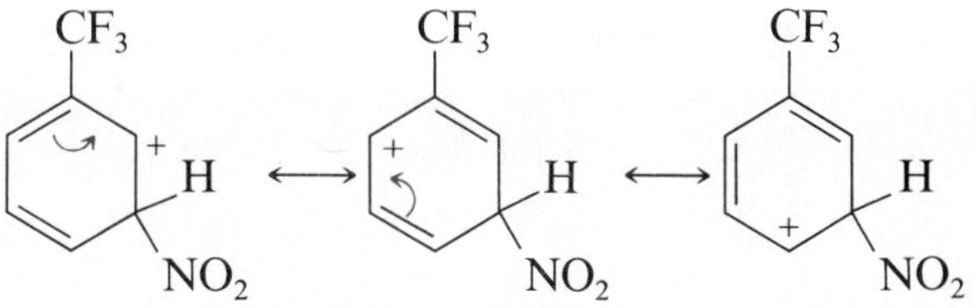

Attack at the meta position of (trifluoromethyl)benzene leads to a more stable intermediate than attack at either ortho or para positions, so meta substitution is observed. However the intermediate from meta attack is still less stable than that formed during nitration of benzene, so it is formed more slowly. In other words, the trifluoromethyl group is a **deactivating meta director.**

As a general rule, **all meta directors are deactivating substituents.** Substituents which destabilize carbocations cause the reaction itself to occur more slowly. The substituent atom attached to the benzene ring generally has a partial positive charge, which has the effect of withdrawing electron density away from the ring thus decreasing the stability of the cyclohexadienyl cation intermediate. The carbon-oxygen bond of benzaldehyde, for example, is highly polarized. Carbon, as the less electronegative element, is the positive end of the dipole. The cyclohexadienyl cation intermediates from nitration of benzaldehyde are therefore

Ortho attack:

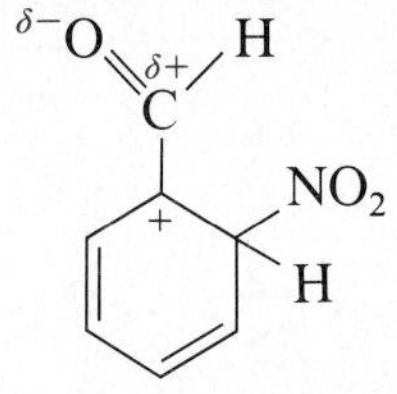

Meta attack:

Para attack:

Unstable because
of adjacent posi-
tively polarized
atoms

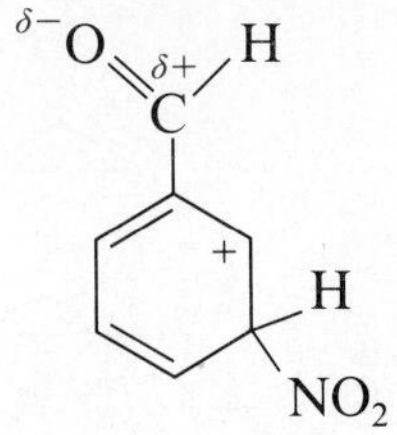

Positively polarized
atoms not adjacent;
most stable inter-
mediate

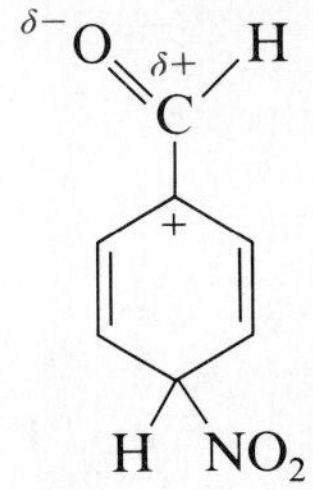

Unstable because
of adjacent positively
polarized atoms

Nitration of benzaldehyde gives the meta-substituted product almost exclusively.

PROBLEM 6.9

Each of the following reactions has been reported in the chemical literature, and the principal organic product has been isolated in good yield. Write a structural formula for the isolated product of each reaction.

(a) Treatment of benzoyl chloride ($C_6H_5\overset{O}{\overset{\|}{C}}Cl$) with chlorine and ferric chloride

(b) Treatment of methyl benzoate ($C_6H_5\overset{O}{\overset{\|}{C}}OCH_3$) with nitric acid and sulfuric acid

(c) Nitration of propiophenone ($C_6H_5\overset{O}{\overset{\|}{C}}CH_2CH_3$)

SAMPLE SOLUTION

(a) Benzoyl chloride has a carbonyl group attached directly to the ring. A $-\overset{O}{\overset{\|}{C}}Cl$ substituent is meta-directing. The reaction conditions, namely, use of chlorine and ferric chloride, are those that introduce a chlorine onto the ring. The product is *m*-chlorobenzoyl chloride.

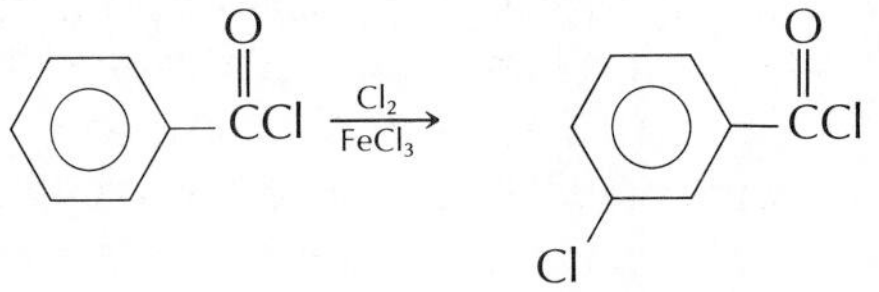

Benzoyl chloride   *m*-Chlorobenzoyl chloride
(isolated in 62% yield)

6.12 MECHANISTIC EXPLANATION OF RATE AND ORIENTATION EFFECTS: HALOGENS

The halogens are unique in that they are **deactivating ortho,para directors.** Halogens are more electronegative than carbon and tend to draw electrons away from the carbon to which they are bonded, in the same manner as a trifluoromethyl group. Thus the ring is deactivated toward electrophilic substitution since the cyclohexadienyl intermediate is less stable than the corresponding intermediate for benzene.

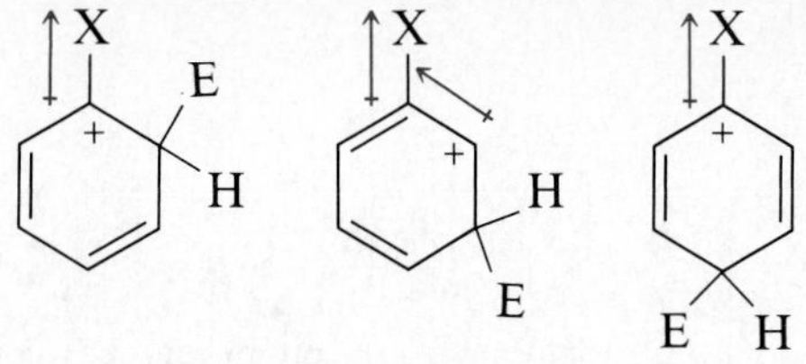

All these ions are less stable when X = F, Cl, Br, or 1 than when X = H

However, like hydroxyl groups and amino groups, halogen substituents possess un-shared electron pairs that can be donated to a positively charged carbon. This electron donation stabilizes the intermediates derived from ortho and from para attack.

Resonance involving the halogen lone pair:

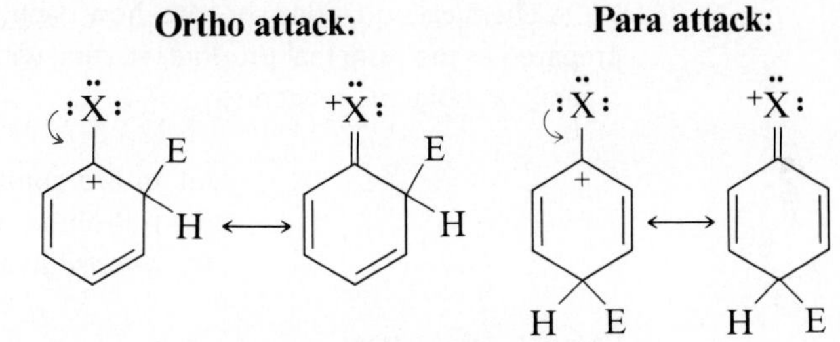

Ortho attack: **Para attack:**

Comparable resonance stabilization of the intermediate leading to meta substitution is not possible; therefore the halogens are ortho,para directors.

6.13 SYNTHESIS OF DISUBSTITUTED AROMATIC COMPOUNDS

Since the position of electrophilic attack on an aromatic ring is controlled by the directing effects of substituents already present, the preparation of disubstituted aromatic compounds requires that careful thought be given to the order of introduction of the two groups.

Compare the independent preparations of *m*-bromoacetophenone and *p*-bromoacetophenone from benzene. Both syntheses require a Friedel-Crafts acylation step and a bromination step, but the major product is determined by the *order* in which the two steps are carried out. When the meta-directing acetyl group is introduced first, the final product is *m*-bromoacetophenone.

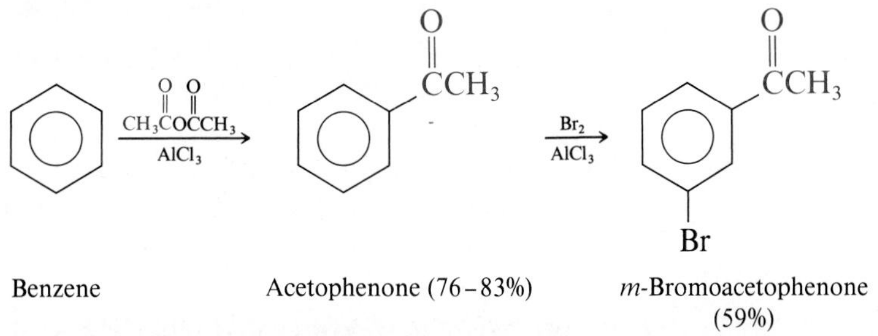

Benzene	Acetophenone (76–83%)	*m*-Bromoacetophenone (59%)

When the ortho,para–directing bromo substituent is introduced first, the final product is *p*-bromoacetophenone (along with some of its ortho isomer, from which it is separated by distillation).

Benzene → $\xrightarrow[\text{FeBr}_3]{\text{Br}_2}$ → Bromobenzene (65–75%) → $\xrightarrow[\text{AlCl}_3]{\text{CH}_3\text{COCCH}_3}$ → p-Bromoacetophenone (69–79%)

In planning a synthesis, the *orientation* of the substituents in the desired product determines the *order* of the reactions used to carry out the scheme. If the substituents are meta in the product, then the meta-directing group must be introduced first. Likewise, if the substituents are ortho or para, the first substituent on the ring must be an ortho,para–directing one.

PROBLEM 6.10

Write chemical equations showing how each of the following compounds could be prepared as the principal product, starting with benzene and using any necessary organic or inorganic reagents.

(a) *m*-Bromonitrobenzene
(b) *p*-Bromonitrobenzene
(c) *p-tert*-Butylnitrobenzene

SAMPLE SOLUTION

(a) The nitro and bromine substituents are situated meta to each other in the desired product, and therefore the meta-directing nitro group must be introduced *first* in the synthesis. Thus nitration of benzene is followed by bromination of nitrobenzene.

Benzene → $\xrightarrow[\text{H}_2\text{SO}_4]{\text{HNO}_3}$ → Nitrobenzene → $\xrightarrow[\text{FeBr}_3]{\text{Br}_2}$ → m-Bromonitrobenzene

---

6.14 AROMATIC SIDE-CHAIN REACTIONS

In this section we shift the focus from reactions that take place on the aromatic ring to reactions in which the ring acts as a substituent.

In Chapter 5 (Section 5.7) you saw how conjugated dienes undergo addition reactions proceeding through formation of resonance-stabilized allylic carbocation intermediates. A carbocation adjacent to a benzene ring is also resonance-stabilized and is called a **benzylic carbocation.**

Benzylic carbocation

Electrophilic additions to double bonds adjacent to an aromatic ring are *regioselective* (Section 5.1.3). The major product results from formation of the more stable benzylic carbocation. This can be seen in the addition of hydrogen chloride to indene. Only a single chloride is formed.

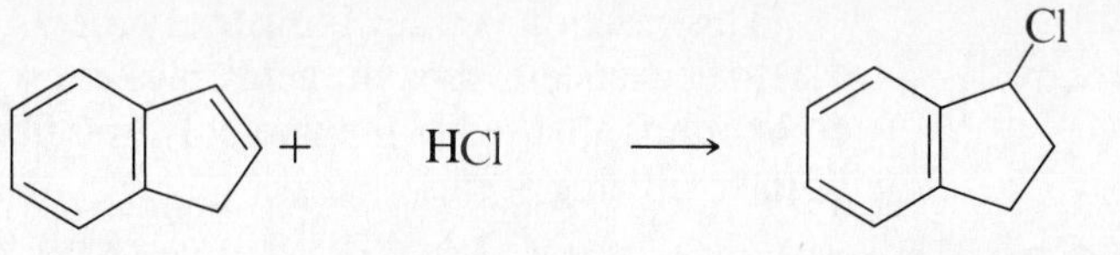

Indene   Hydrogen chloride   1-Chloroindane (75–84%)

Only the benzylic chloride is formed because protonation of the double bond occurs in the direction that gives a secondary carbocation that is benzylic.

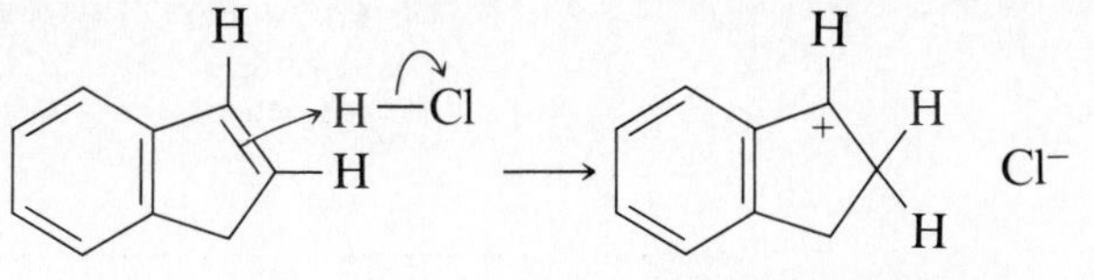

Carbocation that leads to
observed product

Protonation in the opposite direction also gives a secondary carbocation, but one that is not benzylic.

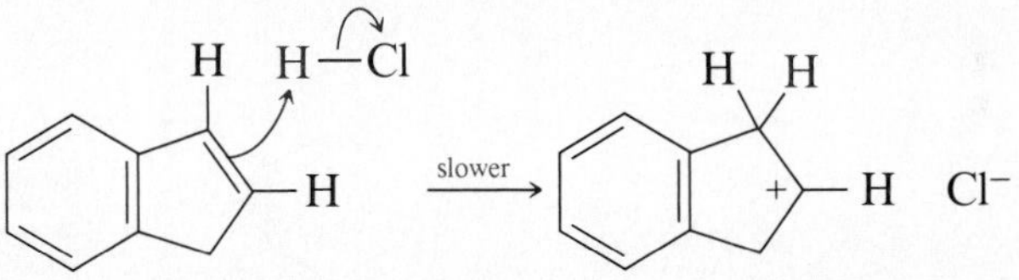

Less stable carbocation

This alternative carbocation does not receive the extra increment of stabilization that its benzylic isomer does and so is formed more slowly. The orientation of addition is controlled by the rate of carbocation formation; the more stable benzylic carbocation is formed faster and is the one that determines the reaction product.

The reaction of alkylbenzenes with oxidizing agents provides a striking example of the effect that a benzene ring has on reactions that take place at the benzylic position. While potassium permanganate cleaves alkenes readily (Section 5.2), it does not react either with benzene or with alkanes.

$$RCH_2CH_2R' \xrightarrow{KMnO_4} \text{no reaction}$$

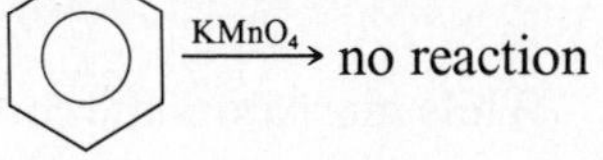

$\xrightarrow{KMnO_4}$ no reaction

On the other hand, an alkyl side chain on a benzene ring is oxidized on being heated with potassium permanganate. The product is benzoic acid or a substituted derivative of benzoic acid.

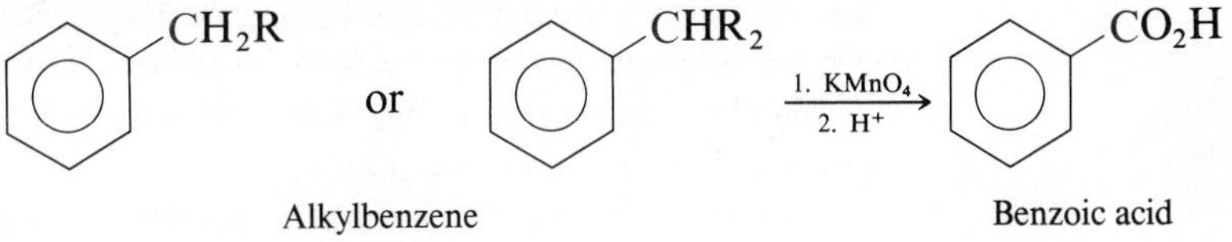

Alkylbenzene   Benzoic acid

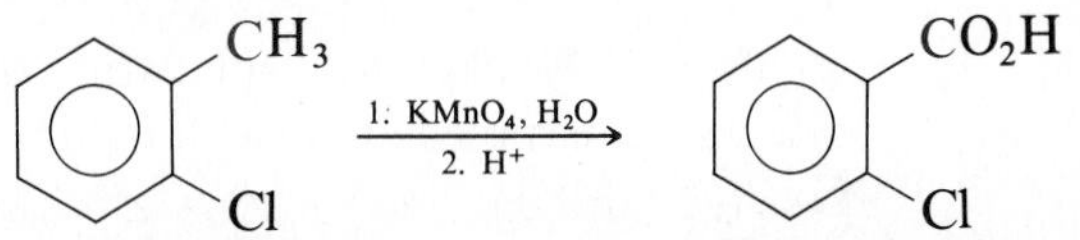

o-Chlorotoluene   o-Chlorobenzenecarboxylic
                                   acid (76–78%)
                                 (o-Chlorobenzoic acid)

This reaction is a key industrial process, being used to convert xylenes (Section 6.5) to benzenedicarboxylic acids, also known commercially as phthalic acids. These compounds are used to prepare polyester fibers. Oxygen in air with a catalyst is used as the oxidizing agent.

$$H_3C-\underset{\text{p-Xylene}}{\bigcirc}-CH_3 \xrightarrow[\text{catalyst}]{\text{air}} HO\overset{O}{\overset{\|}{C}}-\underset{\substack{\text{1,4-Benzenedicarboxylic}\\\text{acid (terephthalic acid)}}}{\bigcirc}-\overset{O}{\overset{\|}{C}}OH$$

Side-chain oxidation of alkylbenzenes is important in certain metabolic processes. One way in which the body gets rid of foreign substances is by oxidation in the liver to compounds more easily excreted in the urine. Toluene, for example, is oxidized to benzoic acid by this process and is eliminated rather readily.

$$\underset{\text{Toluene}}{\bigcirc\!\!\!-CH_3} \xrightarrow[\substack{\text{cytochrome P450}\\\text{(an enzyme in}\\\text{the liver)}}]{O_2} \underset{\text{Benzoic acid}}{\bigcirc\!\!\!-CO_2H}$$

Benzene, with no alkyl side chain, undergoes a different reaction in the presence of these enzymes (see boxed material following Sections 6.5 and 10.14), which converts it to a substance capable of inducing mutations in DNA (deoxyribonucleic acid). This difference in chemical behavior seems to be responsible for the fact that benzene is carcinogenic while toluene is not.

6.15 A GENERAL VIEW OF AROMATICITY. HÜCKEL'S RULE

While most aromatic compounds contain a benzene ring, the presence of such a ring in not a necessary condition for aromaticity. In other words, aromatic compounds exist which do not contain a benzene ring.

Two criteria must be met for a hydrocarbon to be aromatic. First, the π-electron system must have a *continuous cyclic array of parallel p orbitals.* Figure 6.3a (page 160) illustrates the p orbitals of benzene, for example. As you learned in Section 4.1.2, a π bond results from the "side-by-side" overlap of adjacent parallel p orbitals. For a molecule to be aromatic, the "ring" of p orbitals must not be broken with a saturated (sp^3) carbon or other "insulator."

The second criterion is that the number of π electrons must equal $4n + 2$, where n is an integer. This statement is known as **Hückel's rule,** and was first described in the 1930s by E. Hückel. An explanation of the origins of Hückel's rule would require a discussion of molecular orbital theory which is beyond the scope of an introductory text.

The most common case of Hückel's rule is when $n = 1$, that is, when the molecule has six π electrons. Benzene, of course, fits this example and is predicted to be aromatic.

PROBLEM 6.11

Determine which, if any, of the following molecules is aromatic. Explain your reasoning.

(a)

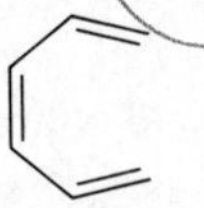

(b)

(c) +

SAMPLE SOLUTION

(a) Although the structure shown has six π electrons from three double bonds, the molecule is not cyclic. A cyclic array of p orbitals is a requirement for aromaticity. The molecule, 1,3,5-hexatriene, is not aromatic.

A number of aromatic compounds and ions are known that satisfy Hückel's rule with a value of n other than 1. Several of these are shown in Figure 6.4. As you will see in the next section, Hückel's rule also predicts aromaticity in molecules that are not hydrocarbons.

Two compounds which early chemists thought might be stabilized by delocalization of their π electrons were cyclobutadiene and cyclooctatetraene.

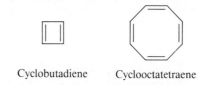

Cyclobutadiene   Cyclooctatetraene

However, the complete absence of compounds based on these ring systems contrasted starkly with the abundance of compounds based on the benzene nucleus.

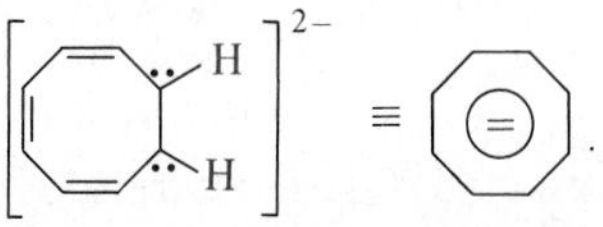

Cyclooctatetraene dianion
($n = 2$; 10 π electrons)

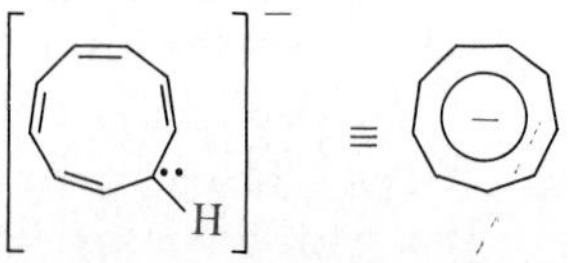

Cyclononatetraenide anion
($n = 2$; 10 π electrons)

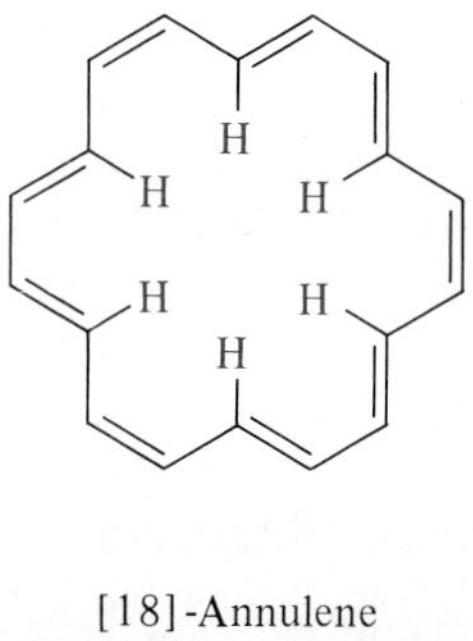

[18]-Annulene
($n = 4$; 18 π electrons)

FIGURE 6.4 Aromatic compounds and ions that satisfy Hückel's rule with values of n other than 1.

133 pm

146 pm

Molecular geometry of cyclooctatetraene.

Structural studies have confirmed the absence of appreciable π-electron delocalization in cyclooctatetraene. It is a *nonplanar* hydrocarbon with four short carbon-carbon bond distances and four longer carbon-carbon bond distances. In other words, cyclooctatetraene is best represented as a cyclic polyene having alternating single and double bonds, as shown in Figure 6.5.

Cyclobutadiene itself has escaped chemical characterization for more than 100 years. Studies of highly substituted derivatives suggest that the ring is best represented not as a square but as a rectangle having two longer single bonds and two shorter double bonds. The instability of cyclobutadiene supports the idea, described earlier by Hückel's rule, that a continuous cyclic array of parallel p orbitals is a necessary, but not a sufficient, condition for aromaticity. The number of π electrons must also satisfy the $4n + 2$ rule, and cyclobutadiene, having four π electrons, does not.

6.16 HETEROCYCLIC AROMATIC COMPOUNDS

Cyclic compounds that contain at least one atom other than carbon within their ring are called **heterocyclic compounds.** Nitrogen and oxygen are the most common heteroatoms, although heterocyclic compounds containing other atoms such as sulfur are known. Many heterocyclic compounds are aromatic, as they fit the criteria for aromaticity described in the preceding section. Examples of aromatic heterocyclic compounds include pyrrole, furan, and pyridine.

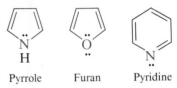

Pyrrole   Furan   Pyridine

In Figure 6.6 you can see that all three of these heterocycles are planar and have a continuous cyclic array of p orbitals. As pictured in Figure 6.6a and b, both pyrrole and furan utilize four π electrons from two double bonds and an unshared electron pair from the heteroatom to make up the six π electrons necessary for aromaticity.

Pyridine is aromatic because, as seen in Figure 6.6c, it has six π electrons in its ring exclusive of the nitrogen unshared pair. Therefore, these electrons do not interact with the π system and remain a nitrogen lone pair in an sp^2 hybridized orbital.

A number of aromatic heterocyclic compounds, some of which contain the rings just described, are important in living systems. Several of these are shown in Figure 6.7.

6.17 SUMMARY

An *aromatic* compound is one which is substantially more stable than expected on the basis of structural formulas that restrict electron pairs to regions between two nuclei. Benzene is an aromatic hydrocarbon. Neither of the two Kekulé formulas for benzene adequately describes its structure or properties. Benzene is said to be a *resonance hybrid* of these two Kekulé forms (Section 6.2).

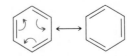

alternatively written as

Resonance description using the two Kekulé forms of benzene

Circle-in-a-ring symbol for benzene

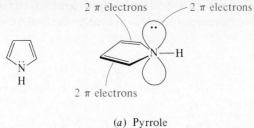

2 π electrons   2 π electrons

(a) Pyrrole

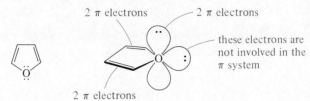

2 π electrons   2 π electrons

these electrons are
not involved in the
π system

2 π electrons

(b) Furan

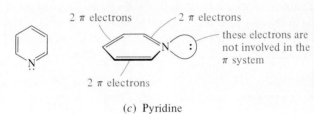

2 π electrons   2 π electrons

these electrons are
not involved in the
π system

2 π electrons

(c) Pyridine

FIGURE 6.6 Electron distribution and aromaticity. (a) Pyrrole has six π electrons. (b) Furan has six π electrons plus an unshared pair in an oxygen sp^2 orbital, which is perpendicular to the π system and does not interact with it. (c) Pyridine has six π electrons plus an unshared pair in a nitrogen sp^2 orbital.

Pyridoxine
(Vitamin B_6)

Nicotinamide
(The vitamin niacin)

FIGURE 6.7 Several important aromatic heterocyclic compounds.

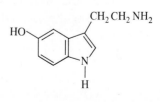

Furfural
(Isolated from corncobs)

Serotonin
(Brain neurotransmitter)

Disubstituted benzene derivatives are named as *ortho, meta,* and *para* according to the relative positions of the substituents (Section 6.5).

Ortho   Meta   Para

On reaction with electrophilic reagents, aromatic substances undergo substitution rather than addition (Section 6.6).

$$ArH \ + \ \ E^+ \ \longrightarrow \ \ ArE \ \ + \ H^+$$

Arene   Electrophile   Product of   Proton
                   electrophilic
                 aromatic substitution

Substitution occurs by attack of the electrophile on the π electrons of the aromatic ring in the rate-determining step to form a cyclohexadienyl cation intermediate. Loss of a proton from this intermediate restores the aromaticity of the ring and yields the product of *electrophilic aromatic substitution.*

Benzene   Electrophile   Cyclohexadienyl   Product of   Proton
                          cation intermediate   electrophilic
                                             aromatic
                                             substitution

Table 6.3 presents some typical examples of electrophilic aromatic substitution reactions; it illustrates conditions for carrying out the nitration, sulfonation, halogenation, and Friedel-Crafts alkylation and acylation of aromatic substances (Sections 6.6 to 6.8).

Some of the aromatic starting materials in the table bear substituents on the ring, and these substituents influence both the *rate* at which reaction occurs and the *orientation* of substitution (Sections 6.9). Substituents are classified as activating or deactivating according to whether they cause the ring to react more rapidly or less rapidly than benzene toward electrophilic aromatic substitution. Substituents are arranged into three major categories:

1. *Activating and ortho,para–directing* (Section 6.10). These substituents stabilize the cyclohexadienyl cation formed in the reaction. They include —N̈R$_2$, —ÖR, —R, —Ar, and related species. The most strongly activating members of this group are bonded to the ring by a nitrogen or oxygen atom that bears an unshared pair of electrons.

2. *Deactivating and meta-directing* (Section 6.11). These include —CF$_3$, —C̈R (with O double-bonded), —C≡N, —NO$_2$, and related species. All the ring positions are deactivated, but because the meta positions are deactivated less than the ortho and para, meta substitution is favored.

3. *Deactivating and ortho,para–directing* (Section 6.12). The halogens are the most prominent members of this class. They withdraw electron density from

TABLE 6.3
Representative Electrophilic Aromatic Substitution Reactions

Reaction and comments	General equation and specific example
Nitration The active electrophile in the nitration of benzene and its derivatives is the nitronium cation ($:\overset{..}{O}=\overset{+}{N}=\overset{..}{O}:$). It is generated by reaction of nitric acid and sulfuric acid. Very reactive arenes—those that bear strongly activating substituents—undergo nitration in nitric acid alone.	$ArH + HNO_3 \xrightarrow{H_2SO_4} ArNO_2 + H_2O$ Arene Nitric acid Nitroarene Water $F-\text{⬡} \xrightarrow[H_2SO_4]{HNO_3} F-\text{⬡}-NO_2$ Fluorobenzene p-Fluoronitrobenzene (80%)
Sulfonation Sulfonic acids are formed when aromatic compounds are treated with sources of sulfur trioxide. These sources can be concentrated sulfuric acid (for very reactive arenes) or solutions, called *oleum,* of sulfur trioxide in sulfuric acid (for benzene and arenes less reactive than benzene).	$ArH + SO_3 \longrightarrow ArSO_3H$ Arene Sulfur trioxide Arenesulfonic acid 1,2,4,5-Tetramethylbenzene $\xrightarrow[H_2SO_4]{SO_3}$ 2,3,5,6-Tetramethylbenzenesulfonic acid (94%)
Halogenation Chlorination and bromination of arenes is carried out by treatment with the appropriate halogen in the presence of a Lewis acid catalyst. Very reactive arenes undergo halogenation in the absence of a catalyst.	$ArH + X_2 \xrightarrow{FeX_3} ArX + HX$ Arene Halogen Aryl halide Hydrogen halide $HO-\text{⬡} \xrightarrow{Br_2} HO-\text{⬡}-Br$ Phenol p-Bromophenol (80–84%)
Friedel-Crafts alkylation Carbocations, usually generated from an alkyl halide and aluminum chloride, attack the aromatic ring to yield alkylbenzenes.	$ArH + RX \xrightarrow{AlCl_3} ArR + HX$ Arene Alkyl halide Alkylarene Hydrogen halide Benzene + Cyclopentyl bromide $\xrightarrow{AlCl_3}$ Cyclopentylbenzene (54%)
Friedel-Crafts acylation Acyl cations (acylium ions) generated by treating an acyl chloride or acid anhydride with aluminum chloride attack aromatic rings to yield ketones.	$ArH + RCCl \xrightarrow{AlCl_3} ArCR + HCl$ Arene Acyl chloride Ketone Hydrogen chloride or $ArH + RCOCR \xrightarrow{AlCl_3} ArCR + RCOH$ Arene Acid anhydride Ketone Carboxylic acid $CH_3O-\text{⬡} \xrightarrow[AlCl_3]{CH_3CO\,CCH_3} CH_3O-\text{⬡}-CCH_3$ Anisole p-Methoxyacetophenone (90–94%)

the ring, making halobenzenes less reactive than benzene. Lone pair electron donation stabilizes the cyclohexadienyl cations corresponding to attack at the ortho and para positions more than those formed by attack at the meta positions, giving rise to the observed orientation.

Carbocations adjacent to a benzene ring are resonance-stabilized and are called *benzylic carbocations.* Electrophilic additions to double bonds adjacent to an aromatic ring exhibit regioselectivity resulting from formation of the more stable benzylic carbocation (Section 6.14).

$$C_6H_5CH\!=\!CHCH_3 + HBr \longrightarrow \overset{\overset{\displaystyle Br}{\displaystyle |}}{C_6H_5CHCH_2CH_3}$$
$$\text{(formed via } C_6H_5\overset{+}{C}HCH_2CH_3)$$

Alkyl side chains on a benzene ring are oxidized on being heated with potassium permanganate, providing a useful synthesis of substituted benzoic acids.

Two general criteria must be met for a molecule to be aromatic (Section 6.15). First, it must contain a continuous cyclic array of parallel p orbitals. Second, according to Hückel's rule, the number of π electrons must equal $4n + 2$, where n is an integer. Aromatic compounds which contain one or more atoms other than carbon in their rings are called *heterocyclic aromatic compounds* (Section 6.16).

ADDITIONAL PROBLEMS

Aromaticity

6.12 Each of the following statements incorrectly describes some aspect of benzene's structure. Write a corrected version of each statement.

    (a) Benzene exists as a ring of alternating double and single bonds.

    (b) The most stable conformation of the benzene ring is the chair form.

    (c) All the carbon atoms of benzene are sp hybridized.

6.13 Draw the Kekulé resonance structures that contribute to the structures of the aromatic compounds shown.

       Biphenyl   Naphthalene   Anthracene

6.14 Explain why an aromatic ring cannot contain an sp^3 hybridized carbon atom.

6.15 List the possible numbers of π electrons allowed by Hückel's rule for an aromatic molecule up to $n = 5$.

6.16 Which of the following species is aromatic? Assume a planar shape for each.

    (a)

    (b)

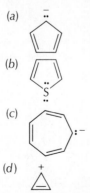

    (c)

    (d)

6.17 Identify the heterocyclic component of each compound in Figure 6.7 as being a pyridine, pyrrole, or furan ring.

Nomenclature

6.18 Write the structures and give a correct systematic name for all the isomeric:
- (a) Nitrotoluenes
- (b) Dichlorobenzoic acids
- (c) Tribromophenols

6.19 How many monobromo derivatives are possible for (you do not need to name them):
- (a) Anthracene
- (b) Naphthalene
- (c) Phenanthrene

6.20 Give a systematic name for each of the following compounds, identifying the C_6H_5 group as a phenyl group.

(a) $CH_3CHCH_2CH_2CH_3$
     |
     C_6H_5

2-phenyl pentane

(b) $CH_3CH_2CHCH_2OH$
     |
     C_6H_5

2 phenyl 1-butanol

(c) $C_6H_5CH{=}CHCH_3$

1-phenyl-1 propene

(d) $C_6H_5-\bigcirc-OH$

4-phenyl-1 cyclohexano l

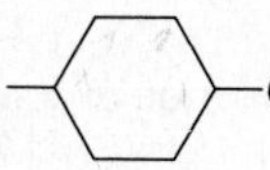

6.21 Write the structure corresponding to each of the following:
- (a) p-Chlorophenol
- (b) m-Nitroacetophenone
- (c) 2-Ethyl-1-phenyl-1-butene
- (d) 1,2-Diphenylcyclohexene
- (e) p-Diisopropylbenzene
- (f) 2,4,6-Tribromoaniline
- (g) 3,5-Dibromobenzoic acid

6.22 Give a correct systematic name for each of the following aromatic compounds.

(a) 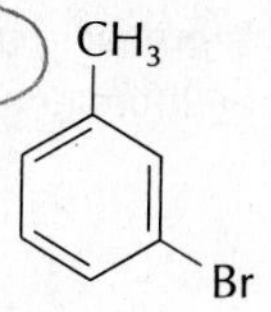 CH_3 ... Br

(b) 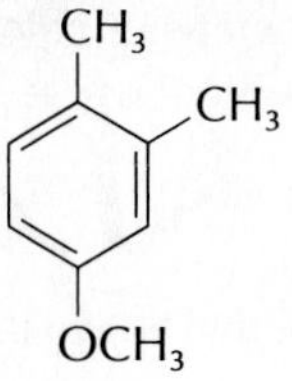 CH_3, CH_3, OCH_3

(c) 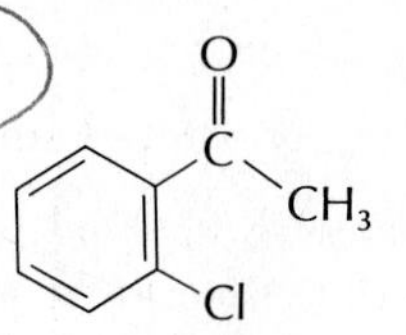 C=O, CH_3, Cl

(d) 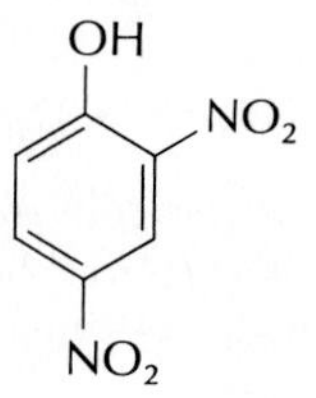 OH, NO_2, NO_2

Substitution Reactions

6.23 Give reagents for effecting each of the following reactions and write the principal products. If an ortho,para mixture is expected, show both. If the meta isomer is the expected major product, write only that isomer.

(a) Nitration of benzene

(b) Nitration of the product of (a)

(c) Bromination of toluene

(d) Bromination of (trifluoromethyl)benzene

(e) Sulfonation of anisole

(f) Sulfonation of acetanilide ($C_6H_5NHCCH_3$, with $\overset{O}{\overset{\|}{}}$ on the C)

(g) Chlorination of bromobenzene

6.24 Give the major product or products of each of the following Friedel-Crafts reactions.

(a) Alkylation of benzene with benzyl chloride ($C_6H_5CH_2Cl$)

(b) Alkylation of anisole with benzyl chloride

(c) Acylation of benzene with benzoyl chloride ($C_6H_5\overset{O}{\overset{\|}{C}}Cl$)

(d) Acylation of the product of (a) with benzoyl chloride

6.25 Write the structure of the product from monobromination of the following compound with Br_2 and $FeBr_3$.

6.26 What combination of acyl chloride or acid anhydride and arene would you choose to prepare the following compound by a Friedel-Crafts acylation reaction?

$$C_6H_5\overset{O}{\overset{\|}{C}}CH_2C_6H_5$$

Substitution Mechanisms

6.27 Write the structures of the electrophilic species present in each of the types of electrophilic aromatic substitution reactions: halogenation, nitration, sulfonation, and Friedel-Crafts alkylation and acylation.

6.28 In each of the following pairs of compounds choose which one will react faster with the indicated reagent and write a chemical equation for the faster reaction:

(a) Toluene or chlorobenzene with nitric acid and sulfuric acid

(b) Fluorobenzene or (trifluoromethyl)benzene with benzyl chloride and aluminum chloride

(c) Methyl benzoate ($C_6H_5\overset{O}{\overset{\|}{C}}OCH_3$) or phenyl acetate ($C_6H_5O\overset{O}{\overset{\|}{C}}CH_3$) with bromine in the presence of ferric bromide

(d) Acetanilide ($C_6H_5NH\overset{O}{\overset{\|}{C}}CH_3$) or nitrobenzene with sulfur trioxide in sulfuric acid

6.29 Write a step-by-step mechanism describing the reaction of benzene with chlorine in the presence of ferric chloride ($FeCl_3$). Be sure to write all the resonance structures contributing to the intermediate in the reaction.

6.30 Write structural formulas for the cyclohexadienyl cations formed from aniline ($C_6H_5NH_2$) during:

(a) Ortho bromination (four resonance structures)

(b) Meta bromination (three resonance structures)

(c) Para bromination (four resonance structures)

6.31 Write a structural formula for the most stable cyclohexadienyl cation intermediate formed in both the following reactions. Is the cyclohexadienyl cation more or less stable than the corresponding intermediate formed by electrophilic attack on benzene?

(a) Nitration of isopropylbenzene

(b) Bromination of nitrobenzene

6.32 Identify each of the following aromatic substituents as being activating and ortho,para–directing or deactivating and meta-directing:

(a) —S̈H (thiol)

(b) —N̈(CH$_3$)$_3$ (trimethylammonium)

(c) —S̈(CH$_3$)$_2$ (dimethylsulfonium)

Aromatic Side-Chain Reactions

6.33 Give the structure of the major product of each of the following reactions.

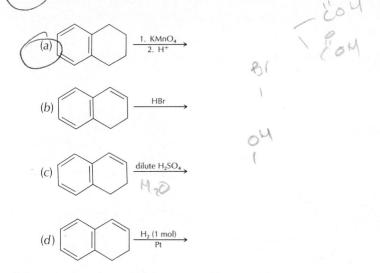

6.34 An aromatic dicarboxylic acid called o-phthalic acid is used commercially in the synthesis of plasticizers. o-Phthalic acid is obtained by oxidation of o-xylene with air in the presence of a catalyst. Write a chemical equation for this process.

Synthesis of Aromatic Compounds

6.35 Write equations showing how you could prepare each of the following from benzene and any necessary organic or inorganic reagents. If an ortho,para mixture is formed in any step of your synthesis, assume that you can separate the two isomers.

(a) Isopropylbenzene

(b) p-Isopropylbenzenesulfonic acid

(c) m-Chloroacetophenone

(d) p-Chloroacetophenone

6.36 Two students wish to prepare m-chloronitrobenzene beginning with benzene. One student carries out the chlorination step first; the other student begins with the nitration step. Write equations for both these reaction schemes. Which one gives the correct product?

6.37 Beginning with toluene and any other necessary organic or inorganic reagents, outline reaction schemes which will prepare:

(a) m-Bromobenzoic acid

(b) p-Bromobenzoic acid

6.38 In an attempt to prepare the compound shown, three synthetic reaction schemes were attempted. Only one of these worked; which one was it? What are the products from the other two pathways?

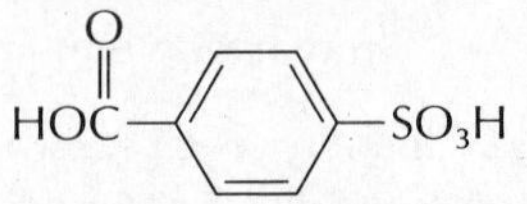

$$HOC-\text{(ring)}-SO_3H$$ with $\overset{O}{\underset{\parallel}{}}$ on the HOC

I: $C_6H_6 \xrightarrow[SO_3]{H_2SO_4} \xrightarrow[AlCl_3]{CH_3CH_2Br} \xrightarrow[\text{2. }H^+]{\text{1. }KMnO_4}$

II: $C_6H_6 \xrightarrow[AlCl_3]{CH_3CH_2Br} \xrightarrow[SO_3]{H_2SO_4} \xrightarrow[\text{2. }H^+]{\text{1. }KMnO_4}$

III: $C_6H_6 \xrightarrow[AlCl_3]{CH_3CH_2Br} \xrightarrow[\text{2. }H^+]{\text{1. }KMnO_4} \xrightarrow[SO_3]{H_2SO_4}$

CHAPTER 7

STEREOCHEMISTRY

This chapter focuses on the spatial relationships and arrangements of atoms and molecules. Upon completion of this chapter you should be able to:

- Identify a chiral molecule by locating a chiral carbon atom.
- Explain what is meant by the term *enantiomer*.
- Explain what is meant by the term *diastereomer*.
- Draw perspective (wedge-and-dash) representations of chiral molecules.
- Draw Fischer projections of chiral molecules.
- Describe how a plane of symmetry relates to whether or not a molecule is chiral.
- Explain optical activity as a property of chiral molecules.
- Specify the absolute configuration of a molecule using the *R-S* notational system.
- Specify alkene stereochemistry using the *E-Z* notational system.
- Explain why addition reactions to achiral alkenes give racemic mixtures of products.
- Explain the meaning of the term *meso isomer*.
- Give the number of stereoisomers possible for a molecule having more than one chiral center.
- Describe how a racemic mixture may be resolved into individual enantiomers.

We live in a three-dimensional world; in fact, a dictionary definition of three-dimensional is "lifelike." Molecules are three-dimensional, as we saw in Chapters 1 and 2 when we discussed bonding and molecular shapes. The chemical changes that take place all around us, especially those in living organisms, are influenced by the three-dimensional relationships between molecules. The study of the spatial arrangement of atoms and molecules is known as **stereochemistry.** You were introduced to stereochemistry previously in Chapter 2, where the conformations of alkanes and cycloal-kanes were discussed, and to cis-trans stereoisomers in ring systems (Section 2.15) and alkenes (Section 4.1.4). This chapter describes another element of stereochemis-try, one based on mirror-image relationships between molecules.

7.1 MIRROR IMAGES AND CHIRALITY

The concept of mirror images is fundamental to our study of stereochemistry. Con-sider what you see when a dinner fork is held in front of a mirror. The **mirror image** is what you observe in the mirror. Two ways of viewing this are shown in Figure 7.1. Figure 7.1b is viewed along the edge of the mirror. The object in front of the mirror is on the right; the mirror image is on the left.

Now consider whether the object and its mirror image are superimposable. Does every part of the mirror image exactly match the original object? It does, so we say the object (the fork) is superimposable on its mirror image. Other objects that are super-imposable on their mirror images include a spoon, a shovel, a thumbtack, and a plain (undecorated) T-shirt.

Now consider what you see when you hold your left hand in front of a mirror; the mirror image is your right hand, as shown in Figure 7.2. Next try to superimpose your left and right hands by sliding them on top of one another. You cannot, of course. Putting your hands palm to palm is *not* superposition; they must lie on top of one another.

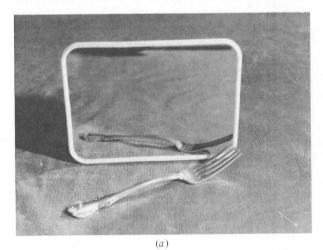

(a)

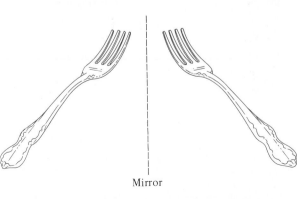

Mirror

(b)

FIGURE 7.1 (a) A dinner fork and its mirror image; (b) a second view, looking down the edge of the mirror.

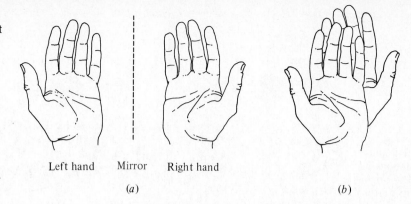

FIGURE 7.2 *(a)* An object (a left hand) and its mirror image (a right hand). *(b)* Hands are chiral; the mirror images are not superimposable. *(Used by permission from Chemistry General, Organic, Biological, by Jacqueline I. Kroschwitz and Melvin Winokur, McGraw-Hill Book Company, New York, 1985.)*

Left hand   Mirror   Right hand

(a)   *(b)*

An object that is *not* superimposable on its mirror image is **chiral,** from the Greek *cheiros,* meaning "hand." An object that *is* superimposable on its mirror image is *not* chiral; it is **achiral.** Thus hands are chiral; dinner forks are achiral.

PROBLEM 7.1
Identify each of the following as chiral or achiral.

    (a) A glove   *Chiral*
    (b) A shoe   *Chiral*
    (c) An undecorated cup and saucer   *achiral*

7.2 MOLECULAR CHIRALITY. ENANTIOMERS

Now let us consider chirality as it applies to chemical compounds. Figure 7.3*a* illustrates the structure of dichloromethane (CH_2Cl_2) using a ball-and-stick representation. Figure 7.3*b* represents the mirror image of the molecule in *(a)*. By sliding the mirror image across the page, you can see that the two are the *same;* the object and its mirror image are identical to each other and are superimposable. Dichloromethane is an *achiral molecule.*

Next consider the mirror images of bromochlorofluoromethane (CHBrClF) shown in Figure 7.4. If you attempt to superimpose the mirror image *(b)* on the original molecule *(a)*, you will be unsuccessful; the groups will not match (as drawn in

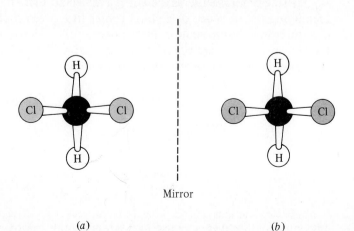

FIGURE 7.3 Ball-and-stick representations of CH_2Cl_2. *(a)* The original molecule; *(b)* its mirror image.

Mirror

(a)   *(b)*

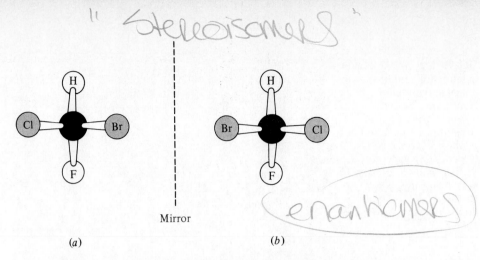

"Stereoisomers"

enantiomers

FIGURE 7.4 Ball-and-stick representations of CHBrClF. (a) The original molecule; (b) its mirror image.

Mirror

(a)   (b)

the figure, the Br and Cl will not coincide). Bromochlorofluoromethane is a *chiral molecule,* as the mirror images are nonsuperimposable.

The two mirror images of bromochlorofluoromethane differ only in the arrangement of their atoms in space, and thus they are *stereoisomers* of each other. Stereoisomers that are related as an object (or molecule) and its *nonsuperimposable mirror image* are called **enantiomers.**

An important point to recognize about enantiomers is that most physical properties such as density and melting and boiling points are the same. A significant difference between enantiomers is a physical property known as *optical activity,* to be discussed in Section 7.6.

7.3 CHIRAL CENTERS IN CHIRAL MOLECULES

The chiral molecule just cited, bromochlorofluoromethane, has four different atoms (Br, Cl, F, H) bonded to the carbon atom. In general, molecules of the type

$$w-\overset{\displaystyle x}{\underset{\displaystyle z}{C}}-y$$

are chiral when w, x, y, and z are different substituents. A carbon atom that bears four different substituents is variously referred to as a **chiral center,** a **chiral carbon atom,** or an **asymmetric carbon atom.***

If a single chiral center is present in a molecule, then the molecule is chiral. Thus identifying the presence of a chiral center in a given molecule is a rapid way to determine that the molecule is chiral. For example, C-2 is a chiral center in 2-butanol; the four substituents are H, OH, CH_3, and CH_2CH_3. 2-Propanol on the other hand, is achiral; none of its carbons bears four different substituents.

$$CH_3CH_2-\overset{\displaystyle OH}{\underset{\displaystyle H}{C}}-CH_3 \qquad CH_3-\overset{\displaystyle OH}{\underset{\displaystyle H}{C}}-CH_3$$

Chiral center

2-Butanol   2-Propanol
(chiral)   (achiral)

* Recently, chemists have noted that it is the molecule, not a single carbon atom, that is chiral. The terms *stereogenic center* or *stereocenter* have been proposed to replace "chiral center." The term chiral center is still commonly employed, however, and will be used in this text.

TABLE 7.1

Some Naturally Occurring Chiral Molecules Having One Chiral Center

Compound name	Structural formula†	Where found

Lactic acid	$\overset{\displaystyle O}{\overset{\displaystyle \|}{CH_3 \overset{*}{C}HCOH}}$
	  $\|$
	  OH
		In muscles; can also be obtained from milk

Nicotine	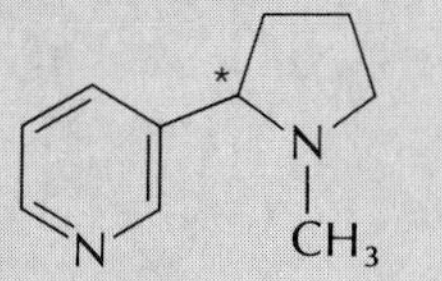	Tobacco

Malic acid	$HO_2C\overset{*}{C}HCH_2CO_2H$
	  $\|$
	  OH
		Apples and other fruits

Glyceraldehyde	$\overset{\displaystyle O}{\overset{\displaystyle \|}{HOCH_2 \overset{*}{C}HCH}}$
	  $\|$
	  OH
		Carbohydrate formed during energy-producing breakdown of sugars in the body

α-Phellandrene	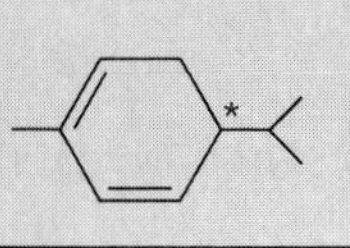	Component of eucalyptus oil; used in perfumes

† An asterisk (*) indicates location of the chiral center.

Table 7.1 gives examples of some naturally occurring chiral molecules characterized by the presence of a single chiral center.

PROBLEM 7.2

Examine the following molecules to determine if any have chiral centers:

  (a) 2-Bromopentane   (c) 4-Ethyl-4-methyloctane
  (b) 3-Bromopentane   (d) 1,2-Dibromobutane

SAMPLE SOLUTION

(a) A chiral carbon has four different substituents. In 2-bromopentane, C-2 satisfies this requirement, and 2-bromopentane is a chiral molecule.

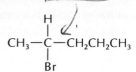

$$CH_3-\overset{\displaystyle H}{\underset{\displaystyle Br}{\overset{\displaystyle |}{\underset{\displaystyle |}{C}}}}-CH_2CH_2CH_3$$

7.4 THREE-DIMENSIONAL REPRESENTATIONS OF MOLECULES. PERSPECTIVE VIEWS AND FISCHER PROJECTIONS

Being able to visualize the spatial relationships of the groups attached to a chiral carbon is a difficult task. In addition, three-dimensional information must then be described on a sheet of paper, which is two-dimensional.

Practicing with molecular models can be very helpful in learning to manipulate three-dimensional molecular formulas. The ball-and-stick representations of Figures 7.3 and 7.4 illustrate how molecular models of these molecules would appear.

Two representations are commonly used by chemists to draw the three-dimensional structure of a molecule: **perspective views** and **Fischer projections.** Perspective views are drawings having bonds represented by wedges and dashes, and were first introduced in Chapter 1 (Figure 1.4). Bonds going toward you are represented by wedges; those projecting away from you are drawn using dashed lines. The molecule can also be drawn with two bonds in the plane of the paper. The structure of dichloromethane, shown in Figure 7.3, becomes

$$Cl-C-Cl \qquad \text{the same as} \qquad \begin{array}{c} Cl \quad Cl \\ C \\ H \quad H \end{array}$$

PROBLEM 7.3

Draw perspective views of the enantiomers of:

      (a) Bromochlorofluoromethane, shown in Figure 7.4

      (b) The chiral molecules found in Problem 7.2

SAMPLE SOLUTION

(a) The bonds pointing "back" are represented by dashed lines, those pointing "forward" by wedges:

$$\begin{array}{ccc} & H & \\ Cl-C-Br & \vdots & Br-C-Cl \\ & F & & F \end{array}$$

Fischer projections were introduced in the late nineteenth century by a German chemist, Emil Fischer. Fischer was responsible for many of the early discoveries in carbohydrate and protein chemistry and was awarded the Nobel Prize in Chemistry in 1902.

A **Fischer projection** is drawn by orienting a chiral molecule as shown in Figure 7.5. If we flatten the molecule by projecting it onto the plane of the page, a cross is formed with the chiral carbon at its center. The horizontal line represents bonds from the chiral center that are pointing toward you; the vertical line represents bonds that are pointing away from you. The chiral carbon itself is not explicitly shown.

PROBLEM 7.4

Repeat Problem 7.3 using Fischer projections.

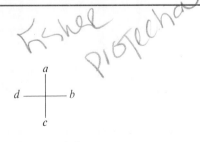

FIGURE 7.5 The Fischer projection formula of a chiral molecule is derived from a particular orientation of the bonds to the chiral center.

These equivalent three-dimensional orientations

are represented by the Fischer projection formula

SAMPLE SOLUTION

(a) The perspective drawings in the sample solution are in the correct orientation for Fischer projections. The projections are

$$
\begin{array}{c@{\;}c@{\;}c}
& \text{H} & \\
\text{Cl}\!\!-\!\!\!\!&\!\!\!\!|\!\!\!\!&\!\!\!\!-\!\!\text{Br} \\
& \text{F} &
\end{array}
\qquad
\begin{array}{c@{\;}c@{\;}c}
& \text{H} & \\
\text{Br}\!\!-\!\!\!\!&\!\!\!\!|\!\!\!\!&\!\!\!\!-\!\!\text{Cl} \\
& \text{F} &
\end{array}
$$

You can see from the sample solution that drawing the mirror image of a Fischer projection is an easy task—the groups on the horizontal line switch positions. Use caution in altering the orientation of a Fischer projection, however. Rotating a Fischer projection 90° changes the spatial arrangement of groups.

$$
\begin{array}{c@{\;}c@{\;}c}
& \text{H} & \\
\text{Cl}\!\!-\!\!\!\!&\!\!\!\!|\!\!\!\!&\!\!\!\!-\!\!\text{Br} \\
& \text{F} &
\end{array}
\qquad
\begin{array}{c}
\text{not the} \\
\text{same as}
\end{array}
\qquad
\begin{array}{c@{\;}c@{\;}c}
& \text{Cl} & \\
\text{F}\!\!-\!\!\!\!&\!\!\!\!|\!\!\!\!&\!\!\!\!-\!\!\text{H} \\
& \text{Br} &
\end{array}
$$

Fischer projections are widely used when several chiral centers are present in the molecule. You will see numerous examples of Fischer projections in the chapter on carbohydrates (Chapter 15).

7.5 SYMMETRY IN ACHIRAL STRUCTURES

Certain features related to molecular symmetry can often enable us to determine, by inspection, if a molecule is chiral or achiral.

A **plane of symmetry** bisects a molecule in such a way that one-half of the molecule is the mirror image of the other half. Planes of symmetry are shown in Figure 7.6 for two achiral molecules, 2-propanol and dichloromethane. There is one plane of symmetry in 2-propanol, while dichloromethane has two. Any molecule

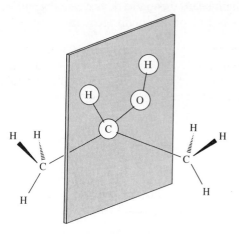

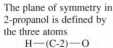

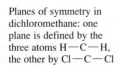

FIGURE 7.6 Planes of symmetry in the achiral molecules 2-propanol and dichloromethane.

The plane of symmetry in 2-propanol is defined by the three atoms
H—(C-2)—O

Planes of symmetry in dichloromethane: one plane is defined by the three atoms H—C—H, the other by Cl—C—Cl

having a plane of symmetry is *achiral*. A chiral molecule cannot have a plane of symmetry.

A reasonable question at this point would be to wonder why we should worry about planes of symmetry when chiral carbons have four different groups attached to them; locating these groups might seem easier than locating a symmetry plane. We will see later, in Section 7.12, that a molecule having two or more chiral centers may be chiral *or* achiral and locating a plane of symmetry becomes important.

7.6 PROPERTIES OF CHIRAL MOLECULES. OPTICAL ACTIVITY

The foundations of organic stereochemistry were laid by Jacobus van't Hoff and Charles LeBel, independently of each other, in 1874. They concluded that constitutionally identical molecules could differ in their arrangements of atoms in space. These molecules differing in the arrangement of their atoms in space differed from one another with respect to a physical property called optical activity.

Optical activity is the ability of a molecule to rotate the plane of polarized light. The individual enantiomers of a chiral substance exhibit optical activity and are said to be *optically active*. To understand what this means, let us first consider some characteristics of visible light and then see how optical activity is measured.

Visible light is a form of electromagnetic radiation. As with all electromagnetic radiation, a beam of light is composed of waves. Imagine that the waves which make up a beam of light are something we can see. If we were to stand in the path of a light beam looking at the source, we would see that the light beam comprises many waves which vibrate perpendicular to their direction of propagation.

Waves of a light beam
looking at the source

Passing the light beam through a device called a *polarizer* allows transmission of only those light waves all of which vibrate in the same plane. This is called **plane-polarized light**.

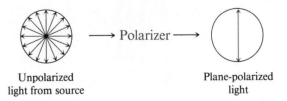

Unpolarized
light from source

Plane-polarized
light

(Polaroid sunglasses reduce glare by the same phenomenon. Reflected sunlight is slightly polarized, and the sunglass lenses, which are polarizers, prevent the glare from passing through.)

A solution containing one enantiomer of a chiral substance is then placed in the path of the beam of plane-polarized light. The plane of polarization will undergo a rotation of an angle α (alpha) either to the right or to the left as viewed looking toward the source.

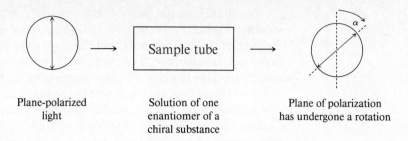

Plane-polarized
light

Solution of one
enantiomer of a
chiral substance

Plane of polarization
has undergone a rotation

This rotation of plane-polarized light is what is meant when a substance is said to be optically active. The angle of rotation is the observed **optical rotation** and is detected by viewing the polarized light through a second polarizer, called the *analyzer*. The laboratory device used to measure optical activity is called a **polarimeter,** illustrated schematically in Figure 7.7.

Optical activity is a physical property, just as melting point, boiling point, and density are. The optical activity of a substance is expressed as the *specific rotation* and is given the symbol [α]. The temperature (in degrees Celsius) and the wavelength of light (usually the so-called sodium D line having a wavelength of 589 nm) are indicated as superscripts and subscripts, respectively. Rotation to the right (clockwise) when viewing the source is taken as positive (+), while rotation to the left is taken as negative (−).* The specific rotation of a substance is therefore cited in the form $[\alpha]_D^{25}$ +15°.

The specific rotation [α] may be calculated from the observed rotation α according to the expression

$$[\alpha] = \frac{100\alpha}{cl}$$

where *c* is the concentration of the sample in grams per 100 mL of solution and *l* is the length of the polarimeter tube in decimeters (1 dm = 10 cm).

PROBLEM 7.5
Cholesterol (Section 17.5), when isolated from natural sources, is obtained as a single enantiomer. The observed rotation of a 0.3-g sample of cholesterol in 15 mL of chloroform solution contained in a 10-cm polarimeter tube is −0.78°. Calculate the specific rotation of cholesterol.

---

* An older designation of optical rotation to the right and left was *d* and *l*, from dextrorotatory (Latin *dextro-*, "to the right") and levorotatory (Latin *levo*, "to the left").

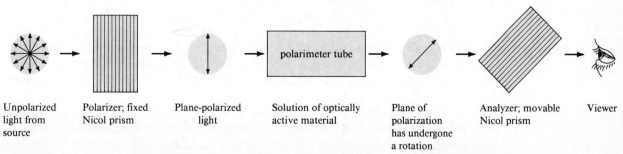

Unpolarized
light from
source

Polarizer; fixed
Nicol prism

Plane-polarized
light

Solution of optically
active material

Plane of
polarization
has undergone
a rotation

Analyzer; movable
Nicol prism

Viewer

FIGURE 7.7 Measurement of optical activity using a polarimeter.

Our discussion has focused on the optical rotation of one enantiomer of a chiral substance. What if we determine the optical rotation of the *other* enantiomer, that is, the mirror image of the substance? The angle of rotation will be the *same,* but it will be *in the opposite direction.* For example, if one enantiomer of a chiral substance has a specific rotation $[\alpha]$ of $+37°$, the other enantiomer will have $[\alpha] = -37°$.

A mixture of *equal amounts* of the two enantiomers of a chiral substance has *no* optical rotation $(\alpha = 0)$, because the positive and negative rotations of the individual enantiomers cancel each other out. Mixtures containing equal quantities of the enantiomers of a chiral substance are called **racemic mixtures** and are *optically inactive.* Achiral substances are also optically inactive, as optical activity is a physical property possessed only by chiral molecules. In order to be optically active, a substance must be chiral and one enantiomer must be present in greater amounts than the other.

7.7 ABSOLUTE CONFIGURATIONS

Imagine that you reach into a box containing several pairs of gloves, with the goal of choosing one for your right hand and one for your left. How can you tell when you have found a pair? Your hands become a point of reference, since both they and the gloves are chiral and a right glove will only fit on your right hand and a left glove on your left.

The scenario just described is one of determining the absolute configuration of an object, in this case a glove. **Absolute configuration** refers to the precise spatial arrangement of an object. We can determine the absolute configuration of a glove, i.e., whether it is for a right or left hand, by comparing it to our hands, whose absolute configuration we know.

While the problem is easily solved for gloves, the situation is not nearly as simple for a molecule. Neither the sign nor the magnitude of optical rotation by itself provides any information concerning the absolute configuration of a substance. Thus, one of the structures shown below is $(+)$-2-butanol and the other is $(-)$-2-butanol, but in the absence of additional information we cannot tell which is which.

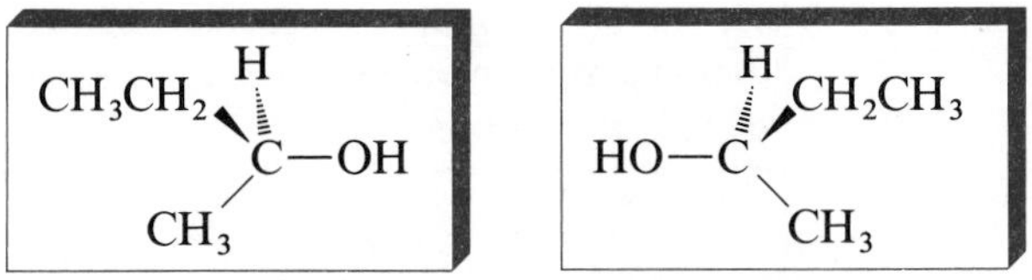

Enantiomers of 2-butanol

While today the absolute configurations of many chiral compounds are known, none were known prior to 1951. In that year the true three-dimensional structure of a salt of $(+)$-tartaric acid was determined by x-ray crystallography. Once its absolute configuration had been determined, the absolute configurations of many other compounds whose configurations had been related to $(+)$-tartaric acid were also revealed. In fact, $(+)$-2-butanol is the stereoisomer shown above on the left, and $(-)$-2-butanol is the one on the right.

Once we know the absolute configuration of a molecule, we are then left with the difficulty of how to uniquely describe that configuration in words without having to always draw the molecular structure. Since we cannot call molecules "right- and left-handed" as we can gloves, another system is needed.

7.8 THE *R-S* NOTATIONAL SYSTEM OF ABSOLUTE CONFIGURATION

In Chapter 2 and elsewhere you have learned how a system of nomenclature (IUPAC) permits structural information to be communicated without the necessity of writing a constitutional formula for each compound. It is reasonable, therefore, to have a notational system that simply and unambiguously describes the absolute configuration of a molecule.

The system used for specifying absolute configurations was developed in 1956 by two English chemists, R. S. Cahn and Sir Christopher Ingold, in collaboration with a colleague in Switzerland, Vladimir Prelog. The key feature of their system is a priority ranking, or *precedence,* of substituents based on atomic number. We can illustrate the method by considering the enantiomers of 2-bromo-2-chloro-1,1,1-trifluoroethane (halothane), a compound used in hospitals as an anesthetic.

Enantiomers

$$F_3C \blacktriangleright \underset{\underset{Cl}{|}}{\overset{\overset{H}{|}}{C}} \blacktriangleleft Br \qquad \text{and} \qquad Br \blacktriangleright \underset{\underset{Cl}{|}}{\overset{\overset{H}{|}}{C}} \blacktriangleleft CF_3$$

$$\text{A} \qquad\qquad\qquad\qquad \text{B}$$

2-Bromo-2-chloro-1,1,1-trifluoroethane
(halothane)

The absolute configuration of each enantiomer may be specified according to the following set of rules:

1. Identify the atoms attached to the chiral carbon, and rank them in order of *decreasing atomic number.* This sequence determines the precedence, or priority ranking, of the substituents. For a substituent containing several atoms, it is the atom attached to the chiral center that determines the ranking. The substituents attached to the chiral carbon in halothane are H, Cl, Br, and CF_3. In order of decreasing atomic number, the atoms attached to the chiral center are

$$Br \quad > Cl > C > \quad H$$

   Highest   Lowest

2. Orient the molecule so that the lowest-ranking substituent points away from you. In other words, the bond to the lowest-ranking substituent should be pointed back, or "below" the page (dashed bond). Hydrogen is the lowest-ranking substituent on the chiral center in the example and is already in the correct orientation.

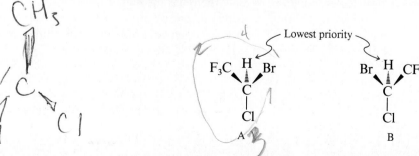

3. Next depict the three remaining substituents as they appear to you when the molecule is viewed from this perspective.

(2)   (4)   (4)   (2)

F_3C Br   Br CF_3

Cl   Cl
(3)   (3)

[( ) = priority ranking; 4 highest]

A   B

This view can be thought of as being similar to sitting behind a three-spoked steering wheel in a car; the lowest-ranked substituent is the steering column, and the other substituents are the spokes projecting from the center.

4. If the order of *decreasing precedence* of the three highest-ranking substituents ($Br > Cl > CF_3$) appears in a *clockwise* sense, as in A, the absolute configuration of the molecule is R (from the Latin *rectus,* "right"). If the order appears in a counterclockwise sense, as in B, the absolute configuration is S (from the Latin *sinister,* "left").

F_3C H Br   Br H CF_3

C   C

Cl   Cl

Clockwise (R)   Counterclockwise (S)

A   B

(R)-2-Bromo-2-chloro-   (S)-2-Bromo-2-chloro-
1,1,1-trifluoroethane   1,1,1-trifluoroethane

Recall that molecules A and B are the two enantiomers of halothane, and notice that one has the R absolute configuration and the other S. It is a general rule that *the enantiomer of a substance which has the R absolute configuration is S, and vice versa.*
Next let us consider the absolute configurations of the enantiomers of 2-butanol.

$$CH_3—\overset{\overset{\displaystyle H}{|}}{\underset{\underset{\displaystyle OH}{|}}{C}}—CH_2CH_3$$

The substituents attached to the chiral center are H, OH, CH_3, and CH_2CH_3. The highest- and lowest-ranking substituents can readily be identified as OH and H, respectively. The methyl and ethyl groups both have carbon atoms attached to the chiral center, however. Which, then, has the higher priority?

Substituent atoms are evaluated one by one, working outward from the point of attachment to the chiral center. Precedence (priority) is determined at the first point of difference. Thus $—CH_2CH_3$ [$—C(C,H,H)$] outranks $—CH_3$ [$—C(H,H,H)$].

$$—\overset{\overset{\displaystyle H}{|}}{\underset{\underset{\displaystyle H}{|}}{C}}—C \quad \text{higher priority than} \quad —\overset{\overset{\displaystyle H}{|}}{\underset{\underset{\displaystyle H}{|}}{C}}—H$$

Following the procedure described earlier, we arrive at the following absolute configurations for the enantiomers of 2-butanol:

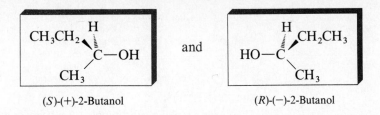

(S)-(+)-2-Butanol   and   (R)-(−)-2-Butanol

PROBLEM 7.6

Identify the higher-priority group in each of the following pairs:

(a) —CH₂CH₂CH₃ or —CH(CH₃)₂
(b) —C(CH₃)₃ or —CH₂OH
(c) —CH(CH₃)₂ or —CH₂CH₂OH
(d) —CH₂F or —CH(CH₃)₂

SAMPLE SOLUTION

(a) Beginning at the point of attachment, each "layer" of atoms is evaluated until a difference is found. Thus —CH(CH₃)₂ [—C(C,C,H)] outranks —CH₂CH₂CH₃ [—C(C,H,H)]

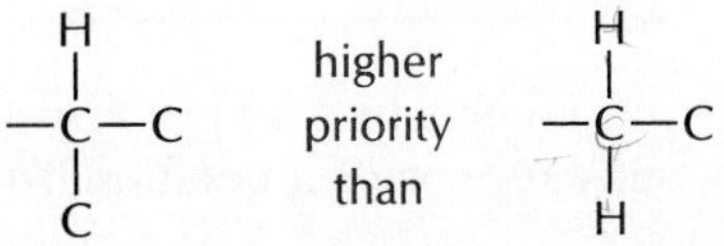

$$\underset{\overset{|}{C}}{\overset{\overset{H}{|}}{-C-C}} \qquad \begin{array}{c} \text{higher} \\ \text{priority} \\ \text{than} \end{array} \qquad \underset{\overset{|}{H}}{\overset{\overset{H}{|}}{-C-C}}$$

Substituent groups that contain multiple bonds may also be assigned a priority ranking. An atom that is multiply bonded to another is considered to be replicated as a substituent on that atom:

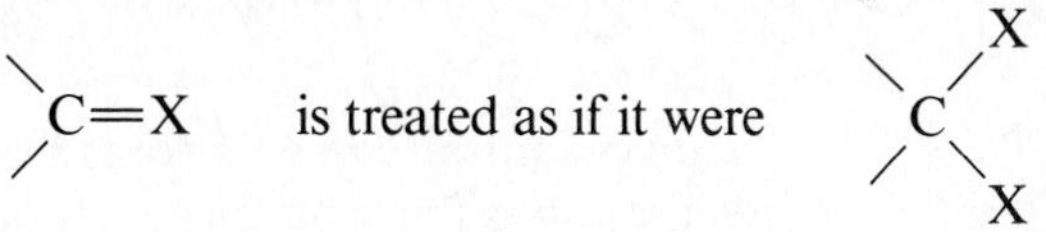

$$\overset{}{\underset{}{>}}C=X \qquad \text{is treated as if it were} \qquad \overset{X}{\underset{X}{>C<}}$$

Thus a carbonyl group has a higher priority than an alcohol group. For example,

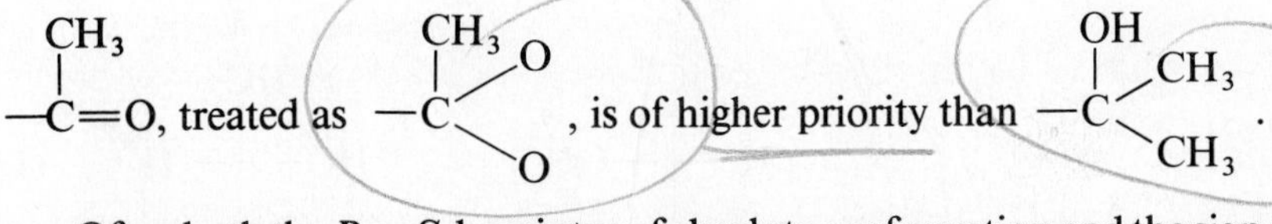

$$\underset{\overset{|}{C}}{\overset{\overset{CH_3}{|}}{-C}}=O, \text{ treated as } \quad -C\overset{\overset{CH_3}{\diagup}\overset{O}{}}{\underset{O}{\diagdown}} \text{, is of higher priority than } -C\overset{\overset{OH}{|}}{\underset{CH_3}{\diagdown CH_3}}.$$

Often both the *R* or *S* descriptor of absolute configuration and the sign of rotation are incorporated into the name of the compound, as in (R)-(−)-2-butanol and (S)-(+)-2-butanol. It is important for you to realize, however, that the sign of rotation of a substance *cannot* be predicted from the absolute configuration. Optical rotation is a physical property that must be measured in the laboratory, as described in Section 7.6.

PROBLEM 7.7

Assign absolute configurations as *R* or *S* to each of the following compounds:

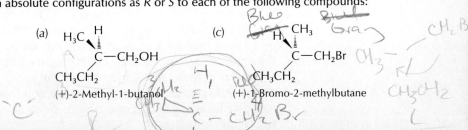

(a) (+)-2-Methyl-1-butanol

(c) (+)-1-Bromo-2-methylbutane

(b)

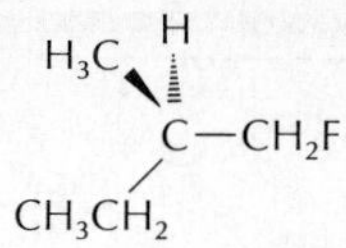

H_3C H

C—CH_2F

CH_3CH_2

(+)-1-Fluoro-2-methylbutane

(d) H_3C H

C—CH=CH_2

HO

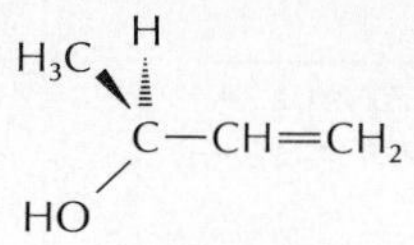

(+)-3-Buten-2-ol

SAMPLE SOLUTION

(a) The highest-ranking substituent at the chiral center of 2-methyl-1-butanol is CH_2OH; the lowest is H. Of the remaining two, ethyl outranks methyl.

$$\text{Order of precedence: } CH_2OH > CH_3CH_2 > CH_3 > H$$

The lowest-ranking substituent (hydrogen) at the chiral center points away from us, so the molecule is oriented properly as drawn. The three highest-ranking substituents trace a clockwise path from $CH_2OH \rightarrow CH_3CH_2 \rightarrow CH_3$.

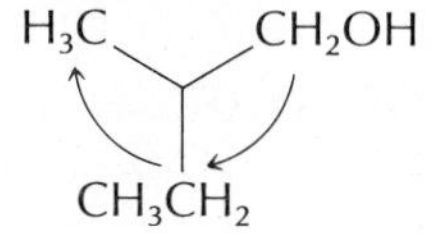

H_3C CH_2OH

CH_3CH_2

Therefore, this compound has the R configuration. It is (R)-(+)-2-methyl-1-butanol.

Biochemists and biologists generally describe the absolute configuration of a substance using a notational system based on the enantiomers of glyceraldehyde.

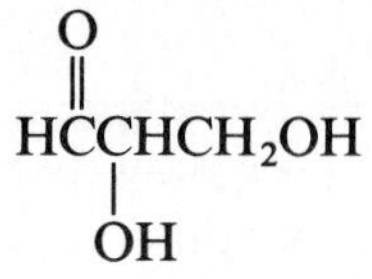

O
‖
HCCHCH$_2$OH
|
OH

Glyceraldehyde

Compounds are described as being either D or L depending on whether their configuration is the same as (+, d) or (−, l) glyceraldehyde, respectively.

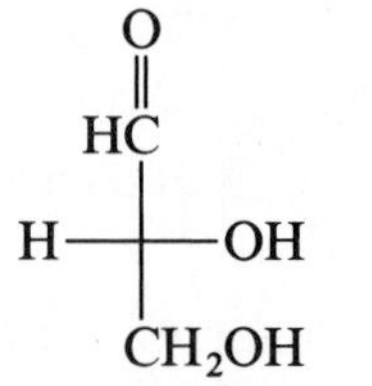

O
‖
HC

H——OH

CH_2OH

D-(+)-Glyceraldehyde

O
‖
HC

HO——H

CH_2OH

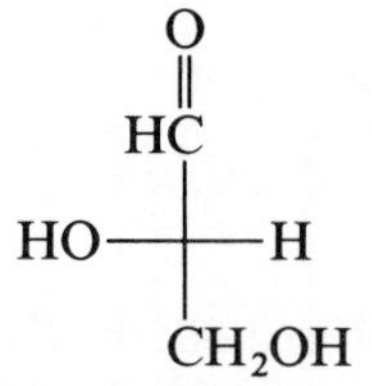

L-(−)-Glyceraldehyde

(Fischer projections)

The D-L notational system will be described more fully in Chapter 15 (Section 15.2).

7.9 NAMING STEREOISOMERIC ALKENES BY THE *E-Z* SYSTEM

In Chapter 4 (Section 4.1.4) you learned that certain alkenes can exist as pairs of stereoisomers due to the restricted rotation about a carbon-carbon double bond. When the substituents on either end of the double bond are the same or are structurally similar to each other, it is a simple matter to apply the cis and trans descriptors to adequately specify the configuration about the double bond. Oleic acid, a material obtained from olive oil, has a cis double bond, for example. Cinnamaldehyde, responsible for the characteristic odor of cinnamon, has a trans double bond.

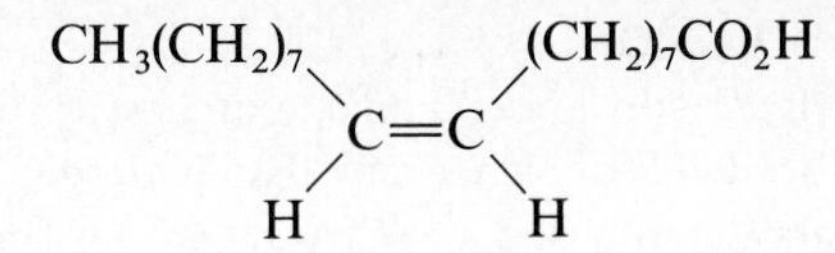

$$CH_3(CH_2)_7 \diagdown \qquad (CH_2)_7CO_2H$$
$$C=C$$
$$H \diagup \qquad \diagdown H$$

Oleic acid

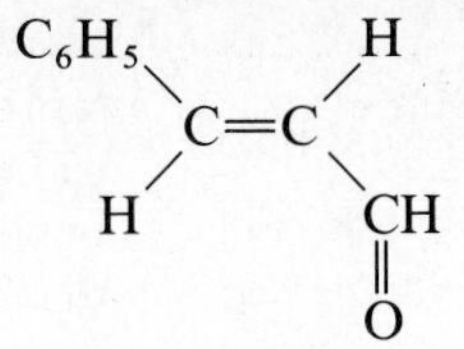

$$C_6H_5 \diagdown \qquad H$$
$$C=C$$
$$H \diagup \qquad \diagdown CH$$
$$\parallel$$
$$O$$

Cinnamaldehyde

However, cis and trans become ambiguous when the similarities or differences between substituents are less clear. For example, it is not at all obvious whether the alkene

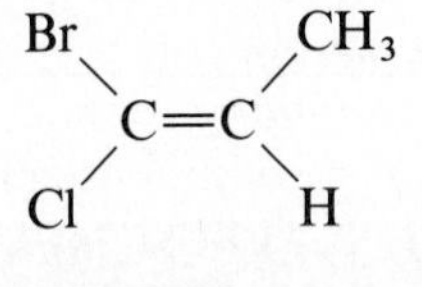

$$Br \diagdown \qquad CH_3$$
$$C=C$$
$$Cl \diagup \qquad \diagdown H$$

should be described as cis or trans. Fortunately, a completely unambiguous system for specifying double bond stereochemistry has been developed, using the priority ranking of substituents described in the preceding section.

Consider the 1-bromo-1-chloropropene stereoisomer just shown. The substituents attached to C-1 are bromine and chlorine. According to the precedence rules described in Section 7.8, bromine has the higher priority or ranking. Of the substituents attached to C-2 (H and CH_3), the methyl group is the higher-ranked. When the higher-ranked substituents are on the *same* side of the double bond, the configuration of the double bond is Z. The Z descriptor stands for the German word *zusammen*, which means "together."

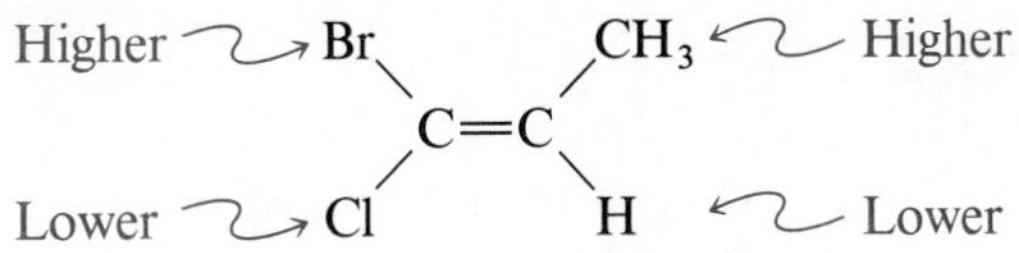

Higher → Br CH_3 ← Higher
$$C=C$$
Lower → Cl   H ← Lower

(*Z*)-1-Bromo-1-chloropropene

When higher-ranked substituents are on opposite sides of the double bond, the double bond has the *E* configuration. The symbol *E* stands for the German word *entgegen*, which means "opposite."

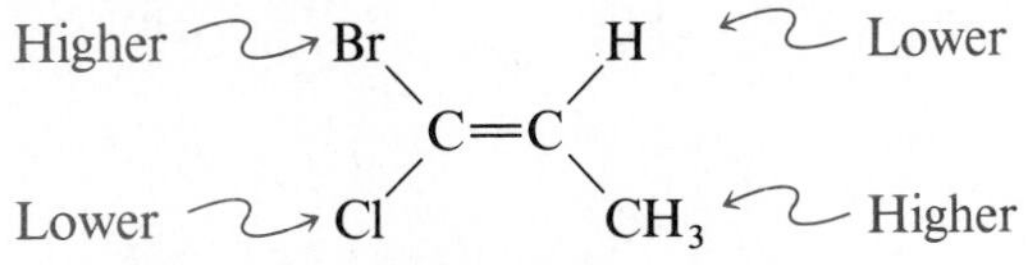

Higher → Br   H ← Lower
$$C=C$$
Lower → Cl CH_3 ← Higher

(*E*)-1-Bromo-1-chloropropene

PROBLEM 7.8
Determine the configuration of each of the following alkenes as Z or E as appropriate.

(a)
$$H_3C \diagdown \qquad CH_2OH$$
$$C=C$$
$$H \diagup \qquad \diagdown CH_3$$

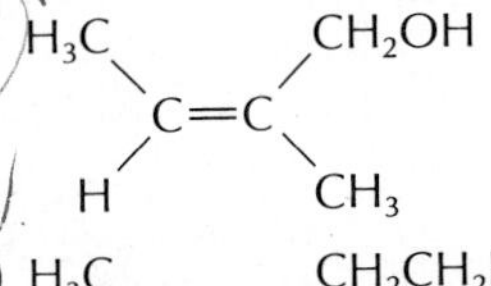

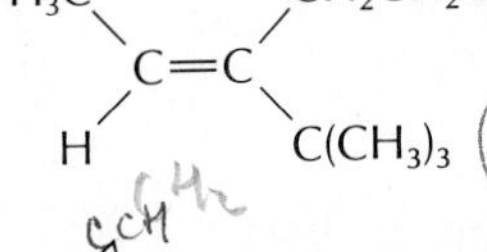

(b)
$$H_3C \diagdown \qquad CH_2CH_2F$$
$$C=C$$
$$H \diagup \qquad \diagdown CH_2CH_2CH_2CH_3$$

(c)
$$H_3C \diagdown \qquad CH_2CH_2OH$$
$$C=C$$
$$H \diagup \qquad \diagdown C(CH_3)_3$$

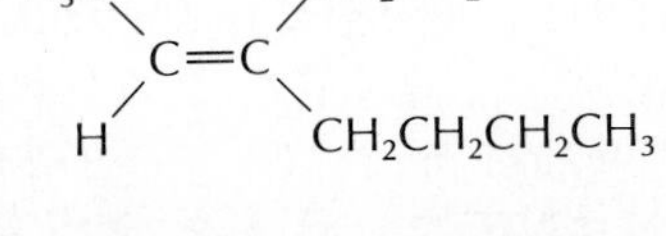

(d)
$$\diagdown \qquad \qquad H$$
$$C=C$$
$$CH_3CH_2 \diagup \qquad \diagdown CH_3$$

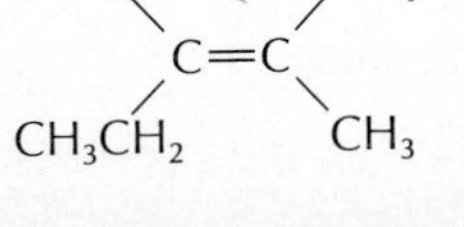

SAMPLE SOLUTION

(a) There is a methyl group and a hydrogen as substituents on one of the doubly bonded carbons. Methyl is of higher rank than hydrogen. The other carbon atom of the double bond bears a methyl and a —CH$_2$OH group. The —CH$_2$OH group is of higher priority than methyl.

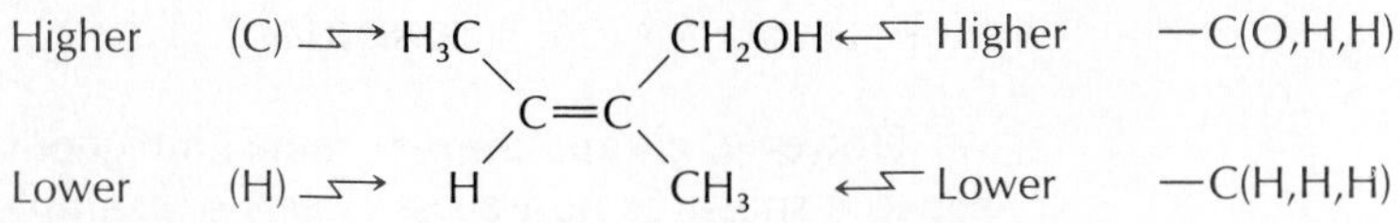

Higher (C) ⟶ H$_3$C   CH$_2$OH ⟵ Higher   —C(O,H,H)

                C=C

Lower (H) ⟶ H   CH$_3$ ⟵ Lower   —C(H,H,H)

Higher-ranked substituents are on the same side of the double bond; the configuration is Z.

7.10 STEREOCHEMISTRY OF CHEMICAL REACTIONS THAT PRODUCE CHIRAL CENTERS

Many of the reactions we have encountered in earlier chapters can produce a chiral product from an achiral starting material. A large number of the reactions of alkenes fall into this category. For example, the reaction of 2-butene with hydrogen bromide converts the achiral alkene into a product that contains a chiral center.

$$CH_3CH=CHCH_3 \xrightarrow{HBr} CH_3\underset{\underset{Br}{|}}{C}HCH_2CH_3$$

(E)- or (Z)-2-butene   2-Bromobutane
(achiral)   (chiral)

In this and related reactions, the chiral product is formed as a *racemic mixture* (Section 7.6) and is *optically inactive*. Remember, for a substance to be optically active, not only must it be chiral but one enantiomer must be present in excess of the other.

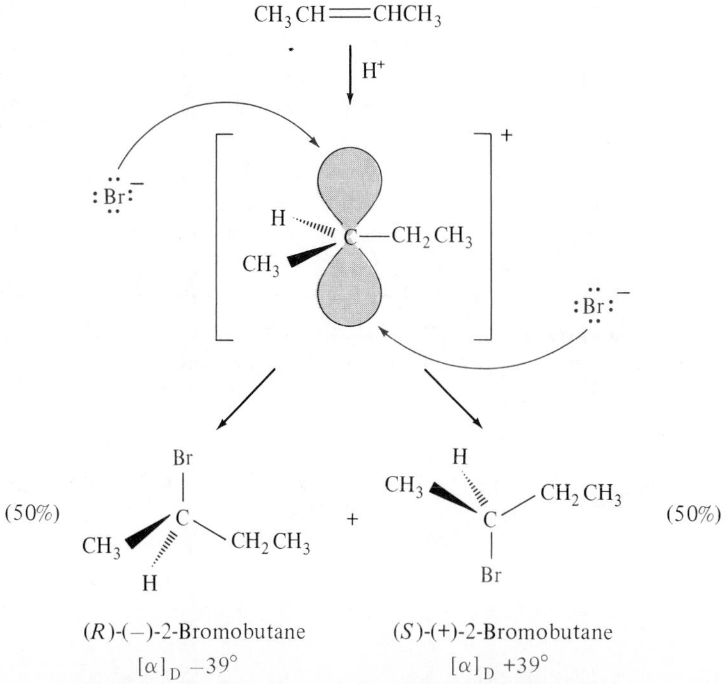

FIGURE 7.8 Ionic addition of hydrogen bromide to either stereoisomer of 2-butene proceeds by way of an achiral carbocation, which leads to equal quantities of (R)- and (S)-2-bromobutane.

(R)-(−)-2-Bromobutane [α]$_D$ −39°   (S)-(+)-2-Bromobutane [α]$_D$ +39°

To understand the reason why a mixture containing equal amounts of enantiomers is produced, let us consider the reaction shown in more detail. In Chapter 5 you saw that in the addition of hydrogen bromide to an alkene such as 2-butene a carbocation is an intermediate. The bonds to the positively charged carbon are coplanar and define a plane of symmetry. This causes the carbocation to be achiral. Once formed, the carbocation may be captured by a bromide ion with equal ease from the top face or the bottom face, as depicted in Figure 7.8. The product, 2-bromobutane, is chiral. However, the two enantiomers are formed in equal amounts, and the product is an optically inactive racemic mixture.

It is a general principle that **optically active products cannot be formed when optically inactive substrates react with optically inactive reagents.** We will discuss further the relationship between stereochemistry and chemical reactions in the next chapter.

7.11 CHIRALITY AND LIVING SYSTEMS

A reactant that is achiral or is a racemic mixture of a chiral substance can give optically active products *only* if it is treated with an optically active reagent or if the reaction is catalyzed by an optically active substance. The best examples of these phenomena are found in living systems. Most biochemical reactions are catalyzed by **enzymes.** Enzymes are chiral and exist as a single enantiomer; they provide an asymmetric environment in which chemical reactions can take place. Thus, in most enzyme-catalyzed reactions only one enantiomer of the product is formed. The enzyme *fumarase,* for example, catalyzes the hydration of fumaric acid, which is achiral, to malic acid in apples and other fruits. Only the *S* enantiomer of malic acid is formed in this reaction.

Fumaric acid $+ H_2O \underset{}{\overset{\text{fumarase}}{\rightleftarrows}}$ (S)-(−)-Malic acid

The presence of optical activity was viewed by Louis Pasteur as perhaps *the* unique criterion of life. He remarked in 1860 that the difference between the chemistry of living and dead matter is "the molecular asymmetry of organic natural products."

More recently, scientists have suggested that perhaps the most valid and sensitive criterion for the recognition of extraterrestrial life would be the detection of optically active compounds. Amino acids, which are molecules that combine together to form proteins and other essential components of living systems, were isolated in minute amounts from a meteorite sample in 1969. However, experiments showed them to be present as a racemic mixture, unlike the optically active amino acids found in living organisms on earth. The most reasonable conclusion is that the amino acids in the meteorite were formed "abiologically," that is, in the absence of a living organism.

7.12 MOLECULES WITH TWO OR MORE CHIRAL CENTERS

So far this chapter has focused on molecules containing a single chiral carbon atom. We will now expand our discussion to include molecules containing more than one chiral center. Consider, for example, 3-bromo-2-butanol.

STEREOCHEMISTRY AND ODOR

In living systems, the enantiomers of a substance can have striking differences in properties. For example, the enantiomeric forms of carvone have distinctly different odors. (R)-$(-)$-Carvone is the principal component and has the characteristic odor of spearmint oil, whereas (S)-$(+)$-carvone is the principal component of caraway seed oil and has the familiar odor of seeded rye bread.

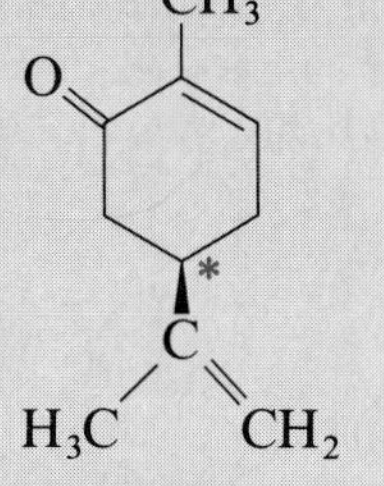

(R)-(−)-Carvone   (S)-(+)-Carvone
(from spearmint oil)   (from caraway seed oil)

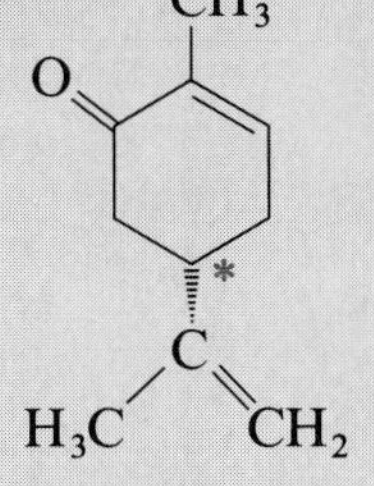

The reason for the difference in odor between (R)- and (S)-carvone results from their different behavior toward receptor sites in the nose. It is believed that volatile molecules occupy only those receptor sites that have the proper shape to accommodate them. These receptor sites are themselves chiral, so one enantiomer may fit one kind of receptor site while the other enantiomer fits a different kind of receptor. An analogy can be drawn to the fact that your right hand fits into a right glove and not a left one, as was discussed in Section 7.7. The receptor site (the glove) can accommodate one enantiomer of a chiral object (your hand) but not the other.

Fragrance chemists have found other examples of odor differences between enantiomers. The (S)-$(-)$ enantiomer of hydroxydihydrocitronellal, for example, has a lily-of-the-valley odor, while the (R)-$(+)$ enantiomer has a minty odor.

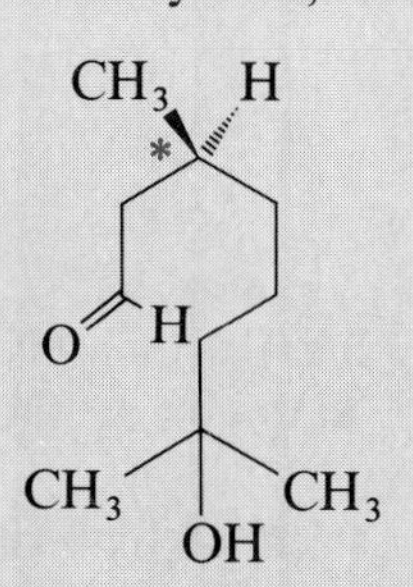

(S)-(−)-Hydroxydihydro-   (R)-(+)-Hydroxydihydro-
citronellal   citronellal

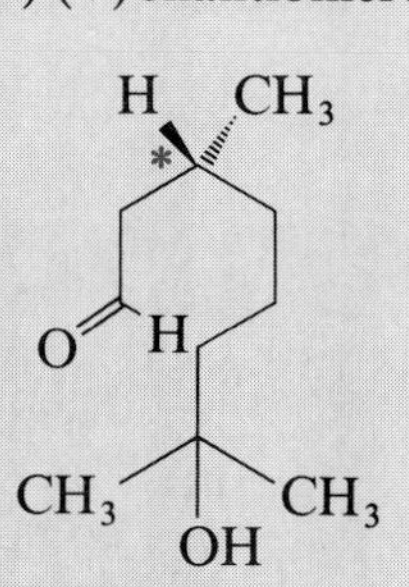

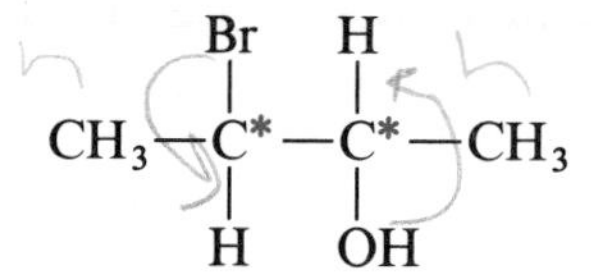

3-Bromo-2-butanol

This molecule contains two chiral centers (C*). How many stereoisomers are possible? The answer can be found by applying the R-S notational system discussed in Section 7.8. The absolute configuration at C-2 may be R or S. Likewise, C-3 may

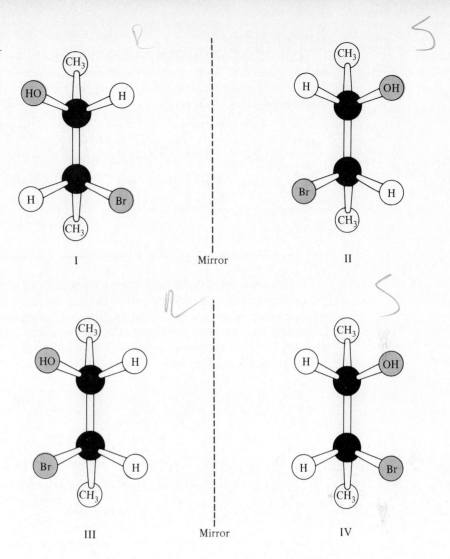

FIGURE 7.9 Ball-and-stick representations of the stereoisomeric 3-bromo-2-butanols.

have either the *R* or the *S* configuration. Thus there are *four* combinations of these two chiral centers.

(2*R*,3*R*)	(compound I)	(2*S*,3*S*)	(compound II)
(2*R*,3*S*)	(compound III)	(2*S*,3*R*)	(compound IV)

Figure 7.9 illustrates these four stereoisomers using ball-and-stick models. Stereoisomers I and II are enantiomers of each other; the enantiomer of (*R,R*) is (*S,S*). Likewise stereoisomers III and IV are enantiomers of each other, the enantiomer of (*R,S*) being (*S,R*).

How then are stereoisomers I and II related to stereoisomers III and IV? Stereoisomer I is *not* a mirror image of III or IV; thus it is not an enantiomer of either one. Stereoisomers that are not related as an object and its mirror image are called **diastereomers.** That is, *diastereomers are stereoisomers that are not enantiomers.* Therefore stereoisomers I and II are diastereomers of III and IV.

PROBLEM 7.9

Fischer projections (Section 7.4) are frequently used to represent the structure of molecules having more than one chiral center. Draw Fischer projections for each of the stereoisomers of 3-bromo-2-butanol.

SAMPLE SOLUTION

Using the ball-and-stick drawings in Figure 7.9 as a guide, first draw perspective views of each stereoisomer. The bonds pointing "down" or away from you are the vertical bonds in the Fischer projection; the bonds pointing "up" or toward you are horizontal. Thus stereoisomer I may be drawn as

$$CH_3$$
$$HO \blacktriangleright C \blacktriangleleft H$$
$$H \blacktriangleright C \blacktriangleleft Br$$
$$CH_3$$

Perspective view

$$CH_3$$
$$HO \!-\!|\!-\! H$$
$$H \!-\!|\!-\! Br$$
$$CH_3$$

Fischer projection

You have learned that enantiomers have equal optical rotations of opposite sign, yet their chemical and physical properties are the same in the absence of a chiral environment, such as a biological organism. Diastereomers, on the other hand, frequently have markedly different physical and chemical properties. For example, the (2R, 3R) stereoisomer of 3-amino-2-butanol is a liquid, while the (2R, 3S) diastereomer is a crystalline solid.

(2R,3R)-3-Amino-2-butanol
(liquid)

(2R,3S)-3-Amino-2-butanol
(solid, mp 49°C)

Diastereomers

PROBLEM 7.10

One other stereoisomer of 3-amino-2-butanol is a crystalline solid. Which one?

Now consider what happens with a molecule which has two chiral centers that are equivalently substituted, as does 2,3-butanediol.

$$\begin{array}{cc} OH & H \\ | & | \\ CH_3-C^*-C^*-CH_3 \\ | & | \\ H & OH \end{array}$$

2,3-Butanediol

Only *three*, not four, stereoisomeric 2,3-butanediols exist. The (2R, 3R) and (2S, 3S) forms, shown in Figure 7.10a, are chiral and enantiomers of each other; each is optically active.

A third combination of chiral centers, (2R, 3S), is *achiral*, however. It is identical to and superimposable on its (2S, 3R) mirror image. Thus this stereoisomer, shown in Figure 7.10b, is optically inactive. Molecules that have chiral centers but which are themselves achiral are called **meso isomers**. The meso isomer of 2,3-butanediol is a *diastereomer* of each of the enantiomers.

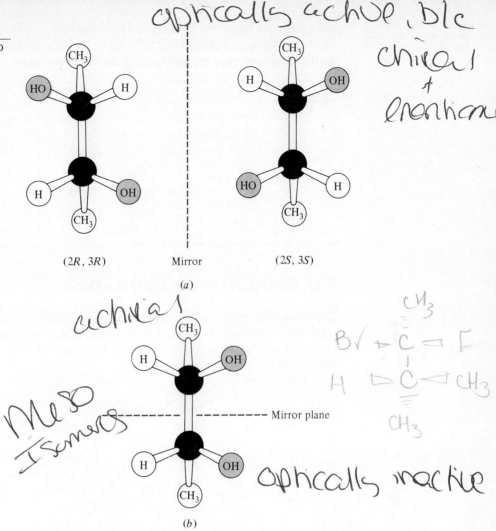

optically active, ble
chiral
+
enantiomers

(2R, 3R)   Mirror   (2S, 3S)

(a)

achiral

Meso Isomers

------- Mirror plane

optically inactive

$Br \rightarrow C \leftarrow F$ (with CH_3)

$H \triangleright C \triangleleft CH_3$ (with CH_3)

(b)

FIGURE 7.10 Ball-and-stick representations of the 2,3-butanediol stereoisomers. (a) The chiral enatiomers; (b) achiral meso isomer.

Notice that one-half of the meso isomer is a reflection of the other. In other words, the meso isomer has a plane of symmetry, indicated by the dashed line in Figure 7.10b. As noted earlier (Section 7.5), a molecule with a plane of symmetry is achiral.

PROBLEM 7.11
Repeat Problem 7.9 for the three stereoisomers of 2,3-butanediol.

In the preceding examples you have seen that a molecule with two chiral centers may have a *maximum* of four stereoisomers. The actual number is less than four if a meso compound is present. As a general rule, *the maximum number of stereoisomers for a particular constitution is 2^n, where n is the number of chiral centers.*

Many naturally occurring substances contain several chiral centers. The carbohydrates, which will be discussed in Chapter 15, are typical. For example, one class of carbohydrates, called aldohexoses, has the constitution

$$HOCH_2\overset{*}{C}H-\overset{*}{C}H-\overset{*}{C}H-\overset{*}{C}H-\overset{O}{\underset{H}{C}}$$
$$\qquad\quad |\quad\ \ |\quad\ \ |\quad\ \ |$$
$$\qquad\quad OH\ \ OH\ OH\ OH$$

An aldohexose

Since there are four chiral centers and no possibility of meso isomers, there are 2^4, or 16, stereoisomeric aldohexoses. All 16 are known, having been isolated either as natural products or as the products of chemical synthesis.

PROBLEM 7.12

2-Ketohexoses are a second type of carbohydrate; they have the constitution shown below. How many stereoisomeric 2-ketohexoses are possible?

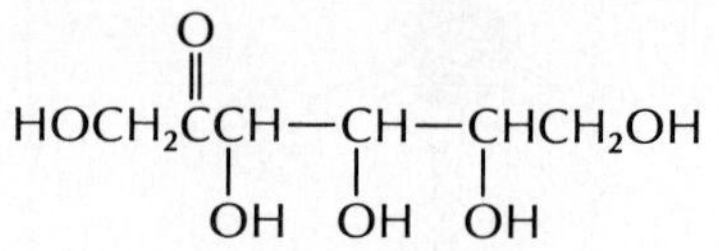

$$\underset{\substack{\text{A 2-ketohexose}}}{\overset{\displaystyle \overset{O}{\underset{\|}{}}}{HOCH_2\overset{O}{\underset{\|}{C}}CH-CH-CHCH_2OH}}$$

OH   OH   OH

A 2-ketohexose

7.13 RESOLUTION OF ENANTIOMERS

The separation of a racemic mixture into its enantiomeric components is called **resolution.** The first resolution, that of tartaric acid, was carried out by Louis Pasteur in 1848. Tartaric acid is present in grapes and is almost always found as its dextrorotatory (2R,3R) stereoisomer.

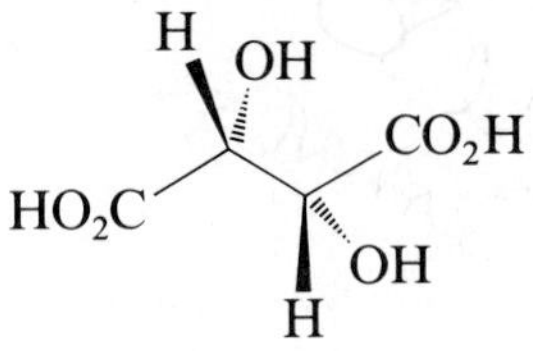

(2R,3R)-Tartaric acid
(mp 170°C
$[\alpha]_D + 12°$)

Occasionally, an optically inactive sample of tartaric acid was obtained. Pasteur noticed that the sodium ammonium salt of optically inactive tartaric acid was a mixture of two mirror image crystal forms. With microscope and tweezers, Pasteur carefully separated the two. He found that one kind of crystal (in aqueous solution) was dextrorotatory, while its mirror image rotated the plane of polarized light an equal amount but was levorotatory.

Although Pasteur was not then able to provide a structural explanation—that had to wait for van't Hoff and LeBel a quarter of a century later—he correctly deduced that the enantiomeric quality of the crystals of the sodium ammonium tartrates must be a consequence of enantiomeric molecules. The rare form of tartaric acid was optically inactive because it contained equal amounts of (+)-tartaric acid and (−)-tartaric acid. It had earlier been called *racemic acid* (from Latin *racemus,* "a bunch of grapes"), a name which subsequently gave rise to our present term for a mixture of enantiomers.

Pasteur's technique of separating enantiomers is not only laborious but requires that the crystal forms of enantiomers be distinguishable. This happens very rarely. Consequently, alternative and more general approaches to optical resolution of enantiomers have been developed. Most of these are based on a strategy of temporarily converting the enantiomers of a racemic mixture to diastereomeric derivatives, which may have different physical properties. The diastereomers are separated, and the enantiomeric starting materials are then regenerated.

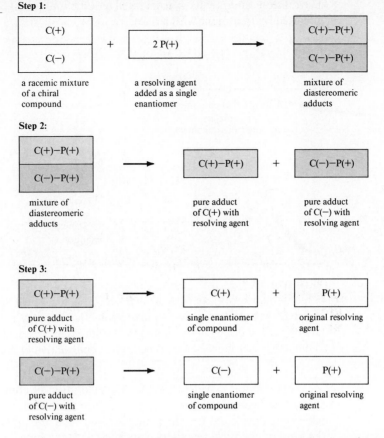

FIGURE 7.11 Diagram illustrating general procedure followed in resolution of a chiral substance into its individual enantiomeric components.

Figure 7.11 illustrates this strategy. Let us say we have a mixture of enantiomers which, for simplicity, we label as C(+) and C(−). Further, let us say that C(+) and C(−) bear some functional group that can combine with a reagent P to yield adducts C(+)-P and C(−)-P. If reagent P is chiral and only one enantiomer of P, say P(+), is added to a racemic mixture of C(+) and C(−), as shown in step 1, then the products of the reaction are C(+)-P(+) and C(−)-P(+). These products are *not* mirror images; they are diastereomers. Diastereomers can have different physical properties, and this difference in physical properties can serve as a means of separating them. In step 2, these diastereomers are separated, usually by recrystallization from a suitable solvent. In step 3, an appropriate chemical reaction is employed to remove the resolving agent [P(+)] from the separated adducts, thereby liberating the enantiomers and regenerating the resolving agent.

Whenever possible, the chemical reactions involved in the formation of diastereomers and their conversion to separate enantiomers are simple acid-base reactions. For example, naturally occurring (S)-(−)-malic acid is often used to resolve alkyl-substituted derivatives of ammonia called *amines*. One such amine that has been resolved in this way is 1-phenylethylamine. Amines are bases, and malic acid is an acid. Proton transfer from (S)-(−)-malic acid to a racemic mixture of (R)- and (S)-1-phenylethylamine gives a mixture of diastereomeric salts.

$$\overset{*}{C_6H_5}CHNH_2 + HO_2CCH_2\overset{*}{C}HCO_2H \longrightarrow \overset{*}{C_6H_5}CHNH_3^+ \quad {}^-O_2CCH_2\overset{*}{C}HCO_2H$$

with substituents CH_3 and OH below the respective starred carbons; on the product side CH_3 and OH.

1-Phenylethyl-
amine
(racemic mixture)

(S)-(−)-Malic
acid
(resolving agent)

1-Phenylethylammonium
(S)-malate
(mixture of diastereomeric salts)

The diastereomeric salts are separated and the individual enantiomers of the amine liberated by treatment with a base.

$$\overset{*}{C_6H_5CHNH_3^+} \quad \overset{*}{^-O_2CCH_2CHCO_2H} + \; 2OH^- \longrightarrow$$
$$\quad\;\; | \qquad\qquad\qquad\quad\; |$$
$$\quad\;\; CH_3 \qquad\qquad\qquad\;\; OH$$

1-Phenylethylammonium   Hydroxide
(S)-malate
(a single diastereomer)

$$\overset{*}{C_6H_5CHNH_2} + \; ^-O_2CCH_2\overset{*}{CHCO_2^-} + 2H_2O$$
$$\quad\;\; | \qquad\qquad\qquad\quad\; |$$
$$\quad\;\; CH_3 \qquad\qquad\qquad\; OH$$

1-Phenylethyl-   (S)-(−)-   Water
amine   Malic acid
(a single   (recovered
enantiomer)   resolving agent)

PROBLEM 7.13

In the resolution of 1-phenylethylamine using (−)-malic acid, the compound obtained by recrystallization of the mixture of diastereomeric salts is (R)-1-phenylethylammonium (S)-malate. The other component of the mixture is more soluble and remains in solution in the recrystallization solvent. What is the configuration of the more soluble salt?

This method is widely used for the resolution of chiral amines and carboxylic acids. Analogous methods based on the formation and separation of diastereomers have been developed for other functional groups; the precise approach depends on the kind of chemical reactivity associated with the functional groups present in the molecule.

7.14 SUMMARY

This chapter has introduced the three-dimensional spatial arrangement of atoms and groups in molecules, known as *stereochemistry*. Table 7.2 summarizes some basic definitions relating to molecular structure and shows where considerations of stereochemistry become important in classifying isomeric structures. Constitutional isomers were introduced in Section 1.8.

An *achiral* object or molecule is identical with and superimposable on its mirror image. An object or molecule is *chiral* if it cannot be superimposed on its mirror image (Section 7.1). The nonsuperimposable mirror images of a molecule are called *enantiomers* (Section 7.2). Chiral molecules contain a carbon atom that bears four different substituents, called a *chiral, or stereogenic, center* (Section 7.3).

The three dimensional structure of a molecule may be represented by ball-and-stick models, perspective views (wedge-and dash representations), or Fischer projections (Section 7.4).

Ball-and stick model   Perspective view   Fischer projection

TABLE 7.2
Basic Definitions Relating to Molecular Structure

Isomers are different compounds that have the same molecular formula. Isomers may be either constitutional isomers or stereoisomers.

Definition	Example
Constitutional isomers (Section 1.8) are isomers that differ in the order in which their atoms are connected.	There are three constitutionally isomeric compounds of molecular formula C_3H_8O: 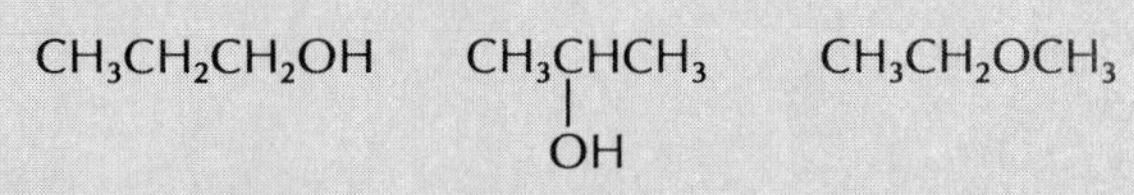 $CH_3CH_2CH_2OH$ CH_3CHCH_3 $CH_3CH_2OCH_3$ $\quad\quad\quad\quad\quad\quad\quad\quad$OH 1-Propanol $\quad$ 2-Propanol $\quad$ Ethyl methyl ether
Stereoisomers are isomers that have the same constitution but differ in the arrangement of their atoms in space.	
Enantiomers (Section 7.2) are stereoisomers that are related as an object and its nonsuperimposable mirror image.	The two enantiomeric forms of 2-chlorobutane are 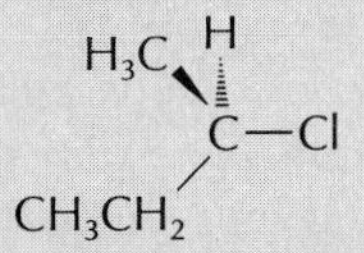 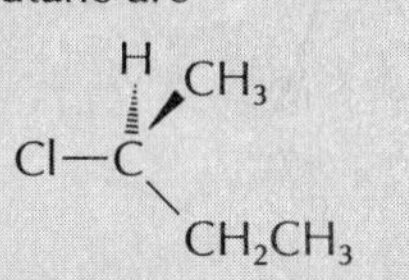 (R)-(−)-2-Chlorobutane $\quad\quad$ (S)-(+)-2-Chlorobutane
Diastereomers (Section 7.12) are stereoisomers that are not enantiomers.	Two stereoisomers of 2,3-dibromobutane are shown below. They are not mirror images; they are diastereomers. *meso*-2,3-Dibromobutane $\quad\quad$ (2S,3S)-2,3-Dibromobutane The cis-trans isomers of a cycloalkane or an alkene are diastereomers of each other. and $\quad$ are diastereomers and $\quad$ are diastereomers

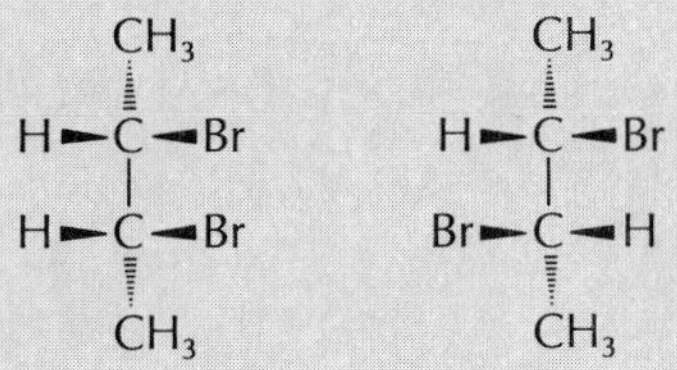

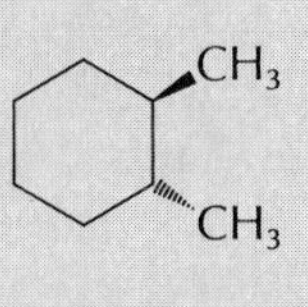

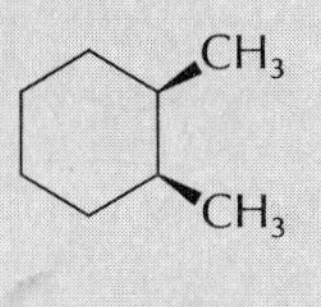

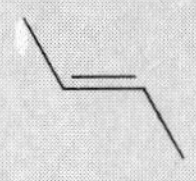

Any structure that has a plane of symmetry is achiral (Section 7.5). A molecule may be achiral because it has no chiral centers or because it is a meso isomer (Section 7.12). *Meso isomers* are those molecules which have two or more chiral centers but are achiral.

Diastereomers (Section 7.12) are stereoisomers that are not enantiomers. The stereoisomers of 2,3-dibromobutane shown in Table 7.2 are diastereomers of each other.

The *absolute configuration* of a molecule is a precise description of the spatial arrangement of its atoms. The absolute configuration of a molecule may be specified using the *R-S* notational system (Section 7.8). Table 7.2 identifies the enantiomers of 2-chlorobutane according to this system.

The priority ranking system developed for the assignment of absolute configurations may also be used to specify the configuration of *stereoisomeric alkenes* (Section 7.9). The higher-priority groups are on the same side in the (*Z*) stereoisomer; they are on opposite sides in the (*E*) stereoisomer.

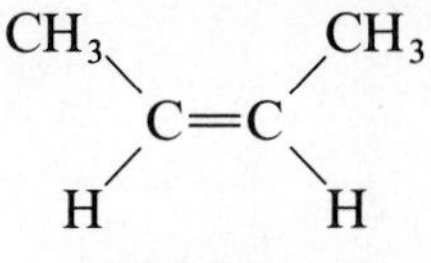

(*Z*)-2-Butene 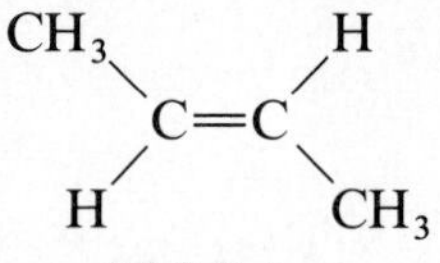   (*E*)-2-Butene

When a structure contains more than one chiral center, the maximum possible number of stereoisomers is 2^n, where *n* is equal to the number of chiral centers in the molecule. The actual number of stereoisomers is less than 2^n when meso isomers are present (Section 7.12).

Optical activity is the ability to rotate plane-polarized light and is a physical property exhibited by chiral molecules (Section 7.6). Enantiomeric forms of the same molecule rotate plane-polarized light an equal amount but in opposite directions. A mixture which contains equal quantities of enantiomers is optically inactive and is a *racemic mixture*. Achiral molecules are also optically inactive.

The individual enantiomers of a racemic mixture may often be separated by a procedure called *resolution* (Section 7.13).

ADDITIONAL PROBLEMS

Chiral Centers and Enantiomers

7.14 Which of the isomeric halides having the molecular formula $C_5H_{11}Br$ are chiral? Which are achiral? Label each chiral center with an asterisk.

7.15 Draw both enantiomers of each chiral stereoisomer identified in Problem 7.14 using perspective (wedge-and-dash) drawings.

7.16 Repeat Problem 7.15 using Fischer projections.

7.17 Identify the relationship in each of the following pairs of molecules. Do the drawings represent compounds which are constitutional isomers or stereoisomers, or are they identical? If they are stereoisomers, are they enantiomers or diastereomers?

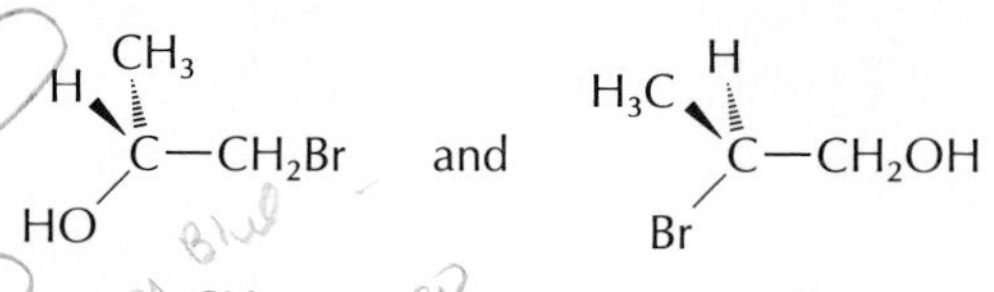

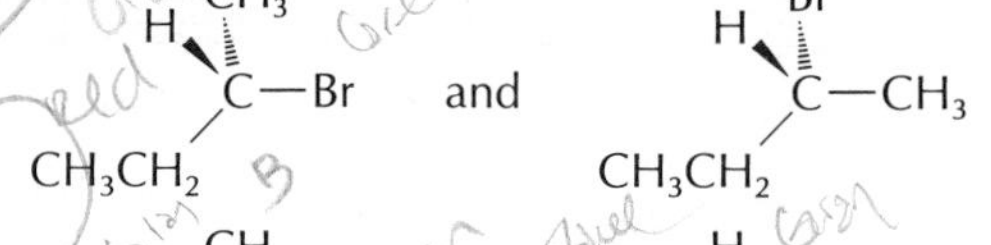

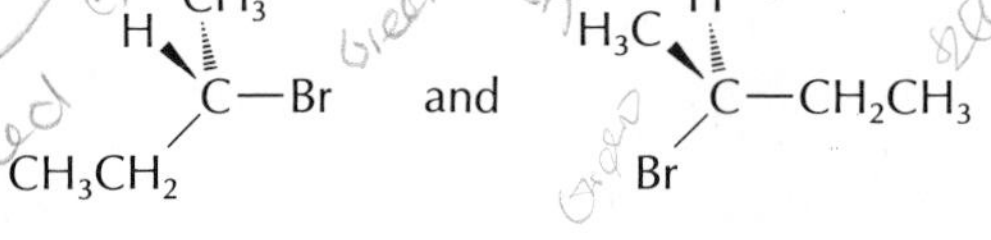

7.18 Redraw each of the molecules in Problem 7.17 using Fischer projections with C-1 at the top and a vertical arrangement of the carbon chain.

Molecules Having Two or More Chiral Centers

7.19 Identify any planes of symmetry in each of the following compounds. Which of the compounds is chiral? Which is achiral?

(a) *cis*-1,2-Dichlorocyclopropane

(b) *trans*-1,2-Dichlorocyclopropane

(c) *cis*-2-Chlorocyclopropanol

(d) *trans*-2-Chlorocyclopropanol

7.20 Write the structures of all the isomers, including stereoisomers, of trichlorocyclopropane. Which of these is (are) chiral?

7.21 Using Fischer projections, draw all the stereoisomers of 2,3-dibromopentane. Specify the relationship between pairs of compounds as being enantiomeric or diastereomeric.

7.22 2,4-Dibromopentane has fewer stereoisomers than 2,3-dibromopentane. Explain.

7.23 One of the stereoisomers of 1,3-dimethylcyclohexane is a meso isomer. Identify which one (cis or trans?).

7.24 Cholesterol is an important molecule in the formation of steroid hormones in our bodies, yet an excess in the blood has been implicated as a contributing factor to certain types of heart disease. Cholesterol has eight chiral centers. How many stereoisomers are possible having the same constitution as cholesterol, assuming there are no meso isomers?

7.25 Identify the chiral centers in each of the following naturally occurring substances.

(a)

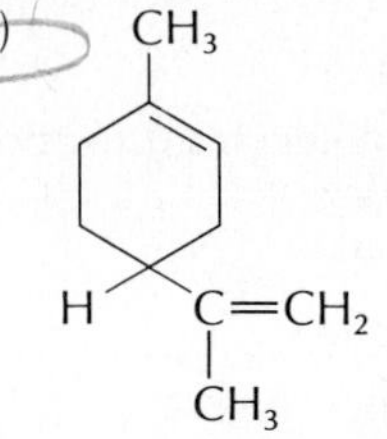

Limonene (a constituent of lemon oil)

(b)

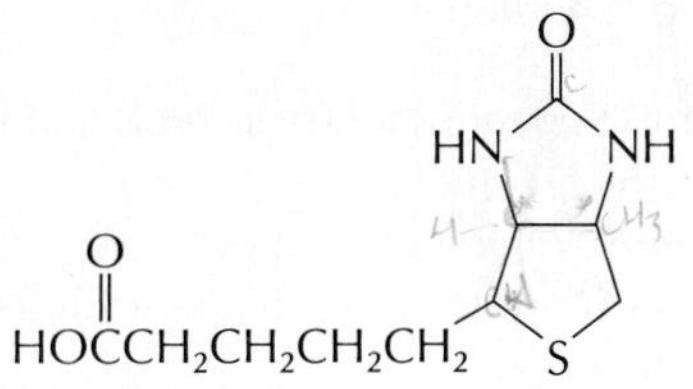

Biotin (a nutrient essential for normal growth)

(c)

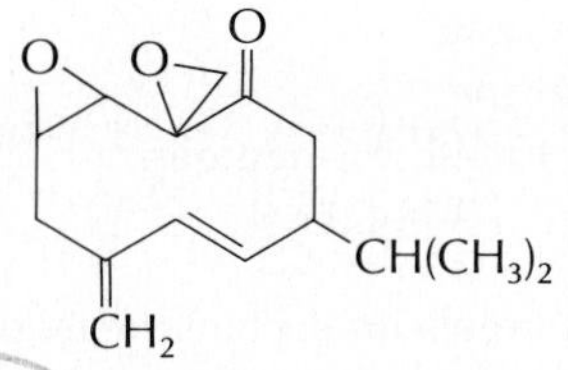

Periplanone B (sex attractant of the American cockroach)

(d)

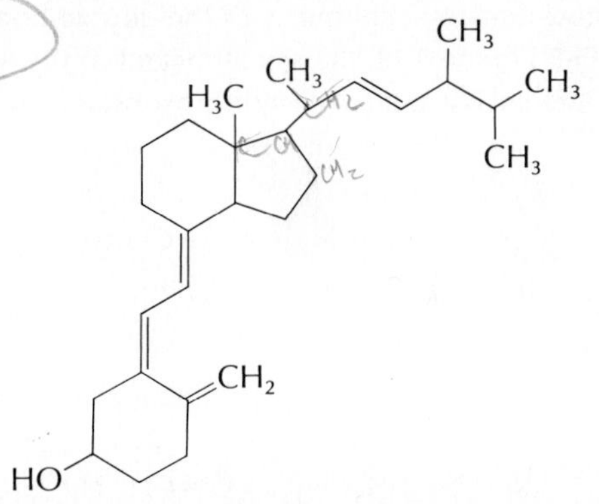

Calciferol (a hormone, also called vitamin D_2, which is involved in calcium deposition in bones)

7.26 How many stereoisomers have the constitution $CH_3CH{=}CHCHCH_3$?

$$\overset{|}{OH}$$

7.27 Carbohydrates having four carbons and an aldehyde group are called aldotetroses and have the general formula

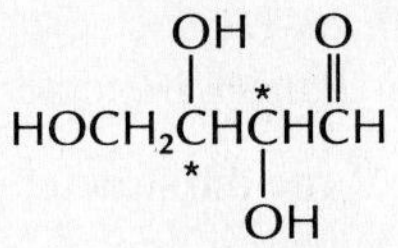

$$
\begin{array}{ccc}
& OH & O \\
& | & \| \\
HOCH_2CHCHCH \\
& * & | \\
& & OH
\end{array}
$$

Draw all the aldotetrose stereoisomers using Fischer projections. Are there any meso isomers?

Absolute Configuration of Chiral Molecules

7.28 Using R-S descriptors, write all the possible combinations for a molecule with three chiral centers.

7.29 Specify the absolute configurations of the enantiomers in Problem 7.15 as R or S.

7.30 Specify the absolute configuration of each molecule in Problem 7.17 using the R-S notational system.

7.31 Specify the absolute configuration of each chiral center in Problem 7.21 using the R-S notational system.

7.32 Draw each of the following molecules using both a perspective view and a Fischer projection.
   (a) (S)-2-Pentanol
   (b) (R)-3-Chloro-2-methylpentane
   (c) (2S,3S)-2-Bromo-3-pentanol
   (d) (R)-3-Bromo-1-pentene

7.33 Draw the R and S enantiomers of a chiral alkene having the molecular formula C_6H_{12}.

Naming Stereoisomeric Alkenes Using the E-Z System

7.34 Write structural formulas for each of the following, clearly showing the stereochemistry around the double bond.
   (a) (Z)-3-Methyl-2-hexene
   (b) (E)-3-Chloro-2-hexene
   (c) (E)-1,2-Dibromo-3-methyl-2-heptene
   (d) (Z)-4-Ethyl-3-methyl-3-heptene

7.35 Write a structural formula for each of the following naturally occurring compounds. Clearly show the stereochemistry of the double bonds.
   (a) (E)-6-Nonen-1-ol, the sex attractant of the Mediterranean fruit fly.
   (b) Geraniol is a naturally occurring substance present in the fragrant oil of many plants. Geraniol is the E isomer of

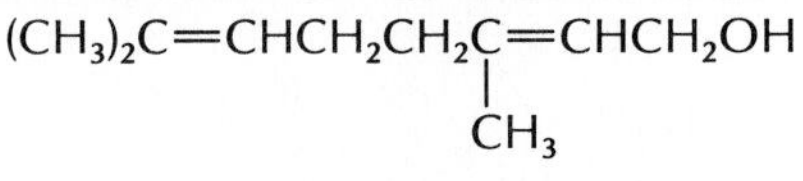

$$
\begin{array}{c}
(CH_3)_2C\!=\!CHCH_2CH_2C\!=\!CHCH_2OH \\
| \\
CH_3
\end{array}
$$

   (c) The sex attractant of the codling moth is the (2Z,6E) stereoisomer of

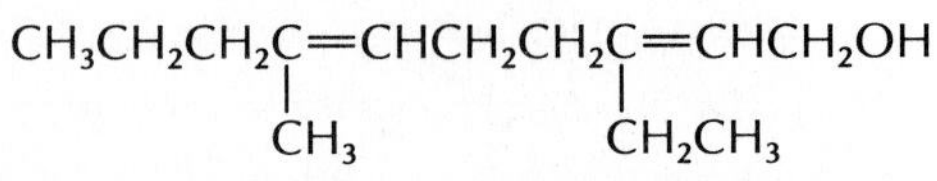

$$
\begin{array}{c}
CH_3CH_2CH_2C\!=\!CHCH_2CH_2C\!=\!CHCH_2OH \\
\qquad | \qquad\qquad\qquad\quad | \\
\qquad CH_3 \qquad\qquad\qquad CH_2CH_3
\end{array}
$$

(d) The sex pheromone of the honeybee is the E stereoisomer of

$$CH_3\overset{\overset{\displaystyle O}{\|}}{C}(CH_2)_4CH_2CH=CHCO_2H$$

Miscellaneous Problems

7.36 An aqueous solution containing 10 g of fructose (a sugar used to sweeten candy and other foods) was diluted to 500 mL with water and placed in a polarimeter tube 20 cm long. The measured optical rotation was −5.20°. Calculate the specific rotation of fructose.

7.37 In Section 7.13 the first resolution experiment, carried out by Louis Pasteur, was described. As noted previously, the tartaric acid isolated from grapes is almost always the (2R,3R) stereoisomer.

     (a) There are two other stereoisomeric tartaric acids. Write their structures and specify the configuration at their chiral centers.

     (b) Could the unusual, optically inactive form of tartaric acid studied by Pasteur have been *meso*-tartaric acid?

7.38 Write a mechanism for the acid-catalyzed hydration of 1-butene which explains why the product is obtained as a racemic mixture.

7.39 How many stereoisomeric products are obtained when (S)-3-chloro-1-butene reacts with HCl by Markovnikov addition? Are these products enantiomers or diastereomers of each other?

CHAPTER 8

NUCLEOPHILIC SUBSTITUTION REACTIONS

This chapter has a mechanistic emphasis and focuses on reactions which are nucleophilic substitutions. Upon completion of this chapter you should be able to:

- Identify a reaction as being a nucleophilic substitution and predict the products.
- Predict the relative reactivities of alkyl halides bearing different halogen atoms.
- Explain what is meant by the phrase "bimolecular nucleophilic substitution."
- Predict the stereochemical result of an S_N2 reaction.
- Predict the effect of steric crowding on an S_N2 reaction.
- Explain what is meant by the phrase "unimolecular nucleophilic substitution."
- Explain how carbocation stability affects the rate of an S_N1 reaction.
- Predict the stereochemical result of an S_N1 reaction.
- Predict whether elimination or substitution will predominate in the reaction of a specific alkyl halide with a Lewis base.

Several million different organic molecules are known, and these are capable, at least in theory, of undergoing an almost countless number of chemical reactions with each other. To study these reactions individually would, of course, be impossible.

Two organizational devices make the problem manageable, however. First, we classify organic molecules according to the *functional groups* they contain. Identifying a functional group reduces the problem from one of having to deal with millions of different molecules to one involving a few dozen families of structurally related compounds. Second, most chemical reactions belong to one or another of a relatively few *reaction types* such as addition, elimination, and substitution. Reaction types may be further subdivided according to mechanism. The present chapter describes a reaction type of *alkyl halides* called *nucleophilic substitution,* which is extremely useful in synthetic organic chemistry. Our goal in presenting the material is to show

how a large number of seemingly diverse reactions bear much in common when viewed from a mechanistic perspective.

8.1 A REVIEW OF REACTION MECHANISMS

Reaction mechanisms were first described in Chapter 3 (Section 3.11). A **reaction mechanism** is a structural description of the individual reaction steps during the conversion of reactants to products. The conversion of alcohols to alkyl halides (Section 3.11) is an example of a *substitution reaction.* In a substitution reaction, one atom or group of atoms substitutes for, or replaces, another atom or group of atoms.

$$R—X + Y \longrightarrow R—Y + X$$

This chapter will explore in more detail substitution reactions of this type in which Y is a nucleophile and RX is an alkyl chloride, bromide, or iodide.

A different type of substitution reaction was encountered in Chapter 6: *electrophilic aromatic substitution.* In these reactions an electron-seeking atom or group replaces a hydrogen on one or more positions of a benzene ring.

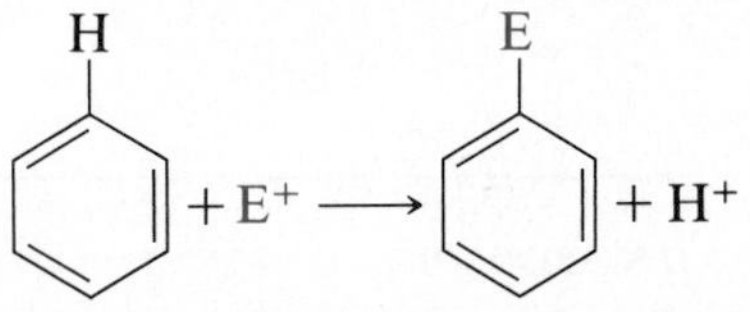

The preparation and reactions of alkenes involve two additional types of reaction mechanisms: eliminations and additions. Alkenes are generally prepared from alcohols or alkyl halides by *elimination reactions* (Section 4.1.6). The dehydrohalogenation of an alkyl halide may proceed either by an *E2 mechanism* (Section 4.1.10) in the presence of a strong base or by an *E1 mechanism* in the absence of a strong base (Section 4.1.11).

The characteristic reactions of alkenes (Chapter 5) are *additions.*

Numerous reagents undergo *electrophilic additions* to alkenes (Sections 5.1.2 to 5.1.7). Hydrogen bromide is also able to undergo *free radical addition* to an alkene (Section 5.1.8).

8.2 NUCLEOPHILIC SUBSTITUTION REACTIONS

The main focus of this chapter will be nucleophilic substitution reactions. These are typical reactions of alkyl halides and may be generalized by the equation

$$R\ddot{X}: + Y:^- \longrightarrow RY + :\ddot{X}:^-$$

| Alkyl halide | Lewis base | Product of nucleophilic substitution | Halide anion |

Recall that a Lewis base (Section 3.13) is a species which can act as an electron-pair donor, and must have at least one unshared pair of electrons. The species may be negatively charged, as shown in the general equation, or may be neutral. A hydroxide ion ($H\ddot{O}:^-$) and water ($H\ddot{O}H$) are both examples of Lewis bases.

As noted above, nucleophilic substitution reactions of the type described in this chapter are typical reactions of *alkyl* halides. Aryl and vinyl halides do *not* undergo these reactions. Although the alkyl halide may contain an aryl or vinyl group elsewhere in the molecule, the carbon undergoing substitution must be an alkyl carbon. That is, it must be sp^3 hybridized.

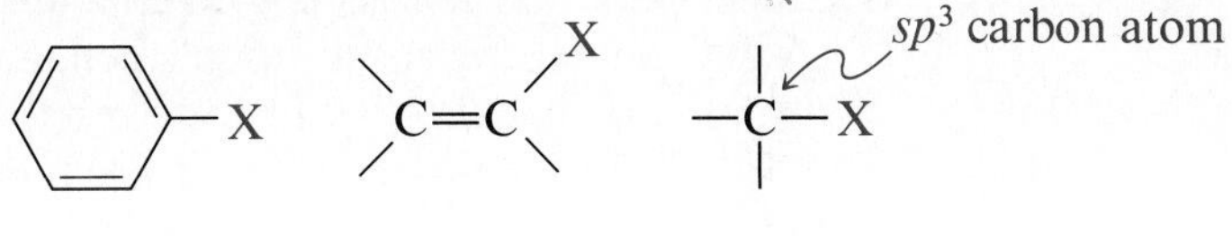

An aryl halide

A vinyl halide

An alkyl halide

sp^3 carbon atom

PROBLEM 8.1

Classify each of the following molecules as being an alkyl, aryl, or vinyl halide.

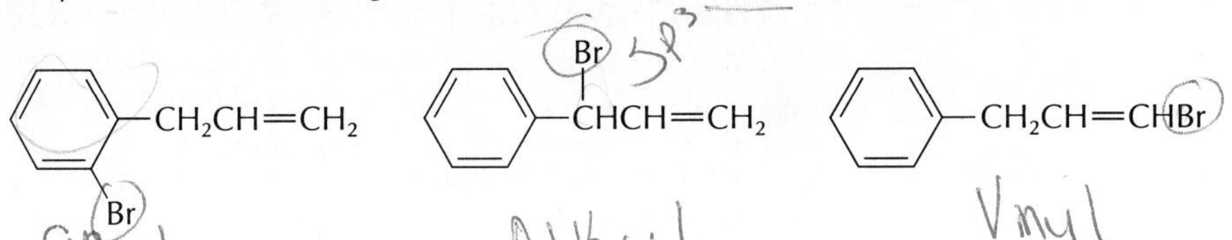

Nucleophilic substitution reactions have four essential components: a **nucleophile** (the Lewis base) which reacts with the **substrate** (the alkyl halide) to give the **product** and a **leaving group** (the halide ion which was displaced).

$$\text{Nucleophile} + \text{substrate} \longrightarrow \text{product} + \text{leaving group}$$

By identifying these four components, all of which are present in a nucleophilic substitution reaction, you can readily focus on the changes taking place.

PROBLEM 8.2

Identify the nucleophile, substrate, product, and leaving group in the nucleophilic substitution reaction

$$CH_3Br + HO^- \longrightarrow CH_3OH + Br^-$$

| Methyl bromide | Hydroxide ion | Methyl alcohol | Bromide ion |

A variety of functional groups may be introduced into an organic molecule by using a nucleophilic substitution reaction and choosing the appropriate nucleophile.

Several examples of nucleophilic substitution reactions are listed in Table 8.1. The nucleophiles shown are all anions, which are generally used as their lithium, sodium, or potassium salts.

PROBLEM 8.3

Write the structure of the principal organic product formed in the reaction of bromoethane with each of the following compounds:

(a) NaOH (sodium hydroxide)
(b) KOC(CH₃)₃ (potassium *tert*-butoxide)
(c) KCN (potassium cyanide)
(d) NaSH (sodium hydrogen sulfide)

[handwritten annotations:]
$HO^- + CH_3CH_2Br \rightarrow CH_3CH_2OH + Br^-$
$(CH_3)_3CO^- + CH_3CH_2Br \rightarrow CH_3COCH_2CH_2 + Br^-$
$C\equiv N: + CH_3CH_2Br \rightarrow NCH_3CH_2 + Br^-$
$HS^- \rightarrow CH_3CH_2SH + Br^-$

SAMPLE SOLUTION

(a) The nucleophile in sodium hydroxide is the negatively charged hydroxide ion. The reaction that occurs is nucleophilic substitution of bromide by hydroxide. The product is ethanol (ethyl alcohol).

[handwritten: nucleophile subst.   leaving]

$$HO^- + CH_3CH_2Br \longrightarrow CH_3CH_2OH + Br^-$$

| Hydroxide ion (nucleophile) | Bromoethane (substrate) | Ethanol *[handwritten: prod.]* (product) | Bromide ion (leaving group) |

Iodide (I⁻) salts are frequently used to prepare alkyl iodides. The replacement of one halide by another is best accomplished when iodide is the nucleophile and chloride or bromide the leaving group. Using acetone as the solvent, the reaction favors product formation because the chloride or bromide salts precipitate as they are formed. Sodium iodide is soluble in acetone; sodium chloride and sodium bromide are not.

$$CH_2=CHCH_2Cl + NaI \xrightarrow{\text{acetone}} CH_2=CHCH_2I + NaCl(s)$$

| Allyl chloride | Sodium iodide | Allyl iodide (77%) | Sodium chloride |

$$\underset{\underset{Br}{|}}{CH_3CHCH_3} + NaI \xrightarrow{\text{acetone}} \underset{\underset{I}{|}}{CH_3CHCH_3} + NaBr(s)$$

| Isopropyl bromide | Sodium iodide | Isopropyl iodide (63%) | Sodium bromide |

These reactions are good examples of *Le Chatelier's principle* (Section 5.1.5). The reaction equilibrium shifts to the right (forms alkyl iodide) to compensate for the precipitation of sodium chloride or sodium bromide.

With this introduction to nucleophilic substitution reactions you can begin to see the importance of alkyl halides in organic chemistry. In earlier chapters you learned about the preparation of alkyl halides from alcohols and the free radical halogenation of alkanes (Chapter 3), and by addition of hydrogen halides to alkenes (Chapter 5). As you now can see, alkyl halides may then become starting materials for a variety of other types of organic compounds by replacement of the halide with an appropriate nucleophile.

TABLE 8.1

Representative Functional Group Transformations by Nucleophilic Substitution Reactions of Alkyl Halides.

Nucleophile and comments	General equation and specific example
Hydroxide ion (HO⁻) The oxygen atom of a metal hydroxide acts as a nucleophile to replace the halogen of an alkyl halide. The product is an *alcohol*.	$HO:^- + R-X \longrightarrow R-OH + X^-$ Hydroxide ion Alkyl halide Alcohol Halide ion $KOH + CH_3CH_2CH_2Br \longrightarrow CH_3CH_2CH_2OH + KBr$ Potassium hydroxide 1-Bromopropane 1-Propanol Potassium bromide
Alkoxide ion (RO:⁻) The oxygen atom of a metal alkoxide acts as a nucleophile to replace the halogen of an alkyl halide. The product is an *ether*.	$R'O:^- + R-X \longrightarrow R'OR + X^-$ Alkoxide ion Alkyl halide Ether Halide ion $(CH_3)_2CHCH_2ONa + CH_3CH_2Br \xrightarrow{\text{isobutyl alcohol}}$ Sodium isobutoxide Ethyl bromide $(CH_3)_2CHCH_2OCH_2CH_3 + NaBr$ Ethyl isobutyl ether (66%) Sodium bromide

Hydrogen sulfide ion (HS:⁻) Use of hydrogen sulfide as a nucleophile permits the conversion of alkyl halides to compounds of the type RSH. These compounds are the sulfur analogs of alcohols and are known as *thiols*.	$HS:^- + R-X \longrightarrow RSH + X^-$
	
	 Hydrogen sulfide ion Alkyl halide Thiol Halide ion
	
	 $KSH + CH_3CH(CH_2)_6CH_3 \xrightarrow{\text{ethanol water}}$
	          $|$
	         Br
	
	 Potassium hydrogen sulfide 2-Bromononane
	
	 $CH_3CH(CH_2)_6CH_3 + KBr$
	     $|$
	    SH
	
	 2-Nonanethiol (74%) Potassium bromide

Cyanide ion (:C≡N:) The negatively charged carbon atom of cyanide ion is usually the site of its nucleophilic character. Use of cyanide ion as a nucleophile permits the extension of a carbon chain by carbon-carbon bond formation. The product is an *alkyl cyanide* or *nitrile*.	$:N\equiv C:^- + R-X \longrightarrow RC\equiv N: + X^-$ Cyanide ion Alkyl halide Alkyl cyanide Halide ion 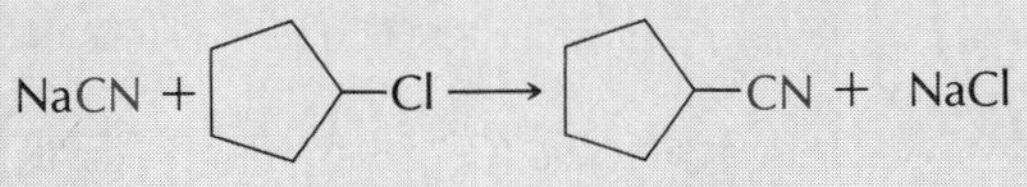 $NaCN + \text{(cyclopentyl)}-Cl \longrightarrow \text{(cyclopentyl)}-CN + NaCl$ Sodium cyanide Cyclopentyl chloride Cyclopentyl cyanide (70%) Sodium chloride

TABLE 8.1 (*continued*)

Nucleophile and comments	General equation and specific example
Alkynide ion (RC≡C:⁻) The conjugate bases of terminal alkynes and acetylene (Section 5.4) are good nucleophiles and react with primary or methyl alkyl halides to form carbon-carbon bonds.	$RC{\equiv}CNa + R'CH_2X \longrightarrow RC{\equiv}CCH_2R' + NaX$ Sodium Alkynide · · · Alkyl halide · · · Alkyne · · · Sodium halide $CH_3CH_2C{\equiv}C{:}^-Na^+ + CH_3I \longrightarrow CH_3CH_2C{\equiv}CCH_3 + NaI$ Sodium 1-butynide · · · Methyl iodide · · · 2-Pentyne · · · Sodium iodide

8.3 RELATIVE REACTIVITY OF HALIDE LEAVING GROUPS

Among alkyl halides, alkyl iodides undergo nucleophilic substitution at the fastest rate, alkyl fluorides at the slowest.

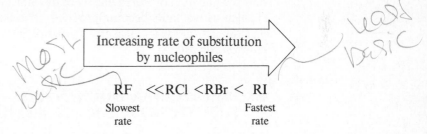

Increasing rate of substitution by nucleophiles

$$RF \ll RCl < RBr < RI$$

Slowest rate · · · · · · · · · · · · Fastest rate

(handwritten: most basic) · · · *(handwritten: least basic)*

As in the elimination reactions of alkyl halides (Section 4.1.10), iodide is the best leaving group because it has the weakest carbon-halogen bond. Alkyl iodides are several times more reactive than bromides and from 50 to 100 times more reactive than chlorides.

PROBLEM 8.4

A single organic product was obtained when 1-bromo-3-chloropropane was allowed to react with one molar equivalent of sodium cyanide. What was this product?

Alkyl fluorides are rarely used as substrates in nucleophilic substitution reactions because they are several thousand times less reactive than alkyl chlorides. Fluoride is the poorest leaving group because it forms the strongest bond to carbon of any of the halogens. Teflon, the Du Pont trade name for polytetrafluoroethylene, has the general formula

$$\left(\begin{array}{cc} F & F \\ | & | \\ -C-C- \\ | & | \\ F & F \end{array}\right)_n$$

In addition of its nonstick properties, Teflon is chemically inert to most laboratory reagents. This inertness can be explained by the strength, and thus the lack of reactivity, associated with the carbon-fluorine bond.

Leaving group ability is also related to basicity. A strongly basic anion is usually a poorer leaving group than a weakly basic one. Fluoride is the most basic and the poorest leaving group among the halide anions, iodide the least basic and the best leaving group.

8.4 MECHANISMS OF NUCLEOPHILIC SUBSTITUTION REACTIONS. A PREVIEW

Nucleophilic substitution reactions of alkyl halides occur by one of two general mechanisms: S_N2, which stands for **substitution nucleophilic bimolecular**, and S_N1, which means **substitution nucleophilic unimolecular.** The terms bimolecular and unimolecular were first used to describe the mechanism of dehydrohalogenation in Section 4.1.10. We will explore further what these terms mean in the sections that follow.

The energy changes that take place during a nucleophilic substitution reaction will be described using potential energy diagrams. A **potential energy diagram** shows how the potential or stored energy of the molecules involved in a chemical reaction changes as reactants become products. The horizontal axis in a potential energy diagram is an arbitrary measure of progress from reactant to product and is called the *reaction coordinate.*

8.5 BIMOLECULAR NUCLEOPHILIC SUBSTITUTION: THE S_N2 MECHANISM

The reaction of methyl bromide with sodium hydroxide to form methyl alcohol is an example of a nucleophilic substitution reaction.

$$CH_3Br + HO^- \longrightarrow CH_3OH + Br^-$$

| Methyl bromide (substrate) | Hydroxide ion (nucleophile) | Methyl alcohol (product) | Bromide ion (leaving group) |

The rate of this reaction is observed to be directly proportional to the concentration of *both* reactants, methyl bromide and sodium hydroxide. This proportionality can be summarized with a *rate equation* (Section 4.1.10):

$$\text{Rate} = k[\text{substrate}][\text{nucleophile}]$$

where k is a proportionality constant called the *rate constant.* If the concentration of either reactant is doubled, the rate will double. If the concentration of both reactants is doubled, the rate will increase by a factor of 4.

The reaction of methyl bromide with sodium hydroxide is said to exhibit second-order kinetic behavior which is interpreted in terms of a *bimolecular* mechanism. The terms *second-order* and *bimolecular* were described earlier in the discussion of bimolecular elimination reactions (Section 4.1.10).

The substitution reaction described above as well as all of those cited in Table 8.1 are examples of **bimolecular nucleophilic substitution,** or S_N2, reactions. An S_N2 reaction is a *concerted* process; the reaction occurs in a single step. E2 reactions (bimolecular eliminations) are another example of concerted reactions.

The energy changes that occur during an S_N2 reaction are illustrated by the energy diagram shown in Figure 8.1. The left end of the x axis represents reactants,

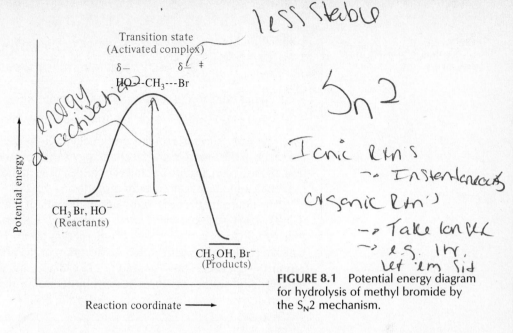

less stable

Sn2

Ionic Rxn's
→ Instenteneous
Organic Rxn's
→ Take long rxc
→ e.g. 1hr.
let 'em sit

FIGURE 8.1 Potential energy diagram for hydrolysis of methyl bromide by the S_N2 mechanism.

and the right end represents products. The point at the top of the curve is the *transition state,* and the species present at that point is called an *activated complex.* As described in Section 4.1.10, the activated complex is shown with dashed lines representing "partial bonds," which are bonds in the process of being formed or broken.

$$HO^- + CH_3Br \longrightarrow [HO\text{---}CH_3\text{---}Br] \longrightarrow HOCH_3 + Br^-$$

| Hydroxide ion | Methyl bromide | Activated complex | Methyl alcohol | Bromide ion |

An unshared electron pair on oxygen becomes partially bonded to carbon as the transition state is approached and serves to "push off" bromide from carbon. The carbon-bromine bond breaks, with the pair of electrons in that bond becoming an unshared electron pair of the bromide ion. In the next section we will explore further the nature of the activated complex in an S_N2 reaction.

NUCLEOPHILIC SUBSTITUTION REACTIONS AND CANCER

Chemicals which can cause cancer in either laboratory animals or humans are called *carcinogens* (see boxed material following Section 6.4). Many different types of chemicals may act as carcinogens in addition to the polycyclic aromatic hydrocarbons described in Chapter 6. Among these are certain *alkylating agents,* compounds which can act as substrates in nucleophilic substitution reactions. For example, chloromethyl methyl ether and bis(2-chloroethyl)amine are both active carcinogens.

$$ClCH_2OCH_3 \qquad (ClCH_2CH_2)_2NH$$

| Chloromethyl methyl ether | Bis(2-chloroethyl)-amine |

One theory of *carcinogenesis,* or how these compounds interact with cells, is that cellular nucleophiles displace the leaving group (chloride in these compounds) to form a *bound adduct.*

$$RCH_2Cl \ + \quad Nu: \quad \longrightarrow \quad Nu—CH_2R$$

Carcinogen   Nucleophile   Bound adduct

Cellular nucleophiles may include functional groups such as $—\overset{..}{N}H_2$, $—\overset{..}{\underset{..}{S}}H$, and $—\overset{..}{\underset{..}{O}}H$, all of which are abundant in the molecules which make up living systems, such as proteins and nucleic acids. The bound adduct changes the shape of the biomolecule by changing its molecular composition. A change in shape will often lead to a change in the function of the molecule, and this can, in turn, lead to a cancer-causing mutation of the cell.

An interesting point can be made about the carcinogenic amine $(ClCH_2CH_2)_2NH$, known as *nitrogen mustard.* This compound is also used to treat certain forms of cancer as a *chemotherapy* agent. The same mechanism by which a cell becomes cancerous—disruption of the normal functioning of the cell through a mutation—is used to kill cancer cells. Because cancer cells are rapidly dividing, they are more susceptible to mutation-causing chemicals such as nitrogen mustard than normal cells in the body.

8.6 STEREOCHEMISTRY OF S_N2 REACTIONS

The transition state is the highest energy point on the reaction pathway. This is true for all chemical reactions, not only those proceeding by the S_N2 mechanism. Learning more about the activated complex present at the transition state will help us understand the mechanism of the reaction. What is the structure of the activated complex? In particular, what is the spatial relationship between the nucleophile and the leaving group as reactants pass over the transition state on their way to products?

A variety of stereochemical studies reveal that S_N2 reactions proceed by **inversion of configuration.** The nucleophile attacks carbon from the side opposite the bond to the leaving group.

$$Nu:^- \ + \quad \overset{w}{\underset{y}{\overset{x}{C}}}—LG \longrightarrow Nu\text{------}\overset{x \; w}{\underset{y}{C}}\text{------}\overset{\delta^-}{LG} \longrightarrow Nu—\overset{w}{\underset{y}{C}}\overset{x}{} \ + :LG$$

Notice that the groups w, x, and y have flipped from the left to the right, much as an umbrella would flip inside out in a windstorm.

For the specific case of the reaction between methyl bromide and the hydroxide ion, the activated complex is as shown.

$$HO^- \ + \quad \overset{H}{\underset{H}{\overset{H}{C}}}—Br \longrightarrow \left[HO\text{------}\overset{H \; H}{\underset{H}{C}}\text{------}Br \right] \longrightarrow HO—\overset{H}{\underset{H}{C}}\overset{H}{} \ + \quad Br^-$$

| Hydroxide ion | Methyl bromide | Activated complex | Methyl alcohol | Bromide ion |

The carbon-oxygen bond is partially formed at the same time the carbon-bromine bond is partially broken. The three hydrogens remain attached to the carbon by full covalent bonds and for an instant are spread out from their normal tetrahedral angles in a planar arrangement. As the reaction continues toward product, the normal tetrahedral angles of carbon are regained.

This picture of the S_N2 mechanism is based on many experiments using optically active alkyl halides as substrates. By comparing the relative configurations of the reactants and products in S_N2 reactions, it was determined that the product configuration was always opposite that of the reactant. In other words, *inversion of configuration* had taken place. For example, reaction of (S)-$(+)$-2-bromooctane with the hydroxide ion gives the 2-octanol enantiomer having a configuration *opposite* to that of the starting alkyl halide.

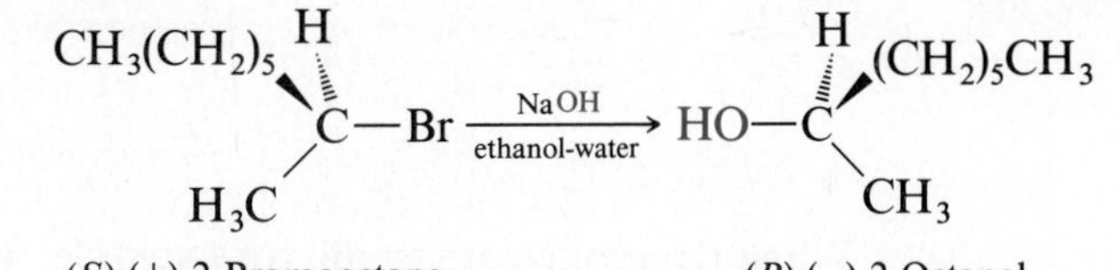

$$CH_3(CH_2)_5 \overset{H}{\underset{H_3C}{\overset{|}{C}}}{-}Br \xrightarrow[\text{ethanol-water}]{NaOH} HO{-}\overset{H}{\underset{CH_3}{\overset{(CH_2)_5CH_3}{C}}}$$

(S)-$(+)$-2-Bromooctane (R)-$(-)$-2-Octanol

Nucleophilic substitution in this case had occurred with inversion of configuration, consistent with the transition state representation shown.

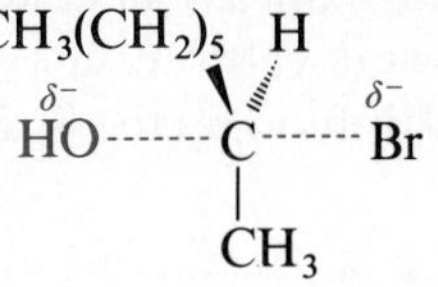

$$\overset{CH_3(CH_2)_5}{\underset{CH_3}{\overset{\delta^-}{HO}\cdots\overset{H}{\underset{|}{C}}\cdots\overset{\delta^-}{Br}}}$$

PROBLEM 8.5

Would you expect the 2-octanol formed by S_N2 hydrolysis of $(-)$-2-bromooctane to be optically active? If so, what will be its absolute configuration and sign of rotation? What about the 2-octanol formed by hydrolysis of racemic 2-bromooctane?

It is generally true of S_N2 reactions that nucleophiles attack carbon from the side opposite the bond to the leaving group.

8.7 STERIC EFFECTS IN S_N2 REACTIONS

In Section 8.3 you saw that alkyl halides undergo nucleophilic substitution reactions at different rates if the halide is changed; alkyl iodides react the fastest and alkyl fluorides the slowest. The rate of an S_N2 reaction also depends on the degree of substitution at the carbon that bears the leaving group.

In general, S_N2 reactions occur fastest when the carbon that bears the leaving group is least crowded. The crowding around a carbon by bulky groups is called **steric hindrance.** We may summarize this dependence of rate on substrate structure as follows:

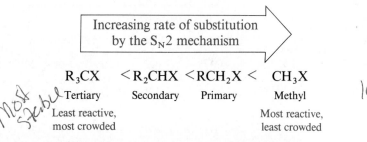

This trend is easily explained by our picture of how an S_N2 reaction occur shown in Figure 8.2. Alkyl substituents directly attached to the carbon that b leaving group block approach of the nucleophile and slow the rate of s

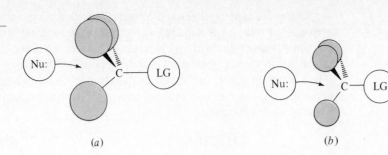

(a) (b)

FIGURE 8.2 (*a*) Bulky groups hinder attack by a nucleophile on a tertiary alkyl halide. (*b*) Smaller groups such as hydrogen do not hinder attack by the nucleophile.

When the groups are small, for example, when two hydrogens are attached as in a primary alkyl halide, no such hindrance to attack is present.

As a general statement, the **least sterically hindered (least crowded) alkyl halides react the fastest** in an S_N2 process. All substitutions of methyl and primary halides by anionic nucleophiles occur by the S_N2 mechanism. Most substitutions of secondary substrates also occur by an S_N2 process, but tertiary substrates are practically inert to substitution by the S_N2 mechanism.

PROBLEM 8.6

Identify the compound in each of the following pairs that reacts with sodium iodide in acetone by the S_N2 mechanism at the faster rate:

(*a*) 1-Chlorohexane or cyclohexyl chloride
(*b*) 1-Bromopentane or 3-bromopentane
(*c*) 2-Chloropentane or 2-fluoropentane
(*d*) 2-Bromo-2-methylhexane or 2-bromo-5-methylhexane

SAMPLE SOLUTION

(*a*) Compare the structures of the two chlorides. 1-Chlorohexane is a primary alkyl chloride; cyclohexyl chloride is secondary. Primary alkyl halides are less crowded at the site of substitution than secondary ones and react faster in substitution by the S_N2 mechanism. 1-Chlorohexane is more reactive.

$CH_3CH_2CH_2CH_2CH_2CH_2Cl$

1-Chlorohexane
(primary, more reactive)

Cyclohexyl chloride
(secondary, less reactive)

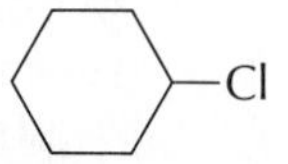

Before proceeding, let us summarize what we know about S_N2 reactions:

Kinetics:	Second-order; therefore we infer the reaction is bimolecular
Stereochemistry:	Inversion of configuration at the carbon which bears the leaving group
Substrate reactivity:	Least crowded substrates (methyl, primary) are most reactive

As you will see in the next several sections, a second type of nucleophilic substitution called S_N1 exists which differs from the S_N2 mechanism in a number of ways.

8.8 THE UNIMOLECULAR (S_N1) MECHANISM OF NUCLEOPHILIC SUBSTITUTION

As we have just seen, tertiary alkyl halides are practically inert to substitution by the S_N2 mechanism because of steric hindrance at the reaction site. You might wonder if tertiary alkyl halides undergo nucleophilic substitution at all. You will see in this section that they do, but by a mechanism different from S_N2.

tert-Butyl bromide reacts with water to form *tert*-butyl alcohol.

$$(CH_3)_3CBr + H_2O \longrightarrow (CH_3)_3COH + \quad HBr$$

| *tert*-Butyl bromide | Water | *tert*-Butyl alcohol | Hydrogen bromide |

Unlike reactions of primary and secondary alkyl halides, the rate of this reaction does *not* depend on the concentration of the nucleophile but is proportional *only* to the concentration of the substrate *tert*-butyl bromide.

$$\text{Rate} = k[\text{substrate}]$$

Because the reaction rate depends on only one of the reactants, the reaction is said to exhibit first-order kinetics and is assumed to be a *unimolecular process* (Section 4.1.11). The symbol for a **unimolecular nucleophilic substitution is S_N1.** Elimination reactions that proceed by the E1 mechanism are also unimolecular processes (Section 4.1.11).

The rate of S_N1 reactions in general depends only on the concentration of the substrate. If the nucleophile concentration does not affect the rate of the substitution reaction, how then does the substitution process occur? An S_N1 reaction is a two-step process. The first step is the **slow step** of the reaction. The word *slow* is used only in a relative sense, comparing the rate of the first reaction step to that of the second. The reaction can be very fast and be over in a few seconds, yet it will still have a slow step. Since a reaction can only proceed as fast as the slowest step, the slow step of a process is also called the **rate-determining step.**

The first step of an S_N1 reaction is dissociation of the alkyl halide to form a *carbocation intermediate* (Section 3.12).

$$(CH_3)_3C \overset{\frown}{-} Br \underset{}{\overset{\text{slow}}{\rightleftharpoons}} (CH_3)_3C^+ + \quad Br^-$$

| *tert*-Butyl bromide | *tert*-Butyl cation | Bromide ion |

Once generated, the carbocation is rapidly captured by water.

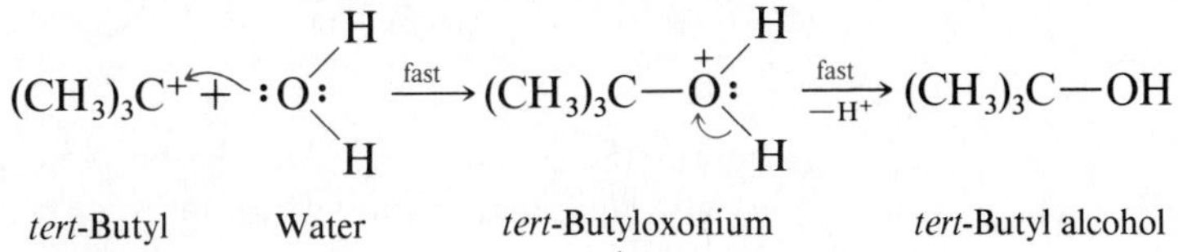

$$(CH_3)_3C^+ \overset{\frown}{+} :\overset{\displaystyle H}{\underset{\displaystyle H}{O}}: \xrightarrow{\text{fast}} (CH_3)_3C - \overset{\displaystyle H}{\underset{\displaystyle H}{\overset{+}{O}}}: \xrightarrow[-H^+]{\text{fast}} (CH_3)_3C - OH$$

| *tert*-Butyl cation | Water | *tert*-Butyloxonium ion | *tert*-Butyl alcohol |

Because the nucleophile is a solvent molecule, H_2O, this process is called a **solvolysis reaction.** When the solvent is water, as in this case, the reaction is known as **hydrolysis.**

We may illustrate the energy changes during an S_N1 process with the potential energy diagram shown in Figure 8.3. You can see that the shape is quite different than the energy diagram for an S_N2 reaction (Figure 8.1). Each step of the S_N1 reaction

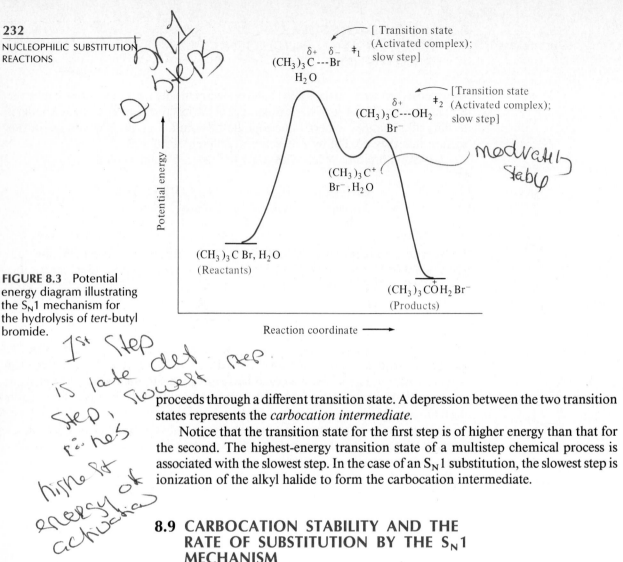

[handwritten annotations: "SN1", "2 step", "1st step is late det step, rates higher energy of activation", "flowest step", "moderately stable"]

FIGURE 8.3 Potential energy diagram illustrating the S_N1 mechanism for the hydrolysis of *tert*-butyl bromide.

proceeds through a different transition state. A depression between the two transition states represents the *carbocation intermediate.*

Notice that the transition state for the first step is of higher energy than that for the second. The highest-energy transition state of a multistep chemical process is associated with the slowest step. In the case of an S_N1 substitution, the slowest step is ionization of the alkyl halide to form the carbocation intermediate.

8.9 CARBOCATION STABILITY AND THE RATE OF SUBSTITUTION BY THE S_N1 MECHANISM

The stability of the intermediate carbocation is the most important factor in determining the rate of an S_N1 substitution reaction. A more stable carbocation is formed faster than a less stable one. In Chapter 3 you learned that the stability order for carbocations (Section 3.12) is

$$\overset{+}{C}H_3 \; < R\overset{+}{C}H_2 < R_2\overset{+}{C}H < \; R_3\overset{+}{C}$$

Least stable   Most stable

The trend in S_N1 reactivity is exactly the same as that for carbocation stability. Thus tertiary alkyl halides are good candidates for reaction by the S_N1 mechanism because they form stable tertiary carbocations. Tertiary alkyl halides are the least reactive by the S_N2 mechanism because they are the most sterically crowded. Substrates which are the most reactive by the S_N1 process are least reactive by S_N2, and vice versa.

S_N1 reactivity: Methyl < primary < secondary < tertiary
S_N2 reactivity: Tertiary < secondary < primary < methyl

Least   Most
reactive   reactive

PROBLEM 8.7

Identify the compound in each of the following pairs that reacts at the faster rate in an S_N1 reaction:

(a) Isopropyl bromide or isobutyl bromide (1-bromo-2-methylpropane)
(b) Cyclopentyl iodide or 1-methylcyclopentyl iodide
(c) Cyclopentyl bromide or 1-bromo-2,2-dimethylpropane
(d) *tert*-Butyl chloride or *tert*-butyl iodide

SAMPLE SOLUTION

(a) Isopropyl bromide, $(CH_3)_2CHBr$, is a secondary alkyl halide, while isobutyl bromide, $(CH_3)_2CHCH_2Br$, is primary. Since the rate-determining step in an S_N1 reaction is carbocation formation and since secondary carbocations are more stable than primary carbocations, isopropyl bromide is more reactive in nucleophilic substitution by the S_N1 mechanism than is isobutyl bromide.

Compounds which are capable of forming an allylic carbocation (Section 5.7) or a benzylic carbocation (Section 6.14), both of which are resonance-stabilized, undergo substitution by the S_N1 mechanism.

$$CH_2{=}CH{-}\overset{+}{C}H_2 \qquad \text{Allyl cation} \qquad \text{Benzyl cation}$$

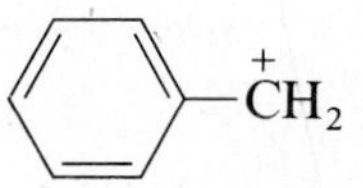

Allyl cation   Benzyl cation

Hydrolysis of 3-chloro-3-methyl-1-butene, for example, proceeds rapidly and gives a mixture of two allylic alcohols.

$$(CH_3)_2\underset{\underset{Cl}{|}}{C}CH{=}CH_2 \xrightarrow[Na_2CO_3]{H_2O} (CH_3)_2\underset{\underset{OH}{|}}{C}CH{=}CH_2 + (CH_3)_2C{=}CHCH_2OH$$

3-Chloro-3-
methyl-1-butene

2-Methyl-3-buten-
2-ol (85%)

3-Methyl-2-buten-
1-ol (15%)

Both alcohols are formed from the same carbocation. Water may react with the carbocation to give either a primary alcohol or a secondary alcohol.

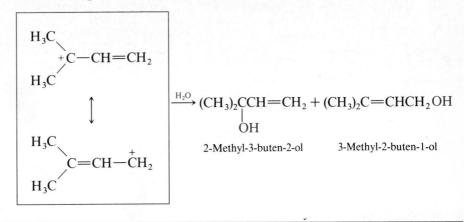

PROBLEM 8.8

A second compound of molecular formula C_5H_9Cl undergoes hydrolysis to give the *same* mixture of alcohols as the preceding equation. Write the structure of this compound.

8.10 STEREOCHEMISTRY OF S_N1 REACTIONS

While bimolecular nucleophilic substitution reactions (S_N2) proceed with complete inversion of configuration at the substrate carbon (Section 8.6), the situation is somewhat less clear-cut for unimolecular (S_N1) substitutions. When the leaving group is attached to the chiral center of an optically active halide, ionization leads to a carbocation intermediate that is *achiral.* It is achiral because the three bonds to the positively charged carbon lie in the same plane, and this plane is a plane of symmetry for the carbocation. Recall from the previous chapter (Section 7.5) that a molecule (or intermediate as in this case) having a plane of symmetry is achiral.

Figure 8.4 shows how a symmetrical carbocation should react with a nucleophile to form a product which is a racemic mixture (Section 7.6). Attack at the top face of the carbocation should proceed at the same rate as attack from the bottom. Attack from the top face gives rise to *inversion of configuration,* as the product has the opposite configuration as the starting material. Attack from the bottom face leads to *retention of configuration.* We would expect the product of substitution by the S_N1 mechanism to be formed as a racemic mixture and to be optically inactive. This outcome is rarely observed in practice, however.

As an example, consider the hydrolysis of 2-bromooctane in aqueous ethanol, a reaction that proceeds by the S_N1 mechanism. When optically active substrate is used, the product 2-octanol is formed with 66 percent inversion of configuration.

$$CH_3(CH_2)_5 \overset{CH_3 \;\; H}{\underset{}{C}}{-}Br \xrightarrow[\text{ethanol}]{H_2O} HO{-}\overset{H\;\;CH_3}{\underset{(CH_2)_5CH_3}{C}} + CH_3(CH_2)_5\overset{CH_3\;\;H}{\underset{}{C}}{-}OH$$

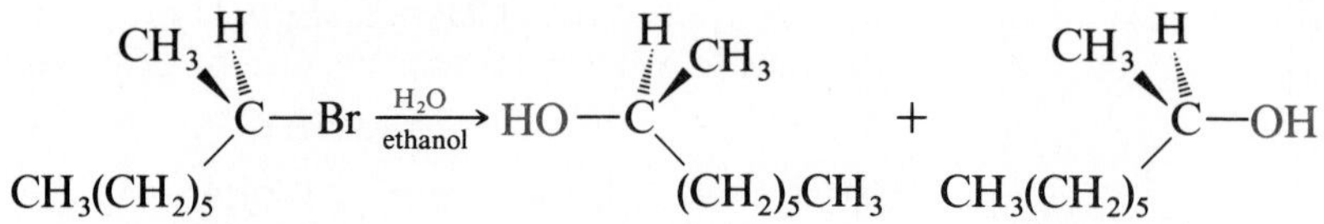

(R)-(−)-2-Bromooctane   (S)-(+)-2-Octanol   (R)-(−)-2-Octanol

66% net inversion corresponds
to 83% *S*, 17% *R*.

As a general rule, an S_N1 reaction of an optically active substrate proceeds with *partial loss of optical activity;* that is, the products are partially racemic. Inversion of configuration predominates, but it is accompanied by some retention as well.

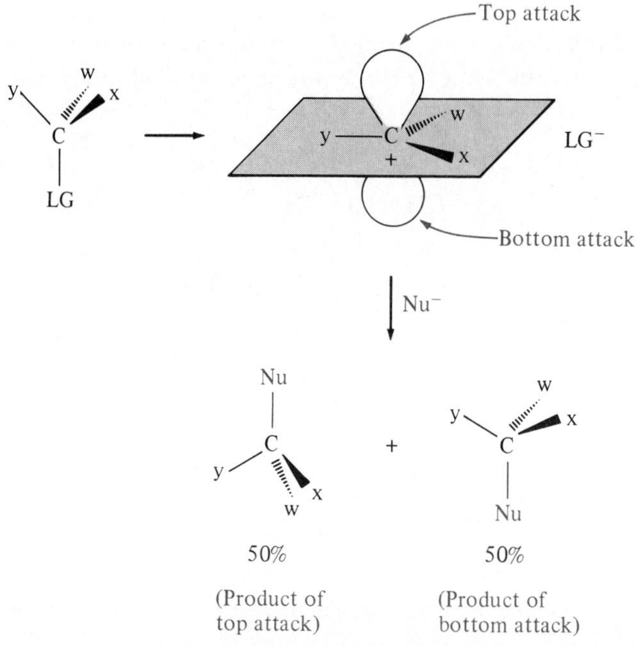

FIGURE 8.4 Formation of a racemic mixture by nucleophilic substitution via a carbocation intermediate.

Top attack

Bottom attack

Nu^-

LG^-

50%

(Product of top attack)

50%

(Product of bottom attack)

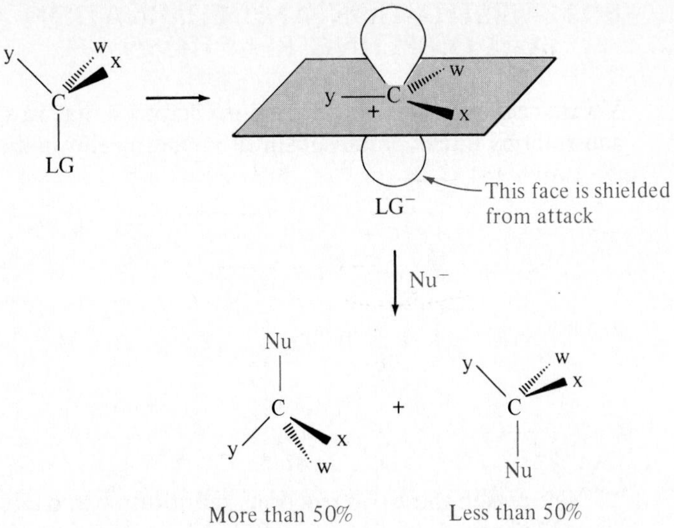

FIGURE 8.5 Inversion of configuration predominates in S$_N$1 reactions because one face of the carbocation is shielded by the leaving group.

We can explain this general result by recognizing that the carbocation intermediate is not completely free of its halide ion counterpart when it is attacked by the nucleophile. As depicted in Figure 8.5, the halide ion leaving group shields one face of the carbocation, making attack by the nucleophile on the opposite face more likely.

To summarize, S$_N$1 reactions are substitutions characterized by:

Kinetics:	First-order; we infer the reaction is unimolecular
Stereochemistry:	Partial (occasionally complete) racemization
Substrate reactivity:	Substrates forming most stable carbocation (tertiary) are most reactive

A comparison of S$_N$1 and S$_N$2 reactions is summarized in Table 8.2.

TABLE 8.2
Comparison of S$_N$1 and S$_N$2 Reactions

S$_N$1 reactions	Factor	S$_N$2 reactions
First-order; rate depends on concentration of substrate	Kinetics	Second-order; rate depends on concentration of substrate and nucleophile
Partial (occasionally complete) racemization	Stereochemistry	Complete inversion of configuration
Alkyl halides which form most stable carbocations (tertiary) are most reactive	Substrate reactivity	Least crowded alkyl halides (methyl, primary) are most reactive
Unimolecular process; substrate dissociates in slow step to form carbocation intermediate; nucleophile captures intermediate in second, fast step	Mechanism	Bimolecular process, nucleophile displaces leaving group by attack at side opposite to bond to leaving group in single, concerted reaction step

8.11 SUBSTITUTION AND ELIMINATION AS COMPETING REACTIONS

You have seen in this chapter and in Chapter 4 that an alkyl halide and a Lewis base can react by nucleophilic substitution and by elimination.

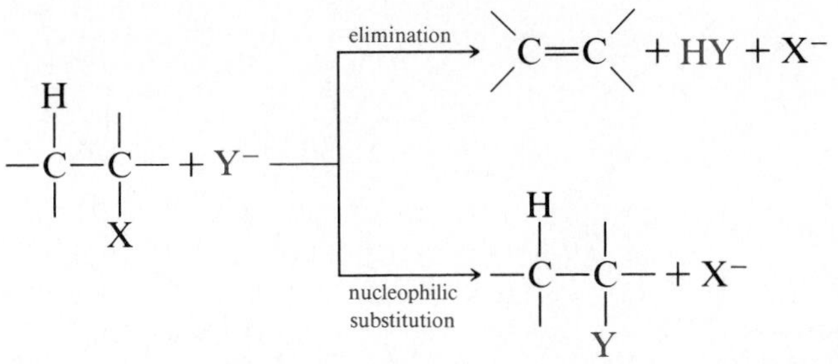

The reason these two reactions, substitution and elimination, compete with one another is that a Lewis base can act as a *nucleophile* or as a *base* in a chemical reaction.

Nucleophilicity (nucleophilic strength) and **basicity** (base strength) are related to each other but do not mean the same thing. Nucleophilicity is a kinetic parameter and is a measure of how rapidly a species will react with an alkyl halide in a substitution process. Basicity is a thermodynamic parameter and refers to the ability of an electron-pair donor to abstract a proton from a suitable donor in an acid-base reaction. As a general rule, strong bases are also good nucleophiles.

How, then, can we predict whether substitution or elimination will be the principal reaction observed with a particular combination of reactants? The important factors to consider are the *structure of the alkyl halide* and the *nature of the Lewis base.*

8.11.1 Primary Alkyl Halides

The predominant reaction is substitution by the S_N2 mechanism, regardless of the basicity of the nucleophile. Thus substitution predominates even with strong bases such as hydroxide and alkoxides. For example,

$$CH_3CH_2CH_2Br \xrightarrow[\text{CH}_3\text{CH}_2\text{OH, 55°C}]{\text{NaOCH}_2\text{CH}_3} CH_3CH_2CH_2OCH_2CH_3 + CH_3CH=CH_2$$

Propyl bromide — Ethyl propyl ether (91%; product of nucleophilic substitution) — Propene (9%; product of elimination)

Increasing steric hindrance near a primary site—as in isobutyl bromide, for example—decreases the rate of substitution and brings about an increase in the proportion of elimination.

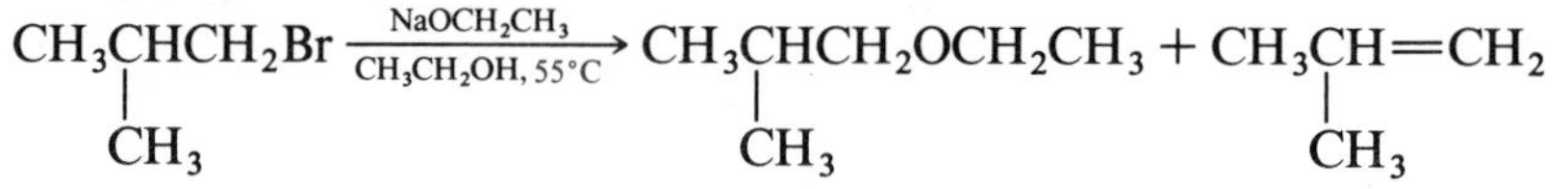

$$CH_3CHCH_2Br \xrightarrow[\text{CH}_3\text{CH}_2\text{OH, 55°C}]{\text{NaOCH}_2\text{CH}_3} CH_3CHCH_2OCH_2CH_3 + CH_3CH=CH_2$$
$$\text{CH}_3 \qquad\qquad \text{CH}_3 \qquad\qquad \text{CH}_3$$

Isobutyl bromide — Ethyl isobutyl ether (40%) — 2-Methylpropene (60%)

8.11.2 Secondary Alkyl Halides

Crowding is more pronounced in secondary substrates, making the rate of substitution by the S_N2 mechanism much slower. The result is that secondary alkyl halides yield mainly the product of elimination on reaction with alkoxide bases. Reaction by the E2 mechanism is faster than reaction by the S_N2 mechanism.

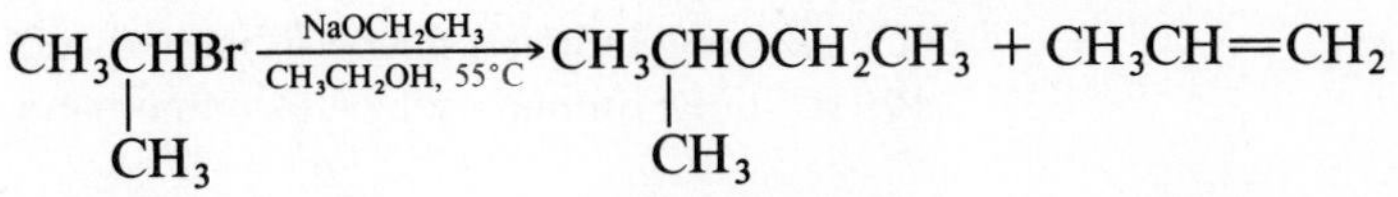

$$CH_3\underset{\underset{CH_3}{|}}{CH}Br \xrightarrow[CH_3CH_2OH,\ 55°C]{NaOCH_2CH_3} CH_3\underset{\underset{CH_3}{|}}{CH}OCH_2CH_3 + CH_3CH{=}CH_2$$

| Isopropyl bromide | Ethyl isopropyl ether (13%) | Propene (87%) |

While the E2 reaction is favored with strong bases such as alkoxides and hydroxide, anions which are *less* basic will tend to act as nucleophiles in an S_N2 reaction. Thus the cyanide ion, which is a much weaker base than hydroxide, gives mainly the product of substitution with secondary alkyl halides.

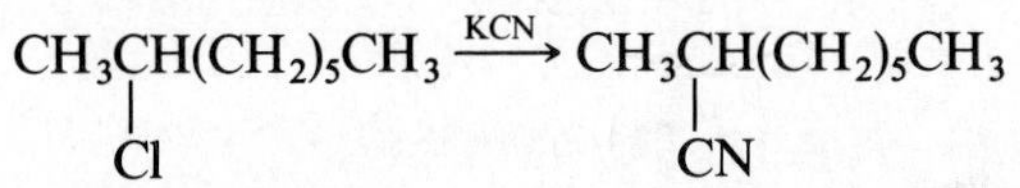

$$CH_3\underset{\underset{Cl}{|}}{CH}(CH_2)_5CH_3 \xrightarrow{KCN} CH_3\underset{\underset{CN}{|}}{CH}(CH_2)_5CH_3$$

| 2-Chlorooctane | 2-Cyanooctane (70%) |

8.11.3 Tertiary Alkyl Halides

In the presence of strong bases, substitution by the only mechanism available (S_N1) is too slow to compete with the rapid rate of E2 elimination.

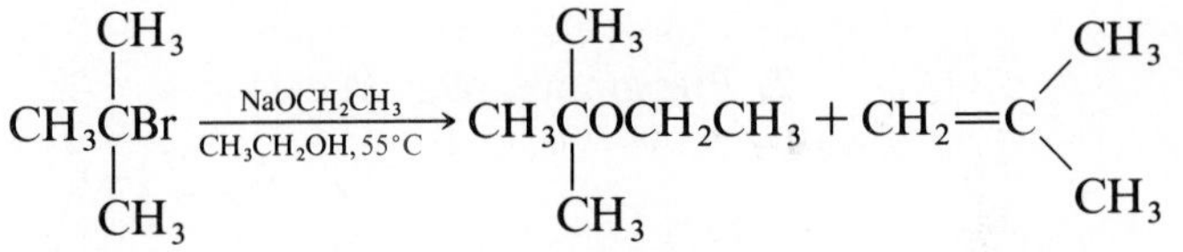

$$CH_3\underset{\underset{CH_3}{|}}{\overset{\overset{CH_3}{|}}{C}}Br \xrightarrow[CH_3CH_2OH,\ 55°C]{NaOCH_2CH_3} CH_3\underset{\underset{CH_3}{|}}{\overset{\overset{CH_3}{|}}{C}}OCH_2CH_3 + CH_2{=}C\underset{CH_3}{\overset{CH_3}{\diagup}}$$

| *tert*-Butyl bromide | *tert*-Butyl ethyl ether (7%) | 2-Methyl-propene (93%) |

Lastly, substitution by the S_N1 mechanism (solvolysis) will predominate with tertiary alkyl halides in the absence of a strong base. A typical medium for solvolysis is water, alcohol (methanol or ethanol), or a mixture of the two. In the solvolysis of the tertiary bromide 2-bromo-2-methylbutane, for example, the ratio of substitution to elimination is 64 : 36 in pure ethanol but falls to 1 : 99 in the presence of 2 *M* sodium ethoxide.

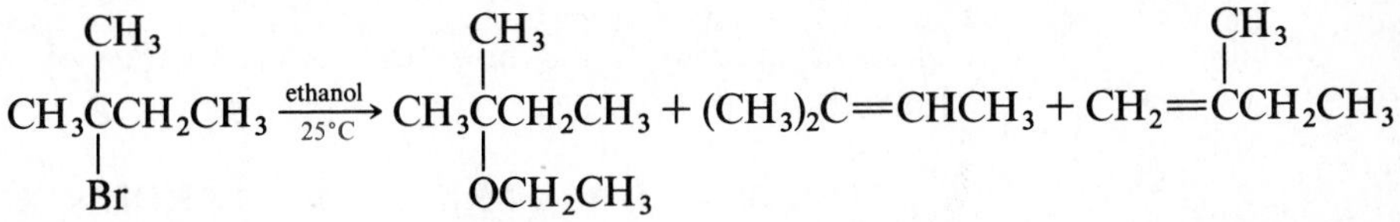

$$CH_3\underset{\underset{Br}{|}}{\overset{\overset{CH_3}{|}}{C}}CH_2CH_3 \xrightarrow[25°C]{ethanol} CH_3\underset{\underset{OCH_2CH_3}{|}}{\overset{\overset{CH_3}{|}}{C}}CH_2CH_3 + (CH_3)_2C{=}CHCH_3 + CH_2{=}\overset{\overset{CH_3}{|}}{C}CH_2CH_3$$

| 2-Bromo-2-methyl-butane | 2-Ethoxy-2-methylbutane (Major product in absence of sodium ethoxide) | 2-Methyl-2-butene | 2-Methyl-1-butene |

(Alkene mixture is major product in presence of sodium ethoxide)

PROBLEM 8.9

Predict the principal organic product of each of the following reactions:

(a) Cyclohexyl bromide and potassium ethoxide

(b) Ethyl bromide and ⬡—OK

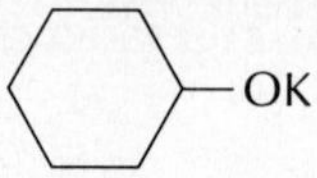

(c) *sec*-Butyl bromide solvolysis in methanol

(d) *sec*-Butyl bromide solvolysis in methanol containing 2 *M* sodium methoxide

SAMPLE SOLUTION

(a) Cyclohexyl bromide is a secondary halide and reacts with alkoxide bases by elimination rather than substitution. The principal organic products are cyclohexene and ethanol.

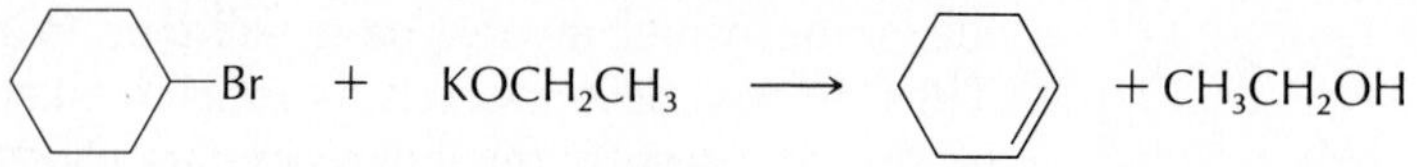

$$\text{⬡—Br} \quad + \quad \text{KOCH}_2\text{CH}_3 \quad \longrightarrow \quad \text{⬡} \quad + \text{CH}_3\text{CH}_2\text{OH}$$

| Cyclohexyl bromide | Potassium ethoxide | Cyclohexene | Ethanol |

8.12 SUMMARY

Nucleophilic substitution reactions play an important role in functional group transformations (Section 8.2). Among the synthetically useful processes accomplished by nucleophilic substitution are:

   1. Preparation of alkyl iodides.

$$RX + I^- \xrightarrow{\text{acetone}} RI + X^-$$

   2. Preparation of alkyl cyanides.

$$RX + CN^- \longrightarrow RCN + X^-$$

   3. Preparation of ethers.

$$RX + R'O^- \longrightarrow ROR' + X^-$$

   4. Preparation of thiols and thioethers.

$$RX + R'S^- \longrightarrow RSR' + X^-$$

   5. Preparation of alkynes.

$$RX + R'C\equiv C^- \longrightarrow RC\equiv CR' + X^-$$

The preparation of alcohols from alkyl halides by nucleophilic substitution is feasible,

$$RX + HO^- \longrightarrow ROH + X^-$$

but not often used because alkyl halides are typically prepared from alcohols (Chapter 3), rather than vice versa.

   Among alkyl halide substrates, the order of leaving group reactivity is (Section 8.3)

$$I^- > Br^- > Cl^- > F^-$$

Nucleophilic substitution reactions can be classified by two mechanisms:

1. **S$_N$2** *(substitution nucleophilic bimolecular)*
2. **S$_N$1** *(substitution nucleophilic unimolecular)*

The S$_N$2 mechanism (Section 8.5) is *concerted;* the reaction occurs in a single bimolecular step. The nucleophile attacks the substrate carbon from the side opposite the bond to the leaving group.

$$\overset{\delta^-}{Y}\text{-----}\overset{|}{\underset{|}{C}}\text{-----}\overset{\delta^-}{X}$$

Substitution by the S$_N$2 mechanism takes place with *inversion of configuration* (Section 8.6). The rate of S$_N$2 reactions is sensitive to steric effects (Section 8.7). The less crowded the carbon bearing the leaving group, the faster it reacts with good nucleophiles. The S$_N$2 order of reactivity for alkyl halides is

$$CH_3 > \text{primary} > \text{secondary} > \text{tertiary}$$

Most reactive — Least reactive

The S$_N$1 mechanism (Section 8.8) is a *two-step process* in which the slow, rate-determining step is carbocation formation.

$$RX \xrightarrow{\text{slow}} R^+ + X^-$$

$$R^+ + Y^- \xrightarrow{\text{fast}} RY$$

The rate of reaction is dictated by carbocation stability (Section 8.9). The more stable the carbocation, the faster RX will react by the S$_N$1 reaction. Therefore, the order of alkyl halide reactivity in an S$_N$1 reaction is

$$\text{Tertiary} > \text{secondary} > \text{primary} > CH_3$$

Most reactive — Least reactive

Alkyl halides which can form an allyl or benzyl cation upon ionization readily undergo S$_N$1 substitution.

The rate of an S$_N$1 reaction depends only on the concentration of the substrate RX. It is independent of the concentration as well as the nature of the nucleophile. The nucleophile plays no role in an S$_N$1 reaction until after the rate-determining step. The presence of a planar carbocation intermediate suggests that optically active alkyl halides should yield racemic products by the S$_N$1 mechanism. Racemization, however, is usually incomplete, and partial net inversion of configuration is normally observed (Section 8.10).

When an alkyl halide is capable of undergoing elimination, this can compete with nucleophilic substitution (Section 8.11). Substitution predominates with primary alkyl halides. Secondary alkyl halides undergo substitution effectively with bases weaker than hydroxide. The normal reaction of a secondary alkyl halide with a base as strong as or stronger than hydroxide is elimination. Elimination predominates when tertiary halides react with any anion. Solvolysis reactions in the absence of added anions cause nucelophilic substitution by the S$_N$1 mechanism to be the major reaction of tertiary alkyl halides.

ADDITIONAL PROBLEMS

Predicting Products of Substitution and Elimination Reactions

8.10 Write the structure of the principal organic product to be expected from the reaction of 1-bromopropane with each of the following:

(a) Sodium iodide in acetone

(b) Sodium acetate ($CH_3\overset{\displaystyle O}{\overset{\displaystyle \|}{C}}ONa$)

(c) Sodium ethoxide (CH_3CH_2ONa)

(d) Sodium cyanide (NaCN)

(e) Sodium azide (NaN_3)

(f) Sodium hydrogen sulfide (NaSH)

(g) Sodium methanethiolate ($NaSCH_3$)

8.11 Repeat Problem 8.10 for 2-bromopropane as the substrate.

8.12 Each of the reagents in Problem 8.10 converts 2-bromo-2-methylpropane to the same product. What is this product?

8.13 Write the structure of the major product, including stereochemistry, to be expected from reaction of (R)-2-bromopentane with each of the following reagents:

(a) Sodium iodide in acetone

(b) Sodium cyanide

(c) Solvolysis in ethanol

(d) Sodium ethoxide

8.14 Identify the product in each of the following reactions:

(a) $ClCH_2CH_2CHCH_2CH_3 \xrightarrow[\text{acetone}]{\text{NaI (1.0 equiv)}} C_5H_{10}ClI$
  with Cl substituent on the middle carbon

(b) $BrCH_2CH_2Br + NaSCH_2CH_2SNa \longrightarrow C_4H_8S_2$

(c) $ClCH_2CH_2CH_2CH_2Cl + Na_2S \longrightarrow C_4H_8S$

8.15 What two stereoisomeric substitution products would you expect from the hydrolysis of cis-1,4-dimethylcyclohexyl bromide? From trans-1,4-dimethylcyclohexyl bromide?

8.16 Each of the following nucleophilic substitution reactions has been reported in the chemical literature. Some of them involve reactants that are somewhat more complex than those we have dealt with to this point. Nevertheless, you should be able to predict the product based on analogy to what you know about nucleophilic substitution in simple systems.

(a) $BrCH_2\overset{\displaystyle O}{\overset{\displaystyle \|}{C}}OCH_2CH_3 \xrightarrow[\text{acetone}]{\text{NaI}}$

(b) $O_2N\text{—}\langle\text{—}\rangle\text{—}CH_2Cl \xrightarrow[\text{acetone}]{\text{NaI}}$

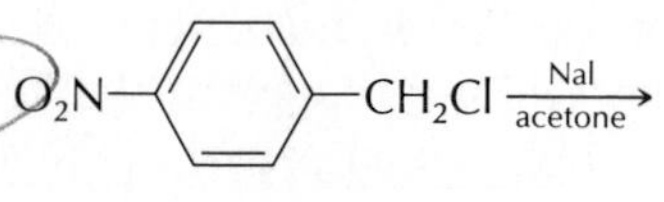

(c) $O_2N\text{—}\langle\text{—}\rangle\text{—}CH_2Cl \xrightarrow[\text{acetic acid}]{CH_3\overset{\displaystyle O}{\overset{\displaystyle \|}{C}}ONa}$

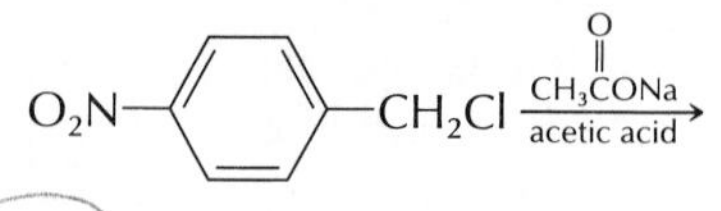

(d) $CH_3CH_2OCH_2CH_2Br \xrightarrow[\text{ethanol-water}]{\text{NaCN}}$

S_N2 and S_N1 Mechanisms

8.17 Arrange the isomers of molecular formula C_4H_9Cl in order of decreasing rate of reaction with sodium iodide in acetone by the S_N2 mechanism.

8.18 In each of the following indicate which S_N2 reaction will occur faster. Explain your reasoning.

(a) $CH_3CH_2CH_2CH_2Br$ or $CH_3CH_2CH_2CH_2I$ with sodium cyanide

(b) 1-Bromobutane or 2-bromobutane with sodium iodide in acetone

(c) 1-Chloro-2,2-dimethylbutane or 1-chlorohexane with sodium iodide in acetone

8.19 Indicate which compound from each of the following pairs will undergo S_N1 solvolysis in warm ethanol at the faster rate. Explain your reasoning.

(a) 1-Bromo-2-methylpropane or 2-bromobutane

(b) 1-Chlorocyclohexene or 3-chlorocyclohexene

(c) Chlorocyclohexane or 3-chlorocyclohexene

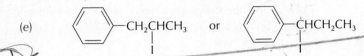

(d) CH$_3$—⟨⟩—Br   or   ⟨⟩—CH$_2$Br

(e) ⟨⟩—CH$_2$CHCH$_3$ or   ⟨⟩—CHCH$_2$CH$_3$

8.20 Give the mechanistic symbol (S_N1, S_N2, E1, or E2) most consistent with each of the following statements.

(a) Methyl halides react with sodium ethoxide in ethanol only by this mechanism.

(b) Unhindered primary halides react with sodium ethoxide in ethanol mainly by this mechanism.

(c) When cyclohexyl bromide is treated with sodium ethoxide in ethanol, the major product is formed by this mechanism.

(d) The principal substitution product obtained by solvolysis of *tert*-butyl bromide in ethanol arises by this mechanism.

(e) These reaction mechanisms proceed in a single step.

(f) This substitution mechanism involves a carbocation intermediate.

8.21 The ratio of elimination to substitution is exactly the same (26 percent elimination) for 2-bromo-2-methylbutane and 2-iodo-2-methylbutane in 80% ethanol/20% water at 25°C.

(a) By what mechanism does substitution most likely occur in these compounds under these conditions?

(b) By what mechanism does elimination most likely occur in these compounds under these conditions?

(c) Which substrate undergoes substitution faster?

(d) Which substrate undergoes elimination faster?

(e) What two substitution products are formed from each substrate?

(f) What two elimination products are formed from each substrate?

(g) Why do you suppose the ratio of elimination to substitution is the same for the two substrates?

Synthesis Using Nucleophilic Substitution Reactions

8.22 Select the combination of alkyl bromide and potassium alkoxide that would be the most effective in the syntheses of the following ethers.

(a) $CH_3OC(CH_3)_3$

(b) 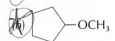—OCH$_3$

(c) (R)-$CH_3CH_2CHCH_2OCH(CH_3)_2$
$\qquad\qquad\;\; |$
$\qquad\qquad\; CH_3$

(d) $(CH_3)_3CCH_2OCH_2CH_3$

8.23 A second combination of reactants (alkyl bromide and potassium alkoxide) can be written for each of the ethers in Problem 8.22; however, for parts (a) to (c), ether formation is not the major reaction for these combinations. Write equations which show the major product formed in each of these cases.

8.24 Outline an efficient synthesis of each of the compounds shown below from the indicated starting material and any necessary organic or inorganic reagents. In most cases more than one step is required.

   (a) Ethanethiol (CH_3CH_2SH) from ethyl alcohol
   (b) Cyclopentyl cyanide from cyclopentane
   (c) Cyclopentyl cyanide from cyclopentene
   (d) Cyclopentyl cyanide from cyclopentanol
   (e) $NCCH_2CH_2CN$ from ethyl alcohol
   (f) Isobutyl iodide from isobutyl chloride

8.25 Describe the preparation of each of the following alkynes from acetylene using an alkylation reaction (Section 5.4).

   (a) 1-Hexyne
   (b) 2-Hexyne
   (c) 3-Hexyne
   (d) 4-Methyl-1-pentyne

SPECTROSCOPY

This chapter will focus on the role of spectroscopy in the determination of chemical structures. Upon completion of this chapter you should be able to:

- Explain what is meant by the wave and particle character of electromagnetic radiation.
- Explain how the frequency and wavelength of electromagnetic radiation are related.
- Identify the functional group present in a molecule on the basis of its infrared spectrum.
- Explain how conjugation affects the absorption of ultraviolet or visible light by a molecule.
- Explain what is meant by chemical shift in a proton nuclear magnetic resonance (^{1}H nmr) spectrum.
- Explain how spin-spin splitting affects the appearance of a ^{1}H nmr spectrum.
- Interpret a ^{1}H nmr spectrum in terms of:
    The number of signals
    The chemical shift of each signal
    The relative intensity of each signal
    The multiplicity of each signal
- Predict the number of signals in the ^{13}C nmr spectrum of a specific compound.
- Explain the origin of the molecular ion in a mass spectrum.

In most of the chapters of this text we have been and will be discussing the chemical and physical properties of the various classes of organic molecules. A discussion of organic chemistry would not be complete, however, without a brief mention of the tools chemists use to determine the structure of organic molecules.

Prior to the 1950s, the structure of a chemical substance was determined by using information obtained from chemical reactions. This information included the identification of functional groups by chemical tests, along with degradation experiments in which the substance was cleaved into smaller, more readily identified fragments. Typical of this approach was determining the location of the double bond in an alkene by ozonolysis (Section 5.2). After considering all the available chemical evidence, the chemist proposed a candidate structure consistent with the observations.

Qualitative tests and chemical degradation as structural probes have been to a large degree replaced in present-day organic chemistry by instrumental methods of **spectroscopic analysis.** These methods are all based on the absorption of energy by a molecule, and all examine how a molecule responds to that absorption of energy. This chapter will focus on **infrared (ir) spectroscopy, ultraviolet-visible (uv-vis) spectroscopy, nuclear magnetic resonance (nmr) spectroscopy, and mass spectrometry.** The discussion of these techniques will emphasize the role of spectroscopy in structure determination, touching only briefly on the theoretical foundations of the methods.

9.1 ELECTROMAGNETIC RADIATION

Our discussion of spectroscopy begins with a brief description of electromagnetic radiation. **Electromagnetic radiation** includes forms of radiation such as visible light, radiowaves, microwaves, and x-rays. Electromagnetic radiation is said to have a dual nature; it has the properties of both waves and particles. While some of the properties of electromagnetic radiation are best understood by considering radiation as a stream of particles, other properties are best described by viewing radiation as a wave traveling at the speed of light.

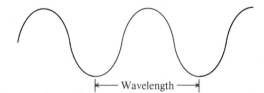

The **wavelength** λ (lambda) is the length of one complete cycle measured from any point on one wave to the same point on the next wave. It is most easily seen as the distance between two successive crests or two successive troughs. The **frequency** v (nu) of a wave is the number of waves that passes a given point per unit of time, usually seconds(s). The SI unit of frequency is s^{-1}, sometimes cited as **hertz, Hz** and informally spoken of as "cycles per second."

The wavelength and frequency of a light wave are related by the equation

$$\lambda \times v = c$$

where c is the speed of light ($c = 3.0 \times 10^8$ m/s). Thus the **frequency of radiation is inversely proportional to its wavelength.** That is, the shorter the wavelength, the greater the frequency, and vice versa.

PROBLEM 9.1

(a) Violet light has a wavelength of 4×10^{-7} m. Calculate the frequency of this light.
(b) Red light has a wavelength of 8×10^{-7} m. Is the frequency of red light greater or less than violet light? Calculate the frequency of red light.

SAMPLE SOLUTION
(a) By rearranging the equation which relates the wavelength and frequency of radiation, frequency is given by

$$v = \frac{c}{\lambda}$$

Substituting the value for the wavelength of violet light,

$$v = \frac{3.0 \times 10^8 \text{ m/s}}{4 \times 10^{-7} \text{ m}}$$

$$= 7.5 \times 10^{14} \text{ s}^{-1}$$

The frequency of violet light is 7.5×10^{14} Hz.

Electromagnetic radiation also has the properties of particles. The particles of light are called *photons.* In 1900 the German physicist Max Planck related the energy of a photon to its frequency. The *energy of a photon is directly proportional to the frequency of the radiation.* Mathematically this relationship can be expressed as

$$E = hv$$

where h represents the constant of proportionality, called *Planck's constant.* The relationship between frequency and energy means that radiation of higher frequency possesses more energy than lower-frequency radiation.

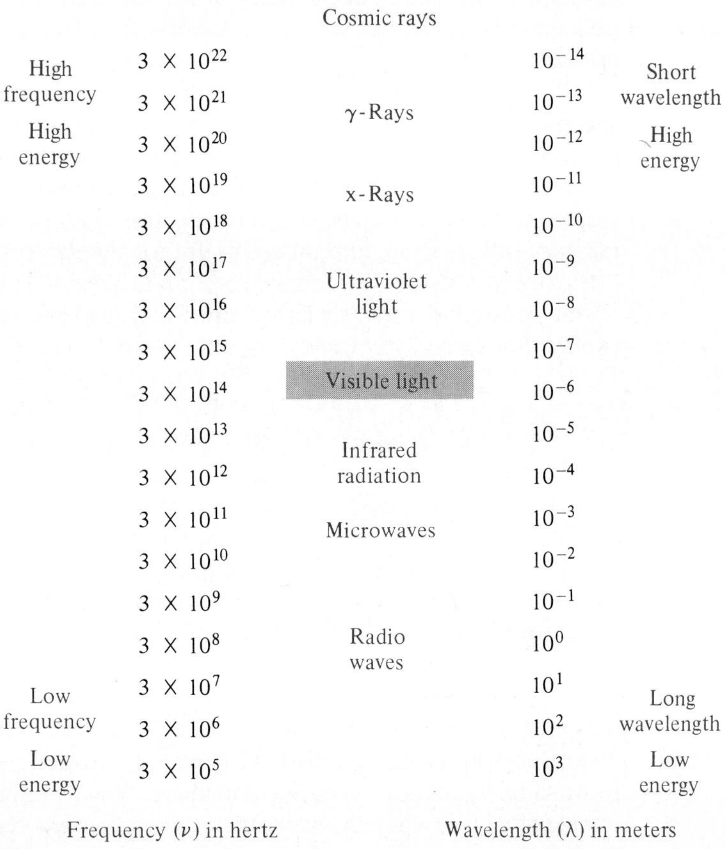

	Frequency (v) in hertz		Wavelength (λ) in meters	
		Cosmic rays		
High frequency	3×10^{22}		10^{-14}	Short wavelength
	3×10^{21}	γ-Rays	10^{-13}	
High energy	3×10^{20}		10^{-12}	High energy
	3×10^{19}	x-Rays	10^{-11}	
	3×10^{18}		10^{-10}	
	3×10^{17}	Ultraviolet light	10^{-9}	
	3×10^{16}		10^{-8}	
	3×10^{15}		10^{-7}	
	3×10^{14}	Visible light	10^{-6}	
	3×10^{13}		10^{-5}	
	3×10^{12}	Infrared radiation	10^{-4}	
	3×10^{11}		10^{-3}	
	3×10^{10}	Microwaves	10^{-2}	
	3×10^9		10^{-1}	
	3×10^8	Radio waves	10^0	
Low frequency	3×10^7		10^1	Long wavelength
	3×10^6		10^2	
Low energy	3×10^5		10^3	Low energy

FIGURE 9.1 The electromagnetic spectrum.

PROBLEM 9.2
Which is of higher energy, a photon of violet light or a photon of red light? (Refer to
Problem 9.1 for the wavelengths of red and violet light.)

Figure 9.1 illustrates the range of photon energies, called the **electromagnetic spectrum.** Depending on its source, a photon can have a vast amount of energy. Cosmic rays and x-rays are streams of very high energy photons. Ultraviolet radiation present in sunlight has enough energy to damage the skin, and excessive exposure to intense sunlight is a cause of certain forms of skin cancer. Radio waves, on the other hand, are of low energy. We are bombarded by them constantly with no ill effects.

9.2 GENERAL PRINCIPLES OF SPECTROSCOPIC ANALYSIS

When the molecules of a chemical compound are exposed to electromagnetic radiation, they may absorb photons of radiation energy. Molecules of different structure are selective with respect to the frequencies of radiation they absorb. The particular energies absorbed by a molecule provide a sensitive indicator of its structure. Detecting the absorption of radiation energy by molecules is the basis of *spectroscopic analysis.*

An instrument called a **spectrometer,** or **spectrophotometer,** measures a sample's absorption of electromagnetic radiation. A sample is placed in a compartment through which the radiation passes. The absorption of electromagnetic radiation by the sample is sensed by a detector as the frequency of radiation is varied. The intensity of radiation measured at the detector is compared with that at the source. When a frequency is reached at which the sample absorbs radiation, the detector senses a decrease in intensity.

The graphical representation of the molecule's absorption pattern is called a **spectrum.** A spectrum consists of a series of peaks at particular frequencies; its interpretation can provide information about the structure of the sample molecule.

With this general background, let us examine briefly three important spectroscopic methods: ir, uv-vis, and nmr spectroscopy. Each method provides complementary information, and all are useful for the determination of molecular structures. We will discuss nmr spectroscopy at greater length than ir or uv-vis, as it normally provides more structural information to an organic chemist than the other two methods.

9.3 INFRARED SPECTROSCOPY

Prior to the introduction of nmr spectroscopy, infrared (ir) spectroscopy was the instrumental method most often used for structure determination of organic compounds. IR still retains an important place in the chemist's inventory because of its usefulness in *identifying the presence of certain functional groups within a molecule.*

Infrared radiation comprises the portion of the electromagnetic spectrum (Figure 9.1) in which wavelengths range from approximately 10^{-4} to 10^{-6} m; it lies between microwaves and visible light. The portion of the infrared region most useful for structural determination lies between 2.5 μm (micrometer; 1 μm $= 10^{-6}$ m) and 16 μm, and has been expanded in Figure 9.2. Infrared spectra are generally recorded in units of frequency called **wavenumbers.** A wavenumber is the reciprocal of the wavelength $(1/\lambda)$ when the wavelength is expressed in centimeters. Thus the units of

2.5 × 10⁻⁶ m
2.5 μm
4000 cm⁻¹

16 × 10⁻⁶ m
16 μm
625 cm⁻¹

High-energy end

Low-energy end

wavenumbers are cm⁻¹. Wavenumbers are directly proportional to energy, and infrared spectra span the range from 4000 to 625 cm⁻¹. Thus 4000 cm⁻¹ is the high-energy end of the scale, and 625 cm⁻¹ is the low-energy end.

When a molecule absorbs a photon of infrared radiation, its energy increases, and this increased energy can be considered to be associated with bond vibrations, especially vibrations involving the *stretching and bending of bonds.* The vibrations for hydrogen cyanide, HCN, are shown in Figure 9.3.

How can the absorption of infrared radiation provide information about the structure of an unknown substance? Most common functional groups absorb infrared energy at characteristic frequencies in the infrared spectrum. The presence of a peak at a particular frequency can be used to determine whether an unknown molecule contains a certain functional group.

Stretching vibrations of many functional groups including C—H, O—H, and C=O are found in the region from 4000 to 1600 cm⁻¹. Table 9.1 lists the frequencies (in wavenumbers) of the characteristic absorptions of several functional groups commonly found in organic compounds.

A typical infrared spectrum, that of the ketone 2-hexanone, is shown in Figure 9.4. The spectrum appears as a series of absorption peaks (pointing down on the page) of varying shape and intensity. The data in Table 9.1 reveal that the absorption by a (C=O) group of a ketone appears in the range of 1710 to 1750 cm⁻¹, and we therefore assign the strong absorption at 1720 cm⁻¹ to the carbonyl group. Carbonyl groups are among the structural units most readily identified by infrared spectroscopy.

Although the infrared spectrum of an organic compound usually contains a number of peaks, frequently only the peak associated with the functional group is assigned. One other area of the spectrum of 2-hexanone is worth noting, however. The group of peaks around 3000 cm⁻¹ is seen in most organic molecules, because it is in this region that absorption due to carbon-hydrogen stretching occurs.

The region of an infrared spectrum from 1400 to 625 cm⁻¹ is where the variations from compound to compound are the greatest. This part of the spectrum is called the **fingerprint region,** as no two chemical compounds have the same exact pattern of peaks, just as no two persons have identical fingerprint patterns. The peaks in the fingerprint region are often difficult to assign, but the pattern they form can be useful when comparing the infrared spectrum of a sample to that of a reference compound in the same way that criminals are identified by fingerprint comparisons.

Figure 9.5 is the infrared spectrum of a second compound, 2-hexanol. The most noteworthy feature of this spectrum is the broad band centered at 3400 cm⁻¹ due to the hydroxyl (O—H) functional group.

Other than by peak-by-peak comparison with a reference spectrum, there is no reliable way to determine the complete structure of an unknown substance from its

H — C ≡ N

H — C ≡ N

H — C ≡ N

FIGURE 9.3 Stretching and bending vibrations of hydrogen cyanide, HCN.

C — H stretch:
3312 cm⁻¹

C — H bend:
712 cm⁻¹

C ≡ N stretch:
2089 cm⁻¹

TABLE 9.1

Infrared Absorption Frequencies of Some Common Structural Units

Structural unit	Frequency, cm^{-1}	Structural unit	Frequency, cm^{-1}
Single bonds		**Double bonds**	
Stretching vibrations:			
—O—H (alcohols)	3200–3600	$\diagdown$C=C$\diagup$	1620–1680
—O—H (carboxylic acids)	2500–3600 3350–3500	$\diagdown$C=O (carbonyl compounds):	
$\diagdown$N—H			
		Aldehydes and ketones	1710–1750
sp C—H	3310–3320	Carboxylic acids	1700–1725
sp^2 C—H	3000–3100	Acid anhydrides	1800–1850 and 1740–1790
sp^3 C—H	2850–2950	Acyl halides	1770–1815
		Esters	1730–1750
sp^2 C—O	1200	Amides	1680–1700
sp^3 C—O	1025–1200		
		Triple bonds	
		—C≡C—	2100–2200
		—C≡N	2240–2280

infrared spectrum alone. Later in this chapter you will see how nuclear magnetic resonance spectroscopy assists in this task. As stated earlier, the primary information that can be gained from an infrared spectrum is an indication of the functional group or groups present in a molecule.

PROBLEM 9.3

Which one of the following compounds is most consistent with the infrared spectrum given in Figure 9.6?

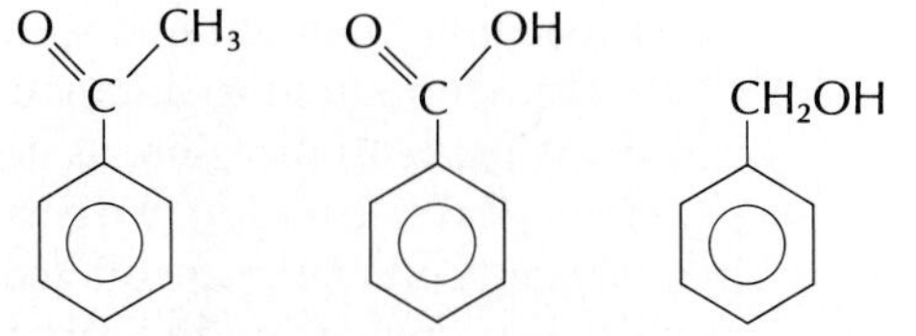

Acetophenone   Benzoic acid   Benzyl alcohol

9.4 ULTRAVIOLET-VISIBLE (UV-VIS) SPECTROSCOPY

The portion of the electromagnetic spectrum (Figure 9.1) that lies just beyond the infrared region is visible light. The frequency range of visible light is between 12,500 and 25,000 cm^{-1}. As wave numbers are directly proportional to energy, visible light is approximately 10 times more energetic than infrared radiation. Red light is the low-energy end of the visible region; violet light is the high-energy end. Ultraviolet radiation lies beyond the violet end of visible light; it encompasses the region from

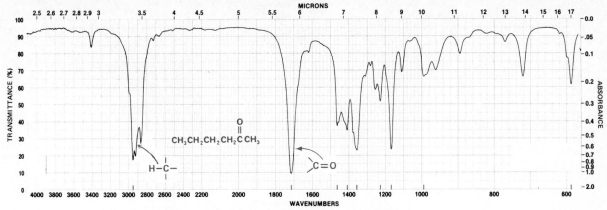

FIGURE 9.4   The infrared spectrum of 2-hexanone.

25,000 to 50,000 cm^{-1}. Positions of absorption in ultraviolet-visible spectroscopy are usually expressed as wavelengths in **nanometers, nm** (1 nm = 10^{-9} m). Thus, the visible region corresponds to 800 to 400 nm and the ultraviolet region to 400 to 200 nm.

Electron excitation occurs when a molecule absorbs a photon having energy in the visible or ultraviolet region. An electron in the molecule undergoes a transition from the usual ground state to the next highest energy vacant orbital, an excited state. Thus uv-vis spectroscopy probes the electron distribution in a molecule and is particularly useful when conjugated π-electron systems are present.

Figure 9.7 shows the ultraviolet spectrum of a conjugated diene, *cis,trans*-1,3-cyclooctadiene. Note there are fewer features than in an infrared spectrum. The peak corresponds to the excitation of a π electron of the conjugated diene system. As is typical of most uv spectra, the absorption peak is rather broad. The wavelength at which absorption is a maximum is referred to as the λ_{max} of the sample. In this case λ_{max} is 230 nm.

Information about the π-electron framework of a molecule can be learned from both the λ_{max} and the intensity of the absorption peak. The peak intensity is expressed by a unit called *molar absorptivity* or *extinction coefficient,* abbreviated ϵ_{max}. Table 9.2 lists the λ_{max} and ϵ_{max} for ethylene, a conjugated diene, a conjugated triene, and a conjugated polyene. The data show that as the number of double bonds in conjugation with one another increases, λ_{max} shifts to a longer wavelength and ϵ_{max} increases.

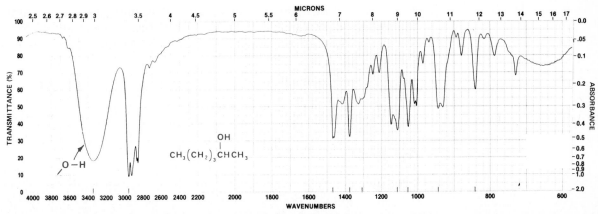

FIGURE 9.5   The infrared spectrum of 2-hexanol.

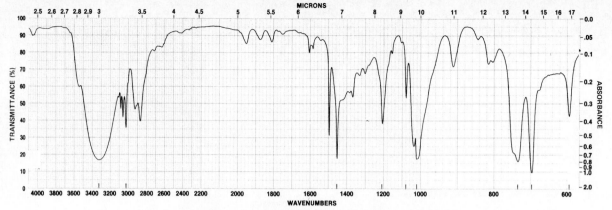

FIGURE 9.6 The infrared spectrum of the unknown compound in Problem 9.3.

Chromophores are the structural units of conjugated compounds that absorb electromagnetic radiation in the ultraviolet or visible region. The chromophore is responsible for the color in certain compounds. Lycopene, for example, contributes to the red color of tomatoes and paprika. The chromophore in lycopene is a conjugated system of 11 double bonds. Lycopene absorbs the blue-green fraction of visible light (λ_{max} = 505 nm) and appears red.

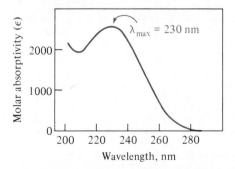

The concept of chromophores is at the heart of the chemistry of dyes. Dyes isolated from plant materials have been known for thousands of years. The blue dye indigo (Figure 9.8*a*) was used in India as long as 4000 years ago. For many centuries clothing has been dyed red with alizarin. In this century synthetic dyes have largely replaced natural ones in clothing and processed food manufacturing. Two examples of synthetic dyes are FD&C Yellow No. 3 (a food coloring) and para red (a clothing dye), shown in Figure 9.8*b*.

9.5 NUCLEAR MAGNETIC RESONANCE SPECTROSCOPY

To the organic chemist, nmr spectroscopy is one of the most valuable tools available for determining molecular structures. You have seen in this chapter how bond vibrations are affected when a molecule absorbs infrared radiation and how absorp-

FIGURE 9.7 The ultraviolet spectrum of *cis, trans*-1,3-cyclooctadiene. *(Adapted from R. Isaksson, J. Roschester, J. Sandstrom, and L.G. Wistrand, Journal of the American Chemical Society, 1985, vol. 107, p. 4074; used by permission of the American Chemical Society.)*

TABLE 9.2

The Effect of Conjugation on λ_{max} and Peak Intensity (ϵ_{max}).

	λ_{max}, nm	ϵ_{max}
Ethylene, $CH_2=CH_2$	175	15,000
1,3-Butadiene, $CH_2=CH-CH=CH_2$	217	21,000
1,3,5-Hexatriene, $CH_2=CH-CH=CH-CH=CH_2$	258	35,000
β-Carotene (11 double bonds),	465	125,000

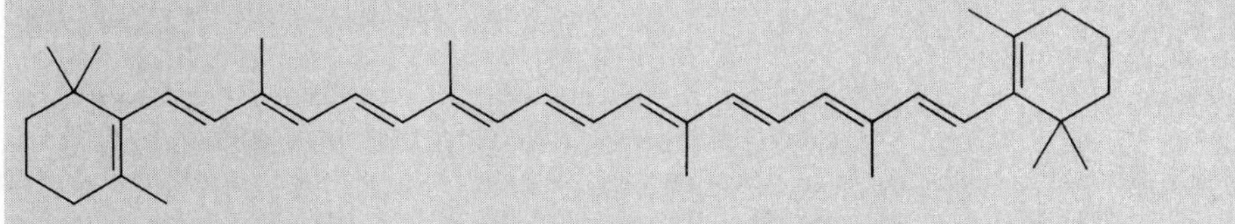

tion of visible or ultraviolet light results in electron excitation. What process is involved in the technique known as nuclear magnetic resonance (nmr) spectroscopy?

NMR spectroscopy is based on transitions between **nuclear spin states.** To understand how this can provide structural information, we must first discuss briefly the concept of nuclear spin. The nuclei of certain atoms possess a property called spin; those atoms of greatest interest to organic chemists are protium (^1H) and carbon-13 (^{13}C). The protium isotope is 99.9 percent of the natural abundance of hydrogen, while carbon-13 constitutes only about 1% of natural carbon. Thus proton nmr (^1H nmr) spectroscopy has been the type of nmr spectroscopy most widely used by organic chemists. Although our discussion of nmr spectroscopy will focus on ^1H nmr, a brief description of ^{13}C nmr spectroscopy is included in Section 9.12.

Like an electron, a proton has two (nuclear) spin states ($+\frac{1}{2}$ and $-\frac{1}{2}$) which are of equal energy. When a proton is placed in a strong magnetic field, the nuclear spin states are no longer of equal energy. For a very powerful magnet having a field strength of about 14,100 gauss (G), the energy difference between the upper and lower spin state is about 6×10^{-6} kcal/mol. This amount of energy corresponds to

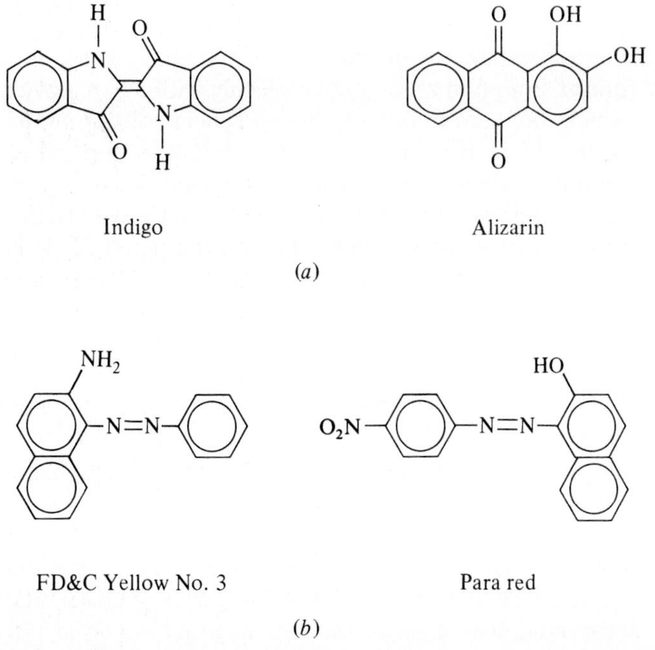

Indigo   Alizarin

(a)

FD&C Yellow No. 3   Para red

(b)

FIGURE 9.8 (a) Natural dyes indigo (blue) and alizarin (red); (b) synthetic dyes FD&C Yellow No. 3 (food coloring) and para red (clothing).

FIGURE 9.9 An external magnetic field causes the two nuclear spin states to have different energies. When the magnetic field strength H_0 is 14,100 G, the energy difference is equivalent to the energy of a photon of radiofrequency radiation (60 MHz).

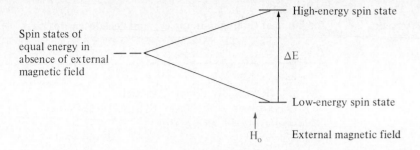

radiation having a frequency of 60×10^6 Hz (60 MHz), which is in the radio frequency region of the electromagnetic spectrum (Figure 9.1).

A nucleus in the lower-energy spin state can absorb radio frequency energy from an external source to "jump" to the higher-energy state, as shown in Figure 9.9. When the frequency of the radiation matches the energy difference between the nuclear spin states, we say the spin states and the radiation are in **resonance** with each other and the radiation is absorbed by the sample. This absorption of energy is the basis of nmr spectroscopy.

9.6 NUCLEAR SHIELDING AND CHEMICAL SHIFT

Our discussion of transitions between the two spin states of a proton has assumed that the proton was "isolated" from any neighboring atoms. In real molecules, of course, a proton is bonded to another atom by a two-electron covalent bond. These electrons, indeed all the electrons in a molecule, affect the magnetic environment of the proton. Let us examine the difference between a bare proton and a proton in an organic compound.

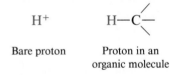

Let us compare the magnetic field strength necessary to bring about resonance for a proton bound in an organic molecule with a bare proton. In an nmr experiment in which the radio frequency energy is held constant at 60 MHz, the bare proton absorbs energy at a magnetic field strength of approximately 14,100 G. The bound proton requires a higher field strength, however. The electrons in the molecule **shield** the proton from the external magnetic field. We say that the nmr signal of a bound proton appears at a higher field than the signal of the bare proton, and we call this difference a **chemical shift**. A chemical shift is the *change in the resonance position of a nucleus which is brought about by its molecular environment.*

9.7 HOW CHEMICAL SHIFT IS MEASURED

Chemists compare the shielding of protons in organic molecules by specifying their chemical shifts relative to a standard substance. This substance is tetramethylsilane, $(CH_3)_4Si$, abbreviated as TMS. The protons of TMS are more shielded than those of almost all other organic compounds. In a solution containing TMS, all the relevant signals appear to the left of the TMS peak. The orientation of the spectrum on the chart is adjusted electronically so that the TMS peak coincides with the zero grid line.

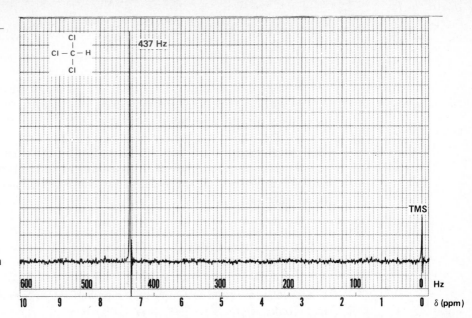

FIGURE 9.10 The proton magnetic resonance (^{1}H nmr) spectrum of chloroform ($CHCl_3$).

In other words, the chemical shift of TMS is zero. Peak positions are measured in frequency units (hertz) downfield, or to the left, of the TMS peak.

Figure 9.10 is the 60-MHz nmr spectrum of chloroform ($CHCl_3$) containing a few drops of TMS. The signal due to the proton in chloroform appears 437 Hz downfield from the TMS peak. The chemical shift of a peak, in δ (delta) units, is reported in parts per million (ppm) from the TMS peak.

$$\text{Chemical shift } (\delta) = \frac{\text{position of signal} - \text{position of TMS peak}}{\text{spectrometer frequency}} \times 10^6$$

Thus the chemical shift for the proton in chloroform is

$$\delta = \frac{437 \text{ Hz} - 0 \text{ Hz}}{60 \times 10^6 \text{ Hz}} \times 10^6 = 7.28 \text{ ppm}$$

Nuclear magnetic resonance spectra are recorded on chart paper that is calibrated in both parts per million and hertz, and both refer to the TMS peak as the zero point. All chemical shifts (δ values) in this text will be reported in parts per million.

PROBLEM 9.4

Calculate the chemical shifts in parts per million for each of the following compounds, given their shifts in hertz measured downfield from tetramethylsilane on a 60-MHz field strength spectrometer.

(a) Bromoform ($CHBr_3$), 413 Hz
(b) Iodoform (CHI_3), 322 Hz
(c) Methyl chloride (CH_3Cl), 184 Hz

SAMPLE SOLUTION

(a) The chemical shift in bromoform is calculated from the equation given above:

$$\delta = \frac{413 \text{ Hz (CHBr}_3) - 0 \text{ Hz (TMS)}}{60 \times 10^6 \text{ Hz}} \times 10^6 = 6.88 \text{ ppm}$$

Although chloroform is a good solvent for many organic compounds, it is not widely used as a medium for measuring nmr spectra; its own nmr signal could potentially obscure a signal in the sample. Chloroform-d (CDCl$_3$) is used instead. Because the magnetic properties of deuterium, the mass-2 isotope of hydrogen (D = ^{2}H), are different from those of protium (^{1}H), CDCl$_3$ does not give an nmr signal under the conditions employed for measuring proton (^{1}H) nmr spectra. Thus CDCl$_3$ exhibits no peaks in the spectrum that would obscure those of an organic molecule being examined.

9.8 CHEMICAL SHIFT AND MOLECULAR STRUCTURE

What makes nmr spectroscopy such a powerful tool for structure determination is that *protons in different chemical environments have different chemical shifts.* This is because protons in different environments experience different amounts of shielding. We can correlate the chemical shifts observed in an nmr spectrum with the kind of protons in a molecule with the aid of Table 9.3.

TABLE 9.3

Chemical Shifts of Representative Types of Protons

Type of proton	Chemical shift (δ), ppm*	Type of proton	Chemical shift (δ), ppm*
H—C—R	0.9–1.8	H—C—NR	2.2–2.9
H—C—C≡C	1.6–2.6	H—C—Cl	3.1–4.1
H—C—C(=O)—	2.1–2.5	H—C—Br	2.7–4.1
H—C≡C—	2.5	H—C—O	3.3–3.7
H—C—Ar	2.3–2.8	H—NR	1–3†
H—C=C⟨	4.5–6.5	H—OR	0.5–5†
H—Ar	6.5–8.5	H—OAr	6–8†
H—C(=O)	9–10	H—OC(=O)	10–13†

  * Approximate values relative to tetramethylsilane; other groups within the molecule can cause a proton signal to appear outside the range cited.

  † The chemical shifts of protons bonded to nitrogen and oxygen are temperature- and concentration-dependent.

9.9 INTERPRETING PROTON (^{1}H) NMR SPECTRA

We have now reached the point of being able to ask the question, "How can we use the information provided in an nmr spectrum to arrive at the structure of a compound?" There are four essential features of a ^{1}H nmr spectrum that should be considered:

1. The *number of signals*
2. The *chemical shift* (δ) of each signal
3. The *intensity of the signals* as measured by the area under each peak
4. The *multiplicity,* or number of peaks into which each signal is split

In the sections which follow you will see how each of these features of a ^{1}H nmr spectrum is used to provide structural information. Rather than beginning with the interpetation of the spectrum of an unknown compound, we will first correlate the structures of several compounds with their ^{1}H nmr spectra in terms of the features listed above.

9.9.1 Number of Signals

The number of signals in an nmr spectrum equals the *number of nonequivalent types of hydrogens in the molecule.* Recall from the previous section that nonequivalent protons experience different degrees of shielding and so have different chemical shifts.

Consider the molecule chloromethyl methyl ether as an example.

$$ClCH_2OCH_3$$

Chloromethyl methyl ether

This molecule has two different "types" of protons, that is, protons in two different environments. These are the methylene (CH_2) protons and the protons of the methyl (CH_3) group. The ^{1}H nmr spectrum of chloromethyl methyl ether is expected to have two signals, and this is confirmed by Figure 9.11. Remember that the signal at 0 ppm is due to TMS; it does not arise from the sample.

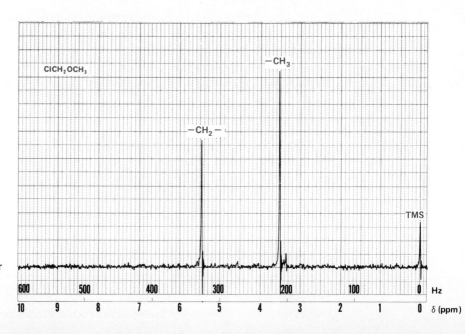

FIGURE 9.11 The ^{1}H nmr spectrum of chloromethyl methyl ether, $ClCH_2OCH_3$.

PROBLEM 9.5

Predict the number of signals that would be expected in the ^1H nmr spectrum of each of the following compounds:

(a) Ethanol, CH_3CH_2OH
(b) Diethyl ether, $CH_3CH_2OCH_2CH_3$
(c) tert-Butyl bromide, $(CH_3)_3CBr$
(d) 2-Chloropropane, CH_3CHCH_3
$\qquad\qquad\qquad\qquad\quad |$
$\qquad\qquad\qquad\qquad\quad Cl$

SAMPLE SOLUTION

(a) Ethanol has three "types" of protons: the protons of the methyl (CH_3) group, the protons of the methylene (CH_2) group, and the hydroxyl proton. The nmr spectrum of ethanol has three signals.

9.9.2 Chemical Shifts

Which peak is due to which set of protons? The approximate chemical shifts of different types of protons can be predicted by using "correlation tables" such as Table 9.3. The methyl group in chloromethyl methyl ether $ClCH_2OCH_3$ is attached to an electronegative atom (oxygen), and the methyl protons appear at $\delta = 3.5$ ppm in the spectrum. This chemical shift is consistent with the data in Table 9.3 for this type of proton (3.3 to 3.7 ppm). In general, electronegative substituents decrease the shielding of protons on the carbon to which they are attached. The methylene group is adjacent to two electronegative atoms (chlorine and oxygen), and thus the methylene protons are even less shielded and appear farther to the left (downfield) in the spectrum at $\delta = 5.5$ ppm.

PROBLEM 9.6

Using the data in Table 9.3, predict the approximate chemical shift for each signal in the compounds listed in Problem 9.5.

SAMPLE SOLUTION

(a) The signal for the methyl protons of ethanol would be found between $\delta = 0.9$ and 1.8 ppm. The methylene protons, because of the adjacent oxygen atom, would be expected to absorb in the region $\delta = 3.3$ to 3.7 ppm. The chemical shift of the hydroxyl proton may vary but is expected to be in the region $\delta = 0.5$ to 5 ppm. The actual chemical shifts for ethanol are $\delta = 1.2$ ppm (CH_3), 3.6 ppm (CH_2), and 4.0 ppm (OH).

9.9.3 Signal Intensities

Another piece of information that may be used to assign the peaks of chloromethyl methyl ether is the signal intensity. The intensity of an nmr peak is measured by the area enclosed by the peak, and the *area is proportional to the number of equivalent protons responsible for that signal* (peak areas are measured electronically). Thus the ratio of the areas of the two peaks in the spectrum of chloromethyl methyl ether is 2:3 for the CH_2 and CH_3 protons, respectively. Remember that peak areas give *relative, not absolute*, numbers of protons. Thus a 2:3 intensity ratio could indicate a 4:6 or 6:9 ratio of protons just as well as a 2:3 ratio.

PROBLEM 9.7

Predict the relative intensities of the signals for each of the compounds given in Problem 9.5.

SAMPLE SOLUTION

(a) The intensity of each signal is determined by the number of equivalent protons giving rise to that signal. Thus the three signals of ethanol have intensities in the ratio $3:2:1$ from the CH_3, CH_2, and OH peaks, respectively.

So far we have discussed three of the features of nmr spectra: the number of signals, their chemical shift, and peak intensities. The remaining topic, signal multiplicity, is described in the next two sections.

9.10 SPIN-SPIN SPLITTING IN NMR SPECTROSCOPY

The signals in the nmr spectrum of chloromethyl methyl ether (Figure 9.11) are both single sharp absorptions called **singlets**. It is quite common to see nmr spectra, however, in which the signal due to a particular type of proton is not a singlet but instead appears as a collection of peaks. The signal may be split into two peaks (a **doublet**), three peaks (a **triplet**), four peaks (a **quartet**), and so on. For example, consider the ^1H nmr spectrum of 1,1-dichloroethane Cl_2CHCH_3 shown in Figure 9.12. The spectrum consists of two absorptions of intensities $1:3$ resulting from the methine ($-\overset{|}{\underset{|}{C}}-H$) and methyl protons, respectively. The methyl protons appear as a *doublet* centered at $\delta = 2.0$ ppm, and the signal for the methine proton is a *quartet* centered at $\delta = 5.9$ ppm.

There is a simple rule that allows us to predict splitting patterns in ^1H nmr spectroscopy. The *multiplicity of the signal for a particular proton is equal to the number of protons that are vicinal to it, n, plus 1 (n + 1)*. By vicinal we mean attached to the carbon adjacent to the carbon bearing the proton being observed. Thus the three methyl protons of 1,1-dichloroethane and the methine proton are vicinal to each other. The methine signal is split into a quartet (four peaks) by the three methyl protons, and the methyl signal is split into a doublet (two peaks) by the single methine proton.

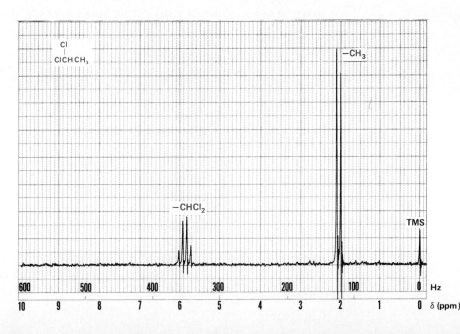

FIGURE 9.12 The ^1H nmr spectrum of 1,1-dichloroethane.

This proton splits the
signal for the methyl
protons into a doublet

$$\underset{\substack{| \\ Cl}}{\overset{\substack{Cl \\ |}}{H-C}} - \overset{\substack{H \\ |}}{\underset{\substack{| \\ H}}{C}} H$$

These three protons
split the signal for
the methine proton
into a quartet.

The physical basis for peak splitting involves interactions between the nuclear spins of the vicinal protons, hence the phrase "spin-spin splitting." We will not discuss the physical phenomenon further, however, but rather focus on the use of splitting in the interpretation of ^{1}H nmr spectra.

The splitting which is typically observed in ^{1}H nmr spectra is between vicinal hydrogens, as we saw for 1,1-dichloroethane. The hydrogens are separated by three bonds; hydrogens separated by four or more bonds do not normally give rise to observable splitting. Also, splitting is generally only observed for protons bound to carbon. Thus hydroxyl protons (O—H) and amine protons (N—H) are usually observed as singlets in ^{1}H nmr spectra.

Lastly, we must note that only hydrogens which are nonequivalent will split each other. Thus ethane, for example, shows only a single sharp peak in its ^{1}H nmr spectrum. Even though the protons of one methyl group are vicinal to those of the other, they do not split each other because they are equivalent.

PROBLEM 9.8

Describe the appearance of the ^{1}H nmr spectrum of each of the following compounds. How many signals would you expect to find, what will be their relative intensities, and into how many peaks will each signal be split?

(a) 1,2-Dichloroethane   (d) 1,2,2-Trichloropropane
(b) 1,1,1-Trichloroethane   (e) 1,1,1,2-Tetrachloropropane
(c) 1,1,2-Trichloroethane

SAMPLE SOLUTION

(a) All the protons of 1,2-dichloroethane (ClCH$_2$CH$_2$Cl) are equivalent and thus have the same chemical shift. Protons that have the same chemical shift do not split each other, so the ^{1}H nmr spectrum of 1,2-dichloroethane consists of a single sharp peak.

9.11 PATTERNS OF SPIN-SPIN SPLITTING

At first glance splitting may seem to complicate the interpretation of nmr spectra. In fact, it makes structure determination easier by providing much additional information. Spin-spin splitting tells us how many protons are vicinal to a proton responsible for a particular signal. With practice, you can pick out characteristic patterns of peaks and associate them with particular structural types.

One of the most common patterns observed in ^{1}H nmr spectra is that of the ethyl group. In compounds of the type CH$_3$CH$_2$X, especially where X is an electronegative atom or group, the ethyl group appears as two signals, a triplet and a quartet. A good example is ethyl bromide CH$_3$CH$_2$Br, shown in Figure 9.13. The methylene proton signal is split into a quartet by coupling with the methyl protons. The signal for the methyl protons is a triplet because of vicinal coupling to the two protons of the adjacent methylene group.

FIGURE 9.13 The ^{1}H nmr spectrum of ethyl bromide, showing the characteristic triplet-quartet pattern of an ethyl group.

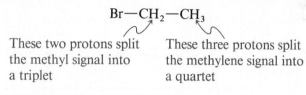

$$Br—CH_2—CH_3$$

These two protons split   These three protons split
the methyl signal into   the methylene signal into
a triplet   a quartet

PROBLEM 9.9

Describe the splitting, if any, which would be observed for each signal in the ^{1}H nmr spectrum of the compounds listed in Problem 9.5.

SAMPLE SOLUTION

(a) The methyl absorption of ethanol would appear as a triplet arising from splitting by the two protons of the adjacent methylene group. The methylene absorption would be a quartet, being split by the three methyl protons. Recall that splitting is usually only observed for protons attached to carbon, and thus the hydroxyl proton appears as a singlet.

$$CH_3—CH_2—OH \qquad \text{This proton is a singlet}$$

These protons   These protons
appear as a   appear as a
triplet   quartet

To summarize the important points about the use of spin-spin coupling in the interpretation of ^{1}H nmr spectra:

1. Splitting is normally observed only between nonequivalent vicinal protons bound to carbon.
2. The multiplicity of the signal is equal to the number of vicinal nonequivalent protons plus 1.

PROBLEM 9.10

To which one of the following compounds does the nmr spectrum of Figure 9.14 correspond?

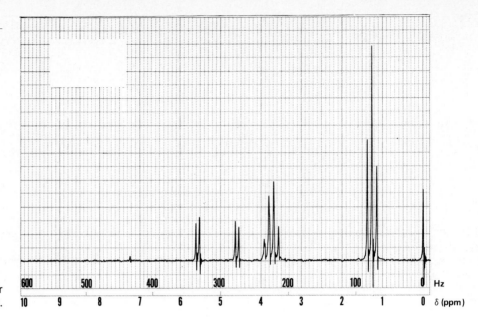

FIGURE 9.14 The ^{1}H nmr spectrum for Problem 9.10.

$$ClCH_2C(OCH_2CH_3)_2 \qquad Cl_2CHCH(OCH_2CH_3)_2 \qquad CH_3CH_2OCHCHOCH_2CH_3$$

with Cl substituents as shown in the structures:

ClCH₂C(OCH₂CH₃)₂ with Cl below the central carbon;

CH₃CH₂OCHCHOCH₂CH₃ with Cl above the first CH and Cl below the second CH.

9.12 CARBON-13 NUCLEAR MAGNETIC RESONANCE

As mentioned in Section 9.5, certain nuclei other than the proton possess the property of nuclear spin, and modern nmr spectrometers allow the study of these nuclei. Biochemists, for example, make frequent use of ^{31}P nmr for the study of cellular metabolism and nucleic acid structure. When we use ^{1}H nmr as a tool for identifying organic compounds, the structure of the carbon skeleton is inferred based on information obtained about the hydrogens attached to those carbons. What about observing the carbons directly?

The major isotope of carbon (98.9 percent ^{12}C) has no nuclear spin and therefore is incapable of giving an nmr signal. The mass 13 isotope, ^{13}C (1.1 percent of natural carbon), has a nuclear spin and is suitable for nmr study. While ^{1}H nmr has been available to chemists for over 30 years, routine ^{13}C nmr has been in use only for the last 10 to 15 years. Why? The problem is *sensitivity*. For a given amount of sample, the ^{13}C nmr signal is 10,000 times weaker than the ^{1}H nmr signal.

The solution to the sensitivity problem has evolved with the advent of high-speed computers and very powerful superconducting magnets built into the nmr instruments. A ^{13}C nmr instrument is able to collect many thousands of spectra and to average the data with its built-in computer, producing a composite spectrum in which the peaks are many times more prominent than in a single spectrum.

Although a complete discussion of ^{13}C nmr spectroscopy is beyond the scope of this book, a brief look at the technique is informative. Figure 9.15 shows both the ^{1}H nmr spectrum and the ^{13}C nmr spectrum for 1-chloropentane. The two spectra are obviously quite different. While the proton spectrum (Figure 9.15*a*) contains a large, ill-defined peak because all of the protons have very similar chemical shifts, each carbon gives a separate sharp signal in the ^{13}C nmr spectrum (Figure 9.15*b*). Notice

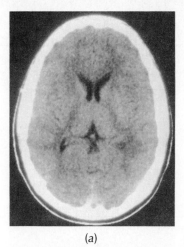 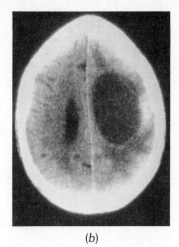

(a)   (b)

A magnetic resonance image showing a two-dimensional cross
section of (a) a normal brain and (b) a brain containing a tumor, as
indicated by the dark shadow. (*Photographs courtesy of NIH.*)

the widely separated chemical shifts of the five peaks, one for each of the carbon
atoms in 1-chloropentane; they cover a range of over 30 ppm compared with less
than 3 ppm for the ^{1}H nmr signals for the same compound. In general, proton signals
span a range of about 12 ppm; ^{13}C signals span a range of over 200 ppm.

MAGNETIC RESONANCE IMAGING

Within the last several years physicians in large hospitals have been using a new
diagnostic technique called *magnetic resonance imaging*, MRI. MRI in actuality is nuclear magnetic resonance, but the word *nuclear* was dropped from the
name so as to avoid confusion with nuclear medicine which involves radioac-
tive isotopes. Nuclear magnetic resonance is a way of studying the chemical
environment of a particular nucleus, but radioactivity is not involved in any
way.

MRI is felt by many in the medical community to be a significant advance
over the widely used computerized tomography, CT (or CAT), scan. CT scans
employ potentially harmful ionizing radiation (x-rays), whereas MRI does not.
Also, MRI is able to distinguish between healthy and diseased tissues without
the need for ingesting contrast agents which can sometimes provoke an allergic
reaction with certain patients.

The basis for MRI is a very advanced form of proton (^{1}H) nmr. The
behavior of water molecules, which of course contain two protons each, is
observed in the MRI technique. Since every cell in our body contains significant amounts of water, every tissue and organ can be visualized. The protons in
water molecules of diseased cells behave differently in the nmr image than
protons in normal, healthy cells.

Advanced computer methods can record two-dimensional images of thin
cross-sectional slices of the portion of the body being studied. A collection of
these slices gives the physician a three-dimensional view of the organ of interest.

MRI is certainly a diagnostic technique which will increase in popularity in
the near future. It has even been suggested that hospitals will begin replacing
aging CT scanners with MRI units.

TABLE 9.4
Chemical Shifts of Representative Carbons

Type of carbon	Chemical shift (δ), ppm*	Type of carbon	Chemical shift (δ), ppm*
RCH_3	0–35	$\diagdown C=C \diagup$	100–150
R_2CH_2	15–40		
R_3CH	25–50	⬡	110–175
RCH_2NH_2	35–50		
RCH_2OH	50–65	$\diagdown C=O$	190–220
$-C\equiv C-$	65–90		

* Approximate values relative to tetramethylsilane.

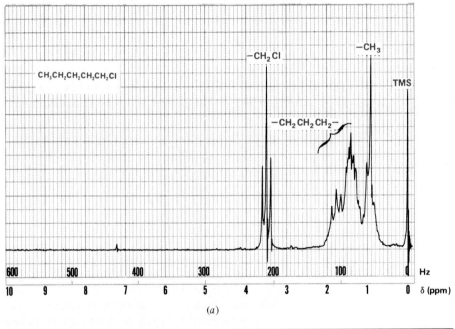

FIGURE 9.15 The (a) ^1H nmr spectrum and (b) ^{13}C nmr spectrum of 1-chloropentane. (*^{13}C spectrum is taken from Carbon-13 NMR Spectra: A Collection of Assigned, Coded, and Indexed Spectra, by LeRoy F. Johnson and William C. Jankowski, Wiley-Interscience, New York, 1972. Reprinted by permission of John Wiley & Sons, Inc.*)

Chemical shifts in ^{1}H nmr are measured relative to the protons of tetramethylsilane (TMS); chemical shifts of the carbon atoms in a ^{13}C nmr spectrum are measured relative to the carbons of TMS. Table 9.4 lists typical chemical shift ranges for some representative types of carbon atoms.

PROBLEM 9.11

How many signals would be observed in the ^{13}C nmr spectra of each of the following compounds?

(a) Propylbenzene
(b) Isopropylbenzene
(c) 1,3,5-Trimethylbenzene

SAMPLE SOLUTION

(a) The number of carbon signals is equal to the number of nonequivalent carbon atoms in the molecule. Although propylbenzene contains nine carbon atoms, there are only seven different ones. The two positions ortho to the side chain are equivalent, as are the two meta positions.

Propylbenzene

9.13 MASS SPECTROMETRY

Mass spectrometry is the instrumental method of structure determination that is the most different from the others in the group discussed in this chapter. It does not depend on the selective absorption of particular frequencies of electromagnetic radiation, but rather examines what happens to a molecule when it is bombarded with high-energy electrons. If an electron having an energy of about 10 electron volts (10 eV = 230.5 kcal/mol) collides with an organic molecule, the energy transferred as a result of that collision is sufficient to dislodge one of the molecule's electrons.

$$A:B + e^- \longrightarrow A \overset{+}{\cdot} B + 2e^-$$

We say the molecule AB has been ionized by **electron impact.** The species that results is called the **molecular ion, M$^+$.** The molecular ion is positively charged and has an odd number of electrons — it is a **cation radical.** The molecular ion has the *same mass as the neutral molecule* less the negligible mass of a single electron.

While energies of about 10 eV are required, energies of about 70 eV are used in most mass spectrometers. Electrons this energetic not only bring about the ionization of a molecule but impart a large amount of energy to the molecular ion. The molecular ion dissipates this excess energy by dissociating into smaller fragments. Dissociation of a cation radical produces a neutral fragment and a positively charged one. The molecular ion A $\overset{+}{\cdot}$ B can dissociate in several ways. Among them are the fragmentations

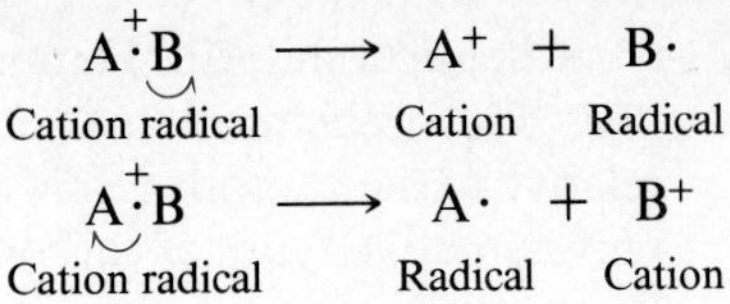

$$\underset{\text{Cation radical}}{A \overset{+}{\cdot} B} \longrightarrow \underset{\text{Cation}}{A^+} + \underset{\text{Radical}}{B \cdot}$$

$$\underset{\text{Cation radical}}{A \overset{+}{\cdot} B} \longrightarrow \underset{\text{Radical}}{A \cdot} + \underset{\text{Cation}}{B^+}$$

Ionization and fragmentation produce a mixture of particles, some neutral and some positively charged. The mass-to-charge ratio (m/z) of each of the positive ions is measured. As the charge is almost always 1, the value of m/z is equal to the mass of the ion. A **mass spectrum** is the distribution of the positive ions formed by fragmentation of the molecular ion. The **fragmentation pattern** is characteristic of a particular compound.

The most common format for displaying mass spectra is a bar graph on which relative ion intensity is plotted versus m/z. Figure 9.16 shows the mass spectrum of benzene in bar graph form. The most intense peak in the mass spectrum is called the **base peak** and is assigned a relative intensity of 100. Peak intensities are reported relative to the base peak. The base peak in the mass spectrum of benzene corresponds to the molecular ion (M^+) at $m/z = 78$.

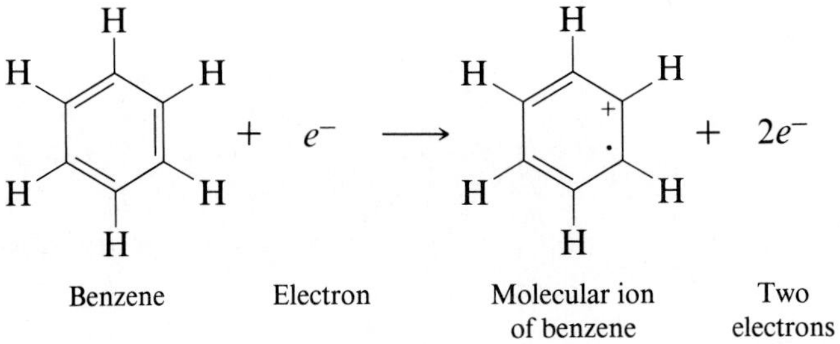

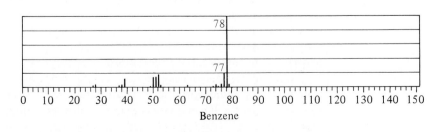

Benzene does not undergo extensive fragmentation; none of the fragment ions in its mass spectrum are as abundant as the molecular ion.

Many of the fragmentation processes in mass spectrometry proceed to form a stable carbocation, and the principles developed in earlier chapters (Sections 3.12 and 6.14) are applicable. Alkylbenzenes of the type $C_6H_5CH_2R$ undergo cleavage of the bond to the benzylic carbon to give $m/z = 91$ as the base peak. The mass spectrum in Figure 9.17 and the fragmentation diagram shown below illustrate this for propylbenzene.

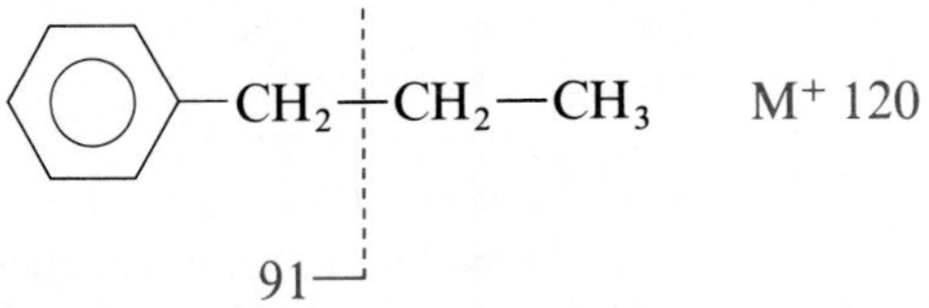

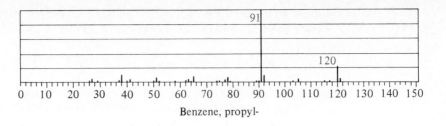

FIGURE 9.17 The mass spectrum of propylbenzene. [Reproduced with permission from "The NIST/EPA/MSDC Mass Spectral Database," PC Version 2.0, National Institute of Standards and Technology (1988).]

Benzene, propyl-

PROBLEM 9.12

The three compounds shown are isomers and exhibit molecular ion peaks (M$^+$) at $m/z = 134$. The base peak appears at $m/z = 105$ for one of the compounds and at $m/z = 119$ for the other two. Match the compounds with the appropriate m/z value for their base peaks.

The mass spectra of thousands of compounds have been measured by mass spectrometry and have provided a vast library of fragmentation patterns to aid in structure determination. Rapid computer searching of these libraries has enabled them to be used in such diverse areas as hospital clinical laboratories and in the analysis of toxic waste dumps.

9.14 MOLECULAR FORMULA AS A CLUE TO STRUCTURE

Chemists are often confronted with the problem of identifying the structure of an unknown compound. Sometimes the unknown can be shown to be a sample of a compound previously reported in the chemical literature; on other occasions, the unknown may be truly that—a compound never before encountered by anyone. An arsenal of powerful techniques is available to simplify the task of structure determination. It should be pointed out, however, that the molecular formula of a molecule provides more information than might be apparent at first glance.

Consider, for example, a substance with the molecular formula C_7H_{16}. We know immediately that the compound is an alkane because its molecular formula corresponds to the general formula for that class of compounds, C_nH_{2n+2}, where $n = 7$.

What about a substance with the molecular formula C_7H_{14}? This compound cannot be an alkane but may be either a cycloalkane or an alkene because both these classes of hydrocarbons correspond to the general molecular formula C_nH_{2n}. Any time a ring or a double bond is present in an organic molecule, its molecular formula has two fewer hydrogen atoms than that of an alkane with the same number of carbons.

Various names have been given to this relationship between molecular formulas and classes of hydrocarbons. It is sometimes referred to as an *index of hydrogen deficiency* and sometimes as *elements of unsaturation* or *sites of unsaturation*. We will use the term *sum of double bonds and rings* because it is more descriptive than the others. In the interests of economy of space, however, the full term sum of double bonds and rings will be abbreviated as SODAR.

$$\text{Sum of double bonds and rings (SODAR)} = \tfrac{1}{2}[C_nH_{2n+2} - C_nH_x]$$

where C_nH_x is the molecular formula of the compound.

Thus, a molecule that has a molecular formula of C_7H_{14} has a SODAR of 1.

$$\text{SODAR} = \tfrac{1}{2}[C_7H_{16} - C_7H_{14}]$$
$$\text{SODAR} = \tfrac{1}{2}[2] = 1$$

Thus, the structure has one ring or one double bond. A molecule of molecular formula C_7H_{12} has four fewer hydrogens than the corresponding alkane; it has a SODAR of 2 and can have two rings, two double bonds, one ring and one double bond, or one triple bond.

A halogen substituent, like hydrogen, is monovalent, and when present in a molecular formula is treated as if it were hydrogen for the purposes of computing the number of double bonds and rings. Oxygen atoms have no effect on the relationship between molecular formulas, double bonds, and rings. They are ignored when computing the SODAR of a substance.

How does one distinguish between rings and double bonds? This additional piece of structural information is revealed by catalytic hydrogenation experiments in which the amount of hydrogen that reacts is measured exactly. Each of a molecule's double bonds consumes one molar equivalent of hydrogen, while rings do not undergo hydrogenation. A substance with a SODAR of 5 that takes up 3 moles of hydrogen must therefore have two rings.

PROBLEM 9.13

How many rings are present in each of the following compounds? Each consumes 2 moles of hydrogen on catalytic hydrogenation.

(a) $C_{10}H_{18}$ (d) C_8H_8O
(b) C_8H_8 (e) $C_8H_{10}O_2$
(c) $C_8H_8Cl_2$ (f) C_8H_9ClO

SAMPLE SOLUTION

(a) The molecular formula $C_{10}H_{18}$ contains four fewer hydrogens than the alkane having the same number of carbon atoms ($C_{10}H_{22}$). Therefore, the SODAR of this compound is 2. Since it consumes two molar equivalents of hydrogen on catalytic hydrogenation, it must have two double bonds and no rings.

9.15 SUMMARY

Structure determination in modern-day organic chemistry relies heavily on instrumental methods. Three of the methods discussed in this chapter are based on detecting the absorption of electromagnetic radiation by molecules.

Electromagnetic radiation (Section 9.1) includes forms of radiation such as visible light, radio waves, and x-rays. Electromagnetic radiation has the properties of both waves and particles. The *wavelength* λ and *frequency* ν of electromagnetic radiation are related by the equation

$$\lambda \times \nu = c$$

where c = speed of light = 3.0×10^8 m/s. The particles of radiation are called *photons*. The energy of a photon is directly proportional to the frequency of the radiation. The common methods of spectroscopic analysis are:

1. *Infrared spectroscopy* (Section 9.3). This method probes molecular structure by examining transitions between vibrational energy levels using electromagnetic radiation in the 625 to 4000 cm^{-1} range. It is useful for determining the presence of certain *functional groups* based on their characteristic absorption frequencies.

2. *Ultraviolet-visible spectroscopy* (Section 9.4). Transitions between electronic energy levels involving electromagnetic radiation in the 200- to 800-nm range form the basis of uv-vis spectroscopy. The absorption peaks tend to be broad but are sometimes useful in indicating the presence of particular *conjugated π-electron* systems within a molecule.

3. *^1H nuclear magnetic resonance spectroscopy* (Sections 9.5 to 9.11). In the presence of an external magnetic field, the nuclear spin states of a proton have slightly different energies. Protons absorb electromagnetic radiation in the radio frequency region, causing the nuclear spin to "flip" from the low-energy state to the high-energy state. The energy depends on the extent to which a nucleus is shielded from the external magnetic field, called the *chemical shift* (Section 9.6). Protons in different environments within a molecule have different chemical shifts. The *number of signals* (Section 9.9) in a ^{1}H nmr spectrum reveals the number of chemical shift nonequivalent protons in a molecule; the *integrated areas,* or *peak intensities,* tell us their relative ratios; their *chemical shifts* indicate the kind of environment surrounding the proton; and the *splitting pattern* (Sections 9.10 and 9.11) is related to the number of protons on adjacent carbons.

4. *^{13}C nuclear magnetic resonance spectroscopy* (Section 9.12). Using special techniques for signal enhancement, high-quality ^{13}C nmr spectra may be obtained, and these provide a useful complement to proton spectra. In many substances a separate signal is observed for each carbon atom.

5. *Mass spectrometry* (Section 9.13). Mass spectrometry exploits the information obtained when a molecule is ionized by electron impact and then dissociates to smaller fragments. Positive ions are separated and detected according to their mass-to-charge ratio.

ADDITIONAL PROBLEMS

Electromagnetic Radiation; Infrared and Ultraviolet Spectra

9.14 (a) Microwave ovens heat food by using radiation having a frequency of 1.5×10^{10} Hz. What is the wavelength of this radiation? The speed of light is 3.0×10^8 m/s.

(b) The frequency and wavelength of sound waves follow the same inverse relationship as electromagnetic radiation. If high C on the musical scale (1024 Hz) has a wavelength of 32.4 cm, what is the speed of sound in air?

9.15 An infrared spectrum exhibits a broad band in the region between 3000 and 3500 cm^{-1} and a strong peak at 1710 cm^{-1}. Choose which of the following substances best fits the data and explain your reasoning.

$$C_6H_5CH_2CH_2OH \qquad C_6H_5CH_2\overset{\displaystyle O}{\overset{\displaystyle \|}{C}}OH \qquad C_6H_5CH_2\overset{\displaystyle O}{\overset{\displaystyle \|}{C}}CH_3$$

9.16 Choose which of the following compounds has the longest wavelength uv absorption (λ_{max}) and explain your reasoning.

$$CH_3CH=CHCH_2CH_3 \qquad CH_2=CHCH=CHCH_3$$

¹H NMR Spectra

9.17 Complete the following table relating to ¹H nmr spectra by supplying the missing data.

Spectrometer frequency	Chemical shift	
	ppm	Hz
(a) 60 MHz	_____	366
(b) 220 MHz	4.35	_____
(c) _____	3.50	210

9.18 How many signals would you expect to find in the ¹H nmr spectrum of each of the following compounds? Ignore splitting effects for this problem.

(a) 1-Bromobutane
(b) 2,2-Dibromobutane
(c) 1,4-Dibromobutane
(d) 1,1,4-Tribromobutane
(e) 1,1,1-Tribromobutane

9.19 Describe the appearance of the ¹H nmr spectrum of each of the following compounds. How many signals would you expect to find, and into how many peaks will each signal be split?

(a) $CH_3CH_2OCH_3$
(b) $CH_3CH_2OCH_2CH_3$
(c) $ClCH_2CH_2OCH_3$
(d) $ClCH_2C(CH_3)_2$
    $\overset{|}{Cl}$
(e) $ClCH_2CH_2CH_2Cl$
(f) $Cl_2CHCH_2\overset{O}{\underset{\|}{C}}CH_2CH_3$
(g) $CH_3OCH_2CH_2OCH_3$

9.20 Using the data in Table 9.3, predict the *approximate* chemical shift expected for each of the protons in Problem 9.19.

9.21 One of the compounds in Problem 9.19 exhibits strong infrared absorption at 1710 cm⁻¹. Identify the compound.

9.22 Each of the following compounds is characterized by a ¹H nmr spectrum that consists of only a single peak having the chemical shift indicated. Identify each compound.

(a) C_8H_{18}, $\delta = 0.9$ ppm
(b) C_5H_{10}, $\delta = 1.5$ ppm
(c) C_8H_8, $\delta = 5.8$ ppm
(d) C_4H_9Br, $\delta = 1.8$ ppm

9.23 Deduce the structure of each of the following compounds on the basis of their ¹H nmr spectra and molecular formulas:

(a) C_8H_{10}: $\delta = 1.2$ ppm (triplet, 3H)
$\delta = 2.6$ ppm (quartet, 2H)
$\delta = 7.1$ ppm (broad singlet, 5H)
(b) $C_{10}H_{14}$: $\delta = 1.3$ ppm (singlet, 9H)
$\delta = 7.0$ to 7.5 ppm (multiplet, 5H)
(c) C_6H_{14}: $\delta = 0.8$ ppm (doublet, 12 H)
$\delta = 1.4$ ppm (heptet, 2H)
(d) $C_4H_6Cl_4$: $\delta = 3.9$ ppm (doublet, 4H)
$\delta = 4.6$ ppm (triplet, 2H)
(e) $C_4H_6Cl_2$: $\delta = 2.2$ ppm (singlet, 3H)
$\delta = 4.1$ ppm (doublet, 2H)
$\delta = 5.7$ ppm (triplet, 1H)

9.24 From among the isomeric compounds of molecular formula C_4H_9Cl, choose the one having a ^1H nmr spectrum that:

(a) Contains only a single peak

(b) Has several peaks including a doublet at $\delta = 3.4$ ppm

(c) Has several peaks including a triplet at $\delta = 3.5$ ppm

(d) Has several peaks including two distinct three-proton signals, one of them a triplet at $\delta = 1.0$ ppm and the other a doublet at $\delta = 1.5$ ppm

Combined Spectral Problems

9.25 Identify each of the following compounds based on the ir and ^1H nmr information provided:

(a) $C_{10}H_{12}O$:   ir: 1710 cm^{-1}

    ^1H nmr: $\delta = 1.0$ ppm (triplet, 3H)

    $\delta = 2.4$ ppm (quartet, 2H)

    $\delta = 3.6$ ppm (singlet, 2H)

    $\delta = 7.2$ ppm (singlet, 5H)

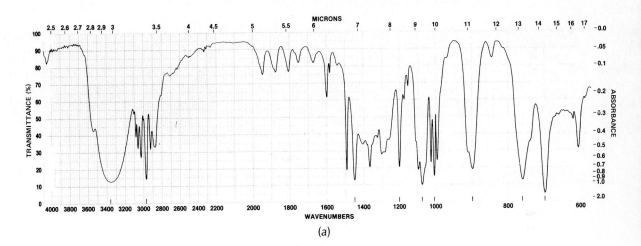

(a)

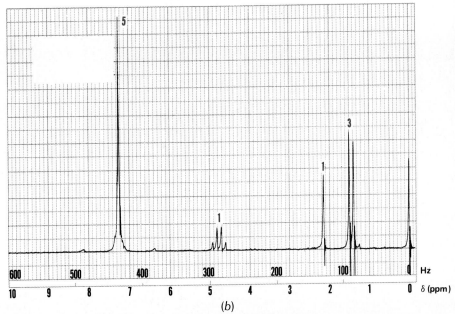

(b)

FIGURE 9.18   (a) The infrared and (b) ^1H nmr spectra of compound A (Problem 9.26). The numbers beside the signals correspond to their relative areas.

(b) $C_6H_{14}O_2$:   ir: 3400 cm^{-1}

^{1}H nmr: $\delta = 1.2$ ppm (singlet, 12 H)

$\delta = 2.0$ ppm (singlet, 2H)

(c) C_4H_7NO: ir: 2240 cm^{-1}

3400 cm^{-1} (broad)

^{1}H nmr: $\delta = 1.65$ ppm (singlet, 6H)

$\delta = 3.7$ ppm (singlet, 1H)

9.26 Compound A ($C_8H_{10}O$) has the infrared and ^{1}H nmr spectra presented in Figure 9.18. What is the structure of compound A?

9.27 Compound B ($C_5H_{10}O$) has the infrared and ^{1}H nmr spectra shown in Figure 9.19. What is the structure of compound B?

9.28 From among the compounds chlorobenzene, o-dichlorobenzene, and p-dichloro-benzene, choose the one that:

(a) Gives the simplest ^{1}H nmr spectrum

(b) Gives the simplest ^{13}C nmr spectrum (fewest number of signals)

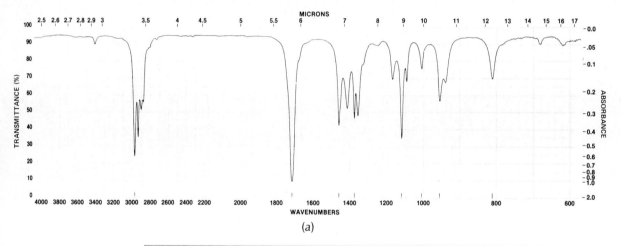

(a)

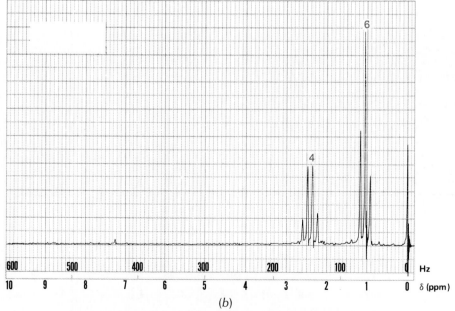

(b)

FIGURE 9.19 (a) The infrared and (b) ^{1}H nmr spectra of compound B (Problem 9.27). The numbers beside the signals correspond to their relative areas.

(c) Has three peaks in its ^{13}C nmr spectrum

(d) Has four peaks in its ^{13}C nmr spectrum

9.29 What is the structure of the compound having a molecular formula $C_5H_{12}O$ which best fits the following spectroscopic information:

       ir: Strong peak at 1150 cm^{-1}

             No peak between 3000 and 3500 cm^{-1}

  ^1H nmr: 3H singlet ($\delta = 3.4$ ppm); 6H doublet ($\delta = 1.0$ ppm); plus other peaks

  ^{13}C nmr: Four signals

9.30 Give the number of signals expected in the ^{13}C nmr spectrum of the substance

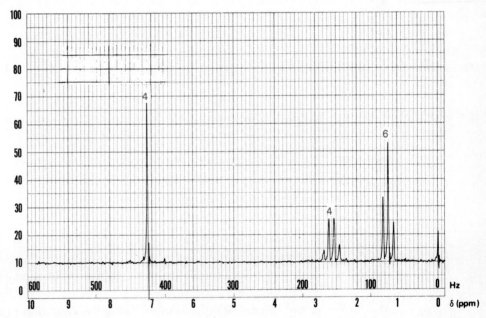

9.31 Figure 9.20 presents the ^1H nmr spectrum of a hydrocarbon, compound C. The mass spectrum of compound C exhibited a molecular ion at $m/z = 134$. What is compound C?

FIGURE 9.20 The ^1H nmr spectrum of compound C (Problem 9.31). The numbers beside the signals correspond to their relative areas.

CHAPTER 10

ALCOHOLS, ETHERS, AND PHENOLS

LEARNING OBJECTIVES

This chapter focuses on the chemical and physical properties of three classes of oxygen-containing organic compounds: alcohols, ethers, and phenols. Upon completion of this chapter you should be able to:

- Explain the progression of oxidation states of common organic substances.
- Describe the preparation of alcohols by the reduction of aldehydes and ketones.
- Explain how alcohols can act as Brönsted acids.
- Describe how alkoxide ions may be formed from alcohols.
- Write a chemical equation describing the preparation of an alkyl ether from a primary alcohol.
- Write a chemical equation describing the oxidation of a primary alcohol to either an aldehyde or a carboxylic acid.
- Write a chemical equation describing the oxidation of a secondary alcohol to a ketone.
- Give a systematic name for an ether or an epoxide.
- Write a chemical equation describing the preparation of an ether by the Williamson method.
- Write chemical equations for the preparation of an epoxide by two methods.
- Predict the product of the reaction of an epoxide with a nucleophile.
- Draw the structure and give a systematic name for a thiol.
- Give a systematic name for a phenol derivative.
- Explain the acidity of phenol derivatives.
- Write a chemical equation describing the formation of a quinone by the oxidation of a 1,4-benzenediol.

The next several chapters will explore the chemistry of various oxygen-containing functional groups: alcohols, ethers, phenols, aldehydes, ketones, carboxylic acids, and derivatives of carboxylic acids.

ROH   ROR′   ArOH

Alcohol   Ether   Phenol

$$\underset{\text{Aldehyde}}{\overset{\displaystyle\overset{O}{\|}}{RCH}} \qquad \underset{\text{Ketone}}{\overset{\displaystyle\overset{O}{\|}}{RCR'}} \qquad \underset{\text{Carboxylic acid}}{\overset{\displaystyle\overset{O}{\|}}{RCOH}}$$

This chapter is devoted to the chemistry of alcohols, ethers, and phenols, all of which contain a single-bonded oxygen atom. You will see that members of all three of these classes of compounds have important commercial use. In addition, a variety of biologically important substances contain these functional groups.

10.1 ALCOHOLS: A REVIEW

Alcohols were first introduced in Chapter 3, and several of the topics covered there are reviewed in this section. Alcohols are given systematic names (Section 3.4) by replacing the **-e** ending of the corresponding alkane name with **-ol.** The position of the hydroxyl group is indicated by number, choosing the sequence that assigns the lower number to the carbon that bears the hydroxyl group.

$$\underset{\overset{\displaystyle|}{CH_3} \quad \overset{\displaystyle|}{OH}}{CH_3CHCH_2CHCH_2CH_3}$$

5-Methyl-3-hexanol
(*not* 2-methyl-4-hexanol)

Alcohols are classified (Section 3.5) as primary, secondary, or tertiary according to whether the hydroxyl group is attached to a primary, secondary, or tertiary carbon. 5-Methyl-3-hexanol is a secondary alcohol. The carbon that bears the hydroxyl group is secondary because it is directly attached to two other carbons.

PROBLEM 10.1

The compound shown, whose common name is tetrahydrolinalool, has a lavender odor and is used in perfumes. Give an acceptable systematic name for this substance, and classify it as a primary, secondary, or tertiary alcohol.

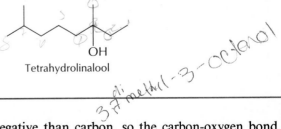

Tetrahydrolinalool

3,7-dimethyl-3-octenol

Oxygen is more electronegative than carbon, so the carbon-oxygen bond of alcohols is polar. In the liquid phase alcohols form a network of *hydrogen bonds* (Section 3.6) between the positively polarized hydrogen and the negatively polarized oxygen of neighboring molecules. One result is that alcohols tend to have much higher boiling points than other compounds of similar molecular weight.

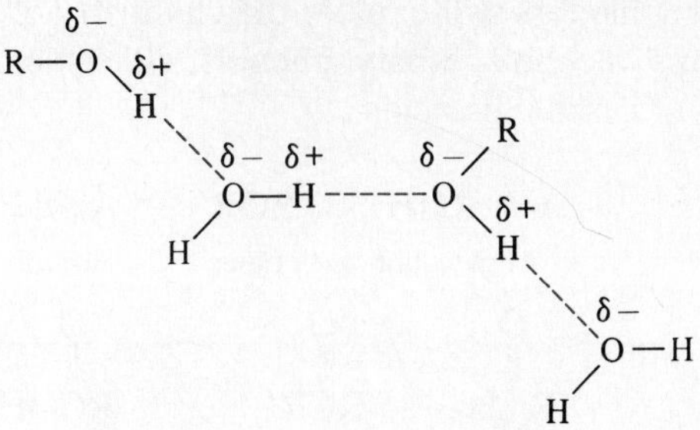

FIGURE 10.1 Hydrogen
bonding between alcohol
and water molecules.

The ability to participate in intermolecular hydrogen bonding not only affects the boiling points of alcohols but also enhances their water solubility. The lower-molecular-weight alcohols, methanol and ethanol, for example, are soluble in water in all proportions. Hydrogen-bonded networks of the type shown in Figure 10.1, in which alcohol and water molecules associate with one another, replace the alcohol-alcohol and water-water hydrogen-bonded networks present in the pure substances.

Higher alcohols become more hydrocarbon-like and less water-soluble. 1-Octanol, for example, dissolves to the extent of only 1 part in 2000 parts of water. Van der Waals forces (Section 2.17) cause 1-octanol molecules to bind to each other through their long alkyl chains, and the energy of hydrogen bonding between water

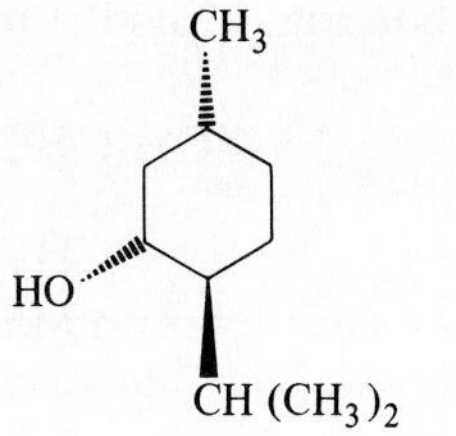

Menthol (obtained from oil of
peppermint and used to flavor
tobacco and food)

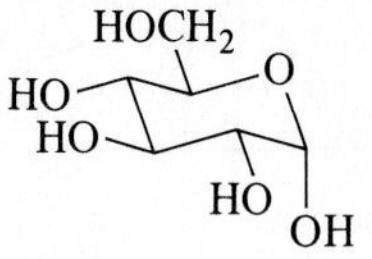

Glucose (a carbohydrate)

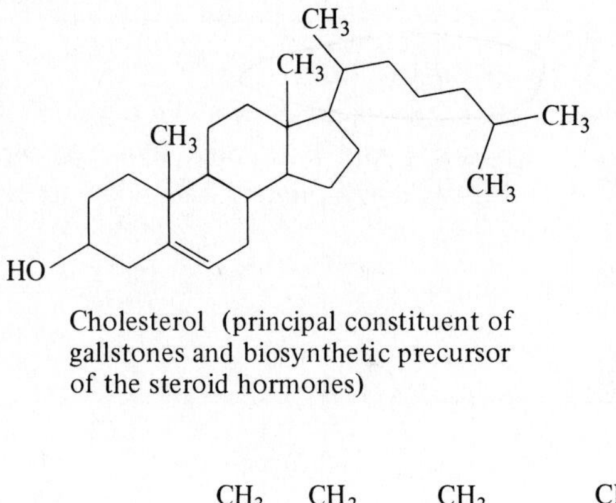

Cholesterol (principal constituent of
gallstones and biosynthetic precursor
of the steroid hormones)

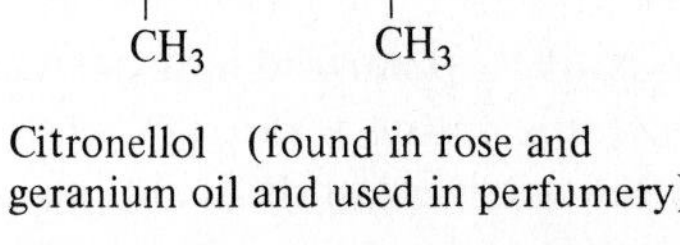

FIGURE 10.2 Some
naturally occurring
alcohols.

Citronellol (found in rose and
geranium oil and used in perfumery)

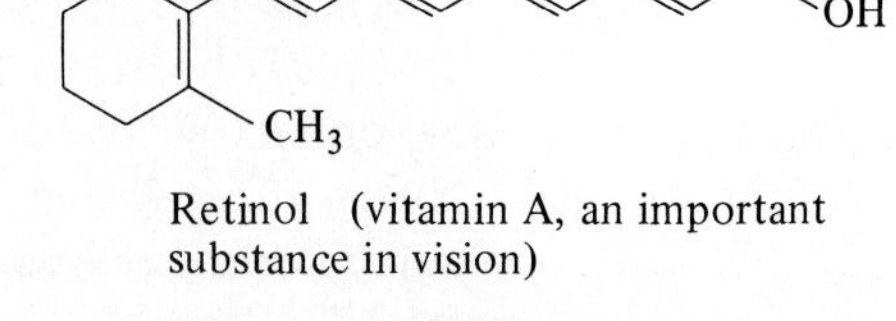

Retinol (vitamin A, an important
substance in vision)

and 1-octanol molecules does not provide enough stabilization to dissociate these
1-octanol aggregates or to disrupt the hydrogen bonds between water molecules.

10.2 NATURAL SOURCES OF ALCOHOLS

Methanol, CH_3OH, is also called **wood alcohol** because it can be prepared by strongly
heating wood in the absence of oxygen, a process known as destructive distillation.
Methanol is very poisonous and can cause blindness or even death if a small amount
is ingested.

  The word *alcohol* to most people refers to ethanol, CH_3CH_2OH, found in alcoholic beverages. When vegetable matter ferments, some of the carbohydrates present
are converted to ethanol and carbon dioxide. Fermentation of barley produces beer;
grapes give wine. Whiskey is the aged distillate of fermented grain. Brandy and
cognac are the aged distillates of fermented grapes and other fruits. The characteristic
flavors, odors, and colors of alcoholic beverages depend on their fruit or grain origin
and how they are aged.

TABLE 10.1
Summary of Reactions Discussed in Earlier Chapters That Yield Alcohols

Reaction (section) and comments	General equation and specific example
Acid-catalyzed hydration of alkenes (Section 5.1.5) The elements of water add to the double bond in accordance with Markovnikov's rule. This reaction is not used frequently for laboratory-scale synthesis of alcohols.	$R_2C{=}CR_2 + H_2O \xrightarrow{H^+} R_2CHCR_2$ (OH) Alkene Water Alcohol $(CH_3)_2C{=}CHCH_3 \xrightarrow[H_2SO_4]{H_2O} CH_3CCH_2CH_3$ (with CH_3 and OH) 2-Methyl-2-butene 2-Methyl-2-butanol (90%)
Hydroboration-oxidation of alkenes (Section 5.1.6) The elements of water add to the double bond with regiochemistry opposite to that of Markovnikov's rule.	$R_2C{=}CR_2 \xrightarrow[2.\ H_2O_2,\ HO^-]{1.\ B_2H_6} R_2CHCR_2$ (OH) Alkene Alcohol $CH_3(CH_2)_7CH{=}CH_2 \xrightarrow[2.\ H_2O_2,\ HO^-]{1.\ B_2H_6} CH_3(CH_2)_7CH_2CH_2OH$ 1-Decene 1-Decanol (93%)
Substitution of alkyl halides (Section 8.2) A reaction useful only with substrates that do not undergo E2 elimination readily. It is rarely used for the synthesis of alcohols since alkyl halides are normally prepared from alcohols.	$RX + HO^- \longrightarrow ROH + X^-$ Alkyl Hydroxide Alcohol Halide halide ion ion [2,4,6-trimethylbenzyl chloride] $\xrightarrow[heat]{H_2O,\ CaO}$ [2,4,6-trimethylbenzyl alcohol] 2,4,6-Trimethylbenzyl chloride 2,4,6-Trimethylbenzyl alcohol (78%)

A wide variety of more complex alcohols occur in nature, and several of these are shown in Figure 10.2.

10.3 REACTIONS THAT LEAD TO ALCOHOLS: A REVIEW AND A PREVIEW

Some of the reactions used for the laboratory synthesis of alcohols have already been discussed and are reviewed in Table 10.1. Additional methods will be introduced in the sections and chapters which follow.

Several methods for preparing alcohols involve **reduction** of carbonyl groups.

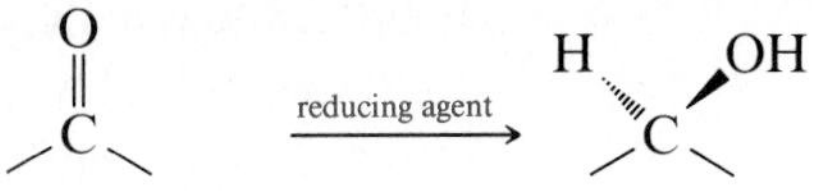

Carbonyl compound   →(reducing agent)→   Alcohol

Before discussing specific reduction methods, let us review the sequence of oxidation states in organic compounds. Among one-carbon compounds, as summarized in Table 10.2, methane and carbon dioxide represent extremes in the oxidation state of carbon. Methane is the most reduced form of carbon, and carbon dioxide is the most oxidized form. In general, *oxidation of a substance increases its oxygen content and reduction decreases it.* Stated in other terms, oxidation at a carbon atom decreases the number of its bonds to hydrogen, and reduction increases the number of its bonds to hydrogen.

The task of choosing the correct reagents for a particular functional group transformation is simplified if one keeps in mind the progression of oxidation states of common organic substances. A change from a lower to a higher oxidation state is an *oxidation reaction.* A reagent that brings about the oxidation of a substance is called an *oxidizing agent.* Conversely, a change from a higher to a lower oxidation state is a *reduction reaction,* brought about with a reagent that is a *reducing agent.*

As an example using the one-carbon compounds in Table 10.2, a reduction reaction would convert formaldehyde (an aldehyde) to methanol (an alcohol). Summarizing the relationship between functional groups:

TABLE 10.2
Oxidation State of Carbon in One-Carbon Compounds

			Number of carbon-oxygen bonds	Number of carbon-hydrogen bonds
Highest oxidation state	Carbon dioxide	$O=C=O$	4	0
	Formic acid	$HCOH$ (with =O)	3	1
	Formaldehyde	HCH (with =O)	2	2
	Methanol	CH_3OH	1	3
Lowest oxidation state	Methane	CH_4	0	4

(increasing oxidation state ↑)

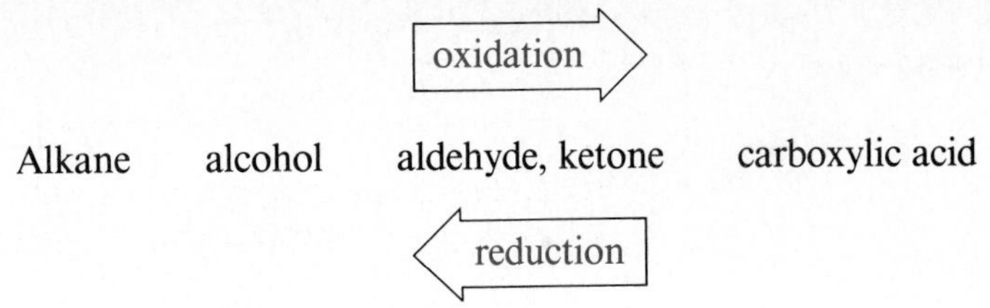

oxidation ⟩

Alkane   alcohol   aldehyde, ketone   carboxylic acid

⟨ reduction

PROBLEM 10.2

One type of breath analyzer used to detect drunk drivers causes the ethanol in the person's breath to undergo the chemical change shown by reaction with a chromium reagent. A color change in the chromium compound from orange to green is used to determine the amount of alcohol present.

$$CH_3CH_2OH \xrightarrow{\text{chromium reagent}} CH_3\overset{\displaystyle O}{\overset{\displaystyle \|}{C}}OH$$

Ethanol   Acetic acid

Classify the chemical change of ethanol as an oxidation or a reduction. Is the chromium reagent an oxidizing agent or a reducing agent?

10.4 PREPARATION OF ALCOHOLS BY REDUCTION OF ALDEHYDES AND KETONES

As described in the preceding section, transforming an aldehyde or ketone into an alcohol requires a process that is a *reduction,* as the number of hydrogens bonded to carbon is greater in the alcohol than in the carbonyl-containing aldehyde or ketone.

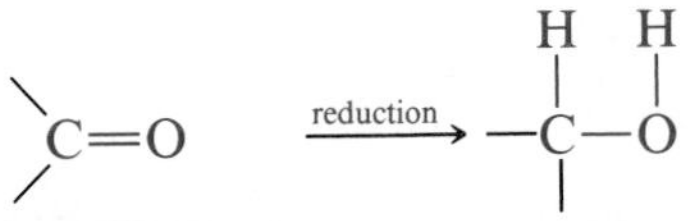

$$\underset{\substack{\text{Carbonyl group} \\ \text{of aldehyde or ketone}}}{\overset{\displaystyle }{\diagdown \!\!\!\diagup C=O}} \xrightarrow{\text{reduction}} \underset{\text{Alcohol}}{\overset{\displaystyle \overset{H}{|}\overset{H}{|}}{-C-O}}$$

Reduction of an aldehyde gives a primary alcohol, while reduction of a ketone gives a secondary alcohol.

$$\underset{\text{Aldehyde}}{R\overset{\displaystyle O}{\overset{\displaystyle \|}{C}}H} \xrightarrow{\text{reducing agent}} \underset{\text{Primary alcohol}}{RCH_2OH}$$

$$\underset{\text{Ketone}}{R\overset{\displaystyle O}{\overset{\displaystyle \|}{C}}R'} \xrightarrow{\text{reducing agent}} \underset{\text{Secondary alcohol}}{R\overset{\displaystyle OH}{\overset{\displaystyle |}{C}}HR'}$$

Hydrogen (H_2) is an effective reducing agent in a process known as *hydrogenation.* The reaction requires a catalyst, and finely divided metals such as platinum, palladium, nickel, and ruthenium are effective for the hydrogenation of aldehydes and ketones. For example,

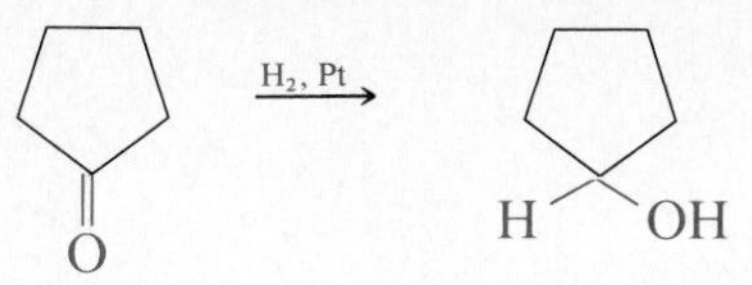

Cyclopentanone   Cyclopentanol (93–95%)

PROBLEM 10.3
Which of the isomeric $C_4H_{10}O$ alcohols can be prepared by hydrogenation of aldehydes?
Which can be prepared by hydrogenation of ketones? Which cannot be prepared by
hydrogenation of a carbonyl compound?

For most laboratory-scale reductions of aldehydes and ketones, catalytic hydrogenation has been replaced by methods based on **metal hydride** reducing agents. The
two reagents that are most commonly used are **sodium borohydride** and **lithium
aluminum hydride.**

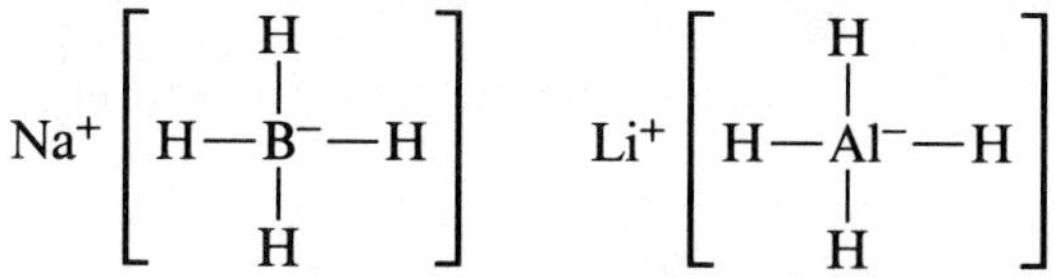

Sodium borohydride   Lithium aluminum hydride
(NaBH$_4$)   (LiAlH$_4$)

Both sodium borohydride and lithium aluminum hydride function as **hydride
donors** when they react with carbonyl compounds. Hydride is the term used to
describe a hydrogen anion, that is, a hydrogen with two electrons, $H:^-$. Boron (or
aluminum) transfers a hydride to the carbon of the carbonyl group, as the carbonyl
oxygen attacks boron or aluminum. In a second step, hydrolysis of the O—B or
O—Al bond gives the alcohol product.

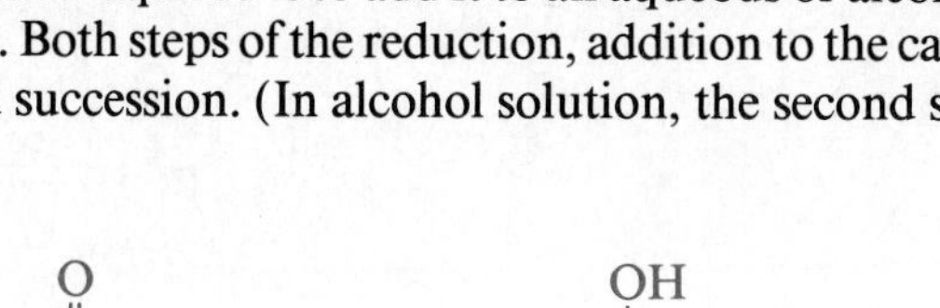

(X = B, Al)

Sodium borohydride is particularly easy to use. It is soluble in water and in
alcohols, and all that is required is to add it to an aqueous or alcoholic solution of an
aldehyde or ketone. Both steps of the reduction, addition to the carbonyl and hydrolysis, occur in rapid succession. (In alcohol solution, the second step is really one of
"alcoholysis.")

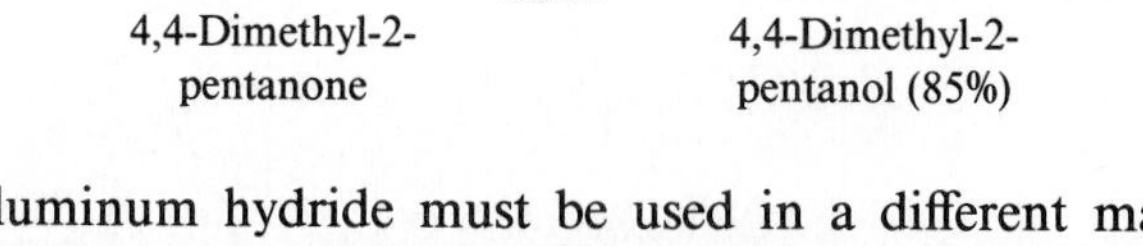

4,4-Dimethyl-2-   4,4-Dimethyl-2-
pentanone   pentanol (85%)

Lithium aluminum hydride must be used in a different manner, however.
LiAlH$_4$ reacts violently with water and alcohols and so must be used in anhydrous
(without water) solvents such as diethyl ether. Following reduction, a *separate* hydrolysis step is required to liberate the alcohol product.

$$CH_3(CH_2)_5CH \xrightarrow[\text{2. H}_2\text{O}]{\text{1. LiAlH}_4, \text{ diethyl ether}} CH_3(CH_2)_5CH_2OH$$

Heptanal   1-Heptanol

Note how the reaction has been written as two separate steps by using a "1" above the arrow and a "2" below it. It is important for you to recognize how that differs from a reaction where two reagents are added at the same time!

Neither sodium borohydride nor lithium aluminum hydride reduces isolated carbon-carbon double bonds. This makes possible the selective reduction of a carbonyl group in a molecule that contains both carbon-carbon and carbon-oxygen double bonds.

$$(CH_3)_2C=CHCH_2CH_2CCH_3 \xrightarrow[\text{2. H}_2\text{O}]{\text{1. LiAlH}_4, \text{ diethyl ether}} (CH_3)_2C=CHCH_2CH_2CHCH_3$$

6-Methyl-5-hepten-2-one   6-Methyl-5-hepten-2-ol
                                            (90%)

Catalytic hydrogenation would not be suitable for this reaction because hydrogen would add to the carbon-carbon double bond faster than to the carbonyl group.

PROBLEM 10.4

What is the product of each of the following reduction reactions?

(a) $(C_6H_5)_2CHCCH_3 \xrightarrow[\text{2. H}_2\text{O}]{\text{1. LiAlH}_4, \text{ diethyl ether}}$

(b) $(CH_3)_2CHCH_2CH \xrightarrow[\text{ethanol}]{\text{H}_2, \text{Pt}}$

(c) 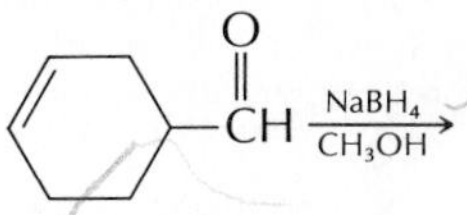 ⬡—CH=O $\xrightarrow[\text{CH}_3\text{OH}]{\text{NaBH}_4}$

(d) 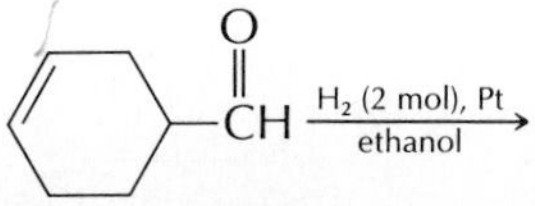 ⬡—CH=O $\xrightarrow[\text{ethanol}]{\text{H}_2 \text{ (2 mol), Pt}}$

SAMPLE SOLUTION

Lithium aluminum hydride converts the ketone to the corresponding secondary alcohol. In this case the product is 1,1-diphenyl-2-propanol.

$$(C_6H_5)_2CHCCH_3 \xrightarrow[\text{2. H}_2\text{O}]{\text{1. LiAlH}_4, \text{ diethyl ether}} (C_6H_5)_2CHCHCH_3$$

1,1-Diphenyl-2-   1,1-Diphenyl-2-
propanone   propanol (84%)

10.5 REACTIONS OF ALCOHOLS: A REVIEW AND A PREVIEW

Alcohols are among the most versatile starting materials for the preparation of a variety of organic compounds. We have already discussed several reactions of alcohols, and these are summarized in Table 10.3.

TABLE 10.3

Summary of Reactions of Alcohols Discussed in Earlier Chapters

Reaction (section) and comments	General equation and specific example
Reaction with hydrogen halides (Section 3.10) The order of alcohol reactivity parallels the order of carbocation stability: 3° > 2° > 1°.* Benzylic alcohols react readily. The order of hydrogen halide reactivity parallels their acidities: HI > HBr > HCl ≫ HF.	ROH + HX ⟶ RX + H_2O Alcohol Hydrogen halide Alkyl halide Water CH_3O —⟨ ⟩—CH_2OH $\xrightarrow{HBr}$ CH_3O —⟨ ⟩—CH_2Br *m*-Methoxybenzyl alcohol *m*-Methoxybenzyl bromide (98%)

Reaction with thionyl chloride (Section 3.14) Thionyl chloride converts alcohols to alkyl chlorides.	ROH + $SOCl_2$ ⟶ RCl + SO_2 + HCl
	 Alcohol Thionyl chloride Alkyl chloride Sulfur dioxide Hydrogen chloride
	
	 $(CH_3)_2C{=}CHCH_2CH_2CHCH_3$ $\xrightarrow[\text{diethyl ether}]{\text{SOCl}_2,\ \text{pyridine}}$
	                  $\mid$
	                  OH
	 6-Methyl-5-hepten-2-ol
	
	 $(CH_3)_2C{=}CHCH_2CH_2CHCH_3$
	                        $\mid$
	                        Cl
	 6-Chloro-2-methyl-2-heptene (67%)

Reaction with phosphorus trihalides (Section 3.14) Phosphorus trihalides convert alcohols to alkyl halides.	$3ROH$ + PX_3 ⟶ $3RX$ + $P(OH)_3$ Alcohol Phosphorus trihalide Alkyl halide Phosphorous acid ⬠—CH_2OH $\xrightarrow{PBr_3}$ ⬠—CH_2Br Cyclopentylmethanol (Bromomethyl)cyclopentane (50%)

Acid-catalyzed dehydration (Section 4.1.7) This is a frequently used procedure for the preparation of alkenes. The order of alcohol reactivity parallels the order of carbocation stability: 3° > 2° > 1°.* Benzylic alcohols react readily.	R_2CCHR_2 $\xrightarrow[\text{heat}]{H^+}$ $R_2C{=}CR_2$ + H_2O
	 $\mid$
	 OH
	 Alcohol Alkene Water
	
	 Br —⟨ ⟩—$CHCH_2CH_3$ $\xrightarrow[\text{heat}]{KHSO_4}$ Br —⟨ ⟩—$CH{=}CHCH_3$
	               $\mid$
	               OH
	 1-(*m*-Bromophenyl)-1-propanol 1-(*m*-Bromophenyl)propene (71%)

* The symbols 3°, 2°, and 1° refer to tertiary, secondary, and primary respectively.

Alcohols undergo reactions involving various combinations of the bond between carbon and oxygen and the O—H bond of the hydroxyl group. For instance, the carbon-oxygen bond of alcohols is cleaved when alcohols are converted to alkyl halides (Section 3.10) or undergo acid-catalyzed dehydration (Section 4.1.7).

$$-\overset{\displaystyle |}{\underset{\displaystyle |}{C}}-O-H$$

This bond is broken when alcohols are
converted to alkyl halides or subjected
to acid-catalyzed dehydration

The ionization of alcohols when they act as weak acids (Section 10.6) is a reaction of the O—H bond.

This bond is broken when alcohols
are converted to alkoxides

$$-\overset{\displaystyle |}{\underset{\displaystyle |}{C}}-O-H$$

Primary and secondary alcohols can exhibit a third reaction type in which a carbon-oxygen double bond is formed by cleavage of both an O—H bond and a C—H bond (Section 10.8).

$$H-\overset{\displaystyle |}{\underset{\displaystyle |}{C}}-O-H \longrightarrow \hspace{1cm} C=O$$

Breaking these two bonds
allows a carbonyl group to
be formed

Carbonyl group of
an aldehyde or ketone

Recall from Section 10.3 that conversion of an alcohol to a carbonyl compound is an *oxidation reaction*. Other reactions of alcohols will be encountered in the chapters on aldehydes and ketones (Chapter 11) and carboxylic acid derivatives (Chapter 13).

10.6 ALCOHOLS AS BRÖNSTED ACIDS

In the presence of strong bases, alcohols can act as proton donors; they are *Brönsted acids* (Section 3.7).

$$B\colon + H-\overset{\displaystyle ..}{\underset{\displaystyle ..}{O}}R \rightleftharpoons B^+ - H + \hspace{0.5cm} {}^-\colon\overset{\displaystyle ..}{\underset{\displaystyle ..}{O}}R$$

Base	Alcohol (acid)	(Conjugate acid)	Alkoxide ion (conjugate base)

Alcohols tend to be slightly weaker acids than water (Table 3.3), with pK_a values in the range 16 to 18. As described in Chapter 3, the position of an acid-base equilibrium favors the formation of the weaker acid and base. Thus, when *tert*-butoxide ion is introduced into aqueous solution, the predominant species present are *tert*-butyl alcohol and hydroxide ion.

$$(CH_3)_3C\overset{\displaystyle ..}{\underset{\displaystyle ..}{O}}\colon^- + H-\overset{\displaystyle ..}{\underset{\displaystyle ..}{O}}-H \rightleftharpoons (CH_3)_3C\overset{\displaystyle ..}{\underset{\displaystyle ..}{O}}H + H\overset{\displaystyle ..}{\underset{\displaystyle ..}{O}}\colon^-$$

tert-Butoxide ion (stronger base)	Water (stronger acid; $K_a \approx 10^{-16}$)	*tert*-Butyl alcohol (weaker acid; $K_a \approx 10^{-18}$)	Hydroxide ion (weaker base)

The most common method for preparation of alkoxides is the reaction between an alkali (group I) metal, usually sodium or potassium, and an alcohol.

$$2CH_3CH_2OH + 2Na \longrightarrow 2Na^+ {}^-OCH_2CH_3 + H_2(g)$$

Ethanol	Sodium	Sodium ethoxide	Hydrogen

As normally carried out, the metal is added to a large excess of the alcohol, and the alkoxide is then used as an alcoholic solution.

Alkoxide ions may also be prepared by reaction of an alcohol with a strong base such as sodium hydride, NaH.

$$\text{ROH} + \text{NaH} \longrightarrow \text{NaOR} + \text{H}_2(g)$$

| Alcohol | Sodium hydride | Sodium alkoxide | Hydrogen |

PROBLEM 10.5

Write equations describing two methods for the preparation of sodium methoxide, $NaOCH_3$.

10.7 CONVERSION OF ALCOHOLS TO ETHERS

Primary alcohols are converted to ethers on heating in the presence of an acid catalyst, usually sulfuric acid.

$$2RCH_2OH \xrightarrow{\text{H}^+,\ \text{heat}} RCH_2OCH_2R + H_2O$$

Primary alcohol   Dialkyl ether   Water

The above reaction is a condensation. A **condensation reaction** is one in which two molecules combine to form a larger one while liberating a small molecule, usually water. In this case two alcohol molecules combine to give an ether and water.

$$2CH_3CH_2CH_2CH_2OH \xrightarrow[130°C]{\text{H}_2\text{SO}_4} CH_3CH_2CH_2CH_2OCH_2CH_2CH_2CH_3 + H_2O$$

1-Butanol   Dibutyl ether (60%)   Water

When applied to the synthesis of ethers, the reaction is effective only with primary alcohols. Elimination to form alkenes (Section 4.1.7) is the principal reaction that occurs at higher temperatures and with secondary and tertiary alcohols.

PROBLEM 10.6

Diethyl ether was the first general anesthetic used in surgery during the nineteenth century. Although the use of diethyl ether in hospitals has diminished due to its flammability, it is still widely used as a laboratory and industrial solvent. Show with a balanced equation how diethyl ether could be prepared by a condensation reaction.

$$\longrightarrow CH_3CH_2OCH_2CH_3$$

Diethyl ether

10.8 OXIDATION OF ALCOHOLS

Recall from Sections 10.3 and 10.5 that alcohols can undergo oxidation to a carbonyl compound. Whether oxidation leads to an aldehyde, a ketone, or a carboxylic acid depends on the alcohol, the oxidizing agent, and the conditions of the reaction.

Primary alcohols may be oxidized either to an aldehyde or to a carboxylic acid.

$$RCH_2OH \xrightarrow{\text{oxidize}} \overset{\displaystyle O}{\overset{\|}{RCH}} \xrightarrow{\text{oxidize}} \overset{\displaystyle O}{\overset{\|}{RCOH}}$$

Primary   Aldehyde   Carboxylic
alcohol   acid

The reagents that are most commonly used for oxidizing alcohols are based on high-oxidation-state transition metals, particularly manganese(VII) and chromium(VI). Vigorous oxidation leads to the formation of a carboxylic acid, but methods are available that permit us to stop the oxidation at the intermediate aldehyde stage.

Oxidation with potassium permanganate converts primary alcohols to carboxylic acids

$$CH_3CH_2CH(CH_2)_4CH_2OH \xrightarrow[\text{H}_2\text{SO}_4,\ \text{H}_2\text{O}]{\text{KMnO}_4} \overset{\displaystyle O}{\overset{\|}{CH_3CH_2CH(CH_2)_4COH}}$$
$$\overset{|}{CH_3} \qquad\qquad\qquad\qquad\qquad \overset{|}{CH_3}$$

6-Methyl-1-octanol   6-Methyloctanoic acid
(66%)

Chromic acid (H_2CrO_4) is a good oxidizing agent and is formed when solutions containing sodium or potassium salts of chromate (CrO_4^{2-}) or dichromate ($Cr_2O_7^{2-}$) are acidified. In most cases carboxylic acids are the major products isolated on treatment of primary alcohols with chromic acid.

$$FCH_2CH_2CH_2OH \xrightarrow[\text{H}_2\text{SO}_4,\ \text{H}_2\text{O}]{\text{K}_2\text{Cr}_2\text{O}_7} \overset{\displaystyle O}{\overset{\|}{FCH_2CH_2COH}}$$

3-Fluoro-1-propanol   3-Fluoropropanoic
acid (74%)

Stopping the oxidation of primary alcohols at the aldehyde stage is best done by using a chromium(VI) reagent in an *anhydrous* medium. When water is present, the carboxylic acid is the major product. One combination effective for the preparation of aldehydes is a chromium trioxide–pyridine complex, $(C_5H_5N)_2CrO_3$, in dichloromethane (CH_2Cl_2). This mixture is called *Collins's reagent.*

$$CH_3CH_2CH(CH_2)_4CH_2OH \xrightarrow[\text{CH}_2\text{Cl}_2]{(\text{C}_5\text{H}_5\text{N})_2\text{CrO}_3} \overset{\displaystyle O}{\overset{\|}{CH_3CH_2CH(CH_2)_4CH}}$$
$$\overset{|}{CH_3} \qquad\qquad\qquad\qquad\qquad\qquad \overset{|}{CH_3}$$

6-Methyl-1-octanol   6-Methyloctanal (69%)

Secondary alcohols are oxidized to ketones by the same reagents that oxidize primary alcohols.

$$\overset{\displaystyle OH}{\overset{|}{RCHR'}} \xrightarrow{\text{oxidize}} \overset{\displaystyle O}{\overset{\|}{RCR'}}$$

Secondary   Ketone
alcohol

The chromium-based reagents are more commonly used than potassium permanganate because they are somewhat milder oxidants and are less prone to decompose the ketone product by overoxidation. Any of the reagents mentioned, including Collins's reagent, may be used for the oxidation of a secondary alcohol.

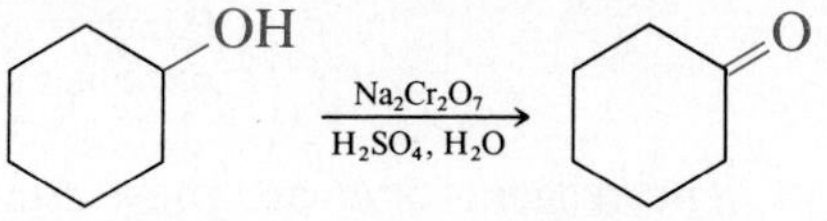

Cyclohexanol Cyclohexanone

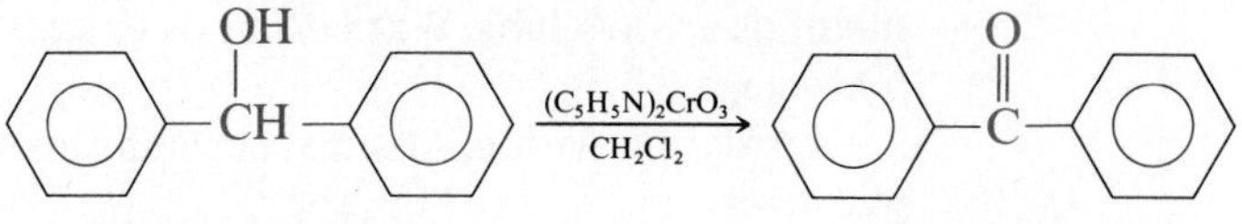

Diphenylmethanol Benzophenone (96%)

Tertiary alcohols have no hydrogen on their hydroxyl-bearing carbon and do not undergo oxidation readily.

$$R—\overset{R'}{\underset{R''}{C}}—OH \xrightarrow{\text{oxidize}} \text{no reaction except under forcing conditions}$$

PROBLEM 10.7

Predict the principal organic product of each of the following reactions:

(a) $ClCH_2CH_2CH_2CH_2OH \xrightarrow[H_2SO_4,\ H_2O]{K_2Cr_2O_7}$

(b) $CH_3\underset{OH}{\underset{|}{C}}HCH_2CH_2CH_2CH_2CH_2CH_3 \xrightarrow[H_2SO_4,\ H_2O]{Na_2Cr_2O_7}$

(c) $CH_3CH_2CH_2CH_2CH_2CH_2CH_2CH_2OH \xrightarrow[CH_2Cl_2]{(C_5H_5N)_2CrO_3}$

(d) 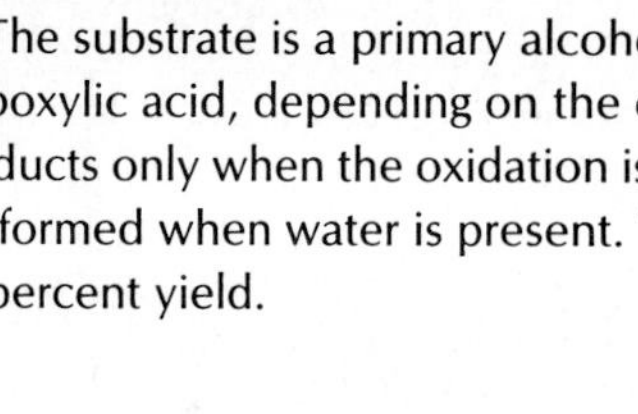

SAMPLE SOLUTION

(a) The substrate is a primary alcohol and can be oxidized either to an aldehyde or to a carboxylic acid, depending on the conditions of the reaction. Aldehydes are the major products only when the oxidation is carried out in anhydrous media; carboxylic acids are formed when water is present. The reaction shown produced the carboxylic acid in 56 percent yield.

$$\underset{\text{4-Chloro-1-butanol}}{ClCH_2CH_2CH_2CH_2OH} \xrightarrow[H_2SO_4,\ H_2O]{K_2Cr_2O_7} \underset{\text{4-Chlorobutanoic acid}}{ClCH_2CH_2CH_2\overset{O}{\overset{\|}{C}}OH}$$

BIOLOGICAL OXIDATION OF ALCOHOLS

Many of the transformations that we describe as laboratory reactions have counterparts that occur in living systems. For example, oxidation of alcohols to carbonyl compounds or the reverse, reduction of carbonyl compounds to alcohols, occur in many biological processes. Ethanol, for example, is oxidized in the liver to acetaldehyde. The enzyme that catalyzes the oxidation of ethanol is called *alcohol dehydrogenase*.

$$CH_3CH_2OH \xrightarrow{\text{alcohol dehydrogenase}} CH_3\overset{\displaystyle O}{\overset{\displaystyle \|}{C}}H$$

Ethanol   Acetaldehyde

The rate of oxidation of ethanol in a particular person is constant, regardless of the concentration of alcohol present. Thus ingestion of ethanol from alcoholic beverages at a rate greater than the rate of oxidation results in a buildup of ethanol in the bloodstream, leading to intoxication.

Acetaldehyde is further oxidized to acetate in a reaction catalyzed by the enzyme *aldehyde dehydrogenase*.

$$CH_3\overset{\displaystyle O}{\overset{\displaystyle \|}{C}}H \xrightarrow{\text{aldehyde dehydrogenase}} CH_3\overset{\displaystyle O}{\overset{\displaystyle \|}{C}}O^-$$

Acetaldehyde   Acetate

Acetate is used by the body in the synthesis of fatty acids and cholesterol (see Chapter 17).

As noted in Section 10.2, methanol (CH_3OH) is a poisonous substance that can cause blindness or death from the ingestion of even small amounts. Methanol is oxidized to formaldehyde in the body by the *same* enzyme system that oxidizes ethanol, alcohol dehydrogenase.

$$CH_3OH \xrightarrow{\text{alcohol dehydrogenase}} H\overset{\displaystyle O}{\overset{\displaystyle \|}{C}}H$$

Methanol   Formaldehyde

Formaldehyde is toxic to the retina of the eye and is the cause of blindness from methanol ingestion. Formaldehyde is further oxidized in the body to formate.

$$H\overset{\displaystyle O}{\overset{\displaystyle \|}{C}}H \xrightarrow{\text{aldehyde dehydrogenase}} H\overset{\displaystyle O}{\overset{\displaystyle \|}{C}}O^-$$

Formaldehyde   Formate

Formate is not utilized rapidly by the body and collects in the blood as formic acid. This lowers the pH of the blood below the normal physiological range, resulting in a fatal condition called *acidosis*.

The treatment for methanol poisoning makes use of the fact that both methanol and ethanol are oxidized in the body by the same enzyme systems.

*Ethers are
Relatively
Inert*

Thus a solution containing ethanol is given to a victim of methanol poisoning. Ethanol is considerably less toxic than methanol, and the ethanol oxidation "ties up" the enzymes, preventing the more harmful methanol oxidation from occurring.

10.9 INTRODUCTION TO ETHERS

Ethers contain a C—O—C unit. In contrast to alcohols, ethers undergo relatively few chemical reactions. Ethers do not have the O—H group present in alcohols, and thus ethers are unable to undergo the deprotonation and oxidation reactions typical of alcohols. This lack of reactivity makes ethers valuable as solvents in a number of synthetically useful chemical reactions.

Epoxides are an important exception to the chemical inertness of ethers. **Epoxides** are ethers in which the C—O—C unit forms a three-membered ring. Because of the strain energy present in such a small ring, epoxides are very reactive substances.

10.10 ETHER NOMENCLATURE

Ethers are described as **symmetrical** or **unsymmetrical,** depending on whether the two groups bonded to oxygen are the same or different.

$$CH_3CH_2OCH_2CH_3 \qquad CH_3CH_2OCH_3$$

Diethyl ether   Ethyl methyl ether
(a symmetrical ether)   (an unsymmetrical ether)

Ethers are named by specifying the two alkyl groups in alphabetical order as separate words, then adding the word *ether* at the end. The prefix *di-* is used in symmetrical ethers, in which both alkyl groups are the same.

Often the common solvent diethyl ether is called just *ether,* or *anesthesia ether.* The latter name derives from its former use an a hospital anesthetic. Diethyl ether is extremely flammable, and its use in hospitals has largely been replaced by less hazardous alternatives such as halothane (Section 7.8).

Cyclic ethers have their ether oxygen as part of a ring. Tetrahydrofuran and tetrahydropyran are cyclic ethers.

Tetrahydrofuran   Tetrahydropyran

In general, the properties of cyclic ethers are very much like those of their noncyclic counterparts. Exceptions, as we noted in the previous section, are the epoxides. At one time epoxides were named as oxides of alkenes. Ethylene oxide and propylene oxide, for example, are the common names of two industrially important epoxides. The industrial use of ethylene oxide is discussed in Section 10.14.

$$H_2C\text{——}CH_2 \qquad CH_2\text{—}CHCH_3$$
$$O \qquad\qquad\qquad O$$

Ethylene oxide   Propylene oxide

The IUPAC system permits two alternative ways to name epoxides. In one of these, epoxides are named as epoxy derivatives of alkanes. According to this system, ethylene oxide becomes epoxyethane and propylene oxide becomes 1,2-epoxypropane. The prefix *epoxy-* always immediately precedes the alkane ending; it is not listed in alphabetical order in the manner of other substituents.

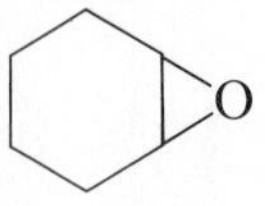

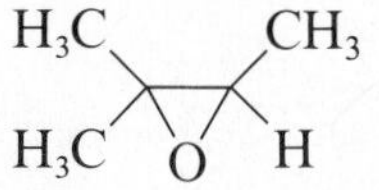

1,2-Epoxycyclohexane   2-Methyl-2,3-epoxybutane

Another way that epoxides are named in the IUPAC system is based on *oxirane* as the parent name of the simplest epoxide, ethylene oxide. Substituents are specified in the usual way and their positions identified by number. Numbering begins at the ring oxygen and proceeds in the direction that gives the lower number to the substituent.

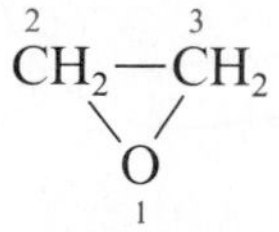

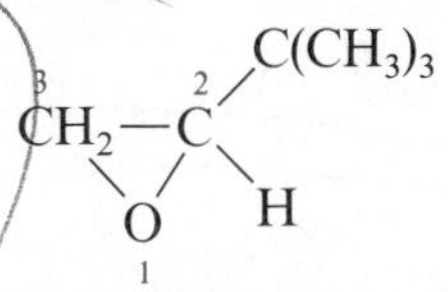

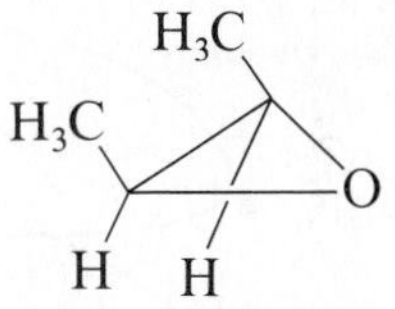

Oxirane   2-*tert*-Butyloxirane   *cis*-2,3-Dimethyloxirane

PROBLEM 10.8
Each of the following ethers has been shown to be or is suspected to be a *mutagen*, which means it can induce genetic mutations in cells. Write the structures of each of these ethers.

(a) Chloromethyl methyl ether
(b) 2-(Chloromethyl)oxirane (also known as epichlorohydrin)
(c) 3,4-Epoxy-1-butene (2-vinyloxirane)

SAMPLE SOLUTION
(a) Chloromethyl methyl ether has a chloromethyl group ($ClCH_2$—) and a methyl group (CH_3—) attached to oxygen. Its structure is $ClCH_2OCH_3$.

10.11 ETHERS IN NATURE

A number of ethers are naturally occurring and have interesting biological properties. For example, monensin is a **polyether antibiotic.** As shown in Figure 10.3, monensin is able to bind a sodium ion with its ether oxygens. The nonpolar alkyl groups of monensin are oriented toward the outside of the monensin–Na^+ complex, thus allowing the complex to penetrate the hydrocarbon-like surface of a cell membrane. This disrupts the normal balance of sodium ions within the cell and interferes with cellular respiration. Small amounts of monensin are added to poultry feed to kill parasites that live in the intestines of chickens.

Two other naturally occurring ethers are illustrated in Figure 10.4. Disparlure is an example of a pheromone. **Pheromones** are compounds that enable an insect to communicate with members of its own or another species.

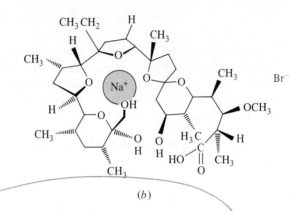

(a)

(b)

FIGURE 10.3 (a) The structure of monensin. (b) The structure of the monensin–sodium bromide complex showing coordination of the sodium ion by oxygen atoms of monensin.

10.12 PREPARATION OF ETHERS

As you saw in Section 10.7, some ethers can be prepared by the reaction of a primary alcohol with sulfuric acid. Another important method used for the preparation of ethers is called the **Williamson ether synthesis.**

Recall from Chapter 8 that an alkoxide ion may act as a nucleophile and react with a substrate such as an alkyl halide. Nucleophilic substitution of an alkyl halide by an alkoxide leads to carbon-oxygen bond formation between the substrate and the nucleophile.

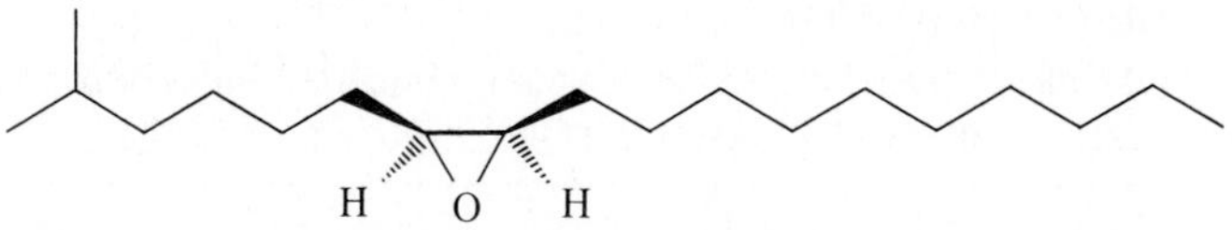

Disparlure (sex attractant of the
female gypsy moth)

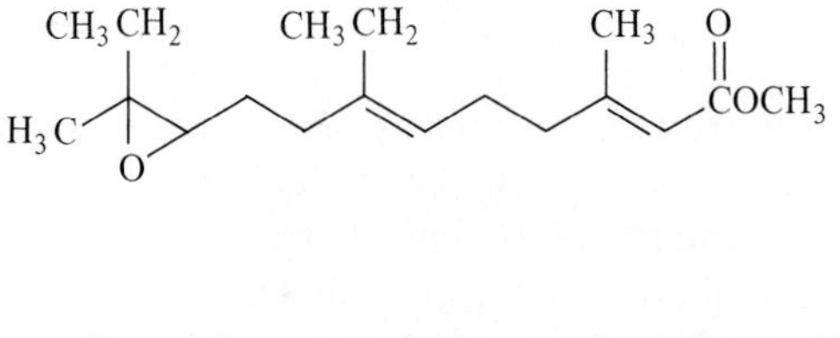

FIGURE 10.4 Two naturally occurring epoxides.

Growth hormone of the cecropia moth

$$\ddot{R}\ddot{O}:^- + R'{-}X \longrightarrow ROR' + X^-$$

Alkoxide   Alkyl   Ether   Halide
ion   halide   ion

Preparation of ethers by the Williamson ether synthesis is most successful when the alkyl halide is one which is reactive toward S_N2 substitution. Methyl halides and primary alkyl halides are the best substrates.

$$CH_3CH_2CH_2CH_2ONa + CH_3CH_2I \longrightarrow CH_3CH_2CH_2CH_2OCH_2CH_3 + NaI$$

Sodium butoxide   Iodoethane   Butyl ethyl ether (71%)   Sodium iodide

---

PROBLEM 10.9

Show how each of the following ethers could be prepared by the Williamson ether synthesis:

(a) $CH_3CH_2OCH_2CH_3$
(b) $CH_3OCH_2CH_2CH_3$ (two ways)
(c) $C_6H_5CH_2OCH_2CH_3$ (two ways)

SAMPLE SOLUTION

(a) Ethers are characterized by a C—O—C unit. One of the carbon-oxygen bonds comes from the alkoxide ion used as a nucleophile, while the other carbon-oxygen bond arises by an S_N2 reaction of the alkoxide with an alkyl halide. The preparation of diethyl ether requires sodium ethoxide as the nucleophile and an ethyl halide as the substrate.

$$CH_3CH_2ONa + CH_3CH_2Br \longrightarrow CH_3CH_2OCH_2CH_3 + NaBr$$

Sodium ethoxide   Ethyl bromide   Diethyl ether   Sodium bromide

---

Secondary and tertiary alkyl halides are poor substrates because they tend to react with alkoxide bases by E2 elimination rather than by S_N2 substitution. Whether the alkoxide base is primary, secondary, or tertiary is much less important than the nature of the alkyl halide. Thus benzyl isopropyl ether is prepared in high yield from benzyl chloride, a primary chloride that is incapable of undergoing elimination, and sodium isopropoxide.

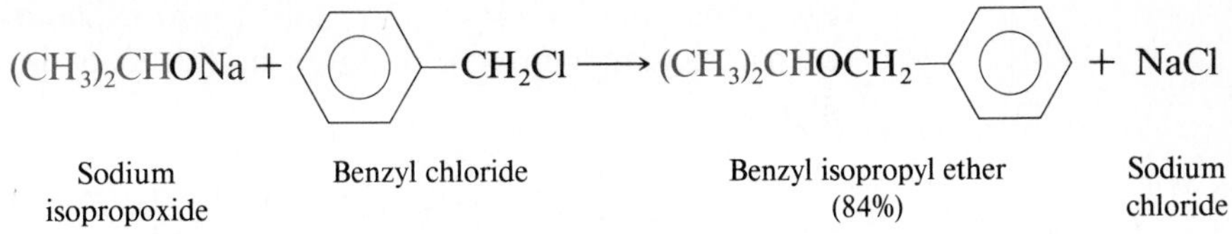

$$(CH_3)_2CHONa + \langle\bigcirc\rangle{-}CH_2Cl \longrightarrow (CH_3)_2CHOCH_2{-}\langle\bigcirc\rangle + NaCl$$

Sodium isopropoxide   Benzyl chloride   Benzyl isopropyl ether (84%)   Sodium chloride

The alternative synthetic route using the sodium salt of benzyl alcohol and an isopropyl halide would be much less effective because of increased competition from the elimination reaction.

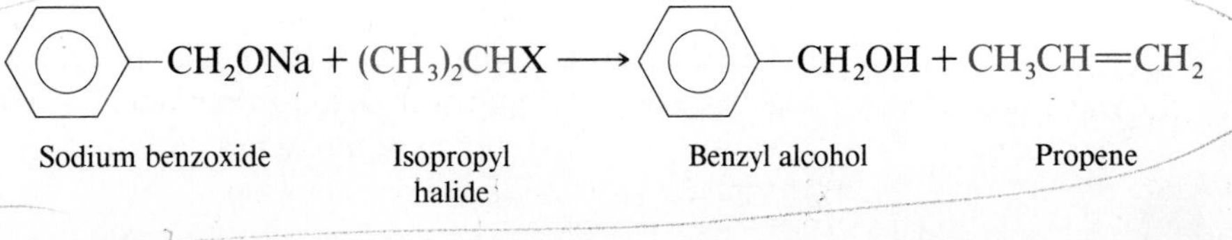

$$\langle\bigcirc\rangle{-}CH_2ONa + (CH_3)_2CHX \longrightarrow \langle\bigcirc\rangle{-}CH_2OH + CH_3CH{=}CH_2$$

Sodium benzoxide   Isopropyl halide   Benzyl alcohol   Propene

10.13 PREPARATION OF EPOXIDES

There are two general methods for the preparation of epoxides: the epoxidation of alkenes and the base-promoted ring closure of halohydrins.

Alkenes react with reagents known as peroxy acids in a process called **epoxidation.**

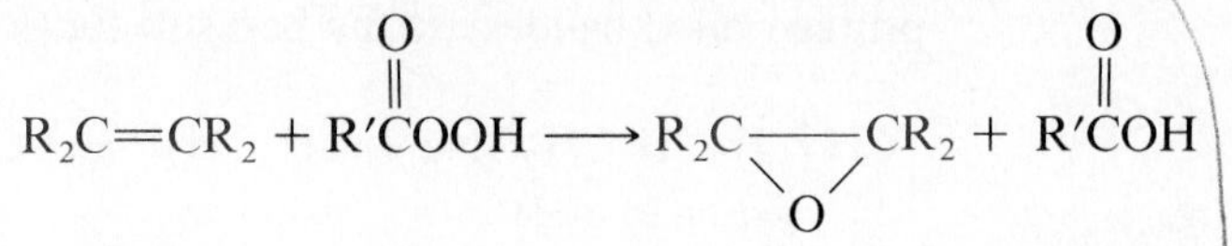

$$R_2C{=}CR_2 + R'COOH \longrightarrow R_2C{-}CR_2 + R'COH$$

| Alkene | Peroxy acid | Epoxide | Carboxylic acid |

Although we will not discuss the mechanism in detail, you can see from the general reaction that an oxygen atom is transferred from the peroxy acid to the alkene to form an epoxide. The reaction occurs in such a way that the stereochemistry of the alkene is retained in the epoxide; groups that are cis in the alkene remain cis in the epoxide, and those that are trans remain trans in the product.

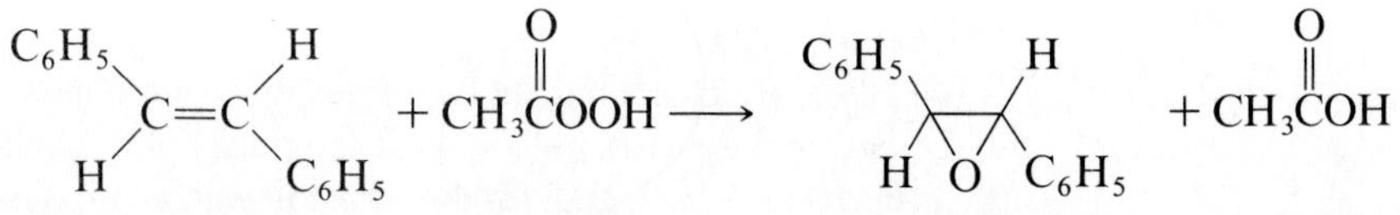

| (E)-1,2-Diphenylethene | Peroxyacetic acid | trans-2,3-Diphenyloxirane (78–83%) | Acetic acid |

PROBLEM 10.10
Figure 10.4 shows the structure of disparlure, the sex attractant of the female gypsy moth. Draw the structure of the alkene which could be used to prepare synthetic disparlure by an epoxidation reaction. Is this alkene the E or Z isomer?

The second method for preparing epoxides involves what amounts to an **intramolecular** (within the same molecule) Williamson ether synthesis. The hydroxyl group and the halide leaving group are on adjacent carbons; that is, they are **vicinal** to each other. Such molecules are called vicinal halohydrins. Treatment with a base such as sodium hydroxide removes the alcohol hydrogen to form the corresponding alkoxide.

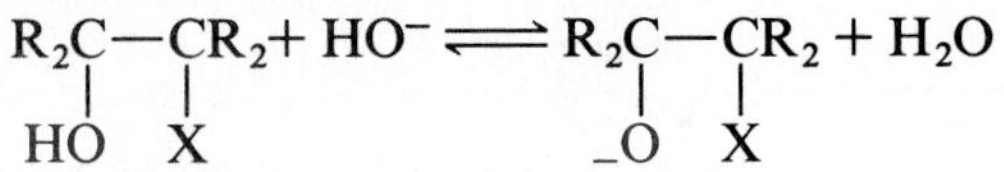

$$R_2C{-}CR_2 + HO^- \rightleftharpoons R_2C{-}CR_2 + H_2O$$

Halohydrin

This alkoxide contains both a nucleophile (the alkoxide oxygen) and a leaving group (the halogen) in close proximity to each other. In a step faster than the reaction of the halohydrin with any external nucleophile, the alkoxide oxygen attacks the carbon that bears the leaving group, giving an epoxide. As in other nucleophilic displacement reactions, the nucleophile approaches carbon from the side opposite the leaving group.

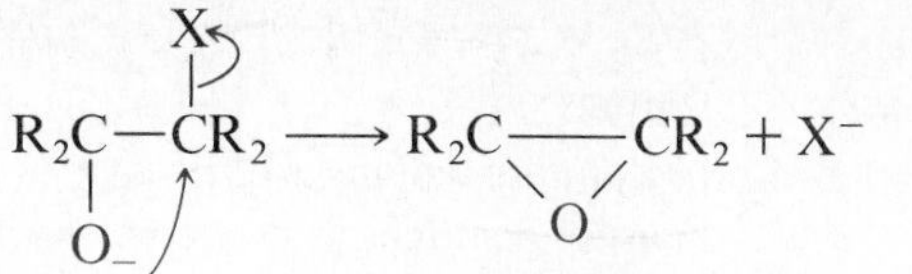

$$R_2C-CR_2 \longrightarrow R_2C\underset{O}{\overset{\displaystyle\triangle}{}}CR_2 + X^-$$

Epoxide

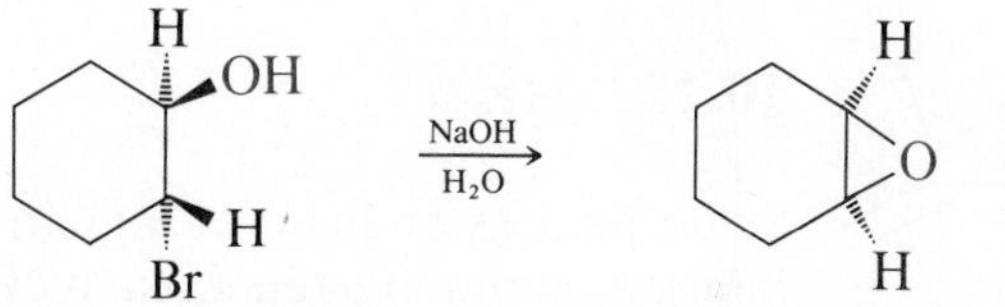

$$\text{trans-2-Bromocyclohexanol} \xrightarrow[H_2O]{NaOH} \text{1,2-Epoxycyclohexane (81\%)}$$

10.14 REACTIONS OF EPOXIDES

The chemical property that most distinguishes epoxides from other ethers is their far greater reactivity toward nucleophilic reagents. Reactions that lead to opening of the three-membered ring relieve the ring strain present in epoxides.

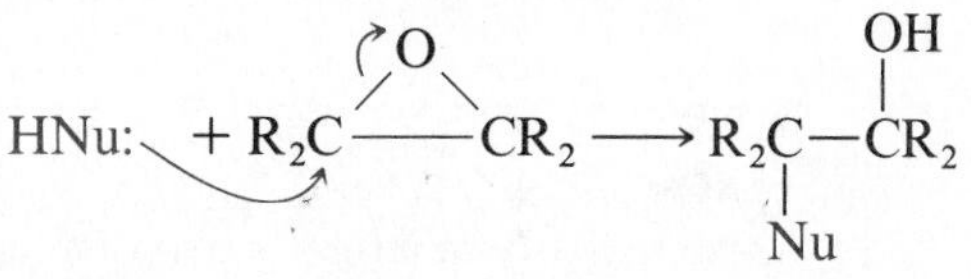

$$HNu: + R_2C-CR_2 \longrightarrow R_2C-CR_2$$

Nucleophile   Epoxide   Product

Ethylene oxide (also known as epoxyethane or oxirane) is a typical epoxide, widely used in the chemical industry (6 *billion* pounds in 1986). It is a synthetic intermediate to a wide variety of familiar end products. For instance, ethylene glycol, used in automotive antifreeze and to make polyester fabrics, is prepared by the hydrolysis of ethylene oxide in dilute sulfuric acid.

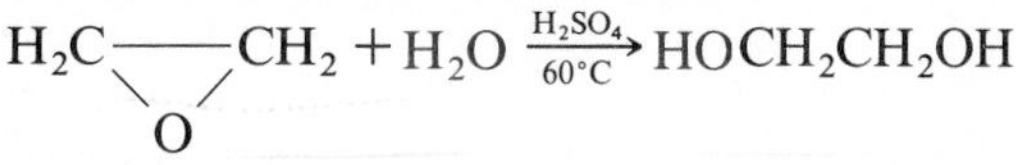

$$H_2C-CH_2 + H_2O \xrightarrow[60°C]{H_2SO_4} HOCH_2CH_2OH$$

Ethylene oxide   Ethylene glycol
(epoxyethane)   (1,2-ethanediol)

Water is the nucleophile in this reaction, which is catalyzed by the sulfuric acid.

Other nucleophiles react with ethylene oxide in an analogous manner to yield 2-substituted derivatives of ethanol. 2-Ethoxyethanol is used as a lacquer thinner and varnish remover as well as an anti-icing additive in aviation fuels. The nucleophile in this reaction is ethanol.

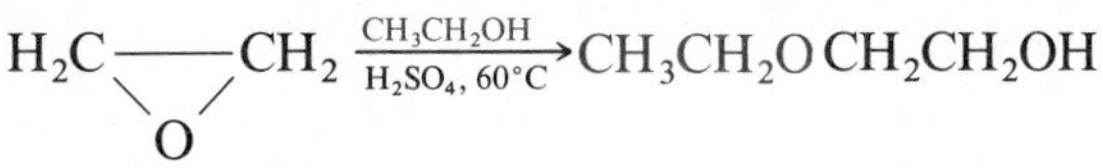

$$H_2C-CH_2 \xrightarrow[H_2SO_4,\,60°C]{CH_3CH_2OH} CH_3CH_2OCH_2CH_2OH$$

Ethylene oxide   2-Ethoxyethanol

Reaction of ethylene oxide with aqueous ammonia (ammonia is the nucleophile) gives 2-aminoethanol, used commercially as a corrosion inhibitor and as an emulsifying agent in some paints.

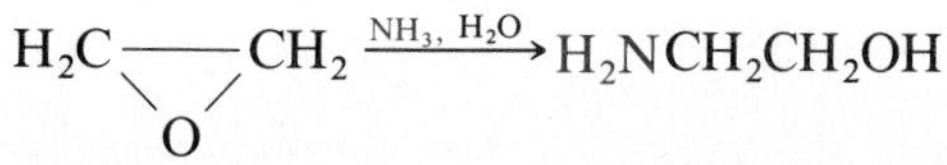

$$H_2C-CH_2 \xrightarrow{NH_3,\,H_2O} H_2NCH_2CH_2OH$$

Ethylene oxide   2-Aminoethanol

PROBLEM 10.11

A commercial material known as *butyl cellosolve,* used as a lacquer solvent and in varnish remover, is made by reaction of ethylene oxide with 1-butanol. Draw the structure of this product.

10.15 THIOLS

Sulfur analogs of alcohols contain the —SH functional group and are known as **thiols.** A common name for a thiol is a *mercaptan,* meaning "mercury-capturing."

$$CH_3SH \qquad CH_3CH_2SH \qquad CH_3CH_2CH_2SH$$

Methanethiol   Ethanethiol   1-Propanethiol

(Methyl mercaptan)   (Ethyl mercaptan)   (Propyl mercaptan)

One characteristic of thiols is their strong, usually disagreeable odor. Since methane (natural gas) and propane (used in liquefied petroleum gas) are odorless compounds, methanethiol and ethanethiol are added in small amounts as odorants to provide warning of a gas leak. Propanethiol is present in onions.

PROBLEM 10.12

3-Methyl-1-butanethiol is part of a mixture of compounds responsible for the odor of a skunk. Draw the structure of this compound.

The natural compound cysteine is a thiol. Cysteine is present in virtually all proteins and enzymes and plays an important structural role in living systems (discussed in Chapter 16).

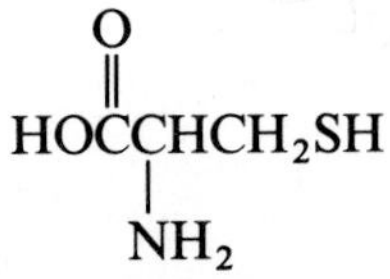

$$\overset{\displaystyle O}{\overset{\displaystyle \|}{HOCCHCH_2SH}}$$

$$\underset{\displaystyle NH_2}{|}$$

Cysteine

The normal function of a thiol group in the body can be disrupted by the presence of toxic metals such as mercury (hence the common name mercaptan). The mercuric or mercury(II) ion (Hg^{2+}) binds to thiol groups present in enzymes, thus inhibiting the biological activity of the enzyme.

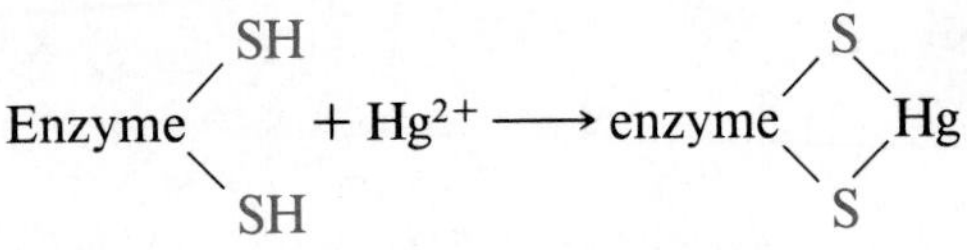

$$Enzyme \overset{\displaystyle SH}{\underset{\displaystyle SH}{<}} + Hg^{2+} \longrightarrow enzyme \overset{\displaystyle S}{\underset{\displaystyle S}{<}} Hg$$

Normal enzyme   Inhibited enzyme

A treatment for mercury poisoning is to give a thiol called *dimercaprol.*

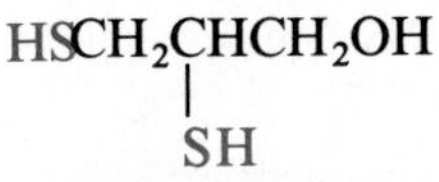

$$HSCH_2CHCH_2OH$$

$$\underset{\displaystyle SH}{|}$$

Dimercaprol

EPOXIDES AND CHEMICAL CARCINOGENESIS

Chemical carcinogens are chemical compounds capable of transforming a normal cell into one which is malignant; in other words, causing cancer. The complex series of steps that occur during this change is called **carcinogenesis.**

Epoxides appear to play a role in the interactions of certain chemical carcinogens with cells. A key step in chemical carcinogenesis is the reaction of a cellular nucleophile with a molecule which can undergo nucleophilic attack. A number of molecules which have been found to be carcinogenic form epoxides in the body; it is these epoxides which are attacked by the cellular nucleophiles. This process of transforming an unreactive molecule into one which is reactive is called **bioactivation.**

One example of a carcinogenic molecule which undergoes bioactivation to form an epoxide is benz[a]pyrene, which was described in Chapter 6. Another example is benzene:

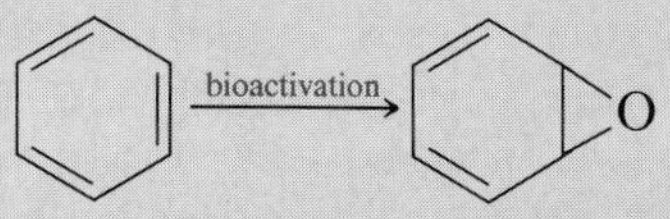

One reason the study of carcinogenesis is so difficult is that metabolic reactions similar to the one shown for benzene are normal parts of cellular metabolism. The biosynthesis of naturally occurring phenols such as tyrosine involve reactions that proceed through epoxide intermediates. And yet these reactions don't cause cancer!

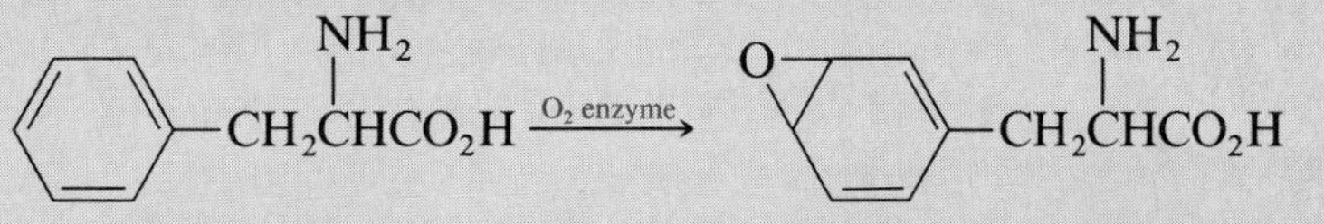

Phenylalanine   Epoxide intermediate

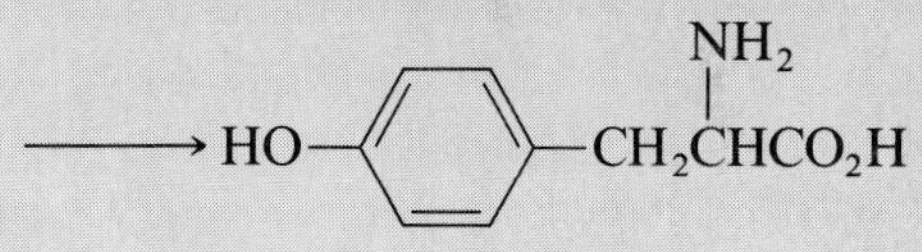

Tyrosine

Dimercaprol competes with the enzymes for the mercury ion, thus allowing the toxic metal to be removed from the cells and excreted from the body. Dimercaprol is also used in the treatment of arsenic poisoning.

10.16 INTRODUCTION TO PHENOLS. NOMENCLATURE

[handwritten: More acidic]

Phenols are compounds that have a hydroxyl group bonded directly to a benzene ring. The parent compound is called simply **phenol.** Several phenol derivatives have common names acceptable in the IUPAC system. For example, the *o-*, *m-*, and *p*-methylphenols are called **cresols.**

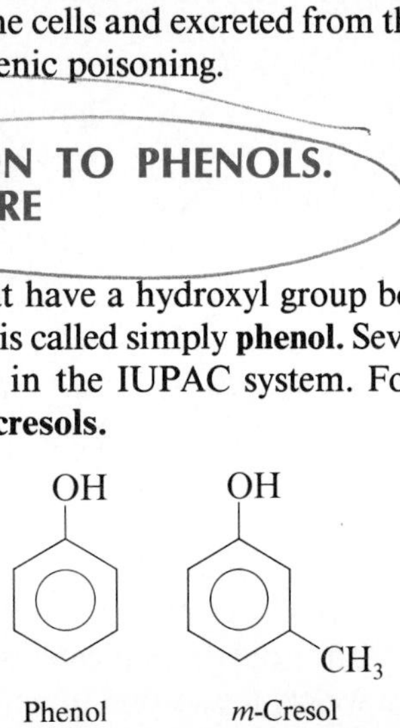

Phenol   *m*-Cresol

Substituted phenols are numbered around the ring beginning with the hydroxyl group and proceeding in the direction that gives the lower number to the next carbon atom bearing a substituent. Thus the example shown is correctly named as 5-chloro-2-methylphenol, *not* 3-chloro-6-methylphenol.

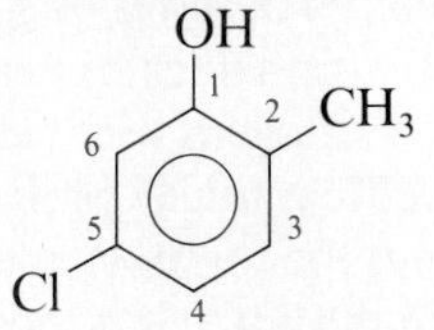

5-Chloro-2-methylphenol

The three dihydroxy derivatives of benzene are named 1,2-, 1,3-, and 1,4-benzenediol, respectively, but each is more familiarly known by its common name (shown in parentheses).

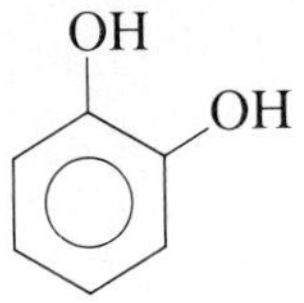

1,2-Benzenediol
(pyrocatechol;
an antioxidant)

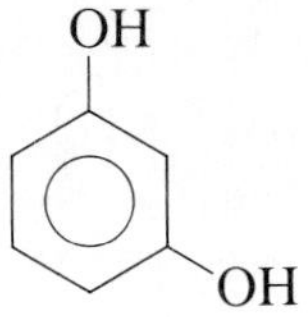

1,3-Benzenediol
(resorcinol;
an adhesive for
wood veneers)

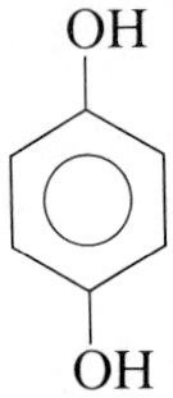

1,4-Benzenediol
(hydroquinone; a
black-and-white photo-
graphic developer)

PROBLEM 10.13

Write a structural formula for each of the following compounds:

(a) 1,2,3-Benzenetriol (pyrogallol, a photographic developer)
(b) 2,4,6-Trinitrophenol (picric acid, an explosive)
(c) 2,4,5-Trichlorophenol (material used in the synthesis of a pesticide)

SAMPLE SOLUTION

(a) Pyrogallol is the common name for 1,2,3-benzenetriol. The three hydroxyl groups occupy adjacent positions on a benzene ring.

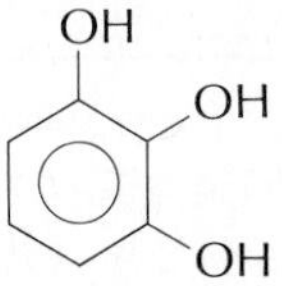

1,2,3-Benzenetriol
(pyrogallol)

10.17 SYNTHETIC AND NATURALLY OCCURRING PHENOL DERIVATIVES

Phenol and the cresols have antiseptic properties and are used commercially in dilute aqueous solution as household disinfectants. Lysol is one brand-name example. Phenol is also used as the starting material for products as varied as the analgesic aspirin and the pesticide 2,4-D. Pentachlorophenol, known commercially as Penta, is a widely used wood preservative.

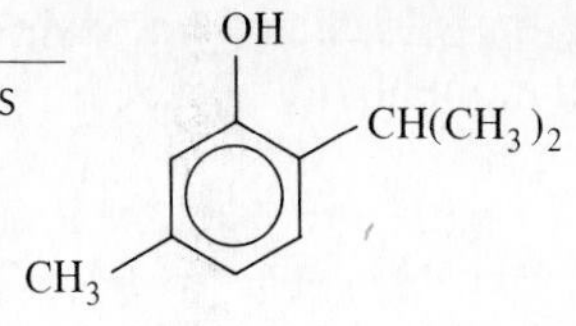

Thymol
(major constituent of oil of thyme)

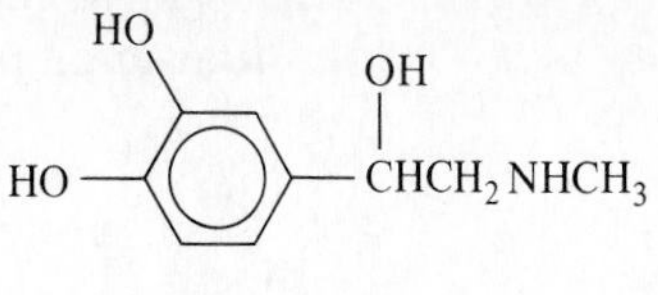

Epinephrine
(Adrenaline; a neurotransmitter
found in mammals)

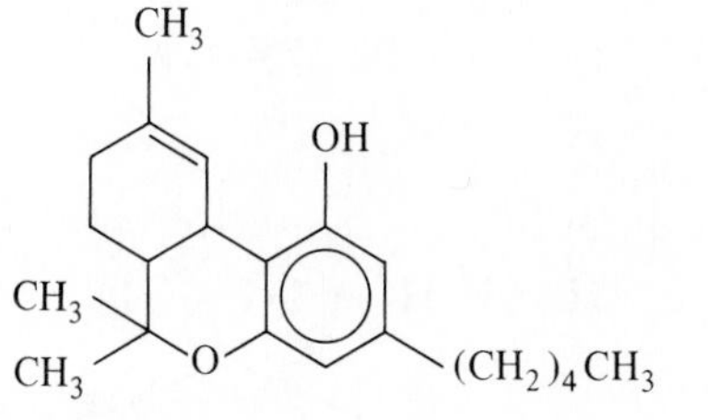

Δ^9-Tetrahydrocannabinol
(active component of marijuana)

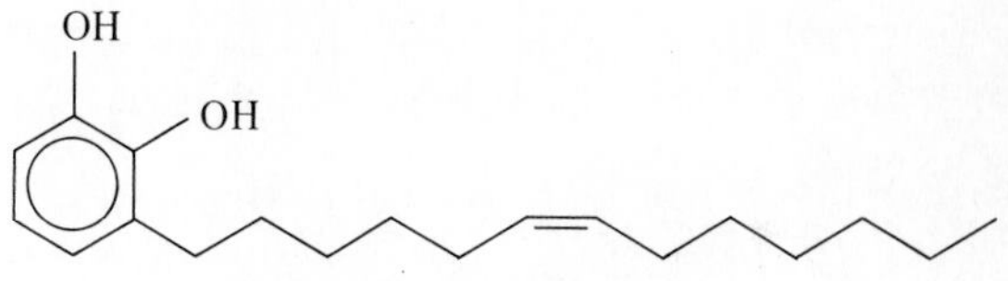

Urushiol (a principal component
of the allergenic oil of poison ivy)

FIGURE 10.5 Some naturally occurring phenols.

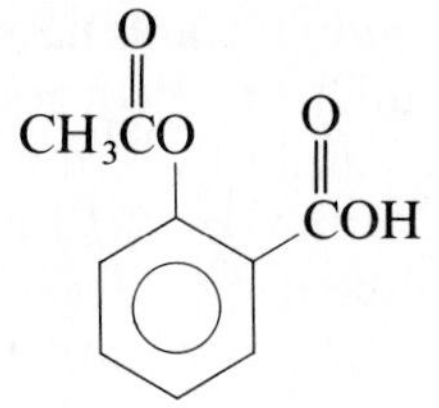

Aspirin
(Acetyl-
salicylic acid)

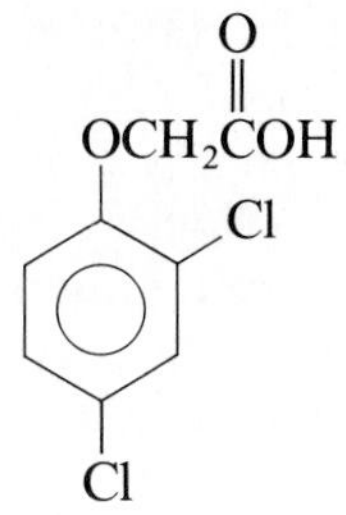

2,4-D
(2,4-Dichloro-
phenoxyacetic acid)

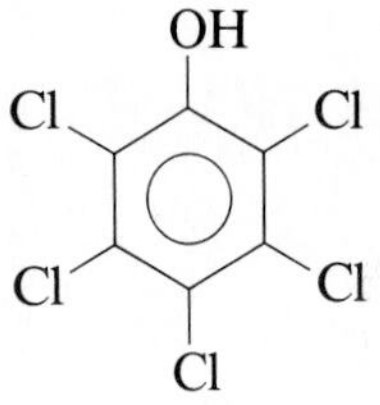

Penta
(Pentachloro-
phenol)

Phenolic compounds are quite common natural products and are found in both plants and animals. As the examples shown in Figure 10.5 illustrate, the phenolic group may be part of molecules which are chemically quite different.

10.18 ACIDITY OF PHENOLS

A prominent characteristic of phenols is their acidity. Phenols are 10^6 to 10^8 times *more* acidic than alcohols. Since both alcohols and phenols contain the hydroxyl group, why is there such a large difference in acidity?

To understand the reason for this large difference in acidity, let us compare the ionization equilibria for phenol and ethanol. The following equation describes the ionization of ethanol:

$$CH_3CH_2OH \rightleftharpoons H^+ + CH_3CH_2O^-$$

Ethanol   Proton   Ethoxide ion
$K_a = 10^{-16}$
($pK_a = 16$)

The negative charge in the ethoxide ion is localized on oxygen. Ionization of phenol generates the phenoxide ion.

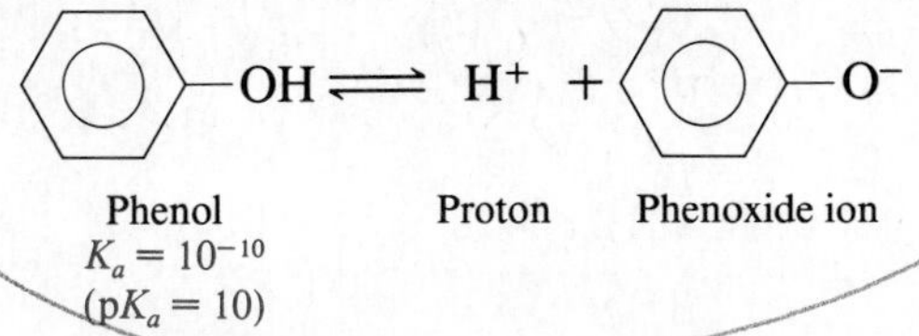

Phenol Proton Phenoxide ion
$K_a = 10^{-10}$
$(pK_a = 10)$

The negative charge in the phenoxide ion is *stabilized* by electron delocalization into the ring as the following resonance structures illustrate.

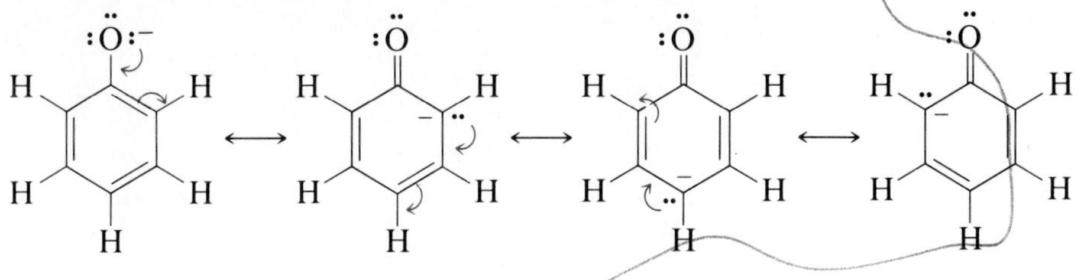

This stabilization of the negative charge of phenoxide ion causes the phenol ionization equilibrium to shift to the right, favoring formation of phenoxide ion. Delocalization of the negative charge is not present in an alkoxide ion, and alcohols are less acidic than phenols.

As Table 10.4 shows, most phenols have ionization constants similar to that of phenol itself. Alkyl substitution produces negligible changes in acidities, as do weakly electronegative groups attached to the ring. Only when the substituent is strongly electron-withdrawing, as is a nitro group (Table 6.2), is a substantial change in acidity noted. The ionization constants of *o*- and *p*-nitrophenol are several hundred times greater than that of phenol. An *o*- or *p*-nitro group greatly stabilizes the phenoxide ion by permitting a portion of the negative charge to be borne by its own oxygens.

Electron delocalization in *o*-nitrophenoxide ion:

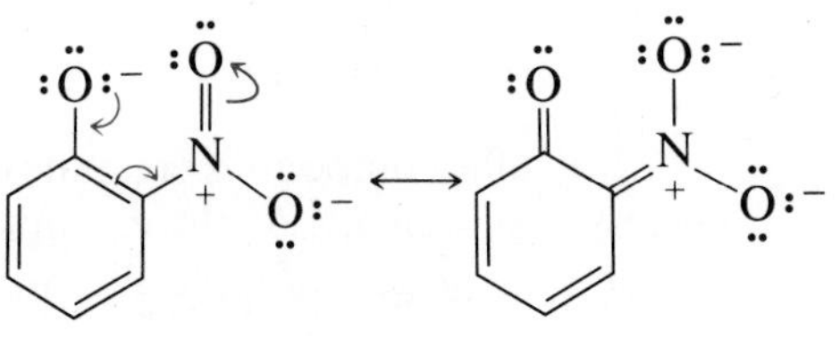

Electron delocalization in *p*-nitrophenoxide ion:

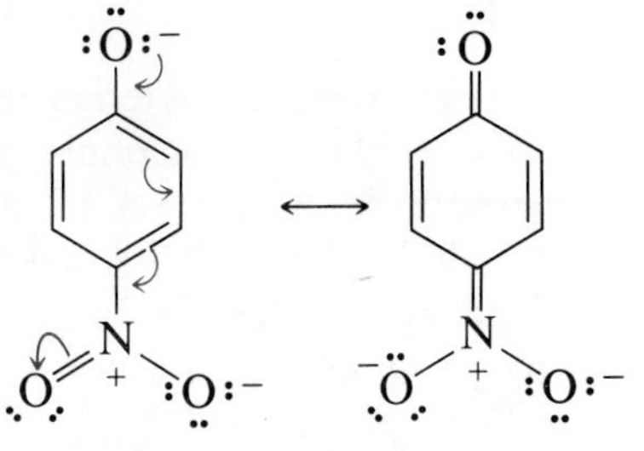

A *m*-nitro group is not directly conjugated to the phenoxide oxygen and thus stabilizes a phenoxide ion to a smaller extent. *m*-Nitrophenol is more acidic than phenol but less acidic than either *o*- or *p*-nitrophenol.

TABLE 10.4

Acidities of Some Substituted Phenols

Compound name	Ionization constant K_a	pK_a
Phenol	1.0×10^{-10}	10.0
o-Cresol	4.7×10^{-11}	10.3
m-Cresol	8.0×10^{-11}	10.1
o-Methoxyphenol	1.0×10^{-10}	10.0
p-Methoxyphenol	6.3×10^{-11}	10.2
o-Nitrophenol	5.9×10^{-8}	7.2
m-Nitrophenol	4.4×10^{-9}	8.4
p-Nitrophenol	6.9×10^{-8}	7.2

A practical result of the greater acidity of phenols compared to alcohols is that phenols react with (and dissolve in) aqueous solutions of strong bases such as sodium hydroxide.

$$ArO-H + \;^-:\ddot{O}H \longrightarrow ArO:^- + H-\ddot{O}H$$

Phenol	Hydroxide ion	Phenoxide ion	Water
Stronger acid	Stronger base	Weaker base	Weaker acid

Thus a phenol can be separated from an alcohol by extraction of an ether solution containing both with an aqueous sodium hydroxide solution. The phenol dissolves in the aqueous layer as its sodium phenoxide salt; the alcohol remains in the ether phase. The phenol may be removed from the separated aqueous phase by acidification, which converts the phenoxide ion to the phenol, followed by extraction of the neutral phenol with ether.

10.19 REACTIONS OF PHENOLS. PREPARATION OF ARYL ETHERS

Aryl ethers can be prepared by the Williamson method (Section 10.12). As described in the preceding section, reaction of a phenol with a base such as sodium hydroxide generates the phenoxide anion. This anion then serves as the nucleophile in a reaction with a suitable alkyl halide.

$$ArO:^- + R-\ddot{X}: \longrightarrow ArO-R + :\ddot{X}:^-$$

Phenoxide anion	Alkyl halide	Alkyl aryl ether	Halide anion

The synthesis may be performed by heating a solution of the phenol and the alkyl halide in the presence of a base.

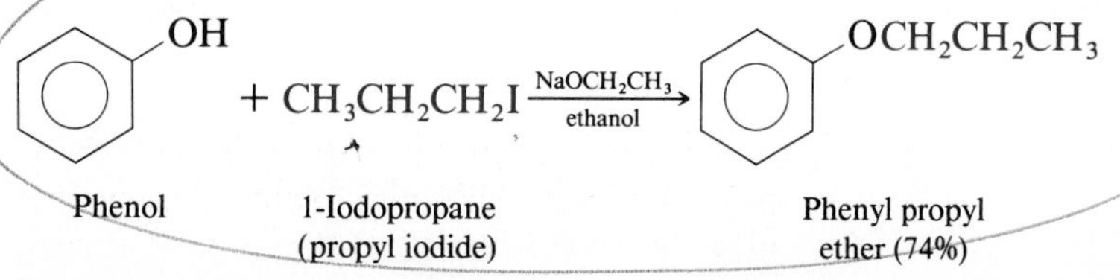

Phenol + $CH_3CH_2CH_2I$ $\xrightarrow[\text{ethanol}]{NaOCH_2CH_3}$ Phenyl propyl ether (74%)

1-Iodopropane (propyl iodide)

ACCIDENTAL FORMATION
OF A DIARYL ETHER:
DIOXIN

Until its use was regulated in 1979, 2,4,5-T (2,4,5-trichlorophenoxyacetic acid) was a widely used herbicide. It is prepared by reaction of 2,4,5-trichlorophenol with chloroacetic acid.

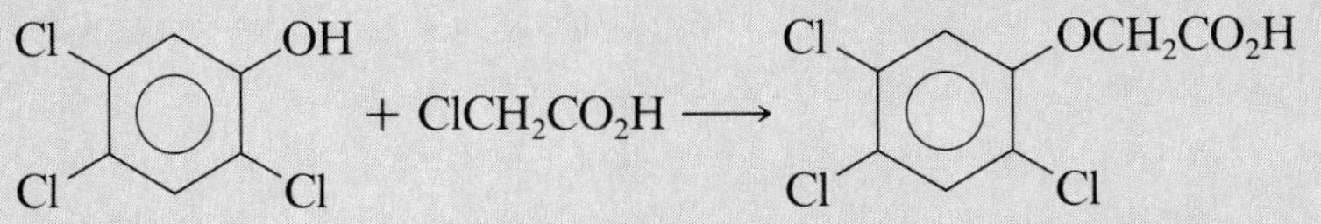

| 2,4,5-Trichlorophenol | Chloroacetic acid | 2,4,5-Trichlorophenoxyacetic acid (2,4,5-T) |

The starting material for the process, 2,4,5-trichlorophenol, is almost always contaminated with small amounts of the diaryl ether 2,3,7,8-tetrachlorodibenzo-*p*-dioxin, better known simply as **dioxin.**

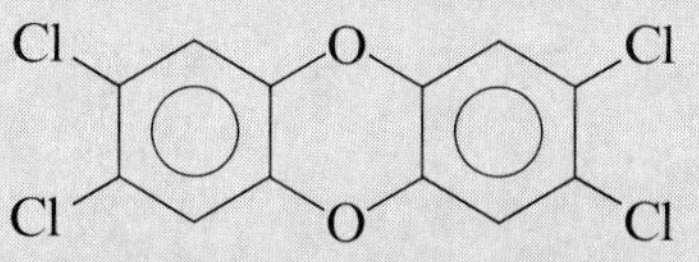

2,3,7,8-Tetrachlorodibenzo-*p*-dioxin
(dioxin)

Dioxin is carried along when 2,4,5-trichlorophenol is converted to 2,4,5-T, and it enters the environment when 2,4,5-T is sprayed on vegetation. Typically, the amount of dioxin present in 2,4,5-T is very small. Agent Orange, a 2,4,5-T – based defoliant used during the Vietnam War, contained about 2 ppm of dioxin.

Tests with animals have revealed that dioxin is one of the most toxic substances known. Toward mice it is about 2000 times more toxic than strychnine and about 150,000 times more toxic than sodium cyanide. Fortunately, however, available evidence indicates that humans are far more resistant to dioxin than are test animals, and so far there have been no human fatalities directly attributable to dioxin. The most prominent symptom seen so far has been a severe skin disorder known as *chloracne.* Yet to be determined is the answer to the question of long-term effects. A connection between exposure to dioxin and the occurrence of a rare type of cancer has been claimed on the basis of a statistical study.

There is another environmental source of dioxin—polychlorinated biphenyls, or PCBs (boxed essay in Section 3.18). When PCBs are strongly heated, as during a fire, one of the decomposition products is dioxin. As a result, firefighters take extra precautions when fighting electrical or other fires in which PCB-containing fluids might be present.

10.20 OXIDATION OF PHENOLS. QUINONES

The oxidation of 1,4-benzenediol derivatives are of use because they yield conjugated dicarbonyl compounds called **quinones.** 1,4-Benzenediol itself, also known as hydroquinone, reacts with mild oxidizing agents such as silver oxide to form *p*-benzoquinone.

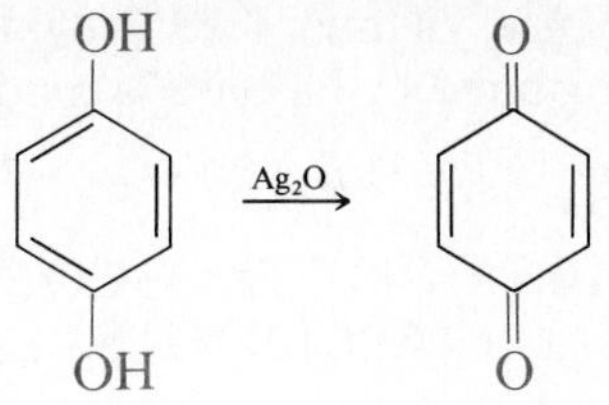

Hydroquinone   p-Benzoquinone
(76–81%)

Hydroquinone has for many years been used as a black-and-white-film developer. Granules of silver bromide (AgBr) activated by exposure to light on the film are reduced to deposits of silver metal (Ag°) on the negative by reaction with hydroquinone.

Quinones are colored; p-benzoquinone for example, is yellow. Many quinones occur naturally and have been used as dyes. Alizarin is a red pigment extracted from the roots of the madder plant. Its preparation from anthracene, a coal tar derivative, in 1868 was a significant step in the development of the synthetic dyestuff industry.

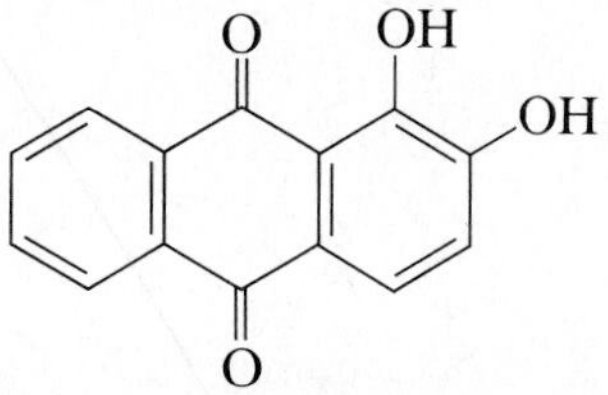

Alizarin

The oxidation of a hydroquinone (1,4-benzenediol derivative) to a quinone is reversible. This ready reversibility is essential to the role that quinones play in cellular respiration, the metabolic process of oxidizing food to carbon dioxide, water, and energy. A key component in this metabolic scheme is **ubiquinone,** also known as **coenzyme Q.**

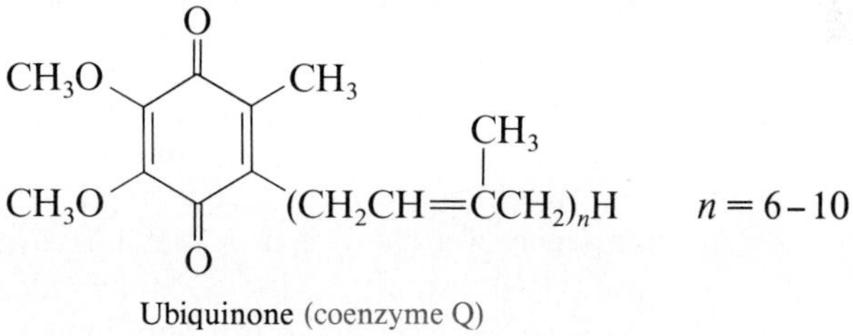

$n = 6-10$

Ubiquinone (coenzyme Q)

The name *ubiquinone* is a shortened form of *ubiquitous quinone,* a term coined to describe the observation that this substance can be found in all cells. The length of its side chain varies among different organisms; the most common form in vertebrates has $n = 10$, while ubiquinones in which $n = 6$ to 9 are found in yeasts and plants.

Another physiologically important quinone is vitamin K. Here K stands for *Koagulation* (Danish), since this substance was first identified as essential for the normal clotting of blood.

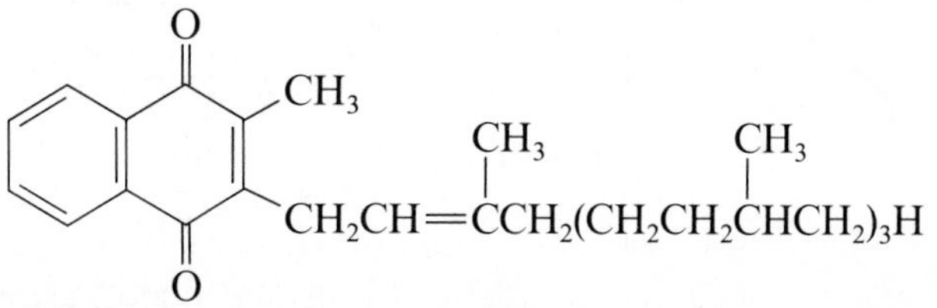

Vitamin K

Some vitamin K is provided in the normal diet, but a large proportion of that required by humans is produced by their intestinal flora.

10.21 SPECTROSCOPIC ANALYSIS OF ALCOHOLS, ETHERS, AND PHENOLS

As discussed in Chapter 9, methods of spectroscopic analysis are widely used by chemists in the determination of molecular structure. Each functional group present in an organic substance exhibits certain characteristic absorptions which aid in identification of that molecule.

Both alcohols and phenols exhibit a band in the infrared spectrum characteristic of the O—H group, appearing in the 3200 to 3650 cm^{-1} region. In addition, the C—O absorption of an alcohol appears in the region between 1025 and 1200 cm^{-1}. For a phenol the absorption occurs around 1200 to 1250 cm^{-1}.

In the ^{1}H nmr spectrum, alcohols exhibit characteristic absorptions due to the hydroxyl proton and any protons on the hydroxyl-bearing carbon.

$$-\overset{|}{\underset{|}{C}}-O-H \qquad H-\overset{|}{\underset{|}{C}}-O-$$

$$\delta = 0.5 - 5 \text{ ppm} \qquad \delta = 3.3 - 4.0 \text{ ppm}$$

The chemical shift of the hydroxyl proton is variable and depends on solvent, temperature, and concentration. The hydroxyl proton of a phenol generally appears in the region from $\delta = 4$ to 12 ppm.

Ethers exhibit the same C—O band in the infrared spectrum as found for alcohols, and the H—C—O unit in the ^{1}H nmr spectrum appears in the same region as for an alcohol. Of course, the hydroxyl proton is absent in an ether!

10.22 SUMMARY

In this chapter we have seen examples of the preparation and reactions of three important classes of organic molecules: alcohols, ethers, and phenols.

Alcohols are most often prepared by reduction using either hydrogen in the presence of a catalyst or a metal hydride reagent such as sodium borohydride or lithium aluminum hydride. Primary alcohols are prepared from aldehydes, secondary alcohols from the corresponding ketone (Section 10.4).

$$\underset{\text{Aldehyde}}{\overset{\displaystyle O}{\overset{\|}{R C H}}} \xrightarrow[\text{or, NaBH}_4, \text{ CH}_3\text{OH}]{\text{H}_2, \text{ catalyst}} \underset{\substack{\text{Primary} \\ \text{alcohol}}}{R C H_2 O H}$$

or 1. LiAlH$_4$
2. H$_2$O

$$\underset{\text{Ketone}}{\overset{\displaystyle O}{\overset{\|}{R C R'}}} \longrightarrow \underset{\substack{\text{Secondary} \\ \text{alcohol}}}{\overset{\displaystyle O H}{\overset{|}{R C H R'}}}$$

Alcohols are weak Brönsted acids (Section 10.6). The range of pK_a values is 16 to 18.

$$\underset{\text{Alcohol}}{\text{ROH}} + \underset{\text{Water}}{\text{H}_2\text{O}} \rightleftharpoons \underset{\substack{\text{Alkoxide} \\ \text{ion}}}{\text{RO}^-} + \underset{\substack{\text{Hydronium} \\ \text{ion}}}{\text{H}_3\text{O}^+}$$

Alkoxide ions may be prepared by reaction of an alcohol with a strong base or by reaction with an alkali metal.

$$ROH + NaH \longrightarrow RO^-Na^+ + H_2(g)$$

$$2ROH + 2Na \longrightarrow 2\ RO^-Na^+ + H_2(g)$$

Primary and secondary alcohols may be used as the starting materials for the preparation of carbonyl compounds (Section 10.8). Whether the product is an aldehyde, a ketone, or a carboxylic acid depends on the alcohols undergoing oxidation and the conditions used. Table 10.5 summarizes these reactions.

Primary alcohols may be converted to dialkyl ethers by heating in the presence of an acid catalyst (Section 10.7). Two molecules of the alcohol combine to form an ether and water in a *condensation* reaction.

$$2RCH_2OH \xrightarrow[\text{heat}]{H^+} RCH_2OCH_2R + H_2O$$
$$\text{Alcohol} \qquad\qquad \text{Dialkyl ether} \qquad \text{Water}$$

$$2(CH_3)_2CHCH_2CH_2OH \xrightarrow[150°C]{H_2SO_4} (CH_3)_2CHCH_2CH_2OCH_2CH_2CH(CH_3)_2$$
$$\text{Isoamyl alcohol} \qquad\qquad\qquad \text{Diisoamyl ether (27\%)}$$

Ethers may also be prepared by the Williamson ether synthesis (Section 10.12). An alkoxide ion displaces a halide leaving group in an S_N2 reaction.

$$RO^- + R'CH_2X \longrightarrow ROCH_2R' + X^-$$
$$\text{Alkoxide} \quad \text{Primary} \qquad\quad \text{Ether} \qquad \text{Halide}$$
$$\text{ion} \qquad \text{alkyl halide} \qquad\qquad\qquad \text{ion}$$

$$(CH_3)_2CHCH_2ONa + CH_3CH_2Br \longrightarrow (CH_3)_2CHCH_2OCH_2CH_3 + NaBr$$
$$\text{Sodium} \qquad\qquad \text{Ethyl} \qquad\qquad \text{Ethyl isobutyl} \qquad \text{Sodium}$$
$$\text{isobutoxide} \qquad\quad \text{bromide} \qquad\qquad \text{ether (66\%)} \qquad \text{bromide}$$

The reaction proceeds best with primary alkyl halides. Elimination, not substitution, is the principal reaction when the alkyl halide is secondary or tertiary.

Epoxides may be prepared from alkenes either by epoxidation or by a base-promoted cyclization of a vicinal halohydrin in what amounts to an intramolecular

TABLE 10.5
Oxidation of Alcohols

Class of alcohol	Desired product	Suitable oxidizing agent(s)
Primary, RCH_2OH	Aldehyde $R\overset{\displaystyle O}{\overset{\displaystyle \|}{C}}H$	$(C_5H_5N)_2CrO_3$ (Collins's reagent)
Primary, RCH_2OH	Carboxylic acid $R\overset{\displaystyle O}{\overset{\displaystyle \|}{C}}OH$	$KMnO_4$ $Na_2Cr_2O_7, H_2SO_4$ H_2CrO_4
Secondary, $R\underset{\displaystyle OH}{\overset{\displaystyle \|}{C}}HR'$	Ketone $R\overset{\displaystyle O}{\overset{\displaystyle \|}{C}}R'$	$(C_5H_5N)_2CrO_3$ $KMnO_4$ $Na_2Cr_2O_7, H_2SO_4$ H_2CrO_4

Williamson reaction (Section 10.13). These reactions are illustrated in the following equations.

Peroxy acid oxidation of alkenes:

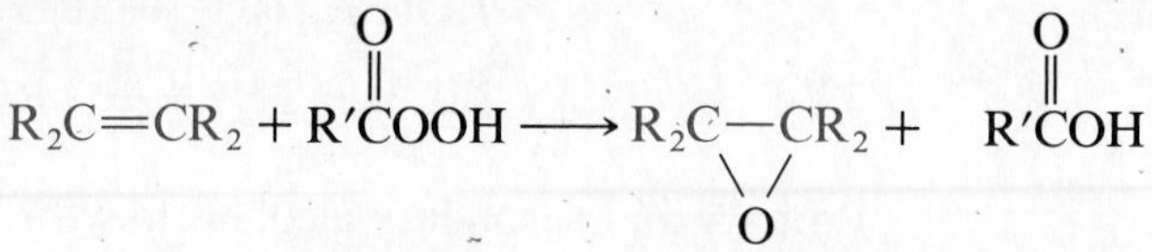

$$R_2C{=}CR_2 + R'\overset{\displaystyle O}{\overset{\|}{C}}OOH \longrightarrow R_2\underset{\underset{\displaystyle O}{\diagdown\diagup}}{C}{-}CR_2 + \quad R'\overset{\displaystyle O}{\overset{\|}{C}}OH$$

Alkene   Peroxy   Epoxide   Carboxylic acid
acid

$$(CH_3)_2C{=}C(CH_3)_2 \xrightarrow{CH_3CO_2OH} \begin{array}{c} H_3C \quad\quad CH_3 \\ \diagdown\diagup \\ C{-}C \\ \diagup \diagdown \quad \diagdown \\ H_3C \quad O \quad CH_3 \end{array}$$

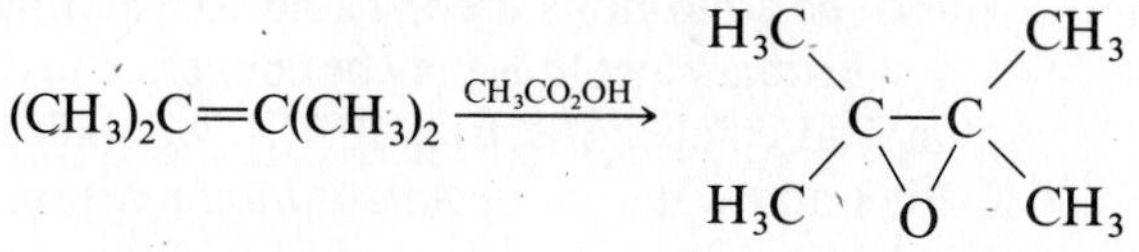

2,3-Dimethyl-2-butene   2,2,3,3-Tetramethyloxirane
(70–80%)

Base-promoted cyclization of vicinal halohydrins:

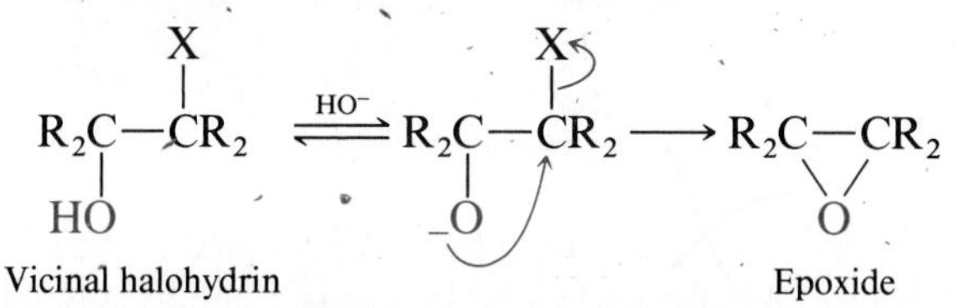

$$\underset{\underset{\displaystyle HO}{|}}{\overset{\overset{\displaystyle X}{|}}{R_2C}}{-}CR_2 \underset{}{\overset{HO^-}{\rightleftharpoons}} \underset{\underset{\displaystyle \_O}{|}}{\overset{\overset{\displaystyle X}{|}}{R_2C}}{-}CR_2 \longrightarrow R_2\underset{\underset{\displaystyle O}{\diagdown\diagup}}{C}{-}CR_2$$

Vicinal halohydrin   Epoxide

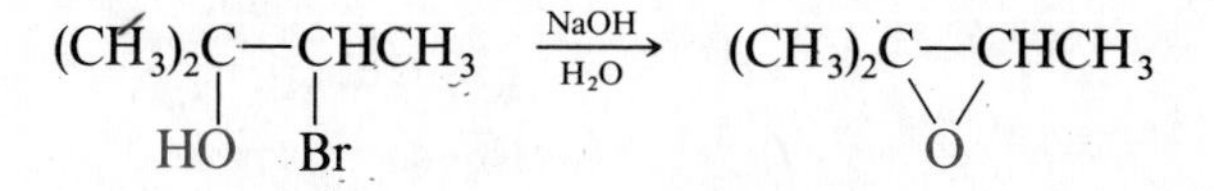

$$(CH_3)_2\underset{\underset{\displaystyle HO}{|}}{C}{-}\underset{\underset{\displaystyle Br}{|}}{C}HCH_3 \xrightarrow[H_2O]{NaOH} (CH_3)_2\underset{\underset{\displaystyle O}{\diagdown\diagup}}{C}{-}CHCH_3$$

3-Bromo-2-methyl-2-butanol   2,2,3-Trimethyloxirane (78%)

Epoxides react with nucleophiles to give products in which the three-membered ring has opened (Section 10.14).

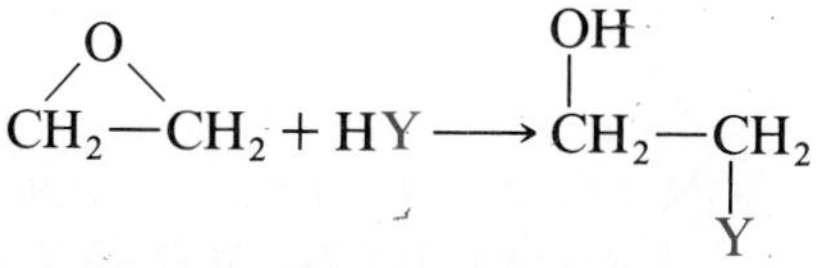

$$\underset{\underset{\displaystyle O}{\diagdown\diagup}}{CH_2}{-}CH_2 + HY \longrightarrow \underset{}{\overset{\overset{\displaystyle OH}{|}}{CH_2}}{-}\underset{\underset{\displaystyle Y}{|}}{C}H_2$$

Phenols are more acidic than alcohols (Section 10.18), as the phenoxide ion is stabilized by delocalization of the negative charge into the aromatic ring. A typical K_a is about 10^{-10} ($pK_a = 10$).

$$ArOH \rightleftharpoons H^+ + \quad ArO^-$$

Phenol   Phenoxide
anion

Phenoxide ions, prepared from phenols in base, react with alkyl halides to give alkyl aryl ethers (Section 10.19).

$$ArO^- + RX \longrightarrow ArOR + X^-$$

Phenoxide   Alkyl   Alkyl   Halide
anion   halide   aryl ether   anion

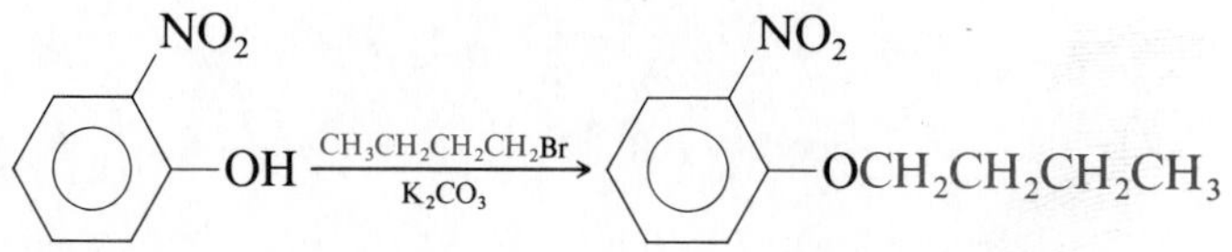

o-Nitrophenol   Butyl o-nitrophenyl
ether (75–80%)

Oxidation of 1,4-benzenediols gives colored compounds known as quinones (Section 10.20).

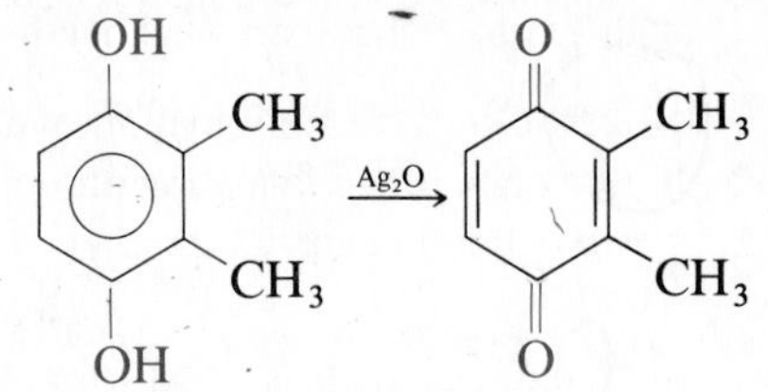

ADDITIONAL PROBLEMS

Alcohols: Preparation and Reactions

10.14 Write a chemical equation, showing all necessary reagents, for the preparation of 1-butanol by each of the following methods:

(a) Hydroboration-oxidation of an alkene
(b) Reduction with lithium aluminum hydride
(c) Catalytic hydrogenation
(d) Nucleophilic substitution of an alkyl halide

10.15 Repeat parts (a) to (c) of Problem 10.14 for the preparation of cyclohexanol.

10.16 Give the structure of the major organic product from the reaction of 1-butanol with each of the following reagents.

(a) $SOCl_2$
(b) $KMnO_4$
(c) $(C_5H_5N)_2CrO_3$ in CH_2Cl_2 (Collins's reagent)
(d) $K_2Cr_2O_7$, H_2SO_4
(e) H_2SO_4, heat
(f) Sodium amide, $NaNH_2$
(g) Potassium metal

10.17 Repeat Problem 10.16 with 2-butanol as the reactant.

10.18 By writing the appropriate chemical equations, show how 1-hexanol could be converted into:

(a) 1-Hexene
(b) 1-Chlorohexane

(c) Hexanal, $CH_3CH_2CH_2CH_2CH_2\overset{\displaystyle O}{\overset{\displaystyle \|}{C}}H$

(d) Hexanoic acid, $CH_3CH_2CH_2CH_2CH_2\overset{\displaystyle O}{\overset{\displaystyle \|}{C}}OH$

10.19 Which of the isomeric $C_5H_{12}O$ alcohols can be prepared by sodium borohydride reduction of a carbonyl compound?

Ethers and Epoxides

10.20 Write the structures of all the constitutionally isomeric ethers of molecular formula $C_5H_{12}O$ and give an acceptable name for each.

10.21 Many ethers, including diethyl ether, are effective as general anesthetics. Because simple ethers are quite flammable, their place in medical practice has been taken by highly halogenated nonflammable ethers. Two such general anesthetic agents are *isoflurane* and *enflurane*. These compounds are isomeric: isoflurane is 1-chloro-2,2,2-trifluoroethyl difluoromethyl ether, while enflurane is 2-chloro-1,1,2-trifluoroethyl difluoromethyl ether. Write the structural formulas of isoflurane and enflurane.

10.22 The octane rating of unleaded gasoline may be boosted by adding a small amount of *tert*-butyl methyl ether (known commercially as MTBE). Write the structure of this substance and outline a method for its preparation that uses the Williamson method starting with an alcohol and an alkyl halide.

10.23 Only one combination of alkyl halide and alkoxide is appropriate for the preparation of each of the following ethers by the Williamson ether synthesis. What is the correct combination in each case?

(a) CH_3CH_2O-⬡ (cyclopentyl)

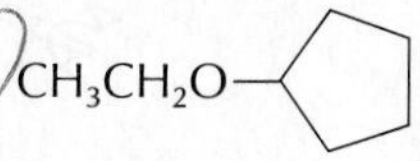

(b) $CH_2=CHCH_2OCH(CH_3)_2$

(c) $(CH_3)_3COCH_2C_6H_5$

10.24 Write the structure of the principal organic products A to D formed from the following sequence of steps:

(a) 2-Methyl-1-butene + H_2O (H^+ cat.) $\longrightarrow$ A

(b) A + NaH $\longrightarrow$ B

(c) $CH_3CH_2CH_2OH + PBr_3 \longrightarrow$ C

(d) B + C $\longrightarrow$ D

10.25 Predict the principal organic product of each of the following reactions. Specify stereochemistry where appropriate.

(a) $CH_3CH=CHCH_2Cl + (CH_3)_3CO^-K^+ \longrightarrow$

(b) $CH_3CH_2I +$ (C with CH_3CH_2, CH_3, C_6H_5, $-O^-K^+$) $\longrightarrow$

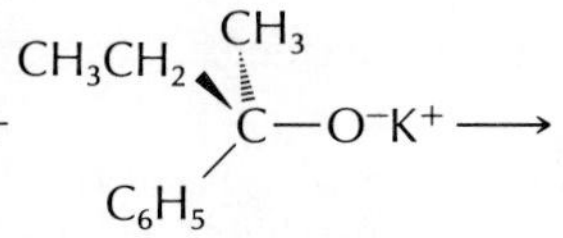

(c) $CH_3CH_2CHCH_2Br \xrightarrow{\text{NaOH}}$
with OH substituent

(d) (phenyl)$C=C$ with H, CH_3, H + (phenyl)$-\overset{O}{\overset{\|}{C}}-COOH \longrightarrow$

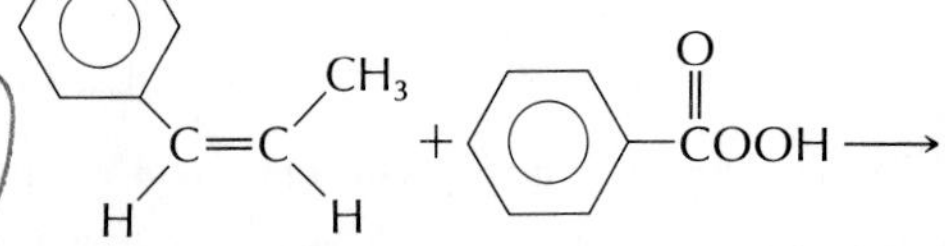

Phenols and Aryl Ethers

10.26 The IUPAC rules permit the use of common names for a number of familiar phenols and aryl ethers. These common names are listed below along with their systematic names. Write the structure of each compound.

(a) *Vanillin* (4-hydroxy-3-methoxybenzaldehyde): a component of vanilla bean oil, which contributes to its characteristic flavor.

(b) *Thymol* (2-isopropyl-5-methylphenol): obtained from oil of thyme.

(c) *Carvacrol* (5-isopropyl-2-methylphenol): present in oil of thyme and marjoram.

(d) *Salicyl alcohol* (o-hydroxybenzyl alcohol): obtained from bark of poplar and willow trees. (*Hint*: Benzyl alcohol has the formula $C_6H_5CH_2OH$.)

10.27 Name each of the following compounds:

(a) benzene ring with OH at top, CH_2CH_3 on right, NO_2 at bottom

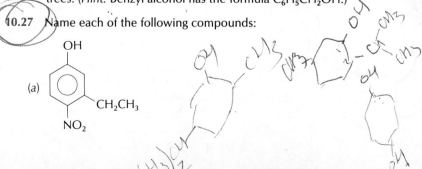

(b) 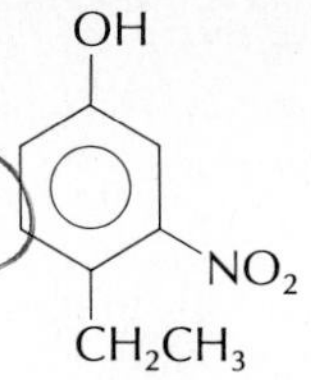 structure: OH on benzene ring, NO$_2$, CH$_2$CH$_3$

(c) 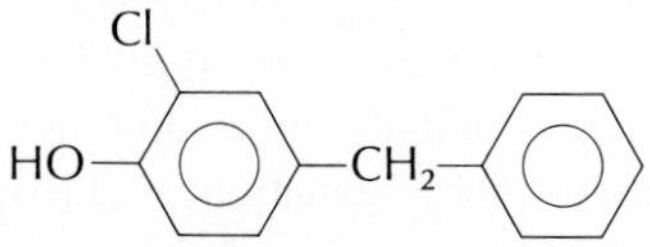 HO— (ring with Cl) —CH$_2$— (ring)

10.28 Fats and oils in food turn rancid by reaction with the oxygen in air. Small amounts of **antioxidants,** compounds which inhibit reaction with oxygen, are added to most prepackaged foods as preservatives. Two phenols, BHA and BHT, are widely used for this purpose. Give acceptable IUPAC names for each of these compounds.

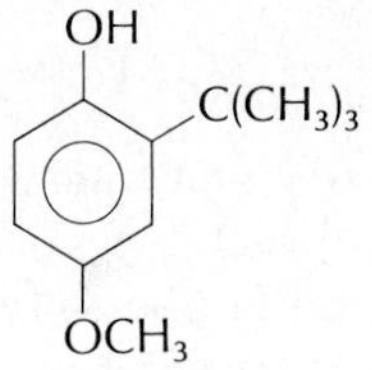

BHA
(Butylated
hydroxy anisole)

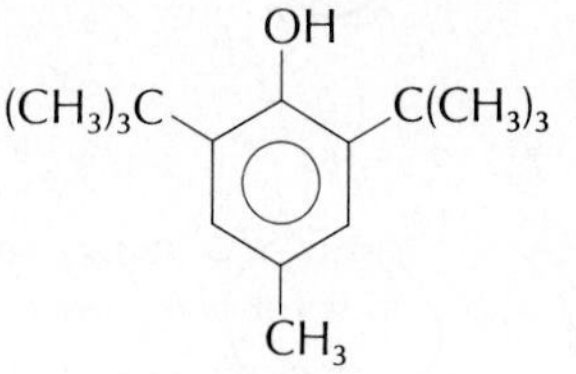

BHT
(Butylated
hydroxy toluene)

10.29 Write a balanced chemical equation for each of the following reactions:
(a) Phenol + sodium hydroxide $\longrightarrow$
(b) Product of (a) + ethyl bromide $\longrightarrow$
(c) m-Cresol + ethylene oxide $\longrightarrow$

Miscellaneous Problems

10.30 The cis isomer of 3-hexen-1-ol (CH$_3$CH$_2$CH=CHCH$_2$CH$_2$OH) has the characteristic odor of green leaves and grass. Suggest a synthesis for this compound from acetylene and any necessary organic or inorganic reagents. (*Hint:* You will find reviewing Sections 5.4 and 5.5 helpful for this problem.)

10.31 Each of the following reactions has been carried out by research chemists and described in a chemical journal. Although the molecules are somewhat more complex than those we are used to seeing, the reaction types are ones we have encountered. Identify the principal organic product in each case.

(a) CH$_3$—(cyclohexane ring with OH and C$_6$H$_5$) $\xrightarrow[\text{heat}]{\text{H}_2\text{SO}_4}$

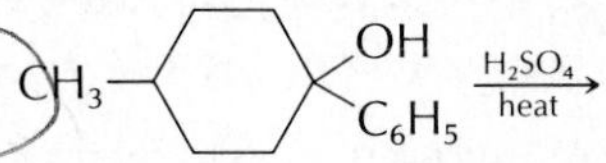

(b) CH$_2$=CHCH=CHCH$_2$CH$_2$CH$_2$OH $\xrightarrow[\text{CH}_2\text{Cl}_2]{\text{(C}_5\text{H}_5\text{N)}_2\text{CrO}_3}$

(c) CH$_3$CHC≡C(CH$_2$)$_3$CH$_3$ $\xrightarrow[\substack{\text{H}_2\text{SO}_4,\ \text{H}_2\text{O} \\ \text{acetone}}]{\text{H}_2\text{CrO}_4}$
   with OH below

(d) CH$_3$CCH$_2$CH=CHCH$_2$CCH$_3$ (with two C=O groups) $\xrightarrow[\text{2. H}_2\text{O}]{\text{1. LiAlH}_4,\ \text{diethyl ether}}$

(e) (benzene ring with OH, Cl, OH) $\xrightarrow[\text{H}_2\text{SO}_4]{\text{K}_2\text{Cr}_2\text{O}_7}$

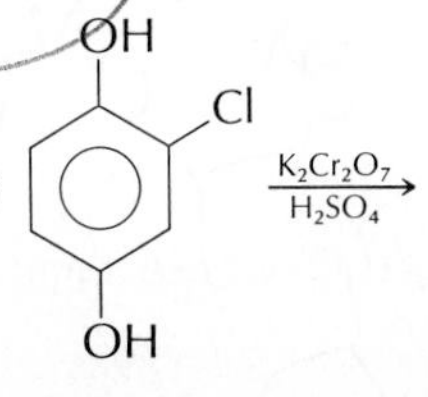

(f)

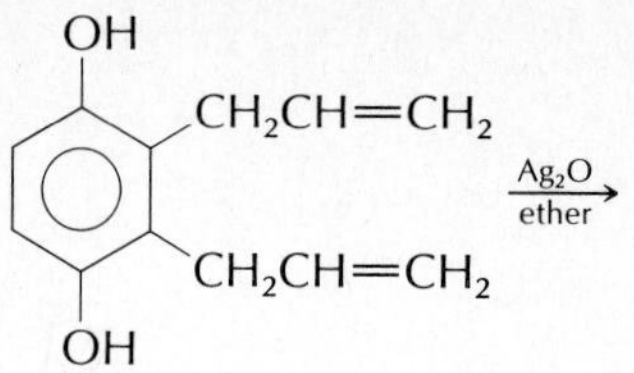

$$\xrightarrow[\text{ether}]{Ag_2O}$$

10.32 Sorbitol is a sweetener often substituted for cane sugar since it is better tolerated by diabetics. It is also an intermediate in the commercial synthesis of vitamin C. Sorbitol is prepared by high-pressure hydrogenation of glucose over a nickel catalyst. What is the structure (including stereochemistry) of sorbitol?

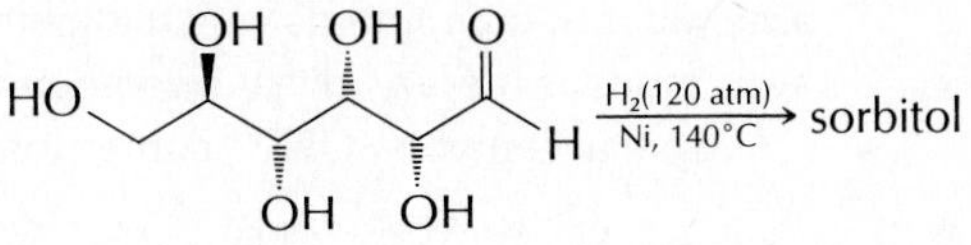

$$\xrightarrow[\text{Ni, 140°C}]{H_2(120\ atm)} \text{sorbitol}$$

Glucose

10.33 R. B. Woodward was the leading organic chemist of the middle part of this century. Known primarily for his achievements in the synthesis of complex natural products, he was awarded the Nobel Prize in Chemistry in 1965. He entered Massachusetts Institute of Technology as a 16-year-old freshman in 1933 and four years later was awarded the Ph.D. While a student there he carried out a synthesis of *estrone,* a female sex hormone. The early stages of Woodward's estrone synthesis required the conversion of *m*-methoxybenzalde-hyde to *m*-methoxybenzyl cyanide, which was accomplished in three steps.

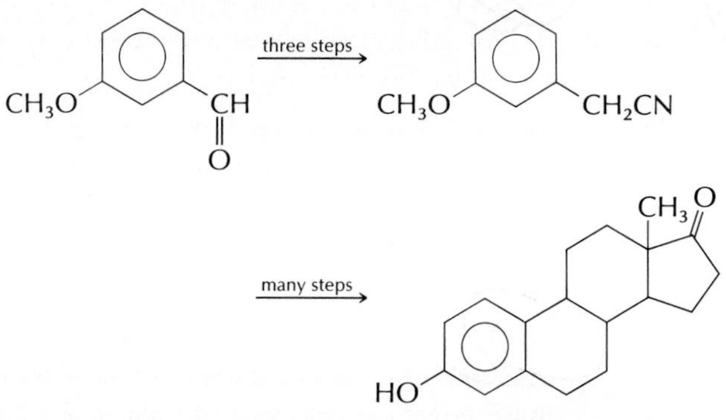

Estrone

Suggest a reasonable three-step sequence, showing all necessary reagents, for the preparation of *m*-methoxybenzyl cyanide from *m*-methoxybenzaldehyde.

10.34 All the following questions pertain to ^1H nmr spectra of isomeric ethers having the molecular formula $C_5H_{12}O$.
  (a) Which one has only singlets in its ^1H nmr spectrum?
  (b) Along with other signals, this ether has a coupled doublet-heptet pattern. None of the protons responsible for this pattern are coupled to protons anywhere else in the molecule. Identify this ether.
  (c) In addition to other signals in its ^1H nmr spectrum, this ether exhibits two signals at relatively low field. One is a singlet; the other is a doublet. What is the structure of this ether?
  (d) In addition to other signals in its ^1H nmr spectrum, this ether exhibits two signals at relatively low field. One is a triplet; the other is a quartet. Which ether is this?

10.35 Identify compound E ($C_{14}H_{14}O$) whose ir and ^1H nmr spectra are shown in Figure 10.6.

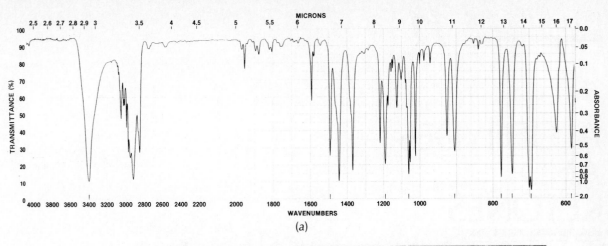

(a)

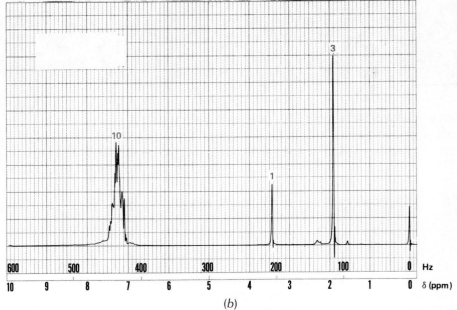

(b)

FIGURE 10.6 The (a) ir and (b) ^{1}H nmr spectra of compound E (Problem 10.35). The numbers beside the signals correspond to their relative areas.

CHAPTER *11*

ALDEHYDES
AND
KETONES

LEARNING OBJECTIVES

This chapter introduces the chemical and physical properties of two related classes of compounds, aldehydes and ketones. A central theme of this chapter will be reactions of the carbonyl group, $\diagdown C = O$. Upon completion of this chapter you should be able to:

- Provide an acceptable IUPAC name for an aldehyde or a ketone.
- Explain how the polarity of the carbonyl group affects physical properties of aldehydes and ketones.
- Describe how the polar nature of the carbonyl group affects nucleophilic additions to aldehydes and ketones.
- Write an equation for the hydration of an aldehyde or a ketone.
- Write a chemical equation for the formation of an acetal.
- Write a chemical equation for the formation of a cyanohydrin.
- Write chemical equations describing the reaction of an aldehyde or a ketone with derivatives of ammonia: primary amines and hydrazine derivatives.
- Show how an organometallic compound can be used to form new carbon-carbon bonds.
- Describe by using chemical equations the reaction of a Grignard reagent with an aldehyde or a ketone.
- Plan the synthesis of a specific alcohol using a Grignard reagent by working the problem backward.
- Explain the process of enolization and draw the enol form of a specific aldehyde or ketone.
- Write a chemical equation for the aldol condensation of an aldehyde.

- Write a chemical equation for the oxidation of an aldehyde.

■ Describe the characteristic features of the infrared or ^1H nmr spectrum of an aldehyde or ketone.

In this chapter we continue our study of oxygen-containing organic compounds. Occupying a central position in this group of compounds are the **aldehydes** and **ketones.** Aldehydes and ketones contain what is probably the most important functional group in organic chemistry, the carbonyl group.

$$\diagdown C = O$$

Carbonyl group

As you will see throughout this chapter, the carbonyl group is found in numerous substances of biological interest, from flavors in fruits to human sex hormones. Also, you will find that aldehydes and ketones are widely used as starting materials for the preparation of other classes of organic compounds.

11.1 NOMENCLATURE

Aldehydes and ketones differ in that aldehydes have at least one hydrogen attached to the carbon atom of the carbonyl group, while ketones have two alkyl and/or aryl groups as substituents on the carbonyl group.

$$\underset{\text{Aldehyde}}{\overset{\displaystyle O}{\underset{R}{\overset{\|}{\underset{}{C}}}}\diagdown H} \qquad \underset{\text{Ketone}}{\overset{\displaystyle O}{\underset{R}{\overset{\|}{\underset{}{C}}}}\diagdown R'}$$

The R groups may be any combination of alkyl or aryl (aromatic) groups.

The base name of both aldehydes and ketones is derived from the longest chain of carbon atoms that *contains the carbonyl group.* An aldehyde is named by replacing the *-e* ending of the corresponding alkane name with *-al.* It is not necessary to number the position of the carbonyl group in an aldehyde because in order to have a hydrogen attached, the carbonyl *must* be at the end of the chain. This carbon is always considered C-1. Substituents are specified by name and position on the chain in the usual way, numbering from the carbonyl group.

$$\underset{\text{Pentanal}}{CH_3CH_2CH_2CH_2\overset{\displaystyle O}{\overset{\|}{C}}H} \qquad \underset{\text{4,4-Dimethylhexanal}}{CH_3CH_2\underset{\underset{\displaystyle CH_3}{|}}{\overset{\overset{\displaystyle CH_3}{|}}{C}}CH_2CH_2\overset{\displaystyle O}{\overset{\|}{C}}H}$$

Certain common names of familiar aldehydes are acceptable as IUPAC names. A few examples include

$$\underset{\substack{\text{Formaldehyde} \\ \text{(methanal)}}}{H\overset{\displaystyle O}{\overset{\|}{C}}H} \qquad \underset{\substack{\text{Acetaldehyde} \\ \text{(ethanal)}}}{CH_3\overset{\displaystyle O}{\overset{\|}{C}}H} \qquad \underset{\substack{\text{Benzaldehyde} \\ \text{(benzenecarbaldehyde)}}}{\text{⬡}-\overset{\displaystyle O}{\overset{\|}{C}}H}$$

Notice that formaldehyde violates the "rule" that aldehydes have one hydrogen attached to the carbonyl: it has two; all the others have one. Formaldehyde is often used in biology labs as a specimen preservative.

PROBLEM 11.1

Certain aldehydes are known by their common names. As the following examples illustrate, they often tell little of the structure of the compound. Provide an alternative IUPAC name for each.

(a) $(CH_3)_2CHCH_2\overset{\displaystyle O}{\overset{\|}{C}}H$ Isovaleraldehyde (found in oranges and lemons)

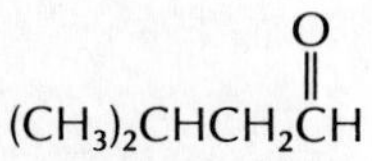

(b) $CCl_3\overset{\displaystyle O}{\overset{\|}{C}}H$ Chloral (used to make a powerful sedative)

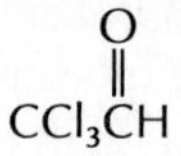

(c) $C_6H_5CH{=}CH\overset{\displaystyle O}{\overset{\|}{C}}H$ Cinnamaldehyde (responsible for the odor of cinnamon)

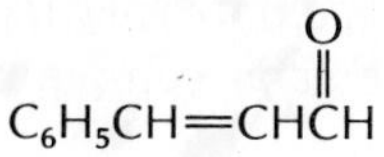

(d) HO—⬡—$\overset{\displaystyle O}{\overset{\|}{C}}H$ Vanillin (a popular food flavoring)
     CH_3O

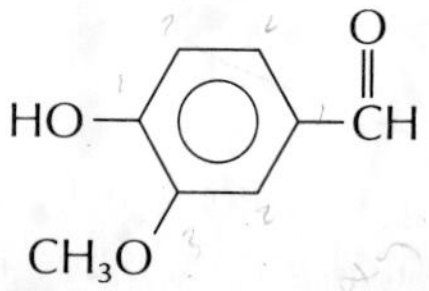

SAMPLE SOLUTION

(a) Writing out the structure reveals the longest chain to be four carbons long. Thus isovaleraldehyde is named as a substituted butanal. The methyl group is attached to C-3, so the name is 3-methylbutanal.

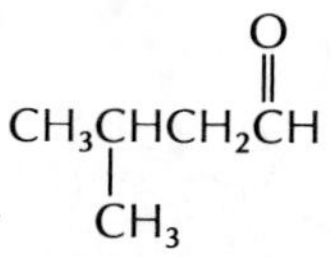

$$CH_3CHCH_2\overset{\displaystyle O}{\overset{\|}{C}}H$$
$$\underset{\displaystyle CH_3}{|}$$

3-Methylbutanal

Ketones are named in a manner similar to aldehydes. The *-e* ending of an alkane is replaced by *-one* in the longest continuous chain containing the carbonyl group. The location of the carbonyl is specified by numbering the chain in the direction that gives the lower number for this group.

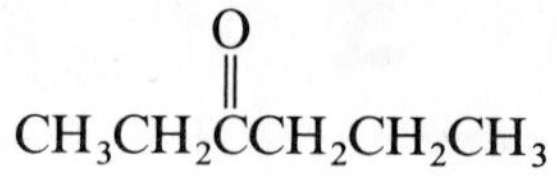

$$CH_3CH_2\overset{\displaystyle O}{\overset{\|}{C}}CH_2CH_2CH_3 \qquad CH_3\underset{\displaystyle \underset{CH_3}{|}}{CH}CH_2\overset{\displaystyle O}{\overset{\|}{C}}CH_3 \qquad CH_3{-}⬡{=}O$$

3-Hexanone	4-Methyl-2-pentanone	4-Methylcyclo-
(*not* 4-hexanone)	(*not* 2-methyl-4-pentanone)	hexanone

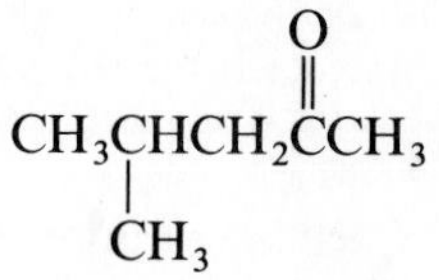

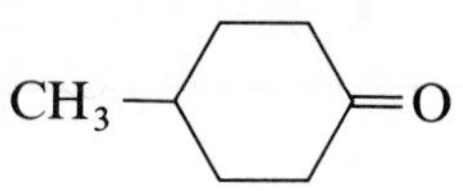

Although systematic names are preferred, the IUPAC rules also permit ketones to be named by separately designating both the groups attached to the carbonyl followed by the word *ketone*. The groups are listed alphabetically.

[handwritten margin notes: CH_3CH–CH_2CH with O; CH_3; "3 methyl butanal"; "Anne"]

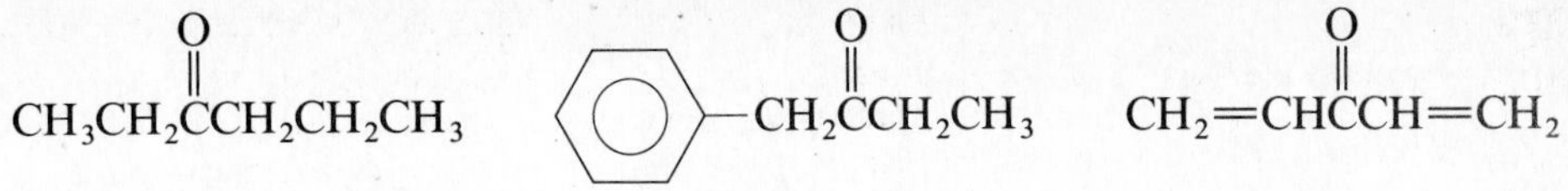

$$CH_3CH_2\overset{\overset{\displaystyle O}{\|}}{C}CH_2CH_2CH_3$$

Ethyl propyl ketone

$$\text{⟨phenyl⟩}-CH_2\overset{\overset{\displaystyle O}{\|}}{C}CH_2CH_3$$

Benzyl ethyl ketone

$$CH_2{=}CH\overset{\overset{\displaystyle O}{\|}}{C}CH{=}CH_2$$

Divinyl ketone

PROBLEM 11.2

As with aldehydes, a number of ketones are known by their common names. For each
of the following give a systematic name and a name using the word *ketone*.

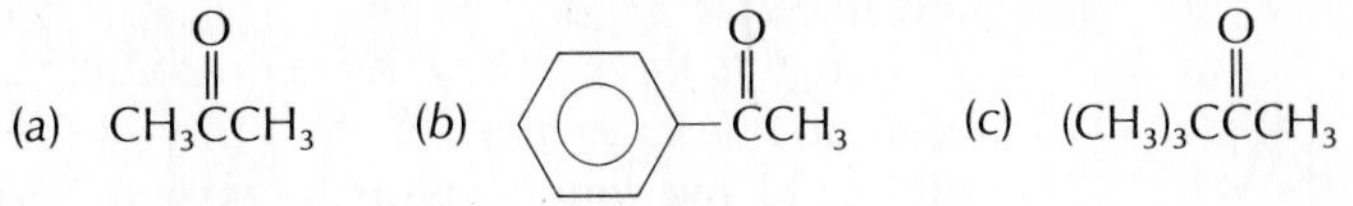

(a) $CH_3\overset{\overset{\displaystyle O}{\|}}{C}CH_3$

Acetone

(b) $\text{⟨phenyl⟩}-\overset{\overset{\displaystyle O}{\|}}{C}CH_3$

Acetophenone

(c) $(CH_3)_3C\overset{\overset{\displaystyle O}{\|}}{C}CH_3$

Pinacolone

SAMPLE SOLUTION

(a) Acetone is a common solvent used as nail polish remover and in organic chemistry
laboratories to rinse glassware. The chain is three carbons long with the carbonyl at C-2;
thus the systematic name for acetone is 2-propanone. Acetone has two methyl groups
attached to the carbonyl and may be called dimethyl ketone.

$$CH_3\overset{\overset{\displaystyle O}{\|}}{C}CH_3$$

2-Propanone
(dimethyl ketone)

Common names often reflect the carbonyl functional group present in the molecule. Two examples in human biochemistry are retin*al*, an aldehyde important in
vision, and progester*one*, a ketone which is a female sex hormone. The structures of
these compounds are shown in Figure 11.1.

11.2 STRUCTURE AND BONDING OF THE CARBONYL GROUP

Two of the more notable aspects of the carbonyl group are its geometry and its
polarity. As you will see in the sections which follow, the polarity of the carbonyl
group plays a significant role in the chemistry of aldehydes and ketones.

The carbonyl group and the atoms attached to it lie in the same plane. Formaldehyde, for example, is a planar molecule. The bond angles involving the carbonyl
group are close to 120° and do not vary much among simple aldehydes and ketones.

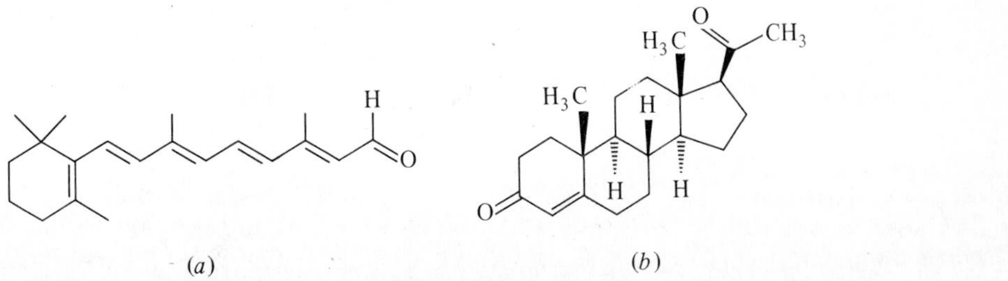

FIGURE 11.1 (a) Retinal
(plays an important role in
the vision process); (b)
progesterone (female sex
hormone).

(a)

(b)

$$121.7° \qquad 121.7°$$

O
‖
C
/ \
H   H
116.5°

Formaldehyde

Using formaldehyde as an example, we can describe the bonding of a carbonyl group according to an sp^2 hybridization model for carbon as shown in Figure 11.2. You have previously seen the sp^2 hybridization bonding model for carbon in the discussions of the structure of carbocations (Section 3.12) and the bonding of alkenes (Section 4.1.2).

Using its three sp^2 hybridized orbitals, carbon forms σ bonds to two hydrogen atoms and an oxygen atom. An unhybridized p orbital on carbon participates in π bonding by overlapping with an oxygen $2p$ orbital.

You learned in Chapter 1 (Section 1.4) that oxygen is more electronegative than carbon. As a result, the electron density in both the σ and π components of the carbon-oxygen double bond is displaced toward oxygen. The carbonyl group is polarized so that carbon is partially positive and oxygen is partially negative.

$$\overset{\delta^+}{\underset{}{\text{C}}}=\overset{\delta^-}{\underset{}{\text{Ö}}}: \qquad \text{also written as} \qquad \text{C}=\overset{\longrightarrow}{\text{Ö}}:$$

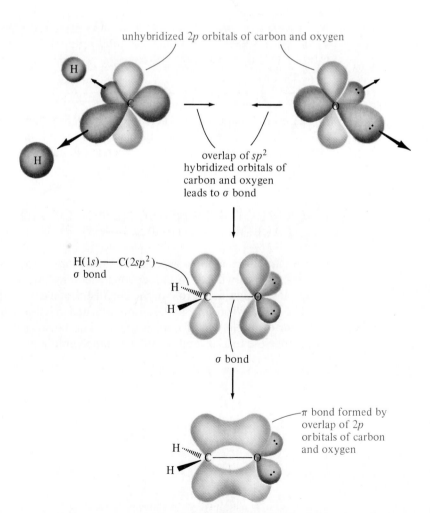

FIGURE 11.2 A description of bonding in formaldehyde based on sp^2 hybridization of carbon and oxygen.

You will see the significance of the polarity of the carbonyl group as we study the reactions of aldehydes and ketones later in this chapter.

11.3 PHYSICAL PROPERTIES

Aldehydes and ketones have higher boiling points than hydrocarbons of similar molecular weight because of intermolecular dipole-dipole attractive forces involving the carbonyl group. This difference can be seen by comparing the boiling points of propanal and acetone to butane. All three compounds have the same molecular weight, but propanal and acetone, both of which contain a carbonyl group, are polar compounds and boil at a higher temperature than butane which is nonpolar.

$$CH_3CH_2CH_2CH_3 \qquad CH_3CH_2CHO \qquad CH_3CCH_3$$

Butane	Propanal	Acetone
(bp −0.4°C)	(bp 49.5°C)	(bp 56.2°C)

Hydrogen bonds between the carbonyl oxygen and water cause aldehydes and ketones to have a higher level of water solubility than hydrocarbons. Acetone, for example, is miscible with water in all proportions. The degree of solubility in water decreases with increasing molecular weight. While more soluble in water than alkanes, aldehydes and ketones tend to be less soluble in water than an alcohol of similar molecular weight. Benzaldehyde, for example, dissolves in water to the extent of only 0.3 g/100 mL, a solubility lower by a factor of 10 than the solubility of benzyl alcohol but four times higher than that of benzene.

PROBLEM 11.3
Briefly explain why benzyl alcohol is more soluble in water than benzaldehyde. (*Hint:* You will find it helpful to review Section 10.1.)

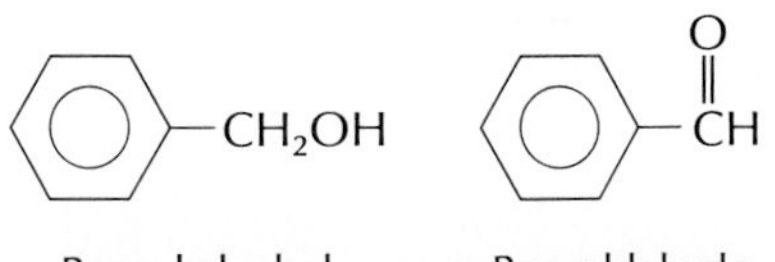

| Benzyl alcohol | Benzaldehyde |

11.4 SOURCES OF ALDEHYDES AND KETONES

Low-molecular-weight aldehydes and ketones are important industrial chemicals. Frequently the methods to prepare these compounds use specialized catalysts to promote the air oxidation of an alcohol or an alkene. Formaldehyde, which is used as the starting material for a number of plastics and as a preservative, is prepared by oxidation of methanol using a silver or iron–molybdenum oxide catalyst.

$$CH_3OH + \tfrac{1}{2}O_2 \xrightarrow[500°C]{catalyst} HCHO + H_2O$$

| Methanol | Oxygen | Formaldehyde | Water |

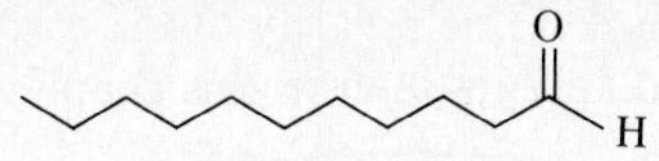

Undecanal
(sex pheromone of greater wax moth)

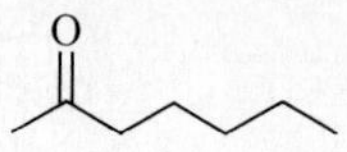

2-Heptanone
(component of alarm pheromone of bees)

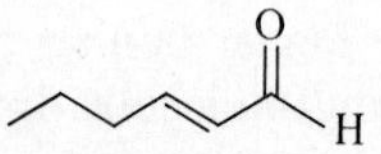

trans-2-Hexenal
(alarm pheromone of myrmicine ant)

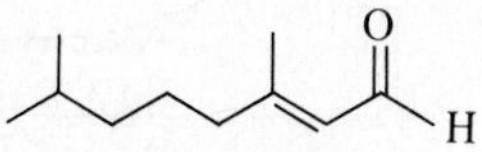

Citral
(present in lemon grass oil)

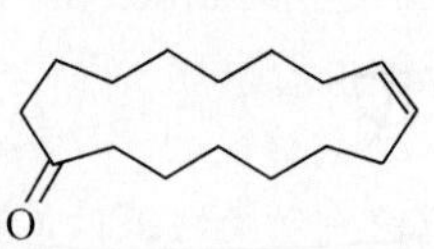

Civetone
(obtained from scent glands of
African civet cat)

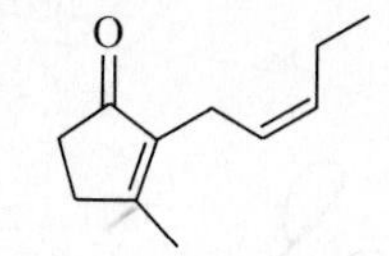

Jasmone
(found in oil of jasmine)

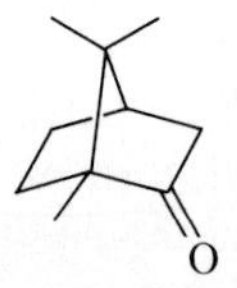

Camphor
(isolated from certain Indonesian
trees: has characteristic odor)

FIGURE 11.3 Some naturally occurring aldehydes and ketones.

Acetaldehyde is made by air oxidation of ethylene in the presence of palladium chloride and cupric chloride.

$$CH_2{=}CH_2 + \tfrac{1}{2}O_2 \xrightarrow[\text{H}_2\text{O}]{\text{PdCl}_2,\ \text{CuCl}_2} CH_3\overset{\displaystyle O}{\overset{\|}{C}}H$$

Ethylene   Oxygen   Acetaldehyde

Acetaldehyde is used as a starting material for a variety of organic compounds including acetic acid, which is used in the manufacture of acetate fibers and plastics.

Many aldehydes and ketones occur naturally in both plants and animals, and several are shown in Figures 11.1 and 11.3.

11.5 PREPARATION OF ALDEHYDES AND KETONES: A REVIEW

You have encountered several methods for preparing aldehydes and ketones in previous chapters. These include cleavage of alkenes by ozonolysis (Chapter 5), Friedel-Crafts acylation of aromatic compounds (Chapter 6), and oxidation of primary and secondary alcohols (Chapter 10). Specific examples of each of these methods are given in Table 11.1.

TABLE 11.1

Summary of Reactions Discussed in Earlier Chapters That Yield Aldehydes and Ketones

Reaction (section) and comments	General equation and specific example

Ozonolysis of alkenes (Section 5.2) This cleavage reaction is more often seen in structural analysis than in synthesis. The substitution pattern around a double bond is revealed by identifying the carbonyl-containing compounds that comprise the product. Hydrolysis of the ozonide intermediate in the presence of zinc permits aldehyde products to be isolated without further oxidation.

$$\underset{\text{Alkene}}{\overset{R}{\underset{R'}{}}C=C\overset{H}{\underset{R''}{}}} \xrightarrow[\text{2. H}_2\text{O, Zn}]{\text{1. O}_3} \underset{\text{Two carbonyl compounds}}{R\overset{O}{\overset{\|}{C}}R' + R''\overset{O}{\overset{\|}{C}}H}$$

$$\underset{\substack{\text{2,6-Dimethyl-}\\\text{2-octene}}}{} \xrightarrow[\text{2. H}_2\text{O, Zn}]{\text{1. O}_3} \underset{\text{Acetone}}{CH_3\overset{O}{\overset{\|}{C}}CH_3} + \underset{\substack{\text{4-Methylhexanal}\\(91\%)}}{H\overset{O}{\overset{\|}{C}}CH_2CH_2\overset{CH_3}{\underset{|}{CH}}CH_2CH_3}$$

Friedel-Crafts acylation of aromatic compounds (Section 6.6) Acyl chlorides and carboxylic acid anhydrides acylate aromatic rings in the presence of aluminum chloride. The reaction is an electrophilic aromatic substitution in which acylium ions are generated and attack the ring.

$$ArH + R\overset{O}{\overset{\|}{C}}Cl \xrightarrow{AlCl_3} Ar\overset{O}{\overset{\|}{C}}R + HCl \quad \text{or}$$

$$ArH + R\overset{O}{\overset{\|}{C}}O\overset{O}{\overset{\|}{C}}R \xrightarrow{AlCl_3} Ar\overset{O}{\overset{\|}{C}}R + RCO_2H$$

$$\underset{\text{Anisole}}{CH_3O-\bigcirc} + \underset{\text{Acetic anhydride}}{CH_3\overset{O}{\overset{\|}{C}}O\overset{O}{\overset{\|}{C}}CH_3} \xrightarrow{AlCl_3} \underset{\substack{p\text{-Methoxyacetophenone}\\(90-94\%)}}{CH_3O-\bigcirc-\overset{O}{\overset{\|}{C}}CH_3}$$

Oxidation of primary alcohols to aldehydes (Section 10.8) Use of the chromium trioxide–pyridine complex (Collins's reagent) in anhydrous media such as dichloromethane permits oxidation of primary alcohols to aldehydes while avoiding overoxidation to the corresponding carboxylic acids.

$$\underset{\text{Primary alcohol}}{RCH_2OH} \xrightarrow[\text{CH}_2\text{Cl}_2]{(C_5H_5N)_2CrO_3} \underset{\text{Aldehyde}}{R\overset{O}{\overset{\|}{C}}H}$$

$$\underset{\text{1-Decanol}}{CH_3(CH_2)_8CH_2OH} \xrightarrow[\text{CH}_2\text{Cl}_2]{(C_5H_5N)_2CrO_3} \underset{\text{Decanal (63–66\%)}}{CH_3(CH_2)_8\overset{O}{\overset{\|}{C}}H}$$

Oxidation of secondary alcohols to ketones (Section 10.8) Many oxidizing agents are available for converting secondary alcohols to ketones. Collins's reagent may be used, as well as other Cr(VI)-based reactions such as chromic acid or potassium dichromate and sulfuric acid. Potassium permanganate may also be used.

$$\underset{\text{Secondary alcohol}}{R\overset{}{\underset{OH}{\underset{|}{CH}}}R'} \xrightarrow{Cr(VI)} \underset{\text{Ketone}}{R\overset{O}{\overset{\|}{C}}R'}$$

$$\underset{\text{1-Phenyl-1-pentanol}}{C_6H_5\overset{}{\underset{OH}{\underset{|}{CH}}}CH_2CH_2CH_2CH_3} \xrightarrow[\substack{\text{acetic acid}\\\text{water}}]{CrO_3} \underset{\substack{\text{1-Phenyl-1-pentanone}\\(93\%)}}{C_6H_5\overset{O}{\overset{\|}{C}}CH_2CH_2CH_2CH_3}$$

315

CHEMICAL COMMUNICATION
IN THE INSECT WORLD:
HOW BUGS BUG BUGS

Several of the compounds shown in Figure 11.3 are described as being phero-mones. **Pheromones** (Greek *pherein,* "to transfer," and *hormon,* "to excite") are compounds, or mixtures of two or more substances, used by members of an animal species to communicate with other members of the same species. *trans-* 2-Hexenal (Figure 11.3), for example, is used by a species of ant to warn other ants of danger and is called an alarm pheromone.

Aldehydes and ketones are not the only classes of compounds which may act as pheromones. Bark beetles, for example, release a mixture of substances which comprise an "aggregation pheromone" into the bark of a tree that causes other beetles to congregate at that site. Two of the substances in this aggregation pheromone are the alcohols shown.

$$\text{and} \qquad \underset{\displaystyle |}{\overset{\displaystyle CH_3}{HOCH_2CHCH}}{=}CH_2$$

Bark beetle aggregation pheromones

While the tree might be able to cope with a few scattered beetles, infestation by a mass of beetles fosters the transmission of a fungal infection at one site which can cause severe damage as it grows and spreads.

Insect pheromones have received considerable study by chemists and biol-ogists in recent years. Why the interest in insect pheromones? Broad-spectrum insecticides such as DDT, while effective in killing insects, have also proven to be harmful to other animal species (see boxed essay at end of Section 3.18 in Chapter 3). These insecticides persist for long periods of time in the environ-ment and make no distinction between beneficial and harmful insects.

The problem is how to effectively control the spread of harmful insects while sparing the beneficial ones. The "ideal" insecticide would be species-specific—that is, it would have an effect on only the insect species whose control is desired and would be harmless to other insects and higher animals.

Insect pheromones appear to be an answer to the problem. A trap with a minute amount of the sex attractant of an insect pest attracts that bug but has no effect on others. Since the compound is not being sprayed, the effect on the environment is negligible. A sex attractant called *disparlure* is now being used in an attempt to control the spread of the gypsy moth.

$$CH_3(CH_2)_{10}CH \overset{\displaystyle }{\underset{\displaystyle O}{\diagdown \diagup}} CH(CH_2)_4CH(CH_3)_2$$

Disparlure

Gypsy moths are pests which have caused extensive defoliation of forests in the eastern United States. Traps baited with disparlure attract males which become confused and are then unable to locate female moths for mating.

Another attractant used for pest control is the synthetic substance *trimed-lure.*

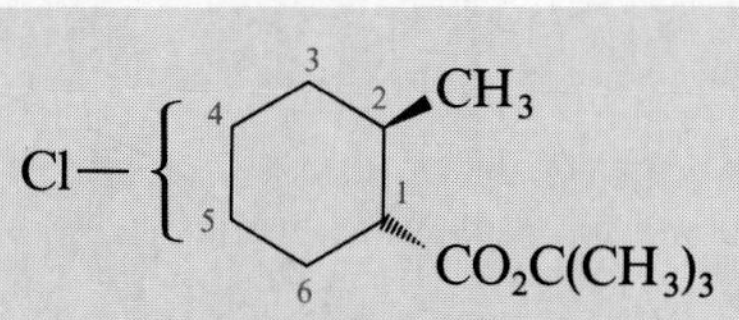

Trimedlure (chlorine
may be on C-4 or C-5)

Trimedlure is a powerful attractant for the Mediterranean fruit fly and is used to detect infestations of this damaging insect. Researchers are finding other commercial uses of insect pheromones for pest control every year.

11.6 REACTIONS OF ALDEHYDES AND KETONES: A REVIEW AND A PREVIEW

You have previously seen one characteristic reaction of aldehydes and ketones, namely, their reduction to alcohols. Aldehydes yield primary alcohols upon reduction, and ketones yield secondary alcohols (Section 10.4).

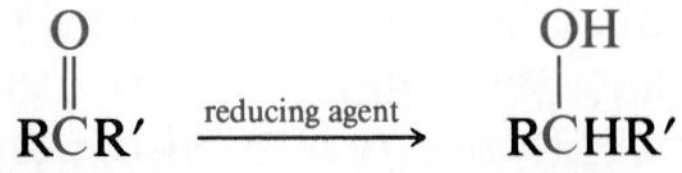

$$\underset{\substack{\text{Aldehyde or} \\ \text{ketone}}}{\overset{\displaystyle O}{\overset{\|}{RCR'}}} \xrightarrow{\text{reducing agent}} \underset{\substack{\text{Primary or} \\ \text{secondary alcohol}}}{\overset{\displaystyle OH}{\overset{|}{RCHR'}}}$$

Reagents effective as reducing agents include hydrogen in the presence of a metal catalyst, sodium borohydride, or lithium aluminum hydride.

The polar nature of the carbonyl group is an important factor in almost all aldehyde and ketone reactions. Nucleophiles attack the electron-deficient carbon of the carbonyl group; electrophiles attack the electron-rich oxygen of the carbonyl.

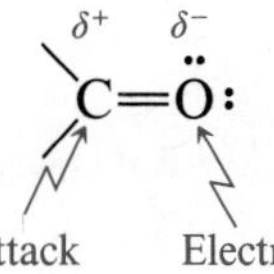

$$\overset{\delta+}{\underset{}{}}C=\overset{\delta-}{\ddot{O}}:$$

Nucleophiles attack   Electrophiles attack
here   here

Many of the reactions presented in the remaining sections of this chapter involve **nucleophilic addition** to the carbonyl group and may be represented by the equation:

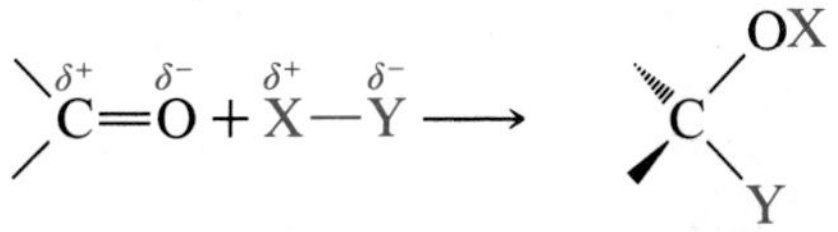

$$\overset{\delta+}{\underset{}{}}C=\overset{\delta-}{O} + \overset{\delta+}{X}-\overset{\delta-}{Y} \longrightarrow \underset{\substack{\text{Product of} \\ \text{nucleophilic addition}}}{\overset{OX}{\underset{Y}{C}}}$$

The mechanistic features of nucleophilic addition to aldehydes and ketones can be illustrated by **hydration,** the addition of a water molecule to the carbonyl group.

11.7 HYDRATION OF ALDEHYDES AND KETONES

Aldehydes and ketones react with water in a rapidly reversible equilibrium process.

$$\underset{\substack{\text{Aldehyde} \\ \text{or ketone}}}{\overset{\overset{\displaystyle O}{\parallel}}{R C R'}} + \underset{\text{Water}}{H_2O} \underset{\text{fast}}{\rightleftharpoons} \underset{\substack{\text{Geminal diol} \\ \text{(hydrate)}}}{\overset{\overset{\displaystyle OH}{|}}{\underset{\underset{\displaystyle OH}{|}}{R C R'}}}$$

The product is called a *geminal diol* because two hydroxyl groups are attached to the same carbon. The term *geminal* describes two functional groups that are attached to the same carbon (see also Section 4.3.3).

Overall the reaction is classified as an **addition reaction**, as the elements of water, $\overset{\delta+}{H}$—$\overset{\delta-}{OH}$, have added to the carbonyl group. The electrophilic proton (H^+) becomes bonded to the negatively polarized carbonyl oxygen, and the nucleophilic hydroxyl to the positively polarized carbon.

The amount of hydrate present at equilibrium varies greatly and is normally much greater for aldehydes than for ketones. Formaldehyde (H_2C=O) exists in aqueous solution almost completely in the hydrate form (99+%), whereas only slightly over 0.1% of acetone, $(CH_3)_2C$=O, is present as the hydrate. The position of equilibrium depends strongly on the nature of the carbonyl group and is influenced by a combination of *electronic* and *steric* effects.

Consider first the electronic effect of substituents on the stabilization of the carbonyl group. Substituents which destabilize the carbonyl group will favor hydrate formation. A striking example is seen in the case of hexafluoroacetone. In contrast to the almost negligible hydration of acetone, hexafluoroacetone is completely hydrated.

$$\underset{\substack{\text{Hexafluoro-} \\ \text{acetone}}}{\overset{\overset{\displaystyle O}{\parallel}}{CF_3 C CF_3}} + \underset{\text{Water}}{H_2O} \rightleftharpoons \underset{\substack{1,1,1,3,3,3- \\ \text{Hexafluoro-2,} \\ \text{2-propanediol}}}{\overset{\overset{\displaystyle OH}{|}}{\underset{\underset{\displaystyle OH}{|}}{CF_3 C CF_3}}} \qquad K_{hydr} = 22,000$$

Carbonyl groups are stabilized by groups that release electrons to the positively polarized carbon. Trifluoromethyl groups, on the other hand, *destabilize* the carbonyl group by withdrawing electrons from it. The less stabilized a carbonyl group is, the greater will be the equilibrium constant for addition to it.

PROBLEM 11.4

Trichloroacetaldehyde, also known as chloral (Problem 11.1*b*), exists as a stable hydrate. It is used in this form as a prescription sleep medication. A Mickey Finn is made by combining chloral hydrate with alcohol. Draw the structure of chloral hydrate and write an equation for its formation.

To understand the role played by steric effects, let us examine the geminal diol product (the hydrate). The carbon that bears the two hydroxyl groups is sp^3 hybrid-

Step 1: Nucleophilic addition of hydroxide ion to the carbonyl group

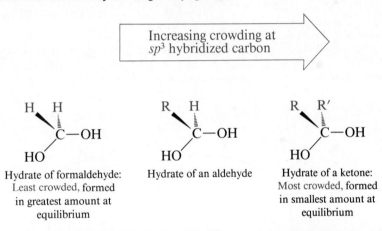

Step 1: (Hydroxide + Aldehyde or ketone → slow)

Step 2: Proton transfer from water to the intermediate formed in the first step

Step 2: (Water + intermediate → fast → Geminal diol + Hydroxide ion)

FIGURE 11.4 Sequence of steps that describes the mechanism of hydration of an aldehyde or ketone under base-catalyzed conditions.

ized. Its substituents are more crowded than they are in the starting aldehyde or ketone. Increased crowding can be better tolerated when the substituents are small hydrogens than when they are large alkyl groups.

Increasing crowding at sp^3 hybridized carbon

Hydrate of formaldehyde: Least crowded, formed in greatest amount at equilibrium

Hydrate of an aldehyde

Hydrate of a ketone: Most crowded, formed in smallest amount at equilibrium

The rate of hydration of aldehydes and ketones is catalyzed by either acids or bases. Under basic conditions, as shown in Figure 11.4, the first step is nucleophilic addition of hydroxide ion to the carbonyl group. The product of the nucleophilic addition step is an alkoxide anion. This anion abstracts a proton from water in the second step to yield the geminal diol product and to regenerate the hydroxide ion.

Three steps are involved in the acid-catalyzed hydration reaction, as shown in Figure 11.5. The first step is protonation of the carbonyl oxygen. Hydrogen ions are electrophilic and are attracted to the electron-rich carbonyl oxygen. The second step is nucleophilic addition of water to the carbon of the carbonyl group. The last step, a proton transfer from the conjugate acid of the geminal diol to a water molecule, regenerates the hydronium ion.

The hydration of aldehydes and ketones illustrates some important general features of nucleophilic addition reactions. Under basic conditions, the first step is *addition of the nucleophile to the positively polarized carbon of the carbonyl group.* Under acidic conditions, however, the first step is *protonation of the negatively polarized oxygen of the carbonyl group.* The protonated species is better able to undergo nucleophilic addition by a weak nucleophile such as water. Protonation of the carbonyl oxygen causes the carbon of the carbonyl to become more positive and thus more electrophilic. This can be respresented by resonance structures:

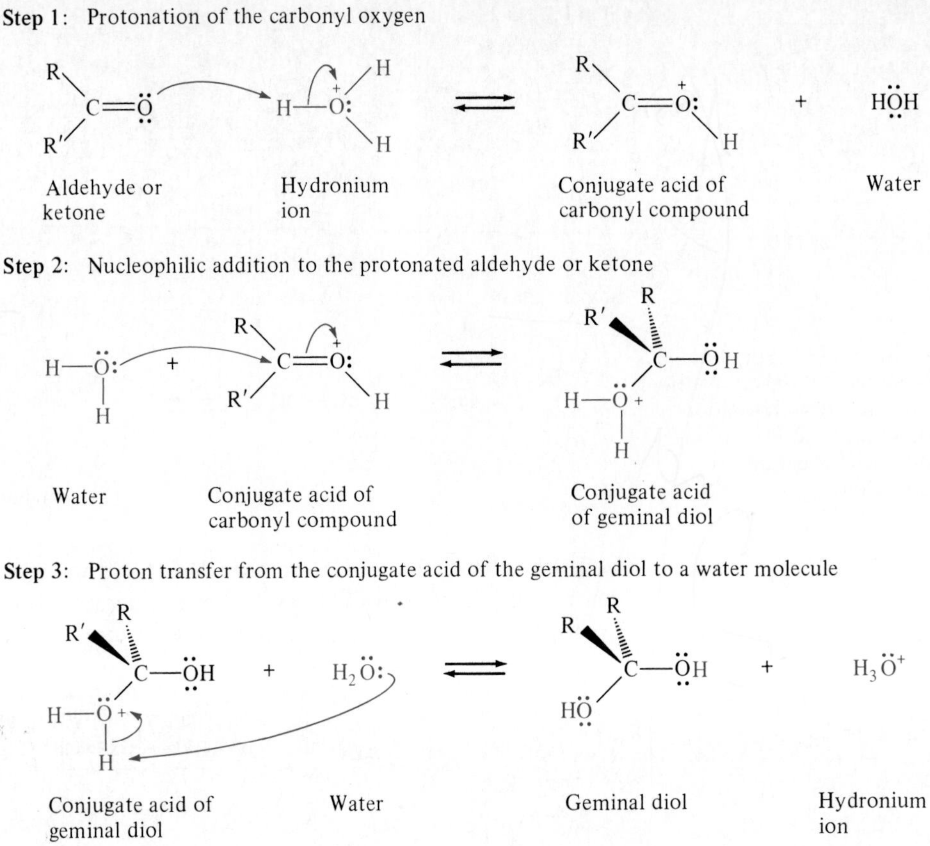

Step 1: Protonation of the carbonyl oxygen

Aldehyde or ketone   Hydronium ion   Conjugate acid of carbonyl compound   Water

Step 2: Nucleophilic addition to the protonated aldehyde or ketone

Water   Conjugate acid of carbonyl compound   Conjugate acid of geminal diol

Step 3: Proton transfer from the conjugate acid of the geminal diol to a water molecule

Conjugate acid of geminal diol   Water   Geminal diol   Hydronium ion

FIGURE 11.5 Sequence of steps that describes the mechanism of hydration of an aldehyde or ketone under acid-catalyzed conditions.

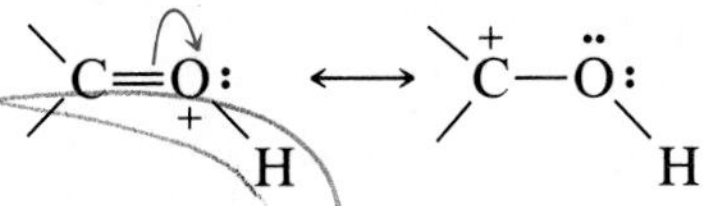

11.8 ACETAL FORMATION

Hydration is only one example of the nucleophilic addition reactions that aldehydes and ketones may undergo. Aldehydes also react with alcohols under conditions of acid catalysis to yield geminal diethers known as **acetals.**

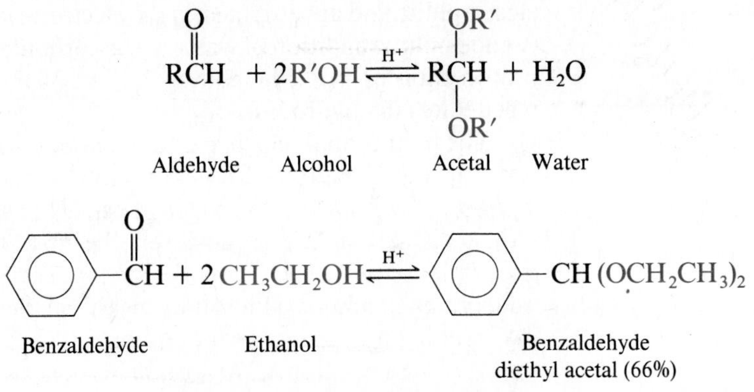

$$\underset{\text{Aldehyde}}{\overset{\overset{\displaystyle O}{\|}}{R\underset{}{C}H}} + \underset{\text{Alcohol}}{2R'OH} \underset{}{\overset{H^+}{\rightleftharpoons}} \underset{\text{Acetal}}{\overset{\overset{\displaystyle OR'}{|}}{R\underset{\underset{\displaystyle OR'}{|}}{C}H}} + \underset{\text{Water}}{H_2O}$$

Benzaldehyde   Ethanol   Benzaldehyde diethyl acetal (66%)

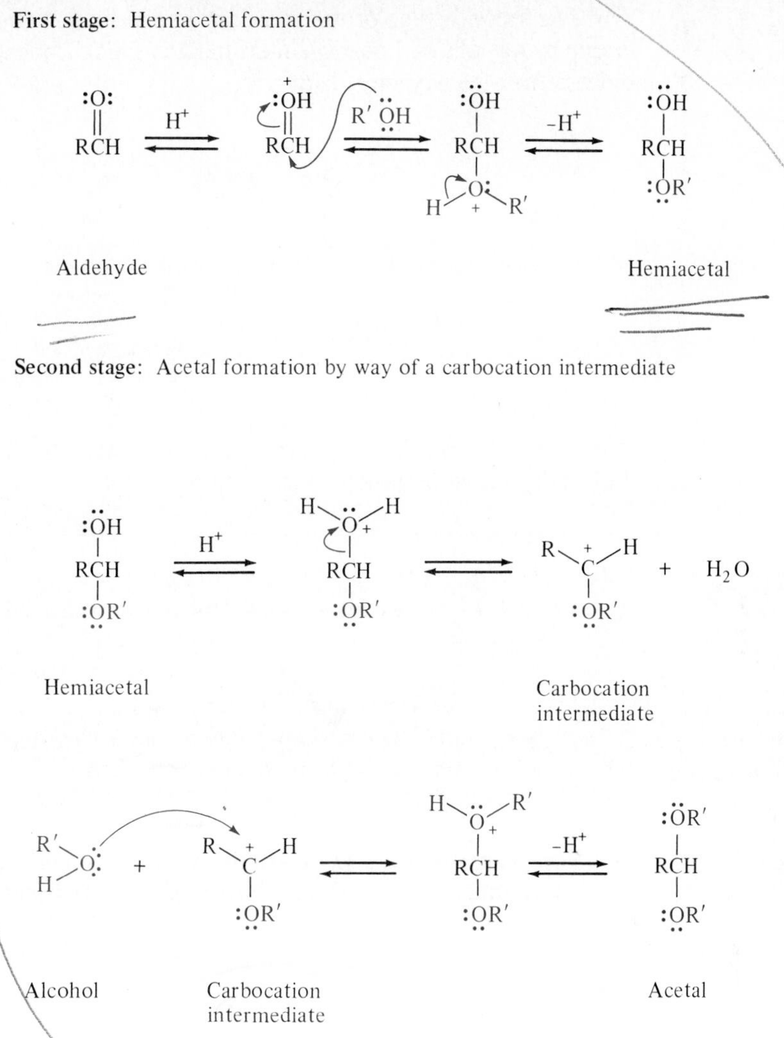

First stage: Hemiacetal formation

$$\underset{\text{Aldehyde}}{\overset{\displaystyle :\!O\!:}{\underset{RCH}{\|}}} \xrightarrow{H^+} \overset{\displaystyle \overset{+}{:}OH}{\underset{RCH}{\|}} \overset{R'\ \ddot{O}H}{\underset{H\ \ddot{O}\!:\!\!\searrow R'}{\rightleftharpoons}} \overset{\displaystyle :\!\ddot{O}H}{\underset{RCH}{\underset{H\ \ddot{O}\!:\!\!\searrow R'}{|}}} \xrightarrow{-H^+} \overset{\displaystyle :\!\ddot{O}H}{\underset{RCH}{\underset{:OR'}{|}}} \quad \underset{\text{Hemiacetal}}{}$$

Second stage: Acetal formation by way of a carbocation intermediate

$$\underset{\text{Hemiacetal}}{\overset{\displaystyle :\!\ddot{O}H}{\underset{RCH}{\underset{:\ddot{O}R'}{|}}}} \xrightarrow{H^+} \overset{H\searrow\ddot{O}\swarrow H}{\underset{RCH}{\underset{:\ddot{O}R'}{|}}} \rightleftharpoons \underset{\text{Carbocation}\atop\text{intermediate}}{\overset{R\searrow\overset{+}{C}\swarrow H}{\underset{:\ddot{O}R'}{|}}} + \ H_2O$$

$$\underset{\text{Alcohol}}{\overset{R'\searrow\ddot{O}\!:}{\underset{H}{}}} + \underset{\text{Carbocation}\atop\text{intermediate}}{\overset{R\searrow\overset{+}{C}\swarrow H}{\underset{:\ddot{O}R'}{|}}} \rightleftharpoons \overset{H\searrow\overset{+}{\ddot{O}}\swarrow R'}{\underset{RCH}{\underset{:OR'}{|}}} \xrightarrow{-H^+} \underset{\text{Acetal}}{\overset{:\ddot{O}R'}{\underset{RCH}{\underset{:OR'}{|}}}}$$

FIGURE 11.6 Mechanism describing the two stages of acetal formation.

Acetal formation is reversible. An equilibrium is established between the reactants, i.e., the carbonyl compound and the alcohol, and the acetal product. The position of equilibrium is favorable for acetal formation from most aldehydes, especially when excess alcohol is present as the reaction solvent. For most ketones the position of equilibrium is unfavorable, and other methods must be used for the preparation of acetals from ketones.*

The overall reaction proceeds in two stages, as shown in Figure 11.6. In the first stage one molecule of the alcohol undergoes acid-catalyzed nucleophilic addition to the aldehyde to yield a **hemiacetal.** This step is entirely analogous to the addition of a molecule of water to the carbonyl group. As described in Section 11.7, protonation of the carbonyl oxygen is the first step of the reaction under acidic conditions. This is followed by nucleophilic addition of an alcohol molecule and subsequent release of a proton to give the hemiacetal.

---

* At one time it was customary to designate the products of addition of alcohols to ketones as *ketals.* This term has been dropped from the IUPAC system of nomenclature, as the term *acetal* is now applied to the products of alcohols with both aldehydes and ketones.

The second stage of the reaction involves conversion of the hemiacetal to an acetal by way of a carbocation intermediate. This carbocation is stabilized by electron release from its oxygen substituent.

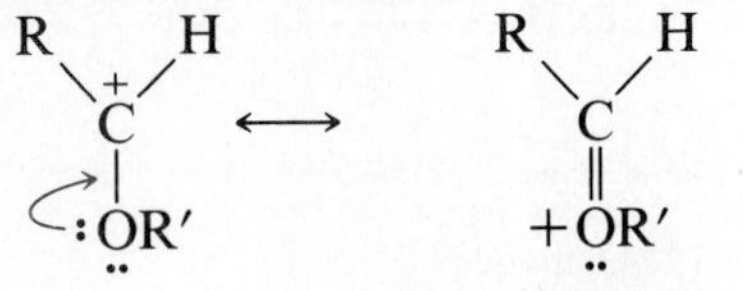

$$
\begin{array}{ccc}
\underset{\underset{\underset{\displaystyle :OR'}{|}}{\overset{\displaystyle +}{C}}}{\overset{\displaystyle R\diagdown \diagup H}{}} & \longleftrightarrow & \underset{\underset{\underset{\displaystyle +OR'}{\|}}{C}}{\overset{\displaystyle R\diagdown \diagup H}{}}
\end{array}
$$

A particularly stable
resonance form
as both carbon and oxygen
have octets of electrons

Nucleophilic capture of the carbocation intermediate by a second molecule of alcohol leads, following proton release, to an acetal.

PROBLEM 11.5

Consider acid-catalyzed acetal formation between acetaldehyde and methanol. Write structural formulas for:

      (a) The hemiacetal intermediate
      (b) The carbocation intermediate (two resonance forms)
      (c) The acetal product

You will encounter hemiacetals and acetals again in Chapter 15, as they are central to understanding the chemistry of carbohydrates.

11.9 CYANOHYDRIN FORMATION

Structurally a cyanohydrin corresponds to the addition of hydrogen cyanide to the carbonyl group of an aldehyde or ketone.

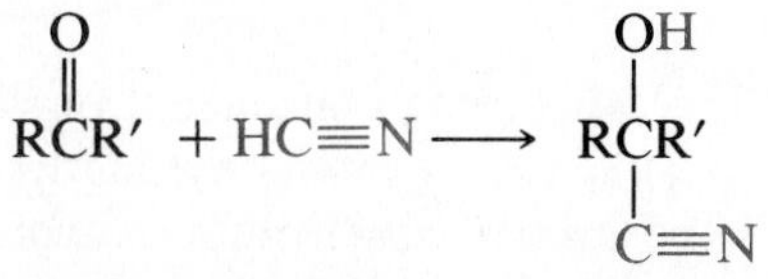

$$
\underset{\text{Aldehyde}\atop\text{or ketone}}{\overset{\displaystyle O \atop \displaystyle \|}{RCR'}} + \underset{\text{Hydrogen}\atop\text{cyanide}}{HC\equiv N} \longrightarrow \underset{\text{Cyanohydrin}}{\overset{\displaystyle OH \atop \displaystyle |}{\underset{\displaystyle | \atop \displaystyle C\equiv N}{RCR'}}}
$$

The reaction occurs in two steps: a nucleophilic addition of a cyanide ion to the carbonyl carbon followed by proton transfer from hydrogen cyanide.

Nucleophilic addition step:

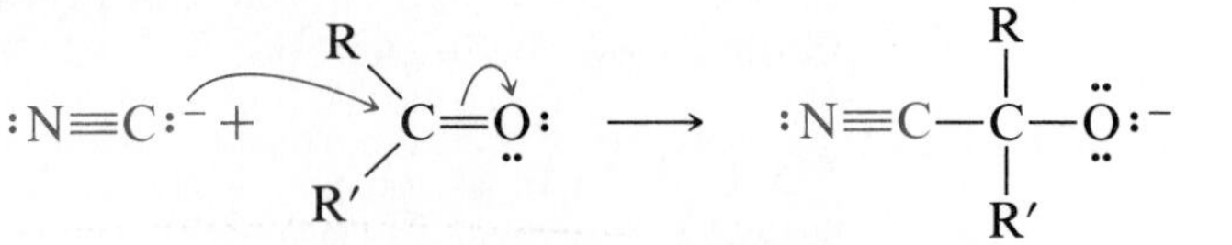

$$
:N\equiv C:^{-} + \underset{R'}{\overset{R}{\diagdown}}C=\overset{..}{\underset{..}{O}}: \longrightarrow :N\equiv C-\underset{R'}{\overset{R}{\underset{|}{\overset{|}{C}}}}-\overset{..}{\underset{..}{O}}:^{-}
$$

    Cyanide ion   Aldehyde or ketone   Conjugate base of cyanohydrin

Proton transfer step:

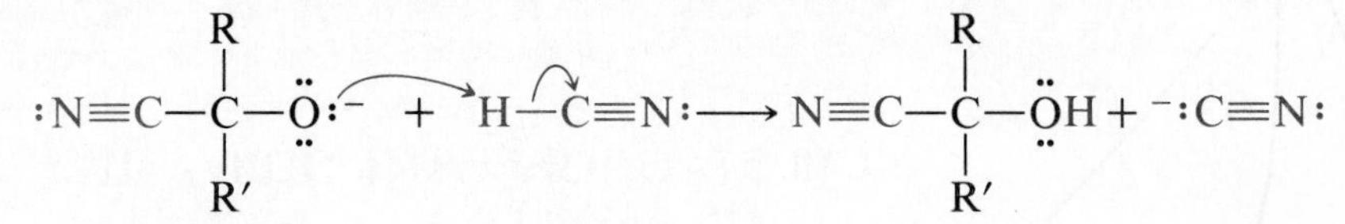

$$:N\equiv C-\underset{\underset{R'}{|}}{\overset{\overset{R}{|}}{C}}-\ddot{O}:^- \quad + \quad H-C\equiv N: \longrightarrow N\equiv C-\underset{\underset{R'}{|}}{\overset{\overset{R}{|}}{C}}-\ddot{O}H \quad + \quad ^-:C\equiv N:$$

| Conjugate base of cyanohydrin | Hydrogen cyanide | Cyanohydrin | Cyanide ion |

Overall, the reaction consumes hydrogen cyanide and is catalytic in cyanide ion. Adding an acid to a solution containing an aldehyde or ketone and sodium or potassium cyanide ensures that sufficient cyanide ion is always present to cause the reaction to proceed at a reasonable rate.

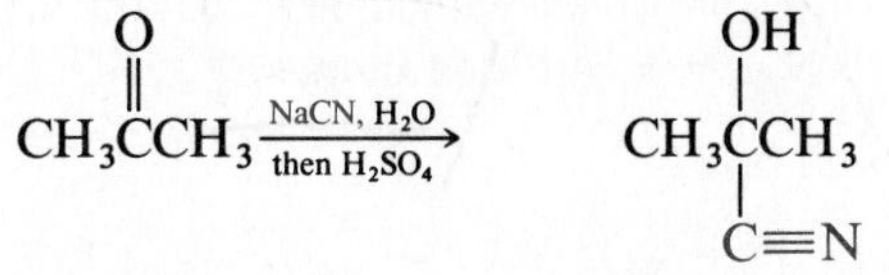

$$CH_3\overset{\overset{O}{\|}}{C}CH_3 \xrightarrow[\text{then } H_2SO_4]{NaCN,\ H_2O} CH_3\underset{\underset{C\equiv N}{|}}{\overset{\overset{OH}{|}}{C}}CH_3$$

| Acetone | 2-Cyano-2-propanol (77–78%) (acetone cyanohydrin) |

Cyanohydrin formation is of synthetic value in that the cyano group may be converted to a carboxylic acid function by hydrolysis (Chapter 12) or to an amine by reduction (Chapter 14).

Some cyanohydrins occur in nature. One species of millipede, for example, stores benzaldehyde cyanohydrin along with an enzyme that catalyzes its cleavage to benzaldehyde and hydrogen cyanide. When attacked, the insect ejects a mixture of the cyanohydrin and the enzyme, repelling the invader by spraying it with poison gas (HCN).

PROBLEM 11.6
Draw the structure of benzaldehyde cyanohydrin.

Another use of a cyanohydrin which has caused some controversy is the anticancer drug Laetrile. Certain fruit pits (peaches and apricots, for example) contain a cyanohydrin derivative called amygdalin.

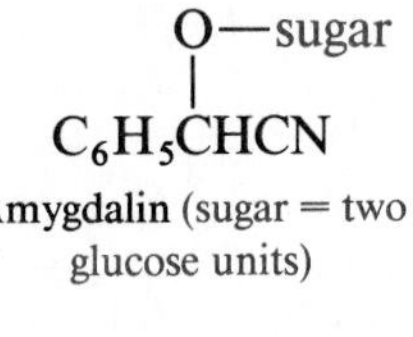

$$C_6H_5\overset{\overset{O-sugar}{|}}{C}HCN$$

Amygdalin (sugar = two glucose units)

Claims have been made, to date unsubstantiated by independent research, that this substance is effective as an anticancer drug. Although not approved for use in the United States by the FDA (Food and Drug Administration), Laetrile, the drug form of amygdalin, has been obtained privately by cancer patients in countries such as Mexico where its application to cancer chemotherapy is permitted. In general, FDA approval of a drug requires that it be shown to be both safe *and* effective. Approval of Laetrile has been withheld because its efficacy has not been demonstrated.

11.10 REACTION WITH DERIVATIVES OF AMMONIA

A number of compounds of the type GNH_2 (where G is one of a variety of groups) react with aldehydes and ketones to give products in which the $\diagdown C{=}O$ functional group is converted to $\diagdown C{=}NG$ and a molecule of water is formed. Primary amines are compounds of the type GNH_2, in which G is an alkyl or aryl group. Primary amines react with aldehydes and ketones to form the corresponding *N*-alkyl- or *N*-aryl-substituted **imines**.

$$GNH_2 + \underset{\substack{\text{Aldehyde} \\ \text{or ketone}}}{\overset{\overset{\displaystyle O}{\|}}{RCR'}} \longrightarrow \underset{\substack{\textit{N}\text{-substituted} \\ \text{imine}}}{\overset{\overset{\displaystyle NG}{\|}}{RCR'}} + \underset{\text{Water}}{H_2O}$$

$$\underset{\text{Primary amine}}{GNH_2}$$

$$\underset{\text{Methylamine}}{CH_3NH_2} + \underset{\text{Benzaldehyde}}{\left\langle\bigcirc\right\rangle\!-\!\overset{\overset{\displaystyle O}{\|}}{CH}} \longrightarrow \underset{\substack{\textit{N}\text{-Benzylidene-} \\ \text{methylamine}}}{\left\langle\bigcirc\right\rangle\!-\!CH{=}NCH_3} + \underset{\text{Water}}{H_2O}$$

Imine formation is a reversible reaction. Both imine formation and imine hydrolysis have been studied extensively because of their relevance to biochemical processes. Many biological reactions involve initial binding of a carbonyl compound to an enzyme or coenzyme by way of imine formation (see the boxed essay on the next page).

Other compounds of the type GNH_2 that react with aldehydes and ketones include aryl derivatives of hydrazine, $ArNHNH_2$ (G = ArNH). The products formed from the reaction of a hydrazine with an aldehyde or a ketone are called **hydrazones**.

$$\underset{\text{Acetophenone}}{\overset{\overset{\displaystyle O}{\|}}{C_6H_5CCH_3}} + \underset{\substack{\text{2,4-Dinitrophenyl-} \\ \text{hydrazine}}}{O_2N\!-\!\left\langle\bigcirc\right\rangle^{\!\!NO_2}\!\!-\!NHNH_2} \longrightarrow \underset{\substack{\text{Acetophenone} \\ \text{2,4-dinitrophenyl-} \\ \text{hydrazone (87-91\%)}}}{\underset{\underset{\displaystyle C_6H_5CCH_3}{\|}}{NNH}\!-\!\left\langle\bigcirc\right\rangle^{\!\!O_2N}\!\!-\!NO_2} + \underset{\text{Water}}{H_2O}$$

In an application to qualitative analysis, an unknown aldehyde or ketone is treated with 2,4-dinitrophenylhydrazine to form a 2,4-dinitrophenylhydrazone derivative. By comparing the melting point of the 2,4-DNP derivative of an unknown aldehyde or ketone to that of one of known structure, the unknown can be identified. This was a common practice in organic analysis until recently, when it was superseded by spectroscopic methods.

325

11.11 REACTIONS THAT
INTRODUCE NEW CARBON-
CARBON BONDS.
ORGANOMETALLIC
COMPOUNDS

We will conclude our discussion of nucleophilic additions with one of the most synthetically useful reactions of aldehydes and ketones, their reaction with Grignard reagents to form alcohols. Let us first take a brief look at Grignard reagents and why they are so valuable to organic chemists engaged in synthesis.

11.11 REACTIONS THAT INTRODUCE NEW CARBON-CARBON BONDS. ORGANOMETALLIC COMPOUNDS

One of the challenges facing an organic chemist is how to construct the carbon skeleton of a desired compound—a new drug, for example. For reasons of availability and cost, one or more reactions that will convert the starting material to a compound having a different number of carbons are often critical to a synthesis.

As you will see in the next several sections, aldehydes and ketones react with species called Grignard reagents to give alcohols having a greater number of carbons than the starting material. Before discussing the Grignard reaction, let us first examine some general aspects of carbon-carbon bond formation in a chemical reaction.

In what types of reactions are new carbon-carbon bonds formed? Friedel-Crafts alkylation and acylation (Section 6.6) are quite useful for this purpose, but these reactions are limited to the synthesis of aromatic compounds. A more general approach would be to have a carbon nucleophile which could attack another carbon as a substrate.

$$R:^- \; + \; R'{-}X \longrightarrow R{-}R' + X^-$$

Nucleophilic substitution using the cyanide ion, CN^- to attack an alkyl halide (Section 8.2), offers one example of this approach. Another example is the synthesis of substituted alkynes by alkylation of an acetylide anion, for example, sodium acetylide, $Na^{+-}C{\equiv}CH$ with an alkyl halide (Section 5.4).

THE CHEMISTRY OF VISION

There are two types of photoreceptor cells in the eye, rods and cones. Cones are sensitive to color but function only in bright light. Rod vision operates even in dim light but only in black, white, and shades of gray, which is why we lose our perception of colors at night. Because rods are larger and easier to work with in the laboratory, most of what we know of the visual process concerns rod vision. The complete process is quite complex but can be illustrated in broad outline as shown in Figure 11.7.

We obtain vitamin A (retinol) in our diet from substances such as β-carotene, found in carrots and other vegetables. Oxidation of vitamin A in the liver forms the aldehyde **retinal.** Following isomerization of one of its double bonds, 11-*cis*-retinal forms an imine by reaction with an amino group of the protein **opsin.** Thus vitamin A becomes covalently bonded to the protein. The imine formed from 11-*cis*-retinal and opsin is called **rhodopsin** or **visual purple;** it is the chemical substance responsible for the visual process. When rhodopsin absorbs a photon of visible light, the cis double bond of the retinal unit undergoes an isomerization which triggers a nerve impulse detected by the brain as a visual image. Following hydrolysis of the imine, the released all-trans retinal is reisomerized to the 11-cis form and recycled.

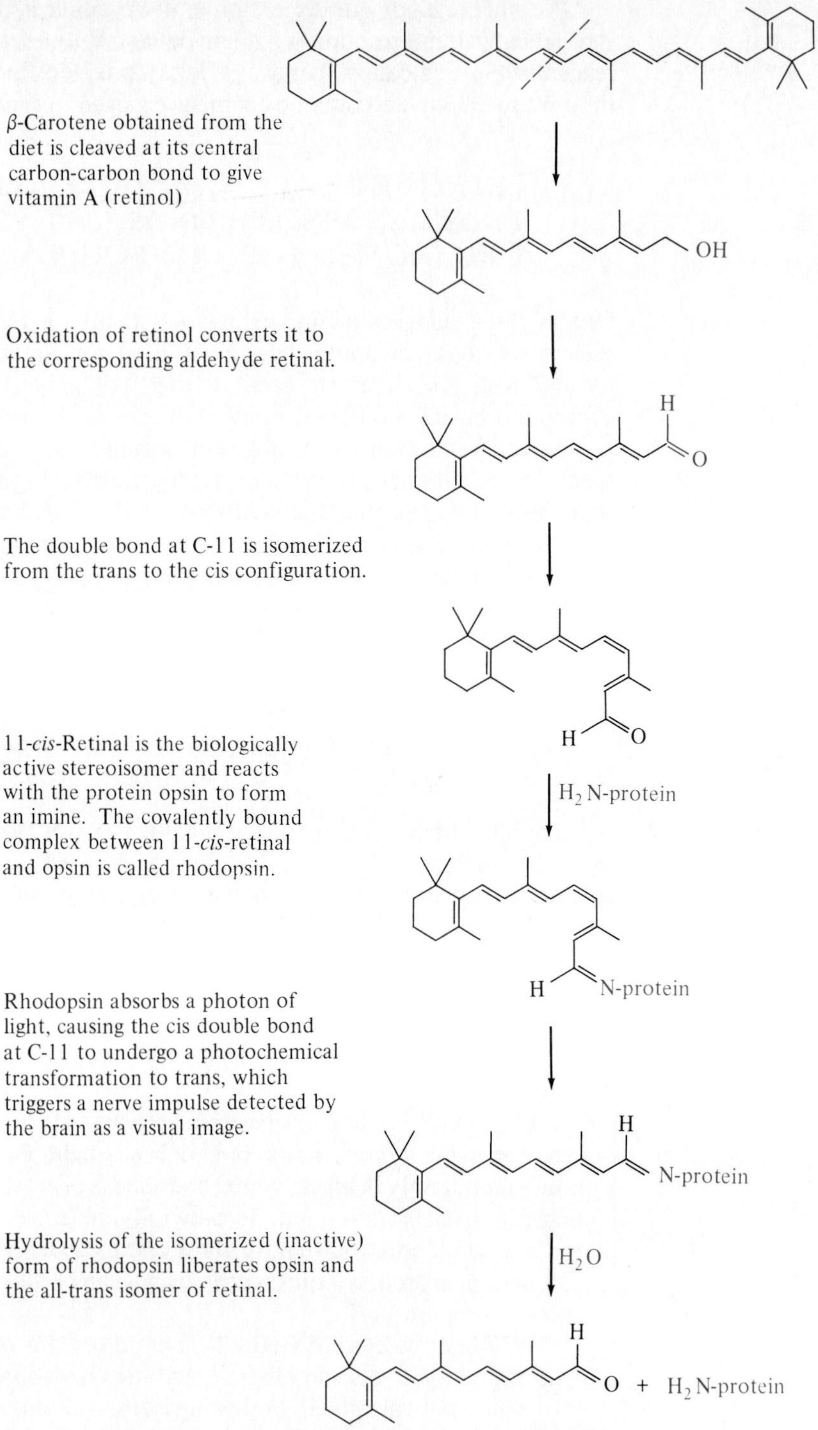

β-Carotene obtained from the diet is cleaved at its central carbon-carbon bond to give vitamin A (retinol)

Oxidation of retinol converts it to the corresponding aldehyde retinal.

The double bond at C-11 is isomerized from the trans to the cis configuration.

11-*cis*-Retinal is the biologically active stereoisomer and reacts with the protein opsin to form an imine. The covalently bound complex between 11-*cis*-retinal and opsin is called rhodopsin.

H$_2$ N-protein

Rhodopsin absorbs a photon of light, causing the cis double bond at C-11 to undergo a photochemical transformation to trans, which triggers a nerve impulse detected by the brain as a visual image.

Hydrolysis of the isomerized (inactive) form of rhodopsin liberates opsin and the all-trans isomer of retinal.

H$_2$O

FIGURE 11.7 Summary diagram illustrating the chemistry of vision.

For a carbon atom to act as a nucleophile, it must have a pair of electrons that can be used to form a bond to a carbon atom of the substrate. In the case of cyanide or an acetylide, the nucleophilic carbon atom bears a full negative charge. A nucleophilic carbon atom may also have a *partial* negative charge, as is the case in compounds in which carbon is bonded to a metal.

When carbon is covalently bonded to an atom more electronegative than itself, such as oxygen, nitrogen, or the halogens, the electron distribution in the bond is polarized so that carbon is slightly positive. Conversely, when carbon is bonded to an atom that is *less* electronegative, such as a metal, the electrons in the bond are more strongly attracted toward carbon. Thus carbon acquires a partial *negative* charge.

$$\overset{|}{\underset{|}{C}} \longleftrightarrow X \qquad \overset{|}{\underset{|}{C}} \longleftrightarrow M$$

X is more electro-   M is less electronegative than carbon   negative than carbon

A species in which carbon is directly bonded to a metal is called an **organometallic compound.** A negatively charged carbon is called a **carbanion,** and organometallic compounds are said to have **carbanionic character.** Although several types of organometallic reagents are commonly used by organic chemists, the most frequently encountered of these is the class of organomagnesium compounds called **Grignard reagents.**

11.12 GRIGNARD REAGENTS

The Grignard reagents are named after the French chemist Victor Grignard. For his studies of organomagnesium compounds Grignard was awarded the 1912 Nobel Prize in Chemistry.

Grignard reagents are prepared directly from organic halides by reaction with magnesium, a group II metal.

$$RX \; + \; Mg \; \xrightarrow{\text{diethyl ether}} \; RMgX$$

Organic   Magnesium   Grignard reagent; an
halide   organomagnesium halide

(R may be methyl or primary, secondary, or tertiary alkyl; it may also be a cycloalkyl, alkenyl, or aryl group)

$$\langle\!\!\!\bigcirc\!\!\!\rangle\!-Br \; + \; Mg \; \xrightarrow[35°C]{\text{diethyl ether}} \; \langle\!\!\!\bigcirc\!\!\!\rangle\!-MgBr$$

Bromobenzene   Magnesium   Phenylmagnesium
                                  bromide (95%)

Anhydrous diethyl ether is the customary solvent for the preparation of Grignard reagents, because it stabilizes the reagent by forming a complex with it. Even more important, ethers such as diethyl ether do not react with Grignard reagents.

The order of halide reactivity is I > Br > Cl > F, with alkyl fluorides being too unreactive to be useful synthetically.

PROBLEM 11.7
Write the structure of the alkylmagnesium halide or arylmagnesium halide formed from each of the following compounds on reaction with magnesium in diethyl ether:

(a) *p*-Bromofluorobenzene   (c) Iodocyclobutane
(b) Allyl chloride (CH_2=CHCH$_2$Cl)   (d) 1-Bromocyclohexene

SAMPLE SOLUTION
(a) Of the two halogen substituents on the aromatic ring, bromine reacts much faster with magnesium than does fluorine. Therefore, fluorine is left intact on the ring, while the carbon-bromine bond is converted to a carbon-magnesium bond.

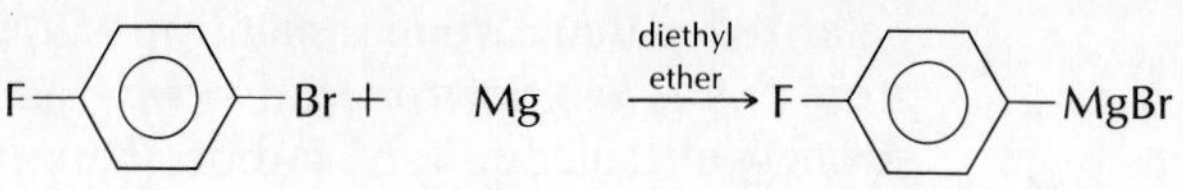

$$F-\!\!\!\bigcirc\!\!\!-Br + Mg \xrightarrow[\text{ether}]{\text{diethyl}} F-\!\!\!\bigcirc\!\!\!-MgBr$$

| *p*-Bromofluoro-benzene | Magnesium | *p*-Fluorophenylmag-nesium bromide |

Grignard reagents are stable species when prepared in suitable solvents such as anhydrous diethyl ether. The solvent must be *anhydrous* (without water), as Grignard reagents react instantly with proton donors such as water or alcohols. Grignard reagents are strongly basic due to the polar nature of their carbon-metal bonds. In the reaction with water or an alcohol, a proton is transferred from the hydroxyl group to the negatively polarized carbon of the organometallic to form a hydrocarbon.

$$\overset{\delta-}{R}-\overset{\delta+}{M} + \overset{\delta+}{H}-\overset{\delta-}{OR'} \longrightarrow RH + R'O^-M^+$$

| Grignard reagent | Water or alcohol | Hydrocarbon | |

$$CH_3CH_2MgCl + H_2O \longrightarrow CH_3CH_3 + MgOHCl$$

| Ethyl magnesium chloride | Water | Ethane (100%) | Hydroxymag-nesium chloride |

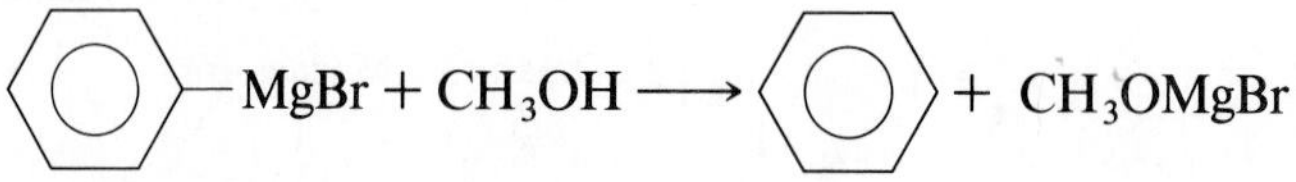

$$\bigcirc\!\!\!-MgBr + CH_3OH \longrightarrow \bigcirc + CH_3OMgBr$$

| Phenylmagnesium bromide | Methanol | Benzene (100%) | Methoxymag-nesium bromide |

11.13 SYNTHESIS OF ALCOHOLS USING GRIGNARD REAGENTS

As noted earlier (Section 11.11), the construction of new carbon-carbon bonds is a major task facing an organic chemist planning a synthesis. Grignard reagents provide a means of carbon-carbon bond formation through their reaction with aldehydes or ketones to produce alcohols.

Recall from Section 11.2 that the carbonyl group is polarized, with the carbon having a partial positive charge. The carbanion-like carbon of the Grignard reagent attacks the carbon of the carbonyl, forming a new carbon-carbon single bond.

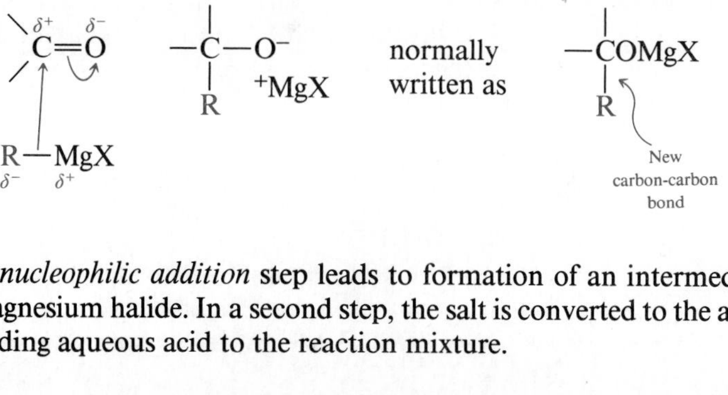

$$\overset{\delta+}{C}=\overset{\delta-}{O} \qquad -\overset{|}{\underset{R}{C}}-O^- \quad ^+MgX \qquad \text{normally written as} \qquad -\overset{|}{\underset{R}{C}}OMgX$$

$$R-\underset{\delta-\quad\delta+}{MgX}$$

New carbon-carbon bond

This *nucleophilic addition* step leads to formation of an intermediate salt, an alkoxymagnesium halide. In a second step, the salt is converted to the alcohol product by adding aqueous acid to the reaction mixture.

$$\underset{\substack{\text{Alkoxymagnesium} \\ \text{halide}}}{R\!-\!\overset{\displaystyle |}{\underset{\displaystyle |}{C}}\!-\!OMgX} + \underset{\substack{\text{Hydronium} \\ \text{ion}}}{H_3O^+} \longrightarrow \underset{\text{Alcohol}}{R\!-\!\overset{\displaystyle |}{\underset{\displaystyle |}{C}}\!-\!OH} + \underset{\substack{\text{Magnesium} \\ \text{ion}}}{Mg^{2+}} + \underset{\substack{\text{Halide} \\ \text{ion}}}{X^-} + \underset{\text{Water}}{H_2O}$$

The type of alcohol produced depends on the carbonyl-containing compound used. Substituents present on the carbonyl group in the reactant stay there—they become substituents on the carbon that bears the hydroxyl group in the product. Thus, formaldehyde reacts with Grignard reagents to yield primary alcohols, aldehydes yield secondary alcohols, and ketones yield tertiary alcohols.

1. **Reaction of a Grignard reagent with formaldehyde:**

$$\underset{\substack{\text{Grignard} \\ \text{reagent}}}{RMgX} + \underset{\text{Formaldehyde}}{\overset{\displaystyle O \atop \displaystyle \|}{H\,C\,H}} \xrightarrow{\substack{\text{diethyl} \\ \text{ether}}} \underset{\substack{\text{Primary alkoxy-} \\ \text{magnesium halide}}}{R\!-\!\overset{\displaystyle H}{\underset{\displaystyle H}{C}}\!-\!OMgX} \xrightarrow{H_3O^+} \underset{\text{Primary alcohol}}{R\!-\!\overset{\displaystyle H}{\underset{\displaystyle H}{C}}\!-\!OH}$$

Cyclohexylmagnesium   Formaldehyde   Cyclohexylmethanol
        chloride   (64–69%)

2. **Reaction of a Grignard reagent with an aldehyde:**

$$\underset{\substack{\text{Grignard} \\ \text{reagent}}}{RMgX} + \underset{\text{Aldehyde}}{\overset{\displaystyle O \atop \displaystyle \|}{R'C\,H}} \xrightarrow{\substack{\text{diethyl} \\ \text{ether}}} \underset{\substack{\text{Secondary} \\ \text{alkoxymagnesium} \\ \text{halide}}}{R\!-\!\overset{\displaystyle H}{\underset{\displaystyle R'}{C}}\!-\!OMgX} \xrightarrow{H_3O^+} \underset{\substack{\text{Secondary} \\ \text{alcohol}}}{R\!-\!\overset{\displaystyle H}{\underset{\displaystyle R'}{C}}\!-\!OH}$$

$$\underset{\substack{\text{Hexylmagnesium} \\ \text{bromide}}}{CH_3(CH_2)_4CH_2MgBr} + \underset{\substack{\text{Ethanal} \\ \text{(acetaldehyde)}}}{\overset{\displaystyle O \atop \displaystyle \|}{CH_3C\,H}} \xrightarrow[\text{2. } H_3O^+]{\text{1. diethyl ether}}$$

$$\underset{\substack{\text{2-Octanol (84\%)}}}{CH_3(CH_2)_4CH_2\underset{\displaystyle OH}{\overset{\displaystyle |}{\underset{\displaystyle |}{C}}}HCH_3}$$

3. **Reaction of a Grignard reagent with a ketone:**

$$\underset{\substack{\text{Grignard} \\ \text{reagent}}}{RMgX} + \underset{\text{Ketone}}{\overset{\displaystyle O \atop \displaystyle \|}{R'C\,R''}} \xrightarrow{\substack{\text{diethyl} \\ \text{ether}}} \underset{\substack{\text{Tertiary} \\ \text{alkoxymagnesium} \\ \text{halide}}}{R\!-\!\overset{\displaystyle R''}{\underset{\displaystyle R'}{C}}\!-\!OMgX} \xrightarrow{H_3O^+} \underset{\substack{\text{Tertiary} \\ \text{alcohol}}}{R\!-\!\overset{\displaystyle R''}{\underset{\displaystyle R'}{C}}\!-\!OH}$$

$$\text{CH}_3\text{MgCl} \; + \; \underset{\text{O}}{\text{cyclopentanone}} \quad \xrightarrow[\text{2. H}_3\text{O}^+]{\text{1. diethyl ether}} \quad \underset{\text{H}_3\text{C} \quad \text{OH}}{\text{methylcyclopentanol}}$$

| Methylmagnesium chloride | Cyclopentanone | 1-Methylcyclopentanol (62%) |

PROBLEM 11.8

Write the structure of the product of the reaction of propylmagnesium bromide with each of the following electrophiles. Assume the reactions are worked up by the addition of dilute aqueous acid in the usual manner.

(a) Formaldehyde $\overset{\text{O}}{\overset{\|}{\text{H}\text{C}\text{H}}}$ (c) Cyclohexanone $\langle\text{cyclohexane ring}\rangle{=}\text{O}$

(b) Benzaldehyde $\text{C}_6\text{H}_5\overset{\text{O}}{\overset{\|}{\text{C}\text{H}}}$ (d) 2-Butanone $\text{CH}_3\overset{\text{O}}{\overset{\|}{\text{C}}}\text{CH}_2\text{CH}_3$

SAMPLE SOLUTION

(a) Grignard reagents react with formaldehyde to give primary alcohols having one more carbon atom than the alkyl halide from which the Grignard reagent was prepared. The product is 1-butanol.

$$\text{CH}_3\text{CH}_2\text{CH}_2{-}\text{MgBr} \quad \xrightarrow[\text{ether}]{\text{diethyl}} \quad \text{CH}_3\text{CH}_2\text{CH}_2 \quad \xrightarrow{\text{H}_3\text{O}^+} \quad \text{CH}_3\text{CH}_2\text{CH}_2\text{CH}_2\text{OH}$$

$$\underset{\text{H}}{\overset{\text{H}}{\text{C}}}{=}\text{O}$$

$$\text{H}{-}\underset{\text{H}}{\overset{|}{\text{C}}}{-}\text{OMgBr}$$

| Propylmagnesium bromide + formaldehyde | | 1-Butanol |

11.14 GRIGNARD REACTIONS IN SYNTHESIS. WORKING BACKWARD

As you saw in the preceding section, Grignard reagents provide a straightforward method for preparing alcohols. What if you are given the structure of the alcohol product and asked to choose the starting materials? Which Grignard reagent would you choose? For that matter, would a Grignard reaction be used at all? Remember, alcohols can also be prepared by reduction of an aldehyde or ketone (Section 10.4). The key to answering these questions is to **work the problem backward.***

Let us consider the preparation of 1-pentanol from carbonyl precursors as an example. One method would be the reduction of pentanal.

$$\text{CH}_3\text{CH}_2\text{CH}_2\text{CH}_2\overset{\text{O}}{\overset{\|}{\text{C}\text{H}}} \quad \xrightarrow[\text{CH}_3\text{OH}]{\text{NaBH}_4} \quad \text{CH}_3\text{CH}_2\text{CH}_2\text{CH}_2\text{CH}_2\text{OH}$$

* The phrase "retrosynthetic analysis" is used to describe this approach to a synthesis problem.

Notice that the starting material has the *same number of carbons* as the product. Reductions do not result in a change in the carbon skeleton going from reactant to product.

Now consider how we could use the Grignard reaction to prepare 1-pentanol. Focus your attention on the hydroxyl-bearing carbon of the target molecule, as this carbon was part of the carbonyl group in the starting material. Mentally disconnect the alkyl group attached to the carbon bearing the hydroxyl group—in essence reversing the carbon-carbon bond formation of the Grignard step. The open arrow in the equation marks it as a retrosynthetic step characterized by a disconnection of chemical bonds.

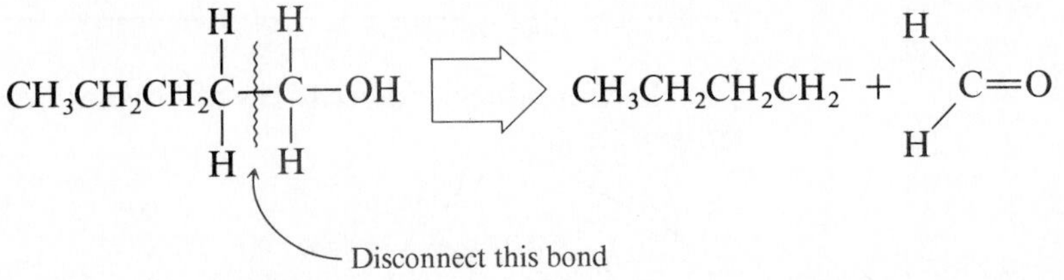

$$CH_3CH_2CH_2C{\overset{\overset{\displaystyle H}{|}}{\underset{\underset{\displaystyle H}{|}}{C}}}{\overset{\overset{\displaystyle H}{|}}{\underset{\underset{\displaystyle H}{|}}{C}}}-OH \implies CH_3CH_2CH_2CH_2{}^- + {\overset{\displaystyle H}{\underset{\displaystyle H}{}}}C{=}O$$

Disconnect this bond

By doing this you have revealed the structures of the Grignard reagent and the carbonyl precursor. 1-Pentanol may be prepared by reaction of formaldehyde and a butylmagnesium halide such as butylmagnesium iodide.

$$CH_3CH_2CH_2CH_2MgI + H_2C{=}O \xrightarrow[\text{2. H}_3\text{O}^+]{\text{1. ether}} CH_3CH_2CH_2CH_2CH_2OH$$

There are two possible combinations of Grignard reagent and aldehyde which will give a secondary alcohol. These can be revealed by disconnecting each of the two alkyl groups of the target alcohol.

1. Disconnect the R—C bond

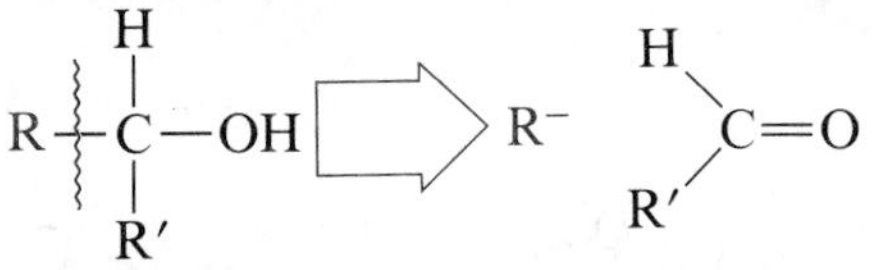

$$R{\overset{\overset{\displaystyle H}{|}}{\underset{\underset{\displaystyle R'}{|}}{C}}}-OH \implies R^- + {\overset{\displaystyle H}{\underset{\displaystyle R'}{}}}C{=}O$$

2. Disconnect the R'—C bond

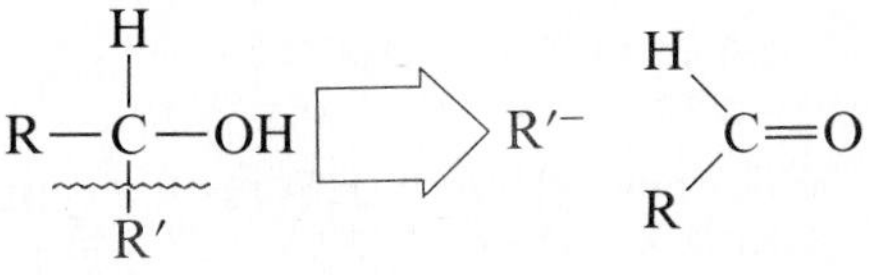

$$R-{\overset{\overset{\displaystyle H}{|}}{\underset{\underset{\displaystyle R'}{|}}{C}}}-OH \implies R'^- + {\overset{\displaystyle H}{\underset{\displaystyle R}{}}}C{=}O$$

PROBLEM 11.9

Suggest how each of the following alcohols might be prepared using a Grignard reagent.

(a) 2-Methyl-1-propanol, $(CH_3)_2CHCH_2OH$

(b) 2-Hexanol, $CH_3\underset{\underset{\displaystyle OH}{|}}{C}HCH_2CH_2CH_2CH_3$ (2 methods)

(c) 1-Phenyl-1-propanol, $C_6H_5\underset{\underset{\displaystyle OH}{|}}{C}HCH_2CH_3$ (2 methods)

SAMPLE SOLUTION

Mentally disconnect the bond between the alkyl group and the hydroxyl-bearing carbon.

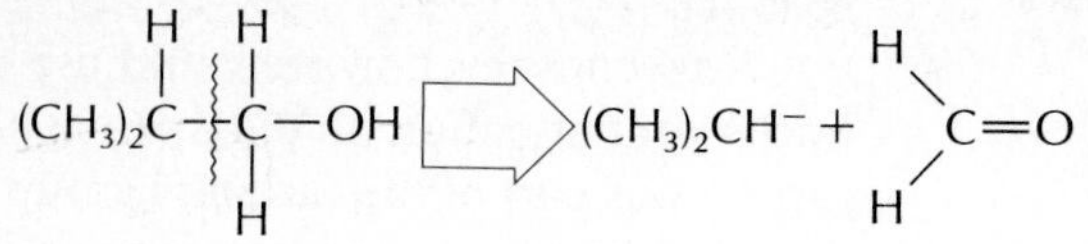

$$(CH_3)_2C\!-\!\!\overset{\displaystyle H}{\underset{\displaystyle H}{\overset{|}{\underset{|}{C}}}}\!\!-OH \implies (CH_3)_2CH^- + \overset{\displaystyle H}{\underset{\displaystyle H}{C}}=O$$

This reveals the appropriate precursors to be formaldehyde and the Grignard reagent derived from 2-bromopropane.

$$(CH_3)_2CHMgBr + H_2C=O \xrightarrow[\text{2. } H_3O^+]{\text{1. diethyl ether}} (CH_3)_2CHCH_2OH$$

There are three combinations of Grignard reagent and ketone that give rise to tertiary alcohols.

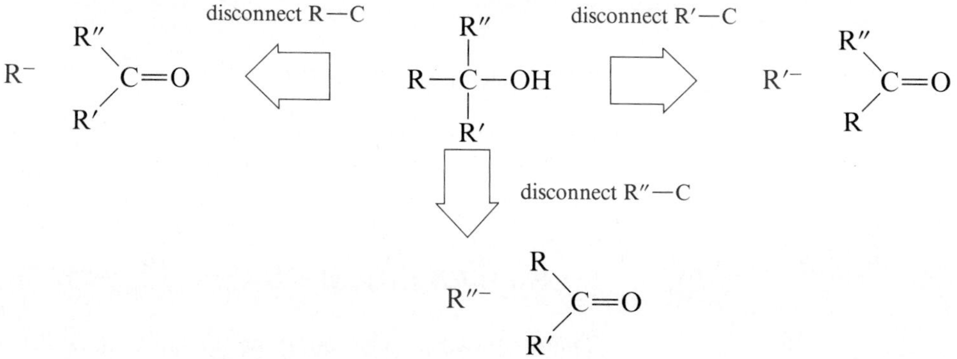

$$R^- \quad \overset{\displaystyle R''}{\underset{\displaystyle R'}{C}}=O \xLeftarrow{\text{disconnect } R-C} R-\overset{\displaystyle R''}{\underset{\displaystyle R'}{\overset{|}{\underset{|}{C}}}}-OH \xRightarrow{\text{disconnect } R'-C} R'^- \quad \overset{\displaystyle R''}{\underset{\displaystyle R}{C}}=O$$

disconnect $R''-C$

$$R''^- \quad \overset{\displaystyle R}{\underset{\displaystyle R'}{C}}=O$$

PROBLEM 11.10

Suggest three ways in which 2-phenyl-2-butanol, $C_6H_5\overset{\displaystyle OH}{\underset{\displaystyle CH_3}{\overset{|}{\underset{|}{C}}}}CH_2CH_3$, can be prepared using Grignard reagents.

11.15 THE α CARBON AND ITS HYDROGENS

While the carbonyl group is the center of reactivity in aldehydes and ketones, the carbon atom adjacent to the carbonyl group is also a site of some interesting and useful chemistry.

The carbon atom adjacent to the carbonyl group is the α carbon atom. A hydrogen connected to the α carbon is an α hydrogen. Carbon atoms two-removed from the carbonyl group are called β carbons; hydrogens attached to β carbons are called β hydrogens.

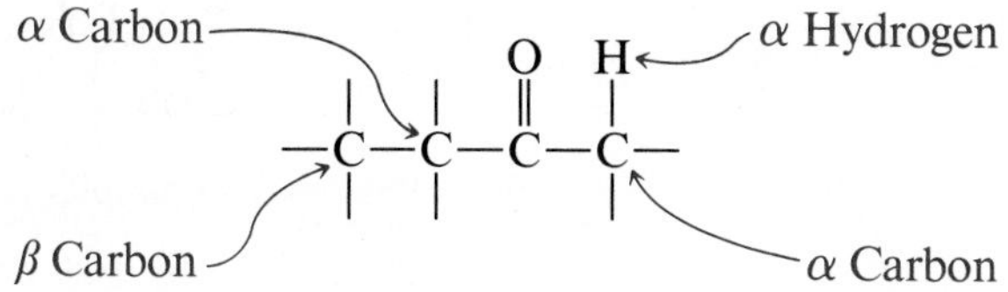

PROBLEM 11.11

How many α hydrogens are there in each of the following?

    (a) 3,3-Dimethyl-2-butanone   (c) Benzyl methyl ketone

    (b) 2,2-Dimethylpropanal   (d) Cyclohexanone

SAMPLE SOLUTION

(a) This ketone has two different α carbons, but only one of them has hydrogen substituents. There are three equivalent α hydrogens. The other nine hydrogens are attached to β carbon atoms.

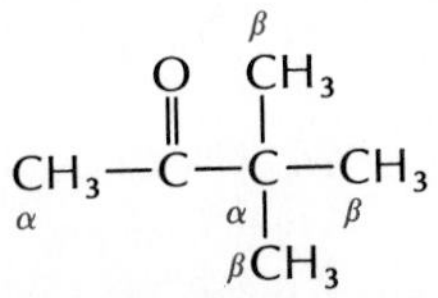

                    3,3-Dimethyl-2-butanone

Other than nucleophilic addition to the carbonyl group, the most important reactions of aldehydes and ketones involve substitution of an α hydrogen. While we will discuss only simple examples, reactions which involve the α carbon and its hydrogens are frequently encountered in biochemistry.

11.16 ENOLS AND ENOLATES. ENOLIZATION

Aldehydes and ketones which have at least one α hydrogen are in equilibrium with an isomer called an **enol.**

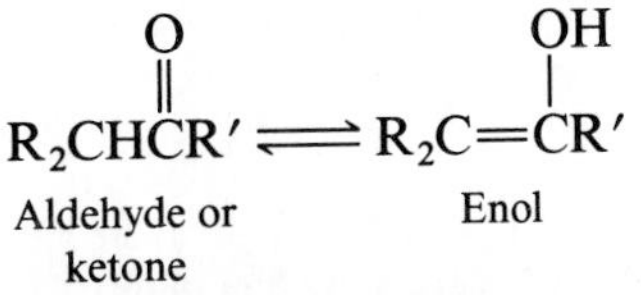

$$R_2CHCR' \rightleftharpoons R_2C{=}CR'$$

      Aldehyde or   Enol
        ketone

This type of equilibrium is called **tautomerism. Tautomers** are isomers which differ by the placement of an atom or group. You should recognize that tautomerism is an *equilibrium* between constitutional isomers and is *not* resonance.

    The amount of enol present at equilibrium is quite small for simple aldehydes and ketones, as shown by these examples.

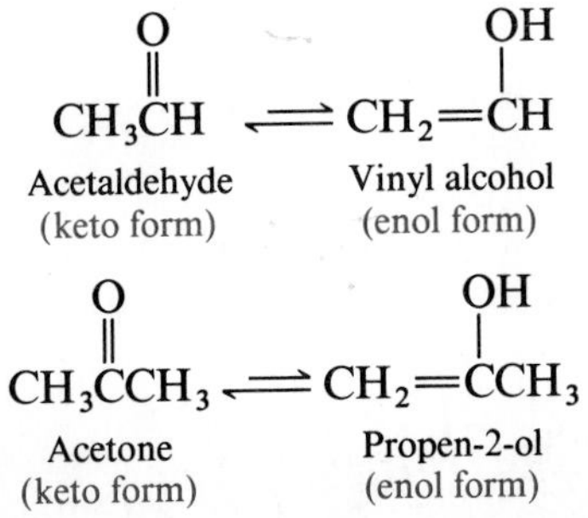

$$CH_3CH \rightleftharpoons CH_2{=}CH \qquad 3 \times 10^{-5}\%\ \text{enol}$$

    Acetaldehyde   Vinyl alcohol
    (keto form)   (enol form)

$$CH_3CCH_3 \rightleftharpoons CH_2{=}CCH_3 \qquad 6 \times 10^{-7}\%\ \text{enol}$$

    Acetone   Propen-2-ol
    (keto form)   (enol form)

PROBLEM 11.12

Write structural formulas corresponding to:

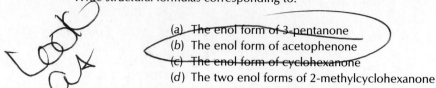

(a) The enol form of 3-pentanone
(b) The enol form of acetophenone
(c) The enol form of cyclohexanone
(d) The two enol forms of 2-methylcyclohexanone

SAMPLE SOLUTION

(a) Remember that it is the α carbon atom that is deprotonated in the enolization process. The ketone 3-pentanone gives a single enol since the two α carbons are equivalent.

$$\underset{\substack{\text{3-Pentanone}\\\text{(keto form)}}}{\overset{\overset{\displaystyle O}{\|}}{CH_3CH_2\overset{\alpha}{C}CH_2CH_3}} \rightleftharpoons \underset{\substack{\text{2-Penten-3-ol}\\\text{(enol form)}}}{\overset{\overset{\displaystyle OH}{|}}{CH_3CH=CCH_2CH_3}}$$

Enolization can be catalyzed by a base such as the hydroxide ion, as shown in Figure 11.8. The base abstracts a proton from the α carbon atom to form the conjugate base of the carbonyl compound. This anion is called an **enolate ion.** The hydroxide ion is a sufficiently strong base to remove an α hydrogen because these protons are far more acidic than would be expected for a simple hydrocarbon.

$$\underset{\displaystyle H}{\overset{\overset{\displaystyle O}{\|}}{R_2CCR'}}$$

This proton is far more acidic than protons attached to simple alkyl groups

Overall reaction:

$$\underset{\text{Aldehyde or ketone}}{\overset{\overset{\displaystyle O}{\|}}{RCH_2CR'}} \quad \underset{}{\overset{HO^-}{\rightleftharpoons}} \quad \underset{\text{Enol}}{\overset{\overset{\displaystyle OH}{|}}{RCH=CR'}}$$

Step 1: A proton is abstracted by hydroxide ion from the α carbon atom of the carbonyl compound.

$$\underset{\substack{\text{Aldehyde}\\\text{or ketone}}}{RCH{-}CR'} + \underset{\substack{\text{Hydroxide}\\\text{ion}}}{:\!\ddot{O}\!:} \rightleftharpoons \underset{\substack{\text{Enolate ion}\\\text{(Conjugate base of}\\\text{carbonyl compound)}}}{RCH{-}CR'} + \underset{\text{Water}}{:\!\ddot{O}\!:}$$

Step 2: A water molecule acts as a Bronsted acid to transfer a proton to the oxygen of the enolate ion.

$$\underset{\substack{\text{Conjugate base of}\\\text{carbonyl compound}}}{RCH=CR'} + \underset{\text{Water}}{:\!\ddot{O}\!:} \rightleftharpoons \underset{\text{Enol}}{RCH=CR'} + \underset{\substack{\text{Hydroxide}\\\text{ion}}}{:\!\ddot{O}\!:}$$

FIGURE 11.8 Sequence of steps that describes the base-catalyzed enolization of an aldehyde or ketone in aqueous solution.

The enolate anion is a resonance-stabilized species, as the negative charge is shared by the α carbon atom and the carbonyl oxygen.

$$\overset{:\ddot{O}:}{\underset{-\,\cdot\cdot}{RCH}-CR'} \longleftrightarrow \overset{:\ddot{O}:^-}{RCH=CR'}$$

Resonance structures
of conjugate base (enolate ion)

It is this delocalization of the negative charge onto the electronegative oxygen that is responsible for the enhanced acidity of aldehydes and ketones. With K_a's (Section 3.7) in the 10^{-16} to 10^{-20} range, aldehydes and ketones are about as acidic as water and alcohols (see Table 3.3).

Protonation of the enolate ion can occur either at the α carbon or at the oxygen. Protonation of the α carbon simply returns the anion to the starting aldehyde or ketone. Protonation at oxygen, as shown in step 2 of Figure 11.8, produces the enol.

You will see an example of how enolate ions are involved in reactions of aldehydes and ketones in the next section when the aldol condensation is discussed.

11.17 THE ALDOL CONDENSATION

As you have just seen, an aldehyde is partially converted to its enolate ion by a base such as a hydroxide ion.

$$\underset{\text{Aldehyde}}{\overset{O}{\overset{\|}{RCH_2CH}}} + \underset{\text{Hydroxide}}{HO^-} \rightleftharpoons \underset{\text{Enolate}}{\overset{O}{\overset{\|}{RCH-CH}}} + \underset{\text{Water}}{H_2O}$$

In a solution that contains both an aldehyde and its enolate ion, the enolate acts as a nucleophile and adds to the carbonyl group of the aldehyde. This addition is exactly analogous to the addition reactions of other nucleophilic reagents to aldehydes and ketones described earlier in this chapter.

$$\overset{O}{\overset{\|}{RCH_2CH}} \\ \underset{-\,\cdot\cdot}{RCH}-C\overset{O}{\underset{H}{\diagdown}} \rightleftharpoons \underset{R}{\overset{O^-}{RCH_2CH}}-\overset{O}{\overset{\|}{CHCH}}$$

The alkoxide formed in this nucleophilic addition step then abstracts a proton from the solvent (usually water or ethanol) to yield the product of **aldol addition.** This product is known as an **aldol** because it contains both an aldehyde function (*ald-*) and a hydroxyl group (*-ol*).

$$\underset{R}{\overset{O^-}{RCH_2CH}}-\overset{O}{\overset{\|}{CHCH}} + H_2O \rightleftharpoons \underset{R}{\overset{OH}{RCH_2CH}}-\overset{O}{\overset{\|}{CHCH}} + HO^-$$

Product of aldol
addition

The significant feature of the aldol addition for you to note is that *carbon-carbon bond formation occurs between the α carbon of one aldehyde and the carbonyl group of another.* We can represent the overall reaction with the following equation:

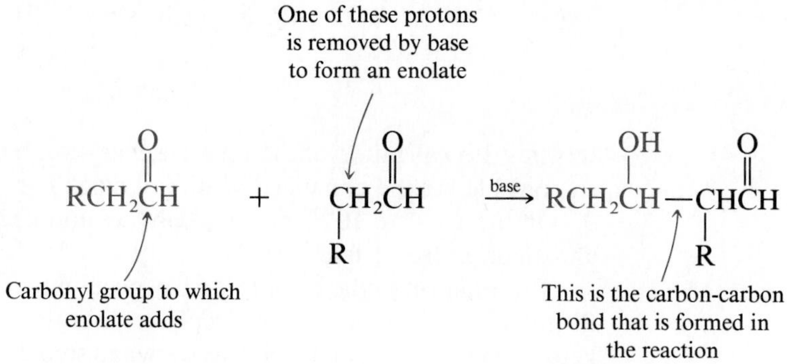

One of these protons
is removed by base
to form an enolate

$$RCH_2CH \;+\; CH_2CH \xrightarrow{\text{base}} RCH_2CH-CHCH$$

Carbonyl group to which
enolate adds

This is the carbon-carbon
bond that is formed in
the reaction

Aldol addition occurs readily with aldehydes.

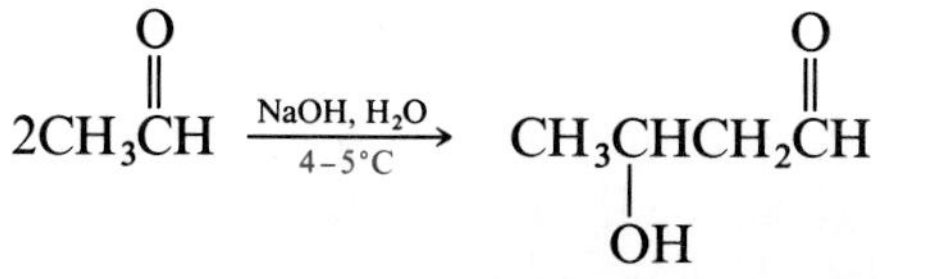

$$2CH_3CH \xrightarrow[4-5^\circ C]{NaOH,\; H_2O} CH_3CHCH_2CH$$

Acetaldehyde   3-Hydroxybutanal (50%)
(acetaldol)

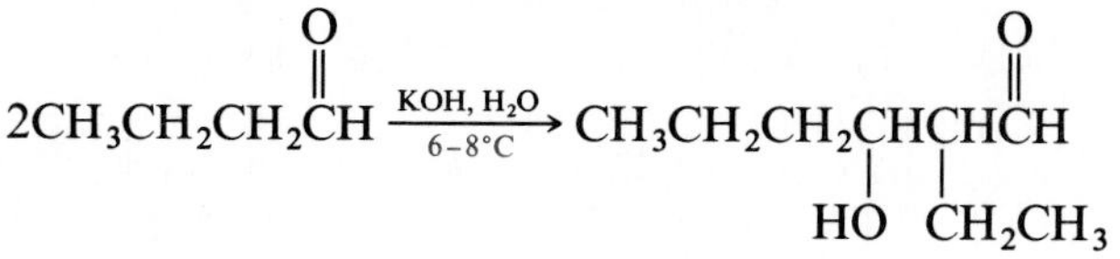

$$2CH_3CH_2CH_2CH \xrightarrow[6-8^\circ C]{KOH,\; H_2O} CH_3CH_2CH_2CHCHCH$$

Butanal   2-Ethyl-3-hydroxyhexanal (75%)

PROBLEM 11.13

Write the structure of the aldol addition product of:

(a) Pentanal   (b) 2-Methylbutanal   (c) 3-Methylbutanal

SAMPLE SOLUTION

(a) A good way to correctly identify the aldol addition product of any aldehyde is to work through the process mechanistically. Remember that the first step is enolate formation and that this *must* involve proton abstraction from the α carbon of a molecule of the starting carbonyl compound.

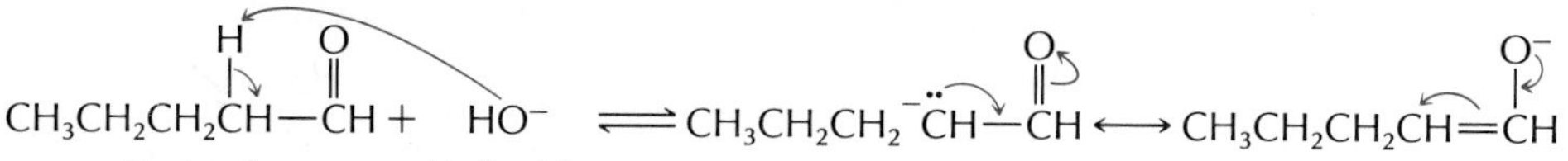

$$CH_3CH_2CH_2CH-CH + HO^- \rightleftharpoons CH_3CH_2CH_2\,\ddot{C}H-CH \longleftrightarrow CH_3CH_2CH_2CH=CH$$

Pentanal   Hydroxide

Now use the negatively charged carbon of the enolate to form a new carbon-carbon bond to the carbonyl group. Proton transfer from the solvent to the alkoxide completes the process.

$$CH_3CH_2CH_2CH_2\overset{\displaystyle O}{\overset{\|}{C}}H \quad :\overset{\displaystyle O}{\overset{\|}{C}}HCH \longrightarrow CH_3CH_2CH_2CH_2\overset{\displaystyle O^-}{\overset{|}{C}}H\overset{\displaystyle O}{\overset{\|}{C}}HCH$$
$$\underset{CH_2CH_2CH_3}{\qquad}\qquad\qquad\qquad\underset{CH_2CH_3CH_3}{\qquad}$$

Pentanal   Enolate of
                    pentanal

$$\xrightarrow{H_2O} CH_3CH_2CH_2CH_2\overset{\displaystyle OH}{\overset{|}{C}}H\overset{\displaystyle O}{\overset{\|}{C}}HCH$$
$$\underset{CH_2CH_2CH_3}{\qquad}$$

3-Hydroxy-2-propylheptanal
(aldol addition product
of pentanal)

Aldol addition is an equilibrium process that lies to the side of products when an aldehyde is the reactant. The equilibrium for a ketone, however, normally lies to the side of starting materials and ketones give poor yields of aldol addition products.

The product of an aldol addition of an aldehyde is a β-hydroxy aldehyde. These compounds undergo dehydration (lose the elements of water) on heating to yield a carbon-carbon double bond which is conjugated with the aldehyde carbonyl. The products of dehydration are called α,β-unsaturated aldehydes.

$$RCH_2\overset{\displaystyle OH}{\overset{|}{\underset{\beta}{C}}}H\overset{\alpha}{C}H\overset{\displaystyle O}{\overset{\|}{C}}H \xrightarrow{heat} RCH_2\overset{\beta}{C}H=\overset{\alpha}{C}\overset{\displaystyle O}{\overset{\|}{C}}H + H_2O$$
$$\underset{R}{\qquad}\qquad\qquad\underset{R}{\qquad}$$

β-Hydroxy aldehyde α,β-Unsaturated   Water
                              aldehyde

If the unsaturated aldehyde is the desired product, the reaction is carried out by heating the starting aldehyde in an aqueous base. Under these conditions, once the aldol addition product is formed, it rapidly loses water to form the α,β-unsaturated aldehyde.

$$2CH_3CH_2CH_2\overset{\displaystyle O}{\overset{\|}{C}}H \xrightarrow[80-100°C]{NaOH, H_2O} CH_3CH_2CH_2CH=\overset{\displaystyle O}{\overset{\|}{C}}CH$$
$$\underset{CH_2CH_3}{\qquad}$$

Butanal   2-Ethyl-2-hexenal (86%)

The overall process by which two molecules of an aldehyde combine to form an α,β-unsaturated aldehyde and a molecule of water is called the **aldol condensation reaction.**

PROBLEM 11.14
Write the structure of the aldol condensation product of each of the aldehydes in Problem 11.13. One of these aldehydes can undergo aldol addition, but not aldol condensation. Which one? Why?

SAMPLE SOLUTION
(a) Dehydration of the product of aldol addition of pentanal introduces the double bond between C-2 and C-3 to give an α,β-unsaturated aldehyde.

$$CH_3CH_2CH_2CH_2\overset{\overset{\displaystyle OH}{|}}{C}H\overset{\overset{\displaystyle O}{\parallel}}{C}H\underset{\underset{\displaystyle CH_2CH_2CH_3}{|}}{C}H \xrightarrow{-H_2O} CH_3CH_2CH_2CH_2CH=\underset{\underset{\displaystyle CH_2CH_2CH_3}{|}}{C}\overset{\overset{\displaystyle O}{\parallel}}{C}H$$

Product of aldol addition of
pentanal (3-hydroxy-2-propyl-
heptanal)

Product of aldol condensation
of pentanal (2-propyl-2-
heptenal)

Aldol condensations and the reverse, called retro–aldol reactions, are very important in the biochemistry of cells. It is by reactions such as these that living organisms break down sugars to smaller molecules and produce the energy that allows the organism to function. The process, known as **glycolysis,** has been studied quite extensively in biochemistry.

11.18 OXIDATION OF ALDEHYDES

Aldehydes are readily oxidized to carboxylic acids by a number of reagents, including those based on Cr(VI) and Mn(VII) (Section 10.8).

$$\underset{\text{Aldehyde}}{R\overset{\overset{\displaystyle O}{\parallel}}{C}H} \xrightarrow{\text{oxidize}} \underset{\text{Carboxylic acid}}{R\overset{\overset{\displaystyle O}{\parallel}}{C}OH}$$

$$\underset{\text{Heptanal}}{CH_3(CH_2)_5\overset{\overset{\displaystyle O}{\parallel}}{C}H} \xrightarrow[\substack{H_2SO_4, H_2O \\ 20°C}]{KMnO_4} \underset{\text{Heptanoic acid (76–78\%)}}{CH_3(CH_2)_5\overset{\overset{\displaystyle O}{\parallel}}{C}OH}$$

$$\underset{\text{Furfural}}{\text{(furan)}\overset{\overset{\displaystyle O}{\parallel}}{C}H} \xrightarrow[H_2SO_4, H_2O]{K_2Cr_2O_7} \underset{\text{Furoic acid (75\%)}}{\text{(furan)}CO_2H}$$

An aldehyde group may also be oxidized by a silver-ammonia complex.

$$R\overset{\overset{\displaystyle O}{\parallel}}{C}H + 2Ag(NH_3)_2{}^+ + 3HO^- \longrightarrow R\overset{\overset{\displaystyle O}{\parallel}}{C}O^- + 2Ag(s) + 2NH_3 + 2H_2O$$

This reaction is known as the **Tollens test,** a qualitative test for aldehydes. Formation of a shiny mirror of silver metal on the walls of a clean test tube is taken as a positive indication of the presence of an aldehyde. The Tollens test was first used in the study of carbohydrates (Chapter 15). Ketones do not react with the Tollens reagent.

11.19 SPECTROSCOPIC ANALYSIS OF ALDEHYDES AND KETONES

Infrared spectroscopy provides perhaps the best way of identifying an aldehyde or ketone, as carbonyl groups are among the easiest functional groups to detect by this method. The carbonyl group gives rise to a strong band in the region 1710 to

1750 cm⁻¹. In addition, aldehydes exhibit weak absorptions due to the carbonyl C—H near 2750 cm⁻¹. These bands are evident in the infrared spectrum of butanal, shown in Figure 11.9.

The ^{1}H nmr spectra of aldehydes exhibit a characteristic absorption in the region $\delta = 9$ to 10 ppm corresponding to the proton of the carbonyl group. Also the α hydrogens of an aldehyde or ketone are typically found in the region $\delta = 2.0$ to 2.5 ppm.

11.20 SUMMARY

Systematic names for both aldehydes and ketones are based on the alkane having the same number of carbons as the longest chain which contains the carbonyl group (Section 11.1). The -e ending is replaced with -al for aldehydes and -one for ketones. The carbonyl group is always C-1 in aldehydes, and numbered from the closer end of the chain in ketones. Substituents are named and located as usual.

$$\underset{\underset{\underset{CH_3}{|}}{CH_3CH_2CHCH}}{\overset{\overset{O}{\|}}{}} \qquad \underset{\underset{\underset{CH_3}{|}}{CH_3CH_2CCH_2CHCH_3}}{\overset{\overset{O}{\|}}{}}$$

2-Methylbutanal   5-Methyl-3-hexanone

The carbon-oxygen double bond of the carbonyl group is *polar* (Section 11.2).

$$\overset{\delta^+ \quad \delta^-}{\diagdown\underset{\diagup}{C}=O}$$

Aldehydes and ketones have higher boiling points than hydrocarbons of similar molecular weight because of intermolecular dipole-dipole attractions between the polar carbonyl group (Section 11.3).

The characteristic reaction of aldehydes and ketones is *nucleophilic addition* to the carbon of the carbonyl (Section 11.6). A summary of the nucleophilic addition reactions introduced in this chapter is presented in Table 11.2. The last entry is an example of the use of an *organometallic compound* called a *Grignard reagent* for the formation of new carbon-carbon single bonds.

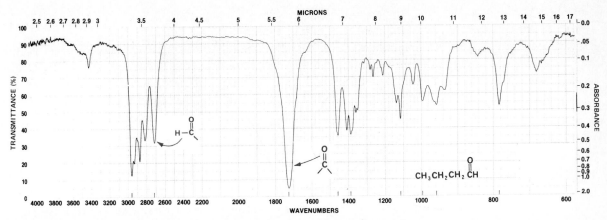

FIGURE 11.9 Infrared spectrum of butanal.

TABLE 11.2

Nucleophilic Addition to Aldehydes and Ketones

Reaction (section) and comments	General equation and typical example
Hydration (Section 11.7) Can be either acid- or base-catalyzed. The elements of water (H—OH) add across the carbon-oxygen double bond.	$\underset{\substack{\text{Aldehyde} \\ \text{or ketone}}}{\overset{\overset{\displaystyle O}{\|}}{RCR'}} + \underset{\text{Water}}{H_2O} \rightleftharpoons \underset{\text{Geminal diol}}{\overset{\displaystyle OH}{\underset{\displaystyle OH}{RCR'}}}$ $\underset{\substack{\text{Chloroacetone} \\ \text{(90\% at equilibrium)}}}{\overset{\overset{\displaystyle O}{\|}}{ClCH_2CCH_3}} \overset{H_2O}{\rightleftharpoons} \underset{\substack{\text{Chloroacetone hydrate} \\ \text{(10\% at equilibrium)}}}{\overset{\displaystyle OH}{\underset{\displaystyle OH}{ClCH_2CCH_3}}}$
Acetal formation (Section 11.8) Reaction is acid-catalyzed. A hemiacetal is an intermediate.	$\underset{\substack{\text{Aldehyde} \\ \text{or ketone}}}{\overset{\overset{\displaystyle O}{\|}}{RCR'}} + \underset{\text{Alcohol}}{2R''OH} \overset{H^+}{\rightleftharpoons} \underset{\text{Acetal}}{\overset{\displaystyle OR''}{\underset{\displaystyle OR''}{RCR'}}} + \underset{\text{Water}}{H_2O}$ *m*-Nitrobenzaldehyde + 2CH$_3$OH $\overset{HCl}{\longrightarrow}$ *m*-Nitrobenzaldehyde dimethyl acetal (76–85%) Methanol
Cyanohydrin formation (Section 11.9) Reaction is base-catalyzed. Cyanohydrins are useful synthetic intermediates; cyano group can be hydrolyzed to —CO$_2$H or reduced to —CH$_2$NH$_2$.	$\underset{\substack{\text{Aldehyde} \\ \text{or ketone}}}{\overset{\overset{\displaystyle O}{\|}}{RCR'}} + \underset{\substack{\text{Hydrogen} \\ \text{cyanide}}}{HCN} \rightleftharpoons \underset{\text{Cyanohydrin}}{\overset{\displaystyle OH}{\underset{\displaystyle CN}{RCR'}}}$ $\underset{\text{3-Pentanone}}{\overset{\overset{\displaystyle O}{\|}}{CH_3CH_2CCH_2CH_3}} \overset{KCN}{\underset{H^+}{\longrightarrow}} \underset{\text{3-Cyano-3-pentanol (75\%)}}{\overset{\displaystyle OH}{\underset{\displaystyle CN}{CH_3CH_2CCH_2CH_3}}}$
Reaction with primary amines (Section 11.10) Isolated product is an imine. Imines are involved in many biological processes involving aldehydes and ketones.	$\underset{\substack{\text{Aldehyde} \\ \text{or ketone}}}{\overset{\overset{\displaystyle O}{\|}}{RCR'}} + \underset{\substack{\text{Primary} \\ \text{amine}}}{R''NH_2} \rightleftharpoons \underset{\text{Imine}}{\overset{\overset{\displaystyle NR''}{\|}}{RCR''}} + \underset{\text{Water}}{H_2O}$ $\underset{\text{2-Methylpropanal}}{\overset{\overset{\displaystyle O}{\|}}{(CH_3)_2CHCH}} + \underset{\textit{tert}\text{-Butylamine}}{(CH_3)_3CNH_2} \longrightarrow \underset{\substack{N\text{-(2-Methyl-1-propylidene)-} \\ \textit{tert}\text{-butylamine (50\%)}}}{(CH_3)_2CHCH=NC(CH_3)_3}$

TABLE 11.2 *(Continued)*

Reaction (section) and comments	General equation and typical example
Reaction with hydrazine and derivatives of hydrazine (Section 11.10) Analogous to the reaction of other amines with aldehydes and ketones. These hydrazones are crystalline solids used as identification derivatives.	$$\underset{\substack{\text{Aldehyde} \\ \text{or ketone}}}{RCR'} + \underset{\text{Hydrazine}}{H_2NNHR''} \longrightarrow \underset{\text{Hydrazone}}{RCR'} + \underset{\text{Water}}{H_2O}$$

2-Dodecanone 2,4-Dinitro-phenylhydrazine 2-Dodecanone 2,4-Dinitrophenylhydrazone (94%)

| **Alcohol synthesis via the reaction of Grignard reagents with carbonyl compounds (Section 11.13)** This is one of the most useful reactions in synthetic organic chemistry. Grignard reagents react with formaldehyde to yield primary alcohols, with aldehydes to give secondary alcohols, and with ketones to form tertiary alcohols. | $$\underset{\substack{\text{Grignard} \\ \text{reagent}}}{RMgX} + \underset{\substack{\text{Aldehyde} \\ \text{or ketone}}}{R'CR''} \xrightarrow[\text{2. } H_3O^+]{\text{1. diethyl ether}} \underset{\text{Alcohol}}{RCOH}$$ |

$$\underset{\substack{\text{Methylmagnesium} \\ \text{iodide}}}{CH_3MgI} + \underset{\text{Butanal}}{CH_3CH_2CH_2CH} \xrightarrow[\text{2. } H_3O^+]{\text{1. diethyl ether}} \underset{\text{2-Pentanol (82\%)}}{CH_3CH_2CH_2CHCH_3}$$

Aldehydes and ketones exist in equilibrium with their corresponding enol forms (Section 11.16). A proton on the α carbon of an aldehyde or ketone is more acidic than an alkyl C—H proton and may be abstracted with a base such as hydroxide ion to give an *enolate ion.*

A summary of the reactions that proceed by way of enol or enolate ion intermediates is presented in Table 11.3.

Aldehydes undergo oxidation to form carboxylic acids upon reaction with mild oxidizing agents (Section 11.18).

ADDITIONAL PROBLEMS

Nomenclature and Physical Properties

11.15 Each of the following aldehydes or ketones is known by a nonsystematic name. Its IUPAC name is provided in parentheses. Write a structural formula for each one.

(a) Pivaldehyde (2,2-dimethylpropanal)

(b) Acrolein (2-propenal)

TABLE 11.3
Reactions of Aldehydes and Ketones That Involve Enol or Enolate Ion Intermediates

Reaction (section) and comments	General equation and typical example
Enolization (Section 11.16) Aldehydes and ketones exist in equilibrium with their enol forms. The rate at which equilibrium is achieved is increased by acidic or basic catalysts. The enol content of simple aldehydes and ketones is quite small.	$R_2CH-\overset{\overset{\displaystyle O}{\|}}{C}R' \rightleftharpoons R_2C=\overset{\overset{\displaystyle OH}{\|}}{C}R'$ Aldehyde or ketone · Enol $(CH_3)_2CH\overset{\overset{\displaystyle O}{\|}}{C}H \rightleftharpoons$ (2-Methyl-1-propen-1-ol structure) 2-Methylpropanal · 2-Methyl-1-propen-1-ol
Enolate ion formation (Section 11.16) An α proton of an aldehyde or ketone is more acidic than most other protons bound to carbon. Aldehydes and ketones are weak acids, with K_a's in the range 10^{-16} to 10^{-20} (pK_a 16 to 20). Their enhanced acidity is due to the electron-withdrawing effect of the carbonyl group and the resonance stabilization of the enolate anion.	$R_2CH\overset{\overset{\displaystyle O}{\|}}{C}R' + HO^- \rightleftharpoons R_2\overset{\cdot\cdot}{\underset{}{C}}\overset{\overset{\displaystyle O}{\|}}{C}R'^- + H_2O$ Aldehyde or ketone · Hydroxide ion · Enolate anion · Water $R_2\overset{\cdot\cdot}{\underset{}{C}}^-\overset{\overset{\displaystyle O}{\|}}{C}R' \longleftrightarrow R_2C=\overset{\overset{\displaystyle O^-}{\|}}{C}R'$ (Principal resonance forms of enolate ion) $CH_3CH_2\overset{\overset{\displaystyle O}{\|}}{C}CH_2CH_3 + HO^- \rightleftharpoons CH_3CH=\overset{\overset{\displaystyle O^-}{\|}}{C}CH_2CH_3 + H_2O$ 3-Pentanone · Hydroxide ion · Enolate anion of 3-pentanone · Water
Aldol condensation (Section 11.17) A reaction of great synthetic value for carbon-carbon bond formation. Nucleophilic addition of an enolate ion to a carbonyl group, followed by dehydration of the β-hydroxy aldehyde, yields an α,β-unsaturated carbonyl compound.	$2RCH_2\overset{\overset{\displaystyle O}{\|}}{C}H \xrightarrow{HO^-} RCH_2C=\overset{\overset{\displaystyle O}{\|}}{C}H + H_2O$ Aldehyde · α,β-Unsaturated aldehyde · Water $CH_3(CH_2)_6\overset{\overset{\displaystyle O}{\|}}{C}H \xrightarrow[CH_3CH_2OH]{NaOCH_2CH_3} CH_3(CH_2)_6CH=C(CH_2)_5CH_3$ with $HC=O$ Octanal · 2-Hexyl-2-decenal (79%)

(c) Deoxybenzoin (benzyl phenyl ketone)
(d) Diacetone alcohol (4-hydroxy-4-methyl-2-pentanone)
(e) Mesityl oxide (4-methyl-3-penten-2-one)
(f) Citral [(E)-3,7-dimethyl-2,6-octadienal]

11.16 Give an acceptable IUPAC name for each of the following aldehydes or ketones.

(a) $(CH_3)_3C\overset{\overset{\displaystyle Cl}{\|}}{C}HCH_2\overset{\overset{\displaystyle O}{\|}}{C}CH_2CH_3$

(b)

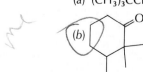

(c) $(CH_3)_2C=CHCH_2\overset{\overset{\displaystyle O}{\|}}{C}H$

(d)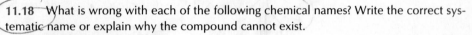

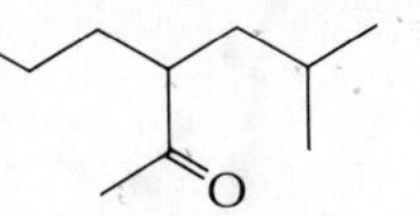

11.17 While the following are all permissible IUPAC names for ketones, each can be named in a different way based on the longest-continuous-chain method. Convert each name into one based on the latter system.

(a) Dibenzyl ketone
(b) Benzyl *tert*-butyl ketone
(c) Ethyl isopropyl ketone
(d) Isobutyl phenyl ketone
(e) Allyl methyl ketone

11.18 What is wrong with each of the following chemical names? Write the correct systematic name or explain why the compound cannot exist.

(a) Cyclohexanal
(b) 2,2-Dimethyl-4-hexanone
(c) 1-Chloro-3-propanone
(d) Trimethylacetaldehyde

11.19 Correctly associate each of the following compounds of similar molecular weight with its boiling point:

Compounds: pentanal; 1-pentanol; hexane
Boiling points, °C: 68.8; 103.4; 138

Aldehyde and Ketone Reactions

11.20 Predict the product of the reaction of propanal with each of the following.

(a) Lithium aluminum hydride, followed by treatment with water
(b) Sodium borohydride in methanol
(c) Hydrogen (nickel catalyst)
(d) Methylmagnesium iodide, followed by dilute acid
(e) Methanol containing a trace of acid
(f) Aniline ($C_6H_5NH_2$)
(g) *p*-Nitrophenylhydrazine
(h) Sodium cyanide with addition of sulfuric acid
(i) Silver-ammonia complex (ammoniacal silver nitrate)
(j) Chromic acid

11.21 Repeat parts (a) to (i) of Problem 11.20 for cyclopentanone.

11.22 (a) Write structural formulas and provide IUPAC names for all the isomeric aldehydes and ketones that have the molecular formula $C_5H_{10}O$. Include stereoisomers.

(b) Which of the isomers in (a) yield chiral alcohols on reaction with sodium borohydride?
(c) Which of the isomers in (a) yield chiral alcohols on reaction with methylmagnesium iodide?

11.23 Write the structure of the principal organic product of each of the following reactions.

(a) 2-Iodopropane with magnesium in diethyl ether
(b) Product of (a) with formaldehyde in ether, followed by dilute acid

(c) Product of (a) with benzaldehyde ($C_6H_5\overset{\overset{\displaystyle O}{\|}}{C}H$) in ether, followed by dilute acid
(d) Product of (a) with cyclopentanone in ether, followed by dilute acid

11.24 Analyze the following structures and determine all the combinations of Grignard reagent and carbonyl compound that will give rise to each.

(a) $CH_3CH_2CHCH_2CH(CH_3)_2$
      |
      OH

(b) $(CH_3)_3CCH_2OH$

(c)

      |
      OH

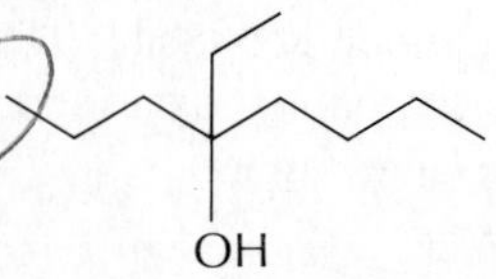

11.25 Which of the compounds in Problem 11.24 could be prepared by reduction of an aldehyde or ketone? Write the reactions which would take place, using lithium aluminum hydride as the reducing agent.

11.26 Supply the structure of the reactant missing from each of the following reactions.

(a) $? + 2CH_3CH_2OH \xrightarrow{H^+} CH_3CH(OCH_2CH_3)_2$

(b)

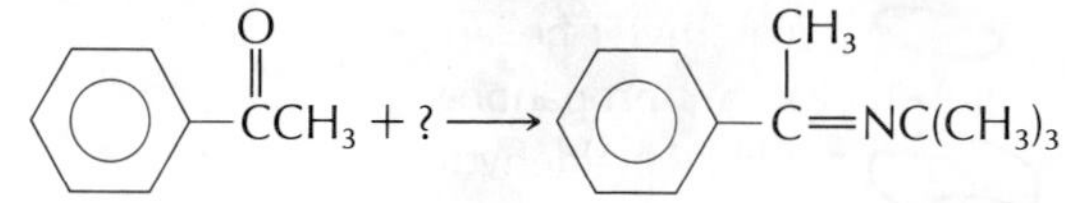

(c) $C_6H_5MgBr + ? \xrightarrow[\text{2. } H_3O^+]{\text{1. diethyl ether}} C_6H_5CCH_2C_6H_5$
                                                          |
                                                          CH_3
   with OH above the central carbon

(handwritten notes at right):
$CH_3 \overset{O}{\overset{\|}{C}} CH_2C_6H_5$
(CH_3) ...

Enols and Enolates

11.27 In each of the following pairs of compounds, choose the one that is able to enolize and write the structure of its enol form.

(a) $(CH_3)_3CCH$ or $(CH_3)_2CHCH$
         ($\overset{O}{\overset{\|}{}}$)

(b) $C_6H_5CC_6H_5$ or $C_6H_5CH_2CCH_2C_6H_5$
         ($\overset{O}{\overset{\|}{}}$)

11.28 Give the structure of the expected organic product in the reaction of propanal with each of the following.

(a) Sodium hydroxide in cold ethanol
(b) Sodium hydroxide in hot ethanol
(c) Product of (b) with sodium borohydride in ethanol

11.29 Butanal is the starting material for the active ingredient in the bug repellent "6 – 12." The product of aldol *addition* of butanal is reduced by catalytic hydrogenation to give the final product which has the formula $C_8H_{18}O_2$. Write a series of equations to outline this procedure.

Synthesis

11.30 The compound known as diphepanol is used as a cough suppressant. The last step in the preparation of diphepanol involves a Grignard reaction. Indicate a plausible set of reactants for this process.

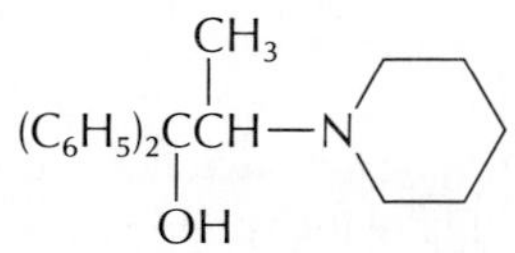

$$ (C_6H_5)_2CCH-N\bigcirc $$
with CH_3 above and OH below the central carbon

Diphepanol, an antitussive (cough suppressant)

11.31 Using 1-bromobutane and any necessary organic or inorganic reagents, suggest efficient methods for the preparation of each of the following alcohols. More than one step may be necessary. (*Hint:* Work backward from the target alcohol.)

(a) 1-Pentanol

(b) 2-Hexanol

(c) 1-Phenyl-1-pentanol

11.32 The fungus responsible for Dutch elm disease is spread by European bark beetles when they burrow into the tree. Other beetles congregate at the site, attracted by the scent of a mixture of chemicals, some emitted by other beetles and some coming from the tree. One of the compounds given off by female bark beetles is 4-methyl-3-heptanol. Suggest an efficient synthesis of this pheromone from alcohols of five or less carbon atoms.

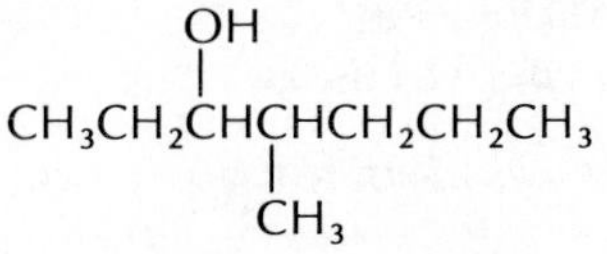

$$\underset{\underset{CH_3}{|}}{CH_3CH_2\overset{\overset{OH}{|}}{C}HCHCH_2CH_2CH_3}$$

4-Methyl-3-heptanol

Miscellaneous Problems

11.33 Addition of phenylmagnesium bromide to 4-*tert*-butylcyclohexanone gives two isomeric alcohols as products. Both alcohols yield the same alkene when subjected to acid-catalyzed dehydration. Suggest reasonable structures for these two alcohols and for the alkene they form on dehydration.

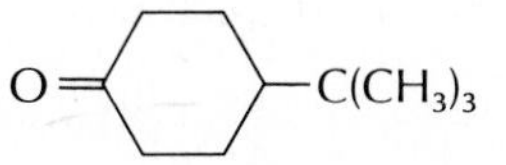

4-*tert*-Butylcyclohexanone

11.34 Compounds that contain both carbonyl and alcohol functional groups are often more stable as cyclic hemiacetals or cyclic acetals than as open-chain compounds. Examples of several of these are shown below. Deduce the structure of the open-chain form of each.

(a)

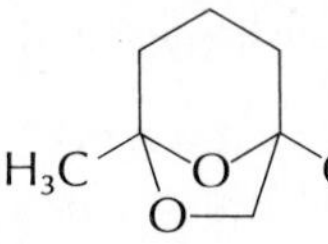

(b)

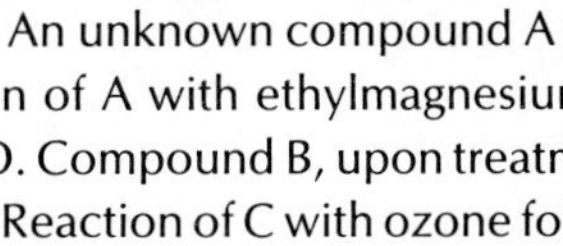

(c)

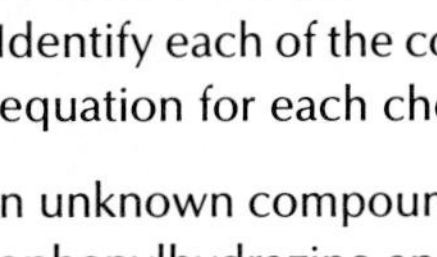

Frontalin (aggregating pheromone of the Southern pine beetle)

11.35 An unknown compound A gave a positive Tollens test (a silver mirror was formed). Reaction of A with ethylmagnesium bromide followed by dilute acid gave compound B, $C_6H_{14}O$. Compound B, upon treatment with concentrated sulfuric acid, gave compound C, C_6H_{12}. Reaction of C with ozone followed by zinc in water gave two products, propanal and acetone. Identify each of the compounds A to C, based on the chemical information given. Write an equation for each chemical reaction described in the problem.

11.36 An unknown compound D did not react with Tollens reagent. Upon reaction with 2,4-dinitrophenylhydrazine an orange solid was obtained. Compound D reacted with sodium cyanide to which acid had been added to give compound E, $C_6H_{11}ON$. Reaction of D with sodium borohydride in methanol gave compound F, an achiral substance which un-

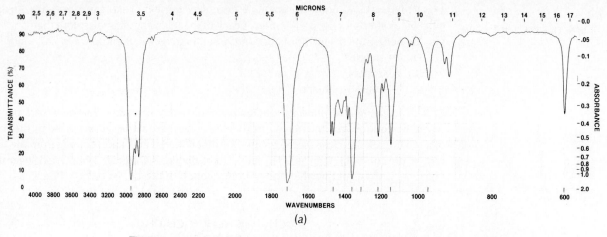

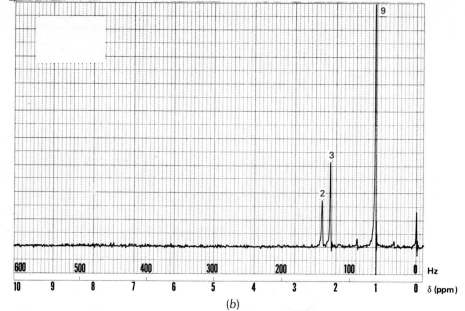

FIGURE 11.10 (a) The infrared and (b) ¹H nmr spectra of compound G (Problem 11.38). The numbers beside the signals correspond to their relative areas.

derwent dehydration with concentrated sulfuric acid to give 2-pentene as the *only* product. Identify compounds D to F and write an equation for each chemical reaction.

11.37 What features would be present in the infrared spectrum of each of the following compounds to allow each one to be distinguished from the others?

2-Pentanone   Pentanal   1-Pentanol

11.38 Identify compound G having the formula $C_7H_{14}O$ on the basis of its infrared and ¹H nmr spectra shown in Figure 11.10.

CHAPTER 12

CARBOXYLIC ACIDS

This chapter describes the properties of one of the most acidic classes of organic compounds, carboxylic acids. Upon completion of this chapter you should be able to:

- Provide an acceptable systematic IUPAC name for a carboxylic acid.
- Use chemical formulas to describe the hydrogen-bonding forces between carboxylic acid molecules in the liquid state and in aqueous solution.
- Write a chemical equation describing the acidity of a carboxylic acid.
- Write a mathematical equation representing the equilibrium constant for ionization of a carboxylic acid.
- Explain the greater acidity of carboxylic acids compared with alcohols by using resonance structures.
- Explain how substituents affect the acidity of a carboxylic acid.
- Write an equation describing the formation of a carboxylate salt from the corresponding carboxylic acid.
- Explain the structural features that enable soaps and detergents to act as cleansing agents.
- Write a chemical equation describing the preparation of a carboxylic acid by carboxylation of a Grignard reagent.
- Write a chemical equation describing the preparation of a carboxylic acid by hydrolysis of a nitrile.
- Write a chemical equation describing the formation of an acyl chloride from a carboxylic acid.
- Write a chemical equation describing the reduction of a carboxylic acid to give a primary alcohol.
- Describe the features of infrared and ^1H nmr spectra that are characteristic of carboxylic acids.

In Chapter 10 we began our discussion of oxygen-containing functional groups with alcohols. We then continued with a description of organic compounds at the next-higher oxidation level—aldehydes and ketones. We now conclude our discussion, in this chapter and the one that follows, with a description of carboxylic acids and certain of their derivatives such as esters and amides.

Carboxylic acids are compounds of the type $\overset{\overset{\displaystyle O}{\displaystyle \|}}{R}COH$. They comprise one of the most frequently encountered classes of organic compounds, commonly found in both biological systems and in the laboratory as synthetic intermediates. Carboxylic acids are important in everything from soap to hormonelike substances which regulate blood pressure and the transmission of pain in our bodies.

12.1 CARBOXYLIC ACID NOMENCLATURE

A number of carboxylic acids are known widely by their common names, and these are considered acceptable alternatives to the systematic names of these compounds. (See Table 12.1; entries 1, 2, and 8 are particularly useful for you to know.) The Latin origins of several of these compounds are described in Section 12.6.

Systematic names for carboxylic acids are derived from the alkane having the same number of carbon atoms as the longest chain containing the carboxyl group. The -e ending of the alkane is replaced by -oic acid, as shown by the first three entries in Table 12.1. Substituents are located by numbering from the carboxyl group. Be

TABLE 12.1
Systematic and Common Names of Some Carboxylic Acids

Entry number	Structural formula	Systematic name	Common name
1	HCO_2H	Methanoic acid	Formic acid
2	CH_3CO_2H	Ethanoic acid	Acetic acid
3	$CH_3(CH_2)_{16}CO_2H$	Octadecanoic acid	Stearic acid
4	$CH_3\underset{\underset{\displaystyle OH}{\displaystyle \|}}{C}HCO_2H$	2-Hydroxypropanoic acid	Lactic acid
5	$C_6H_5\underset{\underset{\displaystyle OH}{\displaystyle \|}}{C}HCO_2H$	2-Hydroxy-2-phenylethanoic acid	Mandelic acid
6	$CH_2{=}CHCO_2H$	Propenoic acid	Acrylic acid
7	$CH_3(CH_2)_7\underset{H}{\overset{}{C}}{=}\underset{H}{\overset{}{C}}(CH_2)_7CO_2H$	(Z)-9-Octadecenoic acid	Oleic acid
8	$C_6H_5{-}CO_2H$	Benzenecarboxylic acid	Benzoic acid
9	2-OH, 1-CO_2H benzene ring	o-Hydroxybenzenecarboxylic acid	Salicylic acid

sure to note that the carboxyl carbon is included when determining the length of the longest carbon chain.

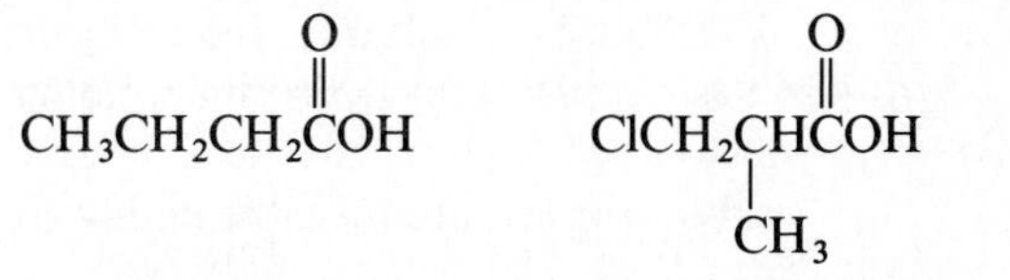

$$CH_3CH_2CH_2\overset{\displaystyle O}{\overset{\displaystyle \|}{C}}OH \qquad ClCH_2\underset{\displaystyle CH_3}{\overset{\displaystyle O}{\overset{\displaystyle \|}{C}}HCOH}$$

Butanoic acid 3-Chloro-2-methylpropanoic acid

You should also be aware of an alternative way of writing the carboxyl group that you will often see in formulas. That is, RCO_2H represents the general formula for a carboxylic acid. Even when written this way, you should recognize the presence of the carbon-oxygen double bond (carbonyl group) in the carboxyl group ($-\overset{\displaystyle O}{\overset{\displaystyle \|}{C}}OH$).

PROBLEM 12.1

Listed below are several carboxylic acids that are known by common names. Give an acceptable systematic (IUPAC) name for each.

(a) $CH_3CH_2CH_2CH_2CH_2CO_2H$ (caproic acid)
(b) $(CH_3)_2CHCH_2CO_2H$ (isovaleric acid)
(c) $(CH_3)_3CCO_2H$ (pivalic acid)
(d) $CH_2{=}\underset{\displaystyle CH_3}{C}CO_2H$ (methacrylic acid)

SAMPLE SOLUTION

(a) The name caproic acid is derived from the Latin word for goat (caper). The carbon chain, including the carboxyl carbon, is unbranched and is six carbon atoms long. As the unbranched alkane with the same number of carbon atoms is hexane, we drop the -e and add -oic acid to get the systematic name hexanoic acid.

12.2 PHYSICAL PROPERTIES

When the melting and boiling points of a carboxylic acid, such as acetic acid, are compared with those of an alcohol of comparable molecular weight (1-propanol), the carboxylic acid both melts and boils at a higher temperature. This indicates the presence of strong attractive forces between carboxylic acid molecules.

Compound (molecular weight)	Melting point, °C	Boiling point, °C
CH_3CO_2H (60 g/mol)	17	118
$CH_3CH_2CH_2OH$ (60 g/mol)	−127	97

A unique hydrogen bonding arrangement, shown in Figure 12.1a, contributes to these attractive forces. The hydroxyl group of one carboxylic acid molecule acts as a proton donor toward the carbonyl oxygen of a second. The second carboxyl group interacts with the first in a likewise manner. The result is that the two carboxylic acid

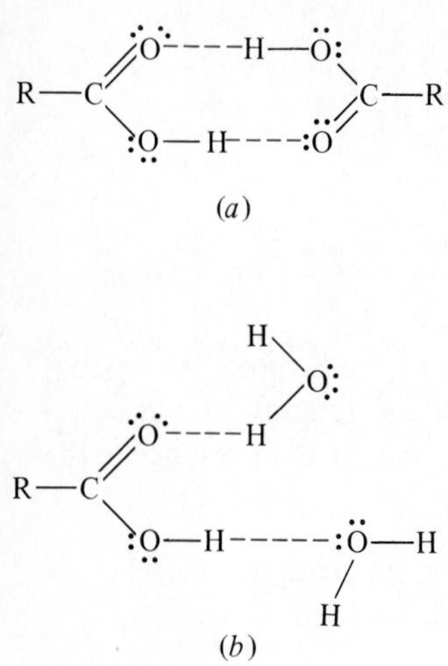

(a)

(b)

FIGURE 12.1 (a) Intermolecular hydrogen bonding between two carboxylic acid molecules to form a dimer. (b) Hydrogen bonding interactions between carboxylic acid and water.

molecules are held together by *two* hydrogen bonds to form a **dimer** (*di* = "two"; *meros* = "parts"). The hydrogen bonding forces are so efficient that some carboxylic acids exist as hydrogen-bonded dimers even in the gas phase.

In aqueous solution, intermolecular association between carboxylic acid molecules is replaced by hydrogen bonding to water (Figure 12.1*b*). The solubility properties of carboxylic acids are similar to those of alcohols. Carboxylic acids of four carbons or less are miscible with water in all proportions.

PROBLEM 12.2

Vinegar is a 5% solution of acetic acid in water. Draw a diagram showing the hydrogen bonding forces present in a vinegar solution.

12.3 ACIDITY OF CARBOXYLIC ACIDS

The acidity of carboxylic acids is their most notable property. As a class, they are among the most acidic organic compounds. With ionization constants (K_a) of about 10^{-5}, carboxylic acids are much stronger acids than water (10^{-16}), alcohols (10^{-18}), or even phenols (10^{-10}).*

PROBLEM 12.3

Acetylsalicylic acid (aspirin) was discovered in 1899, and today is the most widely used analgesic (pain relief medication) in the world. Acetylsalicylic acid has an ionization constant $K_a = 3.3 \times 10^{-4}$. Calculate the pK_a of acetylsalicylic acid.

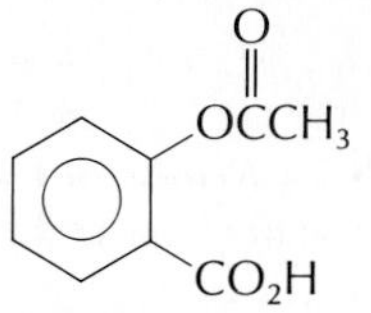

Acetylsalicylic acid (aspirin)

It should be remembered that carboxylic acids are, however, weak acids. A 0.1 *M* solution of acetic acid in water, for example, is only 1.3 percent ionized.

To understand the greater acidity of carboxylic acids compared with water and alcohols, compare the structural changes that accompany the ionization of a representative alcohol (ethanol) and a representative carboxylic acid (acetic acid). The equilibria that define K_a are

Ionization of ethanol

$$CH_3CH_2OH \rightleftharpoons H^+ + CH_3CH_2O^- \qquad K_a = \frac{[H^+][CH_3CH_2O^-]}{[CH_3CH_2OH]} = 10^{-16}$$

Ethanol   Ethoxide ion

Ionization of acetic acid

$$K_a = \frac{[H^+][CH_3CO_2^-]}{[CH_3CO_2H]} = 1.8 \times 10^{-5}$$

Acetic acid   Acetate ion

* You will find it helpful at this point to review the discussion of the acid-base properties of organic molecules and the calculation of ionization constants (Section 3.7).

Ionization of ethanol yields an alkoxide ion in which the negative charge is localized on oxygen. Ionization of acetic acid yields an acetate ion which is stabilized by electron delocalization that permits the negative charge to be shared equally by both oxygens.

This stabilization of the acetate ion increases the extent to which acetic acid undergoes ionization. Thus acetic acid, whose conjugate base (acetate ion) is stabilized, is more acidic than ethanol, whose conjugate base (ethoxide ion) lacks comparable stabilization.

12.4 SUBSTITUENTS AND ACID STRENGTH

Table 12.2 shows that carboxylic acids bearing only alkyl substituents are similar in acidity. Their pK_a's range from 4.7 to 5.1. The slight electron-donating effect of alkyl groups compared to hydrogen is too small to significantly affect the ionization constant of a carboxylic acid.

Electronegative substituents on the other hand, particularly when they are attached to the α carbon atom, significantly *increase* the acidity of a carboxylic acid. As Table 12.2 shows, the monohaloacetic acids are about 100 times more acidic than

TABLE 12.2
Effect of Substituents on Acidity of Carboxylic Acids

Name of acid	Structure	Ionization constant K_a*	pK_a
Standard of comparison			
Acetic acid	CH_3CO_2H	1.8×10^{-5}	4.7
Alkyl substituents have a negligible effect on acidity			
Propanoic acid (propionic acid)	$CH_3CH_2CO_2H$	1.3×10^{-5}	4.9
2-Methylpropanoic acid (isobutyric acid)	$(CH_3)_2CHCO_2H$	1.6×10^{-5}	4.8
2,2-Dimethylpropanoic acid (pivalic acid)	$(CH_3)_3CCO_2H$	0.9×10^{-5}	5.1
Heptanoic acid	$CH_3(CH_2)_5CO_2H$	1.3×10^{-5}	4.9
α-Halogen substituents increase acidity			
Fluoroacetic acid	FCH_2CO_2H	2.5×10^{-3}	2.6
Chloroacetic acid	$ClCH_2CO_2H$	1.4×10^{-3}	2.9
Bromoacetic acid	$BrCH_2CO_2H$	1.4×10^{-3}	2.9
Dichloroacetic acid	Cl_2CHCO_2H	5.0×10^{-2}	1.3
Trichloroacetic acid	Cl_3CCO_2H	1.3×10^{-1}	0.9
Other electronegative groups also increase acidity			
Methoxyacetic acid	$CH_3OCH_2CO_2H$	2.7×10^{-4}	3.6
Cyanoacetic acid	$N{\equiv}CCH_2CO_2H$	3.4×10^{-3}	2.5
Nitroacetic acid	$O_2NCH_2CO_2H$	2.1×10^{-2}	1.7

* In water at 25°C.

acetic acid. Multiple halogen substitution increases the acidity even more; trichloro-acetic acid is almost as acidic as sulfuric acid.

One explanation of the acid-strengthening effect of electronegative atoms or groups suggests that the electron-withdrawing effect of the substituent is transmitted to the carboxylate group through the σ bonds of the molecule. According to this bonding model, the σ electrons in the carbon-chlorine bond of chloroacetate ion, for example, are drawn toward chlorine, leaving the α carbon with a slight positive charge. Because of this positive character, the α carbon attracts electrons from the carboxylate group, thus dispersing the charge and stabilizing the anion. The more stable the anion, the greater the equilibrium constant for its formation.

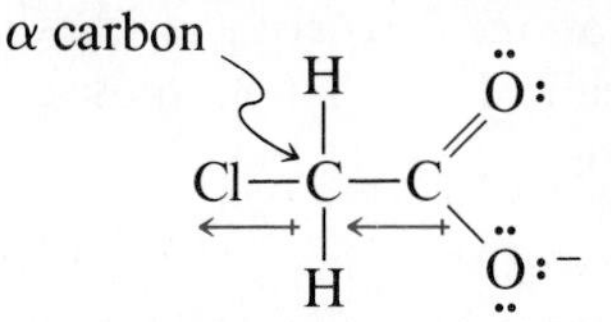

α carbon

The carboxylate anion is stabilized by electron-drawing effect of chlorine

The acid-strengthening effect of a halogen decreases as it becomes more remote from the carboxyl group.

$ClCH_2CO_2H$	$ClCH_2CH_2CO_2H$	$ClCH_2CH_2CH_2CO_2H$
Chloroacetic acid	3-Chloropropanoic acid	4-Chlorobutanoic acid
$K_a = 1.4 \times 10^{-3}$	$K_a = 1.0 \times 10^{-4}$	$K_a = 3.0 \times 10^{-5}$
$pK_a = 2.9$	$pK_a = 4.0$	$pK_a = 4.5$

Although these pK_a values appear numerically very similar, recall (Section 3.7) that pK_a's are logarithmic functions. Therefore, a change in one pK_a unit is equal to a 10-fold change in acidity. Chloroacetic acid ($pK_a = 2.9$), for example, is 14 times more acidic than 3-chloropropanoic acid ($pK_a = 4.0$).

PROBLEM 12.4

Which is the stronger acid in each of the following pairs?

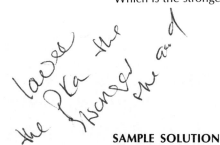

(a) $(CH_3)_3CCH_2CO_2H$ or $(CH_3)_3\overset{+}{N}CH_2CO_2H$
(b) $CH_3CH_2CO_2H$ or $CH_3\underset{\underset{OH}{|}}{CH}CO_2H$

(c) $CH_3\overset{\overset{O}{\|}}{C}CO_2H$ or $(CH_3)_2CHCO_2H$

SAMPLE SOLUTION

(a) The two compounds are viewed as substituted derivatives of acetic acid. The *tert*-butyl substituent has only a small effect on acidity, and the compound $(CH_3)_3CCH_2CO_2H$ is expected to have an acid strength similar to that of acetic acid. A trimethylammonium substituent, on the other hand, is positively charged and is a

powerful electron-withdrawing substituent. The compound $(CH_3)_3\overset{+}{N}CH_2CO_2H$ is expected to be a much stronger acid than $(CH_3)_3CCH_2CO_2H$. The measured ionization constants, shown below, confirm this prediction.

$(CH_3)_3CCH_2CO_2H$	$(CH_3)_3\overset{+}{N}CH_2CO_2H$
Weaker acid	Stronger acid
$K_a = 5 \times 10^{-6}$	$K_a = 1.5 \times 10^{-2}$
$pK_a = 5.30$	$pK_a = 1.83$

Thus $(CH_3)_3\overset{+}{N}CH_2CO_2H$ is 3000 times stronger an acid than $(CH_3)_3CCH_2CO_2H$.

12.5 SALTS OF CARBOXYLIC ACIDS. SOAP

Carboxylic acids react instantly and completely with strong bases such as sodium hydroxide.

$$
\underset{\substack{\text{Carboxylic} \\ \text{acid} \\ \text{(stronger} \\ \text{acid)}}}{R\ddot{C}\overset{\overset{\displaystyle \ddot{O}:}{\|}}{}{\ddot{O}}{-}H} + \underset{\substack{\text{Hydroxide} \\ \text{ion} \\ \text{(stronger} \\ \text{base)}}}{^{-}:\!\ddot{O}H} \longrightarrow \underset{\substack{\text{Water} \\ \text{(weaker} \\ \text{acid)}}}{H{-}\ddot{O}H} + \underset{\substack{\text{Carboxylate} \\ \text{ion} \\ \text{(weaker base)}}}{R\ddot{C}\overset{\overset{\displaystyle \ddot{O}:}{\|}}{}{\ddot{O}}{:}^{-}}
$$

Even weak bases such as sodium bicarbonate ($NaHCO_3$, baking soda) are capable of neutralizing carboxylic acids, as evidenced by the vigorous foaming when baking soda is added to vinegar, which contains acetic acid.

$$
\underset{\substack{\text{Acetic} \\ \text{acid}}}{CH_3\overset{\overset{\displaystyle O}{\|}}{C}OH} + \underset{\substack{\text{Sodium} \\ \text{bicarbonate}}}{NaHCO_3} \longrightarrow \underset{\substack{\text{Sodium} \\ \text{acetate}}}{CH_3\overset{\overset{\displaystyle O}{\|}}{C}O^-Na^+} + \underset{\substack{\text{Carbonic} \\ \text{acid}}}{H_2CO_3}
$$

The carbonic acid that is formed spontaneously dissociates to water and gaseous carbon dioxide which bubbles out of the solution.

$$
\underset{\substack{\text{Carbonic} \\ \text{acid}}}{HO\overset{\overset{\displaystyle O}{\|}}{C}OH} \longrightarrow \underset{\substack{\text{Water}}}{H_2O} + \underset{\substack{\text{Carbon} \\ \text{dioxide}}}{CO_2}
$$

PROBLEM 12.5

Write an ionic equation for the reaction of acetic acid with each of the following and specify whether the equilibrium favors starting materials or products (see Table 3.3 for K_a values).

(a) Sodium ethoxide   (d) Sodium acetylide (pK_a of HC≡CH is 26)
(b) Potassium *tert*-butoxide   (e) Lithium amide
(c) Sodium bromide

SAMPLE SOLUTION

(a) The reaction is an acid-base reaction; ethoxide is the base.

$$
\underset{\substack{\text{Acetic acid} \\ \text{(stronger} \\ \text{acid)}}}{CH_3CO_2H} + \underset{\substack{\text{Ethoxide ion} \\ \text{(stronger} \\ \text{base)}}}{CH_3CH_2O^-} \longrightarrow \underset{\substack{\text{Acetate ion} \\ \text{(weaker} \\ \text{base)}}}{CH_3CO_2^-} + \underset{\substack{\text{Ethanol} \\ \text{(weaker} \\ \text{acid)}}}{CH_3CH_2OH}
$$

Ethanol with a K_a of 10^{-16} (p$K_a = 16$; see Table 3.3) is a much weaker acid than acetic acid. An acid-base equilibrium favors formation of the weaker acid and the weaker base. Thus the position of equilibrium lies well to the right.

The conjugate base of a carboxylic acid is called a **carboxylate ion.** Carboxylate ions are named by replacing the ending *-ic acid* by *-ate.* Thus the conjugate base of acetic acid is the acetate ion. Metal salts of carboxylic acids are named by giving the metal name preceding the carboxylate.

$$CH_3\overset{\displaystyle O}{\overset{\|}{C}}OLi \qquad \overset{\displaystyle O}{\overset{\|}{\underset{}{\bigcirc}}}-CONa$$

Lithium acetate   Sodium benzoate

Sodium benzoate is used as a preservative in a variety of food products such as baked goods and soft drinks.

Most metal carboxylate salts are ionic, and frequently the sodium and potassium salts of carboxylic acids are soluble in water. Carboxylic acids therefore may be extracted from an ether solution into aqueous sodium or potassium hydroxide as their carboxylate salts.

The solubility of salts of unbranched (straight-chain) carboxylic acids having 12 to 18 carbons is the basis for the cleansing action of soaps. Let us consider the sodium salt of stearic acid, sodium stearate, as a representative example of a soap.

Nonpolar (hydrophobic) part

$$\overset{\displaystyle O}{\overset{\|}{C}}O^- \ Na^+$$

Sodium stearate
(sodium octadecanoate)

Polar (hydrophilic) part

Sodium stearate has a polar, ionic carboxylate group at one end of a long hydrocarbon chain. The carboxylate group is **hydrophilic,** or "water-loving," and tends to make the molecule soluble by hydrogen bonding to water molecules. The hydrocarbon chain, on the other hand, is **hydrophobic** ("water-hating"), also called **lipophilic** ("fat-loving"), and tends to associate with other hydrocarbon chains. When sodium stearate is placed in water, it forms a dispersion of spherical aggregates, or collections, called **micelles.** Each micelle is composed of 50 to 100 individual molecules; an idealized representation of a micelle is shown in Figure 12.2.

The polar carboxylate groups of sodium stearate are on the surface of the micelle. There they bind to water molecules and to sodium ions. The nonpolar hydrocarbon chains are directed toward the interior of the micelle, where weak attractive forces bind them together. Micelles tend to repel each other because their surfaces are negatively charged, rather than clustering to form higher aggregates.

Micelle formation is responsible for the cleansing action of soap. The molecules which make up most dirt and grease tend to be nonpolar and repel polar water molecules. Water that contains sodium stearate removes dirt and grease by dissolving it in the hydrocarbon-like interior of the micelles. Since the micelles are in suspension, the dirt and grease are removed when the water is rinsed away.

Sodium stearate is only one example of a soap; sodium and potassium salts of other C_{12} through C_{18} straight-chain carboxylic acids possess similar properties. The carboxylic acids used to make soaps are called *fatty acids.* You will see in the next chapter how soaps are formed from fats and oils.

Detergents are substances, including soaps, that cleanse by micellar action. We usually use the word to refer to *synthetic* detergents, however. One example of a synthetic detergent is sodium lauryl sulfate. This molecule has a long hydrocarbon chain (the hydrophobic or lipophilic end) and a polar sulfate ion (the hydrophilic end). Sodium lauryl sulfate forms soaplike micelles in water.

$$\overset{\displaystyle {}^-O \quad O^-}{\underset{O \quad\ \ O^- \ Na^+}{S}}$$

Sodium lauryl sulfate
(sodium 1-dodecyl sulfate)

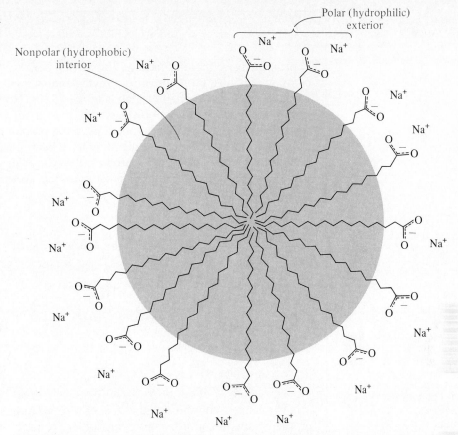

FIGURE 12.2 Idealized representation of a sodium stearate micelle. The shaded interior is the nonpolar, hydrophobic region of the micelle. The exterior surface is the polar, hydrophilic region. A real micelle is likely to be more irregular and to contain voids and channels.

Detergents are designed to be effective in hard water, i.e., water containing calcium or magnesium salts that form insoluble calcium or magnesium carboxylates with soaps. These precipitates rob the soap of its cleansing power and form an unpleasant scum ("bathtub ring"). The calcium and magnesium salts of synthetic detergents such as sodium lauryl sulfate, however, are soluble and retain their micelle-forming ability in water.

12.6 SOURCES OF CARBOXYLIC ACIDS

Many carboxylic acids were first isolated from natural sources and were given names indicative of their origin. Formic acid (Latin *formica,* "ant") was obtained by distilling ants. Since ancient times acetic acid (Latin *acetum,* "vinegar") has been known to be present in wine that has turned sour. Butyric acid (IUPAC: butanoic acid; Latin *butyrum,* "butter") contributes to the odor of rancid butter, and lactic acid (Latin *lac,* "milk") is found in sour milk.

Vinegar is obtained from the fermentation of sugar-containing materials such as cider or wine. The acetic acid produced may be purified by the process known as distillation, giving "white vinegar," a 5% solution of acetic acid in water.

Acetic acid is also a large-scale industrial chemical, with over 3 billion pounds produced in the United States each year. Much of this is used in the production of vinyl acetate for paints and adhesives. One method for its preparation uses carbon monoxide and methanol, both of which are derived from coal.

$$CH_3OH + CO \xrightarrow[\text{heat, pressure}]{\substack{\text{cobalt or rhodium} \\ \text{catalyst}}} CH_3CO_2H$$

Methanol   Carbon   Acetic acid
monoxide

PROSTAGLANDINS

A prominent example of naturally occurring carboxylic acids is the class of compounds called prostaglandins. **Prostaglandins** were first isolated in minute amounts from sheep prostate glands in the 1930s. However, since the 1960s scientists have realized that prostaglandins exist in almost all cells and partici- pate in a variety of regulatory functions in the human body.

More than a dozen prostaglandins have been identified as regulators of biological processes. All are 20-carbon carboxylic acids which contain a cyclo- pentane ring. Two representative prostaglandins, known as PGE_1 and $PGF_{1\alpha}$, are shown.

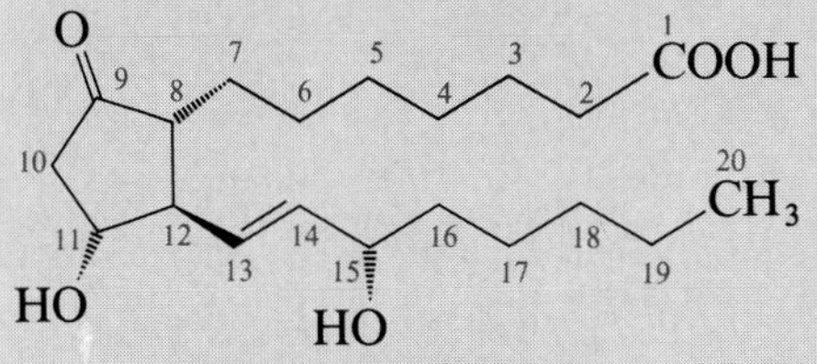

Prostaglandin E₁
(PGE₁)

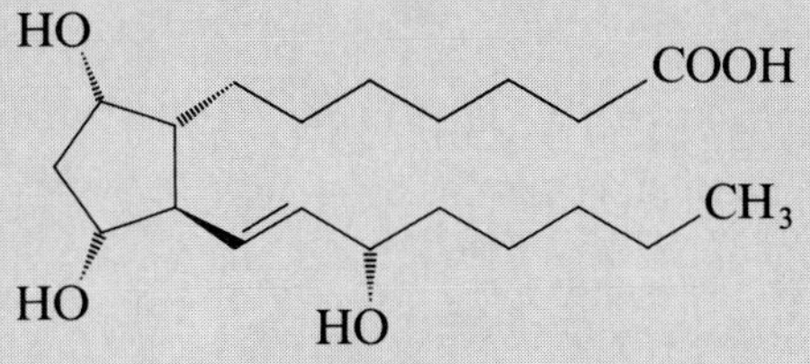

Prostaglandin F₁ₐ
(PGF₁ₐ)

The range of physiological effects of prostaglandins is remarkable. Some prostaglandins relax bronchial tissue, others contract it. PGE_1 dilates blood vessels and lowers blood pressure, and offers promise as a drug to reduce the formation of blood clots.

One result of prostaglandin research has been an answer to the question "How does aspirin work?". Although aspirin has been used as a pain reliever for almost a century, little was known about its mechanism of action. Research now shows that the body makes prostaglandins in response to tissue damage. The consequence of this prostaglandin production is pain and inflammation. Aspirin interferes with prostaglandin biosynthesis, thus lowering prostaglandin levels and reducing pain and inflammation.

Prostaglandins continue to be an important focus of health care research. It is hoped that new drugs, and a greater understanding of human physiology, may develop from this work. In 1982 three European scientists, Sune Berg- strom and Bengt Samuelsson of Sweden and John Vane of Great Britain, shared the Nobel Prize in Physiology or Medicine for their pioneering research in prostaglandins.

The carboxylic acid prepared industrially in greatest amount is 1,4-benzenedi- carboxylic acid (terephthalic acid). It is prepared by side-chain oxidation of *p*-xylene (Section 6.14) and is used in the production of polyester fibers.

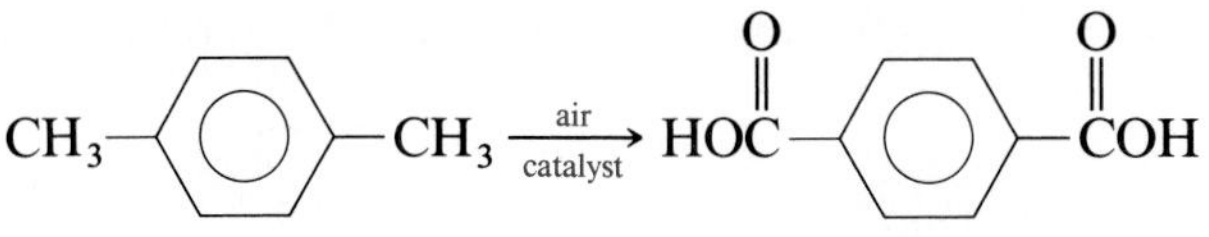

p-Xylene   Terephthalic acid

TABLE 12.3
Summary of Reactions Discussed in Earlier Chapters That Yield Carboxylic Acids

Reaction (section) and comments	General equation and specific example
Oxidative cleavage of alkenes (Section 5.2) Potassium permanganate cleaves alkenes to two carbonyl compounds. If one of the substituents at the double bond is hydrogen, the cleavage product is an aldehyde, which is rapidly oxidized to a carboxylic acid under the reaction conditions.	$RCH{=}CR'_2 \xrightarrow{KMnO_4} RCO_2H + R'_2C{=}O$ Alkene — Carboxylic acid — Carbonyl compound $CH_3CH(CH_2)_3CHCH{=}CH_2 \xrightarrow[H_2O]{KMnO_4} CH_3CH(CH_2)_3CHCO_2H + CO_2$ (substituents CH_3, CH_3) 3,7-Dimethyl-1-octene → 2,6-Dimethylheptanoic acid (45%) + Carbon dioxide
Side-chain oxidation of alkylbenzenes (Section 6.14) A primary or secondary alkyl side chain on an aromatic ring is degraded to a carboxyl group by reaction with a strong oxidizing agent such as potassium permanganate.	$ArCHR_2 \xrightarrow{KMnO_4} ArCO_2H$ Alkylbenzene — Arenecarboxylic acid (ring with CH_3, OCH_3, NO_2) $\xrightarrow[2.\ H^+]{1.\ KMnO_4,\ HO^-}$ (ring with CO_2H, OCH_3, NO_2) 3-Methoxy-4-nitrotoluene → 3-Methoxy-4-nitrobenzoic acid (100%)
Oxidation of primary alcohols (Section 10.8) Potassium permanganate and chromic acid convert primary alcohols to carboxylic acids by way of the corresponding aldehyde.	$RCH_2OH \xrightarrow[K_2Cr_2O_7,\ H_2SO_4]{KMnO_4\ or} RCO_2H$ Primary alcohol — Carboxylic acid $(CH_3)_3CCHC(CH_3)_3 \xrightarrow[H_2O,\ H_2SO_4]{H_2CrO_4} (CH_3)_3CCHC(CH_3)_3$ (substituent CH_2OH) → (substituent CO_2H) 2-tert-Butyl-3,3-dimethyl-1-butanol → 2-tert-Butyl-3,3-dimethylbutanoic acid (82%)
Oxidation of aldehydes (Section 11.18) Aldehydes are particularly sensitive to oxidation and are converted to carboxylic acids by a number of oxidizing agents, including potassium permanganate, chromic acid, and silver oxide.	$RCH{=}O \xrightarrow{oxidizing\ agent} RCO_2H$ Aldehyde — Carboxylic acid (furan ring with $CH{=}O$) $\xrightarrow[H_2SO_4,\ H_2O]{K_2Cr_2O_7}$ (furan ring with CO_2H) Furan-2-carbaldehyde (furfural) → Furan-2-carboxylic acid (furoic acid) (75%)

12.7 PREPARATION OF CARBOXYLIC ACIDS: A REVIEW

You have seen several methods for the preparation of carboxylic acids, all of which are oxidation reactions (Table 12.3). The last two methods, oxidation of primary alcohols and aldehydes, are particularly useful to an organic chemist. A key point to recognize about these two reactions is that they give carboxylic acids having the *same number of carbon atoms* as the starting materials. The hydroxyl-bearing carbon in the primary alcohol becomes the carboxyl carbon of the carboxylic acid upon oxidation. Likewise, the carbonyl carbon of the aldehyde becomes the carboxyl carbon of the carboxylic acid. In the next two sections you will encounter methods that *add one carbon atom* to the carbon skeleton of the starting compound.

12.8 PREPARATION OF CARBOXYLIC ACIDS BY CARBOXYLATION OF GRIGNARD REAGENTS

You saw in the last chapter how alcohols may be prepared from the nucleophilic addition of Grignard reagents to aldehydes and ketones (Section 11.13). Grignard reagents are also used in the preparation of carboxylic acids by a reaction called *carboxylation.* The Grignard reagent reacts with **carbon dioxide** to yield the magnesium salt of a carboxylic acid. Acidification converts the salt to the desired carboxylic acid.

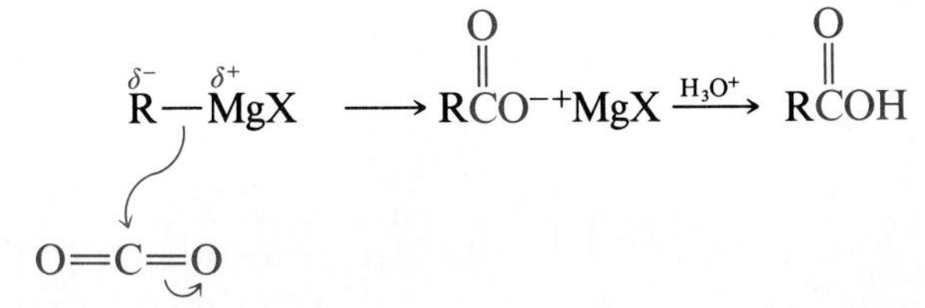

$$\overset{\delta-}{R}-\overset{\delta+}{MgX} \longrightarrow R\overset{O}{\overset{\|}{C}}O^- {}^+MgX \xrightarrow{H_3O^+} R\overset{O}{\overset{\|}{C}}OH$$

$$O=C=O$$

Grignard reagent acts as a nucleophile toward carbon dioxide

Carboxylic acid

$$CH_3CHCH_2CH_3 \underset{\substack{2.\ CO_2 \\ 3.\ H_3O^+}}{\overset{1.\ Mg,\ diethyl\ ether}{\xrightarrow{\hspace{2cm}}}} CH_3CHCH_2CH_3$$
$$\underset{Cl}{|} \qquad\qquad\qquad \underset{CO_2H}{|}$$

2-Chlorobutane

2-Methylbutanoic acid (76–86%)

Generally carboxylations are carried out by simply pouring the ether solution containing the Grignard reagent onto dry ice (solid carbon dioxide). Acidification of the reaction mixture converts the carboxylate salt to the free acid.

Overall, the carboxylation of a Grignard reagent transforms an alkyl or aryl halide to a carboxylic acid having *one more carbon,* the carbon of the carboxyl group.

$$RX \underset{diethyl\ ether}{\overset{Mg}{\xrightarrow{\hspace{1.5cm}}}} RMgX \underset{2.\ H_3O^+}{\overset{1.\ CO_2}{\xrightarrow{\hspace{1.5cm}}}} RCO_2H$$

Alkyl or aryl halide

Grignard reagent

Carboxylic acid

PROBLEM 12.6
Outline procedures for the following conversions, using carboxylation of a Grignard reagent.

(a) 1-Chlorobutane → pentanoic acid
(b) 2-Bromopropane → 2-methylpropanoic acid
(c) Bromobenzene → benzoic acid

SAMPLE SOLUTION

(a) Reaction of the four-carbon alkyl halide 1-chlorobutane with magnesium metal gives the corresponding Grignard reagent. Subsequent reaction with carbon dioxide, followed by acid hydrolysis, yields the desired five-carbon carboxylic acid, pentanoic acid.

$$CH_3CH_2CH_2CH_2Cl \xrightarrow[\substack{2.\ CO_2 \\ 3.\ H_3O^+}]{1.\ Mg,\ ether} CH_3CH_2CH_2CH_2CO_2H$$

1-Chlorobutane   Pentanoic acid

12.9 PREPARATION OF CARBOXYLIC ACIDS BY THE HYDROLYSIS OF NITRILES

A second method for preparing carboxylic acids that adds one carbon atom is by hydrolysis of *nitriles,* Nitriles, also known as alkyl cyanides, are prepared by a nucleophilic substitution reaction (Section 8.2).

$$RX + {}^-:C{\equiv}N: \longrightarrow RC{\equiv}N + X^-$$

Primary or   Cyanide   Nitrile   Halide
secondary   ion   (alkyl   ion
alkyl halide   cyanide)

The reaction is a nucleophilic substitution of the S_N2 type. As you learned in Chapter 8 (Section 8.7), S_N2 reactions are most effective with unhindered primary halides. Also, S_N2 nucleophilic substitutions are only possible with alkyl halide substrates; aryl and vinyl halides do not react.

Once the nitrile group has been introduced into the molecule, it is subjected to a separate hydrolysis step. Usually this is carried out by heating in aqueous acid.

$$RC{\equiv}N + 2H_2O + H^+ \xrightarrow{heat} \overset{\overset{\displaystyle O}{\|}}{R}COH + NH_4^+$$

Nitrile   Water   Carboxylic   Ammonium
                                          acid   ion

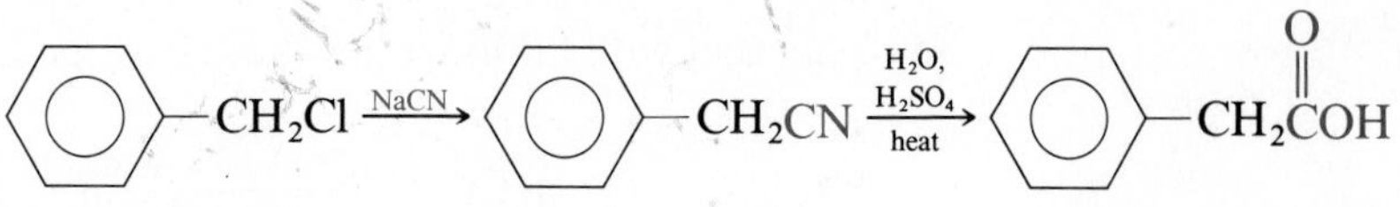

Benzyl chloride   Benzyl cyanide (92%)   Phenylacetic acid (77%)

PROBLEM 12.7

Repeat Problem 12.6 using hydrolysis of a nitrile as a key step. Which of the conversions of Problem 12.6 may not be carried out by this method?

In Chapter 11 (Section 11.9) you learned that aldehydes and ketones may be converted to cyanohydrins by reaction with hydrogen cyanide. Through hydrolysis of the nitrile group, cyanohydrins provide a means of preparing α-hydroxy carboxylic acids.

$$\underset{\text{2-Pentanone}}{CH_3\overset{\overset{\displaystyle O}{\|}}{C}CH_2CH_2CH_3} \xrightarrow[\text{2. H}^+]{\text{1. NaCN}} \underset{\substack{\text{2-Cyano-2-}\\\text{pentanol}}}{CH_3\overset{\overset{\displaystyle OH}{|}}{\underset{\underset{\displaystyle CN}{|}}{C}}CH_2CH_2CH_3} \xrightarrow[\text{heat}]{H_2O,\ HCl} \underset{\substack{\text{2-Hydroxy-2-}\\\text{methylpentanoic}\\\text{acid (60\% from}\\\text{2-pentanone)}}}{CH_3\overset{\overset{\displaystyle OH}{|}}{\underset{\underset{\displaystyle CO_2H}{|}}{C}}CH_2CH_2CH_3}$$

12.10 REACTIONS OF CARBOXYLIC ACIDS

The most apparent chemical property of carboxylic acids, their ability to act as acids, has already been discussed in this chapter (Section 12.3). Another reaction of carboxylic acids is the formation of acyl chlorides by reaction of the carboxylic acid with thionyl chloride, $SOCl_2$.

$$\underset{\substack{\text{Carboxylic}\\\text{acid}}}{RCO_2H} + \underset{\substack{\text{Thionyl}\\\text{chloride}}}{SOCl_2} \longrightarrow \underset{\substack{\text{Acyl}\\\text{chloride}}}{R\overset{\overset{\displaystyle O}{\|}}{C}Cl} + \underset{\substack{\text{Sulfur}\\\text{dioxide}}}{SO_2} + \underset{\substack{\text{Hydrogen}\\\text{chloride}}}{HCl}$$

m-Methoxyphenylacetic acid → *m*-Methoxyphenylacetyl chloride (85%)

Acyl chlorides undergo electrophilic substitution with aromatic compounds in the *Friedel-Crafts acylation reaction* (Section 6.6). In the following chapter you will encounter other uses of acyl chlorides, in the preparation of esters and amides.

As with aldehydes and ketones, reduction of a carboxylic acid may be used to prepare an alcohol. The reduction of a carboxylic acid is more difficult than the reduction of an aldehyde or a ketone, and only the powerful reducing agent lithium aluminum hydride is successful. Both catalytic hydrogenation and sodium borohydride fail to effect the reduction of a carboxylic acid. The product from reduction of a carboxylic acid with lithium aluminum hydride is a primary alcohol.

$$\underset{\substack{\text{Carboxylic}\\\text{acid}}}{RCO_2H} \xrightarrow[\text{2. H}_2O]{\text{1. LiAlH}_4,\ \text{diethyl ether}} \underset{\text{Primary alcohol}}{RCH_2OH}$$

p-(Trifluoromethyl)benzoic acid → *p*-(Trifluoromethyl)benzyl alcohol (96%)

PROBLEM 12.8

Predict the product from the reaction of phenylacetic acid, $C_6H_5CH_2CO_2H$, with each of the following reagents.

(a) KOH   (c) LiAlH$_4$, then H$_2$O

(b) SOCl$_2$ (d) NaHCO$_3$ (sodium bicarbonate)

SAMPLE SOLUTION

(a) Carboxylic acids react with strong bases to yield metal carboxylate salts.

$$
\underset{\substack{\text{Phenylacetic}\\\text{acid}}}{C_6H_5CH_2\overset{\displaystyle O}{\overset{\|}{C}}OH} + \underset{\substack{\text{Potassium}\\\text{hydroxide}}}{KOH} \longrightarrow \underset{\substack{\text{Potassium}\\\text{phenylacetate}}}{C_6H_5CH_2\overset{\displaystyle O}{\overset{\|}{C}}O^-K^+} + \underset{\text{Water}}{H_2O}
$$

In the next chapter we will explore several other reactions of carboxylic acids. These reactions involve formation of compounds which are derivatives of carboxylic acids, principally the classes of molecules called *esters* and *amides*.

12.11 SPECTROSCOPIC PROPERTIES OF CARBOXYLIC ACIDS

Both the infrared and ^1H nmr spectra of carboxylic acids exhibit characteristic absorption patterns. The hydroxyl absorption in the infrared spectrum is a broad band in the 3500 to 2500 cm^{-1} region. The carbonyl group gives rise to a strong band near 1700 cm^{-1}. These features are apparent in the infrared spectrum of 3-methylpentanoic acid, shown in Figure 12.3.

The hydroxyl proton of a carboxyl group is normally the least shielded (appears farthest to the left) of all the protons in a ^1H nmr spectrum. Carboxyl protons appear in the range $\delta = 10$ to 12 ppm, often as a broad peak.

12.12 SUMMARY

Carboxylic acids are compounds of the type $R\overset{\displaystyle O}{\overset{\|}{C}}OH$. Systematic names are derived from the alkane having the same number of carbon atoms as the longest chain containing the carboxyl group (Section 12.1). The -*e* ending of the alkane is replaced by -*oic acid*.

Carboxylic acids have melting and boiling points higher than alcohols of comparable molecular weight because both the carboxyl oxygen and the hydroxyl group can engage in hydrogen-bond formation (Section 12.2).

Carboxylic acids are weak acids, having dissociation constants K_a in the 10^{-4} to 10^{-5} range (Section 12.3). One reason for the greater acidity of carboxylic acids compared with alcohols is the electron delocalization in the carboxylate ion. The negative charge is shared by both oxygens of the anion.

$$
\underset{\text{Carboxylic acid}}{RC\overset{\displaystyle O}{\underset{\displaystyle OH}{}}} \underset{H^-}{\overset{-H^-}{\rightleftharpoons}} \underset{\text{Resonance description of electron}\atop\text{delocalization in carboxylate anion}}{R-C\overset{\displaystyle O}{\underset{\displaystyle O^-}{}} \longleftrightarrow R-C\overset{\displaystyle O^-}{\underset{\displaystyle O}{}}}
$$

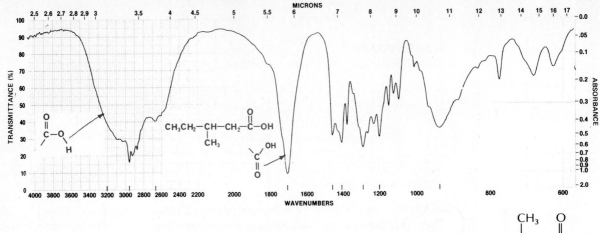

FIGURE 12.3 The infrared spectrum of 3-methylpentanoic acid, $\overset{\quad\ \ \underset{|}{CH_3}\ \ \overset{O}{\overset{||}{}}}{CH_3CH_2CHCH_2COH}$.

Electronegative substituents, especially when located close to the carboxyl group, increase the acidity of carboxylic acids (Section 12.4). Thus trifluoroacetic acid is a rather strong acid when compared to acetic acid.

$$CF_3CO_2H \qquad CH_3CO_2H$$
$$K_a = 5.9 \times 10^{-1} \qquad K_a = 1.8 \times 10^{-5}$$
$$pK_a\ 0.2 \qquad\qquad pK_a\ 4.7$$

Carboxylate salts, $R\overset{O}{\overset{||}{C}}O^- M^+$, of unbranched $C_{12} - C_{18}$ carboxylic acids are used as *soaps* (Section 12.5). The nonpolar hydrocarbon chain dissolves grease in spherical aggregates called *micelles*. The polar carboxylate ions on the micelle surface bind to water molecules and enable the grease to be suspended in the water and removed by rinsing.

Two new methods for the preparation of carboxylic acids were introduced. Both *add one carbon* to the skeleton of the starting material. They are:

1. Carboxylation of Grignard reagents (Section 12.8)
2. Hydrolysis of nitriles (Section 12.9)

Grignard reagents are nucleophilic and add to carbon dioxide to yield carboxylic acids after acidification.

$$RX \xrightarrow[\text{diethyl ether}]{Mg} RMgX \xrightarrow{CO_2} R\overset{O}{\overset{||}{C}}OMgX \xrightarrow{H_3O^+} RCO_2H$$

Primary and secondary alkyl halides undergo nucleophilic substitution by a cyanide ion to form nitriles (alkyl cyanides). Nitriles are converted to carboxylic acids by hydrolysis.

$$\underset{\substack{\text{Primary or}\\\text{secondary}\\\text{alkyl}\\\text{halide}}}{RX} \xrightarrow{CN^-} \underset{\text{Nitrile}}{RCN} \xrightarrow[\text{heat}]{H_2O,\ H^+} \underset{\substack{\text{Carboxylic}\\\text{acid}}}{RCO_2H}$$

Carboxylic acids react with thionyl chloride to form acyl chlorides.

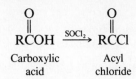

Carboxylic acids also undergo reduction with lithium aluminum hydride to give primary alcohols.

$$RCO_2H \xrightarrow[\text{2. } H_2O]{\text{1. } LiAlH_4} RCH_2OH$$

ADDITIONAL PROBLEMS

Nomenclature, Physical Properties, and Acidity

12.9 Many carboxylic acids are much better known by their common names than by their systematic names. Some of these are given below. Provide a structural formula for each one on the basis of its systematic name.

(a) Tetradecanoic acid (also known as *myristic acid,* it can be obtained from a variety of fats)

(b) 10-Undecenoic acid (also called *undecylenic acid,* it is used, in combination with its zinc salt, to treat fungal infections such as athlete's foot)

(c) 3,5-Dihydroxy-3-methylpentanoic acid (also called *mevalonic acid,* it is an important intermediate in the biosynthesis of terpenes and steroids)

(d) (E)-2-Methyl-2-butenoic acid (also known as *tiglic acid,* it is a constituent of various natural oils)

12.10 Give an acceptable IUPAC name for each of the following carboxylic acids.

(a) $(CH_3)_2CHCO_2H$ (Common name: isobutyric acid)

(b) $CH_3CH_2CH_2CH_2CHCH_2CH_2CO_2H$ (Common name: γ-methylcaprylic acid)
$\quad\quad\quad\quad\quad\quad\quad\quad |$
$\quad\quad\quad\quad\quad\quad\quad\quad CH_3$

4. methyl octa.

(c) CH_3CHCO_2H (Common name: α-chloropropionic acid)
$\quad\quad\quad |$
$\quad\quad\quad Cl$

12.11 Give an acceptable systematic IUPAC name for each of the following:

(a) $CH_3(CH_2)_6CO_2H$

(b) $CH_3(CH_2)_6CO_2^-K^+$

(c) $CH_2{=}CH(CH_2)_5CO_2H$

(d) $H_3C\quad\quad\quad(CH_2)_4CO_2H$
$\quad\quad\quad\searrow\quad\quad\swarrow$
$\quad\quad\quad\quad C{=}C$
$\quad\quad\swarrow\quad\quad\quad\searrow$
$\quad\quad H\quad\quad\quad\quad H$

12.12 The following compounds are used as food preservatives. They are called *antioxidants* and retard oxidation of fats and oils. Write the chemical structure of each.

(a) Sodium benzoate

(b) Calcium propionate (IUPAC name, calcium propanoate)

(c) Potassium sorbate (sorbic acid is $CH_3CH{=}CHCH{=}CHCO_2H$)

12.13 Each of the compounds shown has a molecular weight of 46, yet their boiling points vary over a range of 125°C. Explain this trend with chemical structures.

$$CH_3OCH_3 \quad\quad CH_3CH_2OH \quad\quad HCO_2H$$
$$\text{bp } -24°C \quad\quad\quad \text{bp } 78°C \quad\quad\quad \text{bp } 101°C$$

12.14 Rank the compounds in each group below in order of decreasing acidity (most to least acidic).

(a) Acetic acid, ethane, ethanol

(b) Benzene, benzoic acid, benzyl alcohol, phenol

(c) Acetic acid, ethanol, trifluorocetic acid, 2,2,2-trifluoroethanol

Preparation of Carboxylic Acids

12.15 Write a chemical reaction for the preparation of each of the compounds in Problem 12.12 from the appropriate carboxylic acid.

12.16 Propose methods for preparing butanoic acid from each of the following:

(a) 1-Butanol

(b) Butanal

(c) 1-Butene

(d) 1-Chloropropane (two methods)

12.17 Describe two methods for the preparation of 3-methylhexanoic acid from 2-methyl-1-pentanol.

Reactions of Carboxylic Acids

12.18 Give the product of the reaction of pentanoic acid with each of the following reagents:

(a) Sodium hydroxide

(b) Sodium bicarbonate

(c) Thionyl chloride

(d) Lithium aluminum hydride, then hydrolysis with water

12.19 Show how butanoic acid may be converted to each of the following compounds using any other necessary organic or inorganic reagents.

(a) 1-Butanol

(b) 1-Bromobutane

(c) Pentanoic acid

(d) Butanoyl chloride ($CH_3CH_2CH_2\overset{\overset{\displaystyle O}{\|}}{C}Cl$)

(e) 1-Phenyl-1-butanone ($C_6H_5\overset{\overset{\displaystyle O}{\|}}{C}CH_2CH_2CH_3$)

12.20 Outline a sequence of reactions that will allow pentanal to be converted into:

(a) Pentanoic acid

(b) Hexanoic acid

(c) 2-Hydroxyhexanoic acid

12.21 Unbranched carboxylic acids obtained from natural sources, called fatty acids, usually contain an even number of carbon atoms. To prepare an acid having an odd number of carbons, the necessary conversion is

$$RCO_2H \longrightarrow RCH_2CO_2H$$

Show how this might be done by describing the preparation of nonanoic acid from octanoic acid.

Miscellaneous Problems

12.22 In the presence of the enzyme *aconitase,* the double bond of aconitic acid undergoes hydration. The reaction is reversible and the following equilibrium is established:

Isocitric acid $\underset{}{\overset{H_2O}{\rightleftharpoons}}$
$$\underset{\underset{\displaystyle H}{\diagup}\overset{\overset{\displaystyle HO_2C}{\diagdown}}{C}=\underset{\underset{\displaystyle CH_2CO_2H}{\diagdown}}{\overset{\overset{\displaystyle CO_2H}{\diagup}}{C}}$$
$\underset{}{\overset{H_2O}{\rightleftharpoons}}$ citric acid

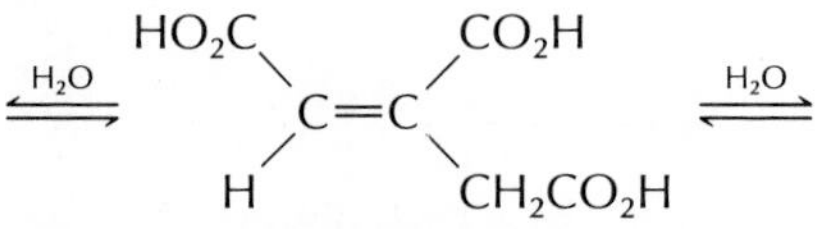

$(C_6H_8O_7)$ Aconitic acid $(C_6H_8O_7)$

(6% at equilibrium)   (4% at equilibrium)   (90% at equilibrium)

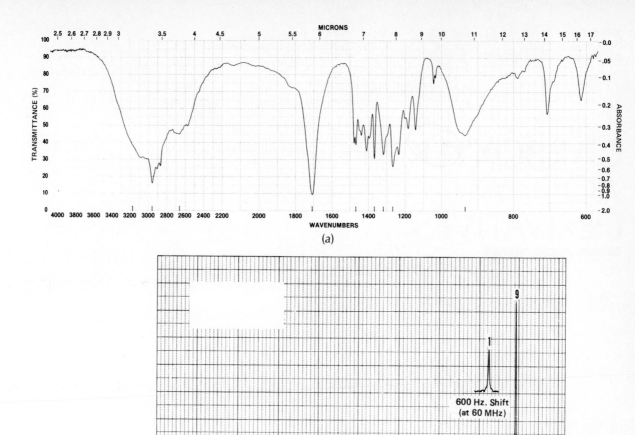

FIGURE 12.4 The (a) infrared and (b) ¹H nmr spectra of compound A, $C_6H_{12}O_2$ (Problem 12.23). The numbers beside the signals correspond to their relative areas.

(a) The major tricarboxylic acid present is *citric acid,* the substance responsible for the tart taste of citrus fruits. Citric acid is achiral. What is its structure?

(b) What must be the constitution of isocitric acid? (Assume no rearrangements accompany hydration.) How many stereoisomers are possible for isocitric acid?

12.23 Identify compound A ($C_6H_{12}O_2$), whose ir and ¹H nmr spectra are shown in Figure 12.4.

12.24 The ¹H nmr spectra of formic acid (HCO_2H), maleic acid (*cis*-$HO_2CCH{=}CHCO_2H$), and malonic acid ($HO_2CCH_2CO_2H$) are similar in that each is characterized by two singlets of equal intensity. Match these compounds with the designations B, C, and D on the basis of the appropriate ¹H nmr chemical shift data.

Compound B: signals at $\delta = 3.2$ and 12.1 ppm

Compound C: signals at $\delta = 6.3$ and 12.4 ppm

Compound D: signals at $\delta = 8.0$ and 11.4 ppm

CHAPTER *13*

CARBOXYLIC ACID DERIVATIVES

LEARNING OBJECTIVES

This chapter focuses on the most common classes of carboxylic acid derivatives: acyl chlorides, anhydrides, esters, and amides. Upon completion of this chapter you should be able to:

- Give an acceptable systematic IUPAC name for a carboxylic acid derivative.
- Write the structure of a carboxylic acid derivative given its systematic name.
- Explain by using chemical equations the general mechanism for a nucleophilic acyl substitution reaction, and describe the role of the tetrahedral intermediate.
- Write a chemical equation for the preparation of an ester by the Fischer esterification method.
- Explain by using chemical equations the mechanism of the Fischer esterification process.
- Write chemical equations for the preparation of esters by using acyl chlorides or acetic anhydride.
- Write chemical equations for the acid-catalyzed and base-promoted hydrolysis of an ester.
- Explain by using chemical equations the reaction of an ester with a Grignard reagent.
- Describe by using chemical equations the preparation of an amide by the reaction of an amine with a carboxylic acid, an acyl chloride, acetic anhydride, or an ester.
- Write a chemical equation for the hydrolysis of an amide under both acidic and basic conditions.
- Describe the characteristic features of the infrared spectra of the various carboxylic acid derivatives.

The previous two chapters described three classes of organic compounds character-ized by the presence of a carbonyl ($\diagdown$ C=O) group: aldehydes, ketones, and carbox-ylic acids. We will now expand our discussion to include the principal classes of carboxylic acid derivatives. Two of these, **esters** and **amides,** are especially important in both organic and biochemistry and will receive particular emphasis.

13.1 NOMENCLATURE OF CARBOXYLIC ACID DERIVATIVES

The four classes of carboxylic acid derivatives that we will encounter are:

$$\overset{O}{\overset{\|}{}}$$

1. Acyl chlorides: RCCl

$$\overset{O}{\overset{\|}{}}\overset{O}{\overset{\|}{}}$$

2. Anhydrides: RCOCR

$$\overset{O}{\overset{\|}{}}$$

3. Esters: RCOR′

$$\overset{O}{\overset{\|}{}}\qquad\overset{O}{\overset{\|}{}}\qquad\overset{O}{\overset{\|}{}}$$

4. Amides: RCNH$_2$, RCNHR′, and RCNR′$_2$

The systematic (IUPAC) names of each class are based on the name of the corresponding carboxylic acid. Each of these compounds possesses an

$$\overset{O}{\overset{\|}{}}\qquad\overset{O}{\overset{\|}{}}$$

acyl group, RC— or ArC—. **Acyl chlorides** are named by replacing the *-ic acid* ending of the carboxylic acid name with *-yl chloride.*

$$\overset{O}{\overset{\|}{}}$$
CH$_3$CCl

$$\overset{O}{\overset{\|}{}}$$
⬡—CCl

Acetyl chloride
(from acetic acid)

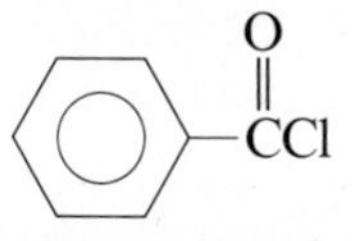

Benzoyl chloride
(from benzoic acid)

Anhydrides are named in a similar manner. The word *acid* is replaced with the word *anhydride.*

$$\overset{O}{\overset{\|}{}}\overset{O}{\overset{\|}{}}$$
CH$_3$COCCH$_3$

$$\overset{O}{\overset{\|}{}}\overset{O}{\overset{\|}{}}$$
⬡—COC—⬡

Acetic anhydride

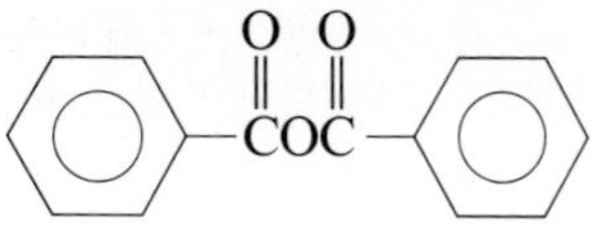

Benzoic anhydride

Esters have both an alkyl or aryl group and an acyl portion, which are named independently.

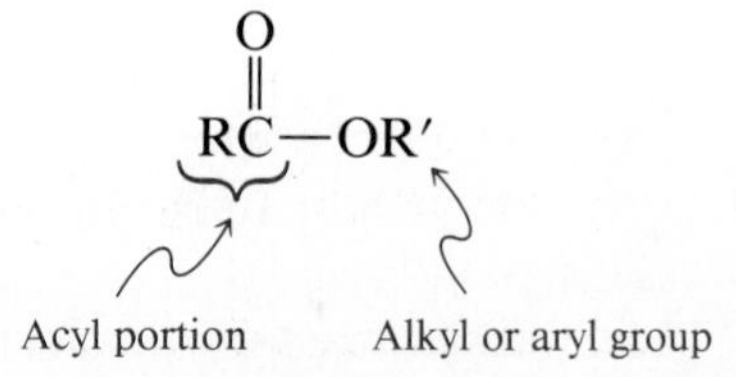

$$\overset{O}{\overset{\|}{}}$$
RC—OR′

Acyl portion   Alkyl or aryl group

The alkyl or aryl group (R′) attached to oxygen is named first, followed by the acyl portion as a separate word. The *-ic* ending of the corresponding acid is replaced by *-ate*.

$$CH_3\overset{\overset{\displaystyle O}{\|}}{C}OCH_2CH_3 \qquad CH_3CH_2\overset{\overset{\displaystyle O}{\|}}{C}OCH_3 \qquad C_6H_5\overset{\overset{\displaystyle O}{\|}}{C}OCH_2CH_2Cl$$

Ethyl acetate   Methyl propanoate   2-Chloroethyl benzoate

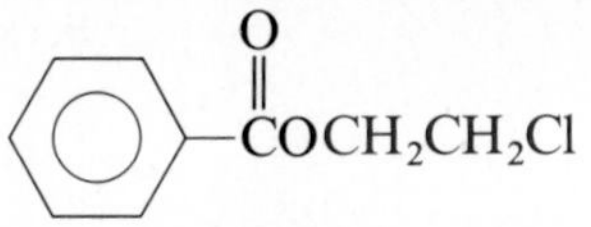

The names of **amides** of the type $R\overset{\overset{\displaystyle O}{\|}}{C}NH_2$ are derived from carboxylic acids by replacing the *-oic* (or *ic*) *acid* ending with *-amide*.

$$CH_3\overset{\overset{\displaystyle O}{\|}}{C}NH_2 \qquad C_6H_5\overset{\overset{\displaystyle O}{\|}}{C}NH_2 \qquad (CH_3)_2CHCH_2\overset{\overset{\displaystyle O}{\|}}{C}NH_2$$

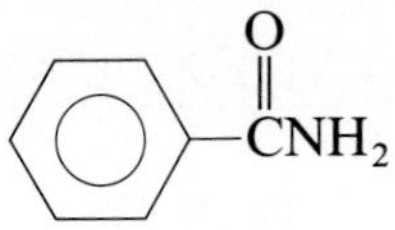

Acetamide   Benzamide   3-Methylbutanamide

Compounds of the type $R\overset{\overset{\displaystyle O}{\|}}{C}NHR'$ and $R\overset{\overset{\displaystyle O}{\|}}{C}NR'_2$ are named as *N*-alkyl and *N,N*-dialkyl substituted derivatives of a parent amide.

$$CH_3\overset{\overset{\displaystyle O}{\|}}{C}NHCH_3 \qquad C_6H_5\overset{\overset{\displaystyle O}{\|}}{C}N(CH_2CH_3)_2 \qquad CH_3CH_2CH_2\overset{\overset{\displaystyle O}{\|}}{C}NCH(CH_3)_2$$

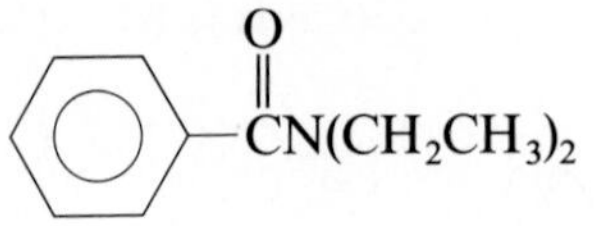

N-Methyl-acetamide   *N,N*-Diethylbenzamide   *N*-Isopropyl-*N*-methylbutanamide

---

PROBLEM 13.1

Write a structural formula for each of the following compounds:

(a) 2-Phenylbutanoyl chloride   (d) 2-Phenylbutyl butanoate
(b) 2-Phenylbutanoic anhydride   (e) 2-Phenylbutanamide
(c) Butyl 2-phenylbutanoate

SAMPLE SOLUTION

(a) When the name of an acyl group is followed by "chloride," it designates an *acyl chloride*. A 2-phenylbutanoyl group is a four-carbon acyl unit that bears a phenyl substituent at C-2.

$$CH_3CH_2\underset{\underset{\displaystyle C_6H_5}{|}}{C}H\overset{\overset{\displaystyle O}{\|}}{C}Cl \qquad \text{2-Phenylbutanoyl chloride}$$

13.2 ACYL TRANSFER REACTIONS. HYDROLYSIS

Acyl transfer reactions are **nucleophilic acyl substitution** reactions that interconvert the various carboxylic acid derivatives.

$$\text{R}\overset{\text{O}}{\overset{\|}{\text{C}}}\text{Y} + \text{HZ}\colon \longrightarrow \text{R}\overset{\text{O}}{\overset{\|}{\text{C}}}\text{Z} + \text{HY}\colon$$

or

$$\text{R}\overset{\text{O}}{\overset{\|}{\text{C}}}\text{Y} + \colon\text{Z}^- \longrightarrow \text{R}\overset{\text{O}}{\overset{\|}{\text{C}}}\text{Z} + \text{Y}\colon^-$$

Such reactions do *not* proceed by either an S_N1 or an S_N2 process. Nucleophilic substitutions of the S_N1 and S_N2 type were examined in Chapter 8 and involve substrates in which the carbon attacked by the nucleophile is sp^3 hybridized. The carbon atom attacked by the nucleophile in nucleophilic acyl substitution reactions is sp^2 hybridized.

To illustrate the mechanism of nucleophilic acyl substitution, let us examine the **hydrolysis** of an acyl derivative. Hydrolysis is the reaction with water, and all carboxylic acid derivatives react with water to give the corresponding carboxylic acid.

$$\underset{\substack{\text{Carboxylic} \\ \text{acid} \\ \text{derivative}}}{\text{R}\overset{\text{O}}{\overset{\|}{\text{C}}}\text{Y}} + \underset{\text{Water}}{\text{H}_2\text{O}} \longrightarrow \underset{\substack{\text{Carboxylic} \\ \text{acid}}}{\text{R}\overset{\text{O}}{\overset{\|}{\text{C}}}\text{OH}} + \underset{\substack{\text{Conjugate} \\ \text{acid of} \\ \text{substituent}}}{\text{HY}}$$

In the hydrolysis reaction shown, the leaving group Y has been replaced by the OH of water. This substitution can be explained by a sequence of steps involving an *addition* of a nucleophile (in this case, H_2O) followed by an *elimination* of a leaving group (in this case, Y). The steps involved in the hydrolysis of an acyl chloride are outlined in Figure 13.1. The nucleophile water *adds* to the carbonyl group of the acyl chloride in the first step, forming a **tetrahedral intermediate.** In the second step of the reaction the tetrahedral intermediate dissociates, *eliminating* the chloride ion and regaining the carbonyl group to form the product carboxylic acid.

PROBLEM 13.2
Acyl chlorides are called *lachrymators* because of the strong irritating effect they have on

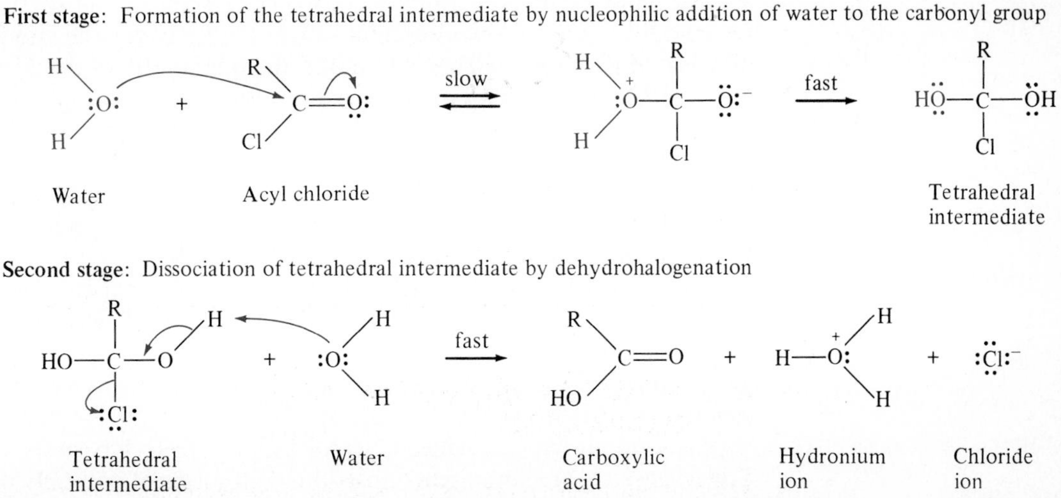

First stage: Formation of the tetrahedral intermediate by nucleophilic addition of water to the carbonyl group

Second stage: Dissociation of tetrahedral intermediate by dehydrohalogenation

FIGURE 13.1 Mechanism of the hydrolysis of an acyl chloride.

the eyes. This effect comes about from the reaction with moisture naturally present in our eyes. Write a balanced equation showing the reaction of benzoyl chloride with water.

The ease of hydrolysis of the various acid derivatives can be compared, and this comparison can be used as a measure of the relative chemical reactivity of these compounds. Using esters as a basis for comparison, acyl chlorides undergo hydrolysis 100 *billion* times faster! In fact, the reaction of an acyl chloride with water can be a violent one, occurring with explosive force and producing large quantities of heat. *Acyl* chlorides also react with nucleophiles many times faster than do *alkyl* chlorides.

Anhydrides are less reactive than acyl chlorides but still undergo hydrolysis about 10 million times faster than esters. Amides, on the other hand, are *less* reactive than esters. Their rate of hydrolysis is less than one one-hundredth that of an ester.

The trend in reactivities can be summarized:

$$\underset{\substack{\text{Acyl}\\\text{chloride}\\\text{Most reactive}}}{\overset{\overset{\displaystyle O}{\parallel}}{RCCl}} > \underset{\text{Anhydride}}{\overset{\overset{\displaystyle O}{\parallel}\ \overset{\displaystyle O}{\parallel}}{RCOCR}} > \underset{\text{Ester}}{\overset{\overset{\displaystyle O}{\parallel}}{RCOR'}} > \underset{\substack{\text{Amide}\\\text{Least}\\\text{reactive}}}{\overset{\overset{\displaystyle O}{\parallel}}{RCNR'_2}}$$

Why are amides the least reactive carboxylic acid derivative? Nitrogen is less electronegative than oxygen and is able to donate an electron pair to the carbonyl group. Thus amides are stabilized by resonance.

$$R-C \overset{\displaystyle O:}{\underset{\displaystyle NR'_2}{}} \longleftrightarrow R-C \overset{\displaystyle \ddot{O}:^-}{\underset{\displaystyle \overset{+}{N}R'_2}{}}$$

This stabilization makes amides less reactive than the other carboxylic acid derivatives toward attack by a nucleophile. Acyl chlorides and anhydrides are stabilized to a much smaller extent than amides or esters and thus are more reactive. One result of this trend is that acyl chlorides and anhydrides, the two most reactive derivatives, are often used to prepare esters and amides, the less reactive, more stable, derivatives.

13.3 NATURAL SOURCES OF ESTERS

A great many esters are naturally occurring substances, and several are shown in Figure 13.2. Esters of low molecular weight are fairly volatile and frequently have pleasing odors. Esters often are found in the fragrant oil of fruits and flowers. For example, the aroma of an orange has been found to contain 30 different esters in addition to 10 carboxylic acids, 34 alcohols, 34 aldehydes and ketones, and 36 hydrocarbons!

Esters are often used in artificially flavored foods, as the ester fragrance may be suggestive of the natural fragrance of the fruit. For example, isoamyl acetate (Figure 13.2) is used to produce a bananalike flavor in foods.

Two of the examples in Figure 13.2 illustrate esters that are *pheromones,* chemicals used by insects to communicate with one another. The Japanese beetle pheromone (Z)-5-tetradecen-4-olide is an example of a cyclic ester — the acyl group and the alcohol portion of the ester are connected in a ring. Cyclic esters are called **lactones.**

Esters of glycerol are called **glycerol triesters, triacylglycerols,** or **triglycerides.** Triacylglycerols are among the most abundant forms of naturally occurring esters. *Tristearin* is one example of a triacylglycerol.

BIOLOGICAL ACYL TRANSFER REACTIONS

In the laboratory an organic chemist uses acyl chlorides and anhydrides to prepare other acyl compounds because of the greater reactivity of these materials. Nucleophilic acyl substitutions are also common in biochemistry, but acyl chlorides are not found in living systems, and the only known example of a naturally occurring carboxylic acid anhydride is *cantharidin,* once believed to be an aphrodisiac, isolated from a beetle.

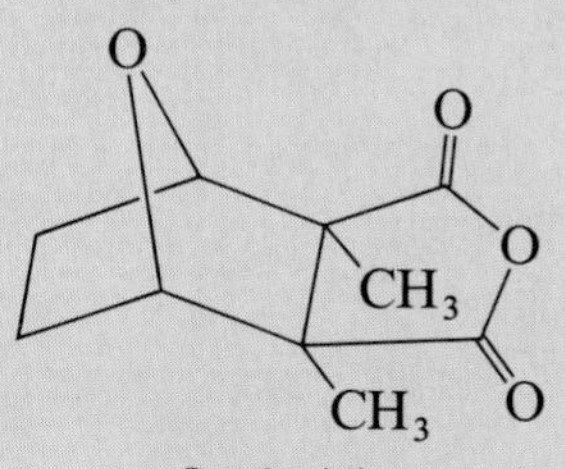

Cantharidin

The kinds of substances involved in many biological acyl transfer reactions are acyl **thioesters,** compounds of the type RCSR'. Thioesters are more reactive than simple esters, RCOR'. One thioester, acetyl coenzyme A, has the relatively complicated structure shown in Figure 17.4a. Acetyl coenzyme A is abbreviated CH_3CSCoA. One of the reactions of acetyl coenzyme A in living systems is the formation of acetate esters by the acyl transfer reaction:

$$\text{CH}_3\text{CSCoA} + \text{ROH} \longrightarrow \text{CH}_3\text{COR} + \text{CoASH}$$

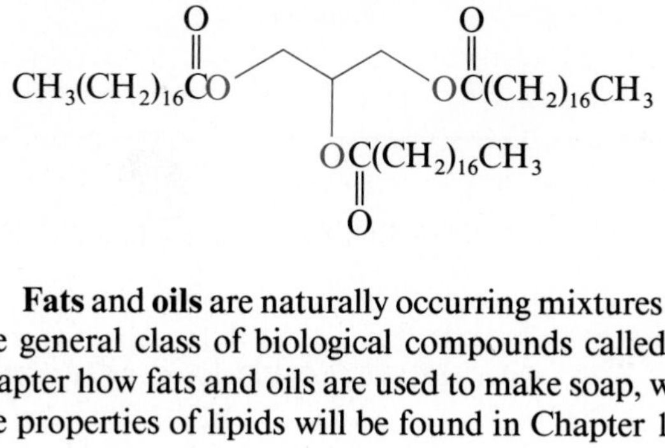

Tristearin, a trioctadecanoyl ester of glycerol found in many animal and vegetable fats (the three carbons and three oxygens of glycerol are shown in color)

Fats and **oils** are naturally occurring mixtures of triacylglycerols. They belong to the general class of biological compounds called *lipids.* You will see later in this chapter how fats and oils are used to make soap, while a more detailed discussion of the properties of lipids will be found in Chapter 17.

13.4 PREPARATION OF ESTERS. FISCHER ESTERIFICATION

A frequently used method for the preparation of esters is the acid-catalyzed reaction of an alcohol with a carboxylic acid.

$(CH_3)_2CHCH_2CH_2O\overset{\overset{\displaystyle O}{\|}}{C}CH_3$

Isoamyl acetate (banana oil)

$CH_3CH_2CH_2CH_2O\overset{\overset{\displaystyle O}{\|}}{C}CH_3$

Butyl acetate (component of apple aroma)

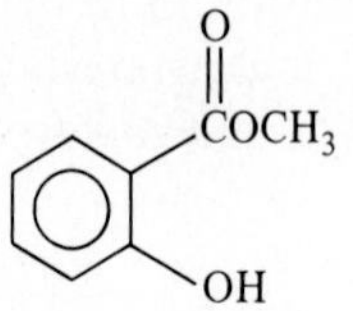

Methyl salicylate
(principal component of oil
of wintergreen)

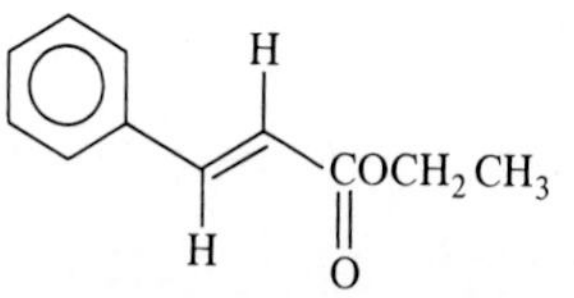

Ethyl cinnamate
(one of the constituents of
the sex pheromone of the
male Oriental fruit moth)

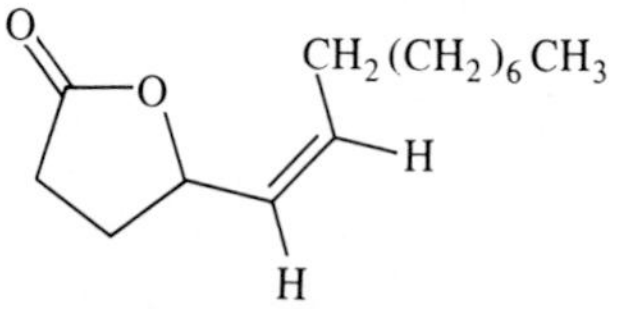

(Z)-5-Tetradecen-4-olide
(sex pheromone of female
Japanese beetle)

FIGURE 13.2 Several
naturally occurring esters.

$$R\overset{\overset{\displaystyle O}{\|}}{C}OH + R'OH \overset{H^+}{\rightleftharpoons} R\overset{\overset{\displaystyle O}{\|}}{C}OR' + H_2O$$

Carboxylic acid   Alcohol   Ester   Water

$$CH_3\overset{\overset{\displaystyle O}{\|}}{C}OH + CH_3CH_2CH_2OH \overset{H_2SO_4}{\rightleftharpoons} CH_3\overset{\overset{\displaystyle O}{\|}}{C}OCH_2CH_2CH_3 + H_2O$$

Acetic acid   1-Propanol   Propyl acetate   Water

PROBLEM 13.3

Write the structure of the ester formed in each of the following reactions:

(a) $CH_3CH_2CH_2CH_2OH + CH_3CH_2\overset{\overset{\displaystyle O}{\|}}{C}OH \overset{H_2SO_4}{\underset{heat}{\longrightarrow}}$

(b) $C_6H_5CH_2OH + (CH_3)_2CHCH_2\overset{\overset{\displaystyle O}{\|}}{C}OH \overset{H_2SO_4}{\underset{heat}{\longrightarrow}}$

(c) $2CH_3OH + HO\overset{\overset{\displaystyle O}{\|}}{C}\text{—}\langle\bigcirc\rangle\text{—}\overset{\overset{\displaystyle O}{\|}}{C}OH \overset{H_2SO_4}{\underset{heat}{\longrightarrow}} C_{10}H_{10}O_4$

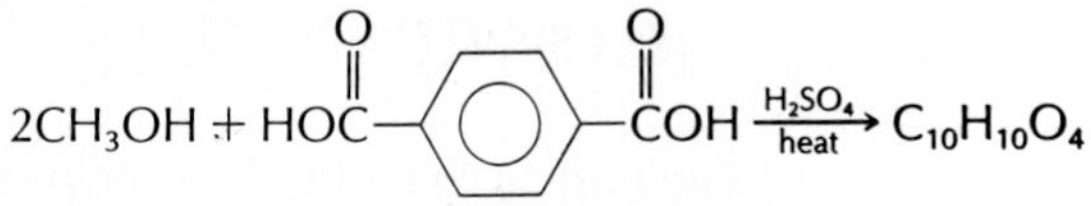

SAMPLE SOLUTION

(a) By analogy to the general equation and to the example cited above, we can write the equation

$$CH_3CH_2CH_2CH_2COH + CH_3CH_2COH \xrightarrow[\text{heat}]{H_2SO_4} CH_3CH_2COCH_2CH_2CH_2CH_3 + H_2O$$

1-Butanol   Propanoic acid   Butyl propanoate   Water

This condensation reaction is called a **Fischer esterification reaction.** It is an equilibrium process, with the position of equilibrium only slightly favoring products.

To be useful as a synthetic procedure for the preparation of esters, reaction conditions must be chosen so that the desired product is present in large amounts at equilibrium. We can understand how this is accomplished by recalling **Le Chatelier's principle** (Section 5.1.5). Simply stated, an equilibrium shifts in the direction necessary to relieve any "stress" applied to it. "Stress" usually means the addition of reactant or the removal of products.

As the following examples illustrate, either of the reactants, the alcohol or the carboxylic acid, may be present in excess to shift the equilibrium toward product formation.

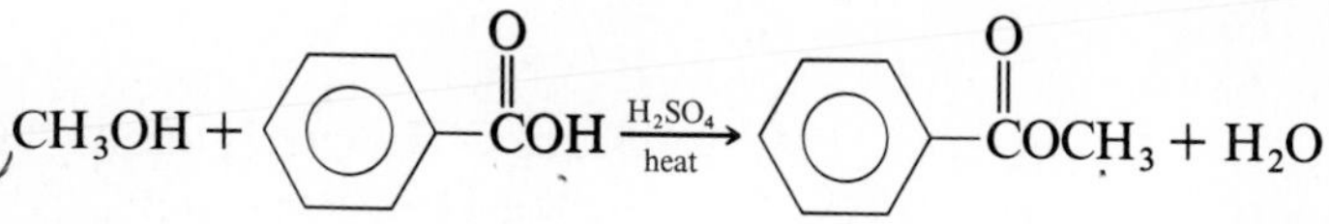

Methanol (0.6 mol) + Benzoic acid (0.1 mol) → Methyl benzoate (isolated in 70% yield based on benzoic acid) + Water

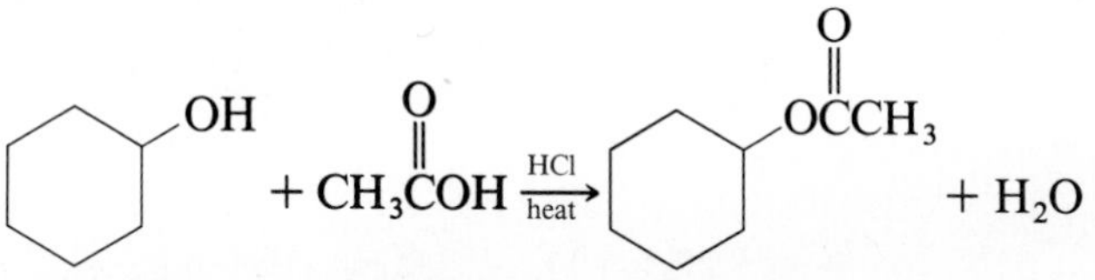

Cyclohexanol (1 mol) + Acetic acid (2.2 mol) → Cyclohexyl acetate (isolated in 53% yield based on cyclohexanol) + Water

Another way to favor the formation of ester is to remove water from the reaction mixture as it is formed. Water is removed by adding benzene as a cosolvent and distilling the benzene-water azeotrope. (When two or more substances distill together as a mixture of constant composition, the mixture is called an *azeotrope.*)

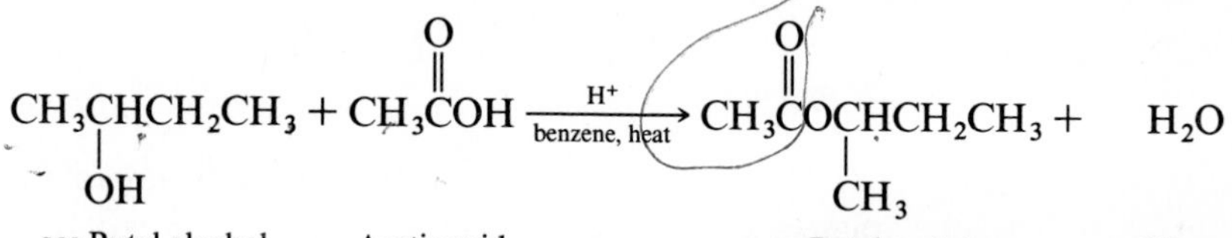

$$CH_3CHCH_2CH_3 + CH_3COH \xrightarrow[\text{benzene, heat}]{H^+} CH_3COCHCH_2CH_3 + H_2O$$

with OH below first, CH₃ below product.

sec-Butyl alcohol (0.20 mol)   Acetic acid (0.25 mol)   sec-Butyl acetate (isolated in 71% yield based on sec-butyl alcohol)   Water (codistills with benzene)

(Handwritten margin notes:) C₆H₅CH₂OCCH₂CH(CH₃)₂ ; Benzyl-3-methyl butanoate ; CH₃ , H₃COCOCCH₃

13.5 MECHANISM OF FISCHER ESTERIFICATION

The mechanism of an acid-catalyzed esterification reaction is worth examining because it illustrates the principles of acyl transfer reaction mechanisms outlined in Section 13.2, especially the role played by the tetrahedral intermediate.

An important point to note is that the ester oxygen (—OR) is the oxygen of the starting alcohol, not the carboxylic acid. The mechanism must account for this observation.

$$R'\overset{\overset{\textstyle O}{\|}}{C}-OH \ + \ RO-H \ \longrightarrow \ R'\overset{\overset{\textstyle O}{\|}}{C}-OR \ + \ H_2O$$

| This oxygen is lost from the acyl group. | This bond is broken in conversion of an alcohol to an ester. | This oxygen is the same one that was part of the starting alcohol. | This oxygen was part of the starting carboxylic acid. |

Consider as an example the formation of methyl benzoate from the reaction of benzoic acid and methanol.

$$C_6H_5\overset{\overset{\textstyle O}{\|}}{C}OH + CH_3OH \overset{H^+}{\rightleftharpoons} C_6H_5\overset{\overset{\textstyle O}{\|}}{C}OCH_3 + H_2O$$

Benzoic   Methanol   Methyl   Water
acid   benzoate

In the first step of the reaction, the acid catalyst protonates benzoic acid at its carbonyl oxygen.

$$C_6H_5C\overset{\ddot{O}:}{\underset{\ddot{O}H}{\diagup}} \overset{H^+}{\rightleftharpoons} C_6H_5C\overset{{}^+\!\ddot{O}H}{\underset{\ddot{O}H}{\diagup}}$$

The protonated species is the conjugate acid of benzoic acid. Protonation of the carbonyl oxygen causes the carbonyl carbon to be more susceptible to attack by a weak nucleophile such as methanol. A similar phenomenon was observed in the discussion of the acid-catalyzed hydration of aldehydes and ketones (Section 11.7).

$$C_6H_5C\overset{{}^+\!\ddot{O}H}{\underset{\ddot{O}H}{\diagup}}\underset{\underset{H\quad CH_3}{\overset{O}{\diagdown}}}{} \overset{slow}{\rightleftharpoons} C_6H_5\underset{\underset{H\quad CH_3}{\overset{+}{O}}}{\overset{\ddot{O}H}{\underset{|}{C}}}-\ddot{O}H \overset{-H^+, fast}{\rightleftharpoons} C_6H_5\underset{:\ddot{O}CH_3}{\overset{\ddot{O}H}{\underset{|}{C}}}-\ddot{O}H$$

Tetrahedral
intermediate

The product of nucleophilic addition of methanol to the carbonyl group is analogous to a hemiacetal formed by addition of methanol to the carbonyl group of an aldehyde or ketone (Section 11.8). The tetrahedral intermediate is not stable, and under the acid-catalyzed conditions of the reaction it undergoes dehydration to form the product, methyl benzoate.

$$\underset{\substack{\text{Tetrahedral}\\\text{intermediate}}}{C_6H_5\overset{\overset{\displaystyle :\ddot{O}H}{|}}{\underset{\underset{\displaystyle :\ddot{O}CH_3}{|}}{C}}-\ddot{O}H} \xrightarrow{\text{H}^+,\text{ fast}} C_6H_5\overset{\overset{\displaystyle :\ddot{O}H}{|}}{\underset{\underset{\displaystyle :\ddot{O}CH_3}{|}}{C}}-\overset{+}{\ddot{O}}\overset{H}{\underset{H}{<}} \rightleftharpoons C_6H_5\overset{\overset{\displaystyle \overset{+}{O}H}{\|}}{\underset{\underset{\displaystyle \ddot{O}CH_3}{|}}{C}} + H_2O$$

$$\Big\Updownarrow -H^+$$

$$\underset{\text{Methyl benzoate}}{C_6H_5\overset{\overset{\displaystyle \ddot{O}:}{\|}}{\underset{\underset{\displaystyle \ddot{O}CH_3}{|}}{C}}}$$

PROBLEM 13.4

Give the structure of the tetrahedral intermediate formed in each reaction of Problem 13.3.

SAMPLE SOLUTION

(a) Following protonation of the carbonyl oxygen, the carboxylic acid undergoes nucleophilic attack by the oxygen of the alcohol.

$$CH_3CH_2\overset{\overset{\displaystyle \ddot{O}:}{\|}}{\underset{\underset{\displaystyle \ddot{O}H}{|}}{C}} \underset{}{\overset{\text{H}^+}{\rightleftharpoons}} CH_3CH_2\overset{\overset{\displaystyle +\ddot{O}H}{\|}}{\underset{\underset{\displaystyle \ddot{O}H}{|}}{C}}$$

$$CH_3CH_2\overset{\overset{\displaystyle \overset{+}{\ddot{O}}H}{\|}}{\underset{\underset{\displaystyle \ddot{O}H}{|}}{C}} + H\ddot{O}CH_2CH_2CH_3 \underset{}{\overset{-\text{H}^+}{\rightleftharpoons}} \underset{\text{Tetrahedral intermediate}}{CH_3CH_2\overset{\overset{\displaystyle :\ddot{O}H}{|}}{\underset{\underset{\displaystyle :\ddot{O}H}{|}}{C}}-OCH_2CH_2CH_2CH_3}$$

The mechanism of acid-catalyzed esterification illustrates the three important elements in acyl transfer reactions:

1. Activation of the carbonyl group by protonation of the carbonyl oxygen
2. Nucleophilic addition to the protonated carbonyl, forming a tetrahedral intermediate.
3. Elimination from the tetrahedral intermediate to restore the carbonyl group

13.6 PREPARATION OF ESTERS. ADDITIONAL METHODS

Fischer esterification is not the only method available for the preparation of esters. Esters are often prepared by using acyl chlorides or anhydrides as the source of the acyl portion of the ester. The primary advantage over Fischer esterification is that ester preparations using acyl chlorides and anhydrides do not involve equilibrium reactions. Thus it is not necessary to use an excess of either reactant, a definite "plus" if the reactants are very expensive.

Acyl chlorides are prepared by reaction of a carboxylic acid with thionyl chloride.

$$\underset{\substack{\text{Carboxylic} \\ \text{acid}}}{\text{RCOH}} + \underset{\substack{\text{Thionyl} \\ \text{chloride}}}{\text{SOCl}_2} \longrightarrow \underset{\substack{\text{Acyl} \\ \text{chloride}}}{\text{RCCl}} + \underset{\substack{\text{Sulfur} \\ \text{dioxide}}}{\text{SO}_2} + \underset{\substack{\text{Hydrogen} \\ \text{chloride}}}{\text{HCl}}$$

$$\underset{\substack{\text{Propanoic} \\ \text{acid}}}{\text{CH}_3\text{CH}_2\text{COH}} \xrightarrow{\text{SOCl}_2} \underset{\substack{\text{Propanoyl} \\ \text{chloride}}}{\text{CH}_3\text{CH}_2\text{CCl}}$$

Reaction of the acyl chloride with an alcohol produces the desired ester.

$$\underset{\substack{\text{Acyl} \\ \text{chloride}}}{\text{RCCl}} + \underset{\text{Alcohol}}{\text{R}'\text{OH}} \longrightarrow \underset{\text{Ester}}{\text{RCOR}'} + \underset{\substack{\text{Hydrogen} \\ \text{chloride}}}{\text{HCl}}$$

The rate of the reaction is increased, and the reaction rendered essentially irreversible, by carrying it out in the presence of a base such as pyridine. Pyridine reacts with hydrogen chloride as it is formed.

$$\underset{\substack{\text{Propanoyl} \\ \text{chloride}}}{\text{CH}_3\text{CH}_2\text{CCl}} + \underset{\text{Methanol}}{\text{CH}_3\text{OH}} + \underset{\text{Pyridine}}{\text{C}_5\text{H}_5\text{N}} \longrightarrow \underset{\substack{\text{Methyl} \\ \text{propanoate}}}{\text{CH}_3\text{CH}_2\text{COCH}_3} + \underset{\substack{\text{Pyridinium} \\ \text{hydrochloride}}}{\text{C}_5\text{H}_5\text{NH}^+\text{Cl}^-}$$

PROBLEM 13.5

Write chemical equations which show how each of the following esters could be prepared by using an acyl chloride in the presence of pyridine.

(a) Ethyl benzoate
(b) Benzyl acetate
(c) Isopropyl 2-methylpropanoate

SAMPLE SOLUTION

(a) From the name of the ester, we can determine that the acyl portion is derived from benzoic acid, and the alkyl group from ethanol. Thus first prepare the acyl chloride, benzoyl chloride. This is then allowed to react with ethanol in the presence of pyridine.

$$\underset{\substack{\text{Benzoic} \\ \text{acid}}}{\text{C}_6\text{H}_5\text{COH}} \xrightarrow{\text{SOCl}_2} \underset{\substack{\text{Benzoyl} \\ \text{chloride}}}{\text{C}_6\text{H}_5\text{CCl}} \xrightarrow[\text{pyridine}]{\text{CH}_3\text{CH}_2\text{OH}} \underset{\text{Ethyl benzoate}}{\text{C}_6\text{H}_5\text{COCH}_2\text{CH}_3}$$

Esters can also be prepared by reaction of an anhydride with an alcohol.

$$\underset{\substack{\text{Acid} \\ \text{Anhydride}}}{\text{RCOCR}} + \underset{\text{Alcohol}}{\text{R}'\text{OH}} \longrightarrow \underset{\text{Ester}}{\text{RCOR}'} + \underset{\substack{\text{Carboxylic} \\ \text{acid}}}{\text{RCOH}}$$

By far the most common use of this reaction is in the preparation of acetate esters. That is, acetic anhydride is allowed to react with an alcohol to give an ester.

$$CH_3\overset{\displaystyle O}{\overset{\|}{C}}O\overset{\displaystyle O}{\overset{\|}{C}}CH_3 + HOCHCH_2CH_3 \longrightarrow CH_3\overset{\displaystyle O}{\overset{\|}{C}}OCHCH_2CH_3 + CH_3\overset{\displaystyle O}{\overset{\|}{C}}OH$$

| Acetic anhydride | sec-Butyl alcohol | sec-Butyl acetate (60%) | Acetic acid |

PROBLEM 13.6

Acetylsalicylic acid is the chemical name for aspirin. Acetylsalicylic acid is the acetate ester of salicylic acid and is made by treating salicylic acid with acetic anhydride. Write a chemical equation for this reaction.

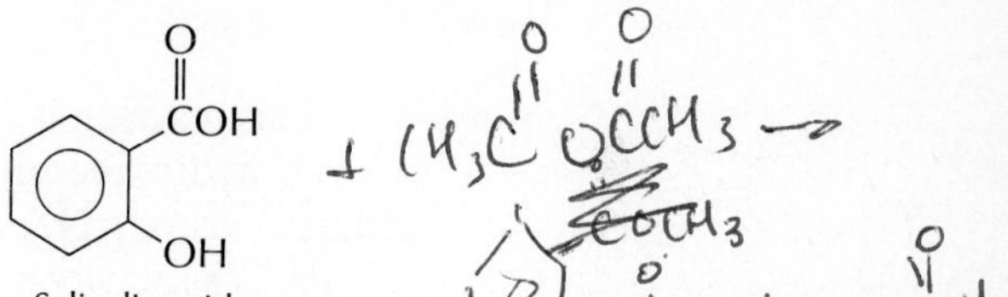

Salicylic acid

As you saw in Section 13.2, anhydrides are less reactive than acyl chlorides. Nevertheless, in spite of this lower reactivity, acetic anhydride is used more frequently than acetyl chloride for the preparation of acetate esters. Acetyl chloride is more expensive, and its high reactivity makes it difficult to handle in large amounts.

13.7 REACTIONS OF ESTERS. HYDROLYSIS

Esters are fairly stable in pure water. They must be since esters are such abundant components of natural products in aqueous environments. However, when an ester is heated with water in the presence of a strong acid or base, *hydrolysis* occurs (Section 13.2). Hydrolysis of an ester gives a carboxylic acid and an alcohol.

$$R\overset{\displaystyle O}{\overset{\|}{C}}OR' + H_2O \underset{}{\overset{H^+}{\rightleftharpoons}} R\overset{\displaystyle O}{\overset{\|}{C}}OH + R'OH$$

| Ester | Water | Carboxylic acid | Alcohol |

The mechanism of acid-catalyzed hydrolysis of an ester is exactly the reverse of that of acid-catalyzed esterification, described in Section 13.5. When esterification is the objective, water is removed from the reaction mixture to encourage ester formation. When ester hydrolysis is the objective, the reaction is carried out in the presence of a large excess of water.

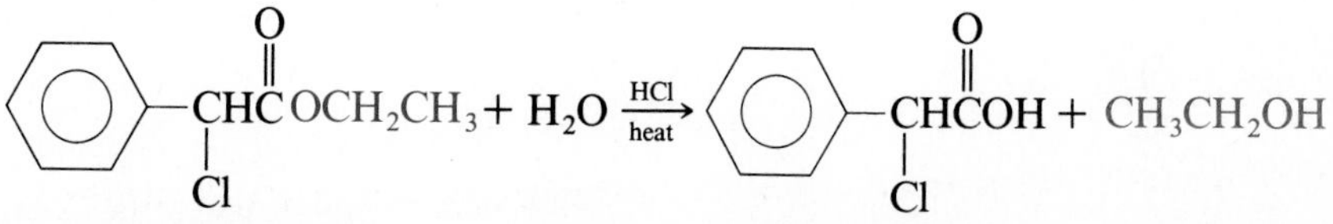

| Ethyl 2-chloro-2-phenylacetate | Water | 2-Chloro-2-phenyl-acetic acid (80%) | Ethanol |

PROBLEM 13.7

One component of beeswax is the ester triacontyl hexadecanoate. Write a balanced equation for the acid-catalyzed hydrolysis of this compound.

$$CH_3(CH_2)_{14}\overset{\overset{\displaystyle O}{\|}}{C}OCH_2(CH_2)_{28}CH_3$$

Triacontyl hexadecanoate

Unlike its acid-catalyzed counterpart, ester hydrolysis in an aqueous base is not reversible.

$$R\overset{\overset{\displaystyle O}{\|}}{C}OR' + NaOH \longrightarrow R\overset{\overset{\displaystyle O}{\|}}{C}O^- Na^+ + R'OH$$

| Ester | Sodium hydroxide | Carboxylate salt | Alcohol |

This reaction is called **base-promoted hydrolysis** rather than base-catalyzed because a hydroxide ion is consumed in the reaction. Since base-promoted hydrolysis of esters is irreversible and is usually faster than hydrolysis in acid, basic conditions are normally used when ester hydrolysis is part of an organic synthesis. The carboxylic acid product exists as a carboxylate salt at the basic pH of the reaction. Acidification of the reaction mixture after hydrolysis is complete converts the carboxylate salt to the free carboxylic acid.

$$CH_2{=}\overset{\overset{\displaystyle O}{\|}}{\underset{\underset{\displaystyle CH_3}{|}}{C}}OCH_3 \xrightarrow[\text{2. H}^+]{\text{1. NaOH, H}_2\text{O, heat}} CH_2{=}\overset{\overset{\displaystyle O}{\|}}{\underset{\underset{\displaystyle CH_3}{|}}{C}}OH + CH_3OH$$

Methyl 2-methyl-
propenoate
(methyl methacrylate)

2-Methyl-
propenoic acid
(methacrylic acid)

Methanol

Base-promoted ester hydrolysis is often called **saponification,** which means "soap-making." Over 2000 years ago, the Phoenicians made soap by heating animal fat with wood ashes. Animal fat contains glycerol triesters, and wood ashes are a source of potassium carbonate, a base. Base-promoted cleavage of the fats produces a mixture of long-chain carboxylic acids as their potassium salts and glycerol.

$$CH_3(CH_2)_x\overset{\overset{\displaystyle O}{\|}}{C}O\diagdown\diagup O\overset{\overset{\displaystyle O}{\|}}{C}(CH_2)_zCH_3$$

$$O\overset{\displaystyle |}{\underset{\underset{\displaystyle O}{\|}}{C}}(CH_2)_yCH_3$$

$$\xrightarrow[\text{heat}]{\text{K}_2\text{CO}_3, \text{H}_2\text{O}} HOCH_2\underset{\underset{\displaystyle OH}{|}}{C}HCH_2OH + \begin{array}{l} KO\overset{\overset{\displaystyle O}{\|}}{C}(CH_2)_xCH_3 \\ KO\overset{\overset{\displaystyle O}{\|}}{C}(CH_2)_yCH_3 \\ KO\overset{\overset{\displaystyle O}{\|}}{C}(CH_2)_zCH_3 \end{array}$$

Glycerol

Potassium
carboxylate salts
(soaps)

PROBLEM 13.8

Trimyristin is obtained from coconut oil and has the molecular formula $C_{45}H_{86}O_6$. On being heated with aqueous sodium hydroxide followed by acidification, trimyristin was converted to glycerol and tetradecanoic acid as the only products. What is the structure of trimyristin?

You learned in Chapter 12 that salts of long-chain carboxylic acids have cleansing properties by forming micelles that dissolve grease (Section 12.5). Soap is still made today by the saponification of fats and oils. The label on a bar of soap will often tell you the origin of the carboxylate salts. For example, if the soap contains "sodium tallowate," it was made from beef tallow (animal fat). The carboxylic acids obtained by saponification of fats and oils are called **fatty acids.**

The mechanism of base-promoted hydrolysis is another example of the addition-elimination process typical of acyl transfer reactions and is outlined in Figure 13.3. In the first step the carbonyl group of the ester undergoes nucleophilic addition by a hydroxide ion. A rapid proton transfer from water in step 2 gives the neutral tetrahedral intermediate.

Step 1: Nucleophilic addition of hydroxide ion to the carbonyl group

| Hydroxide ion | Ester | | Anionic form of tetrahedral intermediate |

Step 2: Proton transfer to anionic form of tetrahedral intermediate

| Anionic form of tetrahedral intermediate | Water | | Tetrahedral intermediate | Hydroxide ion |

Step 3: Dissociation of tetrahedral intermediate

| Hydroxide ion | Tetrahedral intermediate | | Water | Carboxylic acid | Alkoxide ion |

Step 4: Proton transfer steps yield an alcohol and a carboxylate anion

| Alkoxide ion | Water | | Alcohol | Hydroxide ion |

| Carboxylic acid (stronger acid) | Hydroxide ion (stronger base) | | Carboxylate ion (weaker base) | Water (weaker acid) |

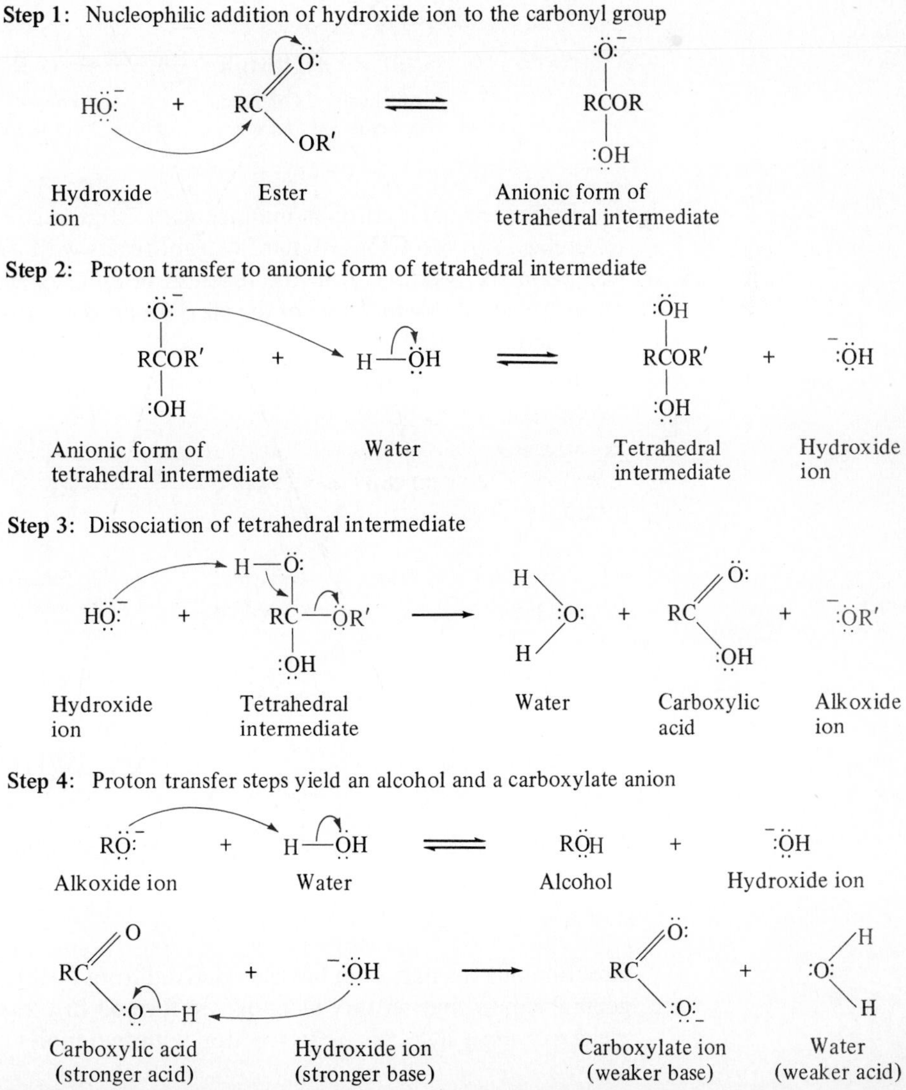

FIGURE 13.3 Sequence of steps that describes the mechanism of base-promoted ester hydrolysis.

Proton removal by a hydroxide ion leads to dissociation of the tetrahedral inter-mediate in step 3, eliminating an alkoxide ion. Lastly, in step 4, rapid proton transfer yields the alcohol and carboxylate ion products. All the steps are reversible except the last one. The equilibrium constant for proton abstraction from the carboxylic acid by hydroxide is so large that step 4 is, for all intents and purposes, irreversible, and this makes the overall reaction irreversible.

PROBLEM 13.9

Using the general mechanism for base-promoted ester hydrolysis as a guide, write an analogous sequence of steps for the saponification of ethyl benzoate.

13.8 REACTIONS OF ESTERS WITH GRIGNARD REAGENTS

In Chapter 11 you learned that Grignard reagents can be used to prepare alcohols by reaction with aldehydes or ketones (Sections 11.13 and 11.14).

$$\underset{\substack{\text{Aldehyde} \\ \text{or ketone}}}{\overset{\overset{\displaystyle O}{\|}}{RCR}} + \underset{\substack{\text{Grignard} \\ \text{reagent}}}{R'MgX} \xrightarrow[\text{2. } H_3O^+]{\text{1. diethyl ether}} \underset{\substack{\text{Primary, secondary,} \\ \text{or tertiary alcohol}}}{\overset{\overset{\displaystyle OH}{|}}{R_2CR'}}$$

Esters react in a similar manner with Grignard reagents with one important difference: *two* moles of Grignard reagent reacts with *each* mole of ester. With one exception the product from the reaction of an ester with a Grignard reagent is a tertiary alcohol in which two of the alkyl (or aryl) groups are the same, coming from the Grignard reagent.

$$\underset{\text{Ester}}{\overset{\overset{\displaystyle O}{\|}}{RCOR'}} + \underset{\substack{\text{Grignard} \\ \text{reagent}}}{2R''MgX} \xrightarrow[\text{2. } H_3O^+]{\text{1. diethyl ether}} \underset{\substack{\text{Tertiary} \\ \text{alcohol}}}{\overset{\overset{\displaystyle OH}{|}}{\underset{\underset{\displaystyle R''}{|}}{RCR''}}} + \underset{\text{Alcohol}}{R'OH}$$

$$2CH_3MgBr + \underset{\substack{\text{Methyl 2-methyl-} \\ \text{propanoate}}}{\overset{\overset{\displaystyle O}{\|}}{(CH_3)_2CHCOCH_3}} \xrightarrow[\text{2. } H_3O^+]{\substack{\text{1. diethyl} \\ \text{ether}}} \underset{\substack{\text{2,3-Dimethyl-} \\ \text{2-butanol (73\%)}}}{\overset{\overset{\displaystyle OH}{|}}{\underset{\underset{\displaystyle CH_3}{|}}{(CH_3)_2CHCCH_3}}} + \underset{\text{Methanol}}{CH_3OH}$$

Methylmag-nesium bromide

Reaction of a formate ester (an ester derived from formic acid) is an exception to the generalization that tertiary alcohols are formed in these reactions. Formate esters yield *secondary* alcohols upon reaction with two moles of a Grignard reagent.

AN ESTER OF AN INORGANIC ACID: NITROGLYCERIN

In 1847 an Italian scientist, Ascanio Sobrero, discovered that glycerol, obtained as a by-product from the manufacture of soap, could be converted into a nitrate triester called nitroglycerin.

$$\underset{\underset{\text{OH}}{|}}{\text{HOCH}_2\text{CHCH}_2\text{OH}} + 3\text{HONO}_2 \longrightarrow \underset{\underset{\text{ONO}_2}{|}}{\text{O}_2\text{NOCH}_2\text{CHCH}_2\text{ONO}_2} + 3\text{H}_2\text{O}$$

Glycerol   Nitric acid   Glycerol trinitrate
(nitroglycerin)

Nitroglycerin is an explosive and extremely unstable liquid, and for years this property made it dangerous to manufacture and use. Alfred Nobel, a Swedish inventor, discovered in 1867 that the explosive power of nitroglycerin could be controlled by absorbing the liquid on an inert powder—producing **dynamite.**

When Nobel died, he left his entire fortune to a foundation that he established in his will. The foundation was to give an award each year, recognizing outstanding accomplishments toward world peace and in the fields of chemistry, physics, literature, and physiology or medicine. These awards are what we know as the Nobel Prizes. The Nobel Peace Prize is presented each year in Oslo, Norway; the other Prizes are presented in Stockholm, Sweden. Since 1969 the Central Bank of Sweden has also awarded a Nobel Memorial Prize in Economic Science. The income from the trust fund established by Alfred Nobel is the source of the monetary award that accompanies each Nobel Prize except the economics award.

Nitroglycerin has another, very different, use. Nitroglycerin is a blood vessel dilator and is used to treat a type of heart pain called angina. Nitroglycerin medication is absorbed rapidly, and the pill, which contains less than a milligram of nitroglycerin, is placed under the tongue for almost immediate relief of the angina symptoms.

$$\underset{\text{Formate}\atop\text{ester}}{\overset{\overset{\text{O}}{\|}}{\text{HCOR}}} + 2R'\text{MgX} \xrightarrow[\text{2. H}_3\text{O}^+]{\text{1. diethyl ether}} \underset{\text{Secondary}\atop\text{alcohol}}{\underset{\underset{\text{R}'}{|}}{\overset{\overset{\text{OH}}{|}}{\text{HCR}'}}}$$

Formate ester   Grignard reagent   Secondary alcohol

PROBLEM 13.10

What combination of ester and Grignard reagent could you use to prepare each of the following alcohols?

(a) 3-Methyl-3-pentanol   (c) $(C_6H_5)_2\text{CHOH}$
(b) 6-Methyl-6-undecanol   (d) $(C_6H_5)_3\text{COH}$

SAMPLE SOLUTION

(a) The carbon that bears the hydroxyl substituent has two ethyl groups and a methyl group attached to it.

$$CH_3CH_2\overset{\overset{\displaystyle CH_3}{|}}{\underset{\underset{\displaystyle OH}{|}}{C}}CH_2CH_3 \qquad \text{3-Methyl-3-pentanol}$$

The two groups that are the same, ethyl in this case, come from the Grignard reagent. Therefore, use an ethyl Grignard reagent and an acetate ester.

$$2CH_3CH_2MgX + CH_3\overset{\overset{\displaystyle O}{\|}}{C}OR \xrightarrow[\text{2. H}_3\text{O}^+]{\text{1. diethyl ether}} CH_3\overset{\overset{\displaystyle OH}{|}}{C}(CH_2CH_3)_2$$

Ethylmagnesium   Acetate   3-Methyl-3-pentanol
halide   ester

Appropriate choices would be, for example, ethylmagnesium bromide and ethyl acetate.

Esters react with Grignard reagents in two stages. The first mole of Grignard reagent reacts with the ester to form a ketone.

$$R\overset{\overset{\displaystyle O}{\|}}{C}OR' + R''MgX \xrightarrow[\text{ether}]{\text{diethyl}} R\overset{\overset{\displaystyle O^-\ ^+MgX}{|}}{\underset{\underset{\displaystyle R''}{|}}{C}}OR' \longrightarrow R\overset{\overset{\displaystyle O}{\|}}{C}R'' + R'OMgX$$

Ester   Grignard   Ketone   Alkoxy mag-
           reagent   nesium halide

The ketone then reacts rapidly with the second mole of Grignard reagent giving, after hydrolysis of the magnesium salt, the alcohol product.

$$R\overset{\overset{\displaystyle O}{\|}}{C}R'' + R''MgX \xrightarrow[\text{2. H}_3\text{O}^+]{\text{1. diethyl ether}} R\overset{\overset{\displaystyle OH}{|}}{\underset{\underset{\displaystyle R''}{|}}{C}}R''$$

Ketone   Grignard   Tertiary
            reagent   alcohol

13.9 POLYESTERS

The fibers used in clothing are all made from polymers, either natural or synthetic. The word **polymer** is from the Greek terms *poly* and *meros,* meaning "many parts." A polymeric chemical has a structural unit which repeats many times in each molecule. For instance, cellulose, which is a polymer of the sugar glucose, is the main component of cotton fiber. The structure of cellulose will be discussed in Chapter 15.

During the 1940s efforts were made to find a cotton substitute which could be made from cheap, readily available raw materials. The answer was **polyester,** a fiber material which we know by trade names such as Dacron or Fortrel. Polyester is a polymer whose repeating units are esters.

To form a polyester, a dicarboxylic acid is condensed with a diol. A typical polyester material is made by reaction of terephthalic acid (*p*-benzenedicarboxylic acid) with ethylene glycol.

$$n\text{HOCH}_2\text{CH}_2\text{OH} + n\text{HO}_2\text{C}-\bigcirc-\text{CO}_2\text{H} \xrightarrow{\text{catalyst}}$$

Ethylene   Terephthalic acid
glycol

$$\left[-\text{OCH}_2\text{CH}_2\text{OC}-\bigcirc-\overset{O}{\overset{\|}{\text{C}}}-\right]_n + n\text{H}_2\text{O}$$

Polyester

Polyesters are used by themselves or as cotton blends in many types of clothing. Polyesters are also made into plastic films (Mylar is an example) and used in consumer goods such as audio cassettes and VCR tapes.

PROBLEM 13.11

Hard plastic materials called polycarbonates are used to make eyeglass lenses, shatter-proof windows, and other glass-substitute products. One type of polycarbonate, called Lexan, is made by reaction of phosgene with a diphenol known by the common name bisphenol-A. Write the structure of the repeating unit of this polymer.

$$\overset{O}{\overset{\|}{\text{ClCCl}}} + \text{HO}-\bigcirc-\overset{\overset{\text{CH}_3}{|}}{\underset{\overset{|}{\text{CH}_3}}{\text{C}}}-\bigcirc-\text{OH} \longrightarrow \text{Lexan}$$

Phosgene   Bisphenol-A

13.10 NATURAL SOURCES OF AMIDES

As with esters, a large number of amides are naturally occurring substances. Several amides which have antibacterial properties and are important as *antibiotics* are shown in Figure 13.4. The penicillins are the most widely used antibiotics. For some types of bacterial infections and for patients who have adverse reactions to penicillin therapy, the cephalosporins and tetracyclines are used.

Amides occur in all biological systems as the backbone of peptides and proteins. We will discuss these types of substances in Chapter 16.

13.11 PREPARATION OF AMIDES

Several methods are available for the preparation of amides. Reaction of carboxylic acids and amines yields amides in a two-step process. The first step is an acid-base reaction in which the acid and the amine combine to form an ammonium salt. On heating, the ammonium carboxylate salt loses water to form an amide.

$$\overset{O}{\overset{\|}{\text{RCOH}}} + \text{R}_2'\text{NH} \longrightarrow \overset{O}{\overset{\|}{\text{RCO}^-}} \text{R}_2'\overset{+}{\text{NH}}_2 \xrightarrow{\text{heat}} \overset{O}{\overset{\|}{\text{RCNR}_2'}} + \text{H}_2\text{O}$$

Carboxylic   Amine   Ammonium   Amide   Water
acid   carboxylate salt

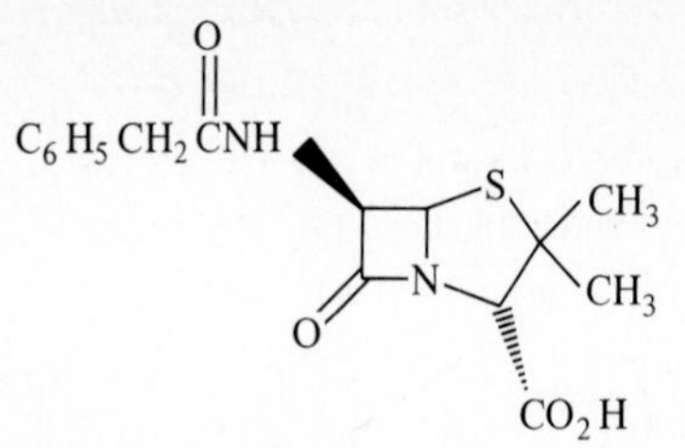

Penicillin G

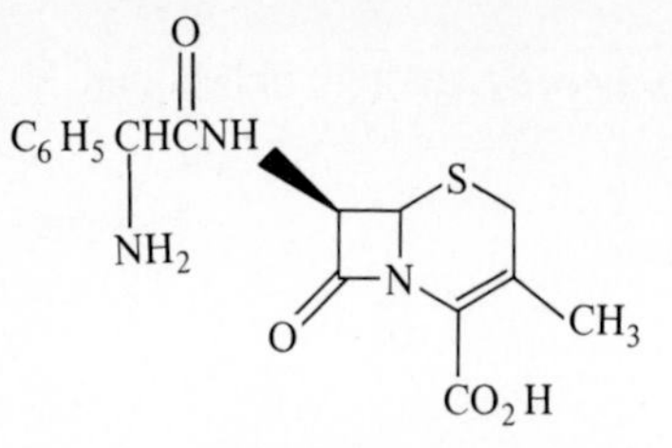

Cephalexin
(Keflex, a cephalosporin)

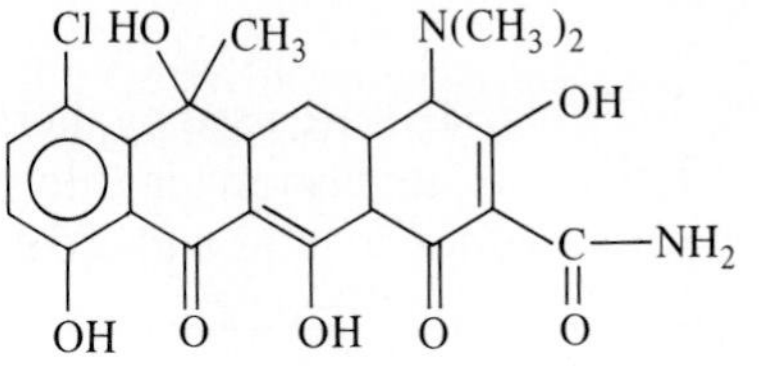

Chlortetracycline
(Aureomycin)

FIGURE 13.4 Amides which are used as antibiotics.

In practice, both steps may be combined in a single operation by heating a carboxylic acid and an amine together.

$$C_6H_5\overset{O}{\overset{\|}{C}}OH + C_6H_5NH_2 \xrightarrow{225°C} C_6H_5\overset{O}{\overset{\|}{C}}NHC_6H_5 + H_2O$$

Benzoic acid   Aniline   N-Phenylbenz-   Water
                                amide (80%)
                                (benzanilide)

Amides may also be prepared by the reaction of ammonia or an amine with an acyl chloride. Two molar equivalents of amine (or ammonia) are required in the reaction with acyl chlorides. One molecule of amine acts as a nucleophile, while the second acts as a Brönsted base and converts HCl formed in the reaction to a salt.

$$2R_2NH + R'\overset{O}{\overset{\|}{C}}Cl \longrightarrow R'\overset{O}{\overset{\|}{C}}NR_2 + R_2\overset{+}{N}H_2\ Cl^-$$

Amine   Acyl   Amide   Hydrochloride
           chloride   salt of amine

$$C_6H_5\overset{O}{\overset{\|}{C}}Cl + 2NH_3 \longrightarrow C_6H_5\overset{O}{\overset{\|}{C}}NH_2 + NH_4Cl$$

Benzoyl   Ammonia   Benzamide   Ammonium
chloride   chloride

The choice of ammonia or an amine determines the substitution at nitrogen in the amide product. This can be summarized as:

Ammonia (NH_3) yields an amide, $R'\overset{O}{\overset{\|}{C}}NH_2$

Primary amines (RNH_2) yield N-substituted amides, $R'\overset{O}{\overset{\|}{C}}NHR$

Secondary amines (R_2NH) yield N,N-disubstituted amides, $R'\overset{O}{\overset{\|}{C}}NR_2$

PROBLEM 13.12

Write an equation showing the preparation of the following amides from the reaction of an acyl chloride with the appropriate amine or ammonia.

$$
\text{(a) } (CH_3)_2CHCNH_2 \qquad \text{(b) } CH_3CNHCH_3 \qquad \text{(c) } C_6H_5CN(CH_3)_2
$$

(with C=O at the carbonyl carbon in each)

SAMPLE SOLUTION

(a) Amides of the type $RCNH_2$ (with C=O) are prepared by reacting an acyl chloride with ammonia.

$$
(CH_3)_2CHCCl + 2NH_3 \longrightarrow (CH_3)_2CHCNH_2 + NH_4Cl
$$

| 2-Methylpropanoyl chloride | Ammonia | 2-Methylpropan- amide | Ammonium chloride |

Two molecules of ammonia are needed because reaction with an acyl chloride produces, in addition to the desired amide, a molecule of hydrogen chloride. An acid-base reaction between hydrogen chloride and ammonia gives the salt ammonium chloride.

Amides in which the acyl portion is derived from acetic acid are called **acetamides**. Acetamides are frequently prepared by reaction of the appropriate amine with acetic anhydride.

$$
CH_3COCCH_3 + 2C_6H_5NH_2 \longrightarrow C_6H_5NHCCH_3 + C_6H_5\overset{+}{N}H_3 \ ^-OCCH_3
$$

| Acetic anhydride | Aniline | N-Phenylacet- amide (acetanilide) | Anilinium acetate |

As in the reaction between amines and acyl chlorides, two moles of amine are used in the preparation of acetamides using acetic anhydride. The second mole of amine forms a salt with the acetic acid produced in the reaction.

Esters may function as acyl transfer agents to the nitrogen of ammonia to form amides.

$$
RCOR' + NH_3 \longrightarrow RCNH_2 + R'OH
$$

| Ester | Ammonia | Amide | Alcohol |

Ammonia is more nucleophilic than water, and thus this reaction can be carried out in aqueous solution.

$$
CH_2{=}\underset{\underset{CH_3}{|}}{C}COCH_3 + NH_3 \xrightarrow{\ H_2O\ } CH_2{=}\underset{\underset{CH_3}{|}}{C}CNH_2 + CH_3OH
$$

| Methyl 2-methyl- propenoate | Ammonia | 2-Methylprop- enamide | Methanol |

Amines, which are substituted derivatives of ammonia, react similarly.

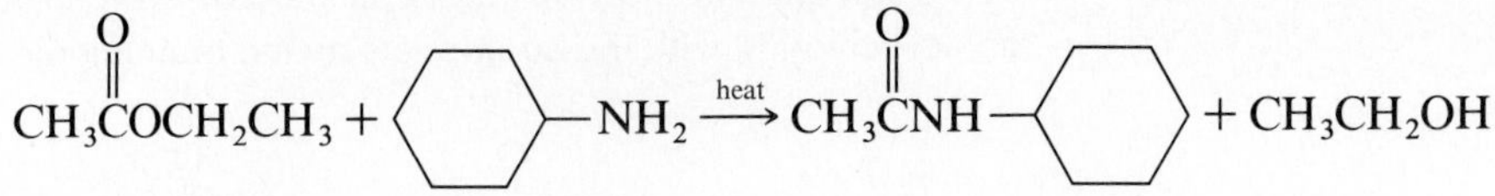

$$CH_3\overset{\overset{\displaystyle O}{\|}}{C}OCH_2CH_3 + \bigcirc\!\!-NH_2 \xrightarrow{\text{heat}} CH_3\overset{\overset{\displaystyle O}{\|}}{C}NH\!\!-\!\!\bigcirc + CH_3CH_2OH$$

| Ethyl acetate | Cyclohexylamine | N-Cyclohexyl-
acetamide | Ethanol |

PROBLEM 13.13

Give the structure of the expected product of the following reaction.

$(CH_3)_2CH\overset{\overset{\displaystyle O}{\|}}{C}NHCH_3$

$$CH_3NH_2 + (CH_3)_2CH\overset{\overset{\displaystyle O}{\|}}{C}OCH_2CH_3 \longrightarrow$$

The amine used to prepare an amide by any of the methods presented in this section must be primary (RNH_2) or secondary (R_2NH). Tertiary amines (R_3N) cannot form amides because they have no proton on nitrogen that can be replaced by an acyl group.

13.12 HYDROLYSIS OF AMIDES

Recall (Section 13.2) that amides are the least reactive of the carboxylic acid derivatives. As a result, the only nucleophilic acyl substitution that amides undergo is hydrolysis. Amides are fairly stable in water, but the amide bond is cleaved on heating in the presence of strong acids or bases.

In acid solution, the products of hydrolysis are the carboxylic acid and the protonated amine, an ammonium ion.

$$R\overset{\overset{\displaystyle O}{\|}}{C}NR'_2 + H_3O^+ \longrightarrow R\overset{\overset{\displaystyle O}{\|}}{C}OH + R'_2\overset{+}{N}H_2$$

| Amide | Hydronium
ion | Carboxylic
acid | Ammonium
ion |

In basic solution, the carboxylic acid is deprotonated, forming the carboxylate ion.

$$R\overset{\overset{\displaystyle O}{\|}}{C}NR'_2 + HO^- \longrightarrow R\overset{\overset{\displaystyle O}{\|}}{C}O^- + R'_2NH$$

| Amide | Hydroxide
ion | Carboxylate
ion | Amine |

The acid-base reactions that occur after the amide bond is broken render the overall hydrolysis irreversible in both cases. The amine product is protonated in acid; the carboxylic acid product is deprotonated in base.

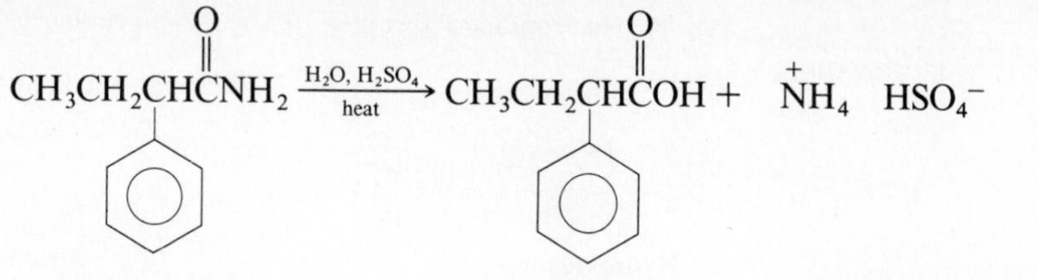

$$\underset{\text{2-Phenylbutanamide}}{CH_3CH_2CHCNH_2} \xrightarrow[\text{heat}]{H_2O, H_2SO_4} \underset{\substack{\text{2-Phenylbutanoic} \\ \text{acid} \\ (88-90\%)}}{CH_3CH_2CHCOH} + \underset{\substack{\text{Ammonium hydrogen} \\ \text{sulfate}}}{\overset{+}{N}H_4 \quad HSO_4^-}$$

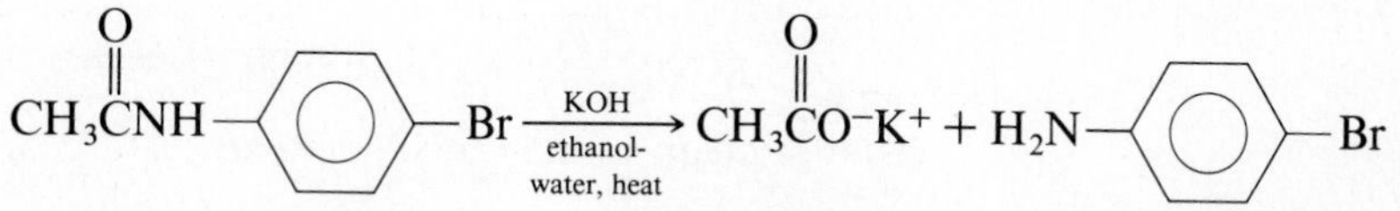

$$\underset{\substack{N\text{-(4-Bromophenyl)acetamide} \\ (p\text{-bromoacetanilide})}}{CH_3CNH \!-\!\! \bigcirc \!\!-\! Br} \xrightarrow[\substack{\text{ethanol-} \\ \text{water, heat}}]{KOH} \underset{\substack{\text{Potassium} \\ \text{acetate}}}{CH_3CO^-K^+} + \underset{p\text{-Bromoaniline (95\%)}}{H_2N \!-\!\! \bigcirc \!\!-\! Br}$$

PROBLEM 13.14

Write a chemical equation for the hydrolysis of each of the following amides, using the conditions indicated.

$$\text{(a)} \quad \underset{}{(CH_3)_2CHCNH_2} \text{ (heat in aqueous hydrochloric acid)}$$

$$\text{(b)} \quad CH_3CNHCH_3 \text{ (heat in aqueous sodium hydroxide)}$$

$$\text{(c)} \quad C_6H_5CNHC_6H_5 \text{ (heat in aqueous hydrochloric acid)}$$

SAMPLE SOLUTION

(a) Hydrolysis under acidic conditions yields the carboxylic acid and the salt of the amine. Thus hydrolysis of 2-methylpropanamide in aqueous hydrochloric acid yields propanoic acid and ammonium chloride

$$\underset{\text{2-Methylpropanamide}}{(CH_3)_2CHCNH_2} + \underset{\text{Water}}{H_2O} \xrightarrow[\text{heat}]{HCl} \underset{\text{2-Methylpropanoic acid}}{(CH_3)_2CHCOH} + \underset{\substack{\text{Ammonium} \\ \text{chloride}}}{\overset{+}{N}H_4 \ Cl^-}$$

The mechanism of amide hydrolysis is similar to the mechanism of ester hydrolysis described in Section 13.7. The steps which describe base-promoted amide hydrolysis are shown in Figure 13.5.

One other important reaction of amides, reduction to give an amine, will be presented in the next chapter.

13.13 POLYAMIDES. NYLON

Polymers of amides, polyamides, can have either a natural or a synthetic origin. Peptides and proteins are natural polyamides found in all living systems (Chapter 16).

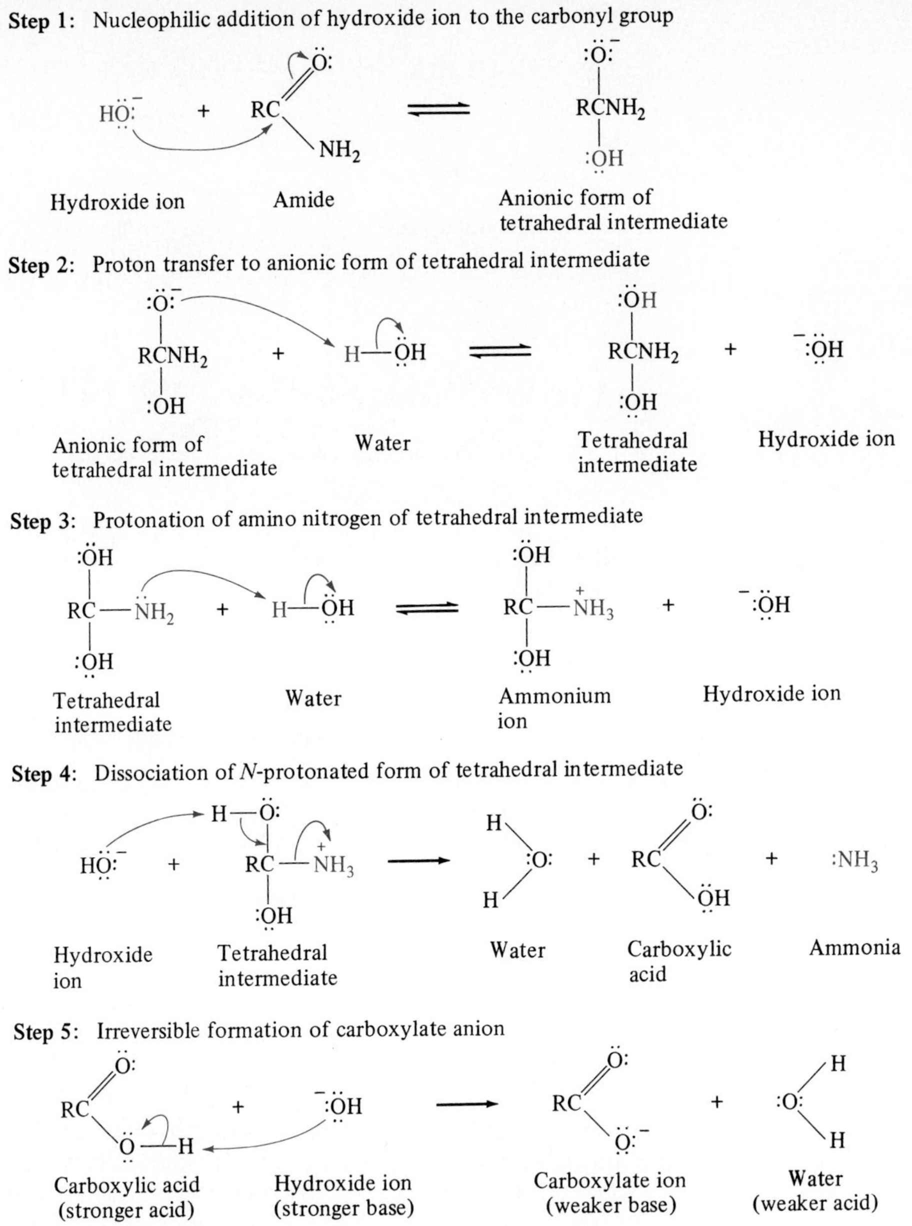

Step 1: Nucleophilic addition of hydroxide ion to the carbonyl group

$$HO^- \; + \; RC\overset{O}{\underset{NH_2}{\big\langle}} \; \rightleftharpoons \; RC\underset{OH}{\overset{O^-}{\underset{|}{\overset{|}{NH_2}}}}$$

Hydroxide ion   Amide   Anionic form of
tetrahedral intermediate

Step 2: Proton transfer to anionic form of tetrahedral intermediate

$$\underset{OH}{\overset{O^-}{\underset{|}{\overset{|}{RCNH_2}}}} \; + \; H{-}OH \; \rightleftharpoons \; \underset{OH}{\overset{OH}{\underset{|}{\overset{|}{RCNH_2}}}} \; + \; {}^-OH$$

Anionic form of   Water   Tetrahedral   Hydroxide ion
tetrahedral intermediate   intermediate

Step 3: Protonation of amino nitrogen of tetrahedral intermediate

$$\underset{OH}{\overset{OH}{\underset{|}{\overset{|}{RC{-}NH_2}}}} \; + \; H{-}OH \; \rightleftharpoons \; \underset{OH}{\overset{OH}{\underset{|}{\overset{|}{RC{-}\overset{+}{N}H_3}}}} \; + \; {}^-OH$$

Tetrahedral   Water   Ammonium   Hydroxide ion
intermediate   ion

Step 4: Dissociation of *N*-protonated form of tetrahedral intermediate

$$HO^- \; + \; \underset{OH}{\overset{H{-}O}{\underset{|}{\overset{|}{RC{-}\overset{+}{N}H_3}}}} \; \longrightarrow \; \overset{H}{\underset{H}{\big\rangle}}O \; + \; RC\overset{O}{\underset{OH}{\big\langle}} \; + \; NH_3$$

Hydroxide   Tetrahedral   Water   Carboxylic   Ammonia
ion   intermediate   acid

Step 5: Irreversible formation of carboxylate anion

$$RC\overset{O}{\underset{O{-}H}{\big\langle}} \; + \; {}^-OH \; \longrightarrow \; RC\overset{O}{\underset{O^-}{\big\langle}} \; + \; \overset{H}{\underset{H}{\big\rangle}}O$$

Carboxylic acid   Hydroxide ion   Carboxylate ion   Water
(stronger acid)   (stronger base)   (weaker base)   (weaker acid)

FIGURE 13.5 Sequence of steps that describes the mechanism of base-promoted amide hydrolysis.

The first synthetic polyamide was made in 1935 by a group of chemists at the Du Pont Company. These chemists condensed a diacid (adipic acid) with a diamine (hexamethylenediamine) to give the material known as nylon.

$$n\mathrm{H_2N(CH_2)_6NH_2} \; + \; n\mathrm{HO_2C(CH_2)_4CO_2H} \longrightarrow$$

Hexamethylenediamine   Adipic acid

$$\left[-\mathrm{NH(CH_2)_6NH\overset{O}{\overset{\|}{C}}(CH_2)_4\overset{O}{\overset{\|}{C}}} \right]_n + n\mathrm{H_2O}$$

Nylon

A second method of nylon preparation involves self-polymerization of caprolactam, a cyclic amide.

Caprolactam   Nylon

Nylon is a popular clothing fiber, as it makes an excellent silk substitute. Nylon can also be fabricated as a plastic and is used for many small machine parts such as gears and bearings.

13.14 SPECTROSCOPIC ANALYSIS OF CARBOXYLIC ACID DERIVATIVES

Infrared spectroscopy has been quite useful in determining structures of carboxylic acid derivatives. The carbonyl band is quite strong, and its position is sensitive to the substituents bound to the carbonyl group. As these examples illustrate, the characteristic absorption frequencies of the various acid derivatives vary considerably. Anhydrides have two carbonyl bands from interaction of the vibrations of the two carbonyl groups.

$$CH_3CCl \qquad CH_3COCCH_3 \qquad CH_3COCH_3 \qquad CH_3CNH_2$$

Acetyl chloride

Acetic anhydride

Methyl acetate

Acetamide

$v_{C=O} = 1822 \text{ cm}^{-1}$

$v_{C=O} = 1748 \text{ cm}^{-1}$ and 1815 cm^{-1}

$v_{C=O} = 1736 \text{ cm}^{-1}$

$v_{C=O} = 1694 \text{ cm}^{-1}$

Chemical shift data in ^1H nmr spectroscopy permit assignment of structure in esters. Consider two isomeric esters, ethyl acetate and methyl propanoate. Both have a methyl singlet and a triplet-quartet pattern for their ethyl group. Notice, however, that there is a significant difference in the chemical shifts of the corresponding signals in each spectrum. The methyl is more shielded ($\delta = 2.0$ ppm) when it is bonded directly to the carbonyl group, as in ethyl acetate, than when it is bonded to the oxygen of methyl propanoate ($\delta = 3.6$ ppm). Likewise, the methylene group is more shielded in methyl propanoate when it is bonded to the carbonyl group ($\delta = 2.3$ ppm) than when bonded to oxygen in ethyl acetate ($\delta = 4.1$ ppm).

Singlet
$\delta = 2.0$ ppm

Quartet
$\delta = 4.1$ ppm

$$CH_3COCH_2CH_3$$

Triplet
$\delta = 1.3$ ppm

Ethyl acetate

Singlet
$\delta = 3.6$ ppm

Quartet
$\delta = 2.3$ ppm

$$CH_2OCCH_2CH_3$$

Triplet
$\delta = 1.2$ ppm

Methyl propanoate

13.15 SUMMARY

The four major classes of carboxylic acids derivatives are acyl chlorides, anhydrides, esters, and amides (Section 13.1). Acyl chlorides are named by replacing the *-ic acid* ending of the corresponding carboxylic acid name with *-yl chloride.* In a similar manner, anhydrides are named by replacing the word *acid* with *anhydride.*

$$\underset{\text{Propanoyl chloride}}{CH_3CH_2\overset{\displaystyle O}{\overset{\|}{C}}Cl} \qquad \underset{\text{Propanoic anhydride}}{CH_3CH_2\overset{\displaystyle O}{\overset{\|}{C}}O\overset{\displaystyle O}{\overset{\|}{C}}CH_2CH_3}$$

The alkyl portion of an ester is named first, followed by the acyl portion in which the -ic ending of the corresponding acid has been replaced by -ate.

$$\underset{\text{Propyl butanoate}}{CH_3CH_2CH_2\overset{\displaystyle O}{\overset{\|}{C}}OCH_2CH_2CH_3}$$

Amides are named by replacing the -oic or -ic ending of the acid name with -amide. Amides substituted on nitrogen are named as N-alkyl or N,N-dialkyl derivatives of the parent amide.

$$\underset{\text{Propanamide}}{CH_3CH_2\overset{\displaystyle O}{\overset{\|}{C}}NH_2} \qquad \underset{\text{N-Phenylpropanamide}}{CH_3CH_2\overset{\displaystyle O}{\overset{\|}{C}}NHC_6H_5}$$

The characteristic reaction of carboxylic acid derivatives is *nucleophilic acyl substitution,* or *acyl transfer* (Section 13.2). This nucleophilic substitution process occurs by a pathway involving addition of a nucleophile to the carbonyl carbon atom to form a *tetrahedral intermediate.* Subsequent elimination of a leaving group gives the acyl derivative product.

$$\underset{\substack{\text{Carboxylic} \\ \text{acid derivative}}}{R\overset{\displaystyle O}{\overset{\|}{C}}-X} + \underset{\text{Nucleophile}}{HY} \rightleftharpoons \underset{\substack{\text{Tetrahedral} \\ \text{intermediate}}}{R\overset{\displaystyle OH}{\overset{|}{\underset{\displaystyle Y}{\underset{|}{C}}}}-X} \rightleftharpoons \underset{\substack{\text{Product of} \\ \text{nucleophilic} \\ \text{acyl substitution}}}{R\overset{\displaystyle O}{\overset{\|}{C}}-Y} + \underset{\substack{\text{Conjugate acid} \\ \text{of leaving} \\ \text{group}}}{HX}$$

Each of the derivatives — acyl chlorides, anhydrides, esters, and amides — may be converted to a carboxylic acid by hydrolysis.

$$\underset{\substack{\text{Carboxylic} \\ \text{acid derivative}}}{R\overset{\displaystyle O}{\overset{\|}{C}}X} + \underset{\text{Water}}{H_2O} \longrightarrow \underset{\substack{\text{Carboxylic} \\ \text{acid}}}{R\overset{\displaystyle O}{\overset{\|}{C}}OH} + \underset{\substack{\text{Conjugate acid} \\ \text{of leaving group}}}{HX}$$

Acyl chlorides and anhydrides are the most reactive acyl derivatives; amides are the least reactive.

Esters may be prepared from the reaction of a carboxylic acid with an alcohol in the presence of an acid catalyst. The reaction is known as *Fischer esterification* (Section 13.4).

$$\underset{\substack{\text{Carboxylic}\\ \text{acid}}}{R\overset{O}{\overset{\|}{C}}OH} + \underset{\text{Alcohol}}{R'OH} \underset{}{\overset{H^+}{\rightleftharpoons}} \underset{\text{Ester}}{R\overset{O}{\overset{\|}{C}}OR'} + \underset{\text{Water}}{H_2O}$$

$$\underset{\substack{\text{Acetic}\\ \text{acid}}}{CH_3\overset{O}{\overset{\|}{C}}OH} + \underset{\text{1-Pentanol}}{CH_3CH_2CH_2CH_2CH_2OH} \overset{H^+}{\rightleftharpoons}$$

$$\underset{\text{1-Pentyl acetate (71\%)}}{CH_3\overset{O}{\overset{\|}{C}}OCH_2CH_2CH_2CH_2CH_3} + \underset{\text{Water}}{H_2O}$$

The mechanism of Fischer esterification illustrates the mechanistic principles of nucleophilic acyl substitution reactions (Section 13.5). Protonation of the carbonyl oxygen by the acid catalyst is followed by nucleophilic attack on the carbonyl carbon by the alcohol oxygen atom. The *tetrahedral intermediate* (shown in the box in the following equation) is not stable, but dissociates to give the ester product and a molecule of water.

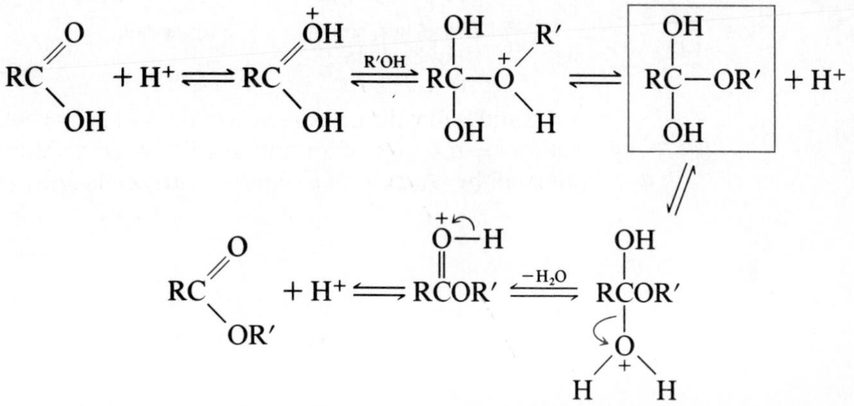

Esters may also be prepared by the reaction of an acyl chloride with an alcohol in the presence of a base such as pyridine (Section 13.6).

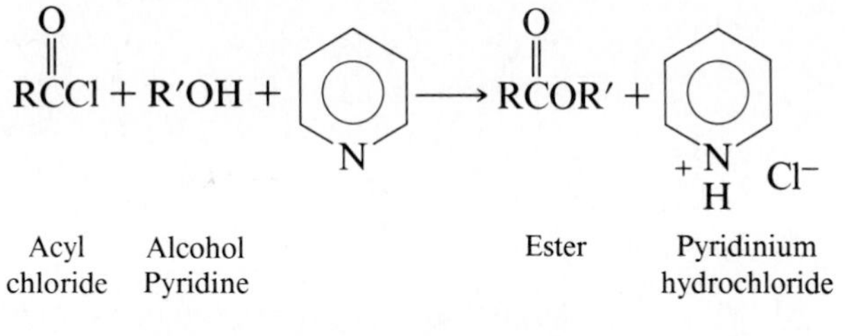

Acetate esters are frequently prepared by reaction of acetic anhydride with an alcohol.

$$\underset{\substack{\text{Acetic}\\ \text{anhydride}}}{2CH_3\overset{O}{\overset{\|}{C}}O\overset{O}{\overset{\|}{C}}CH_3} + \underset{\text{Alcohol}}{ROH} \longrightarrow \underset{\substack{\text{Acetate}\\ \text{ester}}}{CH_3\overset{O}{\overset{\|}{C}}OR} + \underset{\substack{\text{Acetic}\\ \text{acid}}}{CH_3\overset{O}{\overset{\|}{C}}OH}$$

A principal reaction of esters is hydrolysis (Section 13.7). Under acidic conditions, hydrolysis proceeds by a mechanism that is the reverse of Fischer esterification. Base-promoted hydrolysis, also known as *saponification,* is not reversible due to carboxylate salt formation.

Acidic hydrolysis:

$$\underset{\text{}}{\overset{\overset{\displaystyle O}{\|}}{RCOR'}} + H_2O \underset{\longleftarrow}{\overset{H^+}{\rightleftharpoons}} \overset{\overset{\displaystyle O}{\|}}{RCOH} + R'OH$$

Basic hydrolysis:

$$\overset{\overset{\displaystyle O}{\|}}{RCOR'} + HO^- \xrightarrow{H_2O} \overset{\overset{\displaystyle O}{\|}}{RCO^-} + R'OH$$

Alcohols in which two of the substituents are the same may be prepared by reaction of an ester with two moles of Grignard reagent (Section 13.8).

$$2C_6H_5MgBr + \overset{\overset{\displaystyle O}{\|}}{CH_3COCH_2CH_3} \xrightarrow[\text{2. }H_3O^+]{\text{1. diethyl ether}} (C_6H_5)_2\overset{\overset{\displaystyle OH}{|}}{CCH_3}$$

Phenylmagnesium   Ethyl acetate   1,1-Diphenyl-
bromide   ethanol

Amides may be prepared (Section 13.11) by heating an amine with a carboxylic acid or by reaction of an amine with an acyl chloride. Acetamides are frequently prepared by reaction of an amine with acetic anhydride.

$$R_2NH + \overset{\overset{\displaystyle O}{\|}}{R'COH} \xrightarrow{\text{heat}} \overset{\overset{\displaystyle O}{\|}}{R_2NCR'} + H_2O$$

Amine   Carboxylic   Amide   Water
         acid

$$2R_2NH + \overset{\overset{\displaystyle O}{\|}}{R'CCl} \longrightarrow \overset{\overset{\displaystyle O}{\|}}{R_2NCR'} + R_2\overset{+}{N}H_2\ Cl^-$$

Amine   Acyl   Amide   Ammonium salt
       chloride

$$\overset{\overset{\displaystyle O\ \ \ O}{\|\ \ \ \|}}{CH_3COCCH_3} + 2\ RNH_2 \longrightarrow \overset{\overset{\displaystyle O}{\|}}{CH_3CNHR} + \overset{\overset{\displaystyle O}{\|}}{CH_3CO^-}\ H_3\overset{+}{N}R$$

Acetic   Amine   Acetamide   Ammonium
anhydride   acetate salt

Amides undergo irreversible hydrolysis under acidic or basic conditions (Section 13.12). In acid solution, the products of hydrolysis are the carboxylic acid and the protonated amine, an ammonium ion.

$$\overset{\overset{\displaystyle O}{\|}}{RCNR'_2} + H_3O^+ \longrightarrow \overset{\overset{\displaystyle O}{\|}}{RCOH} + R'_2\overset{+}{N}H_2$$

Amide   Hydronium   Carboxylic   Ammonium
        ion   acid   ion

In basic solution, the carboxylic acid is deprotonated, forming the carboxylate ion.

$$\underset{\text{Amide}}{\text{RC}\overset{\displaystyle O}{\overset{\|}{}}\text{NR}_2'} + \underset{\substack{\text{Hydroxide}\\ \text{ion}}}{\text{HO}^-} \longrightarrow \underset{\substack{\text{Carboxylate}\\ \text{ion}}}{\text{RC}\overset{\displaystyle O}{\overset{\|}{}}\text{O}^-} + \underset{\text{Amine}}{\text{R}_2'\text{NH}}$$

Carboxylic acid derivatives exhibit characteristic carbonyl absorption frequencies in the infrared region (Section 13.14). Chemical shift data in ^{1}H nmr spectroscopy may be used to distinguish between ester isomers.

ADDITIONAL PROBLEMS

Nomenclature

13.15 Write a structural formula for each of the following compounds:

(a) m-Chlorobenzoyl chloride

(b) 4-Methylpentanoyl chloride

(c) Trifluoroacetic anhydride

(d) 1-Phenylethyl acetate

(e) Butyl 2-methylbutanoate

(f) N-Ethylbenzamide

(g) N,N-Diphenylacetamide

13.16 Give an acceptable systematic (IUPAC) name for each of the following compounds:

(a) $\underset{\overset{\displaystyle |}{\text{Cl}}}{\text{CH}_3\text{CHCH}_2}\overset{\displaystyle O}{\overset{\|}{\text{C}}}\text{Cl}$

(b) $\text{CH}_3\overset{\displaystyle O}{\overset{\|}{\text{C}}}\text{OCH}_2{-}\bigcirc$

(c) $\text{CH}_3\text{O}\overset{\displaystyle O}{\overset{\|}{\text{C}}}\text{CH}_2{-}\bigcirc$

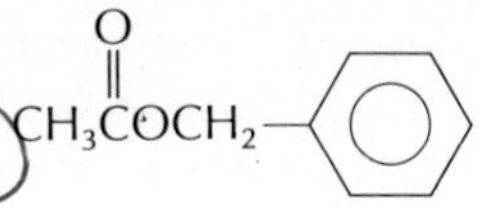

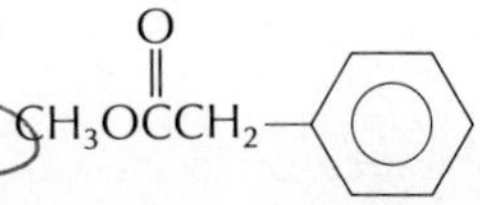

(d) $(\text{CH}_3)_2\text{CHCH}_2\text{CH}_2\overset{\displaystyle O}{\overset{\|}{\text{C}}}\text{NH}_2$

(e) $(\text{CH}_3)_2\text{CHCH}_2\text{CH}_2\overset{\displaystyle O}{\overset{\|}{\text{C}}}\text{NHCH}_3$

(f) $(\text{CH}_3)_2\text{CHCH}_2\text{CH}_2\overset{\displaystyle O}{\overset{\|}{\text{C}}}\text{N(CH}_3)_2$

Preparation and Reaction of Acyl Derivatives

13.17 Write a structural formula for the principal organic product or products of each of the following reactions:

(a) Phenylacetic acid and thionyl chloride

(b) Product of (a) and water

(c) Product of (a) and benzyl alcohol ($\text{C}_6\text{H}_5\text{CH}_2\text{OH}$)

(d) Product of (a) and benzylamine ($\text{C}_6\text{H}_5\text{CH}_2\text{NH}_2$)

(e) Product of (c) and aqueous hydrochloric acid, heat

(f) Product of (c) and aqueous sodium hydroxide, heat

(g) Product of (d) and aqueous hydrochloric acid, heat
(h) Product of (d) and aqueous sodium hydroxide, heat
(i) Acetic anhydride and cyclohexanol
(j) Acetic anhydride and dimethylamine, $(CH_3)_2NH$
(k) Ethyl formate and ethylamine

13.18 Supply the formula of the reactant missing from each of the following equations:

(a) ? + benzoic acid $\xrightarrow{heat}$ benzamide + water
(b) ? + 1-butanol → butyl acetate + acetic acid

(c) ? + H_3O^+ $\xrightarrow{heat}$ HCO_2H + NH_4^+

(d) ? + propanoic acid $\xrightarrow{H^+}$ H_2O + isopropyl propanoate

Acyl Transfer Reaction Mechanisms

13.19 For each of the following acyl transfer reactions write the structure of the tetrahedral intermediate in the process.
(a) Benzoyl chloride and ethanol
(b) Acetic anhydride and ethylamine
(c) Saponification of methyl benzoate
(d) Methyl benzoate and ammonia

13.20 Outline a series of chemical steps that describes the preparation of ethyl acetate by the Fischer esterification reaction.

13.21 Outline a series of chemical steps that describes the mechanism of the base-promoted hydrolysis of the industrial solvent *N,N*-dimethylformamide (DMF).

13.22 One method used by chemists to probe a reaction mechanism is to employ a reagent in which an atom is enriched in an isotope present naturally only in trace amounts. The fate of this isotope can be determined by using a technique known as mass spectrometry. Water enriched in oxygen-18 (^{18}O) has been used to study the hydrolysis of esters. If the base-promoted hydrolysis of ethyl acetate is carried out in ^{18}O labeled HO^- in labeled water, will the ^{18}O isotope be found in the carboxylate ion product or in the alcohol?

Synthesis

13.23 Using ethanol as the source of all the carbon atoms, along with any necessary inorganic reagents, show how you could prepare each of the following:
(a) Acetyl chloride
(b) Ethyl acetate
(c) Acetamide

13.24 Using bromobenzene and any other necessary organic or inorganic reagents, suggest an efficient synthesis of:
(a) 1,1-Diphenyl-1-propanol.
(b) Diphenylmethanol, $(C_6H_5)_2CHOH$

(c) Benzamide, $C_6H_5\overset{\overset{\displaystyle O}{\|}}{C}NH_2$

Miscellaneous Problems

13.25 Cyclic carboxylic acid derivatives usually undergo the same chemical reactions as their counterparts which lack a ring. Thus cyclic esters and amides undergo reactions typical of acyclic esters and amides. With this in mind, give the structure of the organic product of each of the following reactions:

(a) 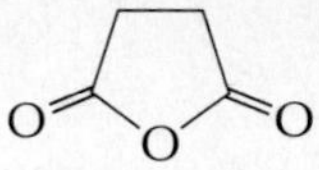 and water

(b) 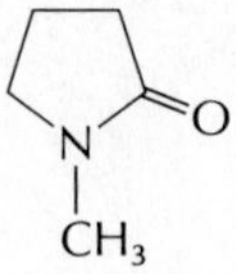 and aqueous hydrochloric acid, heat

(c) 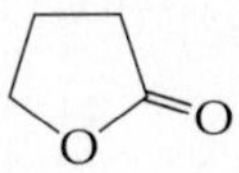 and aqueous sodium hydroxide, heat

(d) 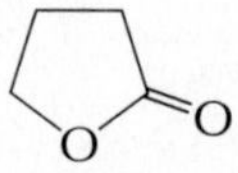 and methylmagnesium bromide (2 mol), then H_3O^+

13.26 The compound having the structure shown was heated with dilute sulfuric acid to give a product having the molecular formula $C_5H_{12}O_3$ in 63 to 71 percent yield. Propose a reasonable structure for this product. What other organic compound is formed in this reaction?

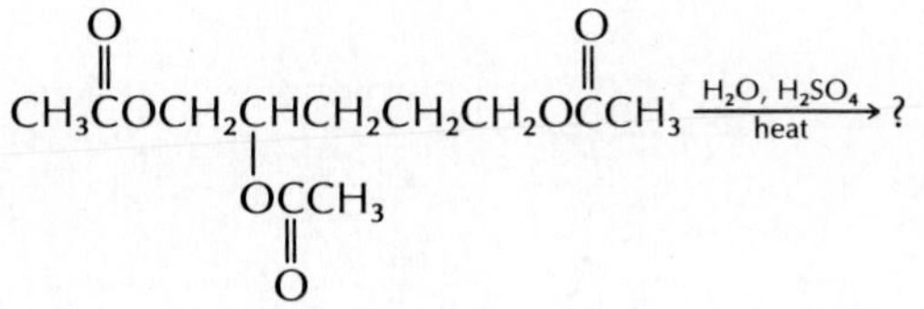

13.27 Novocain is a local anesthetic often used by dentists. Supply the reagents that are missing from the first two steps of the synthesis shown.

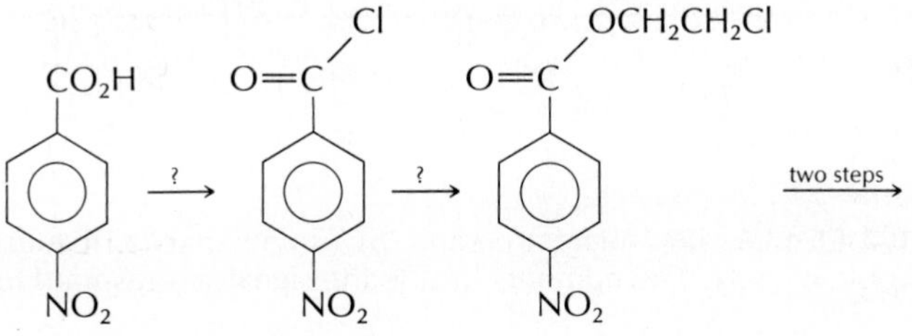

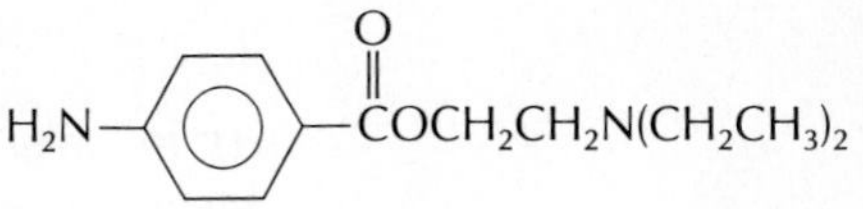

Novocain

13.28 Identify compounds A and B in the following equations:

(a)

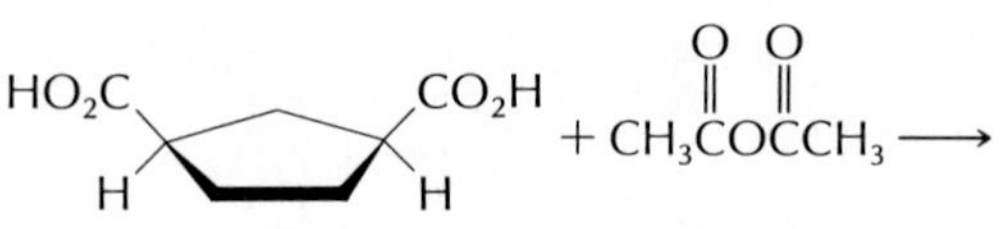

compound A $(C_7H_8O_3)$ + $2CH_3CO_2H$

(b) 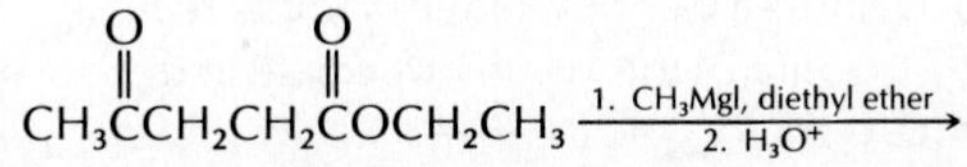 $CH_3\overset{O}{\overset{\|}{C}}CH_2CH_2\overset{O}{\overset{\|}{C}}OCH_2CH_3 \xrightarrow[\text{2. } H_3O^+]{\text{1. } CH_3MgI, \text{ diethyl ether}}$ compound B $(C_6H_{10}O_2)$

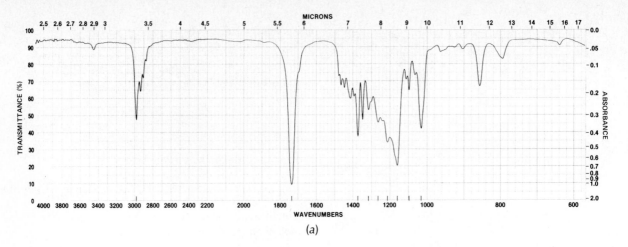

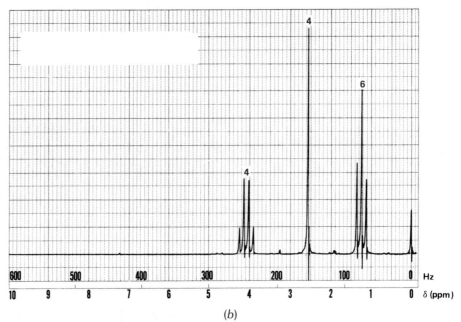

(b)

FIGURE 13.6 (a) The infrared and (b) ^{1}H nmr spectra of compound E ($C_8H_{14}O_4$) in Problem 13.31. The numbers beside the signals correspond to their relative areas.

13.29 When compounds of the type represented by C are allowed to stand in pentane, they are converted to a constitutional isomer.

$$RNHCH_2CH_2O\overset{\overset{\displaystyle O}{\|}}{C}-\text{⬡}-NO_2 \longrightarrow \text{compound D}$$

Compound C

Hydrolysis of either C or D yields $RNHCH_2CH_2OH$ and *p*-nitrobenzoic acid.
  (a) Suggest a reasonable structure for compound D.
  (b) Demonstrate your understanding of the mechanism of this reaction by writing the structure of the key intermediate in the conversion of compound C to compound D.

13.30 A very strong and heat-resistant nylon fabric known by the trade name Nomex is made by the reaction shown between an aromatic diacyl chloride and a diamine. The repeating unit of this polymer has the formula $C_{14}H_{10}N_2O_2$. Write the structure.

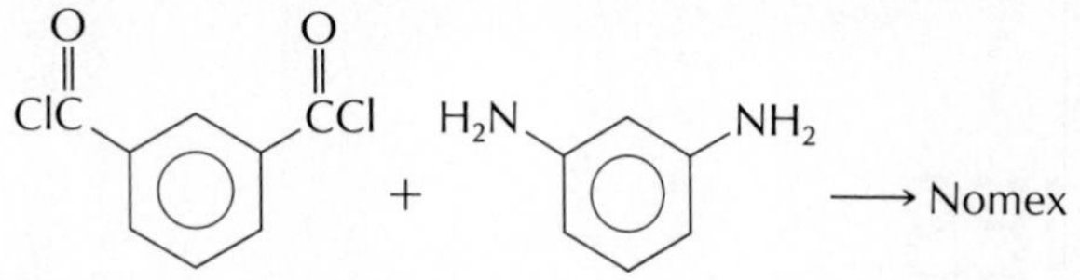

13.31 The infrared and ^{1}H nmr spectra of compound E ($C_8H_{14}O_4$) are shown in Figure 13.6. What is the structure of compound E?

CHAPTER *14*

AMINES

LEARNING OBJECTIVES

This chapter focuses on the chemical and physical properties of *alkylamines* and *arylamines*. Upon completion of this chapter you should be able to:

- Give an acceptable IUPAC name for an alkylamine or arylamine.
- Draw the structure of a primary, secondary, or tertiary amine or a quaternary ammonium salt.
- Describe the bonding of amines by using an orbital hybridization model.
- Explain how the structure of an amine affects its boiling point.
- Write a chemical equation for the reaction of an amine with an acid.
- Explain what is meant by K_b and pK_b for an amine.
- Write a chemical equation describing the preparation of an alkylamine by alkylation of ammonia.
- Write a chemical equation for the preparation of an alkylamine by reduction of an amide.
- Write a chemical equation for the preparation of a primary amine by reduction of a nitrile.
- Write a chemical equation describing the preparation of an arylamine from an aromatic nitro compound.
- Use a chemical equation to describe the reaction of an amine with an alkyl halide.
- Write a series of chemical equations describing the preparation of an alkene from an amine by using the Hofmann elimination.
- Write a chemical equation describing the formation of a diazonium salt from an arylamine.
- Describe with chemical equations the preparation of phenols, aryl halides, and aryl cyanides by using diazonium salts.
- Write a chemical equation for the formation of an azo compound by a diazonium coupling reaction.

Nitrogen-containing compounds are essential to life and are ultimately derived from atmospheric nitrogen. Atmospheric nitrogen is reduced by the process known as *nitrogen fixation* to ammonia, then converted to organic nitrogen compounds.

This chapter describes the chemistry of **amines,** organic derivatives of ammonia. The nitrogen atom of an *alkylamine* is bonded to an sp^3 hybridized carbon. The nitrogen of an *arylamine* is bonded directly to an sp^2 hybridized carbon of an aromatic ring. Cyclic compounds containing at least one atom other than carbon as part of the ring are called *heterocyclic compounds* and when that atom is nitrogen, the compound is called a *heterocyclic amine.*

$$R-\overset{\cdot\cdot}{N}\overset{\diagup}{\diagdown} \qquad Ar-\overset{\cdot\cdot}{N}\overset{\diagup}{\diagdown} \qquad -N\colon$$

| R = alkyl group: alkylamine | Ar = aryl group: arylamine | Heterocyclic amine |

Amines are weak bases and are usually the bases involved in biological acid-base reactions. Amines also frequently are the nucleophiles in biological nucleophilic substitution reactions.

While a large number of amines occur naturally, the laboratory synthesis of amines has reached a high level of development. Many prescription drugs are amines, and chemists in the pharmaceutical industry are constantly improving the methods used for the preparation of amines. This chapter will present several methods of amine synthesis, as well as some characteristic reactions of amines.

14.1 AMINE NOMENCLATURE

Amines are classified as primary, secondary, or tertiary according to the number of carbon atoms directly bonded to the nitrogen atom. The nitrogen atom of a **primary amine** has one substituent, the nitrogen atom of a **secondary amine** is disubstituted, and the nitrogen atom of a **tertiary amine** bears three substituents.

$$R-N\colon\overset{\diagup H}{\diagdown H} \qquad R-N\colon\overset{\diagup R'}{\diagdown H} \qquad R-N\colon\overset{\diagup R'}{\diagdown R''}$$

| Primary amine | Secondary amine | Tertiary amine |

This method of classification differs from that used to classify alcohols and alkyl halides. Alcohols and alkyl halides are classified as primary, secondary, or tertiary according to the degree of substitution at the carbon bearing the functional group. Thus *tert*-butyl alcohol is a tertiary alcohol, while *tert*-butylamine is a primary amine.

$$CH_3-\overset{\overset{\displaystyle CH_3}{|}}{\underset{\underset{\displaystyle CH_3}{|}}{C}}-OH \qquad CH_3-\overset{\overset{\displaystyle CH_3}{|}}{\underset{\underset{\displaystyle CH_3}{|}}{C}}-NH_2$$

| *tert*-Butyl alcohol (tertiary alcohol) | *tert*-Butylamine (primary amine because nitrogen is bonded to one carbon) |

Primary alkylamines may be named by adding the ending -*amine* to the name of the alkyl group that bears the nitrogen. They may also be named as **alkanamines** by naming the alkyl group as an alkane and replacing the -*e* ending with -*amine*.

$CH_3CH_2NH_2$ [cyclohexyl structure]—NH_2 $CH_3CHCH_2CH_2CH_3$
                                                              |
                                                            NH_2

Ethylamine   Cyclohexylamine   1-Methylbutylamine
(ethanamine)   (cyclohexanamine)   (2-pentanamine)

Secondary and tertiary amines are named as *N*-substituted derivatives of primary amines. The parent primary amine is the one with the longest carbon chain.

H_3C CH_3 $N(CH_3)_2$ CH_3
    \   /   /
     N   [cycloheptyl structure] $(CH_3)_2CHCH_2N$
     | \
     H CH_2CH_3

Dimethylamine   *N,N*-Dimethylcycloheptyl-   *N*-Ethyl-*N*-methyliso-
(a secondary   amine   butylamine
  amine)   (a tertiary amine)   (a tertiary amine)

A nitrogen that bears four substituents is positively charged and is called an **ammonium ion.** The anion that is associated with the ammonium ion is also identified in the name. Ammonium salts that have four alkyl groups bonded to nitrogen are called **quaternary ammonium salts.**

$CH_3\overset{+}{N}H_3\ Cl^-$ $(CH_3CH_2)_2\overset{+}{N}H_2\ Br^-$ $C_6H_5CH_2\overset{+}{N}(CH_3)_3\ I^-$

Methylammonium   Diethylammonium   Benzyltrimethyl-
   chloride   bromide   ammonium iodide
                                                      (a quaternary
                                                     ammonium salt)

PROBLEM 14.1
Give an acceptable IUPAC name for each of the following.

(a) [benzene ring]—$CH_2CH_2NH_2$ (c) $CH_3NCH_2CH_2CH_3$
                                                   |
                                                  CH_3

(b) [benzene ring]—$\underset{\underset{NH_2}{|}}{CH}CH_3$ (d) $(CH_3)_3\overset{+}{N}H\ Br^-$

SAMPLE SOLUTION
(a) The amino substituent is bonded to an ethyl group that bears a phenyl substituent at C-2. The compound $C_6H_5CH_2CH_2NH_2$ is named either 2-phenylethylamine or 2-phenylethanamine.

The nitrogen atom of an arylamine is attached to one or more aromatic rings. Arylamines which contain a benzene ring are named as derivatives of **aniline,** $C_6H_5NH_2$. Substituents, including those bonded to nitrogen, are listed in alphabetical order.

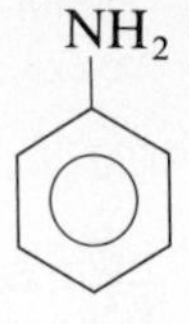

NH₂   NH₂   NHCH₂CH₃

Aniline 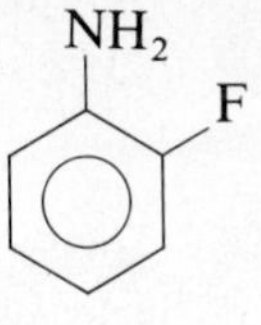   o-Fluoroaniline 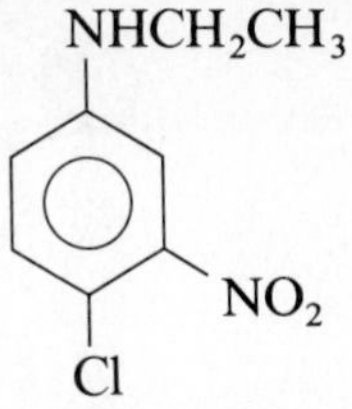   4-Chloro-N-ethyl-3-nitroaniline

---

PROBLEM 14.2

Give an acceptable IUPAC name for each of the following amines.

(a)   (b) 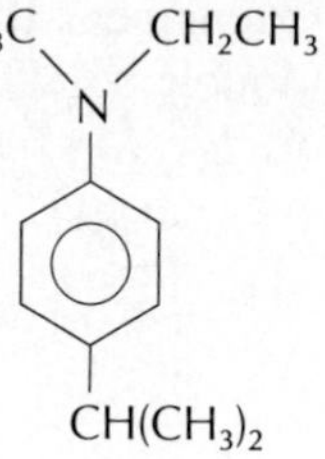   (c)

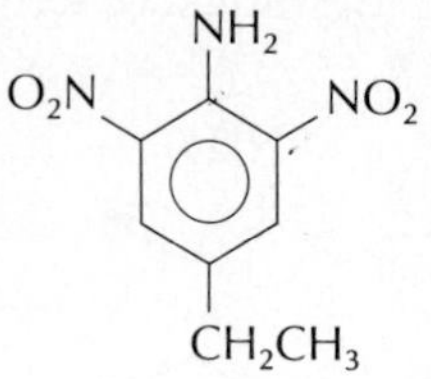

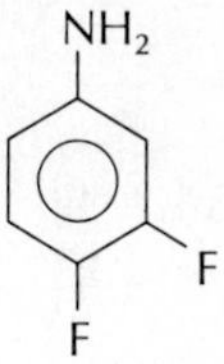

SAMPLE SOLUTION

(a) Naming this compound as a derivative of aniline, there is an ethyl substituent at C-4 and two nitro groups located at C-2 and C-6. The compound is 4-ethyl-2,6-dinitroaniline.

14.2 STRUCTURE AND BONDING

Alkylamines have a pyramidal arrangement of substituents at nitrogen, similar to the structure of ammonia described in Section 1.10. We can illustrate the structure of amines by examining methylamine.

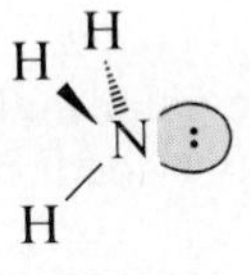

Ammonia 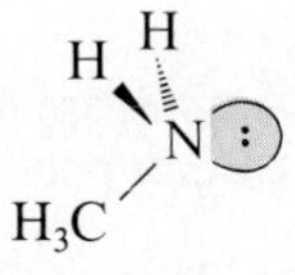   Methylamine

An orbital hybridization description of bonding in methylamine is shown in Figure 14.1. Each of four equivalent sp^3 hybridized orbitals on carbon contains one electron, as shown in (a). A nitrogen atom contributes *five* valence electrons. Three of the sp^3 hybrid orbitals on the nitrogen atom contain one electron each; the fourth sp^3 hybrid orbital contains *two* electrons. As shown in (b), nitrogen forms three σ bonds by using three of its sp^3 hybridized orbitals. The fourth nitrogen orbital contains a **non-bonded,** or **lone, pair** of electrons. This pair of electrons plays a crucial role in the chemical properties of amines—it is the lone pair that gives amines their basic and nucleophilic character.

Aniline, like methylamine, has a pyramidal arrangement of bonds around nitrogen, but its pyramid is somewhat shallower. A somewhat flattened arrangement of bonds to nitrogen permits the unshared electron pair to be partially delocalized into the aromatic π system.

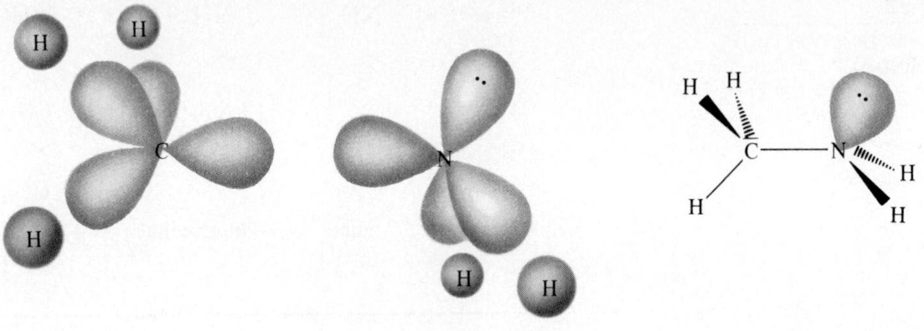

FIGURE 14.1 Orbital hybridization description of bonding in methylamine.

(a)   (b)

Delocalization of the nitrogen lone pair electrons into the aromatic π system strengthens the carbon-nitrogen bond of aniline, shortens it, and gives it "partial double bond character." In resonance terms, this is expressed as

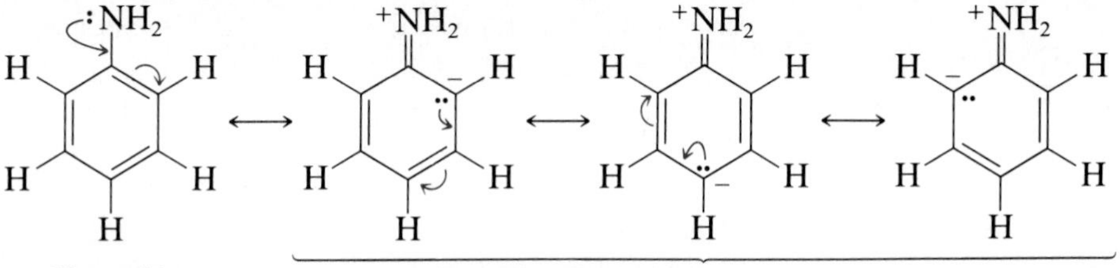

Most stable
Lewis structure
for aniline

Dipolar resonance forms of aniline

Delocalization of the nitrogen lone pair decreases the electron density at nitrogen while increasing it in the aromatic ring system. Aniline's high level of reactivity in electrophilic aromatic substitution reactions (Chapter 6) is one result of this delocalization.

14.3 PHYSICAL PROPERTIES

Amines are more polar than hydrocarbons but less polar than alcohols. This difference in polarity is evident in the trend in boiling points.

$CH_3CH_2CH_3$	$CH_3CH_2NH_2$	CH_3CH_2OH
Propane	Ethylamine	Ethanol
bp $-42°C$	bp $17°C$	bp $78°C$

Amines participate in hydrogen bonding, as shown in Figure 14.2a, and thus have higher boiling points than hydrocarbons of similar molecular weight. However nitrogen is less electronegative than oxygen, and amines are less polar than alcohols. The result is that hydrogen bonding forces between amine molecules are weaker than the hydrogen bonds between alcohol molecules. Thus a lower boiling point is observed for an amine when compared to an alcohol of similar molecular weight.

Among isomeric amines, primary amines have the highest boiling points; tertiary amines the lowest.

$CH_3CH_2CH_2NH_2$	$CH_3CH_2NHCH_3$	$(CH_3)_3N$
Propylamine	N-Methylethylamine	Trimethylamine
(a primary amine)	(a secondary amine)	(a tertiary amine)
bp $50°C$	bp $34°C$	bp $3°C$

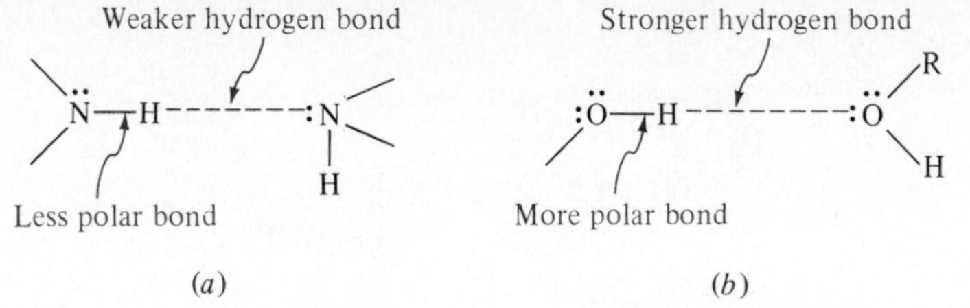

FIGURE 14.2 Hydrogen bonding forces between (a) amine molecules are weaker than hydrogen bonds between (b) alcohol molecules.

Weaker hydrogen bond

Stronger hydrogen bond

Less polar bond

More polar bond

(a)

(b)

PROBLEM 14.3

Using what you know about hydrogen bonding forces, propose an explanation for the trend in boiling points among isomeric amines.

Amines having fewer than six or seven carbon atoms are soluble in water. All amines, even tertiary amines, can act as proton acceptors in hydrogen bonding to water molecules. The basicity of the unshared electron pair on nitrogen enhances the solvation of amines by water, alcohols, and other proton donors. Amines tend to be somewhat more soluble in water than are the corresponding alcohols.

Arylamines resemble alkylamines in their physical properties. Hydrogen bonding affects their boiling points and solubility in water in the same way that it affects alkylamines. For example, aniline is many times more soluble in water than toluene.

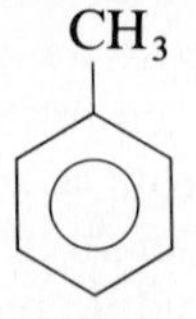

CH_3

NH_2

Toluene
water solubility
(25°C): 0.05 g/100 mL

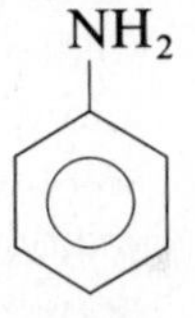

Aniline
water solubility
(25°C): 3.5 g/100 mL

One characteristic physical property of low-molecular-weight amines is their smell. Many common amines are liquids with unpleasant, "fishlike" odors. Two diamines which result from the decay of protein-containing organic matter have names suggestive of their malodorous properties.

$$H_2NCH_2CH_2CH_2CH_2NH_2 \qquad H_2NCH_2CH_2CH_2CH_2CH_2NH_2$$

Putrescine

Cadaverine

14.4 AMINES IN NATURE

The physiological importance of amines has long been recognized. The word *vitamin* was coined in 1912 in the belief that the substances present in our diet that prevented scurvy, rickets, and other diseases were "vital amines." In some cases that belief was confirmed; certain vitamins are amines. Examples include vitamins B_1 (thiamine) and B_6 (a mixture of three compounds), shown in Figure 14.3.

Amines also play active roles in our brain and central nervous system (Figure 14.4). Epinephrine, also called *adrenalin,* is secreted by the adrenal gland when a person is under stress or frightened. This sudden increase in adrenalin prepares the body for "flight or fight." Persons who have a deficiency of dopamine exhibit the uncontrollable shaking of Parkinson's disease. The standard treatment for Parkinson's sufferers is administration of a drug called *L-Dopa* which elevates the dopamine

FIGURE 14.3 Two vitamins which are amines. (a) Thiamine (vitamin B$_1$) deficiency results in beriberi, a disease of the nervous system that, in severe cases, can lead to heart failure. (b) Vitamin B$_6$ is a mixture of three compounds: pyridoxine, pyridoxal, and pyridoxamine. A deficiency of vitamin B$_6$ can lead to disorders of the central nervous system.

(a) Vitamin B$_1$:

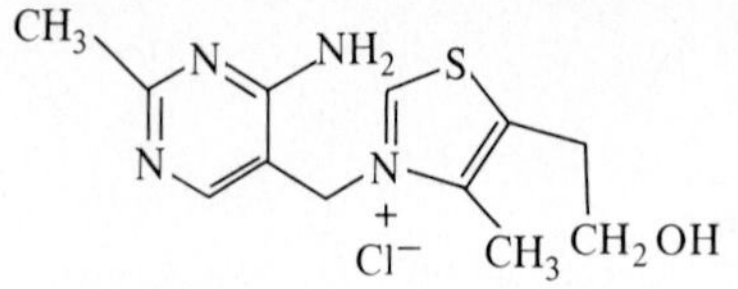

Thiamine

(b) Vitamin B$_6$:

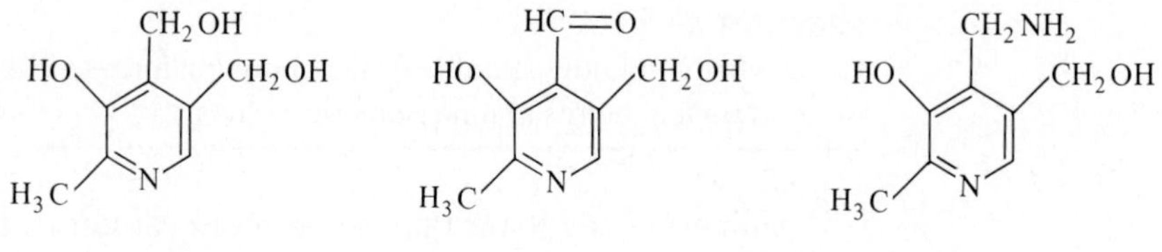

Pyridoxine   Pyridoxal   Pyridoxamine

levels in their brain. Serotonin levels in the brain are believed to be related to some forms of mental illness.

Nitrogen-containing plant materials are called **alkaloids,** or plant bases. Over 5000 different alkaloids are known, and several are shown in Figure 14.5. The analgesic (pain-relieving) properties of morphine have been known since the 1850s. As early as the American Civil War, however, addiction to morphine among injured soldiers was recognized as a major problem. Heroin is the diacetate ester of morphine and was first synthesized in the late 1800s as a cure for morphine addiction. It was soon learned that heroin is *more* addictive than morphine, and today heroin is a major drug abuse problem. Codeine is a less powerful analgesic than morphine or heroin but is also less addictive. Codeine blocks the cough reflex and is often used as an antitussive in cough medications.

14.5 AMINES AS BASES

As a class, amines are the strongest bases of all neutral molecules. Amines react with strong acids by proton transfer to nitrogen to form ammonium salts.

$$R_3N : \ + \ H-X \longrightarrow R_3\overset{+}{N}-H \ X^-$$

Amine   Acid   Ammonium salt

These ammonium salts are ionic and are more soluble in polar solvents than in nonpolar ones. Amines can be extracted from an ether solution into water by shaking

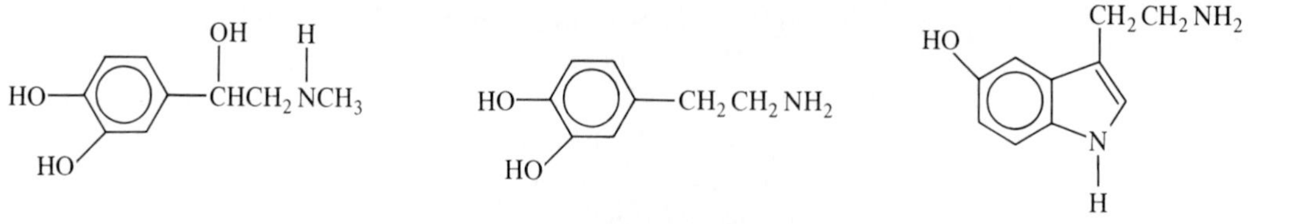

Epinephrine (adrenalin); Hormone secreted by the adrenal gland

Dopamine; A hormone-like substance present in the brain

Serotonin: A hormone synthesized in the pineal gland

FIGURE 14.4 Amines which are active in the brain and central nervous system.

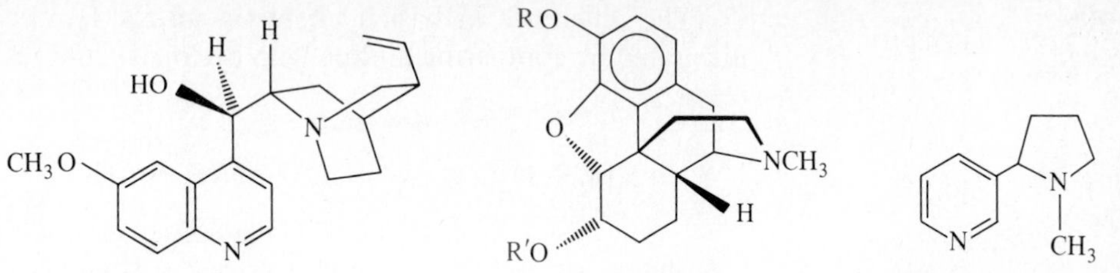

Quinine (alkaloid of
cinchona bark used
to treat malaria)

Alkaloids derived from
the opium poppy:
R = R' = H; morphine

$$R = R' = O\overset{\displaystyle O}{\overset{\displaystyle \|}{C}}CH_3 \text{ ; heroin}$$

R = CH₃ , R' = H; codeine

Nicotine
(toxic alkaloid
present in tobacco)

FIGURE 14.5 Several naturally occurring alkaloids.

with dilute hydrochloric acid. Conversely, adding sodium hydroxide to an aqueous solution of an ammonium salt converts it to the free amine, which can be removed from the aqueous phase by extraction with ether.

Ammonium ion	Hydroxide ion	Amine	Water
(stronger acid)	(stronger base)	(weaker base)	(weaker acid)

The base strength of an amine is related to the equilibrium constant for the reaction:

Amine	Water	Ammonium ion	Hydroxide ion
(weaker base)	(weaker acid)	(stronger acid)	(stronger base)

The basicity constant K_b for an amine is related to K, the equilibrium constant for the reaction, according to the expression

$$K_b = K[H_2O] = \frac{[R_3\overset{+}{N}H][HO^-]}{[R_3N]}$$

and

$$pK_b = -\log K_b$$

Alkylamines are weak bases and have basicity constants in the range of $K_b \approx 10^{-3}$ to 10^{-5}. Arylamines, on the other hand, are several orders of magnitude *less* basic than alkylamines. Arylamines have K_b's in the 10^{-10} range. The sharply decreased basicity of arylamines is because the stabilizing effect of lone pair electron delocalization into the aromatic ring (Section 14.2) is lost on protonation.

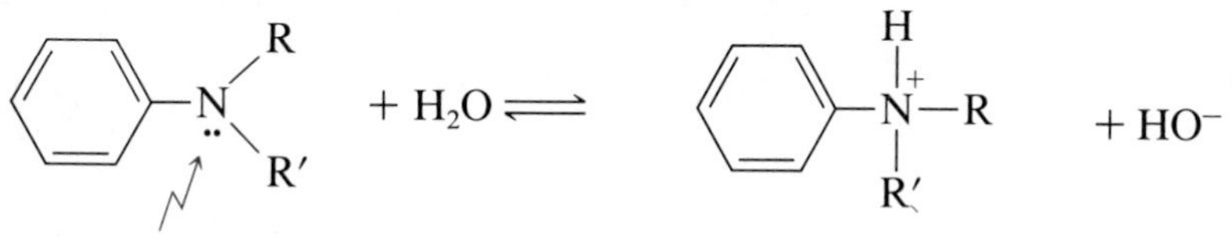

$+ H_2O \rightleftharpoons$

$+ HO^-$

Amine is stabilized by
delocalization of lone
pair into π system of
ring, decreasing the electron
density at nitrogen

Lone pair electrons
transformed to N—H bonded pair

The difference in basicity between an alkylamine and an arylamine can be illustrated by comparing aniline with cyclohexylamine.

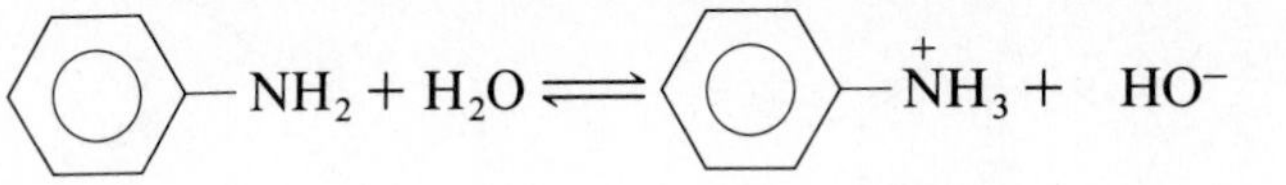

$$\text{Aniline} + H_2O \rightleftharpoons \text{Anilinium ion} + HO^- \qquad (K_b\ 3.8 \times 10^{-10};\ pK_b\ 9.4)$$

| Aniline | Water | Anilinium ion | Hydroxide ion |

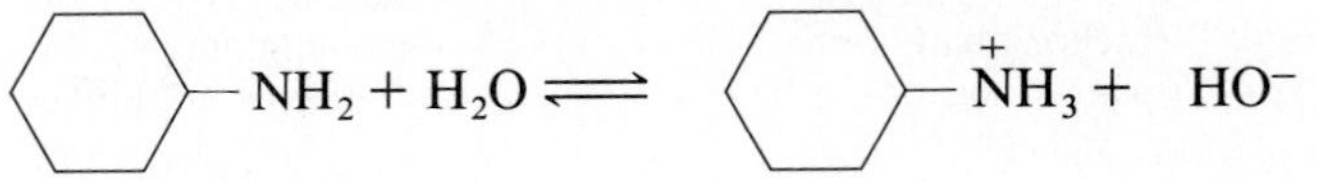

$$\text{Cyclohexylamine} + H_2O \rightleftharpoons \text{Cyclohexylammonium ion} + HO^- \qquad (K_b\ 4.4 \times 10^{-4};\ pK_b\ 3.4)$$

| Cyclohexylamine | Water | Cyclohexylammonium ion | Hydroxide ion |

PROBLEM 14.4
The two amines shown differ by a factor of 40,000 in their K_b values. Which is the stronger base? Why?

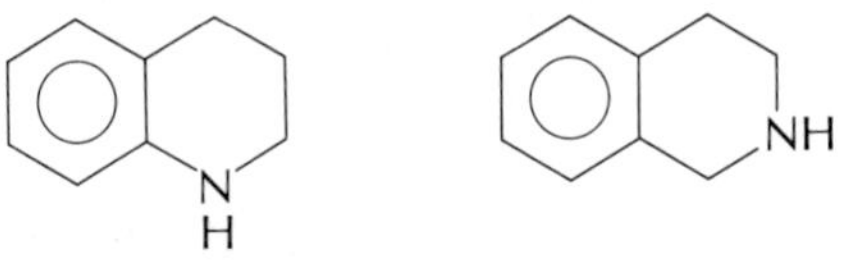

Tetrahydroquinoline   Tetrahydroisoquinoline

Many chemists express the basicity of an amine in terms of the pK_a of its conjugate acid. The stronger the base, the weaker the conjugate acid. Thus the higher the pK_a of an alkyl- or arylammonium ion, the stronger the base.

$$CH_3NH_2 \overset{H^+}{\rightleftharpoons} CH_3\overset{+}{N}H_3 \qquad\qquad C_6H_5NH_2 \overset{H^+}{\rightleftharpoons} C_6H_5\overset{+}{N}H_3$$

| Methylamine $pK_b = 3.4$ | Methylammonium ion $pK_a = 10.6$ | Aniline $pK_b = 9.4$ | Anilinium ion $pK_a = 4.6$ |

The relationship between pK_b and pK_a is a simple one:

$$pK_b\ (\text{base}) + pK_a\ (\text{conjugate acid}) = 14$$

14.6 PREPARATION OF ALKYLAMINES BY ALKYLATION OF AMMONIA

As alkyl derivatives of ammonia, amines are, in principle, capable of being prepared by nucleophilic substitution reactions of alkyl halides with ammonia.

$$2NH_3 + RX \longrightarrow RNH_2 + \overset{+}{N}H_4\ X^-$$

| Ammonia | Alkyl halide | Primary amine | Ammonium halide salt |

This reaction is a bimolecular nucleophilic substitution. As with other nucleophilic substitution reactions of the S_N2 type, primary alkyl halides are the preferred substrates. As a general synthetic method, however, the reaction of ammonia with an alkyl halide is of limited value because the yields of primary amines obtained are typically rather low. The reason is that the primary amine product is itself a nucleophile and competes with ammonia for the alkyl halide.

$$RX + RNH_2 + NH_3 \longrightarrow R_2NH + \overset{+}{N}H_4\ X^-$$

| Alkyl halide | Primary amine | Ammonia | Secondary amine | Ammonium halide salt |

When 1-bromooctane, for example, is allowed to react with ammonia, both the primary amine and the secondary amine are isolated in comparable amounts.

$$CH_3(CH_2)_6CH_2Br \xrightarrow{\text{NH}_3\ (2\ \text{mol})} CH_3(CH_2)_6CH_2NH_2 + [CH_3(CH_2)_6CH_2]_2NH$$

1-Bromooctane (1 mol)   Octylamine (45%)   N,N-Dioctylamine (43%)

In a similar manner, the reaction may continue to form the trialkylamine and even the quaternary ammonium salt.

PROBLEM 14.5

Write equations which show the reaction of N,N-dioctylamine (from the preceding equation) with 1-bromooctane to form tertiary amine and quaternary ammonium salt by-products.

14.7 PREPARATION OF ALKYLAMINES BY REDUCTION

Almost any nitrogen-containing organic compound can be reduced to an amine. The synthesis of amines then becomes a question of the availability of a suitable starting material and the choice of an appropriate reducing agent.

In the last chapter you saw how amides could be prepared by heating an amine with a carboxylic acid or by reaction of an amine with an acyl chloride (Section 13.11). Reduction of the amide carbonyl group to form a new amine is possible by using lithium aluminum hydride.

$$\underset{\text{Amide}}{R\overset{\overset{\displaystyle O}{\|}}{C}NR'_2} \xrightarrow[\text{2. H}_2\text{O}]{\text{1. LiAlH}_4} \underset{\text{Amine}}{RCH_2NR'_2}$$

This reaction is one of the most frequently used methods for the laboratory preparation of amines. For example, reduction of an N-alkyl amide yields a secondary amine in good yield. The by-product contamination described in the preceding section is not a problem in the synthesis of amines by amide reduction.

$$CH_3CH_2CH_2CH_2\overset{\overset{\displaystyle O}{\|}}{C}NHCH_2CH_2CH_2CH_3 \xrightarrow[\text{2. H}_2\text{O}]{\text{1. LiAlH}_4}$$

N-Butylpentanamide

$$CH_3CH_2CH_2CH_2CH_2NHCH_2CH_2CH_2CH_3$$

N-Butyl-1-pentanamine (88%)

Nitriles, compounds of the type $RC\equiv N$, are readily available from primary or secondary alkyl halides by nucleophilic substitution (Section 8.2; see Table 8.1).

$$RX \xrightarrow{\ ^{-}:C\equiv N:\ } RC\equiv N: +\ X^{-}$$

Alkyl   Nitrile   Halide
halide   ion

Catalytic hydrogenation of nitriles converts them to primary alkylamines.

$$RC\equiv N \xrightarrow[\text{catalyst}]{H_2} RCH_2NH_2$$

$$CH_3CH_2CH_2CH_2Br \xrightarrow{NaCN}$$
1-Bromobutane

$$CH_3CH_2CH_2CH_2C\equiv N \xrightarrow[\text{Ni}]{H_2} CH_3CH_2CH_2CH_2CH_2NH_2$$

Pentanenitrile   1-Pentanamine
                                              (56%)

Lithium aluminum hydride reduction is also a very effective method for converting nitriles to amines.

14.8 PREPARATION OF ARYLAMINES

Arylamines are industrial chemicals synthesized as intermediates in the manufacture of food colorings and dyes. The most convenient method for the synthesis of arylamines involves the preparation and reduction of a nitroarene.

$$ArH \longrightarrow ArNO_2 \longrightarrow ArNH_2$$

Arene   Nitroarene   Arylamine

There is no direct means of introducing an amino group onto most aromatic compounds; hence this somewhat indirect method is used. Fortunately, aromatic nitrations are straightforward electrophilic aromatic substitution reactions (Section 6.6). Nitro groups are easily reduced with a variety of reducing agents, including catalytic hydrogenation over a platinum, palladium, or nickel catalyst. Reduction using either iron or tin in hydrochloric acid may also be used.

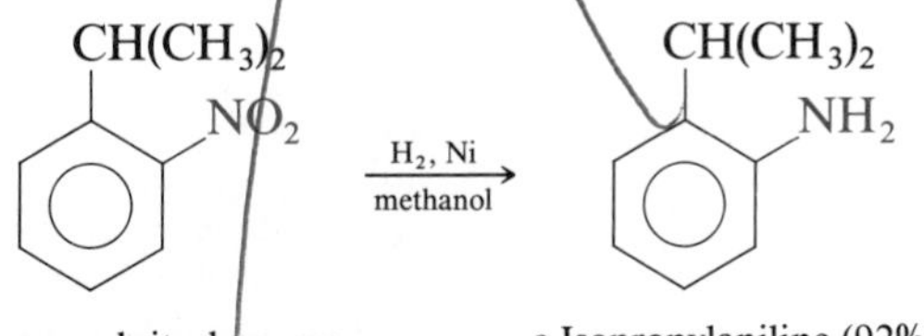

o-Isopropylnitrobenzene $\xrightarrow[\text{methanol}]{H_2,\ Ni}$ o-Isopropylaniline (92%)

$$Cl-\langle\bigcirc\rangle-NO_2 \xrightarrow[\text{2. NaOH}]{\text{1. Fe, HCl}} Cl-\langle\bigcirc\rangle-NH_2$$

p-Chloronitrobenzene $\qquad$ *p*-Chloroaniline (95%)

$$\langle\bigcirc\rangle-\overset{\displaystyle O}{\overset{\displaystyle \|}{C}}CH_3 \xrightarrow[\text{2. NaOH}]{\text{1. Sn, HCl}} \langle\bigcirc\rangle-\overset{\displaystyle O}{\overset{\displaystyle \|}{C}}CH_3$$

O_2N $\qquad\qquad\qquad\qquad$ H_2N

m-Nitroacetophenone $\qquad$ *m*-Aminoacetophenone (82%)

PROBLEM 14.7

Benzocaine is a topical anesthetic found in many ointments. Benzocaine is prepared by a series of reactions that illustrates the strategy used in the synthesis of arylamines. This sequence of reactions is outlined for you; provide reagents to carry out the first and the last steps of the synthesis.

$$\overset{CH_3}{\langle\bigcirc\rangle} \xrightarrow{?} \overset{CH_3}{\underset{NO_2}{\langle\bigcirc\rangle}} \xrightarrow[\text{2. CH}_3\text{CH}_2\text{OH, H}^+]{\text{1. KMnO}_4} \overset{CO_2CH_2CH_3}{\underset{NO_2}{\langle\bigcirc\rangle}} \xrightarrow{?} \overset{CO_2CH_2CH_3}{\underset{NH_2}{\langle\bigcirc\rangle}}$$

Toluene $\qquad\qquad\qquad\qquad\qquad\qquad\qquad\qquad\qquad\qquad\qquad$ Benzocaine

---

14.9 REACTIONS OF AMINES: A REVIEW AND A PREVIEW

Amines exhibit the characteristics of both *bases* and *nucleophiles*. Both the basicity and the nucleophilicity of amines originate in the unshared pair of electrons on nitrogen. When amines act as bases (Section 14.5), they use this electron pair to abstract a proton from a suitable donor.

$$\overset{|}{\underset{|}{\diagdown}}N\!:\ \ H\!-\!X$$

Amine acting as a base

Two examples of amines acting as nucleophiles have been encountered in previous chapters. One example is the reaction of a primary amine with an aldehyde or ketone to form an imine (Section 11.10).

$$R\ddot{N}H_2 \quad \overset{R'}{\underset{R''}{C}}\!=\!O \longrightarrow RNH-\overset{R'}{\underset{R''}{\overset{|}{C}}}-OH \xrightarrow{-H_2O} RN\!=\!\overset{R'}{\underset{R''}{C}}$$

Primary $\quad$ Aldehyde $\qquad\qquad$ Carbinolamine $\qquad\qquad$ Imine
amine $\qquad$ or ketone

$$CH_3NH_2 + C_6H_5\overset{\displaystyle O}{\overset{\displaystyle \|}{C}}H \longrightarrow C_6H_5CH\!=\!NCH_3 + H_2O$$

Methylamine $\quad$ Benzaldehyde $\qquad$ *N*-Benzylidenemethylamine $\quad$ Water
$\qquad\qquad\qquad\qquad\qquad\qquad\qquad$ (70%)

A second example is the formation of an amide by the reaction of an amine with an acyl chloride (Section 13.11).

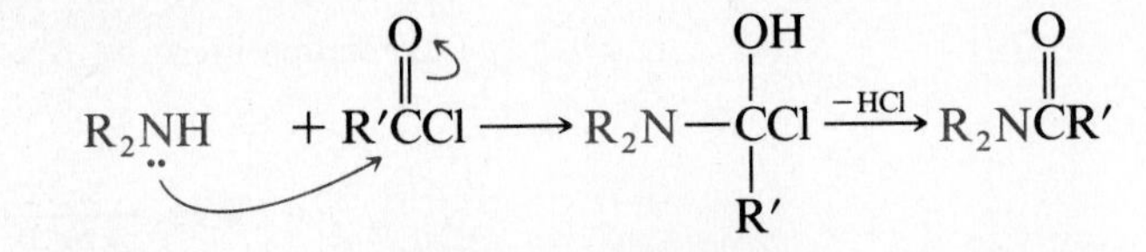

$$R_2NH + R'CCl \longrightarrow R_2N-CCl \xrightarrow{-HCl} R_2NCR'$$

| Primary or secondary amine | Acyl chloride | Tetrahedral intermediate | Amide |

$$CH_3CH_2CH_2CH_2NH_2 + CH_3CH_2CH_2CH_2CCl \longrightarrow$$

Butylamine   Pentanoyl chloride

$$CH_3CH_2CH_2CH_2CNHCH_2CH_2CH_2CH_3$$

N-Butylpentanamide (81%)

In both reactions the first step is the nucleophilic attack of the amine nitrogen atom, using its unshared electron pair, on the positively polarized carbon of the carbonyl group.

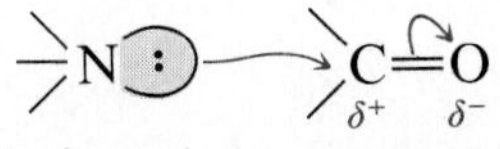

$$\ce{>N: \longrightarrow \overset{\delta+}{C}=\overset{\delta-}{O}}$$

Amine acting as a nucleophile

14.10 REACTION OF AMINES WITH ALKYL HALIDES

Nucleophilic substitution results when amines react with primary alkyl halides.

$$R\ddot{N}H_2 + R'CH_2X \longrightarrow R\overset{+}{N}-CH_2R'\ X^- \longrightarrow R\ddot{N}-CH_2R' + HX$$

| Primary amine | Primary alkyl halide | Ammonium halide salt | Secondary amine | Hydrogen halide |

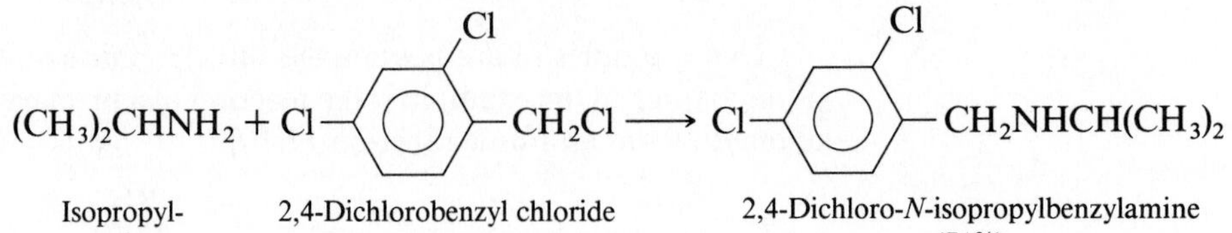

$$(CH_3)_2CHNH_2 + Cl-\underset{Cl}{\bigcirc}-CH_2Cl \longrightarrow Cl-\underset{Cl}{\bigcirc}-CH_2NHCH(CH_3)_2$$

| Isopropyl-amine | 2,4-Dichlorobenzyl chloride | 2,4-Dichloro-*N*-isopropylbenzylamine (71%) |

If the alkyl halide and the amine are not sterically hindered, a second alkylation may follow, converting the secondary amine to a tertiary amine. This process is the same as that discussed for the alkylation of ammonia (Section 14.6). Alkylation need not stop with formation of a tertiary amine; the tertiary amine may itself be alkylated, giving a quaternary ammonia salt.

$$R\ddot{N}H_2 \xrightarrow{R'CH_2X} R\ddot{N}HCH_2R' \xrightarrow{R'CH_2X} R\ddot{N}(CH_2R')_2 \xrightarrow{R'CH_2X} R\overset{+}{N}(CH_2R')_3\ X^-$$

| Primary amine | Secondary amine | Tertiary amine | Quaternary ammonium salt |

Because of its high reactivity toward nucleophilic substitution, methyl iodide is the alkyl halide often used to proceed to the quaternary ammonium salt stage.

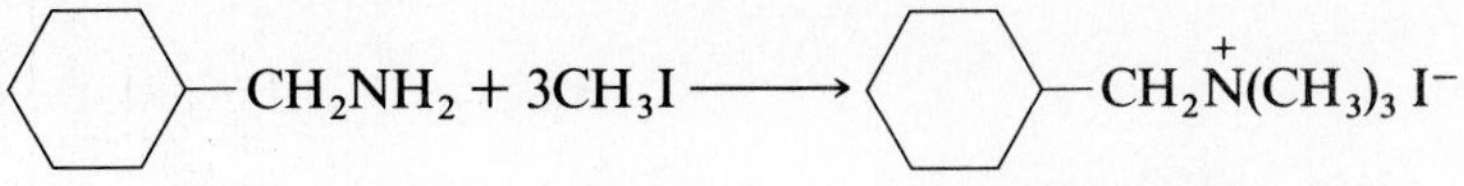

$$\text{(cyclohexyl)}-CH_2NH_2 + 3CH_3I \longrightarrow \text{(cyclohexyl)}-CH_2\overset{+}{N}(CH_3)_3 \ I^-$$

(Cyclohexylmethyl)-	Methyl	(Cyclohexylmethyl)trimethyl-
amine	iodide	ammonium iodide (99%)

PROBLEM 14.8

Choline is an intermediate in formation of acetylchloline, a key substance involved in the transmission of nerve impulses in animals. Synthetic choline is added as a supplement to animal feed. Choline can be prepared by the reaction of trimethylamine with ethylene oxide in water. Choline is a hydroxide salt having the formula $(C_5H_{14}NO)^+ \ (OH)^-$. What is its structure?

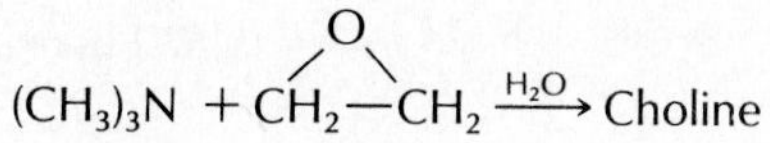

$$(CH_3)_3N \ + \ CH_2\overset{O}{-}CH_2 \xrightarrow{H_2O} Choline$$

Trimethyl-	Ethylene
amine	oxide

As you will see in the next section, quaternary ammonium hydroxide salts are used as substrates in an elimination reaction to form alkenes.

14.11 THE HOFMANN ELIMINATION

The halide anion of quaternary ammonium iodides may be replaced by hydroxide by treatment with an aqueous slurry of silver oxide. Silver iodide precipitates and a solution of the quaternary ammonium hydroxide is formed.

$$2(R_4\overset{+}{N} \ I^-) \ + Ag_2O + H_2O \longrightarrow 2(R_4\overset{+}{N} \ \overset{-}{OH}) + 2AgI$$

Quaternary	Silver	Water	Quaternary	Silver
ammonium iodide	oxide		ammonium	iodide
			hydroxide	

$$\text{(cyclohexyl)}-CH_2\overset{+}{N}(CH_3)_3 \ I^- \xrightarrow[H_2O, \ CH_3OH]{Ag_2O} \text{(cyclohexyl)}-CH_2\overset{+}{N}(CH_3)_3 \ HO^-$$

(Cyclohexylmethyl)trimethyl-	(Cyclohexylmethyl)trimethylammonium
ammonium iodide	hydroxide

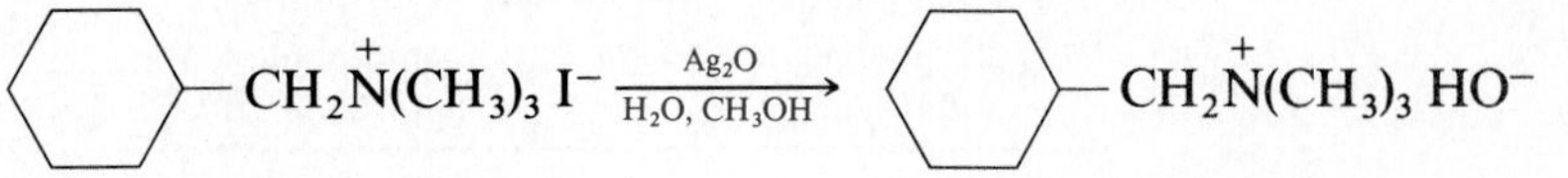

When quaternary ammonium hydroxides are heated, they undergo an elimination reaction to form an alkene and an amine in a process known as the **Hofmann elimination.**

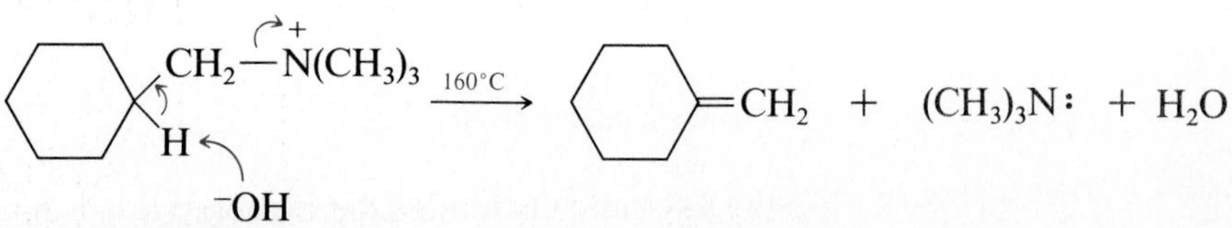

$$\text{(cyclohexyl with } CH_2-\overset{+}{N}(CH_3)_3 \text{ and H, } {}^-OH) \xrightarrow{160°C} \text{(cyclohexyl)}=CH_2 \ + \ (CH_3)_3N\colon \ + H_2O$$

(Cyclohexylmethyl)trimethyl-	Methylenecyclohexane	Trimethylamine	Water
ammonium hydroxide	(69%)		

To prepare an alkene from an amine, the quaternary ammonium salt must first be prepared. This is accomplished by reaction of the amine with an excess of methyl iodide (Section 14.10).

$$RNH_2 + CH_3I \text{ (excess)} \longrightarrow R\overset{+}{N}(CH_3)_3 \; I^-$$

The ammonium hydroxide salt is then prepared as described above and heated to give the desired alkene.

PROBLEM 14.9

Outline the preparation of an alkene from each of the following amines by using the Hofmann elimination.

(a) 3-Pentanamine, CH₃CH₂CHCH₂CH₃
          |
          NH₂

(b) 2-Phenylethylamine, ⬡—CH₂CH₂NH₂

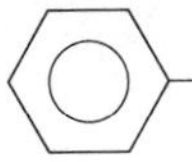

(c) Dicyclohexylamine, (⬡)₂—NH

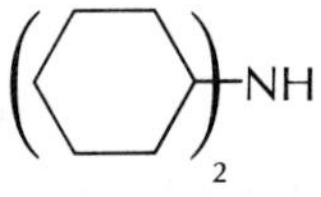

SAMPLE SOLUTION

(a) The quaternary ammonium iodide salt is first prepared by reaction with an excess of methyl iodide. Treatment with an aqueous slurry of silver oxide gives the quaternary ammonium hydroxide salt. The alkene is formed by heating the hydroxide salt.

$$\underset{\substack{| \\ NH_2 \\ \text{3-Pentanamine}}}{CH_3CH_2CHCH_2CH_3} \xrightarrow{CH_3I \text{ (excess)}} \underset{\substack{| \\ +N(CH_3)_3 \; I^- \\ \text{Quaternary} \\ \text{ammonium} \\ \text{iodide salt}}}{CH_3CH_2CHCH_2CH_3}$$

$$\underset{\substack{| \\ +N(CH_3)_3 \; I^- \\ \text{Quaternary} \\ \text{ammonium} \\ \text{iodide salt}}}{CH_3CH_2CHCH_2CH_3} \xrightarrow[\text{H}_2\text{O}]{\text{Ag}_2\text{O}} \underset{\substack{| \\ +N(CH_3)_3HO^- \\ \text{Quaternary} \\ \text{ammonium} \\ \text{hydroxide salt}}}{CH_3CH_2CHCH_2CH_3} \xrightarrow{\text{heat}} \underset{\text{2-Pentene}}{CH_3CH=CHCH_2CH_3} + \underset{\substack{\text{Trimethyl-} \\ \text{amine}}}{N(CH_3)_3}$$

A novel aspect of the Hofmann elimination is its regioselectivity. Elimination in alkyltrimethylammonium hydroxides proceeds in the direction that gives the *less substituted alkene.*

$$\underset{\substack{| \\ +N(CH_3)_3 \; HO^- \\ \textit{sec}\text{-Butyltrimethyl-} \\ \text{ammonium hydroxide}}}{CH_3CHCH_2CH_3} \xrightarrow[\substack{-H_2O \\ -(CH_3)_3N}]{\text{heat}} \underset{\substack{\text{1-Butene (95\%)}}}{CH_2=CHCH_2CH_3} + \underset{\substack{\text{2-Butene (5\%)} \\ \text{(cis and trans)}}}{CH_3CH=CHCH_3}$$

It is the less sterically hindered β hydrogen that is removed by the base in Hofmann elimination reactions. Methyl groups (—CH₃) are deprotonated in preference to methylene groups (—CH$_2$—), and methylene groups are deprotonated in preference to methines (—CH). The regioselectivity of Hofmann eliminations is *opposite* to that predicted by the Zaitsev rule (Section 4.1.7).

PROBLEM 14.10

Give the structure of the alkene formed in major amounts when the hydroxide salts of each of the following quaternary ammonium ions are heated.

(a)
cyclopentane ring with CH$_3$ and N(CH$_3$)$_3$ with + charge

(c) $CH_3CH_2\overset{+}{\underset{\underset{CH_3}{|}}{\overset{\overset{CH_3}{|}}{N}}}CH_2CH_2CH_2CH_3$

(b) $(CH_3)_3CCH_2\underset{\underset{+N(CH_3)_3}{|}}{C}(CH_3)_2$

(d) $CH_3CH_2CH_2\overset{+}{\underset{\underset{CH_3}{|}}{\overset{\overset{CH_3}{|}}{N}}}CH_2CH(CH_3)_2$

SAMPLE SOLUTION

(a) Two alkenes are capable of being formed by elimination, methylenecyclopentane and 1-methylcyclopentene.

(1-Methylcyclopentyl)trimethylammonium hydroxide + HO⁻ heat, −H₂O, −(CH₃)₃N → Methylenecyclopentane (91%) + 1-Methylcyclopentene (9%)

(1-Methylcyclopentyl)-
trimethylammonium
hydroxide

Methylenecyclo-
pentane
(91%)

1-Methyl-
cyclopentene
(9%)

Methylenecyclopentane has the less substituted double bond and is the major product.

14.12 NITROSATION OF AMINES

When solutions of sodium nitrite are acidified, a species called **nitrous acid** forms.

$$:\overset{..}{O}=\overset{..}{N}-\overset{..}{\underset{..}{O}}:^- + H^+ \rightleftharpoons :\overset{..}{O}=\overset{..}{N}-\overset{..}{O}H$$

Nitrite ion
(from sodium
nitrite, NaNO$_2$)

Nitrous acid

Nitrous acid reacts with amines and acts as a **nitrosating agent.** The nitrosation of amines is best illustrated by examining what happens when the amine is secondary. The amine acts as a nucleophile, attacking the nitrogen of the nitrosating agent.

$$R_2\overset{..}{N}\overset{|}{\underset{H}{:}} + \overset{\overset{:\overset{..}{O}}{\|}}{N}-\overset{..}{O}H \longrightarrow R_2\overset{+}{\underset{H}{N}}-\overset{..}{N}=\overset{..}{O}: + H\overset{..}{\underset{..}{O}}:^- \longrightarrow R_2\overset{..}{N}-\overset{..}{N}=\overset{..}{O}: + H_2\overset{..}{O}:$$

Secondary
amine

Nitrous
acid

N-Nitroso-
amine

Water

The product of this reaction is an *N*-nitrosoamine (also called a **nitrosamine**). Nitrosamines have been the object of much research as many of them are carcinogenic in animals. Nitrosamines are found in some foods and in tobacco smoke, as the following examples illustrate.

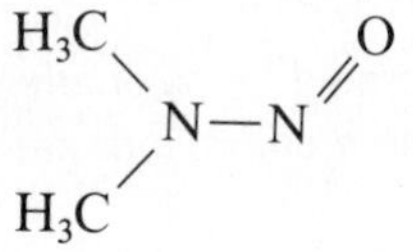

N-Nitrosodimethylamine (formed during tanning of leather; also found in beer and herbicides)

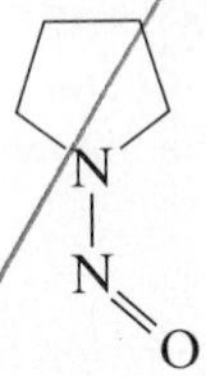

N-Nitrosopyrrolidine (formed when bacon that has been cured with sodium nitrite is fried)

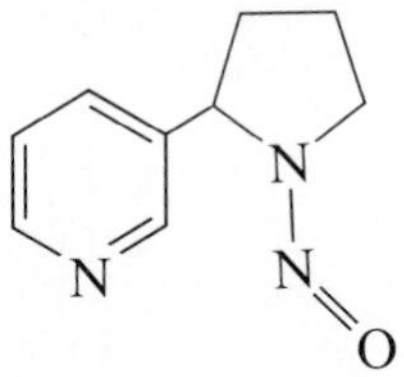

N-Nitrosonornicotine (present in tobacco smoke)

A possible source of nitrosamines in food is the use of sodium nitrite, $NaNO_2$, in prepared meats such as bacon and hot dogs. Trace amounts of sodium nitrite are added to the meat to inhibit growth of the bacterium responsible for botulism. This sodium nitrite also reacts with acid in our stomachs, however, to form nitrous acid which may form nitrosamines in our body. Whether this reaction poses a significant health risk is a subject of debate. Sodium nitrite is approved by the government for use in food because the benefit in preventing botulism poisoning outweighs the potential risk as a source of nitrosamines.

When primary amines are nitrosated, the N-nitroso compounds produced react further to form **diazonium ions.**

$$\underset{\text{Primary amine}}{R\ddot{N}H_2} \xrightarrow{\text{HONO}} \underset{\text{(Not isolable)}}{R\ddot{N}(H)-\ddot{N}=\ddot{O}:} \longrightarrow \underset{\text{Diazonium ion}}{R\overset{+}{N}\equiv N:}$$

Alkyl diazonium ions readily dissociate to carbocations and molecular nitrogen and are too unstable to be useful synthetically.

$$\underset{\text{Alkyl diazonium ion}}{R-\overset{+}{N}\equiv N:} \xrightarrow{\text{fast}} \underset{\text{Carbocation}}{R^+} + \underset{\text{Nitrogen}}{:N\equiv N:}$$

Aryl diazonium ions, on the other hand, play an important role in the synthesis of aromatic compounds.

$$\underset{\text{Aniline}}{C_6H_5NH_2} \xrightarrow[H_2O,\ 0-5°C]{NaNO_2,\ HCl} \underset{\text{Benzenediazonium chloride}}{C_6H_5\overset{+}{N}\equiv N:\ Cl^-}$$

Aryl diazonium ions are considerably more stable than their alkyl counterparts because aryl cations (Ar^+) do not form readily. In cold aqueous solution (0 to 5°C), aryl diazonium salts can be stored for reasonable periods of time and are intermediates in several important reactions that we will explore in the next section.

14.13 REACTIONS OF ARYL DIAZONIUM SALTS

Aryl diazonium ions are used as intermediates in the preparation of several classes of aromatic compounds, summarized in Figure 14.6. In each case a substitution reaction occurs in which the entering group becomes bonded to the ring position from which nitrogen departs. The following examples will illustrate the versatility of aryl diazonium ions in the preparation of ring-substituted aromatic compounds.

One of the most general methods for preparing phenols is by hydrolysis of the corresponding diazonium ion.

$$\overset{+}{Ar}N\equiv N\!: \;+\; H_2O \longrightarrow ArOH + H^+ + :N\equiv N:$$

Aryl diazonium   Water   A phenol   Nitrogen
ion

In this reaction the diazonium salt is prepared by adding an acid (such as sulfuric acid) to a solution of the arylamine in the presence of sodium nitrite. Heating the aqueous solution of the diazonium ion produces the desired phenol.

$$(CH_3)_2CH\!-\!\!\bigcirc\!\!-\!NH_2 \xrightarrow[\text{2. H}_2\text{O, heat}]{\text{1. NaNO}_2,\ \text{H}_2\text{SO}_4,\ \text{H}_2\text{O}} (CH_3)_2CH\!-\!\!\bigcirc\!\!-\!OH$$

p-Isopropylaniline   p-Isopropylphenol (73%)

NH₂ ring with Br $\xrightarrow[\text{2. H}_2\text{O, heat}]{\text{1. NaNO}_2,\ \text{H}_2\text{SO}_4,\ \text{H}_2\text{O, 0–5°C}}$ OH ring with Br
m-Bromoaniline   m-Bromophenol (66%)

PROBLEM 14.11
Outline a synthesis of *m*-bromophenol from benzene.

Aryl iodides may also be prepared using a diazonium salt intermediate. This reaction is especially useful since there is no electrophilic aromatic substitution reaction available for the preparation of aryl iodides. Addition of a solution of potassium iodide to the diazonium salt brings about the reaction.

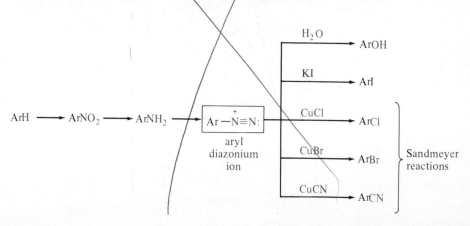

FIGURE 14.6 Flowchart showing the synthetic origin or aryl diazonium ions and their most useful transformations.

$$Ar\overset{+}{-}N\equiv N: + \ I^- \longrightarrow ArI \ + :N\equiv N:$$

Aryl diazonium   Iodide   Aryl   Nitrogen
Ion   ion   iodide

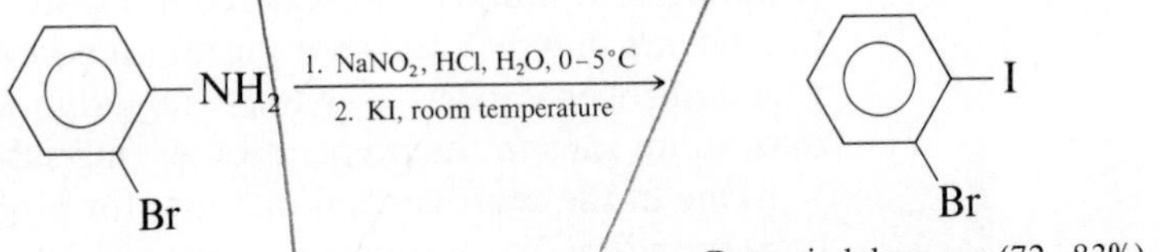

o-Bromoaniline   o-Bromoiodobenzene (72–83%)

Reactions similar to that illustrated above are used in the laboratory synthesis of *thyroxine,* a thyroid hormone used in the treatment of hypothyroidism.

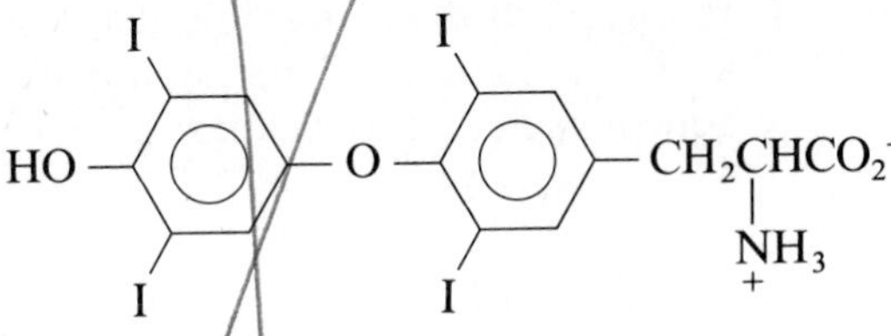

Thyroxine

Aryl chlorides and bromides may also be prepared from aryl diazonium ions. The diazonium ion is treated with copper(I) chloride or bromide, as appropriate:

$$Ar\overset{+}{N}\equiv N: \ \xrightarrow{\text{CuX}} \ ArX \ + :N\equiv N:$$

Aryl diazonium   Aryl chloride   Nitrogen
ion   or bromide

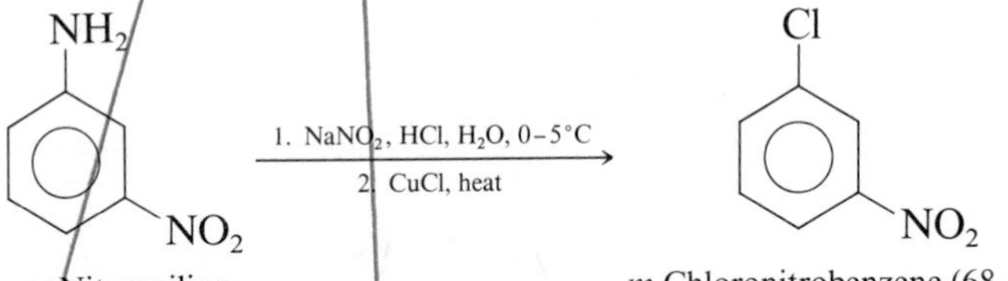

m-Nitroaniline   m-Chloronitrobenzene (68–71%)

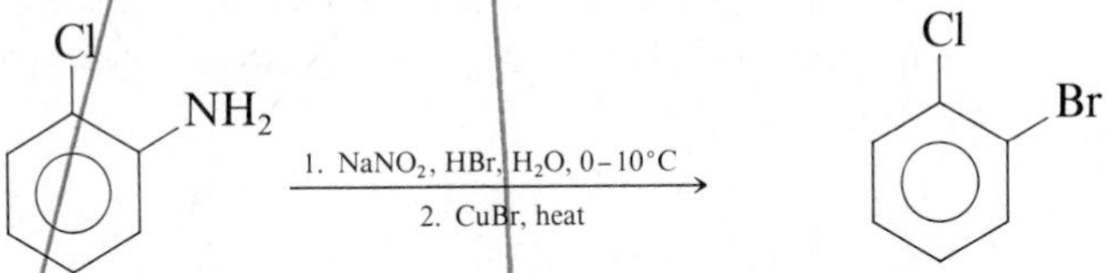

o-Chloroaniline   o-Bromochlorobenzene (89–95%)

Reactions that use copper(I) salts as reagents for replacement of nitrogen in diazonium salts are called **Sandmeyer reactions.** Copper(I) cyanide may also be used to prepare aromatic nitriles (aryl cyanides).

$$Ar\overset{+}{-}N\equiv N: \xrightarrow{\text{CuCN}} ArCN + :N\equiv N:$$

Aryl diazonium   Aryl   Nitrogen
ion   cyanide

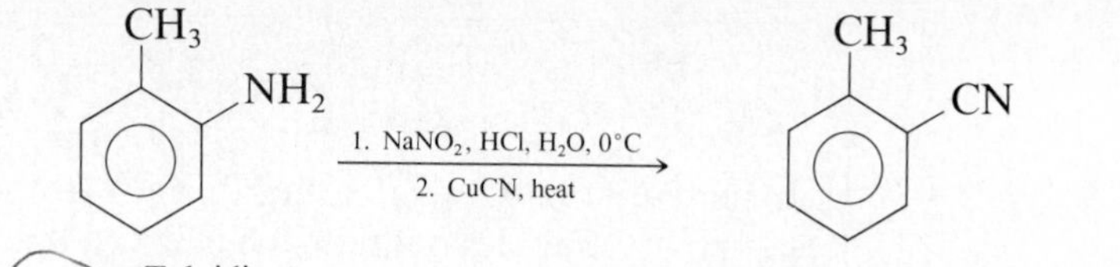

o-Toluidine → o-Methylbenzonitrile (64–70%)

1. NaNO$_2$, HCl, H$_2$O, 0°C
2. CuCN, heat

A reasonable question to ask is: "Why use the Sandmeyer reaction when aromatic halogenation reactions are possible?" (Recall from Chapter 6 that an aromatic ring can be halogenated directly.) The reason the Sandmeyer sequence would be used lies in the use of the *nitro group* as a directing group on an aromatic ring. Halogens are ortho, para-directing groups; nitro directs meta. The nitro group can therefore be used to direct the substitution on an aromatic ring in a meta orientation. The nitro group can then be replaced by a halogen by using the Sandmeyer reaction.

For example, *m*-dibromobenzene is only a minor product when bromobenzene is brominated but can be prepared efficiently from nitrobenzene by way of a diazonium salt. Bromination of nitrobenzene occurs in the meta position, giving *m*-bromonitrobenzene. Reduction of the nitro group gives *m*-bromoaniline. The second bromine is then introduced by the Sandmeyer sequence.

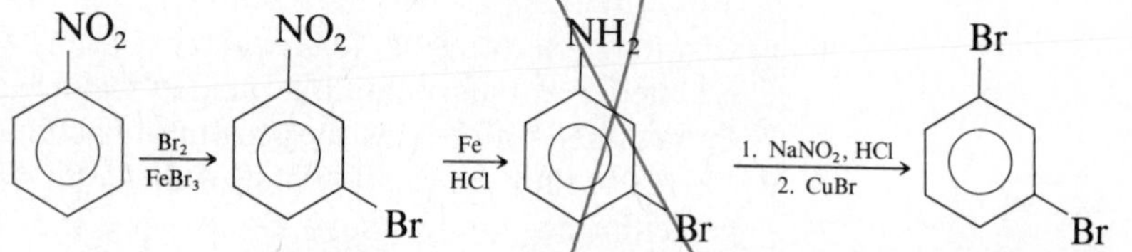

Nitrobenzene →(Br$_2$/FeBr$_3$) *m*-Bromonitrobenzene →(Fe/HCl) *m*-Bromoaniline →(1. NaNO$_2$, HCl 2. CuBr) *m*-Dibromobenzene

PROBLEM 14.12

Outline a method for the preparation of *m*-bromochlorobenzene from nitrobenzene.

14.14 AZO DYES

When aryl diazonium salts react with phenols and aryl amines, nitrogen is not lost. The two aryl groups are joined together by an azo (—N=N—) function in a reaction called **diazonium** or **azo coupling**.

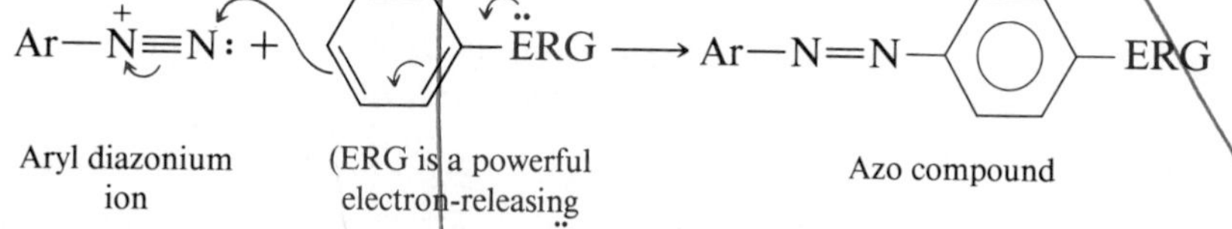

Ar—N≡N: + ⟨ring⟩—ERG ⟶ Ar—N=N—⟨ring⟩—ERG

Aryl diazonium ion — (ERG is a powerful electron-releasing group such as —ÖH or —N̈R$_2$) — Azo compound

The coupling of diazonium ions with phenols or other electron-rich aromatic compounds is a useful commercial reaction, as azo compounds are often highly colored

SULFA DRUGS

In the early 1930s it was learned that an azo dye called *Prontosil* had antibacterial activity. Subsequent research revealed that Prontosil exerts its antibacterial effect by first undergoing degradation to a compound called sulfanilamide.

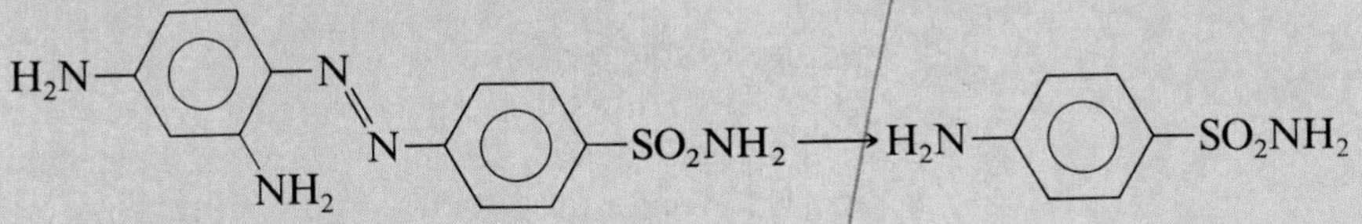

Prontosil   Sulfanilamide

Bacteria require *p*-aminobenzoic acid (PABA) in order to biosynthesize folic acid, an essential growth factor.

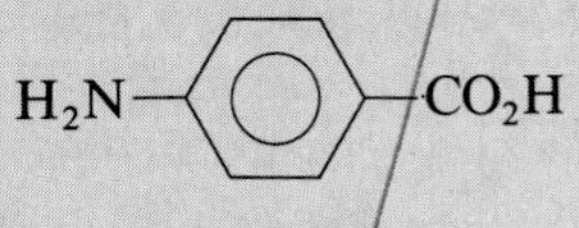

p-Aminobenzoic acid

Structurally, sulfanilamide resembles *p*-aminobenzoic acid, and is mistaken for it by the bacteria so that folic acid biosynthesis is inhibited and bacterial growth is halted. Animals obtain folic acid in their diet, not through biosynthesis, and thus sulfanilamide halts the growth of bacteria without harm to the host.

Sulfanilamide was the first of the **sulfa drugs.** Until the development of penicillin antibiotics, sulfa drugs were the most widely used antibacterial agents. Today they are used as topical agents and in veterinary medicine.

All the sulfa drugs retain the basic structure of sulfanilamide. Thousands of these compounds have been synthesized and tested for antibacterial activity. Some of those which were successful and marketed commercially include

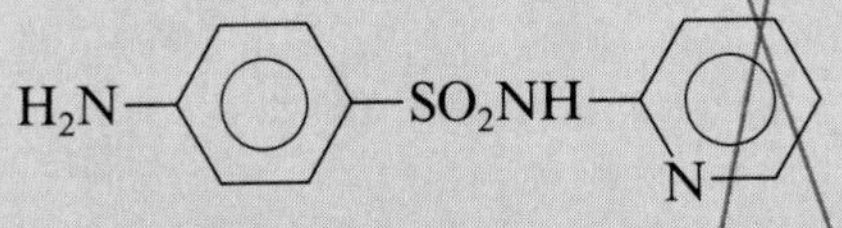

Sulfapyridine 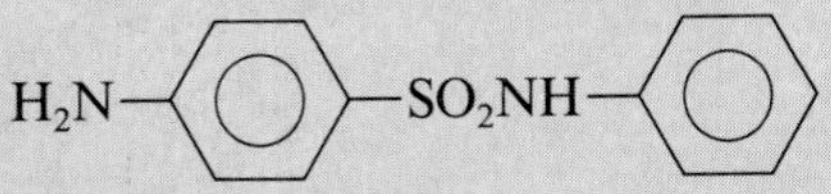   Sulfabenz

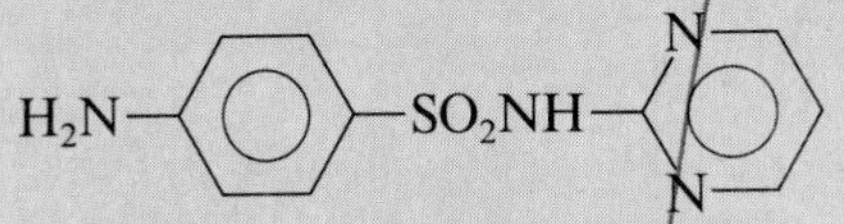

Sulfadiazine 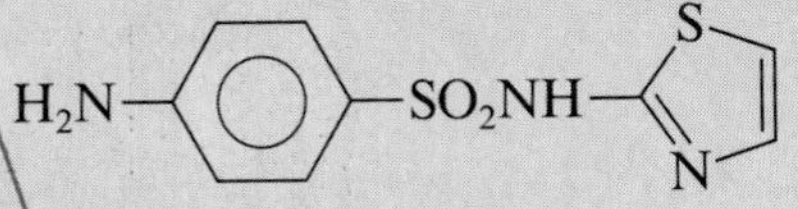   Sulfathiazole

and many of them are used as dyes. Two examples of azo dyes are shown: FD&C Orange No. 1 is used to artificially color foods; para red is used to dye textiles.

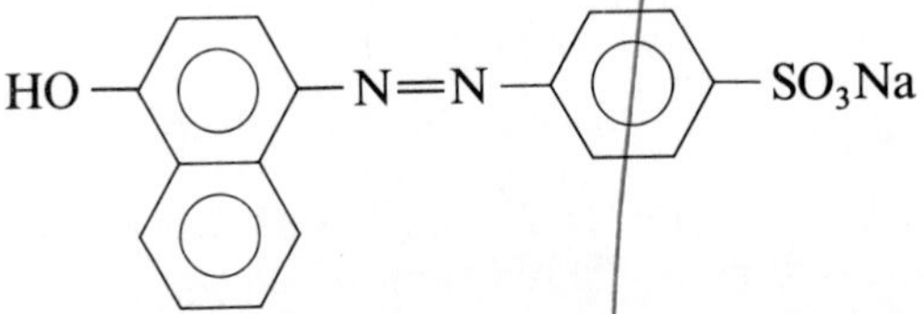

FD&C Orange No. 1 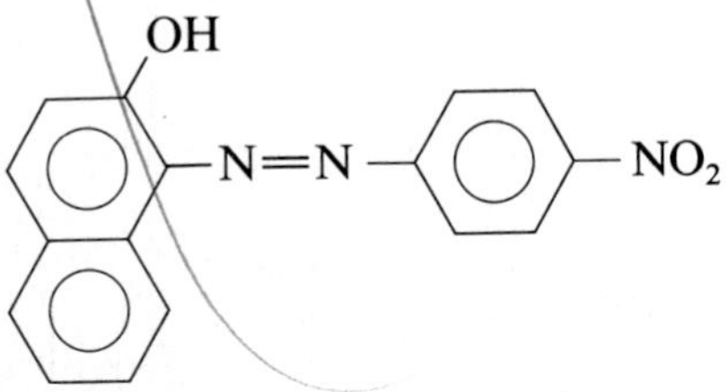   Para red

14.15 SPECTROSCOPIC ANALYSIS OF AMINES

Primary and secondary amines exhibit characteristic absorptions in their infrared spectra due to N—H vibrations. Primary amines exhibit *two* peaks in the range 3000 to 3500 cm^{-1}, while secondary amines exhibit only *one* peak in this region.

Two peaks observed for a primary amine

Two peaks observed for a primary amine $\rightarrow$ $\begin{cases} H \\ H \end{cases}$ $\ddot{N}$—R

One peak observed for a secondary amine $\rightarrow$ H—$\ddot{N}$ $\begin{matrix} R \\ R \end{matrix}$

The segments of the infrared spectra of a primary amine and a secondary amine shown in Figure 14.7 illustrate this difference.

14.16 SUMMARY

Amines are compounds of the type shown, where R, R′, and R″ are alkyl or aryl groups.

Primary amine   Secondary amine   Tertiary amine

Alkylamines are named by adding the ending *-amine* to the name of the alkyl group that bears the nitrogen (Section 14.1). Secondary and tertiary amines are named as *N*-substituted derivatives of primary amines. Arylamines are named as derivatives of aniline.

$$CH_3CH_2CH_2CH_2NH_2$$

Butylamine
(primary amine)

$HNCH_2CH_3$

N-ethylaniline
(secondary amine)

$$(CH_3)_3N$$

Trimethylamine
(tertiary amine)

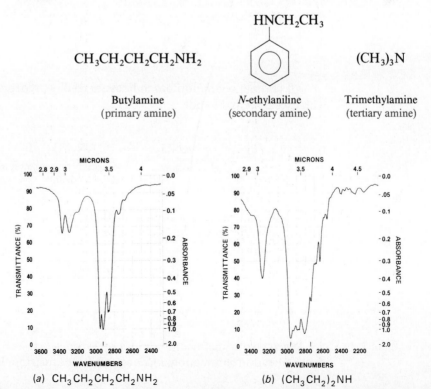

FIGURE 14.7 Portions of the infrared spectra of (*a*) a primary amine, butylamine, and (*b*) a secondary amine, diethylamine. Primary amines exhibit two peaks due to N—H stretching in the 3300 to 3500 cm^{-1} region, while secondary amines show only one.

(*a*) $CH_3CH_2CH_2CH_2NH_2$

(*b*) $(CH_3CH_2)_2NH$

Amines have a pyramidal arrangement of bonds to sp^3 hybridized nitrogen (Section 14.2). The unshared pair of electrons on nitrogen is responsible for the basicity and nucleophilicity of amines.

Alkylamines are weak bases; they can use their unshared electron pair to abstract a proton from a suitable donor (Section 14.5). Basicity constants K_b of alkylamines are in the range 10^{-3} to 10^{-5} (pK_b of 3 to 5).

$$R_3N: + H_2O \rightleftharpoons R_3\overset{+}{N}H + HO^- \qquad K_b = \frac{[R_3\overset{+}{N}H][HO^-]}{[R_3N]}$$

Arylamines are up to 1 million (10^6) times less basic than alkylamines because the nitrogen lone pair of electrons of an arylamine is delocalized into the π system of the aromatic ring.

Primary alkylamines may be prepared by alkylation of ammonia (Section 14.6). A drawback of this method is contamination of the desired primary amine with secondary and tertiary amines and quaternary ammonium salts. Alkylamines may also be prepared by reduction of amides or nitriles (Section 14.7). Methods for the preparation of alkylamines are summarized in Table 14.1.

Arylamines are most often prepared by reduction of the corresponding nitroarene (Section 14.8).

$$ArNO_2 \xrightarrow{\text{reduce}} ArNH_2$$

Nitroarene \qquad\qquad Primary arylamine

Typical reducing agents include iron or tin in hydrochloric acid. Catalytic hydrogenation using a platinum, palladium, or nickel catalyst is also effective.

Amines are alkylated by primary alkyl halides (Section 14.10). The product may be a secondary or tertiary amine or a tetraalkylammonium salt.

$$RNH_2 \xrightarrow{R'CH_2X} RNHCH_2R' \xrightarrow{R'CH_2X} RN(CH_2R')_2 \xrightarrow{R'CH_2X} R\overset{+}{N}(CH_2R')_3 \ X^-$$

Primary amine \qquad Secondary amine \qquad Tertiary amine \qquad Quaternary ammonium salt

Alkyltrimethylammonium iodides are readily prepared by alkylation of amines with excess methyl iodide.

$$RNH_2 \xrightarrow[\text{excess}]{CH_3I} R\overset{+}{N}(CH_3)_3 \ I^-$$

Amine \qquad\qquad Alkyltrimethylammonium iodide

The iodide counterion of an alkyltrimethylammonium iodide may be exchanged for hydroxide by treatment with moist silver oxide.

$$R\overset{+}{N}(CH_3)_3 \ I^- \xrightarrow{Ag_2O, \ H_2O} R\overset{+}{N}(CH_3)_3 \ \overset{-}{O}H$$

Alkyltrimethylammonium iodide \qquad\qquad Alkyltrimethylammonium hydroxide

When alkyltrimethylammonium hydroxides that have β hydrogens are heated, they undergo an elimination, forming an alkene and trimethylamine (Section 14.11).

TABLE 14.1
Preparation of Alkylamines

Reaction (section) and comments	General equation and specific example
Alkylation of ammonia (Section 14.6) Ammonia can act as a nucleophile toward primary and some secondary alkyl halides to give primary amines. Yields tend to be modest because the primary amine is itself a nucleophile and undergoes alkylation. Alkylation of ammonia can lead to a mixture containing a primary amine, a secondary amine, a tertiary amine, and a quaternary ammonium salt.	$RX + 2NH_3 \longrightarrow RNH_2 + NH_4X$ Alkyl halide Ammonia Alkylamine Ammonium halide $C_6H_5CH_2Cl \xrightarrow{NH_3(8\ mol)} C_6H_5CH_2NH_2 + (C_6H_5CH_2)_2NH$ Benzyl chloride Benzylamine Dibenzylamine (1 mol) (53%) (39%)
Reduction of amides (Section 14.7) Lithium aluminum hydride reduces the carbonyl group of an amide to a methylene group. Primary, secondary, or tertiary amines may be prepared by proper choice of the starting amide.	$RCNR'_2 \xrightarrow{reduce} RCH_2NR'_2$ (amide carbonyl O shown above RCNR'_2) Amide Amine $CH_3CNHC(CH_3)_3 \xrightarrow[2.\ H_2O]{1.\ LiAlH_4} CH_3CH_2NHC(CH_3)_3$ N-tert-Butylacetamide N-Ethyl-tert-butylamine (60%)
Reduction of nitriles (Section 14.7) Nitriles are reduced to primary amines by lithium aluminum hydride and by catalytic hydrogenation.	$RC \equiv N \xrightarrow{reduce} RCH_2NH_2$ Nitrile Primary amine $\triangleright\!-CN \xrightarrow[2.\ H_2O]{1.\ LiAlH_4} \triangleright\!-CH_2NH_2$ Cyclopropyl cyanide Cyclopropylmethanamine (75%)

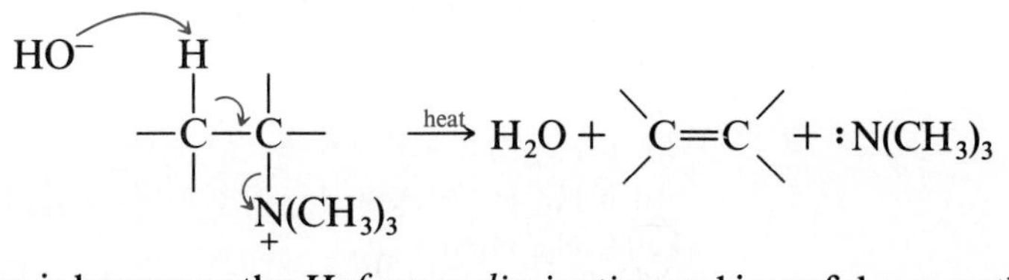

$$HO^- \quad H \qquad \qquad$$

$$-\overset{|}{\underset{|}{C}}-\overset{|}{\underset{\underset{+}{N(CH_3)_3}}{C}}- \xrightarrow{heat} H_2O + \overset{}{\underset{}{C}}=\overset{}{\underset{}{C} + :N(CH_3)_3}$$

This reaction is known as the *Hofmann elimination* and is useful as a method for the preparation of alkenes. The hydroxide ion tends to abstract a proton from the less substituted β carbon to produce the less substituted alkene.

Nitrosation of amines occurs on acidification of a solution containing sodium nitrite in the presence of an amine (Section 14.12). Secondary amines react with nitrosating agents to give *N*-nitrosoamines.

$$R_2NH \xrightarrow{NaNO_2,\ HCl} R_2N-N\overset{O}{\underset{\cdot\cdot}{\diagup\!\!\!\parallel}}$$

Secondary amine   *N*-Nitrosoamine

Primary arylamines yield aryl diazonium salts.

$$ArNH_2 \xrightarrow{NaNO_2,\ HCl} Ar\overset{+}{N} \equiv N:$$

Primary arylamine   Aryl diazonium ion

Aryl diazonium salts are useful synthetic intermediates (Section 14.13). Once formed, the diazonium salt may be treated with an appropriate reagent to give a phenol, aryl halide, or aryl cyanide (Figure 14.6). These reactions using copper(I) salts are called *Sandmeyer reactions.*

Aryl diazonium salts also react with strongly activated aromatic rings to form azo compounds, many of which are highly colored and are used as dyes (Section 14.14).

$$Ar\overset{+}{N}{\equiv}N: \; + \; Ar'H \; \longrightarrow \; ArN{=}NAr' + H^+$$

Aryl diazonium ion	Arylamine or phenol	Azo compound

ADDITIONAL PROBLEMS

Structure, Nomenclature, and Properties

14.13 Write structural formulas for all the amines of molecular formula $C_4H_{11}N$. Give an acceptable name for each one and classify it as a primary, secondary, or tertiary amine.

14.14 Provide a structural formula for each of the following compounds.
(a) Heptylamine
(b) 2-Ethyl-1-butanamine
(c) N-Ethylpentylamine
(d) Dibenzylamine
(e) Tetraethylammonium hydroxide
(f) N-Ethyl-4-methylaniline
(g) 2,4-Dichloroaniline
(h) N,N-Dimethylaniline

14.15 Give the structures and an acceptable name for all the isomers of molecular formula C_7H_9N that contain a benzene ring.

14.16 Name each of the following amines, indicating whether the amine is primary, secondary, or tertiary.
(a) $CH_3CH_2CH_2NHCH_2CH_2CH_3$
(b) $CH_3CH_2CH_2NCH_2CH_2CH_2CH_3$
                     CH_3

(c) $(CH_3)_2CHCH_2CH_2NH_2$
(d) $(CH_3CH_2CH_2)_2NC_6H_5$
(e) $C_6H_5CH_2NHC_6H_5$

14.17 Many naturally occurring nitrogen compounds and many nitrogen-containing drugs are better known by common names than by their systematic names. A few of these are given below. Write a structural formula for each one.
(a) 4-Aminobutanoic acid, better known as *γ-aminobutyric acid,* or *GABA,* involved in metabolic processes occurring in the brain
(b) 2-(3,4,5-Trimethoxyphenyl)ethylamine, better known as *mescaline,* a hallucinogen obtained from the peyote cactus
(c) *trans*-2-Phenylcyclopropylamine, better known as *tranylcypromine,* an antidepressant drug
(d) N-Benzyl-N-methyl-2-propynylamine, better known as *pargyline,* a drug used to treat high blood pressure
(e) 1-Phenyl-2-propanamine, better known as *amphetamine,* a stimulant

14.18 Several amine-containing compounds are used medicinally as local anesthetics. The structure of one of these, lidocaine (also known as Xylocaine), is shown. To increase the water solubility of these compounds when given by injection, they are used as the hydrochloride salt. Draw the structure of lidocaine hydrochloride.

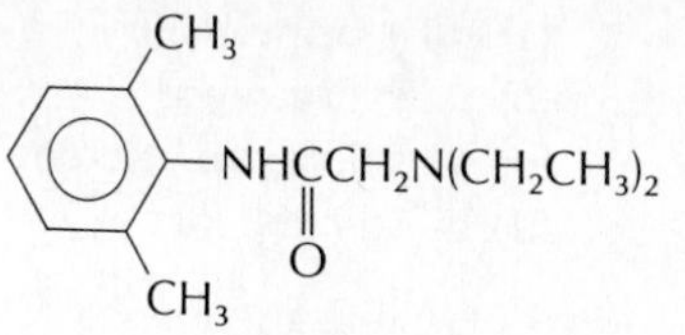

$$CH_3$$

(ring)—$NHCCH_2N(CH_2CH_3)_2$
$\overset{\|}{O}$

$$CH_3$$

Lidocaine

14.19 Arrange the following compounds in order of decreasing basicity.

$$\overset{O}{\overset{\|}{}}$$

$C_6H_5NH_2$ $C_6H_5CH_2NH_2$ $C_6H_5NHCCH_3$

Aniline   Benzylamine   Acetanilide

14.20 *Physostigmine* is an alkaloid obtained from a West African plant; it is used in the treatment of glaucoma. Treatment of physostigmine with methyl iodide gives a single quaternary ammonium salt. What is the structure of this salt?

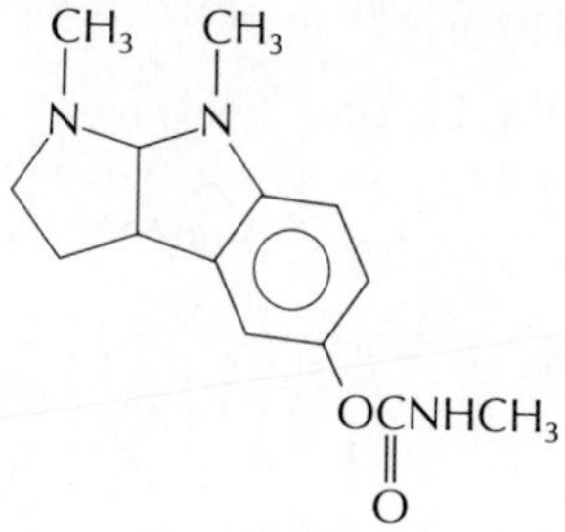

$$CH_3 \quad CH_3$$
$$N \qquad N$$

$$OCNHCH_3$$
$$\overset{\|}{O}$$

Physostigmine

Reactions

14.21 Give the structure of the expected product formed when benzylamine reacts with each of the following reagents:

(a) Hydrogen bromide
(b) Acetic acid
(c) Acetyl chloride
(d) Acetone
(e) Excess methyl iodide

14.22 Repeat Problem 14.21 for aniline.

14.23 The isomeric ring-methylated aniline derivatives are known as toluidines. Thus *p*-methylaniline is called *p*-toluidine. Write the structure of the product formed on reaction of *p*-toluidine with each of the following:

(a) Sodium nitrite, aqueous sulfuric acid, 0 to 5°C
(b) Product of (a), heated in aqueous acid
(c) Product of (a), treated with copper(I) chloride
(d) Product of (a), treated with potassium iodide
(e) Product of (a), treated with copper(I) bromide
(f) Product of (a), treated with copper(I) cyanide
(g) Product of (a), treated with phenol (product is $C_{12}H_{10}N_2O$)

Synthesis

14.24 Outline a sequence of reactions that will convert:

(a) Benzoic acid into *N*-methylbenzylamine
(b) Benzyl bromide ($C_6H_5CH_2Br$) into 2-phenylethylamine

14.25 Outline syntheses of each of the following aromatic compounds from benzene:

(a) o-Isopropylaniline
(b) p-Chloroaniline
(c) m-Chloroiodobenzene
(d) p-Bromophenol

14.26 Outline a sequence of steps that would convert p-nitrotoluene into:
(a) p-Methylaniline
(b) p-Chlorotoluene
(c) p-Cyanotoluene
(d) p-Iodotoluene
(e) p-Methylphenol (p-cresol)

14.27 Identify compounds A to C in the following sequence of reactions:

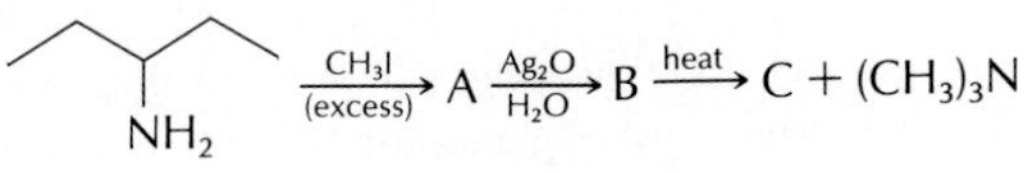

$$\underset{NH_2}{\text{(structure)}} \xrightarrow[\text{(excess)}]{CH_3I} A \xrightarrow[H_2O]{Ag_2O} B \xrightarrow{heat} C + (CH_3)_3N$$

14.28 Write out the sequence of steps that would allow preparation of 1-octene by starting with octanamide.

Miscellaneous Problems

14.29 Provide a reasonable explanation for each of the following observations:
(a) 4-Methylpiperidine has a higher boiling point than N-methylpiperidine.

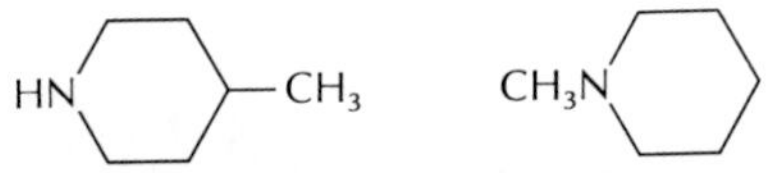

HN⟨ ⟩—CH₃   CH₃N⟨ ⟩

4-Methylpiperidine   N-Methylpiperidine
(bp 129°C)   (bp 106°C)

(b) Two isomeric quaternary ammonium salts are formed in comparable amounts when 4-tert-butyl-N-methylpiperidine is treated with benzyl chloride.

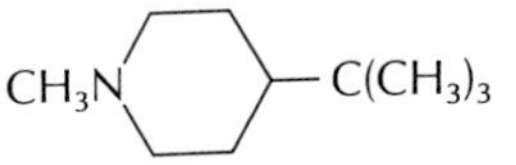

CH₃N⟨ ⟩—C(CH₃)₃   4-tert-Butyl-N-methylpiperidine

14.30 A nitrogen-containing heterocycle of unknown structure (compound D, $C_6H_{13}N$) was treated as shown in the equations given to form compound E. Compound E underwent the indicated reactions to give 2,3-dimethyl-1,3-butadiene and trimethylamine. What are the structures of compounds D and E?

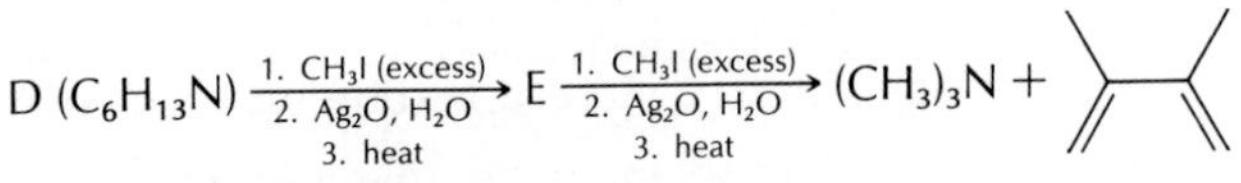

$$D\ (C_6H_{13}N) \xrightarrow[\substack{2.\ Ag_2O,\ H_2O \\ 3.\ heat}]{1.\ CH_3I\ (excess)} E \xrightarrow[\substack{2.\ Ag_2O,\ H_2O \\ 3.\ heat}]{1.\ CH_3I\ (excess)} (CH_3)_3N + \text{(structure)}$$

14.31 When tetramethylammonium hydroxide is heated to 130°C, trimethylamine and methanol are formed. Give an explanation for this result.

14.32 Explain how infrared spectroscopy could be used to distinguish among the following three amines.

⟨ ⟩—NH₂   ⟨ ⟩—NHCH₃   ⟨ ⟩—N(CH₃)₂

14.33 Compound F ($C_8H_{11}N$) shows one peak at 3330 cm^{-1} in its infrared spectrum. The ${}^1H$ nmr spectrum consists of four singlets at $\delta = 7.3$ ppm (5H), 3.8 ppm (2H), 2.4 ppm (3H), and 1.4 ppm (1H). Give a structure for compound F consistent with this data.

CHAPTER 15

CARBOHYDRATES

LEARNING OBJECTIVES

This chapter focuses on one of the major classes of substances common to living systems, carbohydrates. Upon completion of this chapter you should be able to:

- Classify carbohydrates according to their structure.
- Represent the structure of a carbohydrate by using a Fischer projection.
- Classify a carbohydrate as having the D or L configuration from its Fischer projection.
- Construct a Haworth projection of the furanose form of a pentose.
- Construct a Haworth projection of the pyranose form of a hexose.
- Explain how the α and β configurations of a carbohydrate differ.
- Use chemical structures to describe the equilibrium process associated with the term mutarotation.
- Describe the structural features that are characteristic of glycosides.
- Write the chemical structure of a disaccharide.
- Explain the structural difference between cellulose and starch.
- Describe the structural features that characterize reducing sugars.
- Explain the chemical tests used to identify reducing sugars.
- Write a chemical equation describing osazone formation.

At this point we have completed our look at the major functional groups of organic chemistry. In the remaining chapters you will see how organic molecules participate in the chemistry of living systems. As you probably know, an introductory course in **biochemistry** is every bit as lengthy and detailed as one in organic chemistry. Of necessity, our treatment of biochemistry will be brief, with an emphasis on how the principles of organic chemistry apply to the study of living organisms.

Carbohydrates are one of the major classes of substances common to living systems, the others being lipids, proteins, and nucleic acids. Carbohydrates are very familiar to us—many of those which are water-soluble are called *sugars*. Carbohydrates constitute a substantial portion of the food we eat and provide most of the energy that keeps the human engine running. Carbohydrates are structural components of the walls of plant cells and the wood of trees. Genetic information is stored and transferred by way of nucleic acids, complex molecules which have sugar-phosphate backbones. Nucleic acids will be discussed in Chapter 18.

Most of this chapter is devoted to carbohydrate structure. You will see how the principles of stereochemistry can be used to aid in our understanding of this complex subject. We will also look at some chemical reactions of carbohydrates. Most of these reactions are similar to the reactions of alcohols, aldehydes, ketones, and acetals that you have encountered in previous chapters.

15.1 CLASSIFICATION OF CARBOHYDRATES

The word *carbohydrates* is derived from "hydrates of carbon" because the molecular formulas of many carbohydrates correspond to $C_n(H_2O)_m$. It is better to define a carbohydrate as a **polyhydroxy aldehyde** or a **polyhydroxy ketone.**

The basis of carbohydrate classification is the word **saccharide,** from the Latin word for sugar, *saccharum.* A **monosaccharide** is a simple carbohydrate, one that cannot be hydrolyzed to a smaller carbohydrate. Glucose ($C_6H_{12}O_6$), for example, is a monosaccharide. A **disaccharide** gives two monosaccharide molecules on hydrolysis. The two monosaccharides formed by hydrolysis of a disaccharide may be the same or may be different. Sucrose—common table sugar—is a disaccharide that yields one molecule of glucose and one of fructose on hydrolysis.

$$\underset{\text{Sucrose}}{C_{12}H_{22}O_{11}} + \underset{\text{Water}}{H_2O} \longrightarrow \underset{\text{Glucose}}{C_6H_{12}O_6} + \underset{\text{Fructose}}{C_6H_{12}O_6}$$

An **oligosaccharide** (from the Greek *oligos,* "few") yields 3 to 10 monosaccharides on hydrolysis. **Polysaccharides** are hydrolyzed to more than 10 monosaccharides. Cellulose and starch are both polysaccharides that give thousands of glucose molecules when completely hydrolyzed.

Over 200 different monosaccharides are known. They can be grouped according to the number of carbon atoms they contain and whether they are polyhydroxy aldehydes **(aldoses)** or polyhydroxy ketones **(ketoses).** The suffix *-ose* is added to the stem name that defines the number of carbon atoms in the chain. Thus a monosaccharide having four, five, or six carbons, respectively, is called a tetrose, pentose, or hexose. In traditional nomenclature, monosaccharides that are aldehydes are designated by the prefix *aldo-*; those that are ketones are designated by the prefix *keto-*. Thus an aldopentose is a monosaccharide that has five carbons, one of which is the carbonyl group of an aldehyde. Modern systematic nomenclature does not use the *aldo-* and *keto-* prefixes. An aldopentose is simply a *pentose* and a ketopentose is referred to as a *pentulose.*

15.2 GLYCERALDEHYDE AND THE D-L SYSTEM OF STEREOCHEMICAL NOTATION

The German chemist Emil Fischer recognized in the late nineteenth century that stereochemistry was the key to understanding carbohydrate structure. Fischer was responsible for many of the most important fundamental discoveries in the areas of

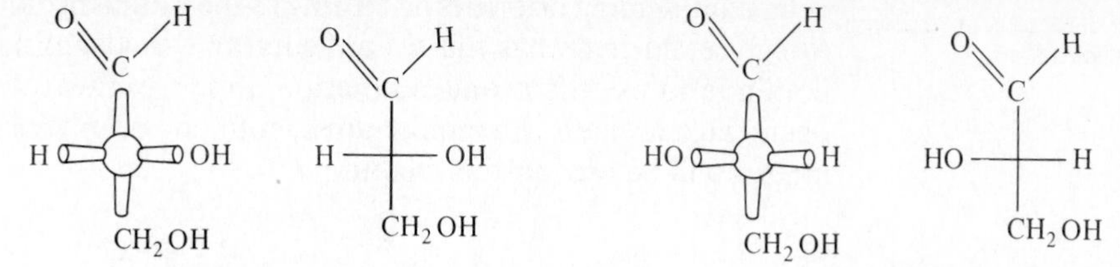

(+)-Glyceraldehyde (D-glyceraldehyde) (−)-Glyceraldehyde (L-glyceraldehyde)

FIGURE 15.1 Three-dimensional representations and Fischer projection formulas of the enantiomers of glyceraldehyde.

carbohydrate and protein chemistry and was awarded the Nobel Prize in Chemistry in 1902. The artistic device used by Fischer to represent stereochemistry in chiral molecules was described in Section 7.4. The representation is known as the *Fischer projection* and is still widely used today, particularly for the study of carbohydrates.

Figure 15.1 illustrates the use of Fischer projections to represent the enantiomers of glyceraldehyde. Glyceraldehyde is an *aldotriose* and is the smallest chiral carbohydrate. The molecule is oriented in the Fischer projection so that the carbonyl group is at the "top." In this orientation the hydroxyl attached to the chiral carbon (C-2) points to the right in (+)-glyceraldehyde and to the left in (−)-glyceraldehyde.

Techniques for determining the absolute configuration of chiral molecules (Section 7.7) were not developed until the 1950s, so it was not possible for Fischer and his contemporaries to know with certainty the absolute configuration of a carbohydrate. A system was adopted based on the arbitrary assumption, later shown to be correct, that the enantiomers of glyceraldehyde have the signs of rotation and absolute configurations shown in Figure 15.1. Two stereochemical descriptors were defined, D and L. The absolute configuration of (+)-glyceraldehyde, as depicted in the figure, was said to be D. The spatial arrangement of atoms in the enantiomer, (−)-glyceraldehyde, was defined as the L configuration. *Compounds having a spatial arrangement of substituents analogous to those of* D-(+)- *and* L-(−)-*glyceraldehyde are said to have the* D *and* L *configurations, respectively.*

PROBLEM 15.1

Identify each of the following Fischer projections as either D- or L-glyceraldehyde.

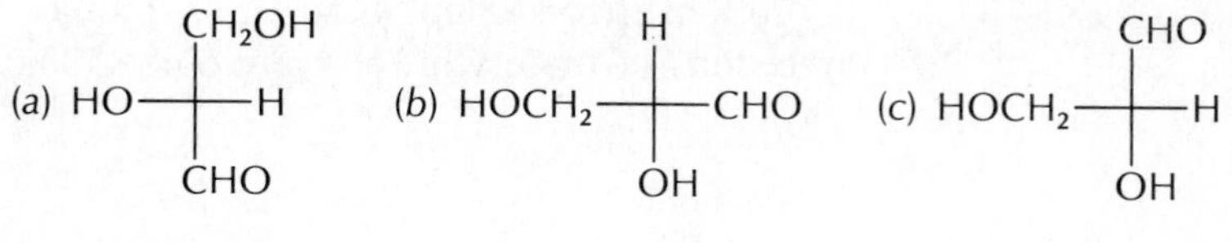

SAMPLE SOLUTION

(a) Redraw the Fischer projection so as to more clearly show the true spatial orientation of the groups. Next, reorient the molecule so that its relationship to the glyceraldehyde enantiomers in Figure 15.1 is apparent.

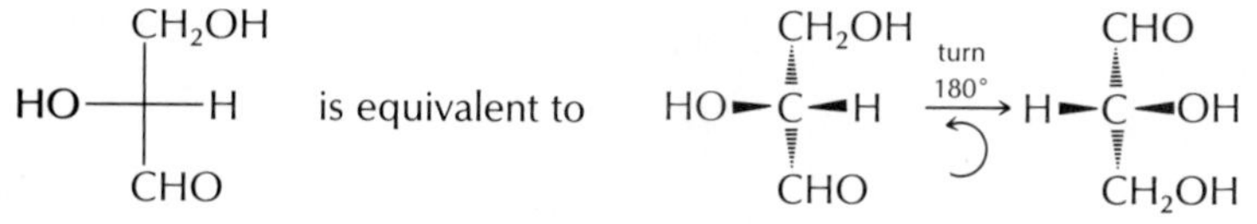

The structure is the same as that of (+)-glyceraldehyde in the figure. It is D-glyceraldehyde.

Why not describe the absolute configuration of carbohydrates by using the *R*-*S* notational system described in Chapter 7? This can certainly be done, as any chiral molecule can be described in terms of the *R* or *S* absolute configurations of its chiral centers. Assigning priorities according to the Cahn-Ingold-Prelog rules (Section 7.8), (+)-glyceraldehyde has the R configuration. Biochemists and biologists, however, continue to use the D and L notation almost exclusively since so many naturally occurring saccharides contain three, four, or even more chiral centers. The D-L system will be used in this chapter.

15.3 THE ALDOTETROSES

The *aldotetroses* are not abundant in nature; however, examination of their structures will illustrate Fischer projection formulas of compounds with more than one chiral center.

The aldotetroses are the four stereoisomers of 2,3,4-trihydroxybutanal. Fischer projection formulas are constructed by orienting the molecule in an eclipsed conformation with the aldehyde group at the top, as shown in Figure 15.2. The four carbon atoms define the main chain of the Fischer projection and are arranged vertically. Horizontal bonds are directed outward, vertical bonds back.

This aldotetrose is called D-erythrose, because the configuration at the highest-numbered chiral center (i.e., farthest from the carbonyl) is analogous to that of D-glyceraldehyde. In other words, the hydroxyl attached to the highest-numbered chiral center points to the right when the aldehyde group is at the top in a Fischer projection. The mirror image of D-erythrose is L-erythrose. The hydroxy group points to the left in a Fischer projection of L-erythrose.

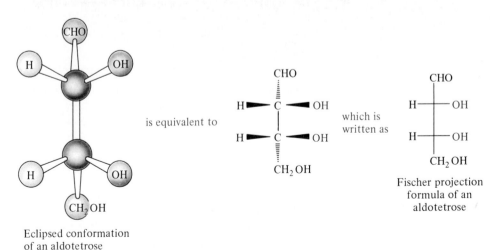

Highest-numbered chiral center has configuration analogous to that of D-glyceraldehyde

D-Erythrose   L-Erythrose

Highest-numbered chiral center has configuration analogous to that of L-glyceraldehyde

Both hydroxyl groups attached to chiral centers are on the same side in Fischer projections of the erythrose enantiomers. The remaining two aldotetrose stereo-

Eclipsed conformation of an aldotetrose

is equivalent to

which is written as

Fischer projection formula of an aldotetrose

FIGURE 15.2   Constructing the Fischer projection of an aldotetrose.

isomers have their hydroxyl groups on opposite sides in Fischer projection formulas. They are *diastereomers* (Section 7.12) of D- and L-erythrose and are called D- and L-threose. The D and L descriptors again specify the configuration of the highest-numbered chiral center. D-Threose and L-threose are enantiomers of each other.

The Fischer projections: (left) CHO at top (C1), HO—H at C2, H—OH at C3, CH₂OH at C4 (C4) labeled D-Threose, with note "Highest-numbered chiral center has configuration analogous to that of D-glyceraldehyde"; (right) CHO at top (C1), H—OH at C2, HO—H at C3, CH₂OH at C4 labeled L-Threose, with note "Highest-numbered chiral center has configuration analogous to that of L-glyceraldehyde"

PROBLEM 15.2

Which of the four aldotetroses just discussed is the following?

15.4 ALDOPENTOSES AND ALDOHEXOSES

The most common saccharides in living systems are the aldopentoses and aldohexoses. Aldopentoses have three chiral centers, and thus there are eight possible stereoisomers ($2^3 = 8$) (Section 7.12). The aldopentose stereoisomers are divided into a set of four D aldopentoses and an enantiomeric set of four L aldopentoses. Fischer projection formulas of the D stereoisomers are given in Figure 15.3.

Among the aldopentoses, D-ribose is a component of many biologically important substances, most notably the ribonucleic acids (Chapter 18). D-Xylose is also very abundant and may be isolated by hydrolysis of the polysaccharides present in corncobs and the wood of trees.

PROBLEM 15.3

L-(+)-Arabinose is a naturally occurring L sugar. It is obtained by acid hydrolysis of the polysaccharide present in mesquite gum. Write a Fischer projection for L-(+)-arabinose.

The aldohexoses include some familiar monosaccharides, as well as one of the most abundant organic compounds on earth, D-glucose. With four chiral centers, 16 stereoisomeric aldohexoses are possible: eight D isomers and eight L isomers. All are known, either as naturally occurring substances or as the products of laboratory synthesis. The eight D aldohexoses are given in Figure 15.3. Remember it is the orientation of the highest-numbered chiral center that identifies a carbohydrate as belonging to the D series. Note that the sign of rotation of a carbohydrate in the D series may be either (+) or (−).

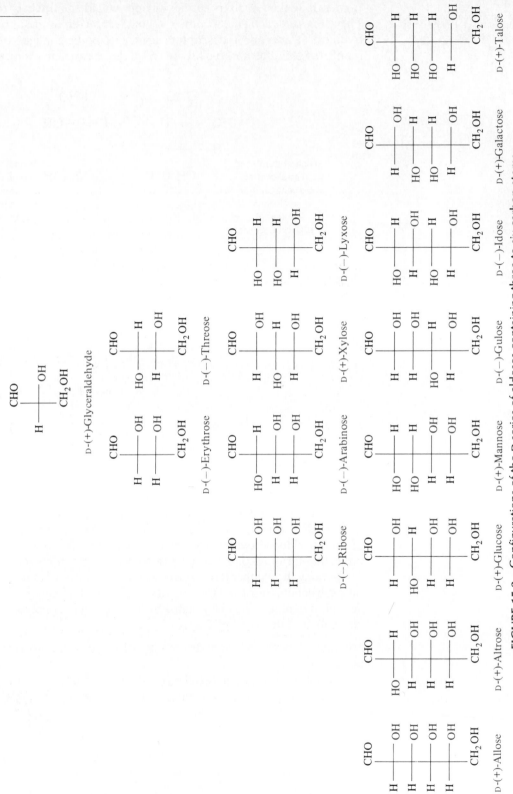

FIGURE 15.3 Configurations of the D series of aldoses containing three to six carbon atoms.

PROBLEM 15.4

Name the following sugar.

```
              CHO
        H ──────── OH
        H ──────── OH
        H ──────── OH
       HO ──────── H
             CH₂OH
```

Of all the monosaccharides, D-(+)-glucose is the best known and the most abundant. Its formation from carbon dioxide, water, and sunlight is the central theme of photosynthesis. Carbohydrate formation by photosynthesis is estimated to be on the order of 100 billion tons per year, a source of energy utilized directly or indirectly by all higher forms of life on the planet. Glucose was isolated from raisins in 1747 and by hydrolysis of starch in 1811. The structure of glucose was determined, in work culminating in 1900, by Emil Fischer.

Glucose is also known as *dextrose.* The term dextrose comes from the optical rotation of D-glucose. An optical rotation in the clockwise (+) direction is dextrorotatory, from the Latin *dextro,* meaning "to the right." *Blood sugar* is primarily glucose. The sugar level in the blood rises several hours after eating a meal; however, excessive sugar in the blood may be an indication of a diabetic condition.

D-(+)-Galactose is a constituent of numerous polysaccharides. It is best obtained by acid hydrolysis of the disaccharide lactose (milk sugar) (Section 15.11). L-(−)-Galactose also occurs naturally and can be obtained by hydrolysis of flaxseed gum and agar. The principal source of D-(+)-mannose is hydrolysis of the polysaccharide of the ivory nut, a large nutlike seed obtained from a South American palm.

15.5 CYCLIC FORMS OF CARBOHYDRATES: FURANOSE FORMS

Aldoses have two types of functionality, an aldehyde group and the hydroxyl groups. Recall from Chapter 11 (Section 11.8) that nucleophilic addition of an alcohol to an aldehyde leads to formation of a hemiacetal. When the hydroxyl and aldehyde functions are part of the same molecule and the hydroxyl is three or four carbons distant from the carbonyl, a *cyclic hemiacetal* is formed. Figure 15.4 illustrates the relationship between two hydroxy aldehydes and their corresponding cyclic hemiacetals.

Cyclic hemiacetal formation is most favorable when the ring that results is five- or six-membered. Five-membered cyclic hemiacetals of carbohydrates are called **furanose** forms; six-membered ones are called **pyranose** forms. The ring carbon that is derived from the carbonyl group, that is, the one that bears two oxygen substituents, is called the **anomeric carbon.**

Aldoses exist almost exclusively as their cyclic hemiacetals; very little of the open-chain (free aldehyde) form is present. To understand the structures of carbohydrates, it is necessary to be able to translate Fischer projection formulas of carbohydrates into their cyclic hemiacetal forms.

Consider, for example, the furanose form of D-ribose, the carbohydrate which forms the backbone of ribonucleic acids (Chapter 18). Hemiacetal formation occurs

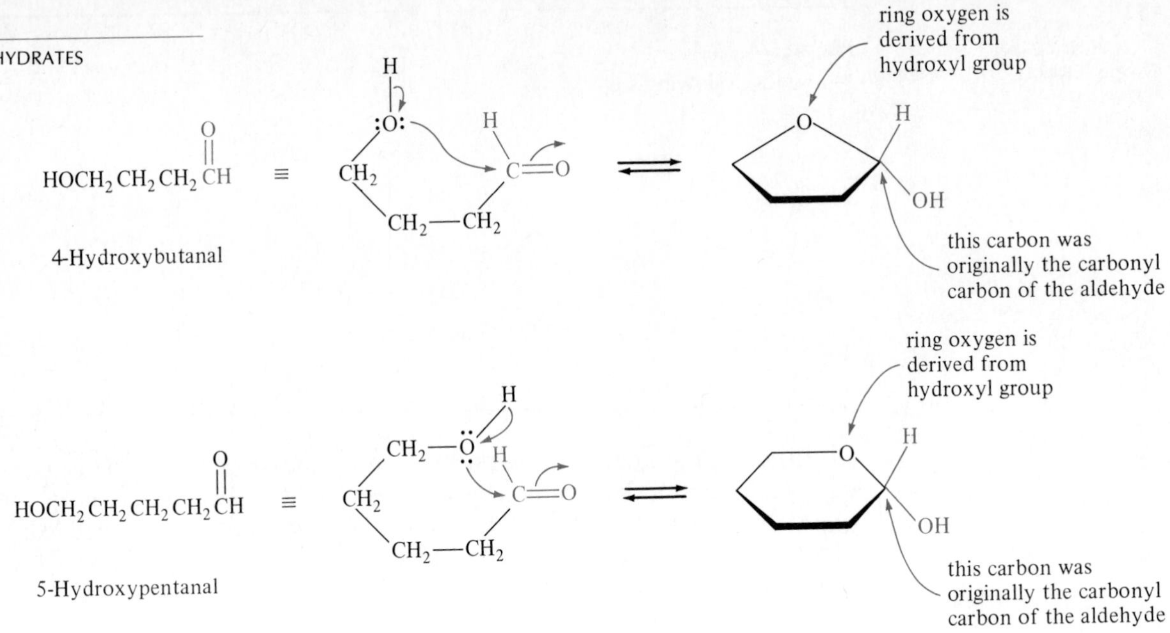

FIGURE 15.4 Cyclic hemiacetal formation in 4-hydroxybutanal and 5-hydroxypentanal.

between the aldehyde group and the hydroxyl at C-4. Furanose ring formation can be visualized by redrawing the Fischer projection in a "coiled" form more suited to cyclization. This is accomplished by rotating the Fischer projection 90° in a clockwise direction.

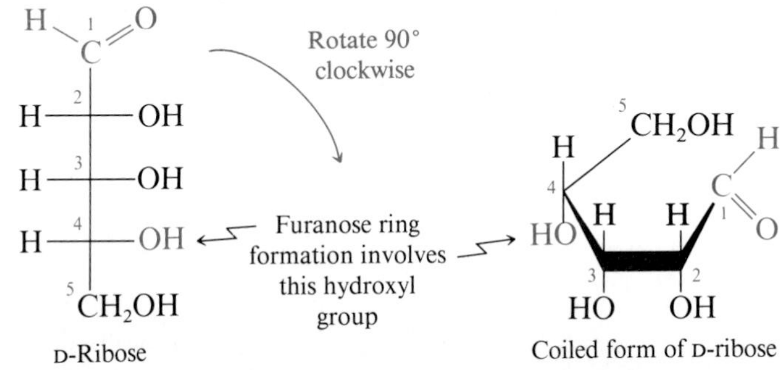

Notice that substituents which are to the right of the carbon chain in the Fischer projection are now "down." The coiled form of D-ribose shown does not have its C-4 hydroxyl group properly oriented for furanose ring formation, however. The coiled form must be redrawn in a conformation that permits cyclic hemiacetal formation by rotating about the C(3)—C(4) bond.

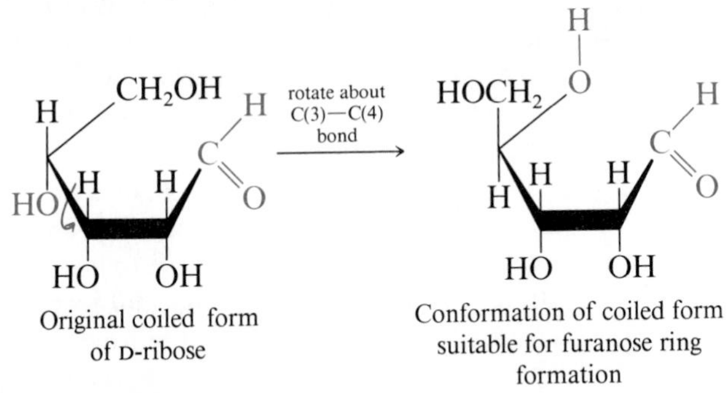

With the C-4 hydroxyl in the proper position, the hemiacetal between this hydroxyl and the carbonyl carbon can form. The anomeric carbon becomes a new chiral center in the hemiacetal structure. Because the carbonyl carbon is planar, attack of the C-4 hydroxyl group can come from the bottom or the top. Thus, the hydroxyl group attached to the anomeric carbon is either up or down in the furanose form. These two structures are stereoisomers. They are diastereomers of each other and are called **anomers.**

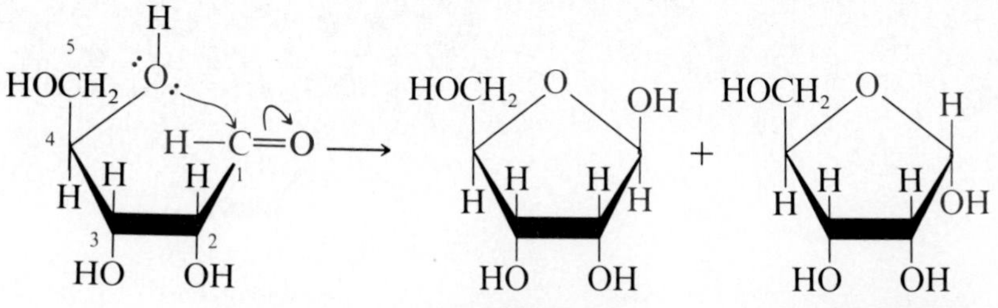

β-D-Ribofuranose
(hydroxyl group at
anomeric carbon is
up)

α-D-Ribofuranose
(hydroxyl group at
anomeric carbon is
down)

Representations of the hemiacetal forms of carbohydrates constructed using planar rings are called **Haworth formulas,** after the British carbohydrate chemist Sir Norman Haworth. Haworth was a corecipient of the 1937 Nobel Prize in Chemistry for his discovery that sugars exist as cyclic hemiacetals and his collaboration on an efficient synthesis of vitamin C.

The two stereoisomeric furanose forms of D-ribose are called α-D-ribofuranose and β-D-ribofuranose. The prefixes α- and β- designate the configuration of the anomeric carbon. When the Haworth formula is drawn as in this section and the hydroxyl group at the anomeric carbon is up in a D-series carbohydrate, then the configuration of the carbohydrate is said to be β. When the anomeric hydroxyl is down, the configuration is α.

PROBLEM 15.5
Write the Haworth formulas corresponding to the furanose forms of each of the following carbohydrates.

(a) D-Xylose   (b) D-Arabinose   (c) L-Arabinose

SAMPLE SOLUTION
(a) The Fischer projection formula of D-xylose is given in Figure 15.3.

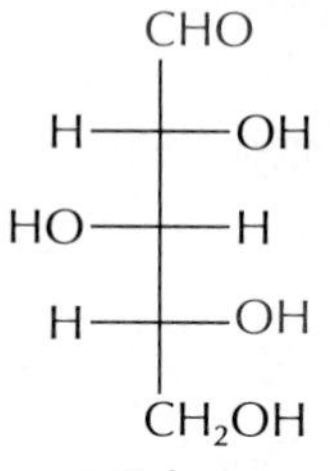

D-Xylose

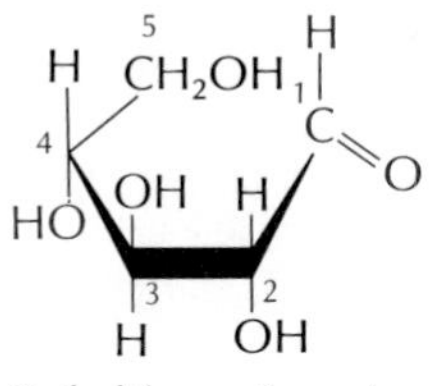

Coiled form of D-xylose

The bond between C-3 and C-4 of the coiled form of D-xylose must be rotated in a counterclockwise sense in order to bring its hydroxyl group into the proper orientation for furanose ring formation.

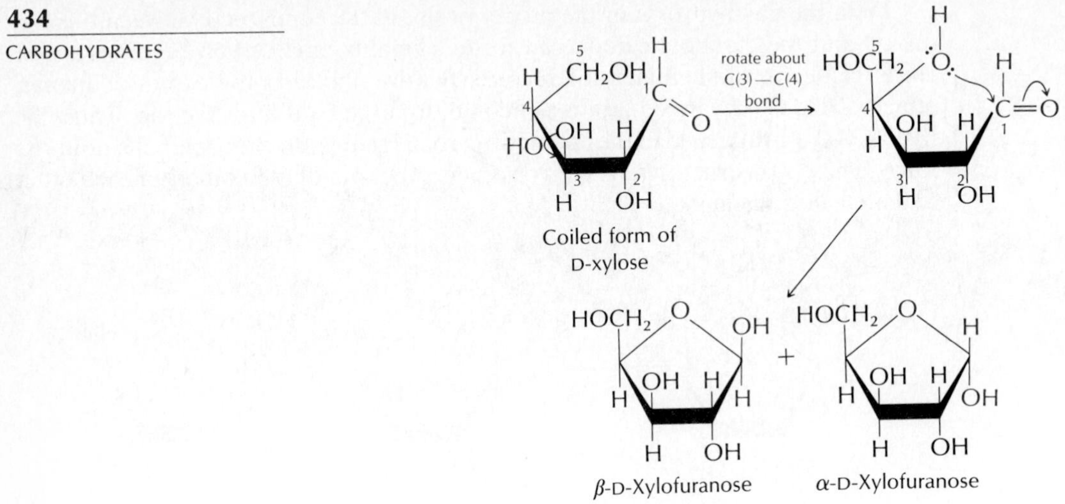

Coiled form of
D-xylose

β-D-Xylofuranose   +   α-D-Xylofuranose

15.6 CYCLIC FORMS OF CARBOHYDRATES: PYRANOSE FORMS

Aldopentoses and aldohexoses are capable of forming six-membered cyclic hemiacetals by addition of the C-5 hydroxyl to the carbonyl group. The **pyranose** form which results may exist in either the α or the β configuration, depending on the orientation of the hydroxyl attached to the anomeric carbon. Haworth representations of the pyranose forms of a carbohydrate are constructed in a manner analogous to furanose rings, described in the preceding section. Construction of the Haworth formulas for the α- and β-pyranose forms of D-glucose is shown in Figure 15.5.

PROBLEM 15.6

D-Ribose exists in two pyranose forms in addition to the furanose forms discussed in Section 15.5. Construct Haworth formulas representing the pyranose forms of D-ribose.

While Haworth formulas are satisfactory for representing *configurational* relationships in pyranose forms, they tell us little about the carbohydrate *conformations*. The six-membered pyranose ring of D-glucose adopts a chair conformation.

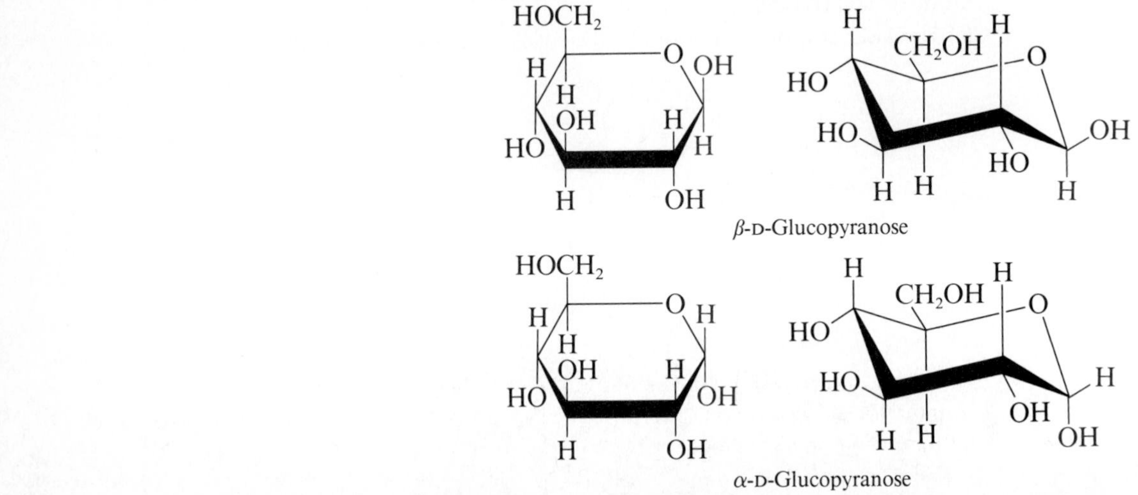

β-D-Glucopyranose

α-D-Glucopyranose

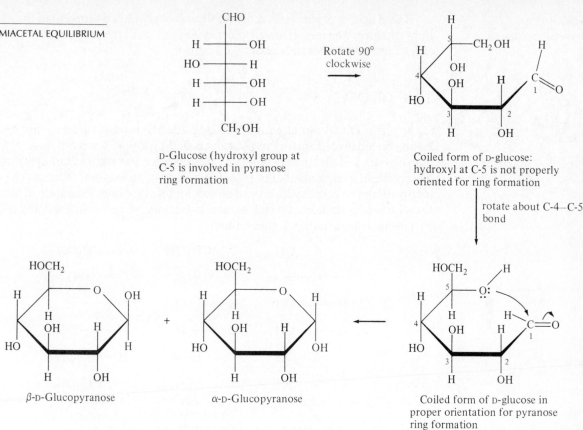

FIGURE 15.5 Haworth formulas for α- and β-pyranose forms of D-glucose.

In the β form of D-glucose, all the ring substituents, including the anomeric hydroxyl, are equatorial. The anomeric hydroxyl is axial in the α isomer; the other substituents remain equatorial.

15.7 HEMIACETAL EQUILIBRIUM

The reactions that give the cyclic hemiacetal forms of carbohydrates are reversible. Thus the furanose and pyranose forms of a carbohydrate in solution are in equilibrium with each other and with the open-chain form. The equilibrium for the pyranose form of D-glucose is

α-D-Glucopyranose
$[\alpha]_D + 112.2°$

Open-chain form
of D-glucose

β-D-Glucopyranose
$[\alpha]_D + 18.7°$

The optical rotations cited for each isomer are those measured immediately after each one is dissolved in water. On standing, the rotation of the solution containing the α isomer decreases from $+112.2°$ to $+52.5°$; the rotation of the solution of the β isomer increases from $+18.7°$ to the same value of $+52.5°$. The change in optical rotation resulting from equilibration of the anomeric forms is called **mutarotation.**

The β isomer of glucose is more stable and constitutes 64 percent of the equilibrium mixture. The less stable α isomer is 36 percent of the mixture, and the open-chain form constitutes only about 0.02 percent.

15.8 KETOSES

Up to this point all our attention has been directed toward aldoses, carbohydrates having an aldehyde function in their open-chain form. Aldoses are more common than ketoses, and their role in biological processes has been more thoroughly studied. Nevertheless, a large number of ketoses are known, and several of them are pivotal intermediates in carbohydrate biosynthesis and metabolism. Examples of some ketoses include D-ribulose, L-xylulose, and D-fructose. Note the -*ulose* suffix (Section 15.1) which characterizes ketose names.

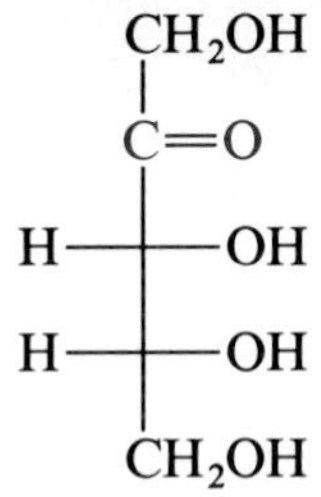

| | D-Ribulose (a key compound in photosynthesis as well as many other aspects of carbohydrate biochemistry) | L-Xylulose (excreted in excessive amounts in the urine of persons afflicted with the mild genetic disorder pentosuria) | D-Fructose (also known as levulose; it is found in honey and is significantly sweeter than table sugar) |

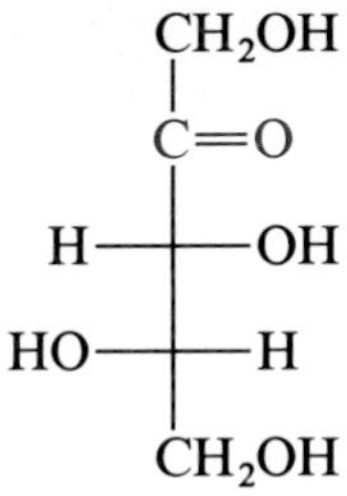

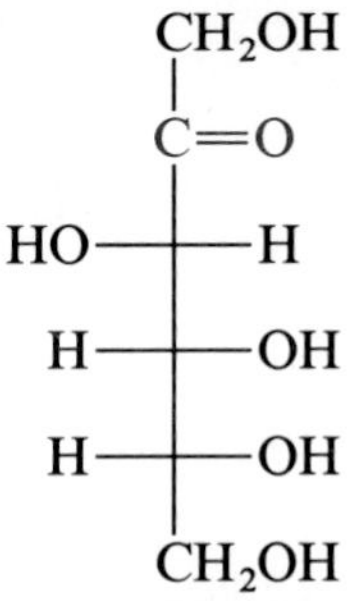

In these three examples the carbonyl group is located at C-2, the most common location for the carbonyl function in ketoses. Fructose is used as a sweetening agent in some foods and soft drinks. Other examples of sweeteners are described in the boxed essay below.

SWEETENERS IN FOOD

Sucrose (table sugar) is the sweetening agent most familiar to us. However, a number of other chemicals, both natural and synthetic, are used by commercial food manufacturers as sweeteners.

Natural sweeteners used in foods include, in addition to sucrose, fructose (Section 15.8) and *invert sugar*. Invert sugar is the mixture of fructose and glucose obtained from hydrolysis of sucrose.

$$\text{Sucrose} + H_2O \longrightarrow \underbrace{\text{glucose} + \text{fructose}}_{\text{Invert sugar}}$$

Both fructose and invert sugar are sweeter than sucrose. Thus less of the sweetener is needed in a food product to achieve the desired sweetness compared to the amount of sucrose that would be necessary.

Another natural sweetener used commercially is sorbitol, produced by catalytic reduction of the corresponding aldohexose, D-glucose.

$$
\begin{array}{c}
\text{CH}_2\text{OH} \\
\text{H} \!-\!\!-\!\!-\! \text{OH} \\
\text{HO} \!-\!\!-\!\!-\! \text{H} \\
\text{H} \!-\!\!-\!\!-\! \text{OH} \\
\text{H} \!-\!\!-\!\!-\! \text{OH} \\
\text{CH}_2\text{OH}
\end{array}
\qquad \text{Sorbitol}
$$

Sorbitol is not broken down by bacteria in the mouth as rapidly as sugar, and thus sorbitol does not promote tooth decay in the same manner as sucrose. Sorbitol is used to sweeten many "sugarless" candies and gum.

Saccharin was the first artificial sweetener, discovered in 1879. It is several hundred times sweeter than sucrose.

Saccharin

Cyclamate

For many decades saccharin was used to sweeten foods used by diabetics or other people who had to limit their intake of sugar for medical reasons. In 1937 a second artificial sweetener was discovered, cyclamate. Although not as sweet as saccharin, cyclamate had the advantage of lacking the bitter aftertaste some people associated with saccharin-sweetened foods and beverages.

By the 1960s artificially sweetened foods were being consumed by a growing segment of the population, primarily for weight control purposes. In 1969, however, research showed that cyclamate caused laboratory animals to develop cancer. Use of cyclamate as a sweetener was banned in 1970, leaving saccharin as the only available artificial sweetener.

In 1977 saccharin was implicated as a cancer-causing agent (carcinogen). Large quantities of saccharin were found to promote bladder cancer in rats. Saccharin too was to be taken off the market; however, Congress overruled the ban and instead required warning labels on saccharin-containing foods and beverages.

Since the early 1980s a new artificial sweetener, aspartame (Section 16.5) has been available. Aspartame, marketed under the trade name Nutrasweet, has been widely accepted in diet beverages and some foods. A disadvantage of aspartame is that it breaks down upon heating, losing its sweetness. Thus diet foods requiring cooking cannot be sweetened with aspartame. Aspartame can also pose medical problems for persons suffering from phenylketonuria, or PKU. Aspartame releases the amino acid phenylalanine into the body. Persons suffering from PKU cannot metabolize phenylalanine, and if phenylalanine levels become too high, permanent brain damage may result.

$$
\underset{\substack{| \\ \text{HO}_2\text{CCH}_2}}{\text{H}_2\text{NCHC}}\overset{\overset{\displaystyle O}{\|}}{}\text{NHCH}\underset{\substack{| \\ \text{CH}_2\text{C}_6\text{H}_5}}{\text{CH}}\overset{\overset{\displaystyle O}{\|}}{\text{C}}\text{OCH}_3
$$

Aspartame

15.9 STRUCTURAL VARIATIONS IN CARBOHYDRATES

Most sugars, including those discussed thus far in this chapter, possess a straight-chain carbon backbone and an oxygen atom on each carbon. Several variations of the general pattern of carbohydrate structure occur in nature. These variations are usually replacement of one or more of the hydroxyl substituents on the main chain by some other atom or group. Examples of these variations include the following.

15.9.1 Deoxy Sugars

In deoxy sugars a hydroxyl group is replaced by hydrogen. Two examples of deoxy sugars are 2-deoxy-D-ribose and L-rhamnose.

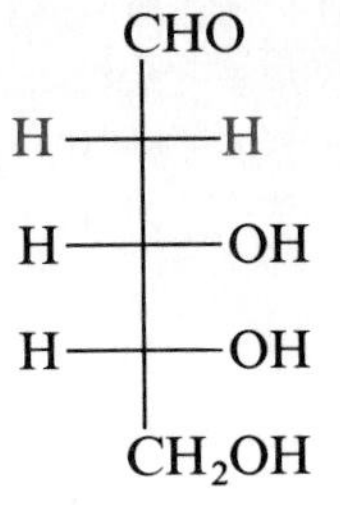

2-Deoxy-D-ribose

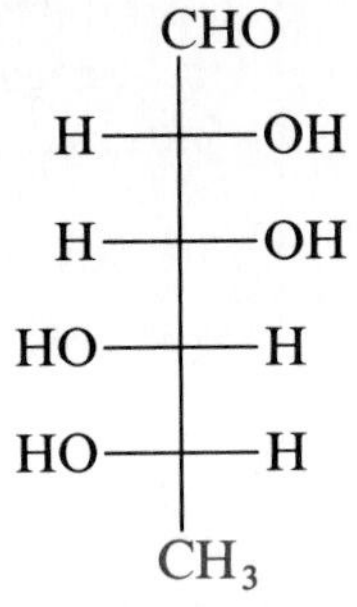

L-Rhamnose
(6-deoxy-L-mannose)

The hydroxyl at C-2 in D-ribose is absent in 2-deoxy-D-ribose. In Chapter 18 you will see how derivatives of 2-deoxy-D-ribose, called *deoxyribonucleotides,* are the fundamental building blocks of deoxyribonucleic acid (DNA), the material responsible for storing genetic information. L-Rhamnose is a compound isolated from a number of plants. Its carbon chain terminates in a methyl group rather than a CH_2OH group.

15.9.2 Amino Sugars

Another structural variation is the replacement of a hydroxyl group in a carbohydrate by an amine group. Such compounds are called *amino sugars.* More than 60 amino sugars are presently known, many of them having been isolated and identified only recently as components of antibiotic substances.

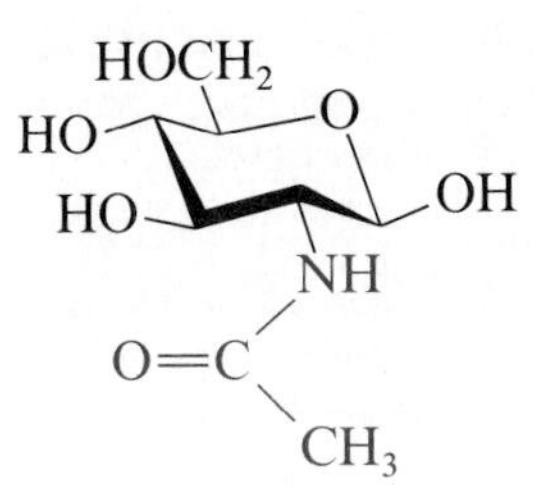

N-Acetyl-D-glucosamine
(shown in its β-pyranose
form; principal component
of the polysaccharide that
is major constituent of
chitin, the substance that
makes up the tough outer
skeleton of arthropods and
insects)

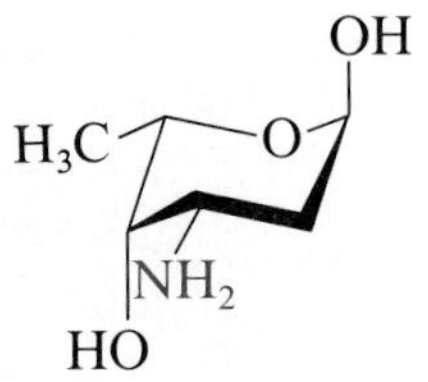

L-Daunosamine (shown in its
α-pyranose form; obtained
from daunomycin and adriamycin,
two powerful drugs used in
cancer chemotherapy)

15.9.3 Branched-Chain Carbohydrates

Carbohydrates that have a carbon substituent attached to the main chain are said to have a *branched chain.* D-Apiose and L-vancosamine are representative branched-chain carbohydrates.

CHO

H——OH

HOCH$_2$——OH

CH$_2$OH

Branching
group

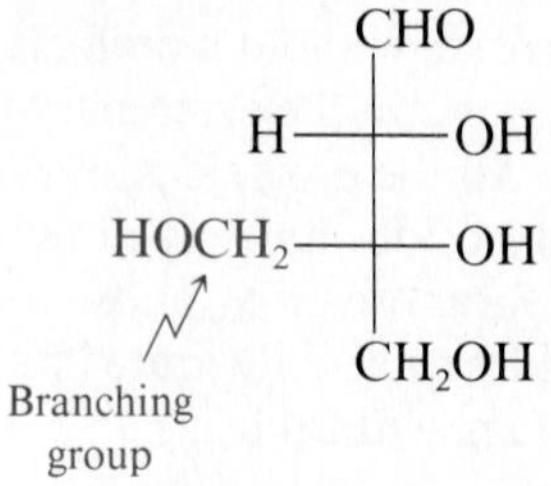

D-Apiose (shown in its
open-chain form; isolated
from parsley and
from the cell wall
polysaccharide of various
marine plants)

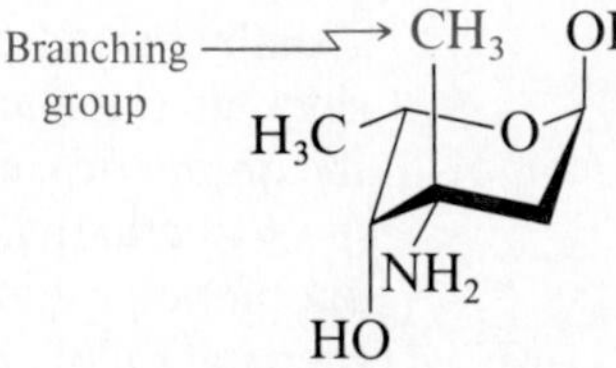

Branching ——→ CH$_3$ OH
group

H$_3$C
O

NH$_2$

HO

L-Vancosamine (shown in its
α-pyranose form; a component
of the antibiotic compound
vancomycin)

15.10 GLYCOSIDES

Glycosides constitute a large class of carbohydrate derivatives characterized by the replacement of the anomeric hydroxyl group by some other substituent. Glycosides are called *O*-glycosides, *N*-glycosides, and *S*-glycosides, according to the atom attached to the anomeric carbon.

HOCH$_2$

HO
O

HO
OC(CH$_3$)$_2$

OH
CN

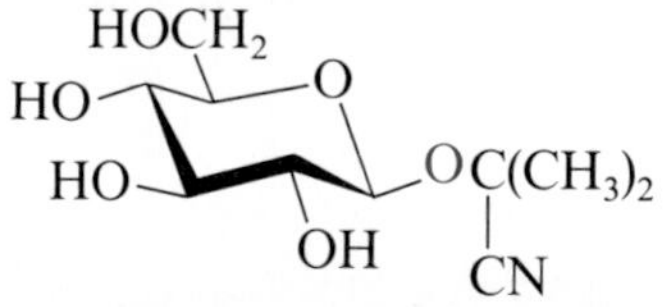

Linamarin
(an *O*-glycoside:
obtained from manioc,
a type of yam widely
distributed in
southeast Asia)

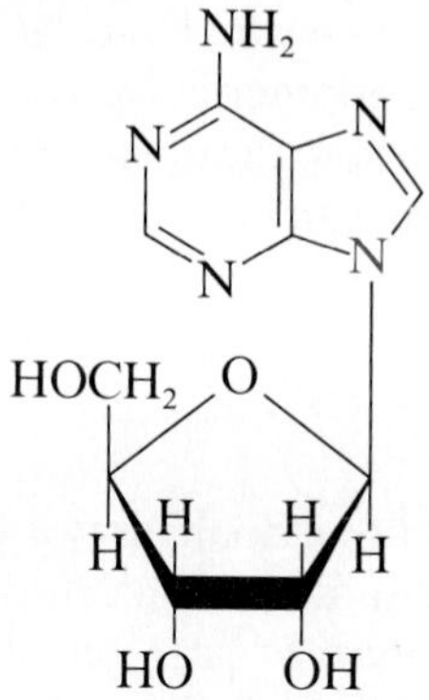

NH$_2$

N
N

N
N

HOCH$_2$ O

H H H H

HO OH

Adenosine
(an *N*-glycoside: also
known as a nucleoside;
adenosine is one of the
fundamental molecules
of biochemistry)

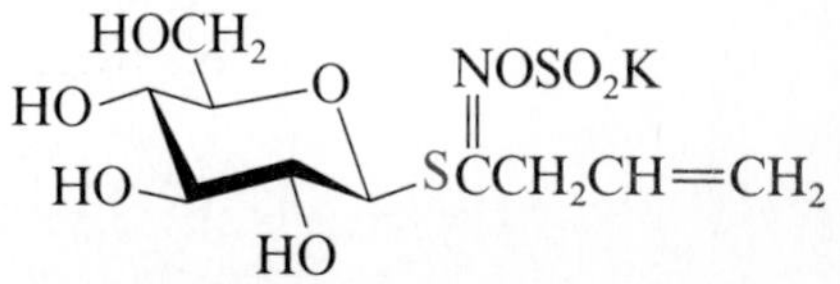

HOCH$_2$

HO
O

HO
NOSO$_2$K

SCCH$_2$CH=CH$_2$

HO

Sinigrin
(an *S*-glycoside:
contributes to the
characteristic flavor
of mustard and
horseradish)

PROBLEM 15.7

Identify by name the carbohydrate component of each of the glycosides shown above.

SAMPLE SOLUTION

Replacement of the group attached to the anomeric carbon with —OH reveals the structure of the carbohydrate. Linamarin is a glycoside formed from the β-pyranose form of D-glucose.

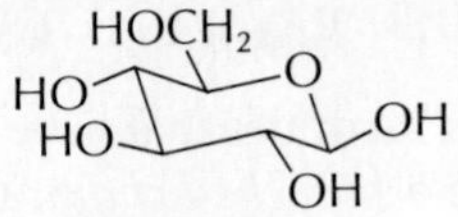

$$\beta\text{-D-Glucopyranose}$$

Usually the word *glycoside* without a prefix is used to mean an *O*-glycoside. Glycosides are classified as α or β in the customary way, according to the configuration at the anomeric carbon. All the glycosides shown above are β-glycosides.

Structurally, *O*-glycosides are mixed acetals; they are ethers that involve the anomeric position of furanose or pyranose forms of carbohydrates. The laboratory preparation of glycosides can be illustrated by the reaction of D-glucose with methanol in the presence of an acid catalyst.

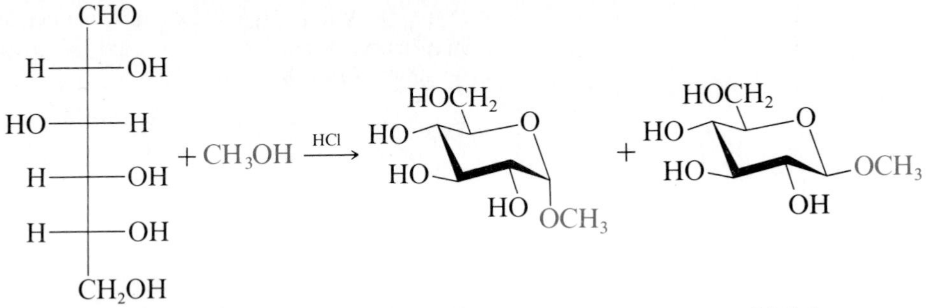

D-Glucose   Methanol

Methyl
α-D-glucopyranoside
(major product; isolated
in 49% yield)

Methyl
β-D-glucopyranoside
(minor product)

Under neutral or basic conditions glycosides are configurationally stable. Unlike the free sugars from which they are derived, glycosides do not exhibit mutarotation. Converting the anomeric hydroxyl to an ether function (hemiacetal → acetal) prevents its reversion to the open-chain form in neutral or basic media. In aqueous acid, acetal formation can be reversed and the glycoside hydrolyzed to an alcohol and the free sugar.

15.11 DISACCHARIDES

Disaccharides are carbohydrates that yield two monosaccharide molecules on hydrolysis. Structurally, disaccharides are glycosides in which the alkoxy group attached to the anomeric carbon is derived from a second sugar molecule.

The best known of all the disaccharides is *sucrose*, common table sugar. Sucrose is a disaccharide in which D-glucose and D-fructose are joined at their anomeric carbons by a glycosidic bond.

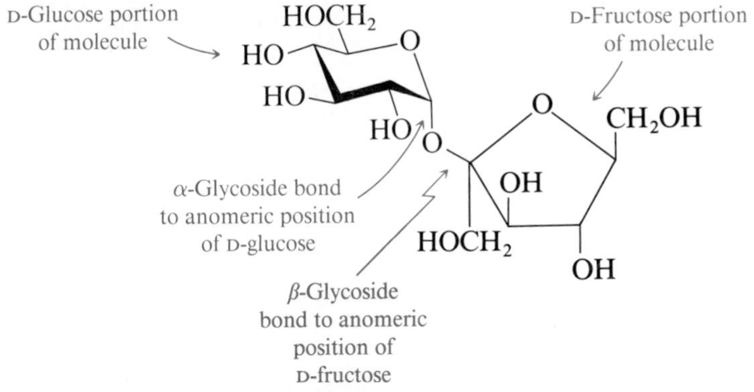

D-Glucose portion
of molecule

D-Fructose portion
of molecule

α-Glycoside bond
to anomeric position
of D-glucose

β-Glycoside
bond to anomeric
position of
D-fructose

Sucrose

Sucrose is obtained primarily from sugarcane and sugar beets. The per capita consumption of sucrose in our diets has been estimated at over 100 pounds per year!

Maltose and cellobiose are two stereoisomeric disaccharides (Figure 15.6). Maltose, or malt sugar, is obtained by the hydrolysis of starch, cellobiose by the hydrolysis of cellulose. The process of brewing beer begins with a malting step that converts the starch in barley and other grains into maltose and glucose. These sugars are then mixed with *hops* (blossoms of the hop plant) and fermented.

Both maltose and cellobiose contain two glucose units joined by a glycosidic bond between C-1 of one glucose molecule and C-4 of a second glucose. How do maltose and cellobiose differ in structure? The configuration at the anomeric carbon (Section 15.5), labeled C-1 in Figure 15.6, is α in maltose and β in cellobiose. In other words, maltose is an α-glycoside and cellobiose is a β-glycoside.

The stereochemistry and points of connection of glycosidic bonds are commonly designated by symbols such as $\alpha(1,4)$ for maltose and $\beta(1,4)$ for cellobiose; α and β designate the stereochemistry at the anomeric position, while the numerals specify the carbon atoms on each of the rings linked by the glycosidic bond.

Lactose, or milk sugar, is another example of a disaccharide. Lactose is found to the extent of 2 to 6 percent in milk. Lactose is composed of a D-galactose unit bonded to a D-glucose unit through a $\beta(1,4)$-glycosidic bond.

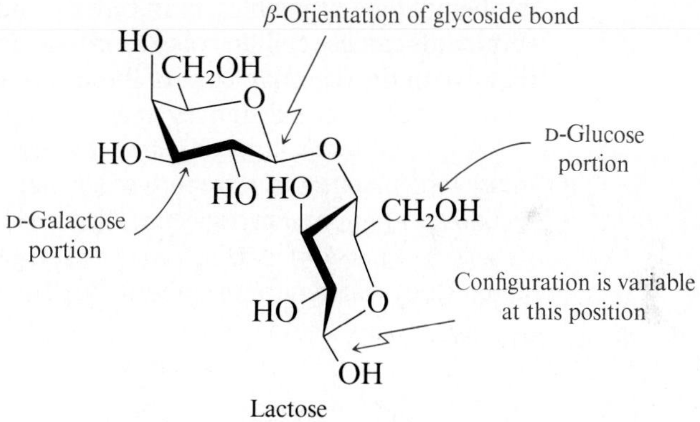

Lactose

Digestion of lactose is facilitated by the enzyme *lactase*. A deficiency of this enzyme makes it difficult to digest lactose and causes abdominal discomfort. Persons having lactose intolerance must either restrict the amount of milk in their diet or add lactase enzyme to the milk before drinking. Milk pretreated with lactase is also available.

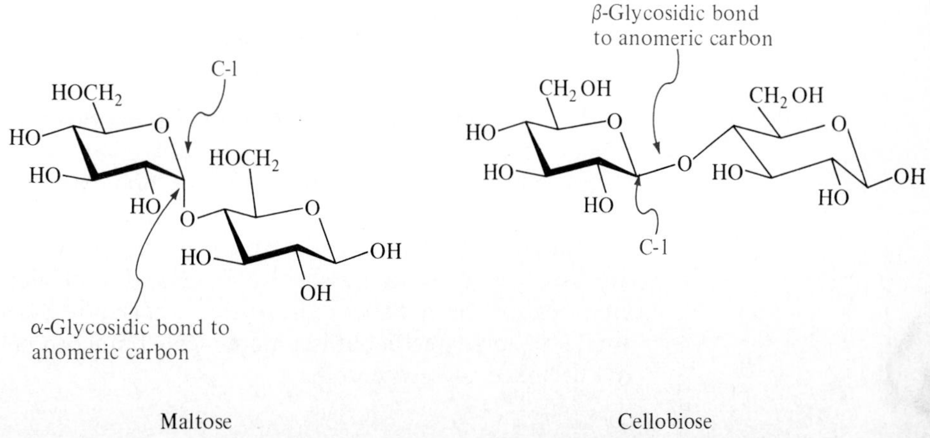

FIGURE 15.6 Two stereoisomeric disaccharides, maltose and cellobiose.

Maltose   Cellobiose

15.12 POLYSACCHARIDES

Cellulose is the principle structural component of vegetable matter. Wood is 30 to 40 percent cellulose, cotton over 90 percent. Photosynthesis in plants is responsible for the formation of over 1 billion tons of cellulose per year.

Structurally, cellulose is a polysaccharide composed of several thousand D-glucose units joined by $\beta(1,4)$-glycosidic linkages.

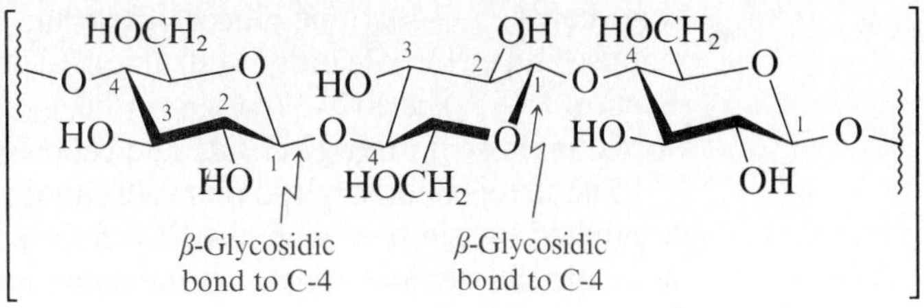

β-Glycosidic bond to C-4   β-Glycosidic bond to C-4

Cellulose

Complete hydrolysis of all the glycosidic bonds of cellulose yields D-glucose. The disaccharide fraction that results from partial hydrolysis is cellobiose (Figure 15.6).

Animals do not possess the enzymes necessary to catalyze the hydrolysis of cellulose and so cannot digest it. Cellulose is the source of *fiber* in our diet. Foods rich in fiber include vegetables, bran cereals, and fiber-enriched breads. Cattle and other ruminants can use cellulose as a food in an indirect way. Colonies of microorganisms that live in the digestive tract of these animals consume cellulose and in the process convert it to other substances that are digestible.

A more direct source of energy for animals is provided by the starches found in many foods. Starch is a mixture of a water-dispersible fraction called **amylose** and a second component, **amylopectin.** Amylose is a polysaccharide made up of about 100 to several thousand D-glucose units joined by $\alpha(1,4)$-glycosidic bonds. A typical starch from wheat contains about 28 percent amylose.

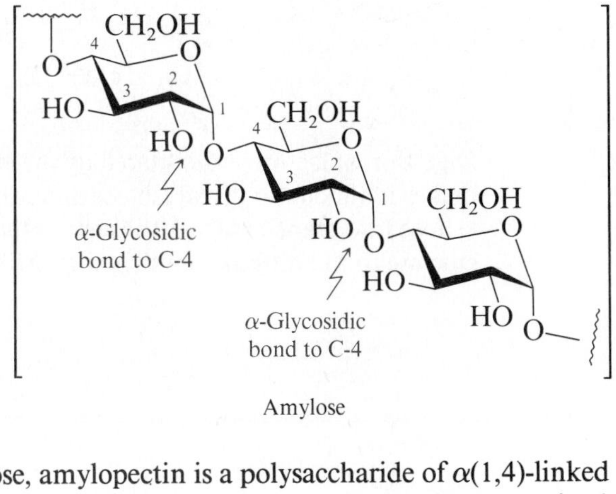

α-Glycosidic bond to C-4   α-Glycosidic bond to C-4

Amylose

Like amylose, amylopectin is a polysaccharide of $\alpha(1,4)$-linked D-glucose units (Figure 15.7). Instead of being a continuous chain of $\alpha(1,4)$ units, however, amylopectin is branched. Short polysaccharide branches of 24 to 30 glucose units are connected to the main chain by an $\alpha(1,6)$ linkage.

Starch is a plant's way of storing glucose to meet its energy needs. Animals can tap that source by eating starchy foods and, with the aid of their α-glycosidase enzymes, hydrolyze starch to glucose. Animals store glucose as **glycogen.** Glycogen is similar to amylopectin but is a more highly branched polysaccharide of $\alpha(1,4)$- and $\alpha(1,6)$-linked D-glucose units.

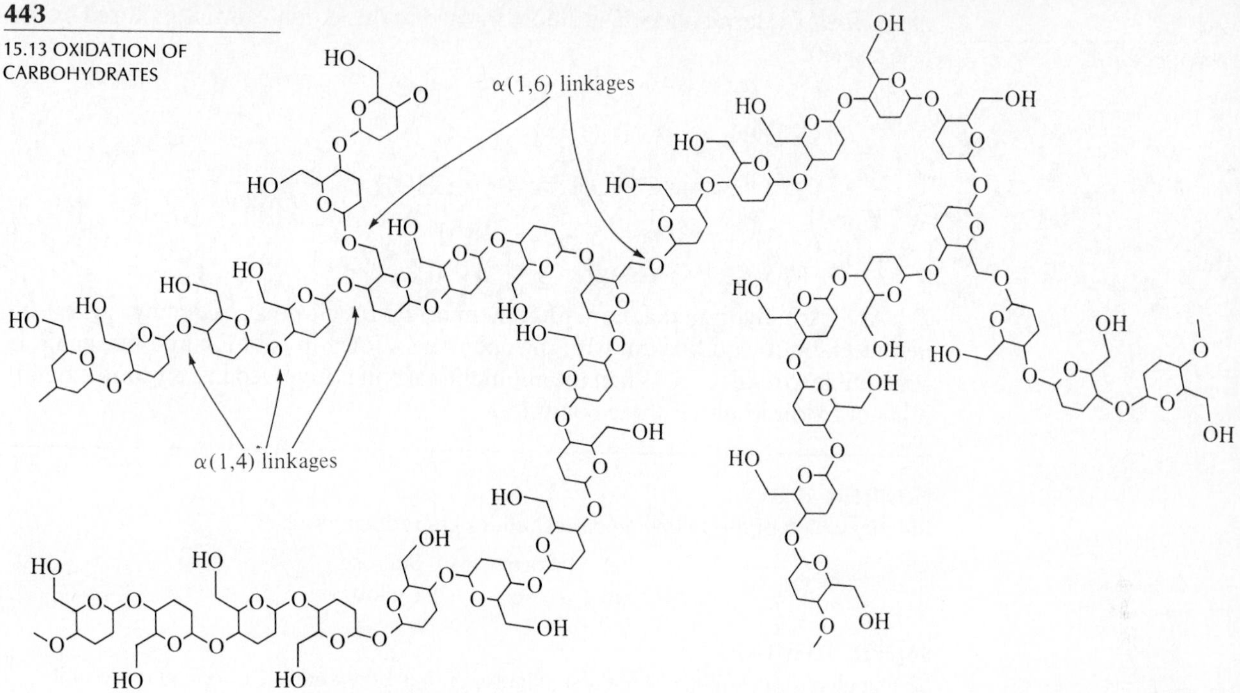

FIGURE 15.7 A portion of the amylopectin structure. α-Glycosidic bonds link C-1 and C-4 of adjacent glucose units. Hydroxyl groups attached to the individual rings have been omitted for clarity.

15.13 OXIDATION OF CARBOHYDRATES

While carbohydrates exist almost entirely as cyclic hemiacetals in aqueous solution, they are in rapid equilibrium with their open-chain forms. The carbonyl functional group of carbohydrates undergoes reactions expected for aldehydes and ketones.

Recall from Chapter 11 (Section 11.18) that a characteristic property of an aldehyde function is its sensitivity to oxidation. Carbohydrates that undergo rapid oxidation are called **reducing sugars**. Oxidizing agents that are used to test for the presence of reducing sugars include *Tollens' reagent* (Section 11.18) and *Benedict's reagent*.

Benedict's reagent is a solution of copper(II) sulfate present as its citrate complex.

$$\underset{\substack{\text{Reducing}\\\text{sugar}}}{\text{RCH}} + \underset{\substack{\text{Copper(II) ion}}}{2\text{Cu}^{2+}} + \underset{\substack{\text{Hydroxide}\\\text{ion}}}{5\text{HO}^-} \longrightarrow \underset{\substack{\text{Carboxy-}\\\text{late anion}}}{\text{RCO}^-} + \underset{\substack{\text{Copper(I)}\\\text{oxide}\\\text{(red ppt)}}}{\text{Cu}_2\text{O}} + \underset{\substack{\text{Water}}}{3\text{H}_2\text{O}}$$

The formation of a red precipitate of copper(I) oxide by reduction of Cu(II) is taken as a positive test for the presence of a reducing sugar. Benedict's reagent is also the basis of a test kit available in drug stores that permits individuals to monitor the glucose levels in their urine.

Aldoses are, of course, reducing sugars since they possess an aldehyde function in their open-chain form. Ketoses, however, are also reducing sugars. Under the basic conditions of the Benedict's and Tollens tests, ketoses equilibrate with aldoses by way of an *enediol* intermediate. The aldose formed in this equilibrium is oxidized by the test reagent.

$$
\underset{\text{Ketose}}{\overset{\displaystyle \text{CH}_2\text{OH}}{\underset{\displaystyle R}{\overset{\displaystyle |}{\underset{\displaystyle |}{\text{C}=\text{O}}}}}}
\;\rightleftharpoons\;
\underset{\text{Enediol}}{\overset{\displaystyle \text{CHOH}}{\underset{\displaystyle R}{\overset{\displaystyle \|}{\underset{\displaystyle |}{\text{C}-\text{OH}}}}}}
\;\rightleftharpoons\;
\underset{\text{Aldose}}{\overset{\displaystyle \text{H}\;\;\text{O}}{\underset{\displaystyle R}{\overset{\displaystyle \diagdown \!\! \diagup}{\underset{\displaystyle |}{\overset{\displaystyle \text{C}}{\underset{\displaystyle |}{\text{CHOH}}}}}}}}
\;\xrightarrow[\text{Tollens' reagent}]{\text{Benedict's or}}\;
\begin{array}{c}\text{positive}\\ \text{test}\end{array}
$$

Any carbohydrate that contains a hemiacetal function is a reducing sugar. The hemiacetal is in equilibrium with the open-chain form in solution and through it is susceptible to oxidation. When the anomeric carbon is involved in a glycosidic bond, it is not oxidized under these conditions.

PROBLEM 15.8

Identify which of the following disaccharides are reducing sugars.

        (a) Maltose   (c) Sucrose
        (b) Lactose   (d) Cellobiose

SAMPLE SOLUTION

(a) The glycosidic linkage of maltose (Figure 15.6) is between C-1 of one glucose unit and C-4 of a second glucose. Therefore maltose contains a free hemiacetal function (C-1 of the second glucose) and is a reducing sugar.

Maltose $\xrightarrow{\text{Cu}^{2+}}$ Open-chain form of maltose   positive test (Cu_2O formed)

15.14 OSAZONE FORMATION

Early research in carbohydrate chemistry was hampered by the physical nature of many sugars. Their water solubility, coupled with a tendency to form noncrystallizable syrups, made isolation and purification difficult. In 1884 Emil Fischer discovered that excess phenylhydrazine reacts with aldoses and ketoses to form crystalline compounds which he called **osazones.** The products contained two adjacent phenylhydrazone units; not only is the aldehyde or ketone function converted to a phenylhydrazone, but so is the alcohol function of the adjacent carbon.

$$
\begin{array}{c}
\text{CHO} \\
\text{H}\!-\!\!\!-\!\text{OH} \\
\text{H}\!-\!\!\!-\!\text{OH} \\
\text{H}\!-\!\!\!-\!\text{OH} \\
\text{CH}_2\text{OH}
\end{array}
\quad\xrightarrow{3\text{C}_6\text{H}_5\text{NHNH}_2}\quad
\begin{array}{c}
\text{CH}=\text{NNHC}_6\text{H}_5 \\
\text{C}=\text{NNHC}_6\text{H}_5 \\
\text{H}\!-\!\!\!-\!\text{OH} \\
\text{H}\!-\!\!\!-\!\text{OH} \\
\text{CH}_2\text{OH}
\end{array}
$$

        D-Ribose   D-Ribose phenylosazone

Three equivalents of phenylhydrazine are required; two are involved in phenylhydrazone formation, while the third undergoes reduction to a molecule of aniline and one of ammonia. The reaction is general for both aldoses and ketoses. It is always limited to adjacent positions on the chain and proceeds no farther than the osazone no matter how much excess phenylhydrazine is present.

Phenylosazones are yellow crystalline solids and are insoluble in water. Carbohydrates are often characterized on the basis of the melting point of their osazone derivative. Osazone formation proved very helpful to Fischer as a tool for interrelating configurations of various sugars. For example, his observation that (+)-glucose, (+)-mannose, and (−)-fructose all gave the same osazone established that all three had the same configuration at C-3, C-4, and C-5.

$$
\begin{array}{cccc}
\text{CHO} & \text{CHO} & \text{CH}_2\text{OH} & \text{CH}=\text{NNHC}_6\text{H}_5 \\
\text{H}-\text{OH} & \text{HO}-\text{H} & \text{C}=\text{O} & \text{C}=\text{NNHC}_6\text{H}_5 \\
\text{HO}-\text{H} & \text{HO}-\text{H} & \text{HO}-\text{H} & \text{HO}-\text{H} \\
\text{H}-\text{OH} & \text{H}-\text{OH} & \text{H}-\text{OH} & \text{H}-\text{OH} \\
\text{H}-\text{OH} & \text{H}-\text{OH} & \text{H}-\text{OH} & \text{H}-\text{OH} \\
\text{CH}_2\text{OH} & \text{CH}_2\text{OH} & \text{CH}_2\text{OH} & \text{CH}_2\text{OH}
\end{array}
$$

D-(+)-Glucose   or   D-(+)-Mannose   or   D-(−)-Fructose $\xrightarrow{3C_6H_5NHNH_2}$ Yellow solid; mp 205°C

Aldoses that differ in configuration only at C-2 yield the same osazone. Ketoses having their carbonyl function at C-2 yield the same osazone as the corresponding aldoses.

PROBLEM 15.9

Name another aldose that reacts with phenylhydrazine to give the same osazone as:

(a) D-Ribose   (b) L-Threose   (c) D-Allose

SAMPLE SOLUTION

(a) Osazone formation in aldoses leads to phenylhydrazone functions at C-1 and C-2. Thus, the aldose that gives the same phenylosazone as D-ribose is the one that has the opposite configuration at C-2, namely, D-arabinose.

$$
\begin{array}{ccc}
\text{CHO} & \text{CHO} & \text{CH}=\text{NNHC}_6\text{H}_5 \\
\text{H}-\text{OH} & \text{HO}-\text{H} & \text{C}=\text{NNHC}_6\text{H}_5 \\
\text{H}-\text{OH} & \text{H}-\text{OH} & \text{H}-\text{OH} \\
\text{H}-\text{OH} & \text{H}-\text{OH} & \text{H}-\text{OH} \\
\text{CH}_2\text{OH} & \text{CH}_2\text{OH} & \text{CH}_2\text{OH}
\end{array}
$$

D-Ribose   or   D-Arabinose $\xrightarrow{3C_6H_5NHNH_2}$

Because the mechanism of osazone formation is somewhat complicated, its elaboration is not appropriate to an introductory treatment of carbohydrate chemistry.

15.15 SUMMARY

Carbohydrates are classified according to their structure (Section 15.1). A *monosaccharide* cannot be hydrolyzed to a smaller carbohydrate; a *disaccharide* gives two monosaccharide molecules on hydrolysis; and so on. Carbohydrates containing an aldehyde group are called *aldoses;* carbohydrates having a ketone group are *ketoses.*

Fischer projection formulas are a convenient way to represent the configurational relationships of carbohydrates (Sections 15.2, 15.3). It is important to remember that vertical bonds are directed away from you, horizontal bonds point toward you.

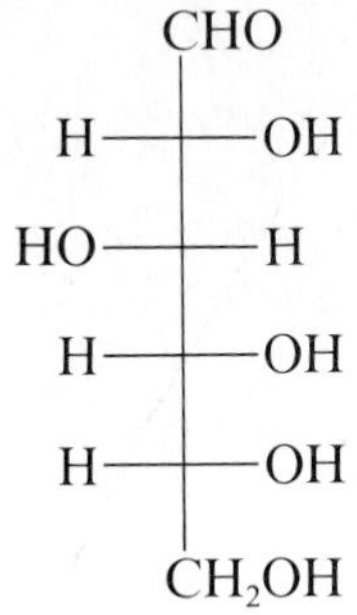

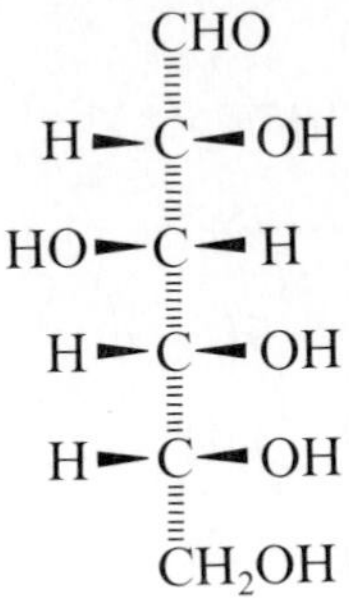

The Fischer
projection of
D-glucose

is equivalent to

When the hydroxyl group at the highest-numbered chiral center is to the right in a Fischer projection, the sugar belongs to the D series; when this hydroxyl is to the left, the compound is an L sugar.

Sugars having five or six carbon atoms are called *pentoses* or *hexoses,* respectively (Section 15.4). Both pentose and hexose sugars exist as cyclic hemiacetals (Sections 15.5, 15.6). Five-membered cyclic hemiacetals are called *furanose* forms, while six-membered ones are called *pyranose* forms.

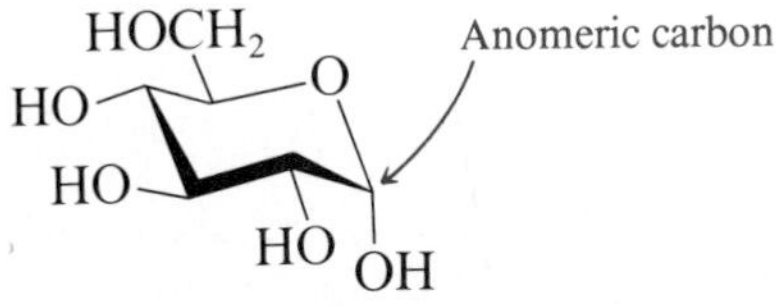

α-Pyranose form of the
aldohexose D-glucose

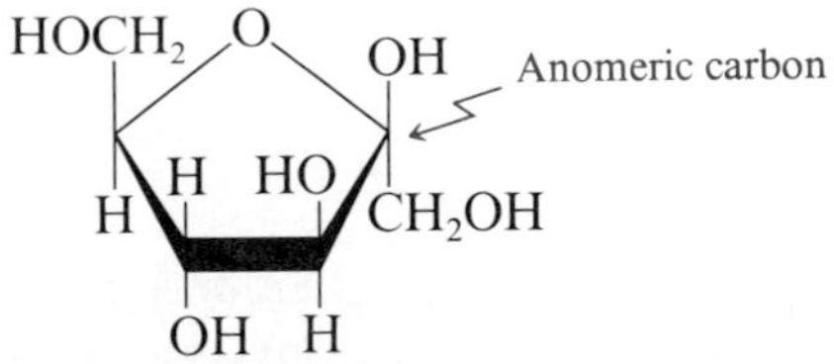

β-Furanose form of the
ketohexose D-fructose

The symbols α and β refer to the configuration at the anomeric carbon. When the direction of the hydroxyl group at this carbon is "down" in a D series carbohydrate, the configuration is α; when the hydroxyl group direction is "up" in the D series, the configuration is β.

The two monosaccharide units of a disaccharide are joined by a glycoside bond (Section 15.11). *Polysaccharides* have many monosaccharide units connected through glycosidic linkages (Section 15.12). Complete hydrolysis of disaccharides and polysaccharides cleaves the glycosidic bonds, yielding the free monosaccharide components.

Carbohydrates undergo chemical reactions characteristic of aldehydes and ketones. Aldoses or ketoses that contain a free hemiacetal function give positive tests with Benedict's or Tollens reagents (Section 15.13). Sugars that react with Benedict's or Tollens reagents are called *reducing sugars.*

Aldoses and ketoses react, by way of their open-chain forms, with three equivalents of phenylhydrazine to yield osazones (Section 15.14). Most osazones crystallize readily and are used to characterize carbohydrates.

ADDITIONAL PROBLEMS

Classification and Structure

15.10 Using Fischer projections, draw the structure of:

(a) A constitutional isomer of glyceraldehyde

(b) An L-ketotetrose

(c) The enantiomer of (b)

15.11 Xylose is a pentose found in many varieties of apples and other fruits. It is also found in both wines and beers, in which it is not fermented by yeast. Refer to the Fischer projection formula of D-(+)-xylose in Figure 15.3 and give structural formulas for:

(a) (−)-Xylose (Fischer projection)

(b) β-D-Xylopyranose

(c) Methyl-β-D-xylopyranoside (a glycoside)

(d) D-Xylose phenylosazone

15.12 (a) Draw the most stable conformation of the β-pyranose form of D-galactose and D-mannose. In what way do each of these differ from β-D-glucopyranose?

(b) Draw the α-pyranose forms of D-galactose and D-mannose, using Haworth formulas.

15.13 Identify the carbohydrates represented by each of the following Fischer projections.

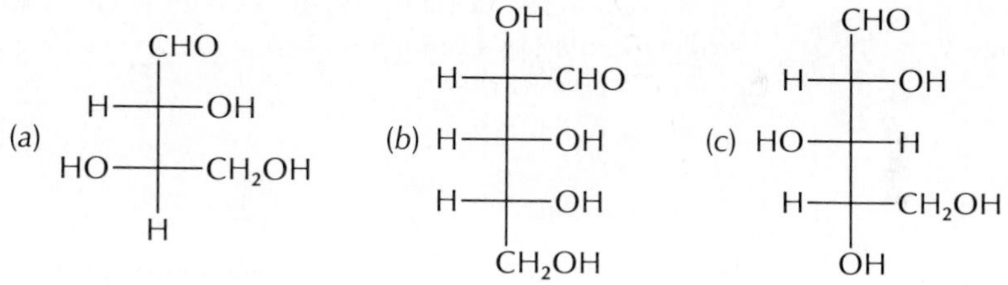

15.14 Disaccharides such as maltose (malt sugar) have a free anomeric hydroxyl group that is not involved in a glycoside bond. While shown in the β orientation in Section 15.11, the α configuration is also known. When dissolved in water, maltose undergoes mutarotation interconverting these two configurations. Draw the structure of the key intermediate in this process.

15.15 Each of the following structures represents the hemiacetal form of a sugar. With the aid of Figure 15.3, identify the sugar from which each is derived. Be sure to specify whether the sugar belongs to the D or the L series.

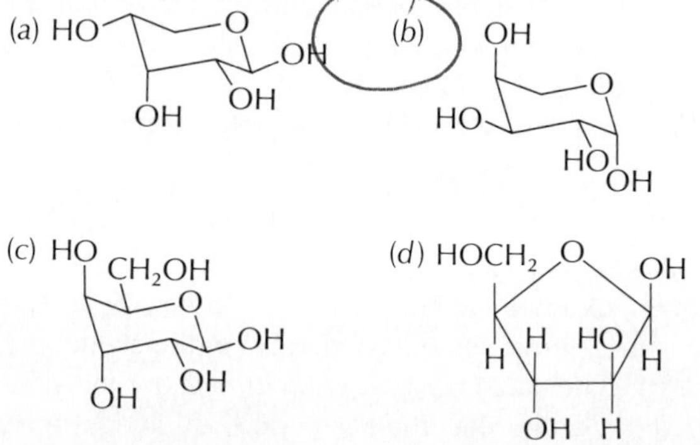

15.16 Trehalose is a naturally occurring disaccharide which on hydrolysis yields two molecules of D-glucose. Trehalose does not reduce Benedict's solution. Based on this information, what can you deduce about the structure of trehalose?

15.17 Write Fischer projection formulas for:
- (a) Cordycepose (3-deoxy-D-ribose): a deoxy sugar isolated by hydrolysis of the antibiotic substance *cordycepin*
- (b) L-Fucose (6-deoxy-L-galactose): obtained from seaweed

Reactions

15.18 Write Fischer projection formulas and name the two aldoses which are capable of existing in equilibrium with the ketose shown.

$$
\begin{array}{c}
CH_2OH \\
| \\
C=O \\
H-\!\!\!-OH \\
H-\!\!\!-OH \\
CH_2OH
\end{array}
$$

15.19 From among the carbohydrates shown in Figure 15.3, choose any that will yield the same phenylosazone as D-xylose.

15.20 The aldehyde function of an aldose is reduced to $-CH_2OH$ by sodium borohydride ($NaBH_4$) by way of the open-chain form. Give the Fischer projection formula of the product formed by sodium borohydride reduction of each of the following.
- (a) D-Erythrose
- (b) D-Arabinose
- (c) D-Glucose
- (d) D-Galactose

15.21 The products formed by reduction of aldoses are called *alditols*. Which of the alditols formed in Problem 15.20 are optically active?
- (a) Erythritol
- (b) Arabinitol
- (c) Glucitol
- (d) Galactitol

15.22 What two alditols are formed by reduction of D-fructose?

15.23 Refer to the Fischer projection formula of the branched-chain carbohydrate D-apiose presented in Section 15.9 for the following:
- (a) How many chiral centers are there in the open-chain form of D-apiose?
- (b) Does D-apiose form an optically active osazone?
- (c) Does D-apiose form an optically active alditol on reduction with sodium borohydride (see Problem 15.20)?
- (d) How many chiral centers are there in the furanose forms of D-apiose?

Miscellaneous Problem

15.24 In 1987 a careful study of the ^{13}C nmr spectrum of D-glucose established its composition in aqueous solution. The α- and β-pyranose forms were 99.7 percent of the mixture, with the α- and β-furanose forms accounting for another 0.29 percent. The free aldehyde was absent; however, 0.0045 percent of the aldehyde hydrate was present.
- (a) Write a Fischer projection formula for the aldehyde hydrate.
- (b) What portion of the glucose exists in a form containing a five-membered ring?

(c) Is the anomeric hydroxyl group cis or trans to the —CH$_2$OH group in the β-pyranose form?

15.25 The specific optical rotations of pure α- and β-D-mannopyranose are $+29.3°$ and $-17.0°$, respectively. When either form is dissolved in water, mutarotation occurs and the observed rotation of the solution changes until a final rotation of $+14.2°$ is observed. Assuming that only α- and β-pyranose forms are present, calculate the percentage of each isomer at equilibrium.

CHAPTER 16

AMINO ACIDS, PEPTIDES AND PROTEINS

LEARNING OBJECTIVES

This chapter focuses on the structure and properties of amino acids and the natural polymers they form, peptides and proteins. Upon completion of this chapter you should be able to:

- Write the structure of an α-amino acid as a zwitterion.
- Explain what is meant by "essential amino acids."
- Portray the structure of naturally occurring amino acids in stereochemical detail.
- Write a chemical equilibrium which expresses the pH dependence of the ionization of an amino acid.
- Explain what is meant by the isoelectric point of an amino acid.
- Write a chemical equation describing the synthesis of an α-amino acid from an α-halo carboxylic acid.
- Write a chemical equation describing the synthesis of an α-amino acid from an aldehyde by the Strecker method.
- Write the chemical formula of a peptide and identify the N-terminal and C-terminal amino acid residues.
- Explain what is meant by the term *peptide bond*.
- Explain what is meant by the primary structure of a peptide.
- Write chemical equations which describe the analysis of the N terminus of a peptide using Sanger's reagent and the Edman degradation.
- Explain how the C terminus of a peptide can be identified using carboxypeptidase-catalyzed hydrolysis.
- Deduce the amino acid sequence of a peptide from the fragments obtained by selective hydrolysis.
- Write a series of chemical equations describing the synthesis of a dipeptide using the appropriate protecting groups.

- Describe the predominant secondary structures of proteins and explain the role played by hydrogen bonds.
- Explain what is meant by the tertiary structure of a protein and how it is stabilized by four types of attractive forces.
- Explain what is meant by the quaternary structure of some proteins.

This chapter is devoted to *amino acids* and the polymers they form, called *peptides* and *proteins*. Proteins are striking in the diversity of roles that they play in living systems: silk, hair, skin, muscle, and connective tissue are proteins. Most biological reactions are catalyzed by proteins called *enzymes*.

As in most aspects of chemistry and biochemistry, structure is the key to function. This chapter will explore the facets of protein structure by first concentrating on their fundamental building blocks, the α-amino acids. Then, after developing the principles of peptide structure and conformation, you will see how the insights gained from these smaller molecules aid our understanding of proteins.

16.1 STRUCTURE OF AMINO ACIDS

Amino acids are carboxylic acids that contain an amine function. The amino acids that are most prevalent in biochemical systems are called α-amino acids because the amino group is attached to the carbon adjacent to the carboxyl group, the α carbon.

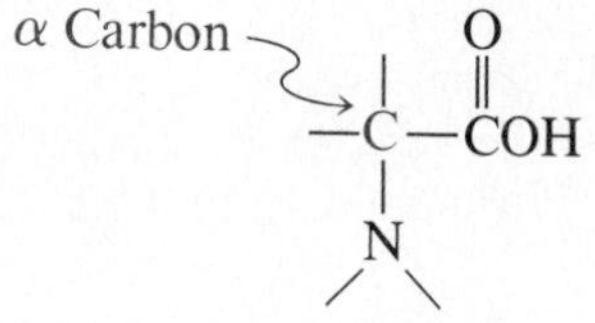

The simplest α-amino acid is glycine, $H_2NCH_2\overset{O}{\overset{\|}{C}}OH$. Glycine is a crystalline solid which does not melt but eventually decomposes at 233°C. It is very soluble in water but practically insoluble in nonpolar organic solvents. These are physical properties that would be expected from an ionic solid, not an organic molecule.

The explanation for these properties is that glycine exists primarily as a **zwitterion,** or **inner salt.**

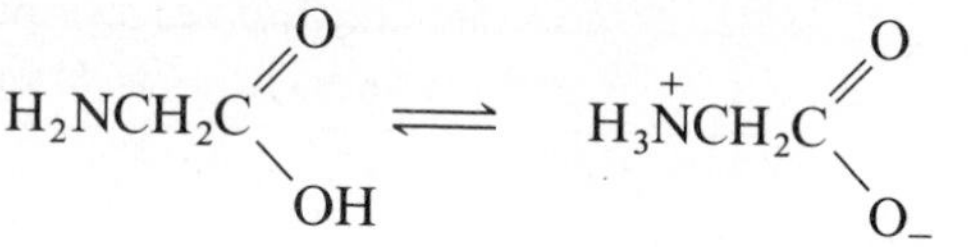

$$H_2NCH_2C\overset{O}{\underset{OH}{\diagup}} \rightleftharpoons H_3\overset{+}{N}CH_2C\overset{O}{\underset{O_-}{\diagup}}$$

Zwitterionic form of glycine

The equilibrium expressed by the equation lies overwhelmingly to the side of the zwitterion.

PROBLEM 16.1

4-Aminobutanoic acid, also known as γ-aminobutyric acid (GABA), is involved in the transmission of nerve impulses. Draw the structure of GABA as a zwitterion.

Most amino acids have only one chiral center

TABLE 16.1
Amino Acids Found in Proteins

Name	Abbreviation	Structural formula†
Amino acids with nonpolar side chains		

Amino acids with nonpolar side chains

Glycine — Gly

$$\underset{\underset{H-CHCO_2^-}{|}}{\overset{+}{\overset{NH_3}{|}}}$$

Alanine — Ala

$$CH_3-\overset{+}{\overset{NH_3}{\underset{|}{CHCO_2^-}}}$$

Valine* — Val

$$(CH_3)_2CH-\overset{+}{\overset{NH_3}{\underset{|}{CHCO_2^-}}}$$

Leucine* — Leu

$$(CH_3)_2CHCH_2-\overset{+}{\overset{NH_3}{\underset{|}{CHCO_2^-}}}$$

Isoleucine* — Ile

$$CH_3CH_2\overset{CH_3}{\underset{|}{CH}}-\overset{+}{\overset{NH_3}{\underset{|}{CHCO_2^-}}}$$

Methionine* — Met

$$CH_3SCH_2CH_2-\overset{+}{\overset{NH_3}{\underset{|}{CHCO_2^-}}}$$

Proline — Pro

$$\text{(cyclic structure with } \overset{+}{NH_2} \text{ and } CO_2^-)$$

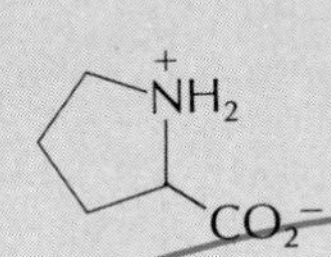

Phenylalanine* — Phe

$$C_6H_5-CH_2-\overset{+}{\overset{NH_3}{\underset{|}{CHCO_2^-}}}$$

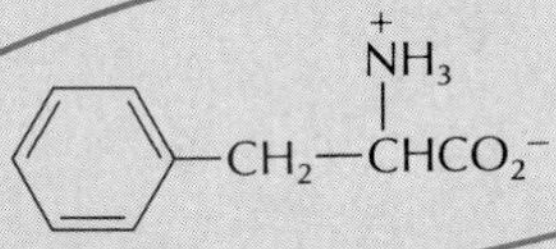

Tryptophan* — Trp

$$\text{(indole)}-CH_2-\overset{+}{\overset{NH_3}{\underset{|}{CHCO_2^-}}}$$

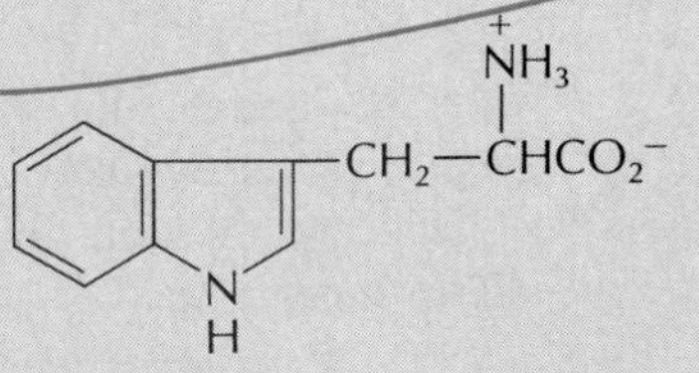

Amino acids with polar, but nonionized side chains

Asparagine — Asn

$$\underset{H_2NCCH_2}{\overset{O}{\|}}-\overset{+}{\overset{NH_3}{\underset{|}{CHCO_2^-}}}$$

Glutamine — Gln

$$\underset{H_2NCCH_2CH_2}{\overset{O}{\|}}-\overset{+}{\overset{NH_3}{\underset{|}{CHCO_2^-}}}$$

Serine — Ser

$$HOCH_2-\overset{+}{\overset{NH_3}{\underset{|}{CHCO_2^-}}}$$

TABLE 16.1 *(continued)*

Name	Abbreviation	Structural formula†
Amino acids with polar, but nonionized side chains		
Threonine*	Thr	$CH_3CH(OH)-CH(\overset{+}{N}H_3)CO_2^-$
Tyrosine	Tyr	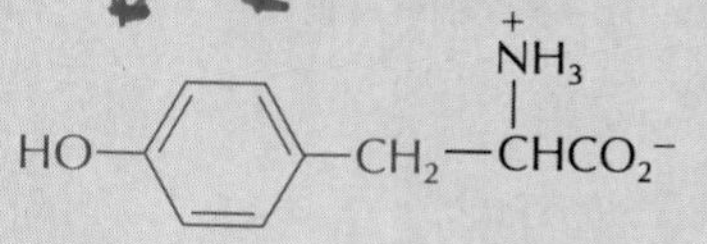 $HO-\!\!\langle\rangle\!\!-CH_2-CH(\overset{+}{N}H_3)CO_2^-$
Cysteine	Cys	$HSCH_2-CH(\overset{+}{N}H_3)CO_2^-$
Amino acids with acidic side chains		
Aspartic acid	Asp	$^-O\overset{O}{\overset{\|}{C}}CH_2-CH(\overset{+}{N}H_3)CO_2^-$
Glutamic acid	Glu	$^-O\overset{O}{\overset{\|}{C}}CH_2CH_2-CH(\overset{+}{N}H_3)CO_2^-$
Amino acids with basic side chains		
Lysine*	Lys	$H_3\overset{+}{N}CH_2CH_2CH_2CH_2-CH(\overset{+}{N}H_3)CO_2^-$
Arginine*	Arg	$H_2N\overset{\overset{+}{N}H_2}{\overset{\|}{C}}NHCH_2CH_2CH_2-CH(\overset{+}{N}H_3)CO_2^-$
Histidine*	His	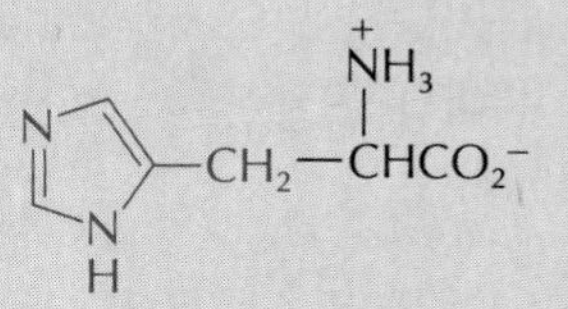 $\underset{H}{N}\!\!\diagup\!\!\diagdown\!\!{N}-CH_2-CH(\overset{+}{N}H_3)CO_2^-$

An asterisk (*) designates an essential amino acid, which must be present in the diet of mammals to ensure normal growth.

† All amino acids are shown in the form present in greatest concentration at pH 7. Asp and Glu exist as anions (have a net negative charge) at this pH, and Lys and Arg exist as cations (have a net positive charge). The remaining amino acids are electrically neutral at pH 7.

There are 20 different α-amino acids normally present in proteins. The structures of these amino acids are shown in Table 16.1. Each amino acid may be represented by a three-letter abbreviation, shown adjacent to the name. All amino acids exhibit the same saltlike character as glycine and exist primarily as zwitterions. All the amino acids from which proteins are derived are α-amino acids, and all but one of these contain a primary amine function ($-NH_2$) and have the general structure

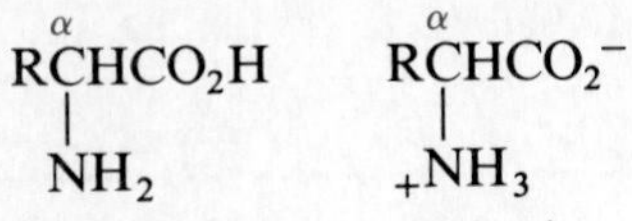

$$\overset{\alpha}{R}CHCO_2H \qquad \overset{\alpha}{R}CHCO_2^-$$
$$| \qquad\qquad\qquad |$$
$$NH_2 \qquad\qquad _+NH_3$$

Nonpolar form   Zwitterion

The exception is proline, a secondary amine in which the amine function is part of a ring.

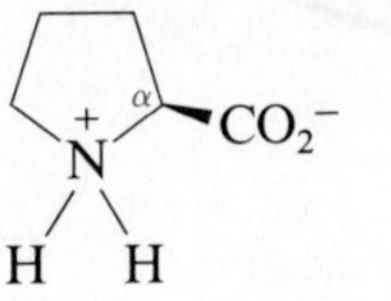

Proline

While humans possess the capacity to biosynthesize some of the amino acids needed for protein formation, there are others which we cannot synthesize from nonprotein components of the diet, and so we must obtain them from proteins in our diet. These amino acids are called **essential amino acids** and are labeled with an asterisk (*) in Table 16.1.

16.2 STEREOCHEMISTRY OF AMINO ACIDS

All the amino acids except glycine are chiral and all have the L configuration at their α-carbon atom when present in proteins. Figure 16.1 compares a typical amino acid L-alanine to L-glyceraldehyde showing their three-dimensional arrangements and their Fischer projections. Recall (Section 15.2) that the configurations of glyceraldehyde form the basis of the D-L system of stereochemical notation.

PROBLEM 16.2
Which of the amino acids in Table 16.1 have more than one chiral center?

The stereochemistry of the other amino acids in Table 16.1 is analogous to that of alanine. The configuration is L when the amine group projects to the left in a Fischer projection oriented with the carboxyl group at the top and the substituent R at the bottom.

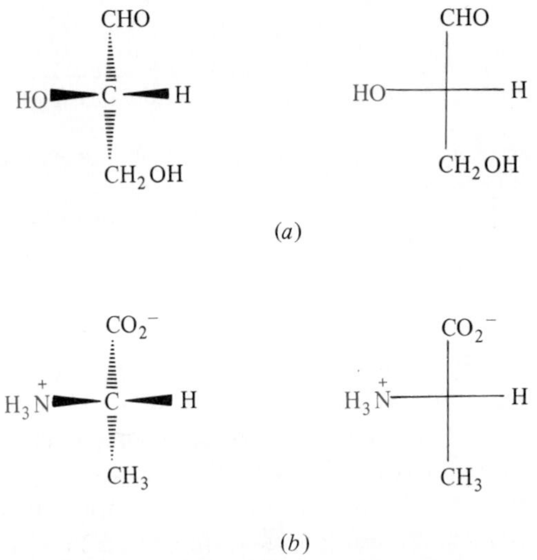

CHO   CHO

HO—C—H   HO——H

CH₂OH   CH₂OH

(a)

CO₂⁻   CO₂⁻

H₃N—C—H   H₃N——H

CH₃   CH₃

(b)

FIGURE 16.1 Three-dimensional representations and Fischer projection formulas for (a) L-glyceraldehyde and (b) L-alanine.

$$H_3\overset{+}{N}\!\!-\!\!\overset{\displaystyle CO_2^-}{\underset{\displaystyle R}{|}}\!\!-\!\!H \qquad \text{L-Configuration of an amino acid}$$

The L configuration of α-amino acids corresponds to S in the Cahn-Ingold-Prelog system (Section 7.8) except in the case of cysteine, where L and R are equivalent. This is because the priority ranking is $\overset{+}{N}H_3 > CO_2^- > R > H$ in all cases except when $R = CH_2SH$ as in cysteine. The group CH_2SH outranks CO_2^-.

16.3 ACID-BASE BEHAVIOR OF AMINO ACIDS

Glycine and the other amino acids are **amphoteric;** they can act as both acids and bases. The acidic functional group is the ammonium ion, $H_3\overset{+}{N}\!-$; the basic functional group is the carboxylate ion, $-CO_2^-$. Consider what happens to glycine in aqueous solution as we change the pH. In a strongly acidic solution (low pH), the predominant species is the protonated form, an ammonium-carboxylic acid. As the pH is raised, the carboxyl group is deprotonated, forming the zwitterion in solutions near neutrality. As the solution becomes more basic, the nitrogen of the zwitterion gives up a proton, forming an amino-carboxylate ion.

$$H_3\overset{+}{N}CH_2C\overset{\displaystyle O}{\underset{\displaystyle OH}{\diagup}} \underset{+H^+}{\overset{-H^+}{\rightleftharpoons}} H_3\overset{+}{N}CH_2C\overset{\displaystyle O}{\underset{\displaystyle O_-}{\diagup}} \underset{+H^+}{\overset{-H^+}{\rightleftharpoons}} H_2NCH_2C\overset{\displaystyle O}{\underset{\displaystyle O_-}{\diagup}}$$

| Species present in strong acid | Zwitterion; predominant species in solutions near neutrality | Species present in strong base |

The pH of an aqueous solution at which the zwitterion is the predominant species is called the **isoelectric point,** or **pI,** of the amino acid. Each amino acid has a characteristic isoelectric point; for glycine pI = 5.97.

PROBLEM 16.3
The isoelectric point of phenylalanine is 5.48. Write the structures of the predominant species present in a solution of phenylalanine at pH:

(a) 2.0   (b) 5.5   (c) 9.0

SAMPLE SOLUTION *acidic*   *basic*
(a) A pH of 2.0 represents an acidic solution; therefore the predominant species is an ammonium-carboxylic acid:

$$C_6H_5CH_2\underset{\underset{\displaystyle +NH_3}{|}}{C}H\overset{\displaystyle O}{\overset{\|}{C}}OH$$

Several of the amino acids in Table 16.1 contain acidic or basic side chains. The presence of these ionizable groups has a dramatic effect on the value of the isoelectric point for that amino acid. Aspartic acid, for example, has a side chain containing a carboxylic acid group and the isoelectric point, pI, = 2.77. Lysine on the other hand,

SEPARATION OF AMINO ACIDS: ELECTROPHORESIS

The ionization of amino acids is the basis of a method used to separate and analyze amino acids called **electrophoresis.** An aqueous mixture of amino acids at a certain pH is placed on a cellulose acetate sheet that is connected to a pair of electrodes. An electric current is applied across the sheet, and each amino acid migrates in a direction determined by its net charge. Positively charged amino acids will migrate toward the negative electrode, and negatively charged amino acids will migrate toward the positive electrode. If the pH of the solution is equal to the isoelectric point of one of the amino acids, that amino acid will not migrate since the zwitterion is electrically neutral.

Figure 16.2 illustrates the separation of a mixture of alanine, lysine, and aspartic acid at a pH of 6. Lysine (pI = 9.7) is positively charged at pH = 6 and migrates to the negative electrode. Aspartic acid (pI = 2.8) is negatively charged at the pH of the experiment and migrates toward the positive electrode. The isoelectric point of alanine is the same as the solution pH (pI = 6), and no migration occurs.

with a side chain containing a basic amine functional group, has an isoelectric point, pI, = 9.74.

16.4 SYNTHESIS OF AMINO ACIDS

There are a variety of methods available for the laboratory synthesis of amino acids. One of the oldest dates back over 100 years and is simply a nucleophilic substitution in which ammonia reacts with an α-halo carboxylic acid.

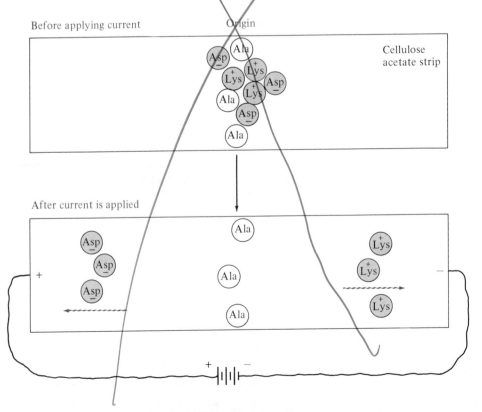

FIGURE 16.2 Separation of amino acids by using electrophoresis. The amino acids are placed in the center of a cellulose acetate strip. A direct current is applied, and the individual amino acids migrate toward an electrode if they have a net charge. (*Used by permission from Chemistry General, Organic, Biological, by Jacqueline I. Kroschwitz and Melvin Winokur, McGraw-Hill Book Company, New York, 1985.*)

$$CH_3CHCO_2H + 2NH_3 \xrightarrow{H_2O} CH_3CHCO_2^- + NH_4Br$$

Br	$_+NH_3$
2-Bromopropanoic acid	Alanine (65–70%)
Ammonia	Ammonium bromide

A second method is called the **Strecker synthesis,** in which an aldehyde is converted to an α-amino acid with one more carbon atom. This is a two-stage procedure in which an α-amino nitrile is an intermediate. The α-amino nitrile is formed by reaction of the aldehyde with ammonia or an ammonium salt and a source of cyanide ion. Hydrolysis of the nitrile group to a carboxylic acid function (Section 12.9) completes the synthesis.

$$\underset{\text{Acetaldehyde}}{CH_3\overset{\overset{O}{\|}}{CH}} \xrightarrow[\text{NaCN}]{NH_4Cl} \underset{\substack{\text{2-Amino-} \\ \text{propanenitrile}}}{CH_3\underset{NH_2}{CH}C\equiv N} \xrightarrow[\text{2. HO}^-]{\text{1. H}_2\text{O, HCl, heat}} \underset{\substack{\text{Alanine} \\ (52-60\%)}}{CH_3\underset{_+NH_3}{CH}CO_2^-}$$

PROBLEM 16.4
Outline the steps in the preparation of valine by the Strecker synthesis (see Table 16.1 for the structure of valine).

Unless a resolution step is introduced into the reaction scheme (Section 7.13), the α-amino acids prepared by the synthetic methods just described are racemic. Optically active amino acids may be obtained by resolving a racemic mixture or by *enantioselective synthesis.* A synthesis is enantioselective if it produces one enantiomer of a chiral compound in an amount greater than its mirror image. Chemists have succeeded in preparing α-amino acids by techniques that are more than 95 percent enantioselective. While this is an impressive feat, we must not lose sight of the fact that the reactions which produce amino acids in living systems do so with 100 percent enantioselectivity.

16.5 PEPTIDES

The principal biochemical reaction of amino acids is the formation of natural polymers called **peptides.** The amino acids are linked together by amide bonds between the amino group of one amino acid and the carboxyl of another. These amide bonds are called **peptide bonds.** The amino acid units of a peptide are known as **amino acid residues.**

Linking two amino acids together gives a **dipeptide;** alanylglycine is a representative dipeptide.

N-terminal amino acid   Peptide bond   C-terminal amino acid

$$H_3\overset{+}{N}CHC\overset{\overset{O}{\|}}{-}NHCH_2CO_2^-$$
$$|$$
$$CH_3$$

Alanylglycine
(Ala-Gly)

By agreement among scientists, peptide structures are written so that the amino group (as $H_3\overset{+}{N}$— or H_2N—) is at the left end of the structure and the carboxyl group (as —CO_2^- or —CO_2H) is at the right. The left and right ends of the peptide are referred to as the **N terminus** (or amino terminus) and the **C terminus** (or carboxyl terminus), respectively. Alanine is the N-terminal amino acid unit or residue in alanylglycine, while glycine is the C-terminal amino acid residue.

The precise order of bonding in a peptide is called its **amino acid sequence.** The amino acid sequence is conveniently specified by using three-letter abbreviations for each amino acid (see Table 16.1) and connecting them by hyphens. The amino acid abbreviations are listed from left to right as they appear in the structure. Thus the first amino acid represents the N terminus, and the last the C terminus. Alanylglycine, for example, is abbreviated Ala-Gly.

The sequence of amino acids in a peptide is crucial in determining the biological function of the molecule. Even for a dipeptide, the sequence makes a difference; Ala-Gly is *not* the same dipeptide as Gly-Ala.

PROBLEM 16.5
Write structural formulas showing the constitution of each of the following dipeptides:

    (a) Gly-Ala   (d) Gly-Glu
    (b) Ala-Phe   (e) Lys-Gly
    (c) Phe-Ala

SAMPLE SOLUTION
(a) Gly-Ala is a constitutional isomer of Ala-Gly. Glycine is the N-terminal amino acid in Gly-Ala; alanine is the C-terminal amino acid.

N-terminal amino acid   C-terminal amino acid

$$H_3\overset{+}{N}CH_2\overset{\overset{\displaystyle O}{\|}}{C}-NHCHCO_2^-$$
$$\underset{\displaystyle CH_3}{|}$$

Glycylalanine
(Gly-Ala)

A dipeptide derivative that is widely used as an artificial sweetener is the C-terminal methyl ester of Asp-Phe. Several years ago a chemist synthesized this compound in connection with research on digestive enzymes. The chemist noted that the new substance had a sweet taste; it is 200 times sweeter than table sugar. The artificial sweetener was given the common name *aspartame* and is marketed under the trade name Nutrasweet. Nutrasweet is found in a variety of diet soft drinks and diet foods (see boxed essay following Section 15.8).

PROBLEM 16.6
Draw the structural formula of Nutrasweet, the C-terminal methyl ester of Asp-Phe.

Up to this point we have dealt with dipeptides. A **tripeptide** contains three amino acid residues, and so on. **Polypeptides** contain many amino acid units. **Proteins** are polypeptides that usually contain 100 to 300 amino acids.

FIGURE 16.3 The
structure of the
pentapeptide leucine
enkephalin. Its amino acid
sequence is
Tyr-Gly-Gly-Phe-Leu,
tyrosine being the
N-terminal and leucine
the C-terminal amino acid.

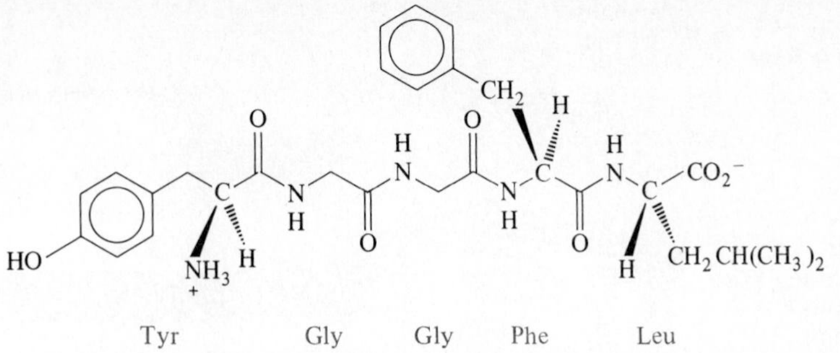

Tyr   Gly   Gly   Phe   Leu

The number of amino acid residues need not be large for a peptide to have significant biological activity. Figure 16.3 shows the structure of *leucine enkephalin*, a pentapeptide. Enkephalins are components of *endorphins*, polypeptides present in the brain that act as the body's own painkillers.

Another "small" peptide with considerable biological activity is *oxytocin*, a nonapeptide whose structure is shown in abbreviated form in Figure 16.4. Oxytocin is a hormone secreted by the pituitary gland that stimulates uterine contractions during childbirth. Oxytocin was the first natural peptide to be synthesized in a laboratory. Vincent du Vigneaud carried out the synthesis of oxytocin at the Cornell Medical College in New York and was awarded the 1955 Nobel Prize in Chemistry for his work.

One structural feature of oxytocin that is worth noting is the presence of a ring in the molecule. The ring is formed by two cysteine residues bonded together through a disulfide bridge.

Disulfide bridges are quite common in the structure of polypeptides and proteins, not only in ring formation but in holding separate polypeptide chains together. For example, *insulin* has two polypeptide chains held together by two disulfide bridges, as depicted in Figure 16.5.

A disulfide bridge forms when the thiol groups of two cysteines undergo oxidation.

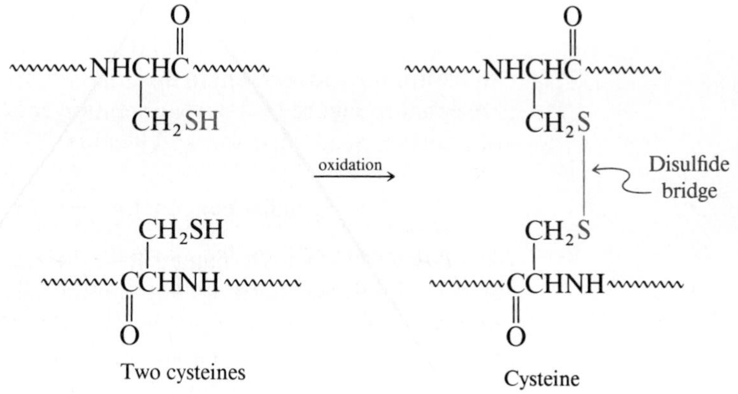

Two cysteines   Cysteine

FIGURE 16.4 The
structure of oxytocin shown
using three-letter amino
acid abbreviations. The
normal carboxyl group at
the C terminus ($-CO_2^-$) is
an amide

$$O$$
$$\|$$
$$(-CNH_2)$$ in oxytocin.

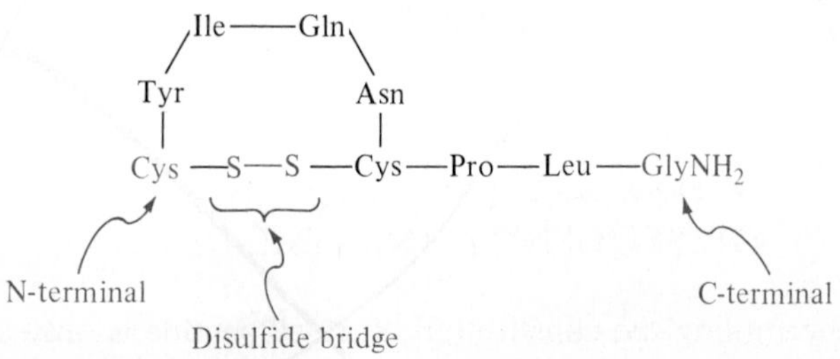

FIGURE 16.5 Depiction of the amino acid sequence in bovine insulin. The two chains, A and B, are joined by the two disulfide bridges. There is also a disulfide bond linking cysteines 6 and 11 in the A chain. Human insulin has threonine and isoleucine at residues 8 and 10, respectively, in the A chain and threonine as the C-terminal amino acid in the B chain.

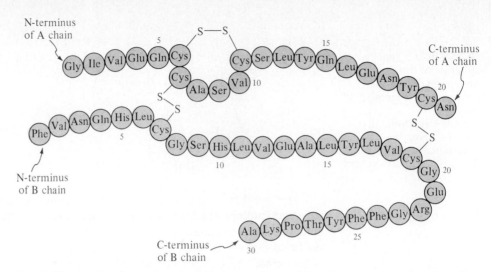

The bridged pair of cysteines is called a cystine residue. In permanent waving, the disulfide bridges of hair are broken, and then remade, using an oxidizing agent, with the hair held in the desired shape.

16.6 PEPTIDE STRUCTURE DETERMINATION: AMINO ACID ANALYSIS

The details of peptide and protein structures are differentiated by chemists and biochemists according to levels. The **primary structure** of a peptide or proteins is *the sequence of amino acids plus any disulfide bridges.* Determining the primary structure of a peptide can be a formidable task, as the number of possible combinations of amino acids is staggering. Given a peptide of unknown structure, how is its amino acid sequence determined?

The first step is to determine which amino acids are present and the relative amounts of each one. The peptide is subjected to acid-catalyzed hydrolysis. Under these conditions each amide bond is cleaved, and a solution is obtained that contains all the amino acids present in the original peptide. For example, hydrolysis of leucine enkephalin (Figure 16.3) gives a solution containing 2 mol of glycine and 1 mol each of tyrosine, phenylalanine, and leucine.

$$\text{Leucine enkephalin} \xrightarrow[\text{heat}]{\text{H}_3\text{O}^+} 2 \text{ Gly} + \text{Tyr} + \text{Phe} + \text{Leu}$$

The amino acids obtained from hydrolysis of an unknown peptide are then separated, and the amount of each amino acid present is determined by reaction with *ninhydrin.* Ninhydrin reacts with an amino acid to give a compound having a violet color. The relative number of moles of each amino acid residue present in the original peptide is proportional to the intensity of the violet color. The separation and identification of the amino acids present in a peptide is usually carried out automatically by using an *amino acid analyzer.*

16.7 PEPTIDE STRUCTURE DETERMINATION: PRINCIPLES OF SEQUENCE ANALYSIS

Determining the constitution of a polypeptide is only the first step in a complete structure determination. The sequence of amino acids in the peptide chain must be determined for the structure to be complete. The number of possible arrangements of amino acids is astonishing. For example, the amino acids obtained from hydrolysis of leucine enkephalin (2 Gly, Tyr, Phe, Leu) could be used to construct 60 different polypeptides! Only one of these, of course, is the correct structure of leucine enkephalin.

Several techniques are used to determine the *amino acid sequence* in a peptide chain. They are:

1. N-terminal group analysis
2. C-terminal group analysis
3. Selective enzyme hydrolysis
4. Partial hydrolysis

Each of these methods will be discussed in the sections which follow.

16.8 END GROUP ANALYSIS: THE N TERMINUS

Identification of the N-terminal amino acid rests on the fact that the N-terminal amino acid has the only α-amino group of the peptide that is *not* involved in an amide bond. Thus the α-amino group of the N-terminal amino acid is capable of acting as a nucleophile.

One method for identifying the N-terminal amino acid involves treatment of the peptide with **Sanger's reagent,** 1-fluoro-2,4-dinitrobenzene.

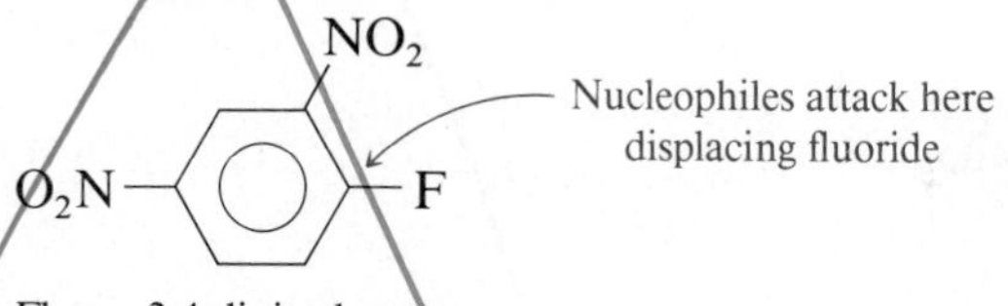

1-Fluoro-2,4-dinitrobenzene

The free amino of the N-terminal amino acid attacks the benzene ring, displacing fluoride. The peptide is then subjected to complete hydrolysis. The N-terminal amino acid is separated and identified as the 2,4-dinitrophenyl derivative. The use of Sanger's reagent is illustrated in Figure 16.6. The technique was developed by Frederick Sanger at Cambridge University (England). Sanger and his coworkers used this technique in research that led to the determination of the amino acid sequence of insulin (Figure 16.5). The determination of the structure of insulin took 10 years to complete, and in 1958 Sanger was awarded the Nobel Prize in Chemistry for his work.

A more modern method for N-terminal analysis of peptides is the **Edman degradation,** developed by the Swedish scientist Pehr Edman. The free amino group of the N-terminal amino acid reacts with phenyl isothiocyanate ($C_6H_5N=C=S$) to give a **phenylthiohydantoin (PTH) derivative.** The most significant feature of the Edman degradation is that following release of the N-terminal amino acid as the PTH derivative, the remainder of the peptide chain remains *unchanged.*

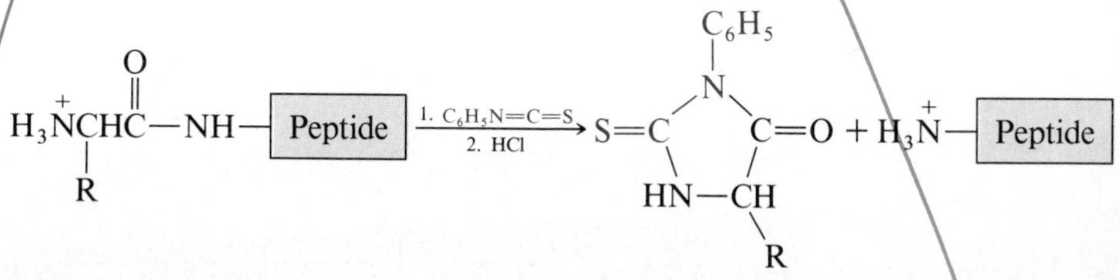

PTH derivative

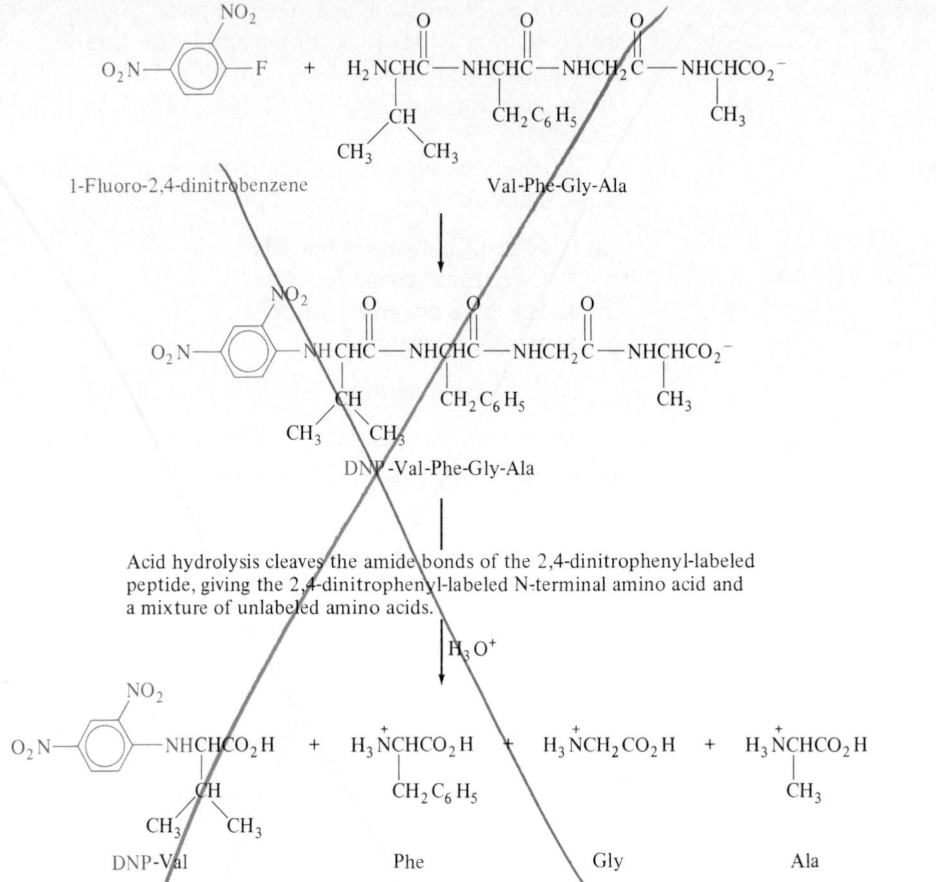

FIGURE 16.6 Use of Sanger's reagent (1-fluoro-2,4-dinitrobenzene) to identify the N-terminal amino acid of a peptide.

1-Fluoro-2,4-dinitrobenzene

Val-Phe-Gly-Ala

DNP-Val-Phe-Gly-Ala

Acid hydrolysis cleaves the amide bonds of the 2,4-dinitrophenyl-labeled peptide, giving the 2,4-dinitrophenyl-labeled N-terminal amino acid and a mixture of unlabeled amino acids.

H_3O^+

DNP-Val   Phe   Gly   Ala

The new peptide has one less amino acid residue than the original and a new N-terminal amino acid, the second in the chain of the original peptide structure. Repeating the Edman procedure allows this newly exposed N terminus to be identified, and so on.

PROBLEM 16.7
Write the structures of the PTH derivatives obtained after the first three cycles of the Edman degradation of leucine enkephalin (Figure 16.3).

SAMPLE SOLUTION
The first Edman degradation cycle gives the PTH derivative of the N-terminal amino acid of the original peptide. In the case of leucine enkephalin, this amino acid is tyrosine.

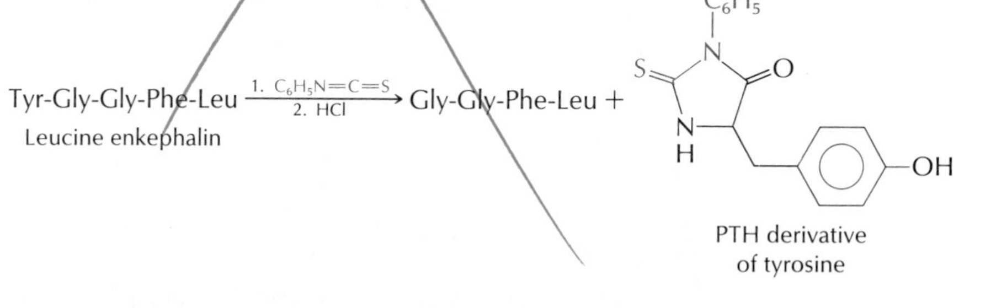

Tyr-Gly-Gly-Phe-Leu $\xrightarrow[\text{2. HCl}]{\text{1. } C_6H_5N=C=S}$ Gly-Gly-Phe-Leu +

Leucine enkephalin

PTH derivative of tyrosine

The Edman degradation procedure has been automated, with the entire operation carried out under computer control. Edman degradations are quite routine for peptide chains up to about 20 amino acid residues. The sequence of 60 amino acid residues has even been determined on a single sample of myoglobin. Myoglobin is a protein having 153 amino acid residues that serves as the oxygen carrier in muscle.

16.9 END GROUP ANALYSIS: THE C TERMINUS

The most widely used method for determining the amino acid at the carboxyl, or C, terminus of a peptide is one based on enzyme-catalyzed hydrolysis. A group of enzymes known as **carboxypeptidases** catalyze the hydrolysis of the peptide bond involving the C-terminal amino acid. By monitoring which amino acid has been released by carboxypeptidase hydrolysis, the identity of the C-terminal amino acid residue of the peptide can be determined.

16.10 SELECTIVE HYDROLYSIS OF PEPTIDES

Carboxypeptidase-catalyzed cleavage of the C-terminal residue of a peptide chain is one example of cleavage of a specific amide bond. Other peptidase enzymes are known which catalyze hydrolysis of specific peptide bonds.

Trypsin, a digestive enzyme present in the intestine, catalyzes only the hydrolysis of peptide bonds involving the carboxyl group of a lysine or arginine residue. In other words, trypsin will cleave a peptide chain to the right (the carboxyl side) of any lysine or arginine residues present. Consider, as an example, the trypsin-catalyzed hydrolysis of the pentapeptide Phe-Arg-Gly-Ala-Gly. Two fragments would be obtained, Phe-Arg and Gly-Ala-Gly.

Trypsin catalyzes
hydrolysis of only this bond

$$\text{Phe-Arg-Gly-Ala-Gly} \xrightarrow[\text{hydrolysis}]{\text{trypsin-catalyzed}} \text{Phe-Arg} + \text{Gly-Ala-Gly}$$

Chymotrypsin, another digestive enzyme, is selective for the hydrolysis of peptide bonds involving the carboxyl group of amino acids with aromatic rings in their side chains, namely, phenylalanine (Phe), tyrosine (Tyr), and tryptophan (Trp).

PROBLEM 16.8

Bradykinin is a peptide present in blood that acts to lower blood pressure. Excess bradykinin, formed in response to the sting of a wasp or other insect, causes severe local pain.

Arg-Pro-Pro-Gly-Phe-Ser-Pro-Phe-Arg   Bradykinin

Write equations describing the enzyme-catalyzed hydrolysis of bradykinin using both trypsin and chymotrypsin.

Nonenzymatic partial hydrolysis of a peptide is carried out in the same manner as the complete hydrolysis by using less strenuous conditions. In partial hydrolysis, some of the peptide bonds are broken, producing peptide fragments of varying lengths. By identifying places where the fragments overlap, the sequence of amino acids in the original peptide can be deduced.

As an example, consider the following fragments obtained from hydrolysis of a pentapeptide: Ala-Gly-Phe + Gly-Phe-Leu + Phe-Leu-Gly. By aligning the fragments where there is overlap, the correct amino acid sequence is obtained.

$$
\text{Hydrolysis fragments} \begin{cases} \text{Ala-Gly-Phe} \\ \phantom{\text{Ala-}}\text{Gly-Phe-Leu} \\ \phantom{\text{Ala-Gly-}}\text{Phe-Leu-Gly} \end{cases}
$$

Original peptide: Ala-Gly-Phe-Leu-Gly

PROBLEM 16.9

Give the structure of the hexapeptide which gave the following fragments upon partial hydrolysis.

Pro-Ala + Leu-Phe-Pro + Pro-Val + Val-Leu-Phe

16.11 THE STRATEGY OF PEPTIDE SYNTHESIS

Chemists and biochemists are interested in peptide synthesis for two reasons. One means of confirming the structure of a natural peptide is to prepare a synthetic peptide having the sequence of amino acids suspected to be present in the natural substance and to then compare the two. Bradykinin, for example (Problem 16.8), was originally believed to be an octapeptide. Laboratory synthesis of bradykinin, however, revealed it to be a nonapeptide having the structure shown.

One other reason for the interest in synthesizing peptides is to learn more about the mechanism of their action. By systematically altering the sequence of amino acids, it is sometimes possible to determine which region of the peptide is most intimately involved in the reactions of that particular substance. Many synthetic peptides have also been prepared in the search for new drugs.

The objective in peptide synthesis may be simply stated: to connect amino acids together in a precise sequence by amide bond formation between them. The problem is that *each* amino acid is *both* an amine and a carboxylic acid. For example, four possible products are expected from the reaction of phenylalanine (Phe) with glycine (Gly).

$$
\overset{+}{\underset{\underset{\text{CH}_2\text{C}_6\text{H}_5}{|}}{\text{H}_3\text{NCHCO}_2^-}} + \overset{+}{\text{H}_3\text{NCH}_2\text{CO}_2^-} \longrightarrow \text{Phe-Gly} + \text{Phe-Phe} + \text{Gly-Phe} + \text{Gly-Gly}
$$

Phenylalanine   Glycine

In other words, each amino acid can react with itself and with each other in two different ways. Remember that a dipeptide A-B is not the same as B-A.

How can we accomplish our goal of forming only one dipeptide product? Assume we wanted to prepare Phe-Gly. This could be done by *protecting* the amino group of phenylalanine *and* the carboxyl group of glycine so that they cannot react under the conditions of peptide formation. This would leave the carboxyl group of phenylalanine and the amino group of glycine free to react. Once the desired peptide bond is formed, the protecting groups are removed to give the desired product.

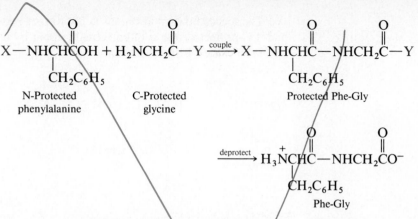

Thus, the synthesis of a dipeptide requires at least five operations:

1. *Protect* the amino group of the N-terminal amino acid.
2. *Protect* the carboxyl group of the C-terminal amino acid.
3. *Couple* the two protected amino acids by amide bond formation between them.
4. *Deprotect* (remove the protecting group) the amino group at the N terminus.
5. *Deprotect* the carboxyl group at the C terminus.

Peptides having more than two amino acid residues are prepared in a manner analogous to the method just outlined for the synthesis of dipeptides.

16.12 PROTECTING GROUPS AND PEPTIDE BOND FORMATION

An amino group can be made less reactive by converting it to an amide. The benzyloxycarbonyl group ($C_6H_5CH_2O\overset{\overset{\displaystyle O}{\|}}{C}$—) is one of the most often used amino-protecting groups. It is attached by acylation of an amino acid with benzyloxycarbonyl chloride.

$$\text{Benzyloxycarbonyl chloride} + \underset{\text{Phenylalanine}}{H_3\overset{+}{N}CHCO^-} \xrightarrow[\text{2. } H^+]{\text{1. NaOH, } H_2O} \text{N-Benzyloxycarbonylphenylalanine (82–87\%)}$$

Just as it is customary to identify individual amino acids by abbreviations, so too with protected amino acids. The abbreviation for a benzyloxycarbonyl group is the letter Z. Thus, *N*-benzyloxycarbonylphenylalanine is represented as

$$\underset{\overset{|}{CH_2C_6H_5}}{ZNHCHCO_2H} \qquad \text{or more simply as Z-Phe}$$

Carboxyl groups are normally protected as esters; benzyl esters are often used.

$$\underset{\text{Glycine}}{H_3\overset{+}{N}CH_2CO_2^-} + \underset{\substack{\text{Benzyl} \\ \text{alcohol}}}{C_6H_5CH_2OH} \xrightarrow[\text{2. } HO^-]{\text{1. } H^+} \underset{\text{Glycine benzyl ester}}{H_2NCH_2CO_2CH_2C_6H_5}$$

The successful formation of a peptide bond between an N-protected amino acid and a C-protected one is often brought about by use of an activating agent, DCCI (*N,N-di*cyclohexyl*c*arbodi*i*mide).

$$
\underset{\substack{\text{Z-Protected} \\ \text{phenylalanine}}}{\text{ZNHCHCOH}} + \underset{\substack{\text{Glycine benzyl} \\ \text{ester}}}{\text{H}_2\text{NCH}_2\text{COCH}_2\text{C}_6\text{H}_5} \xrightarrow{\text{DCCI}}
$$

with CH$_2$C$_6$H$_5$ below the ZNHCHC group.

$$
\underset{\substack{\text{Z-Protected Phe-Gly} \\ \text{benzyl ester}}}{\text{ZNHCHC}-\text{NHCH}_2\text{COCH}_2\text{C}_6\text{H}_5}
$$

with CH$_2$C$_6$H$_5$ below.

The value of using the Z group and the benzyl ester as protecting groups is that removal (deprotection) can be easily accomplished by reactions other than hydrolysis. Treating the Z-protected peptide benzyl ester with hydrogen in the presence of a palladium catalyst removes both protecting groups and gives the desired dipeptide Phe-Gly.

$$
\underset{\substack{\textit{N}\text{-Benzyloxycarbonylphenylalanylglycine} \\ \text{benzyl ester}}}{\text{C}_6\text{H}_5\text{CH}_2\text{OCNHCHCNHCH}_2\text{CO}_2\text{CH}_2\text{C}_6\text{H}_5} \xrightarrow[\text{Pd}]{\text{H}_2}
$$

with CH$_2$C$_6$H$_5$ below.

$$
\underset{\substack{\text{Phenylalanylglycine} \\ (87\%)}}{\overset{+}{\text{H}}_3\text{NCHCNHCH}_2\text{CO}_2^-} + \underset{\text{Toluene}}{2\text{C}_6\text{H}_5\text{CH}_3} + \underset{\substack{\text{Carbon} \\ \text{dioxide}}}{\text{CO}_2}
$$

with CH$_2$C$_6$H$_5$ below.

PROBLEM 16.10
Show the steps involved in the synthesis of Ala-Leu from alanine and leucine using benzyloxycarbonyl and benzyl ester protecting groups and DCCI-promoted peptide bond formation.

In the actual practice of synthesizing large polypeptides, fragments 5 to 10 residues in length are often prepared separately and then combined to give the final compound. Such a method has been used to prepare peptides such as insulin (Figure 16.5), containing 51 amino acid residues.

16.13 SECONDARY STRUCTURES OF PEPTIDES AND PROTEINS

Recall from Section 16.6 that the primary structure of a peptide is its constitution, as dictated by the amino acid sequence. Were the primary structure to be the only consideration, proteins would consist of highly mobile strands of amino acids. This is

Step 1: The Boc-protected amino acid is anchored to the resin. Nucleophilic substitution by the carboxylate anion gives an ester.

Step 2: The Boc protecting group is removed by treatment with hydrochloric acid in dilute acetic acid. After the resin has been washed, the C-terminal amino acid is ready for coupling.

Step 3: The resin-bound C-terminal amino acid is coupled to an N-protected amino acid by using $N.N'$-dicyclohexylcarbodiimide. Excess reagent and $N.N'$-dicyclohexylurea are washed away from the resin after coupling is complete.

Step 4: The Boc protecting group is removed as in step 2. If desired, steps 3 and 4 may be repeated to introduce as many amino acid residues as desired.

Step n: When the peptide is completely assembled, it is removed from the resin by treatment with hydrogen bromide in trifluoroacetic acid.

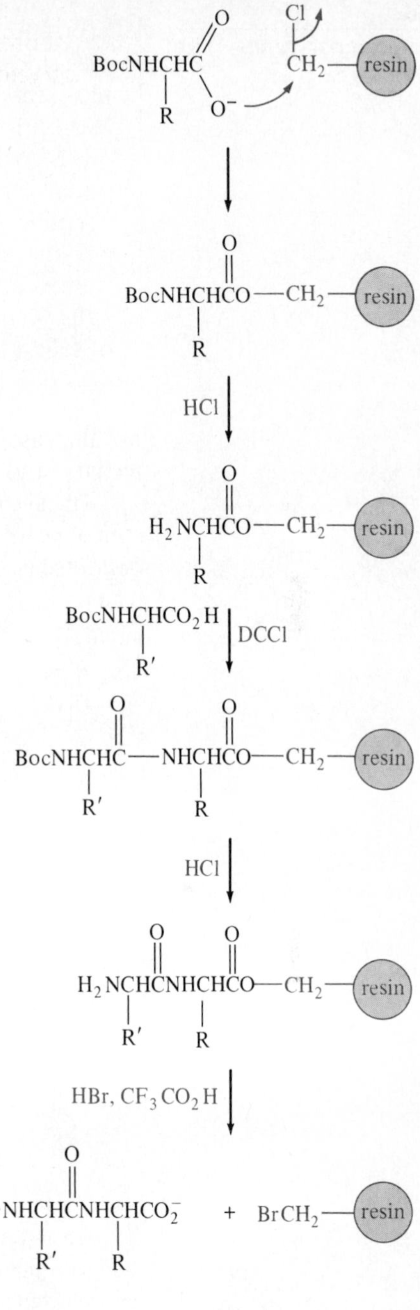

FIGURE 16.7 Peptide synthesis by the Merrifield method. Boc = *tert*-butoxycarbonyl [(CH₃)₃COC—] protecting group.

SOLID-PHASE PEPTIDE SYNTHESIS: THE MERRIFIELD METHOD

In 1962, R. Bruce Merrifield of Rockefeller University (New York) reported the synthesis of the nonapeptide bradykinin by a novel method. The peptide-forming reactions were carried out on the surface of an insoluble polymer called a *solid support*.

The carboxyl end of the desired peptide is bonded to a modified polystyrene resin (Figure 16.7). The new peptide chain is extended toward the N-terminal end by adding amino acid residues one at a time. By anchoring the growing peptide to an insoluble polymer, excess reagents, impurities, and by-products are removed by washing after each operation. When the peptide chain is complete, the C-terminal residue is cleaved from the polymer by treatment with hydrogen bromide in trifluoroacetic acid.

Merrifield was able to automate solid-phase peptide synthesis, and computer-controlled equipment is now available. Using an early version of his "peptide synthesizer," Merrifield and his coworkers were able to synthesize the enzyme ribonuclease in 1969. The 124 amino acid residues of ribonuclease were assembled by using 369 chemical reactions and 11,391 individual steps! In 1984 Merrifield was awarded the Nobel Prize in Chemistry for the development of solid-phase peptide synthesis.

not the case, however. Proteins have a specific shape described by their secondary, tertiary, and quaternary structure.

The **secondary structure** of a peptide or protein is the conformational relationship of nearest-neighbor amino acids in the same or an adjacent molecule. *Hydrogen bonding* between the N—H group of one amino acid residue and the C=O group of

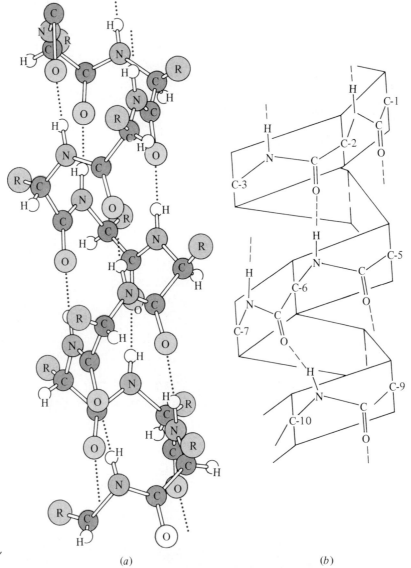

FIGURE 16.8 (a) A protein α-helix. The α-helix is stabilized by hydrogen bonds within the chain between amide protons and carbonyl oxygens. (b) The right-handed helix can be seen when some of the atoms are omitted for clarity. *(Used by permission from Schaum's Outline Series, Theory and Problems of Biochemistry, by Philip W. Kuchel and Gregory B. Ralston, McGraw-Hill Book Company, New York, 1988.)*

(a)   (b)

another is the interaction which plays the greatest role in the secondary structure of a protein.

$$\overset{..}{N}-H\text{--------}:\overset{..}{O}=C$$

Hydrogen bond

Two commonly encountered protein secondary structures are the **α-helix** and the **β-pleated-sheet.** An example of a protein α-helix is shown in Figure 16.8a. The helical structure is stabilized by hydrogen bonds within the chain. A right-handed helical conformation with about 3.6 amino acids per turn permits each carbonyl oxygen to be hydrogen-bonded to an amide proton. The right-handed nature of the helix can be seen in Figure 16.8b.

The α-helix is found in many proteins. The principal protein components of muscle (myosin) and wool (α-keratin), for example, contain high percentages of α-helix. When wool fibers are stretched, these helical regions are elongated by breaking hydrogen bonds. After the stretching force is removed, the hydrogen bonds reform spontaneously and the wool fiber returns to its original shape.

The β-pleated-sheet secondary structure is quite different from the α-helix. Hydrogen bonds form between carbonyl groups and amide protons of adjacent peptide chains, as shown in Figure 16.9. The R group substituents of each chain alternate above and below the plane formed by the hydrogen-bonded amide groups, giving rise to the "pleating." Pleated-sheet structures are usually only stable in proteins which have a large percentage of small R groups, allowing the peptide chains to come close enough for hydrogen bonds to form.

The major protein of silk, called fibroin, is almost entirely a β-pleated-sheet structure. Fibroin has a large number of glycine and alanine residues, both of which

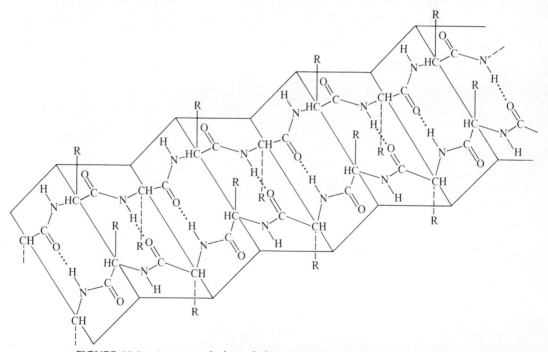

FIGURE 16.9 A protein β-pleated-sheet structure. Hydrogen bonding occurs between amide hydrogen atoms and carbonyl oxygen atoms on adjacent peptide chains. *(Used by permission from Chemistry General, Organic, Biological, by Jacqueline I. Kroschwitz and Melvin Winokur, McGraw-Hill Book Company, New York, 1985.)*

have small R groups (Gly: R = H; Ala: R = CH$_3$). Silk is flexible but resists stretching because the peptide chains of a β-pleated-sheet structure are already in an extended conformation.

16.14 TERTIARY STRUCTURE OF PEPTIDES AND PROTEINS

The folding of an entire polypeptide chain determines the **tertiary structure** of a protein. The way the chain is folded affects both the physical properties of a protein and its biological function. Structural proteins present in skin, hair, tendons, wool, and silk, may have either helical or pleated-sheet secondary structures but in general are elongated in shape. They are classed as **fibrous proteins** and tend to be insoluble in water, as would be expected from their structural role. Figure 16.10 illustrates the ropelike structure of the fibrous protein α-keratin found in wool and hair.

Many other proteins, including most enzymes, operate in aqueous media; some are soluble but most are dispersed as colloids. Proteins of this type adopt an approximately spherical shape and are called **globular proteins.** The globular shape can be seen in the structure of carboxypeptidase A (Figure 16.11), an enzyme that catalyzes removal of the C-terminal amino acid residue from a peptide chain and contains 307 amino acids. Carboxypeptidase A contains a zinc ion (Zn^{2+}) and is called a **metalloenzyme.** The Zn^{2+} ion is essential to the catalytic activity of the enzyme, and its presence influences the tertiary structure.

The tertiary structure of a protein is determined by both bonded and nonbonded interactions between amino acid residues. The bonded interactions include *disulfide bridges* between cysteine residues in the polypeptide chain (Section 16.5) and *ionic attractions* between charged R groups. The nonbonded interactions include *hydrogen bonds* between polar groups of the amino acid side chains and *van der Waals attractions* (Section 2.17), also called *hydrophobic attractions,* between nonpolar side chains. These attractions are summarized in Figure 16.12.

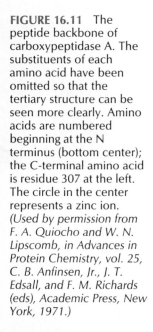

FIGURE 16.10 The stands of polypeptide chains which make up the fibrous protein α-keratin are wound together in a ropelike structure.

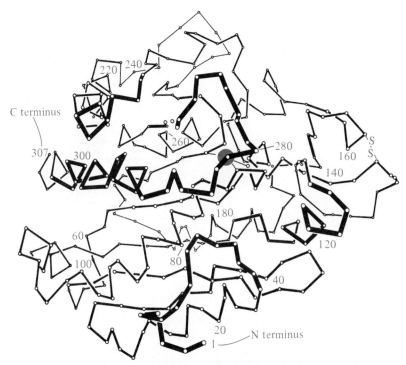

FIGURE 16.11 The peptide backbone of carboxypeptidase A. The substituents of each amino acid have been omitted so that the tertiary structure can be seen more clearly. Amino acids are numbered beginning at the N terminus (bottom center); the C-terminal amino acid is residue 307 at the left. The circle in the center represents a zinc ion. *(Used by permission from F. A. Quiocho and W. N. Lipscomb, in Advances in Protein Chemistry, vol. 25, C. B. Anfinsen, Jr., J. T. Edsall, and F. M. Richards (eds), Academic Press, New York, 1971.)*

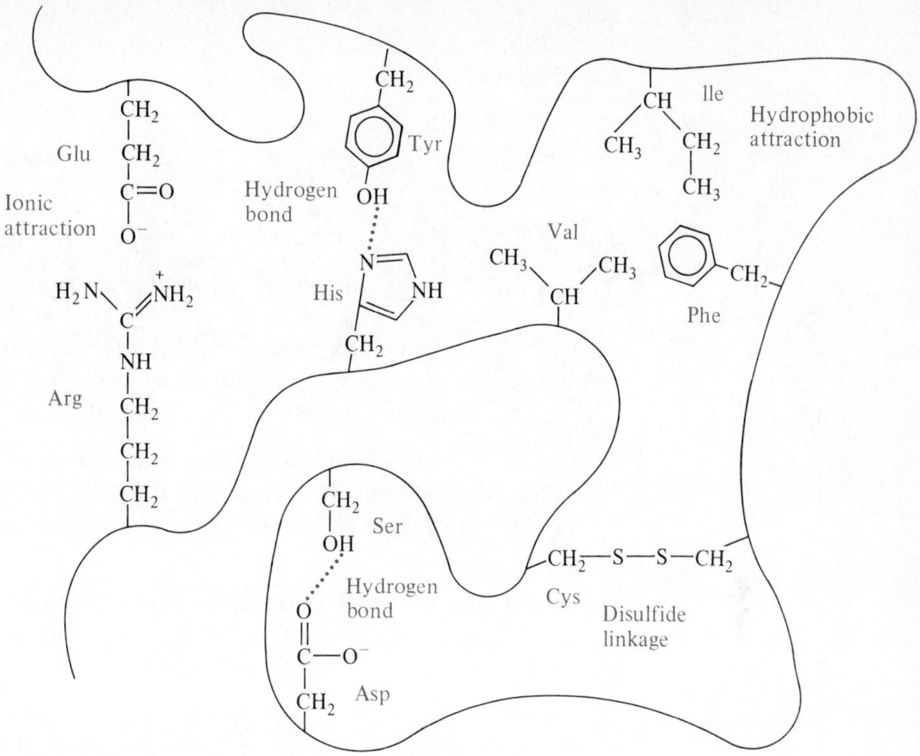

FIGURE 16.12 Schematic representation of the four types of attractions that define the tertiary structure of a protein. *(Used by permission from Chemistry General, Organic, Biological, by Jacqueline I. Kroschwitz and Melvin Winokur, McGraw-Hill Book Company, New York, 1985.)*

16.15 PROTEIN QUATERNARY STRUCTURE. HEMOGLOBIN

Rather than existing as a single polypeptide chain, some proteins are assemblies of two or more chains. The manner in which these subunits are organized is called the **quaternary structure** of the protein.

Hemoglobin, the oxygen-carrying protein of blood, is a large molecule having a molecular weight of about 64,500. Hemoglobin is an assembly of four protein chains and four **heme** units (Figure 16.13). A heme unit is a nonprotein group essential for the biological action of the protein. A heme group coordinates with ferrous iron,

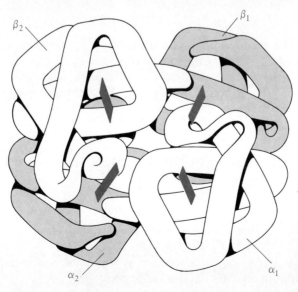

FIGURE 16.13 A schematic representation of hemoglobin showing the four protein chains, labeled α_1, α_2, β_1, and β_2. The four heme units are represented by the colored rectangles. *(Used by permission from Introduction to General, Organic, and Biological Chemistry, by Sally Solomon, McGraw-Hill Book Company, New York, 1987.)*

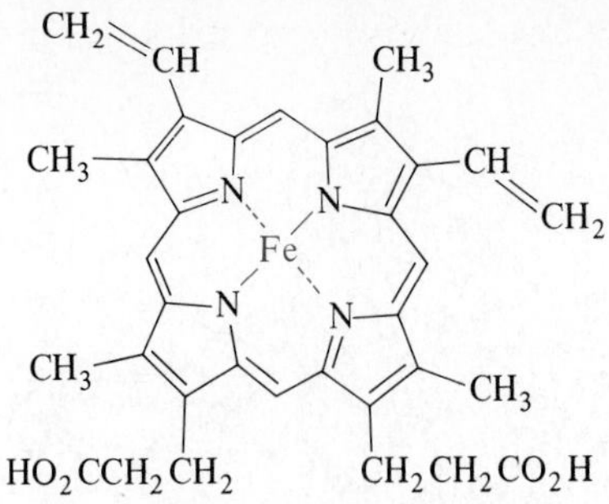

FIGURE 16.14 The structure of heme.

Fe(II), using the four nitrogen atoms of a tetracyclic aromatic substance known as a **porphyrin.** The structure of an individual heme unit is shown in Figure 16.14. Hemoglobin transports oxygen by formation of Fe—O_2 complexes at each of the four heme units.

16.16 SUMMARY

The 20 amino acids which make up all peptides and proteins are all α-amino acids (Section 16.1). Amino acids are *amphoteric,* as they contain both an acidic and a basic functional group. Amino acids exist primarily as *inner salts,* or *zwitterions.*

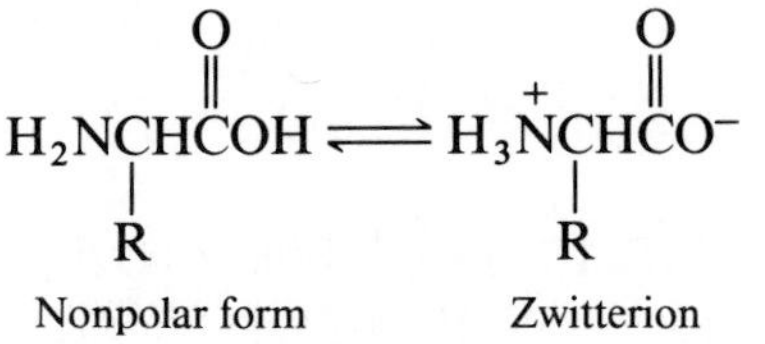

Except for glycine, all the amino acids found in peptides and proteins are chiral and have the L configuration (Section 16.2):

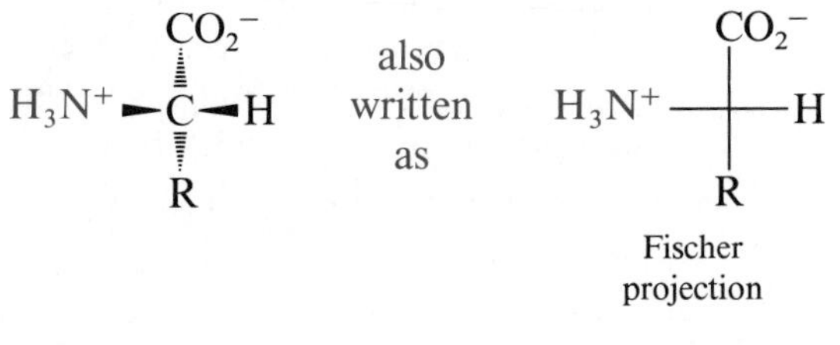

The predominant ionic form of an amino acid present in aqueous solution depends on the pH of the solution (Section 16.3). At low pH, the predominant species is an ammonium-carboxylic acid. As the pH is increased, the carboxyl proton is lost, forming the zwitterion. At high pH, the predominant species is the amino-carboxylate ion.

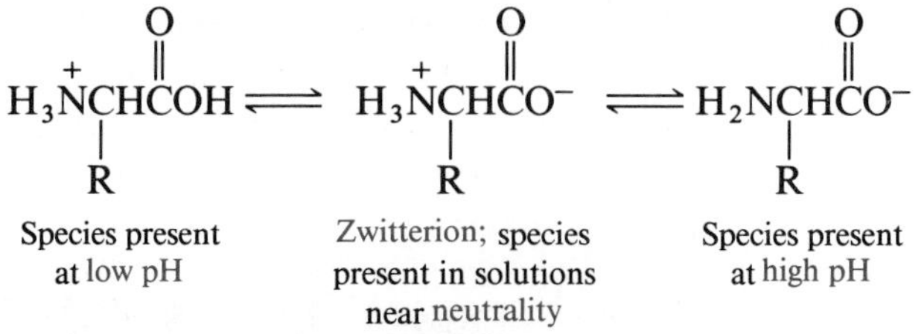

The pH of an aqueous solution at which the concentration of the zwitterion is a maximum is called the *isoelectric point, pI.*

The laboratory synthesis of α-amino acids may be accomplished by (Section 16.4):

1. Reaction of α-halo acids with ammonia:

$$\underset{\overset{|}{Br}}{RCHCO_2H} + 2NH_3 \xrightarrow{H_2O} \underset{\overset{|}{_+NH_3}}{RCHCO_2^-} + NH_4Br$$

2. Reaction of aldehydes with ammonia and cyanide ion followed by hydrolysis of the intermediate α-aminonitrile (Strecker synthesis):

$$\overset{\overset{O}{\|}}{RCH} \xrightarrow[NaCN]{NH_4Cl} \underset{\overset{|}{NH_2}}{RCHC\equiv N} \xrightarrow[\text{2. } HO^-]{\text{1. } H_2O, \text{ HCl, heat}} \underset{\overset{|}{_+NH_3}}{RCHCO_2^-}$$

An amide linkage between two α-amino acids is called a *peptide bond* (Section 16.5). The *primary structure* of a peptide is given by its amino acid sequence plus any disulfide bonds between two cysteine residues (Sections 16.5, 16.6). Peptide structures are written by using three-letter abbreviations to represent each amino acid residue. The *N-terminal amino acid* is written to the left end of the formula; the *C-terminal amino acid* is written at the right end.

$$\underset{\overset{|}{CH_3}}{\overset{+}{H_3}NCHC}\overset{\overset{O}{\|}}{}-NHCH_2C\overset{\overset{O}{\|}}{}-\underset{\overset{|}{CH_2C_6H_5}}{NHCHCO^-}$$

Alanylglycylphenylalanine
(Ala-Gly-Phe)

The constitution of a polypeptide may be determined by complete hydrolysis of the peptide bonds and analysis of the resulting solution of amino acids (Section 16.6). Determination of the amino acid sequence is facilitated by terminal-residue analysis (Section 16.7). The N terminus may be identified by reaction of the peptide with *Sanger's reagent* or by subjecting the peptide to analysis by *Edman degradation* (Section 16.8).

The C terminus is identified by hydrolysis with carboxypeptidase, an enzyme that catalyzes the hydrolysis of the peptide bond to the C-terminal amino acid (Section 16.9). Selective or partial hydrolysis of a peptide may be carried out by using enzymes such as trypsin and chymotrypsin or by nonenzymatic means (Section 16.10). The small fragments obtained by partial hydrolysis provide useful information for determining the overall sequence of the peptide.

Peptide synthesis requires that the number of possible reactions be limited through the use of protecting groups (Section 16.11). Amino-protecting groups include benzyloxycarbonyl (Z) (Section 16.12). Carboxyl groups are protected as benzyl esters. Peptide bond formation between a protected amino acid having a free carboxyl group and a protected amino acid having a free amino group can be accomplished with the aid of *N,N'*-dicyclohexylcarbodiimide (DCCI). Protecting groups can be removed from the peptide product by using hydrogen and a palladium catalyst.

The *secondary structure* of a peptide or protein is the conformational relationship of nearest-neighbor amino acids (Section 16.13). The α-*helix* secondary structure is stabilized by hydrogen bonds between N—H and C=O groups within a single polypeptide chain. The β-*pleated-sheet* secondary structure is stabilized by hydrogen bonds between amino acids on adjacent chains.

The folding of an entire polypeptide chain determines the *tertiary structure* of a protein (Section 16.14). Proteins are classed as either *fibrous* or *globular*. The forces which stabilize the tertiary structure of a protein include disulfide bridges, ionic attractions, hydrogen bonds, and van der Waals (hydrophobic) attractions.

Many proteins consist of two or more chains, and the way in which the various units are assembled into the protein is its *quaternary structure* (Section 16.15). Hemoglobin consists of four polypeptide chains and four *heme* units.

ADDITIONAL PROBLEMS

Amino Acids: Structure, Properties, and Synthesis

16.11 Ten of the amino acids present in proteins are designated as being "essential." Does this statement mean the other amino acids are less important in the construction of proteins? Explain.

16.12 L-Threonine has the D configuration at the carbon atom that bears the hydroxyl group. Write a Fischer projection formula for L-threonine.

16.13 The isoelectric point of isoleucine is 6.02. Write structural formulas for the principal species present when the pH of a solution containing isoleucine is raised from 1 to 6 to 13.

16.14 Give the structure of lysine (isoelectric point of 9.74) in solution at:
(a) pH 1
(b) pH 13

16.15 Give the structure of glutamic acid (isoelectric point of 3.22) in solution at:
(a) pH 1
(b) pH 13

16.16 Outline a series of steps that would allow preparation of phenylalanine:
(a) By the Strecker synthesis
(b) From 2-bromo-3-phenylpropanoic acid

16.17 The penicillin and cephalosporin antibiotics are biosynthesized from amino acid precursors. What amino acids are incorporated into the structure of 6-aminopenicillanic acid?

$H_3\overset{+}{N}$ — [structure] — CH_3, CH_3, CO_2^-, S, N, O 6-Aminopenicillanic acid

Primary Structure of Peptides

16.18 Write the structures and give acceptable names for the dipeptides which can be formed from one molecule each of L-valine and L-alanine.

16.19 If you set out to synthesize the dipeptide Val-Ala from racemic valine and alanine, how many stereoisomers might be obtained?

16.20 Methionine enkephalin is a second peptide present with leucine enkephalin in endorphins. Methionine enkephalin differs from leucine enkephalin only in having methionine instead of leucine as its C-terminal amino acid. What is the amino acid sequence (using three-letter abbreviations) of methionine enkephalin?

16.21 (a) Using three-letter abbreviations for each amino acid residue, give the name of the following peptide.

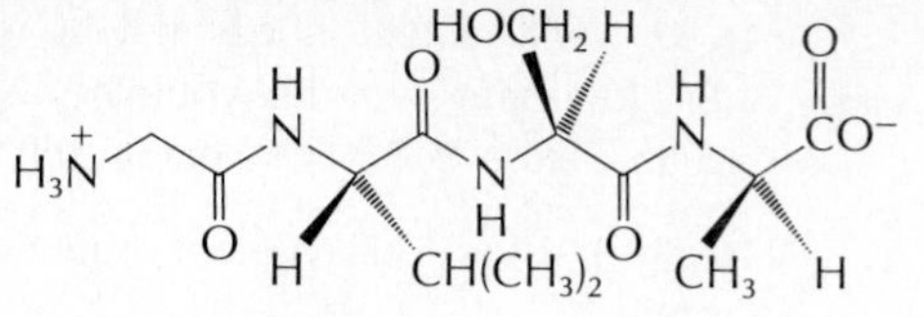

(b) Which, if any, of the amino acid residues in this peptide has the D configuration?

16.22 Complete hydrolysis of a tripeptide gave a solution containing equimolar amounts of alanine, leucine, and proline. Using three-letter abbreviations, list all the possible peptides consistent with these data.

16.23 Treatment of the peptide from Problem 16.22 with carboxypeptidase released proline. Reaction of the peptide with phenyl isothiocyanate followed by hydrochloric acid gave the PTH derivative of leucine. What is the structure of the peptide?

16.24 For the following, refer to the structure of leucine enkephalin, shown in Figure 16.3.
   (a) Give the structure of the derivative formed after reaction with Sanger's reagent and hydrolysis.
   (b) What derivative is obtained after four cycles of the Edman degradation?
   (c) What amino acid is released upon treatment with carboxypeptidase?

16.25 Give the products of the hydrolysis of the peptide shown in the presence of the enzyme trypsin.

Ser-Ala-Arg-Phe-Gly-Ala

16.26 Complete hydrolysis of a tetrapeptide gave equimolar amounts of Leu, Gly, Phe, and Val. One cycle of the Edman degradation gave PTH-Phe. Partial hydrolysis yielded a number of fragments, one of which was a tripeptide which contained Val, Gly, and Phe (not necessarily in that order). Also present in the hydrolysis mixture was a dipeptide containing Leu and Gly. What is the primary structure of the tetrapeptide?

Peptide Synthesis

16.27 Lysine reacts with two equivalents of benzyloxycarbonyl chloride to give a derivative containing two benzyloxycarbonyl groups. What is the structure of this compound?

16.28 Show the steps involved in the synthesis of Val-Ala from valine and alanine using benzyloxycarbonyl and benzyl ester protecting groups and DCCI-promoted peptide bond formation.

16.29 Reaction with hydrogen in the presence of a palladium catalyst is commonly used to remove the benzyloxycarbonyl and benzyl ester protecting groups. Explain what difficulties might arise if hydrolysis were used to remove the protecting groups.

Miscellaneous Problems

16.30 What are the products of each of the following reactions? Your answer should account for all the amino acid residues in the starting peptides.
   (a) Reaction of Leu-Gly-Ser with 1-fluoro-2,4-dinitrobenzene
   (b) Hydrolysis of the compound in (a) in concentrated hydrochloric acid (100°C)
   (c) Treatment of Ile-Glu-Phe with $C_6H_5N{=}C{=}S$, followed by hydrogen chloride
   (d) Reaction of Val-Ser-Ala with benzyloxycarbonyl chloride
   (e) Reaction of the product of (d) with the benzyl ester of valine in the presence of DCCI
   (f) Reaction of the product of (e) with hydrogen and a palladium catalyst.

16.31 Explain the differences between the two commonly encountered types of protein secondary structures.

16.32 As disulfide bridges may be formed by oxidation, they may be cleaved by reduction. How many peptides would be obtained on reduction of the disulfide bridges of bovine insulin (Figure 16.5)? How many sulfhydryl (—SH) groups would each fragment contain?

16.33 List the four types of attractive forces that contribute to the tertiary structure of proteins.

16.34 Are the enzymes trypsin and chymotrypsin more likely to be classified as globular or fibrous proteins?

16.35 *Somatostatin* is a tetradecapeptide of the hypothalamus that inhibits the release of pituitary growth hormone. Its amino acid sequence has been determined by a combination of Edman degradations and enzymatic hydrolysis experiments. On the basis of the following data, deduce the primary structure of somatostatin.

    I.   Edman degradation gave PTH-Ala

    II. Selective hydrolysis gave peptides having the following indicated sequences:

            Phe-Trp
            Thr-Ser-Cys
            Lys-Thr-Phe
            Thr-Phe-Thr-Ser-Cys
            Asn-Phe-Phe-Trp-Lys
            Ala-Gly-Cys-Lys-Asn-Phe

    III. Somatostatin has a disulfide bridge

CHAPTER 17

LIPIDS

LEARNING OBJECTIVES

This chapter focuses on lipids, a major class of substances found in living systems. Upon completion of this chapter you should be able to:

- Explain what physical property enables a substance to be characterized as a lipid.
- Write the chemical structure of a fat or oil and explain the difference between them.
- Explain what is meant by the term *fatty acid*.
- Explain how the iodine number is used to measure the degree of unsaturation of a fat or oil.
- Write the structure of a phospholipid and explain the structural features that enable phospholipid molecules to form cell membranes.
- Write a structural formula typical of a wax.
- Explain the significance of cholesterol in the formation of steroids.
- Describe the function of vitamin D and the bile acids in the body.
- Describe the role of acetyl coenzyme A in the biosynthesis of terpenes.
- Identify the isoprene units in a terpene structure.

Along with proteins, carbohydrates, and nucleic acids, **lipids** form one of the major classes of organic compounds important in living systems. Substances are classed as lipids by their solubility in nonpolar solvents. Lipids play important roles in the chemistry of living systems, from their presence in cell membranes to their action as sex hormones. In addition to a discussion of the various classes of lipids, this chapter will examine how living systems convert simple starting materials to complex natural products of the terpene and steroid class.

17.1 CLASSIFICATION OF LIPIDS

Lipids are naturally occurring substances that are soluble in nonpolar solvents. This is an operational, rather than a structural, distinction. Material from a living organism is shaken with a polar solvent (water or an alcohol-water mixture) and a relatively nonpolar one (diethyl ether, hexane, or dichloromethane). Carbohydrates, proteins, nucleic acids, and related compounds are polar and do not dissolve in the nonpolar solvent—they either dissolve in the aqueous phase or remain behind undissolved. The portion of the natural material that dissolves in the nonpolar solvent is the **lipid fraction.**

The various classes of compounds that make up the lipid fraction will be discussed in the sections which follow. One class of lipids, the prostaglandins, was the subject of a boxed essay in Chapter 12 (Section 12.6).

17.2 FATS AND FATTY ACIDS

Fats are a type of lipid that have a number of functions in living systems, including that of energy storage. While carbohydrates serve as a source of readily available energy, an equal mass of fat delivers over twice the energy. An organism is able to store energy more efficiently in the form of fat because less mass is required compared with storing the same amount of energy in carbohydrate molecules.

The word *fats* usually means both fats and oils. Fats and oils are triesters of glycerol and are called *triacylglycerols,* or *triglycerides* (Section 13.3).

$$
\begin{array}{cc}
\underset{\displaystyle \text{OH}}{\text{HOCH}_2\text{CHCH}_2\text{OH}} &
\underset{\displaystyle \underset{\displaystyle O}{\overset{\displaystyle \text{OC}R''}{\parallel}}}{\overset{\displaystyle O \quad\quad\quad O}{\overset{\displaystyle \parallel \quad\quad\quad \parallel}{\text{R}\text{COCH}_2\text{CHCH}_2\text{OC}R'}}}
\end{array}
$$

Glycerol   Triacylglycerol (triglyceride)

Two typical examples are shown in Figure 17.1. The three acyl groups may all be different, all the same, or one may be different from the other two. In natural fats and oils the acyl groups of triacylglycerols are formed from long-chain carboxylic acids called **fatty acids.** Table 17.1 lists several representative fatty acids. As the examples indicate, most naturally occurring fatty acids possess an even number of carbon atoms and an unbranched carbon chain. The carbon chain may be completely saturated or may incorporate one or more double bonds. Acyl groups containing 14 to 20 carbon atoms are the most abundant in triacylglycerols.

The physical difference between fats and oils is that fats are solids at room temperature (about 25°C), whereas oils are liquids. We often speak of *animal fat* and *vegetable oils,* referring to the fact that fats are more prevalent in animals, while oils are usually of plant origin.

Structurally, fats and oils differ by the number of double bonds in the fatty acid side chains; oils have a greater number of double bonds (are more unsaturated) than fats. These double bonds are generally cis (or *Z*) and interfere with the ability of neighboring molecules to pack together and form a solid. Thus, unsaturation lowers the melting point of the triacylglycerol, giving rise to a liquid oil at room temperature.

Nutrition experts recommend limiting the amount of saturated fat in our diet, and thus vegetable oils have become popular for use in cooking and prepared foods. To make margarine from a vegetable oil and yet have it retain its shape on the dinner

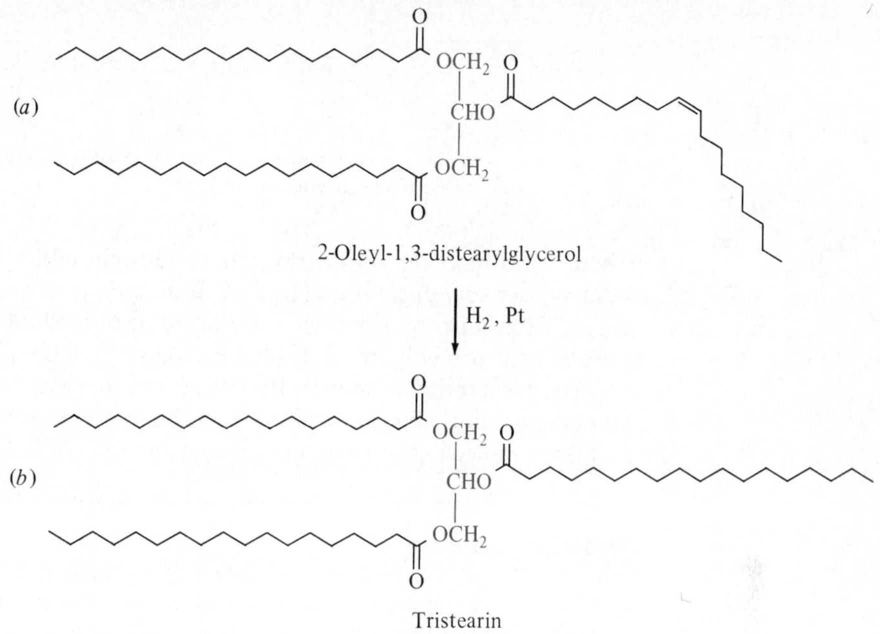

(a)

2-Oleyl-1,3-distearylglycerol

H_2, Pt

(b)

Tristearin

FIGURE 17.1 The structures of two typical triacylgycerols. (a) 2-Oleyl-1,3-distearyl-glycerol is found in cocoa butter. The cis double bond of its 2-acyl group lowers the melting point by interfering with efficient crystal packing. (b) Catalytic hydrogenation converts 2-oleyl-1,3-distearylglycerol to the saturated fat tristearin.

table, the melting point of the oil must be raised above room temperature. Margarines are prepared by catalytic hydrogenation (Figure 17.1) of some, but not all, of the double bonds of the acyl side chains. Thus a margarine label will say that the product contains "partially hydrogenated vegetable oil."

One method used to represent the degree of unsaturation of a fat or oil is the *iodine number,* defined as the number of grams of iodine absorbed per 100 g of fat or oil. Iodine (I_2) itself does not add to alkenes, but the compound iodine chloride (ICl) does and is the iodine derivative used in the analysis. The reaction that occurs with the double bonds of unsaturated triacylglycerols is ·

TABLE 17.1

Some Representative Fatty Acids

Structural formula	Systematic name	Common name
Saturated fatty acids		
$CH_3(CH_2)_{10}COOH$	Dodecanoic acid	Lauric acid
$CH_3(CH_2)_{12}COOH$	Tetradecanoic acid	Myristic acid
$CH_3(CH_2)_{14}COOH$	Hexadecanoic acid	Palmitic acid
$CH_3(CH_2)_{16}COOH$	Octadecanoic acid	Stearic acid
$CH_3(CH_2)_{18}COOH$	Icosanoic acid	Arachidic acid
Unsaturated fatty acids		
$CH_3(CH_2)_7CH{=}CH(CH_2)_7COOH$	(Z)-9-Octadecenoic acid	Oleic acid
$CH_3(CH_2)_4CH{=}CHCH_2CH{=}CH(CH_2)_7COOH$	(9Z, 12Z)-9,12-Octa-decadienoic acid	Linoleic acid
$CH_3CH_2CH{=}CHCH_2CH{=}CHCH_2CH{=}CH(CH_2)_7COOH$	(9Z,12Z,15Z)-9,12,15-Octadeca-trienoic acid	Linolenic acid
$CH_3(CH_2)_4CH{=}CHCH_2CH{=}CHCH_2CH{=}CHCH_2CH{=}CH(CH_2)_3COOH$	(5Z,8Z,11Z,14Z)-5,8,11,14-Icosate-traenoic acid	Arachidonic acid

$$\text{C=C} + \text{ICl} \longrightarrow \overset{\displaystyle I}{\underset{\displaystyle Cl}{-C-C-}}$$

Unsaturated   Addition product
triacylglycerol

There is no reaction with a saturated triacylglycerol. Thus oils with high iodine numbers are very unsaturated (called *polyunsaturated* on food labels). For example, the iodine number of corn oil is 128; the iodine number of butter is 32. A saturated fat such as tristearin (Figure 17.1*b*) has an iodine number of zero.

The preparation of soap by the base-promoted hydrolysis, or *saponification,* of a fat or oil was described in Section 13.7. The products of basic hydrolysis are glycerol and three molecules of soap, the carboxylate salts of fatty acids.

PROBLEM 17.1

Refer to Figure 17.1 and write a chemical equation representing the base-promoted hydrolysis of (*a*) Tristearylglycerol (tristearin); (*b*) 2-Oleyl-1,3-distearylglycerol; (*c*) A second triacylglycerol that gives the same proportion of fatty acid salts as (*b*)

SAMPLE SOLUTION

(*a*) The three acyl groups of tristearin are the same, and thus base-promoted hydrolysis produces 3 mol of stearate ion and 1 mol of the alcohol glycerol.

$$\underset{\text{Tristearin}}{\begin{array}{l} CH_2O_2C(CH_2)_{16}CH_3 \\ | \\ CHO_2C(CH_2)_{16}CH_3 \\ | \\ CH_2O_2C(CH_2)_{16}CH_3 \end{array}} + \underset{\substack{\text{Hydroxide} \\ \text{ion}}}{3HO^-} \longrightarrow \underset{\text{Glycerol}}{\begin{array}{l} CH_2OH \\ | \\ CHOH \\ | \\ CH_2OH \end{array}} + \underset{\text{Stearate ion}}{3CH_3(CH_2)_{16}CO_2^-}$$

17.3 PHOSPHOLIPIDS

Phospholipids are a second type of lipid structurally similar to triacylglycerols. One example of a phospholipid is **phosphatidylcholine,** also called **lecithin.** Lecithin is a diacylglycerol in which the third glycerol hydroxyl has been converted to a phosphate ester. A choline unit $[OCH_2CH_2\overset{+}{N}(CH_3)_3]$ is also attached to the phosphate group.

$$\begin{array}{l} \qquad\qquad \overset{\displaystyle O}{\overset{\displaystyle \|}{CH_2OCR}} \\ \overset{\displaystyle O}{\overset{\displaystyle \|}{R'CO}}\!\!-\!\!\!\!\begin{array}{|c} \;\;\;\;\;\; \\ \hline \\ \hline \end{array}\!\!\!-\!\!H \\ \qquad\qquad CH_2OPO_2^- \\ \qquad\qquad | \\ \qquad\qquad OCH_2CH_2\overset{+}{N}(CH_3)_3 \end{array}$$

Phosphatidylcholine, or lecithin
(R and R' are usually different)

In Chapter 12 you learned how soap molecules associate in water to form spherical micelles. The polar carboxylate "heads" are located on the surface of the micelle,

Water

Hydrophilic "head" groups

Lipophilic "tails"

Hydrophilic "head" groups

Water

FIGURE 17.2 Schematic drawing of a phospholipid bilayer.

where they are solvated by water molecules. Water is excluded from the interior of the micelle by the long-chain alkyl "tails."

Phospholipids such as lecithin also possesses a polar "head group" (the positively charged choline and negatively charged phosphate units) and two nonpolar "tails" (the long-chain acyl groups). Lecithin molecules form a **lipid bilayer,** as shown in Figure 17.2. The nonpolar acyl groups of two opposing layers of lecithin molecules form the interior of the bilayer. The polar groups are solvated by water molecules at the top and bottom outer surfaces.

Phospholipids are the principal components of cell membranes. It is believed that these membranes are composed of lipid bilayers analogous to that shown in Figure 17.2. Nonpolar materials can diffuse through the bilayer from one side to the other relatively easily; polar materials, particularly metal ions such as Na^+, K^+, and Ca^{2+}, cannot. The passage of metal ions through a membrane into and out of a cell is usually assisted by certain proteins present in the lipid bilayer. The metal ion is picked up at one side of the lipid bilayer and delivered at the other, surrounded at all times by a polar environment on its passage through the hydrocarbon-like interior of the membrane. Certain antibiotics such as monensin (Figure 10.3) disrupt the normal functioning of cells by facilitating metal ion transport across cell membranes.

PROBLEM 17.2
Lecithin is used in high-fat commercial food products such as mayonnaise to keep the fat and water from separating into two layers. Explain what structural features of lecithin enable it to be used in this way.

17.4 WAXES

Waxes are a third kind of lipid found in living systems. Waxes are usually mixtures of esters in which both the alkyl and acyl group are unbranched and contain 12 or more carbon atoms. Waxes make up part of the protective coatings of a number of living things, including the leaves of plants, animal fur, and bird feathers. Beeswax, for example, contains the ester triacontyl hexadecanoate as one component of a complex mixture of hydrocarbons, alcohols, and esters.

$$CH_3(CH_2)_{14}\overset{\overset{\displaystyle O}{\|}}{C}OCH_2(CH_2)_{28}CH_3 \qquad \text{Triacontyl hexadecanoate}$$

PROBLEM 17.3
One of the waxes responsible for the shiny coating on holly and rhododendron leaves is oleyl oleate. Using Table 17.1 if necessary, give the systematic name and write the structural formula of this substance.

17.5 STEROIDS. CHOLESTEROL

Steroids form a major class of lipids, having a variety of functions in living systems. Cholesterol is the central compound in any discussion of steroids. It is the most abundant steroid present in humans and the most important one as well since all other steroids arise from it. An average adult has over 200 g of cholesterol; it is found in almost all body tissues with relatively large amounts present in the brain, spinal

HDL, LDL, AND CHOLESTEROL

For a number of years nutritionists and medical professionals have been concerned about the amount of cholesterol consumed in the average person's diet. Evidence has suggested a link between dietary cholesterol and atherosclerosis, a form of heart disease. The aim of reducing dietary cholesterol has been to lower the total cholesterol level in the blood, thus reducing the risk of heart disease.

Research has now shown that the total cholesterol level is not the only risk factor which must be considered. The levels of high-density lipoproteins (HDL) and low-density lipoproteins (LDL) in the bloodstream must also be considered. Some people have termed these "good" cholesterol and "bad" cholesterol.

As a nonpolar lipid, cholesterol has essentially no water solubility. The bloodstream is an aqueous medium, and cholesterol molecules must be held in suspension in order to be transported. This is the function of the lipoproteins —they are water-soluble clusters of protein molecules which suspend and transport lipids throughout the body.

High-density lipoproteins (HDL) are responsible for removing cholesterol from cells and transporting it to the liver, where it is oxidized to form bile acids. The bile acids, in addition to being important in the digestive process, are the predominant means by which cholesterol is excreted from the body. Thus high HDL levels are associated with "good" cholesterol.

Low-density lipoproteins (LDL) carry cholesterol to the cells and are thought to be the main source of arterial cholesterol deposits. High LDL levels in the blood are linked with "bad" cholesterol and a greater tendency toward heart disease and atherosclerosis.

Are there any factors which cause HDL and LDL levels to change? Genetics, which of course we can't control, plays a very important role in our individual susceptibility toward heart disease. There are things we can control, however. HDL levels are increased with regular exercise; LDL levels tend to be higher in persons who smoke or consume large amounts of saturated fats in their diet.

cord, and in gallstones. Cholesterol is the principal constituent of the plaque that builds up on the walls of arteries and restricts the flow of blood in the circulatory disorder known as atherosclerosis.

Cholesterol was isolated in the eighteenth century, but its correct constitution was not determined until 1932 and its stereochemistry not verified until 1955. All steroids are characterized by the tetracyclic ring systems shown in Figure 17.3a. As shown in Figure 17.3b, cholesterol contains this tetracyclic skeleton modified to include an alcohol function at C-3, a double bond at C-5, methyl groups at C-10 and C-13, and a C_8H_{17} side chain at C-17. The name cholesterol is a combination of the Greek words for "bile" (chole) and "solid" (stereos) preceding the alcohol suffix -ol. Cholesterol is found in the body as both the free alcohol and esterified at the C-3 alcohol group by various fatty acids.

17.6 VITAMIN D

A steroid very closely related structurally to cholesterol is its 7-dehydro derivative. 7-Dehydrocholesterol is formed by enzymatic oxidation of cholesterol and has a conjugated diene unit in its B ring. 7-Dehydrocholesterol is present in the tissues of

FIGURE 17.3 (a) The tetracyclic ring system characteristic of steroids. The rings are designated A, B, C, and D as shown. (b) The structure of cholesterol. The numbering system used for steroids is indicated on the formula.

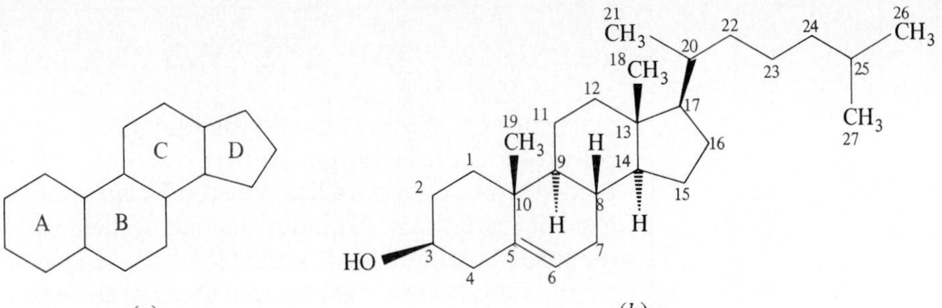

(a)

(b)

the skin, where it is transformed to vitamin D_3 by a sunlight-induced photochemical reaction.

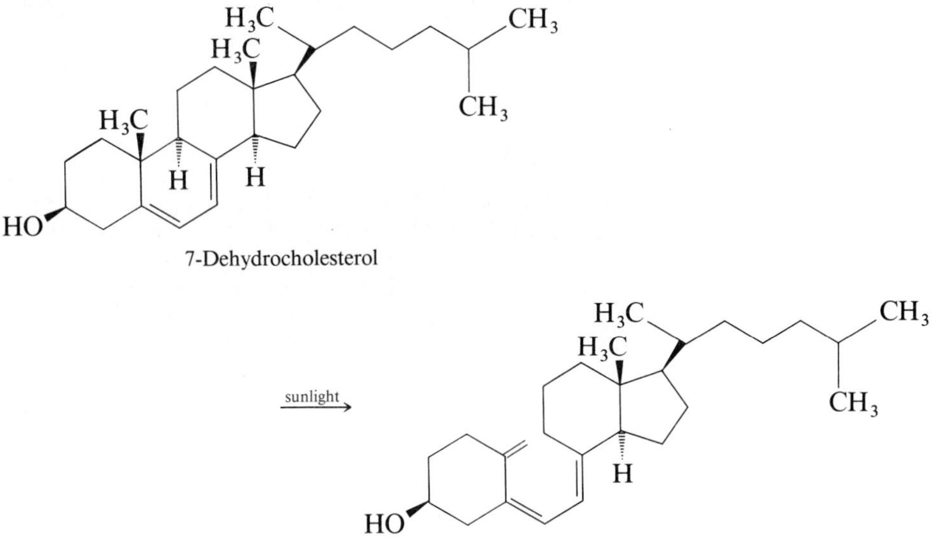

7-Dehydrocholesterol

sunlight

Vitamin D_3

Vitamin D_3 is a key compound in the process by which Ca^{2+} is absorbed from the intestine. Low levels of vitamin D_3 lead to Ca^{2+} concentrations in the body that are insufficient to support proper bone growth, resulting in the disease called *rickets*.

Rickets was once a serious health problem. It was thought to be a dietary deficiency disease because it could be prevented in children be feeding them fish liver oil. Actually, rickets is an environmental disease brought about by a deficiency of sunlight. Where the winter sun is weak, children may not be exposed to enough of its light to convert the 7-dehydrocholesterol in their skin to vitamin D_3 at levels sufficient to promote the growth of strong bones. Fish have adapted to an environment that screens them from sunlight, so they are not directly dependent on a sunlight-induced reaction for their vitamin D_3, and accumulate it by a different process.

While fish liver oil is a good source of vitamin D_3, it is not very palatable. The major dietary source of D-type vitamins at present is *ergosterol*, a steroid found in yeast. Ergosterol is structurally similar to 7-dehydrocholesterol and on irradiation with sunlight or artificial light is converted to vitamin D_2. *Vitamin D_2 has a structure analogous to vitamin D_3 and is comparable with it in preventing rickets. Irradiated ergosterol is added to milk and other foods to ensure that children receive enough vitamin D for their bones to develop properly.

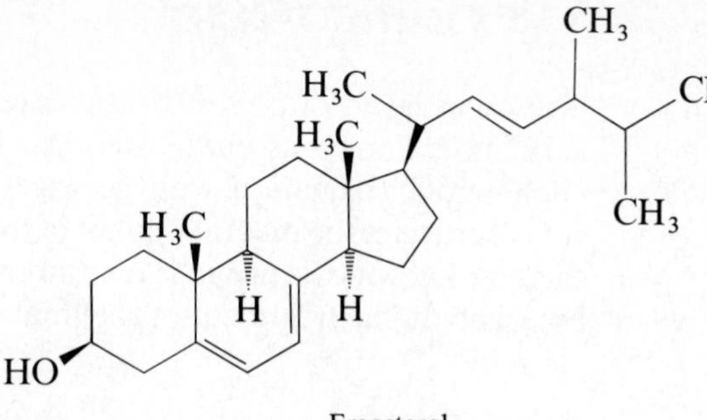

Ergosterol

PROBLEM 17.4

Suggest a reasonable structure for vitamin D_2.

17.7 BILE ACIDS

A significant fraction of the body's cholesterol is used to form **bile acids.** Oxidation in the liver removes a portion of the C_8H_{17} side chain, and additional hydroxyl groups are introduced at various positions on the steroid skeleton. Cholic acid is the most abundant of the bile acids.

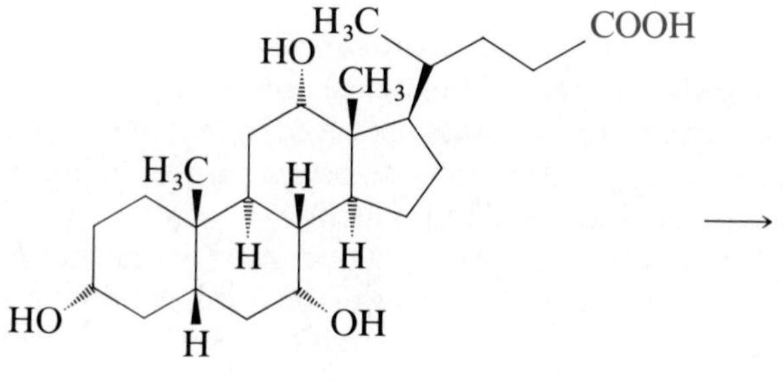

Cholic acid
(a bile acid)

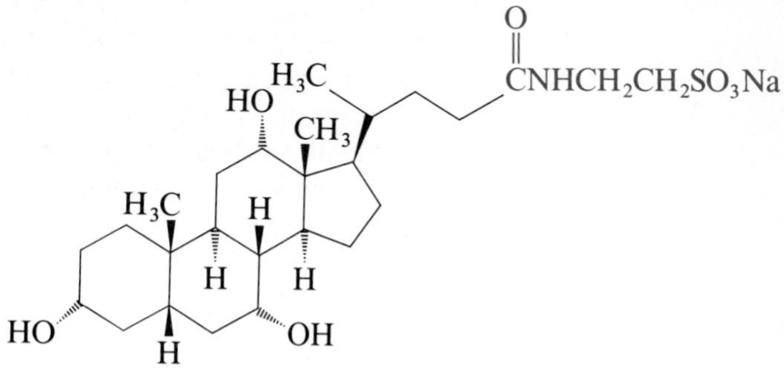

Sodium taurocholate
(a bile salt)

In the form of certain amide derivatives called **bile salts,** bile acids act as emulsifying agents to aid the digestion of fats. Sodium taurocholate is one example of a bile salt. Bile salts have detergent properties similar to those of salts of long-chain fatty acids and promote the transport of lipids through aqueous media.

17.8 CORTICOSTEROIDS

The outer layer, or *cortex,* of the adrenal gland is the source of a large group of substances known as **corticosteroids.** Like the bile acids, they are derived from cholesterol by oxidation, with cleavage of a portion of the alkyl substituent on the D ring. Cortisol is the most abundant of the corticosteroids, while cortisone is probably the best known. Cortisone is commonly prescribed as an anti-inflammatory drug, especially in the treatment of rheumatoid arthritis.

Cortisol   Cortisone

Corticosteroids exhibit a wide range of physiological effects. One important function is to assist in maintaining the proper electrolyte balance in body fluids. Corticosteroids also play a vital regulatory role in the metabolism of carbohydrates.

17.9 SEX HORMONES

Hormones are the chemical messengers of the body; they are secreted by the endocrine glands and regulate biological processes. Corticosteroids, described in the preceding section, are hormones produced by the adrenal glands. The sex glands — testes in males, ovaries in females — secrete a number of hormones which are involved in sexual development and reproduction.

Testosterone is the principal *androgen,* or male sex hormone. Testosterone promotes muscle growth, deepening of the voice, the growth of body hair, and other male secondary sex characteristics. Estradiol is the principal female sex hormone, or *estrogen.* Estradiol is a key substance in the regulation of the menstrual cycle and the reproductive process. It is the hormone most responsible for the development of female secondary sex characteristics.

Testosterone   Estradiol

Testosterone and estradiol are present in the body in only minute amounts, and their isolation and identification required heroic efforts. To obtain 0.01 g of estradiol for study, 4 tons of sow ovaries had to be extracted!

Progesterone is a second female sex hormone. One function of progesterone is to suppress ovulation at certain stages of the menstrual cycle and during pregnancy. Synthetic substances, such as norethindrone, have been developed that are superior to progesterone when taken orally to "turn off" ovulation. By inducing temporary infertility, they form the basis of "the pill" — the oral contraceptive agents that have been in use since the early 1960s.

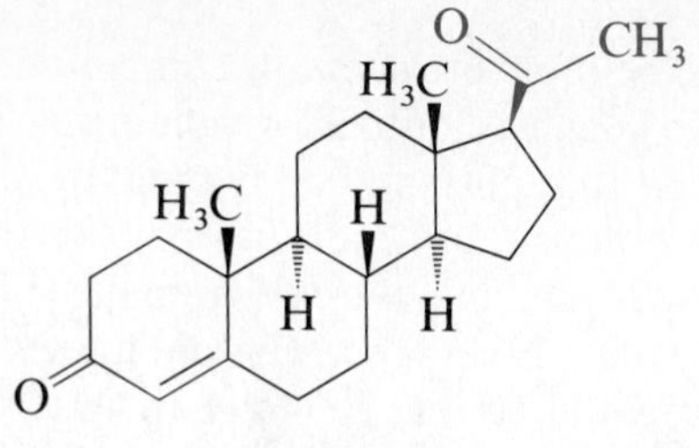

Progesterone

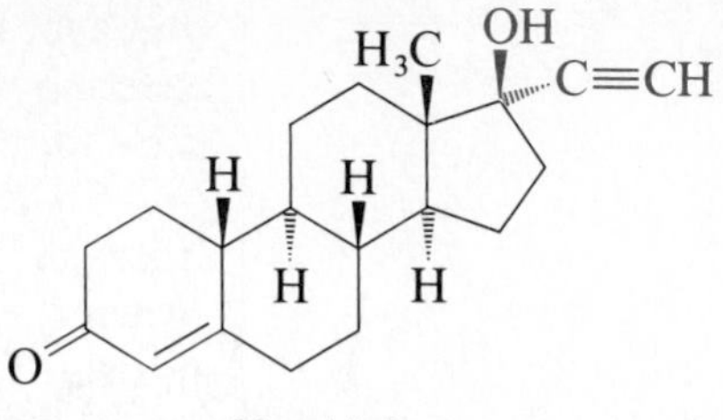

Norethindrone

WHAT ARE ANABOLIC STEROIDS?

As we have seen in this chapter, steroids have a number of functions in human physiology. Cholesterol is a component part of cell membranes and is found in large amounts in the brain. Derivatives of cholic acid assist the digestion of fats in the small intestine. Cortisone and its derivatives are involved in maintaining the electrolyte balance in body fluids. The sex hormones responsible for masculine and feminine characteristics as well as numerous aspects of pregnancy from conception to birth are steroids. The principal male sex hormone, or **androgen**, is testosterone. Testosterone mediates the biochemical processes which promote masculine sexual characteristics such as the deepening of the voice and the growth of body hair.

In addition to being an androgen, testosterone also promotes muscle growth and so is classified as an **anabolic** steroid hormone. Biological chemists distinguish between two major classes of **metabolism: catabolic** and **anabolic** processes. Catabolic processes are degradative pathways in which larger molecules are broken down to smaller ones. Anabolic processes are the reverse; larger molecules are synthesized from smaller ones. While the body mainly stores energy from food in the form of fat, a portion of that energy goes toward the production of muscle from protein. An increase in the amount of testosterone, accompanied by an increase in the amount of food consumed, will cause an increase in the body's muscle mass.

The pharmaceutical industry has developed and studied a number of anabolic steroids for use in veterinary medicine and to promote rehabilitation from injuries that are accompanied by deterioration of muscles. The ideal agent would be one that possessed the anabolic properties of testosterone without its androgenic (masculinizing) effects. Dianabol and stanozolol are among the many synthetic anabolic steroids available by prescription.

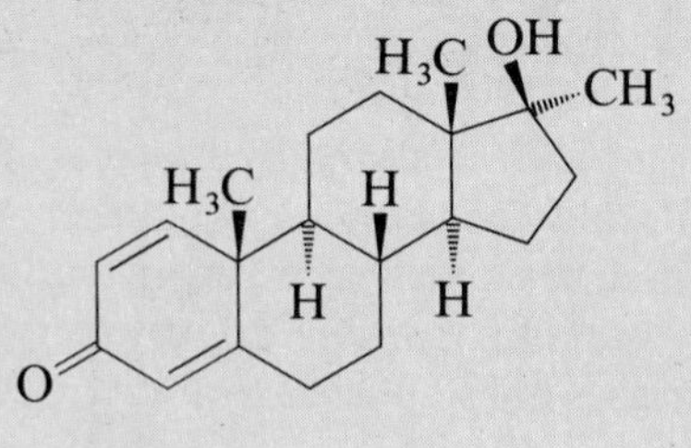

Dianabol

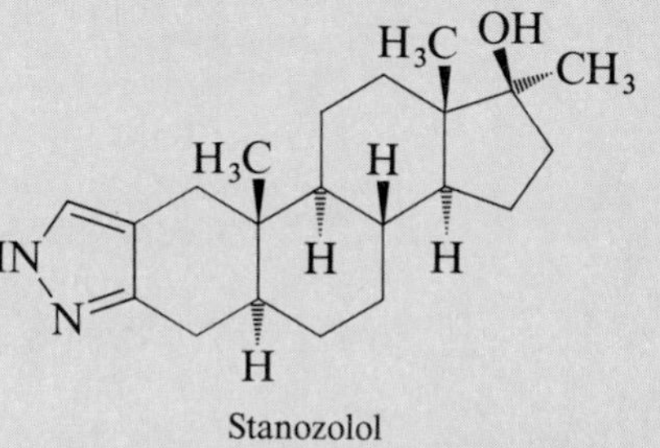

Stanozolol

Sprinter Ben Johnson was stripped of the gold medal he won in the 100-meter dash at the 1988 Olympic Games when tests revealed the presence of stanozolol in his urine. Abuse of anabolic steroids probably exists in all sports but appears to be most prevalent in competitions which place a premium on size and strength such as football, weight lifting, the weight events in track and field (discus and shot put), and body building.

Some scientific studies indicate that the gain in performance obtained through the use of anabolic steroids is small. This seems to be a case, however, where the scientific studies are wrong and the anecdotal evidence of the athletes is right. The scientific studies are done under ethical conditions where patients are treated with "prescription-level" doses of steroids. A 240-pound offensive tackle ("too small" by today's standards) may take several anabolic steroids at a time at 10 to 20 times their prescribed doses in order to weigh the 280 pounds he (or his coach) feels is necessary. The price athletes pay for gains in size and strength can be enormous. The price includes emotional costs (friendships lost because of heightened aggressiveness), sterility, and testicular atrophy (the testes cease to function once the body starts to obtain a sufficient supply of testosterone-like steroids from outside), and an increased risk of premature death from liver cancer or heart disease.

17.10 BIOSYNTHESIS. ACETYL COENZYME A

Many of the complex molecules which make up living systems are the result of **biosynthesis**; the molecule is prepared in the organism by a series of chemical reactions beginning with simple starting materials. Most lipids, for example, are prepared in animals and plants beginning with a type of **acetate** ester. The form in which acetate is utilized in most of its important biochemical reactions is acetyl coenzyme A, shown in Figure 17.4. Acetyl coenzyme A is a **thioester.** It is formed from pyruvic acid by a sequence of steps that is summarized in the following equation:

$$CH_3\overset{O}{\underset{\|}{C}}-\overset{O}{\underset{\|}{C}}OH + CoASH + NAD^+ \longrightarrow$$

Pyruvic acid Coenzyme A Oxidized form of nicotinamide adenine dinucleotide

$$CH_3\overset{O}{\underset{\|}{C}}SCoA + NADH + CO_2 + H^+$$

Acetyl coenzyme A Reduced form of nicotinamide adenine dinucleotide Carbon dioxide Proton

All the individual steps are catalyzed by enzymes. The cofactor NAD^+ is required as an oxidizing agent, and coenzyme A (Figure 17.4b) is the acetyl group acceptor. Coenzyme A is a **thiol;** its chain terminates in a **sulfhydryl** (—SH) group. Acetylation of the sulfhydryl group of coenzyme A gives acetyl coenzyme A.

Because sulfur does not stabilize a carbonyl group by electron donation as well as oxygen does, compounds of the type $R\overset{O}{\underset{\|}{C}}SR'$ are better acyl transfer agents than is $R\overset{O}{\underset{\|}{C}}OR'$. They also contain a greater proportion of enol at equilibrium. Both proper-

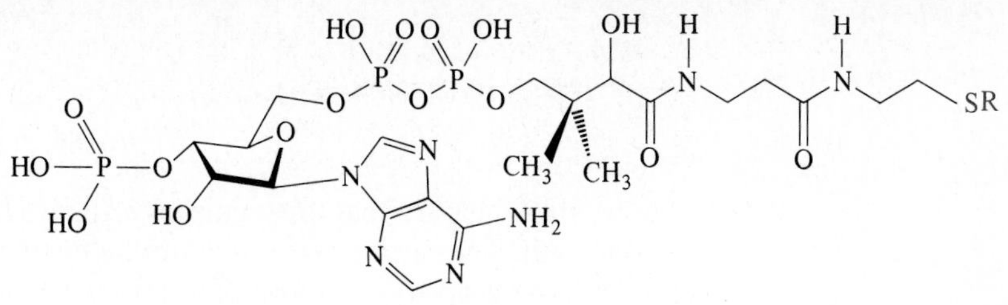

FIGURE 17.4 Structures of (a) acetyl coenzyme A and (b) coenzyme A.

(a) R = $\overset{\overset{\textstyle O}{\|}}{C}CH_3$ Acetyl coenzyme A (abbreviation: $CH_3\overset{\overset{\textstyle O}{\|}}{C}SCoA$)

(b) R = H   Coenzyme A (abbreviation: CoASH)

ties are apparent in the chemistry of acetyl coenzyme A. In some of its reactions acetyl coenzyme A acts as an acyl transfer agent, whereas in others the α carbon atom of the acetyl group is the reactive site.

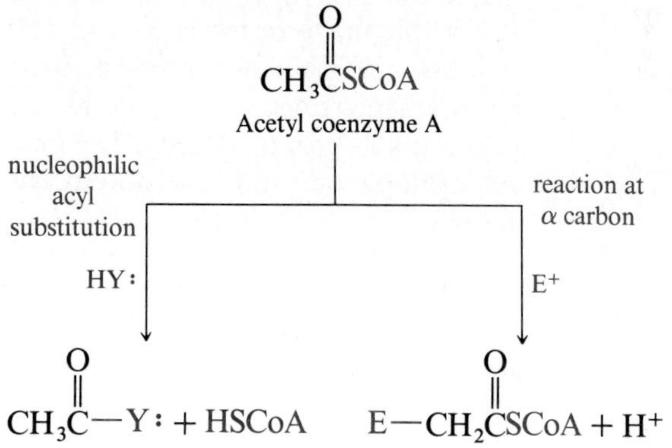

Acetyl coenzyme A is the biosynthetic starting material for the fatty acids which are found in triacylglycerols, phospholipids, and waxes. It is because these compounds are built up by coupling acetate units together that fatty acids have an even number of carbon atoms in an unbranched chain. Cholesterol, and hence all the steroids, are formed through a complex series of transformations beginning with acetate. Most of these reactions have been studied and are understood in some detail. We will look briefly in the following sections at the biosynthesis of one class of lipids called **terpenes**.

17.11 TERPENE BIOSYNTHESIS

The word *essential* as applied to naturally occurring organic substances can have two different meanings. For example, certain amino acids are essential, meaning they are necessary and must be present in our diet because humans lack the ability to biosynthesize them directly.

Essential is also used as the adjective form of the noun *essence*. The mixtures of substances that comprise the fragrant material of plants are called *essential oils* because they contain the essence, i.e., the odor, of the plant. The study of the composition of essential oils is one of the oldest areas of chemical research. Very often, the principal volatile component of an essential oil belongs to the class of chemical substances known as **terpenes**.

Myrcene, a hydrocarbon isolated from bayberry oil, is a typical terpene.

$$(CH_3)_2C{=}CHCH_2CH_2\overset{\overset{\displaystyle CH_2}{\displaystyle \|}}{C}CH{=}CH_2 \equiv$$

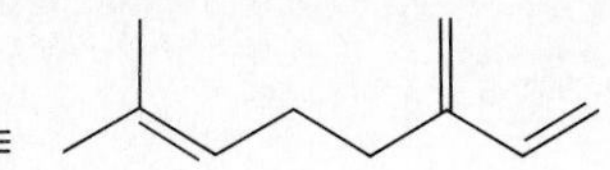

Myrcene

The structural feature that distinguishes terpenes from other natural products is the **isoprene unit.** The carbon skeleton of myrcene (exclusive of its double bonds) corresponds to the head-to-tail union of two isoprene units.

$$CH_2{=}\overset{\overset{\displaystyle CH_3}{\displaystyle |}}{C}{-}CH{=}CH_2 \equiv$$

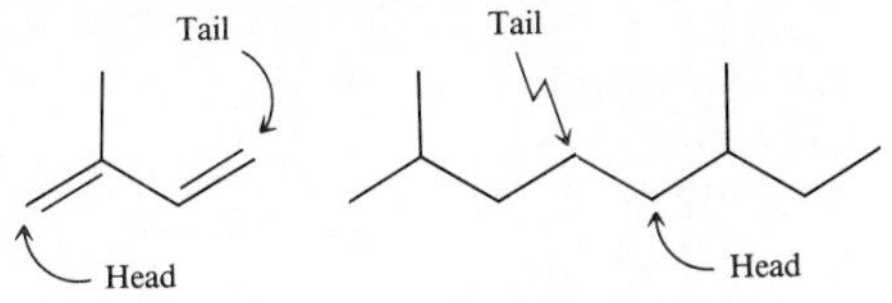

Isoprene
(2-methyl-1,3-butadiene)

Two isoprene units
linked head to tail

Terpenes are often referred to as **isoprenoid** compounds. They are classified according to the number of carbon atoms they contain, as summarized in Table 17.2.

While the term terpene once referred only to hydrocarbons, current usage includes functionally substituted derivatives as well. Figure 17.5 presents the structural formulas for a number of representative terpenes. The isoprene units in some of these are relatively easy to identify. The three isoprene units in the sesquiterpene farnesol, for example, are indicated in structure below. They are joined in a head-to-tail fashion.

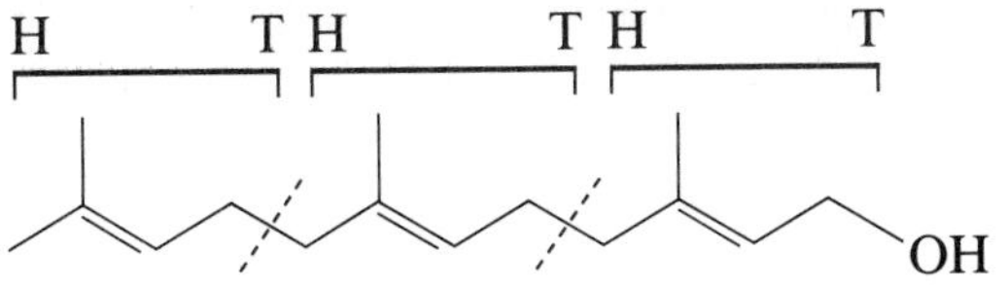

Isoprene units in farnesol
(H = head; T = tail)

Many terpenes contain one or more rings, but these also can be viewed as collections of isoprene units. An example is α-selinene. Like farnesol, it is made up of three isoprene units linked head to tail.

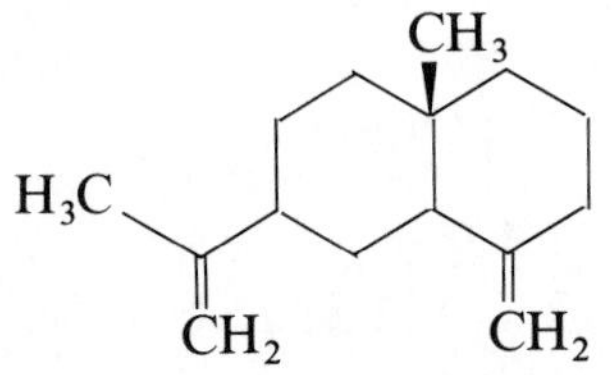

Isoprene units in α-selinene

TABLE 17.2

Classification of Terpenes

Class	Number of carbon atoms
Monoterpene	10
Sesquiterpene	15
Diterpene	20
Triterpene	30
Tetraterpene	40

Monoterpenes

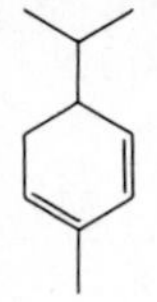

α-Phellandrene
(eucalyptus)

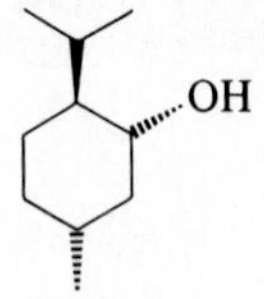

Menthol
(peppermint)

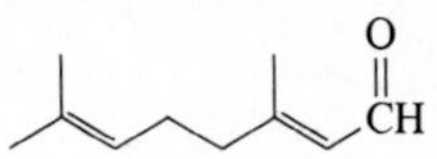

Citral
(lemon grass)

Sesquiterpenes

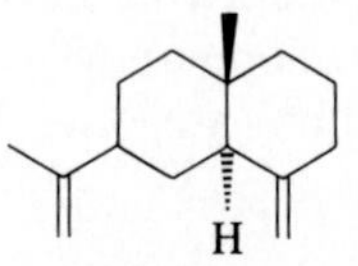

α-Selinene
(celery)

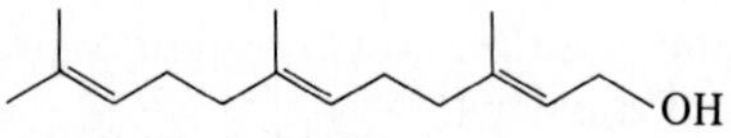

Farnesol
(ambrette)

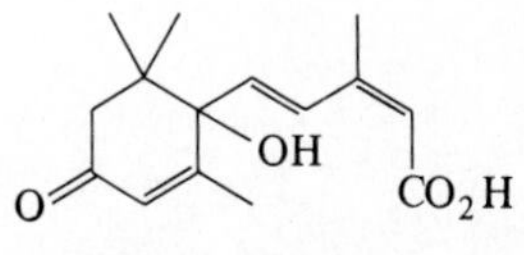

Abscicic acid
(a plant hormone)

Diterpenes

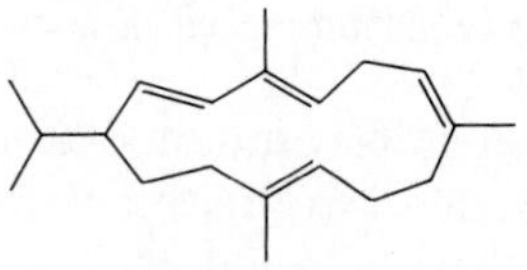

Cembrene
(pine)

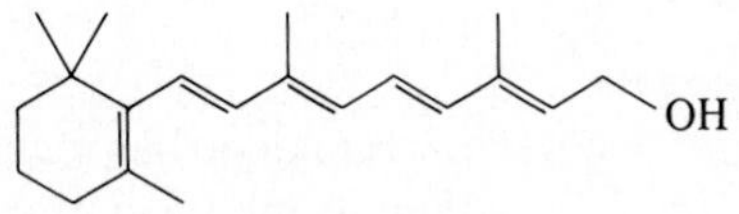

Vitamin A
(present in mammalian tissue and fish oil;
important substance in the chemistry of vision)

Triterpenes

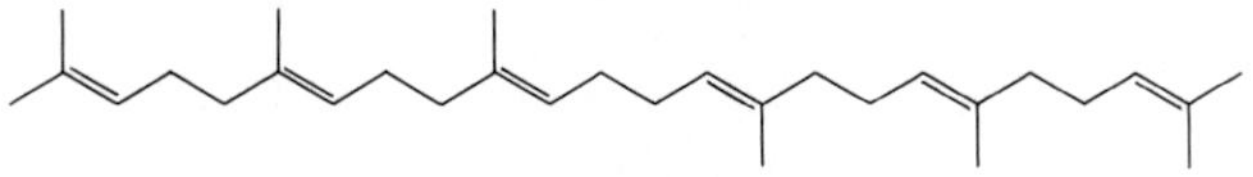

Squalene
(shark liver oil)

Tetraterpenes

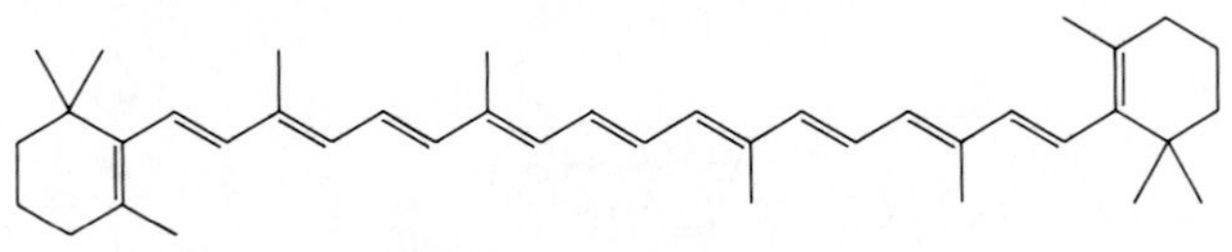

FIGURE 17.5 Some
representative terpenes
and related natural
products.

β-Carotene
(present in carrots and other vegetables;
enzymes in the body cleave β-carotene to vitamin A)

PROBLEM 17.5
Locate the isoprene units in each of the monoterpenes, sesquiterpenes, and diterpenes
shown in Figure 17.5. (In some cases there are two equally correct arrangements.)

SAMPLE SOLUTION
Isoprene units are ⌐⌐⌐ fragments in the carbon skeleton. Functional groups and
multiple bonds are ignored when structures are examined for the presence of isoprene

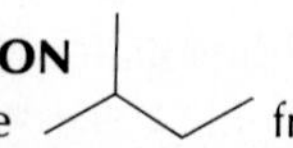

units. The monoterpene α-phellandrene has 10 carbons and thus contains two C_5 isoprene units. There are two equally correct answers:

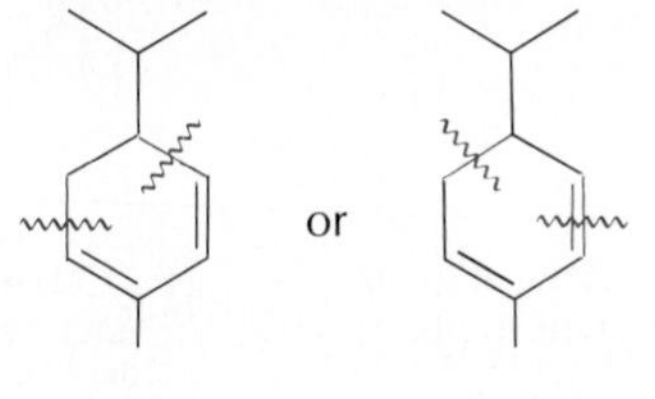

or

Tail-to-tail linkages of isoprene units sometimes occur, especially in the higher terpenes. The C(12)—C(13) bond of squalene unites two C_{15} units in a tail-to-tail manner. Notice, however, that isoprene units are joined head to tail within each C_{15} unit of squalene.

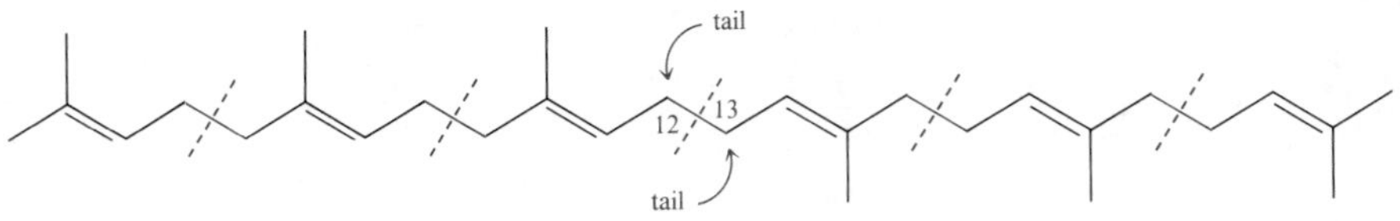

Isoprene units in squalene

Squalene has received a considerable amount of attention, as it is a key intermediate in the biosynthesis of cholesterol from acetate. Squalene undergoes an enzymatic oxidation followed by a cyclization reaction to form the four rings of the steroid skeleton. Several rearrangement steps then lead to the formation of cholesterol.

17.12 ISOPENTENYL PYROPHOSPHATE. THE BIOLOGICAL ISOPRENE UNIT

Isoprenoid compounds such as the terpenes in the previous section are biosynthesized from acetate by a process that involves several stages. The first stage is the formation of mevalonic acid from three molecules of acetic acid.

Acetic acid   Mevalonic acid

In the second stage mevalonic acid is converted to **isopentenyl pyrophosphate.**

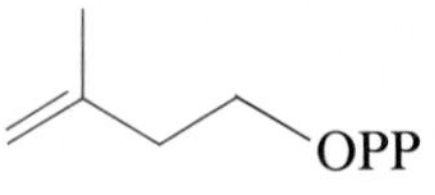

Mevalonic acid   Isopentenyl pyrophosphate

Isopentenyl pyrophosphate is the biological isoprene unit; it contains five carbon atoms connected in the same order as in isoprene. The symbol -OPP is frequently used to represent the pyrophosphate group, as shown in the equation.

Isopentenyl pyrophosphate undergoes an enzyme-catalyzed reaction that converts it, in an equilibrium process, to an isomer, **dimethylallyl pyrophosphate.**

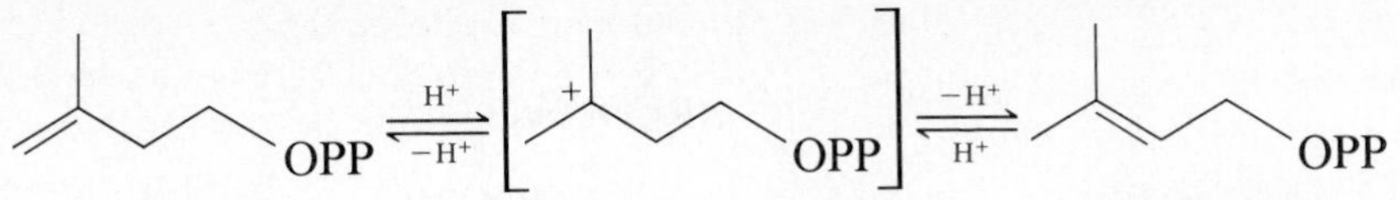

Isopentenyl pyrophosphate   Carbocation intermediate   Dimethylallyl pyrophosphate

Both isopentenyl pyrophosphate and dimethylallyl pyrophosphate are utilized in the biosynthesis of terpenes. This can be illustrated by outlining the formation of geraniol, a naturally occurring monoterpene found in rose oil.

The chemical properties of isopentenyl pyrophosphate and dimethylallyl pyrophosphate are complementary in a way that permits them to react with each other to form a carbon-carbon bond that unites two isoprene units. Using the π electrons of its double bond, isopentenyl pyrophosphate acts as a nucleophile and displaces pyrophosphate from dimethylallyl pyrophosphate.

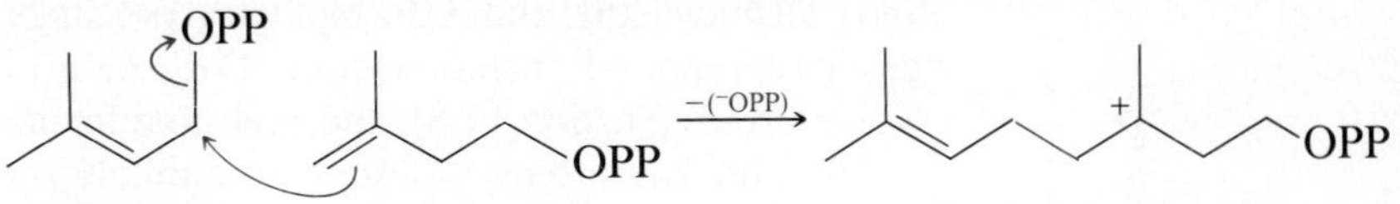

Dimethylallyl pyrophosphate   Isopentenyl pyrophosphate   Ten-carbon carbocation

The tertiary carbocation formed in this step can react according to any of the various reaction pathways available to carbocations. One of these is loss of a proton to give an alkene.

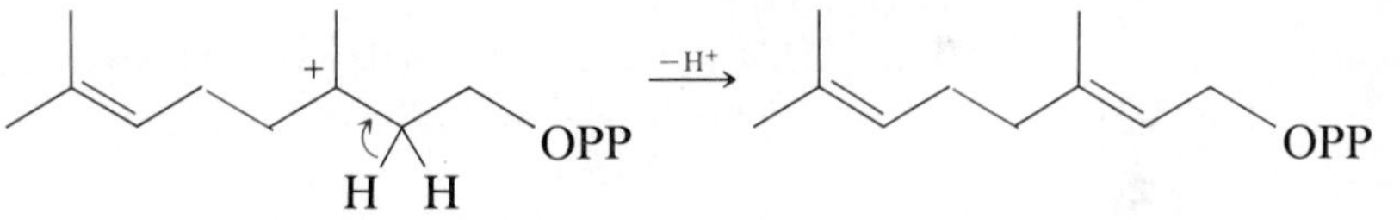

Geranyl pyrophosphate

The product of this reaction is geranyl pyrophosphate. Hydrolysis of the pyrophosphate ester group gives geraniol.

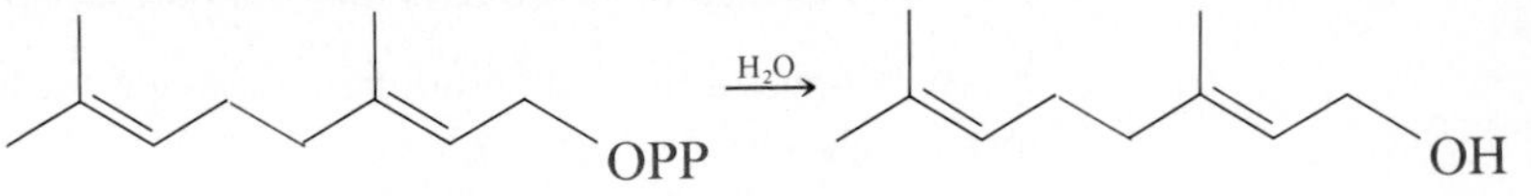

Geranyl pyrophosphate   Geraniol

PROBLEM 17.6
Reaction of geranyl pyrophosphate with isopentenyl pyrophosphate leads to the sesquiterpene farnesol (Figure 17.5). Outline a series of steps for this process.

17.13 SUMMARY

Lipids are naturally occurring substances that are soluble in nonpolar solvents (Section 17.1). *Fats and oils* are glycerol esters of long-chain carboxylic acids (Section 17.2). Typically, the carbon chains of the fatty acids are unbranched and contain an even number of carbon atoms.

$$
\begin{array}{c}
\quad\quad\quad\quad O \\
\quad\quad\quad\quad \parallel \\
\text{RCOCH}_2 \;\; O \\
\quad\quad\quad\;| \quad\; \parallel \\
\quad\quad\quad\text{CHOCR}' \\
\quad\quad\quad\; | \\
\text{R}''\text{COCH}_2 \\
\quad\quad\;\; \parallel \\
\quad\quad\;\; O
\end{array}
$$

Triacylglycerol
(R, R′, and R″ may be the same or different)

Oils are usually of vegetable origin and tend to have one or more double bonds in their fatty acid chains. The degree of unsaturation of a fat or oil is often represented by its *iodine number*. *Phospholipids* such as *lecithin* are structurally similar to triacylglycerols and are the principal components of cell membranes (Section 17.3). *Waxes* are mixtures of substances that usually contain esters of fatty acids and long-chain alcohols (Section 17.4).

Cholesterol is the most prevalant of the *steroids* (Section 17.5). It is found in almost all body tissue and is the biological precursor to the other steroids. Important steroidal compounds include vitamin D (Section 17.6), the bile acids (Section 17.7), corticosteroids (Section 17.8), and male and female sex hormones (Section 17.9).

Most lipids are biosynthesized in animals and plants beginning with *acetate* (Section 17.10). A key intermediate in acetate biosynthesis is *acetyl coenzyme A*. Terpenes and other *isoprenoid* compounds are biosynthesized by way of isopentenyl pyrophosphate (Sections 17.11, 17.12).

ADDITIONAL PROBLEMS

Ester Lipids and Fatty Acids

17.7 What structural features distinguish between each of the following?
  (a) Animal fat and vegetable oil
  (b) A fat and a wax
  (c) A phospholipid and a triacylglycerol

17.8 Phospholipids are more soluble in water than triacylglycerols. Explain.

17.9 By means of chemical equations, show:
  (a) The hydrolysis of trioleylglycerol
  (b) The catalytic hydrogenation of trioleylglycerol

17.10 Compare the iodine number expected for trioleylglycerol with that expected for tristearylglycerol.

17.11 Polar regions of a molecule are often called **hydrophilic** ("water-loving"); nonpolar regions are **hydrophobic** ("water-hating") or **lipophilic** ("lipid-loving"). Identify the hydrophilic and lipophilic parts of lecithin.

17.12 Linolein is a lipid constituent of linseed oil and produces glycerol and linoleic acid upon hydrolysis. Draw the structure of linolein.

17.13 (a) Elaidic acid is a stereoisomer of oleic acid. Write a structural formula for elaidic acid.
  (b) Ricinoleic acid, abundantly present as its glycerol triester in castor oil, is 12-hydroxyoleic acid. Write a structural formula for ricinoleic acid.

17.14 Spermaceti is a wax obtained from the sperm whale. It contains, among other materials, an ester known as *cetyl palmitate,* which is used as an emollient in soaps and cosmetics. The systematic name for cetyl palmitate is *hexadecyl hexadecanoate.*
  (a) Write a structural formula for cetyl palmitate.

(b) Outline a synthesis of cetyl palmitate using hexadecanoic acid as the ultimate source of all its carbon atoms.

Steroids

17.15   Vitamin D is an example of a **fat-soluble vitamin.** Fat-soluble vitamins can be stored in moderate amounts in our body, and thus we are not as dependent on a daily supply of them in our diet. Explain what structural characteristics make vitamin D fat-soluble.

17.16   How many chiral centers are there in cholesterol? What is the number of stereoisomers that can have the cholesterol skeleton?

17.17   In spite of its complexity, cholesterol undergoes reactions with each of the following reagents that are typical of the functional groups it contains. Give the structure of the product formed in each case.

(a) Br_2
(b) H_2 (1 atm), Pt
(c) Acetic anhydride
(d) CrO_3 — pyridine [$(C_5H_5N)_2CrO_3$]

Terpenes and Isoprenoid Biosynthesis

17.18   Identify the isoprene units in β-carotene (Figure 17.5). Which carbons are joined by a tail-to-tail link between isoprene units?

17.19   Identify the isoprene units in each of the following naturally occurring substances:

(a) *Dendrolasin:* A constituent of the defense secretion of a species of ant.

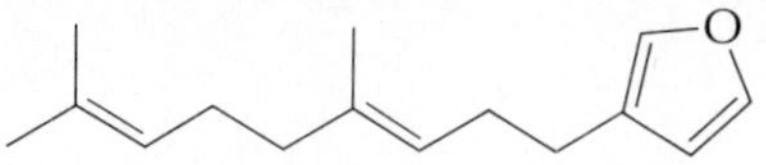

(b) *γ-Bisabolene:* A sesquiterpene found in the essential oils of a large number of plants.

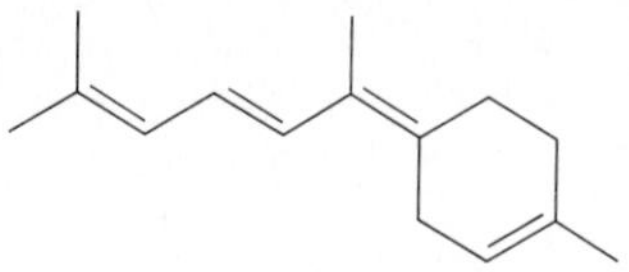

17.20   The isoprenoid compound shown is a scent marker present in the urine of the red fox. Suggest a reasonable synthesis for this substance from 3-methyl-3-buten-1-ol and any necessary organic or inorganic reagents.

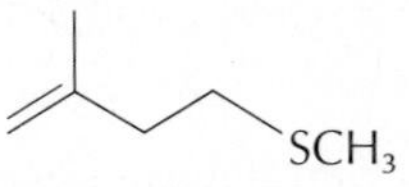

17.21   The ionones are fragrant substances present in the scent of iris and are used in perfume. A mixture of α- and β-ionone can be prepared by treatment of pseudoionone with sulfuric acid.

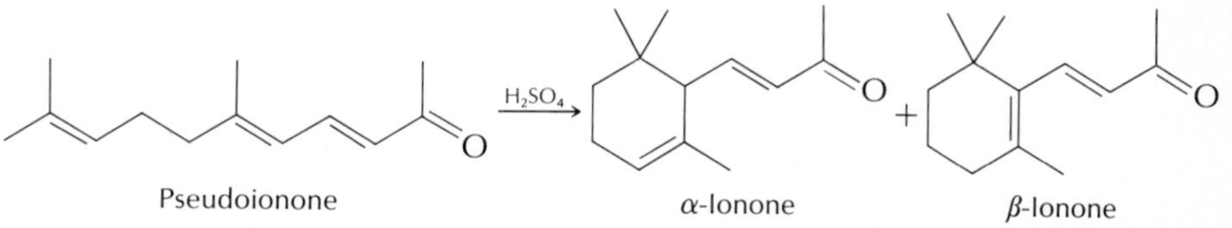

Pseudoionone   α-Ionone   β-Ionone

Write a stepwise mechanism for this reaction.

17.22   Mevalonic acid (Section 17.12) readily forms a lactone (cyclic ester) called mevalonolactone. Write a structural formula for mevalonolactone.

17.23 The monoterpenes α- and β-pinene are obtained from pine trees and are principal constituents of turpentine. Biosynthetically, both are derived from the same carbocation. Write a structural formula for this carbocation.

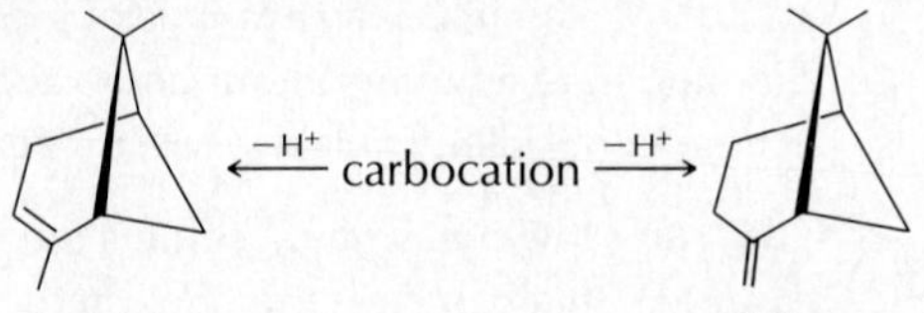

α-Pinene $\qquad$ β-Pinene

NUCLEIC ACIDS

LEARNING OBJECTIVES

This chapter presents a brief look at nucleic acids, the macromolecules responsible for protein biosynthesis and the transfer of genetic information. Upon completion of this chapter you should be able to:

- Distinguish between a purine and a pyrimidine.
- Explain the differences between a nucleoside, a nucleotide, and a nucleic acid.
- Describe how the base sequence of a DNA strand determines the sequence of bases in the complementary strand.
- Explain the roles played by messenger RNA, transfer RNA, and ribosomal RNA in protein biosynthesis.
- Explain what is meant by the term *codon*.

We will conclude our look at biologically significant classes of organic compounds with a brief overview of the nucleic acids. These macromolecules are responsible for protein biosynthesis and the transfer of genetic information. The structure of nucleic acids is quite complex, and our treatment of this subject will be aimed at introducing the molecular components of these important molecules.

18.1 PYRIMIDINES AND PURINES

Nucleic acids were first isolated over 100 years ago and, as their name implies, they are acidic substances present in the nuclei of cells. There are two major kinds of nucleic acids, ribonucleic acid (RNA) and deoxyribonucleic acid (DNA). To understand the complex structure of nucleic acids, we need to first examine some simpler substances, nitrogen-containing aromatic heterocycles called **pyrimidines** and **purines.** The parent substance of each class and the ring numbering system used are shown.

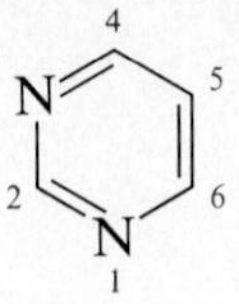

Pyrimidine 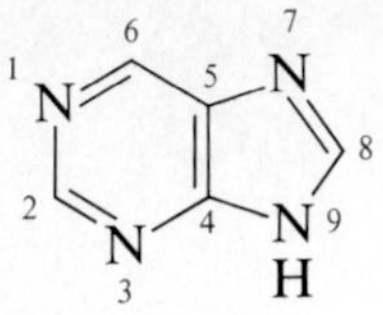   Purine

The pyrimidines that occur in DNA are *cytosine* and *thymine*. RNA also contains cytosine; however, *uracil* is present instead of thymine in RNA.

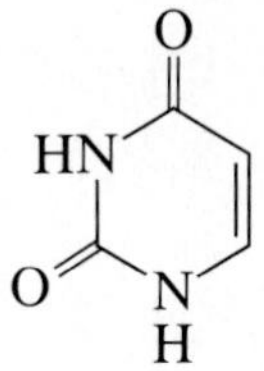

Uracil   Thymine   Cytosine
(occurs in RNA)   (occurs in DNA)   (occurs in both RNA and DNA)

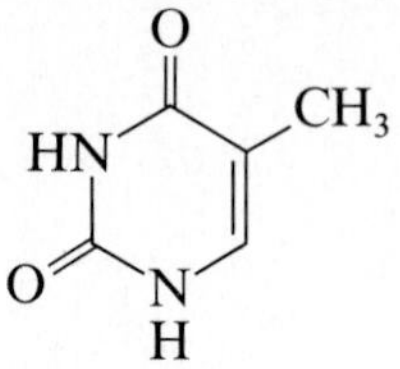

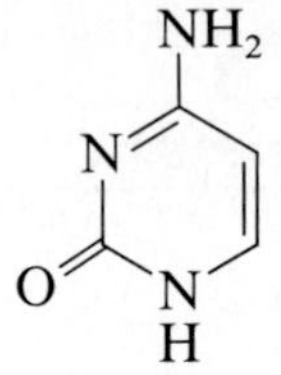

PROBLEM 18.1

5-Fluorouracil, also known as 5-FU, is a drug used in cancer chemotherapy. What is its structure?

The principal purines of both DNA and RNA are adenine and guanine.

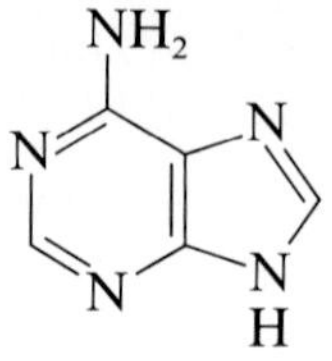

Adenine 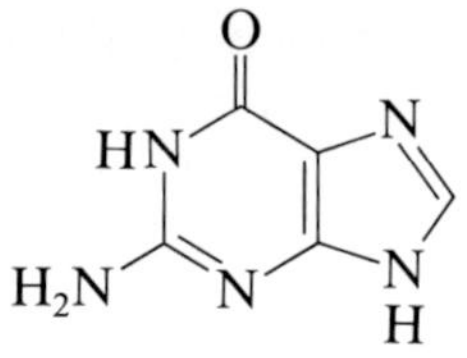   Guanine

The rings of purines and pyrimidines are aromatic and planar. You will see that this flat shape is significant when we consider the structure of nucleic acids.

Pyrimidines and purines occur in substances other than nucleic acids. Two pyrimidine derivatives used as drugs are 5-FU (Problem 18.1) and the antifungal agent flucytosine. Coffee and tea are familiar sources of the natural purine derivative caffeine.

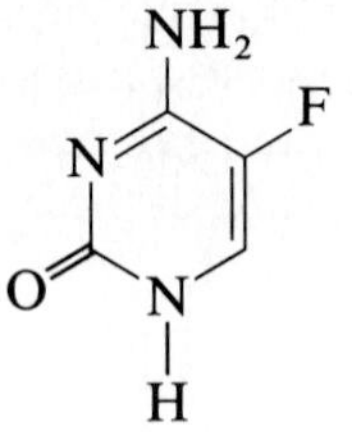

Flucytosine   Caffeine
(a pyrimidine)   (a purine)

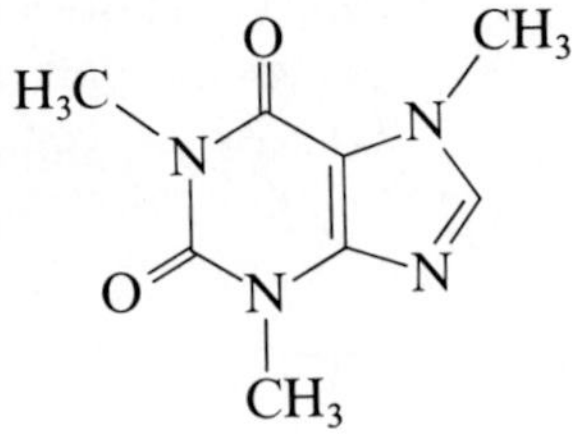

18.2 NUCLEOSIDES

Both RNA and DNA employ a pentose as a fundamental structural unit. In RNA this pentose is D-ribose; in DNA it is 2-deoxy-D-ribose.

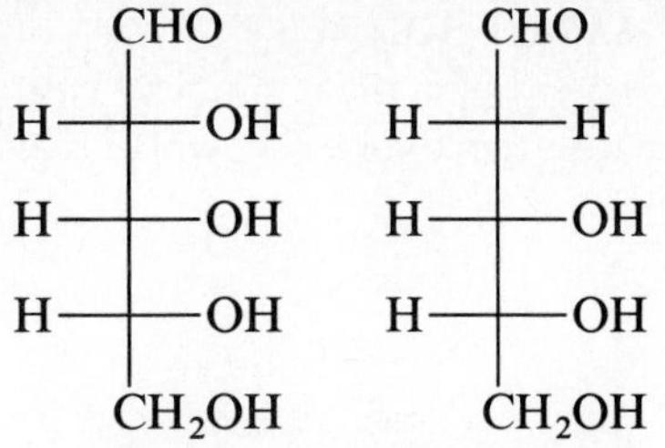

D-Ribose
(in RNA)

2-Deoxy-D-ribose
(in DNA)

In both types of nucleic acids, these sugars exist in their furanose forms (Section 15.5).

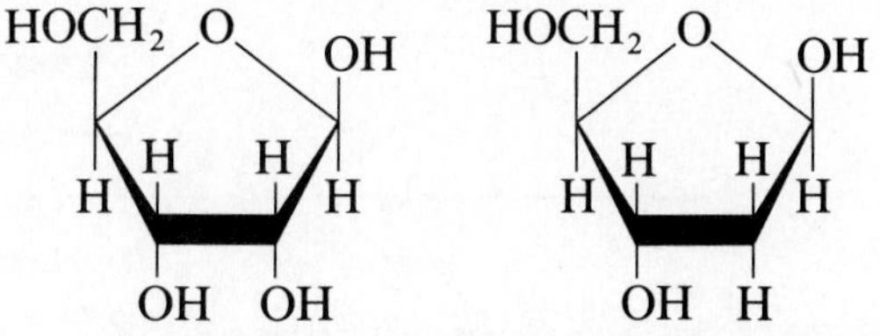

β-D-Ribofuranose

β-D-2-Deoxyribofuranose

In RNA and DNA each pentose occurs as an *N*-glycoside (Section 15.10) in which a pyrimidine or purine substituent is attached to the anomeric carbon. The substituent is referred to as the pyrimidine or purine **base.** The combination of a purine or pyrimidine base with ribose or 2-deoxyribose is called a **nucleoside.**

Two RNA-derived nucleosides are shown. Uridine is a representative pyrimidine nucleoside. Uracil is attached to ribose at N-1 by a β-glycosidic linkage. A typical purine nucleoside is adenosine. The β-glycosidic linkage is between ribose and N-9 of adenine.

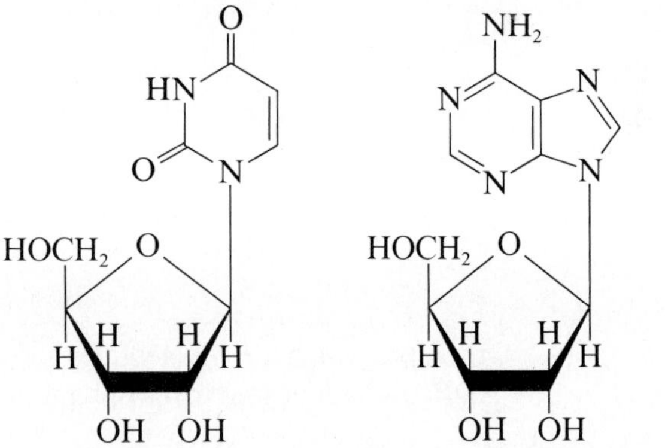

Uridine
(1-β-D-ribofuranosyluracil)

Adenosine
(9-β-D-ribofuranosyladenine)

When naming nucleosides or other nucleic acid fragments, the carbons of the carbohydrate portion are designated as 1′, 2′, 3′, 4′, and 5′ to distinguish them from atoms in the purine or pyrimidine base. Thus, the adenine nucleoside of 2-deoxyribose is called 2′-deoxyadenosine (9-β-D-2′-deoxyribofuranosyladenine).

PROBLEM 18.2

The names of the principal nucleosides obtained from RNA and DNA are given below. Write a structural formula for each one.

(*a*) Thymidine (thymine-derived nucleoside in DNA)
(*b*) Cytidine (cytosine-derived nucleoside in RNA)
(*c*) Guanosine (guanine-derived nucleoside in RNA)

SAMPLE SOLUTION

(a) Thymine is a pyrimidine base present in DNA; its carbohydrate substituent is 2-deoxyribofuranose, which is attached to N-1 of thymine.

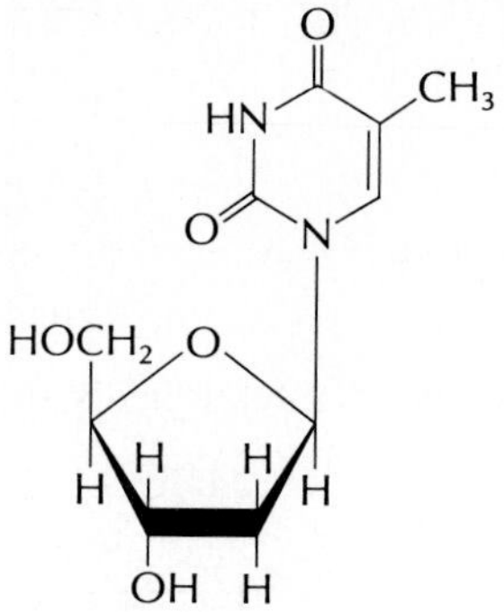

Thymidine

18.3 NUCLEOTIDES

Nucleotides are phosphoric acid esters of nucleosides. The phosphoric acid group is attached at the 5′ position. Adenosine 5′-monophosphate, also called **AMP,** is one example.

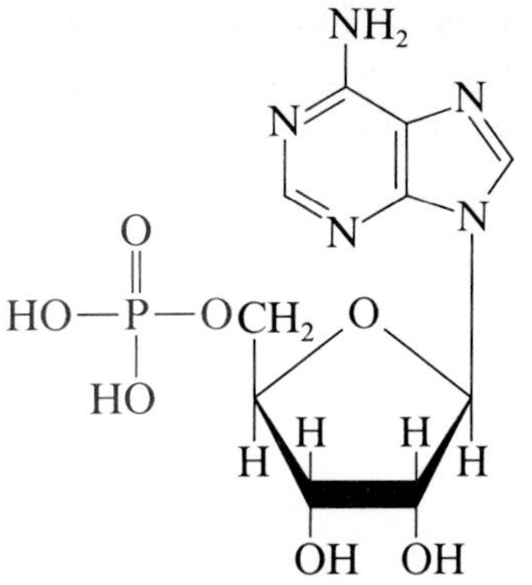

Adenosine 5′-monophosphate (AMP)

Other prominent 5′-nucleotides of adenosine include adenosine disphosphate **(ADP)** and adenosine triphosphate **(ATP).**

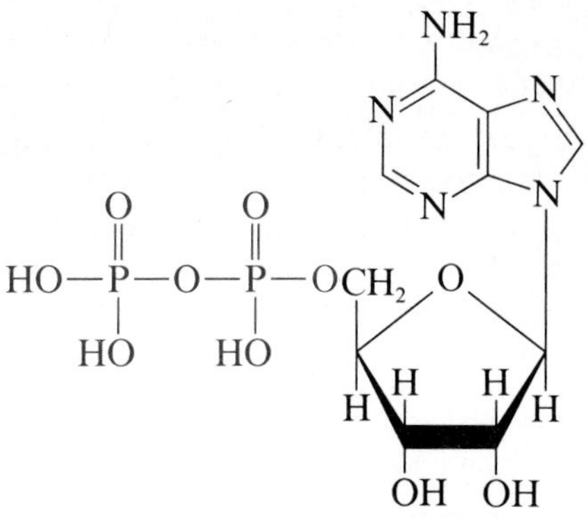

Adenosine diphosphate (ADP)

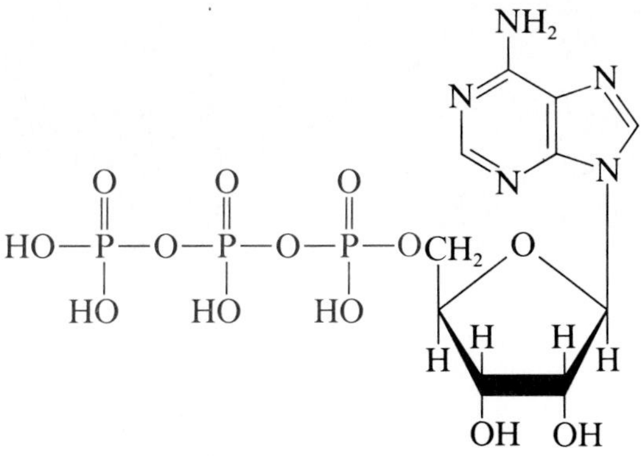

Adenosine triphosphate (ATP)

The formation of AMP, ADP, and ATP from adenosine occurs by a sequence of phosphorylation steps:

$$\text{Adenosine} \xrightarrow[\text{enzymes}]{PO_4{}^{3-}} \text{AMP} \xrightarrow[\text{enzymes}]{PO_4{}^{3-}} \text{ADP} \xrightarrow[\text{enzymes}]{PO_4{}^{3-}} \text{ATP}$$

Each of these steps is endothermic; that is, each is accompanied by the absorption of energy. The energy to drive each step comes from the breakdown of carbohydrates in the body. The body uses ATP as a temporary storage vessel for the chemical energy released when carbohydrates and other foods are broken down to carbon dioxide and water. ATP is described as a "high-energy compound." The high energy is associated with the phosphoric anhydride linkages present in ATP.

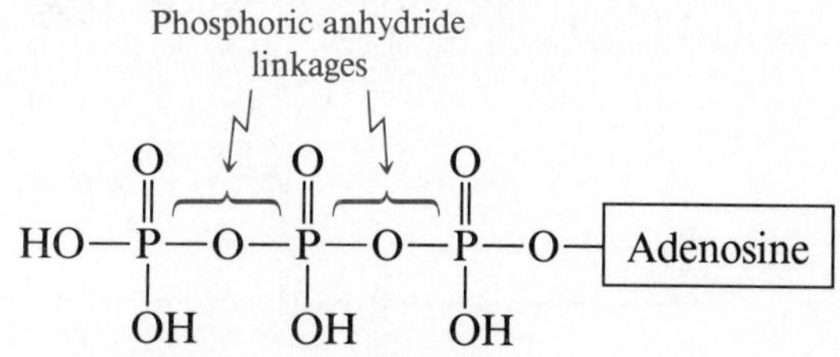

Adenosine triphosphate (ATP)

That energy becomes available to cells when a phosphoric anhydride linkage of ATP undergoes hydrolysis. Phosphoric anhydride hydrolysis is exothermic; conversion of 1 mol of ATP to ADP and phosphate releases 7.3 kcal of energy.

A slightly different kind of nucleotide is adenosine 3'-5'-cyclic monophosphate, also called **cyclic AMP.** Cyclic AMP is a regulator of several biological processes including, for example, both the synthesis and degradation of glycogen (Section 15.12). Cyclic AMP is a cyclic ester of phosphoric acid and adenosine involving the hydroxyl groups at C-3' and C-5'.

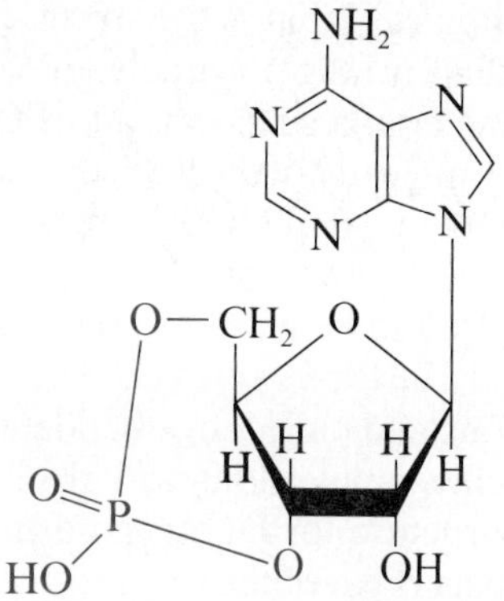

Adenosine 3' – 5'-cyclic monophosphate (cyclic AMP)

18.4 NUCLEIC ACIDS

Nucleic acids are **polynucleotides** in which a phosphate ester unit links the 5' oxygen of one nucleotide to the 3' oxygen of another. Figure 18.1 is a generalized depiction of the structure of a nucleic acid. All nucleic acids have a backbone of alternating sugar and phosphate units. The sugar in ribonucleic acids (RNA) is ribose; the backbone of deoxyribonucleic acids (DNA) contains 2-deoxyribose. The purine and pyrimidine base portions of RNA and DNA differ somewhat (Section 18.1). While both types of nucleic acids contain the purines adenine and guanine, the pyrimidines in RNA are uracil and cytosine, while those in DNA are thymine and cytosine.

Research on nucleic acids progressed slowly until it became evident during the 1940s that they played a role in the transfer of genetic information. It was known that

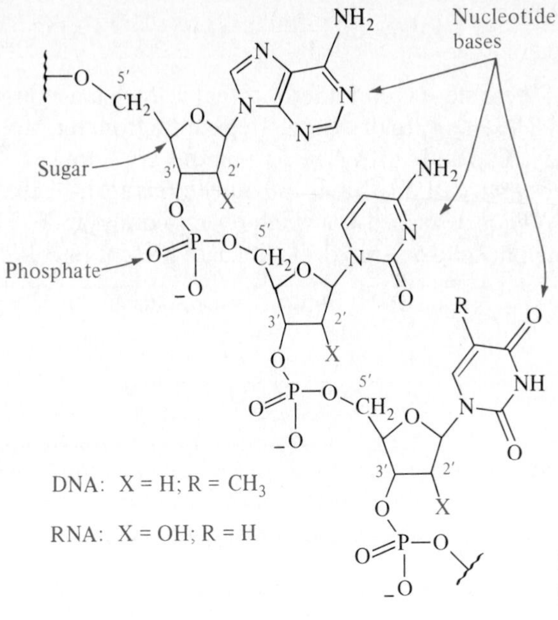

DNA: X = H; R = CH$_3$

RNA: X = OH; R = H

FIGURE 18.1 A portion of a polynucleotide chain.

the genetic information of an organism resides in the chromosomes present in each of its cells and that individual chromosomes are made up of smaller units called *genes*. When it became apparent that genes are DNA, interest in nucleic acids intensified. There was a feeling among scientists that once the structure of DNA was established, the precise way in which DNA carried out its designated role would become more evident. In some respects the problems parallel those of protein chemistry. Knowing that DNA is a polynucleotide is comparable with knowing that proteins are poly-amides. What is the nucleotide sequence (primary structure)? What is the precise shape of the polynucleotide chain (secondary and tertiary structure)? Is the genetic material a single strand of DNA or is it an assembly of two or more strands? The complexity of the problem can be indicated by noting that a typical strand of human DNA contains approximately 100 *million* nucleotides. If the strand were uncoiled, it would be several centimeters long, yet it and many others like it reside in cells too small to be seen with the naked eye.

In 1953 James D. Watson and Francis H. C. Crick at Cambridge University in England pulled together data from biology, biochemistry, chemistry, and x-ray crystallography, along with the insight they gained from molecular models, to propose a structure for DNA and a mechanism for its replication. Their two brief scientific papers paved the way for an explosive growth in our understanding of life processes at the molecular level, the field we now call *molecular biology*. Watson and Crick shared the 1962 Nobel Prize in Physiology or Medicine with Maurice Wilkins. Wilkins, in collaboration with Rosalind Franklin, was responsible for the x-ray crystallographic data.

18.5 STRUCTURE AND REPLICATION OF DNA. THE DOUBLE HELIX

Watson and Crick were aided in their search for the structure of DNA by a discovery that there was a consistent pattern in the composition of DNAs from various sources. While there was a wide variation in the distribution of the bases among species, one-half of the bases in all samples of DNA were purines and the other half were pyrimidines. Furthermore, the ratio of the purine adenine (abbreviated A) to the pyrimidine thymine (T) was always close to 1 : 1. Likewise, the ratio of the purine guanine (G) to the pyrimidine cytosine (C) was also close to 1 : 1. Analysis of human DNA, for example, revealed it to have the following composition:

Purine	Pyrimidine	Base ratio
Adenine (A): 30.3%	Thymine (T): 30.3%	A/T = 1.00
Guanine (G): <u>19.5%</u>	Cytosine (C): <u>19.9%</u>	G/C = 0.98
Total 49.8%	Total 50.2%	

Watson and Crick felt that the constancy in the A/T and G/C ratios was no accident. They proposed that it resulted from complementary structures between A and T and between G and C. Watson and Crick suggested that hydrogen bonds were a key element in their proposed model for the structure of DNA. Consideration of various hydrogen-bonding arrangements revealed that A and T could form the hydrogen-bonded **base pair** shown in Figure 18.2a and that G and C could associate as in Figure 18.2b. The specific base pairings of A to T and of G to C by hydrogen bonds is not only a key element in the Watson-Crick DNA structure but also plays a crucial role in the replication of DNA.

Because each hydrogen-bonded base pair contains one purine and one pyrimidine, A · · · T and G · · · C are approximately the same size. Thus, two nucleic acid chains may be aligned side by side with their bases in the middle, as illustrated in Figure 18.3. The two chains are joined by the network of hydrogen bonds between the paired bases A · · · T and G · · · C. Since x-ray crystallographic data indicated a helical structure, Watson and Crick proposed that the two strands are intertwined as a **double helix** (Figure 18.4).

The Watson-Crick base-pairing model for the structure of DNA holds the key to understanding the process of DNA **replication.** During cell division a cell's DNA is duplicated; the DNA in the new cell is identical to that in the original cell. At one stage of cell division the DNA double helix begins to unwind, separating the two chains. Each strand is then able to form new hydrogen bonds to free nucleotides according to the base pairings A · · · T and G · · · C. Thus each strand is able to serve as a template upon which a new DNA strand is constructed. These new strands are exact complements of the template. As the double helix unravels, each strand becomes one-half of a new and identical DNA double helix. This process is represented schematically in Figure 18.5.

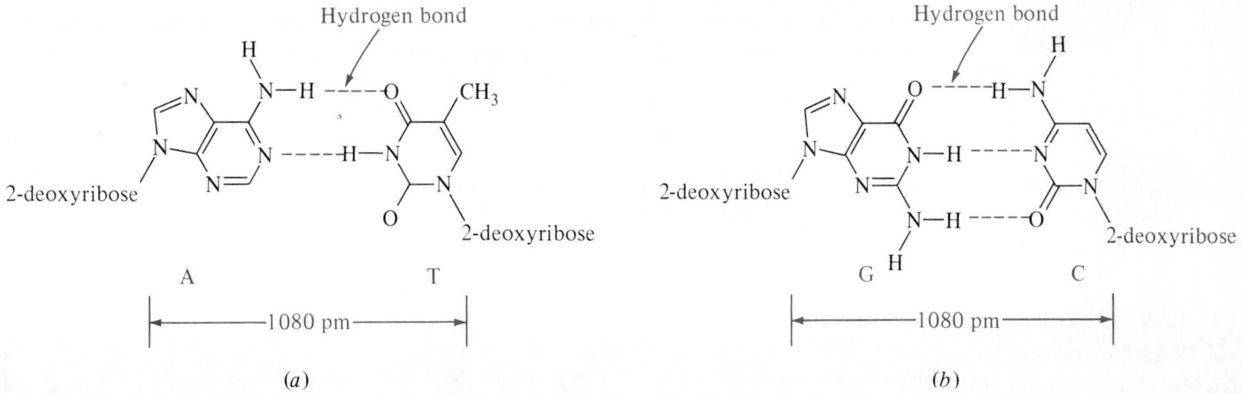

FIGURE 18.2 Base pairings between (a) adenine (A) and thymine (T) and (b) guanine (G) and cytosine (C).

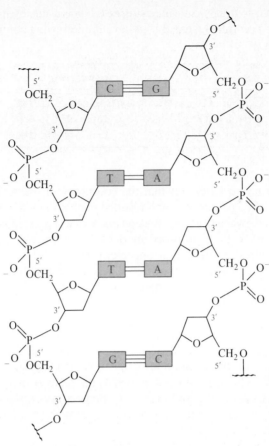

FIGURE 18.3 Hydrogen bonds between complementary bases (A · · · T and G · · · C) permit pairing of two DNA strands. The strands are antiparallel; the 5' end of the left strand is at the top, while that of the right strand is at the bottom.

PROBLEM 18.3

Each of the following represents the base sequence of a portion of one strand of a DNA molecule. What will be the base sequence of the complementary strand in each case?

(a) -T-C-A-G-C-A-   (b) -A-T-G-G-A-C-T-   (c) -G-G-G-C-C-A-G-T-T-

SAMPLE SOLUTION

(a) The complementary base pairs are A · · · T and G · · · C. To find the

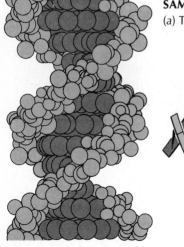

FIGURE 18.4 A space-filling model of a portion of a DNA double helix.

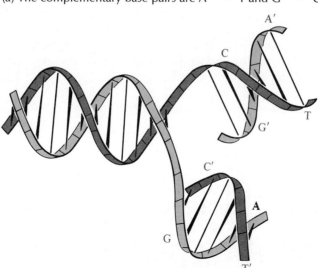

FIGURE 18.5 During DNA replication the double helix unwinds, and each of the original strands serves as a template for the synthesis of its complementary strand.

complementary strand, write the symbol for the complementary base opposite each symbol in the original strand.

Original strand: -T-C-A-G-C-A-
                                         ⋮ ⋮ ⋮ ⋮ ⋮ ⋮
Complementary strand: -A-G-T-C-G-T-

The structural requirements for the pairing of nucleic acid bases are also critical for utilizing genetic information, and in living systems this means protein biosynthesis.

CANCER CHEMOTHERAPY

The treatment of diseases such as cancer with drugs is called *chemotherapy*. A characteristic common to all forms of cancer is the uncontrolled reproduction or growth of cells. The drugs used to control cancer attempt to stop this uncontrolled cell growth by interfering with the DNA replication process.

Anticancer drugs fall into one of several classes. An *antimetabolite* is a compound structurally similar to one normally used by an organism. The drug 5-FU (Problem 18.1), a fluorinated uracil derivative, is an example of an antimetabolite used to treat certain forms of gastrointestinal and breast cancer. The fluorine atom of 5-FU prevents conversion of this uracil derivative into thymine, a compound necessary for DNA synthesis.

A second antimetabolite used to treat uterine cancer and some forms of leukemia is *methotrexate*.

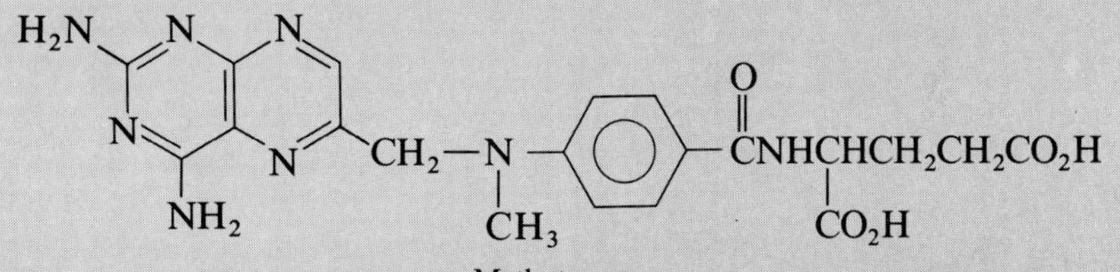

Methotrexate

Methotrexate is an antimetabolite of folic acid. The nutritional requirements of rapidly dividing cancer cells are greater than those of healthy cells. Thus although folic acid is used by all cells, the growth of cancer cells is slowed to a greater extent with methotrexate therapy.

A second important class of cancer chemotherapy agents is the *alkylating agents*. The first alkylating agents used as chemotherapy drugs were the same compounds found to be carcinogens (see boxed essay following Section 8.3). Animal studies during World War II showed that nitrogen mustards destroyed lymphoid tissues. One nitrogen mustard derivative is cyclophosphamide, used to treat lymphomas such as Hodgkin's disease as well as breast, ovarian, and lung cancer.

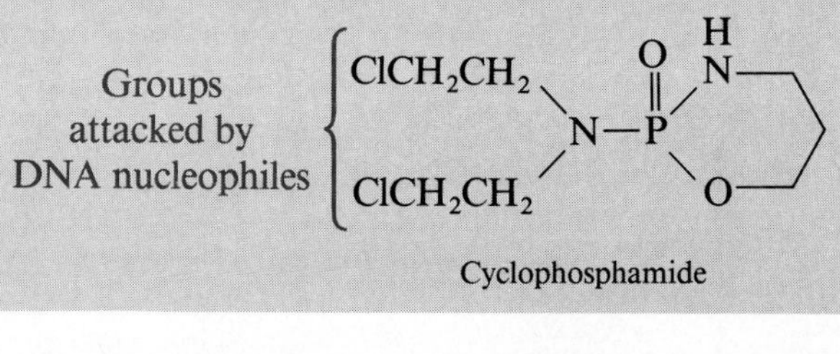

Groups attacked by DNA nucleophiles

Cyclophosphamide

The action of alkylating agents in killing cancer cells relies on the inhibition of DNA replication. The alkylating agent is difunctional; that is, it has two groups that can be attacked by nucleophiles. Cyclophosphamide links the two strands of DNA together by a covalent bridge. Separation of the strands of DNA is a necessary step in replication (Section 18.5). Thus an alkylating agent drug prevents a cancer cell from reproducing. Of course even healthy, rapidly dividing cells are affected by the drug. This is a reason why cancer chemotherapy is associated with side effects such as hair loss and gastrointestinal upset. The hair follicles and the lining of the stomach and intestines have cells that normally reproduce rapidly, and these cells are affected by the anticancer drug to a greater extent than other healthy cells in the body.

AIDS

The explosive growth of our knowledge of nucleic acid chemistry and its role in molecular biology in the 1980s happened to coincide with a challenge to human health that would have defied understanding a generation ago. That challenge is *acquired immune deficiency syndrome,* or **AIDS.** AIDS is a condition in which the body's immune system is devastated by a viral infection to the extent that it can no longer perform its vital function of identifying and destroying invading organisms. AIDS victims die from "opportunistic" infections—diseases that are normally held in check by a healthy immune system but which can become deadly when the immune system is compromised. Within six years of its discovery, AIDS had claimed the lives of 26,000 victims in the United States, and public health officials believe the toll could rise to 200,000 by the early 1990s.

The virus responsible for almost all the AIDS cases in the United States was identified by scientists at the Louis Pasteur Institute in Paris in 1983 and is known as *human immunodeficiency virus-1* (HIV-1). HIV-1 is believed to have originated in Africa fairly recently, where a related virus HIV-2 was discovered in 1986 by the Pasteur Institute group. Both HIV-1 and HIV-2 are classed as **retroviruses** because their genetic material is RNA rather than DNA. HIVs require a host cell to reproduce, and the hosts in humans are the so-called T4 lymphocytes, the cells primarily responsible for inducing the immune system to respond when provoked. The HIV penetrates the cell wall of a T4 lymphocyte and deposits both its RNA and an enzyme called *reverse transcriptase* inside the T4 cell, where the reverse transcriptase catalyzes the formation of a DNA strand which is complementary to the viral RNA. The transcribed DNA then serves as the template for formation of double helical DNA which, with the information it carries for reproduction of the HIV, becomes incorporated into the T4 cell's own genetic material. The viral DNA remains in a latent state until some event, whose nature is not yet known, triggers its activation and induces the host lymphocyte to begin producing copies of the virus which then leave the host to infect other T4 cells. In the course of HIV reproduction, the ability of the T4 lymphocyte to reproduce itself is hampered. As the number of T4 cells decreases so does the body's ability to combat infections.

There is no present known cure for AIDS. The drug most used to treat AIDS patients in the United States is *zidovudine,* also known as azidothymine, or AZT. AZT does not kill the AIDS virus but rather intereferes with its ability to reproduce. Thus AZT controls an AIDS infection but does not cure it.

Zidovudine (AZT)

Zidovudine is taken in 250-mg doses every 4 hours, 24 hours a day. It is a very expensive drug, costing approximately $10,000 per patient per year. Its side effects include bone marrow suppression, leading to low red and white blood cell counts.

The AIDS outbreak has been and continues to be a tragedy on a massive scale. Until a cure is discovered, sustained efforts at preventing its transmission offer our best weapon against the spread of AIDS.

18.6 DNA-DIRECTED PROTEIN BIOSYNTHESIS

Protein biosynthesis is directed by DNA through several types of ribonucleic acids called **messenger RNA (mRNA), transfer RNA (tRNA),** and **ribosomal RNA (rRNA).** Ribosomal RNA is found in ribosomes, the "protein factories" of a cell.

There are two main stages in protein biosynthesis, **transcription** and **translation.** In the **transcription stage** a molecule of mRNA having a nucleotide sequence complementary to one of the strands of a DNA double helix is constructed. A diagram illustrating transcription is presented in Figure 18.6. Ribonucleotides with bases complementary to one of the DNA strands are assembled into a single strand of RNA. Unlike DNA, RNA molecules are single-stranded. Also, thymine does not occur in RNA; the base that pairs with adenine in RNA is uracil.

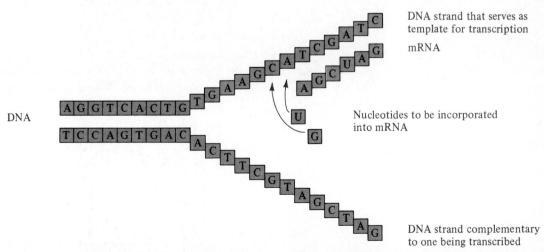

FIGURE 18.6 During transcription a molecule of mRNA is assembled by using one strand of DNA as a template.

TABLE 18.1

The Genetic Code (Messenger RNA Codons)*

Alanine	Arginine	Asparagine	Aspartic acid	Cysteine
GCU GCA	CGU CGA AGA	AAU	GAU	UGU
GCC GCG	CGC CGG AGG	AAC	GAC	UGC
Glutamic acid	Glutamine	Glycine	Histidine	Isoleucine
GAA	CAA	GGU GGA	CAU	AUU AUA
GAG	CAG	GGC GGG	CAC	AUC
Leucine	Lysine	Methionine	Phenylalanine	Proline
UUA CUU CUA	AAA	AUG	UUU	CCU CCA
UUG CUC CUG	AAG		UUC	CCC CCG
Serine	Threonine	Tryptophan	Tyrosine	Valine
UCU UCA AGU	ACU ACA	UGG	UAU	GUU GUA
UCC UCG AGC	ACC ACG		UAC	GUC GUG

* The first letter of each triplet corresponds to the nucleotide nearer the 5′ terminus, the last letter to the nucleotide nearer the 3′ terminus. UAA, UGA, and UAG are not included in the table; they are chain-terminating codons.

In the **translation stage,** the nucleotide sequence of the mRNA is decoded and "read" by ribosomal RNA as an amino acid sequence to be constructed. This **genetic code** is made up of sets of three adjacent nucleotides called **codons.** Read in sets of three nucleotides, the four mRNA bases (A, U, C, G) generate 64 possible "words," more than sufficient to code the 20 amino acids found in proteins. Table 18.1 lists the 64 possible codons of mRNA.

The mechanism of translation makes use of the same complementary base-pairing principle used in replication and transcription. Each of the 20 amino acids is associated with a specific tRNA. Molecules of tRNA are much smaller than DNA and mRNA. They are single-stranded and contain 70 to 90 ribonucleotides arranged in a cloverleaf pattern. Figure 18.7 illustrates phenylalanine tRNA. The characteristic shape of tRNA results from the presence of paired bases in some regions and their

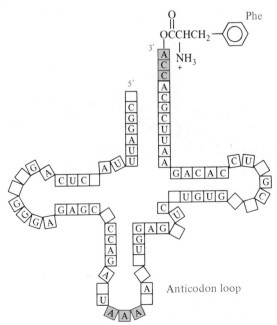

FIGURE 18.7 Phenylalanine tRNA. Transfer RNAs usually contain modified bases, slightly different from those in other RNAs. The anticodon for phenylalanine, AAA, is at the bottom of the figure.

absence in others. The amino acid unique to the particular tRNA is attached, by an ester linkage, to the 3' end of the molecule. At one of the loops of tRNA there is a nucleotide triplet, called the **anticodon,** which is complementary to a codon of mRNA. The codons of mRNA are read by the anticodons of tRNA, and the proper amino acids are transferred in sequence to the growing protein.

PROBLEM 18.4
Give the tRNA anticodon and the amino acid residue specified by each of the following codons.

(a) UCG   (b) GAC   (c) UUU   (d) AGG

SAMPLE SOLUTION
(a) The anticodon is the set of bases complementary to the codon base sequence. Thus the codon UCG has the anticodon AGC. From Table 18.1, UCG is the codon for the amino acid serine.

18.7 SUMMARY

Nucleic acids are responsible for replication of genetic information and they direct protein biosynthesis. *Nucleic acids* are polymeric nucleotides (Section 18.4), *nucleosides* are purine and pyrimidine *N*-glycosides of D-ribose and 2-deoxy-D-ribose (Section 18.2), and *nucleotides* are phosphate esters of nucleosides (Section 18.3). Deoxyribonucleic acid (DNA) exists as a double helix (Section 18.5), in which hydrogen bonds are responsible for complementary base pairing between adenine and thymine and between guanine and cytosine. During cell division the strands unwind and are duplicated. Each strand acts as a template upon which its complement is constructed.

In the *transcription* stage of protein biosynthesis (Section 18.6), a molecule of *messenger RNA* (mRNA) having a nucleotide sequence complementary to that of DNA is assembled. Transcription is followed by *translation,* in which triplets of nucleotides of mRNA called *codons* are recognized by the *transfer RNA* (tRNA) for a particular amino acid and that amino acid is added to the growing peptide chain.

ADDITIONAL PROBLEMS

Structure

18.5   What is the structural distinction between a nucleotide and a nucleoside?

18.6   Nebularine is a toxic nucleoside isolated from a species of mushroom. Its systematic name is 9-β-D-ribofuranosylpurine. Write a structural formula for nebularine.

18.7   The compound vidarabine shows promise as an antiviral agent. Its structure is identical to that of adenosine (Section 18.2) except that D-arabinose (Figure 15.3) replaces D-ribose as the carbohydrate component. Write a structural formula for this substance.

DNA Replication and Protein Biosynthesis

Problems 18.8 to 18.12 refer to the following nucleotide segments of a DNA strand:
    (a) A-A-A-G-G-T-C-C-C-G-T-A
    (b) T-A-C-T-C-G-C-G-G-A-T-G

18.8   Write the nucleotide sequence for the complementary DNA strand of each segment.

18.9   Write the mRNA nucleotide sequence which would arise during transcription of each DNA segment.

18.10   List the codons present in each mRNA segment.

18.11   Determine the amino acid sequence that would be formed from each mRNA nucleotide segment.

18.12   What are the anticodons corresponding to each nucleotide segment in Problem 18.10?

18.13   Using Table 18.1, write a different mRNA nucleotide sequence that would code for each of the amino acid sequences in Problem 18.11.

18.14   Write the DNA nucleotide sequences that would serve as templates for the codons in Problem 18.13.

18.15   In persons who suffer from the genetic disorder sickle cell anemia, a glutamic acid residue in one of hemoglobin's protein chains has been replaced by valine. Referring to the codons in Table 18.1, suggest what base substitution could account for this amino acid change.

18.16   The structure of the pentapeptide leucine enkephalin was shown in Figure 16.3. What set of mRNA codons could produce this substance? Can more than one set of codons give the same compound?

18.17   What sequence of DNA bases would serve as a template for the mRNA codons in Problem 18.16?

Miscellaneous Problem

18.18   The generally accepted theory of how carcinogens act is that they interfere in some manner with the DNA replication process. One class of carcinogenic compounds acts as alkylating agents that react to change the structure of the purine and pyrimidine bases in the DNA molecule. An example is O-methylguanine. Explain how this change could lead to a base-pairing error in DNA replication.

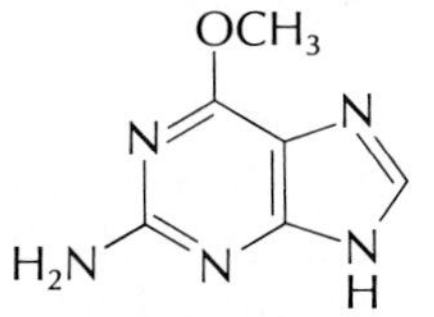

O-Methylguanine

ANSWERS TO IN-TEXT PROBLEMS

Problems are of two types, in-text problems that appear within the body of each chapter and end-of-chapter problems. This appendix gives brief answers to all the in-text problems. More detailed discussions of in-text problems as well as detailed solutions to all the end-of-chapter problems are provided in a separate *Study Guide*. Answers to part (*a*) of those in-text problems with multiple parts have been provided in the form of a sample solution within each chapter and are not repeated here.

CHAPTER 1

1.1 4

1.2 All the third row elements have a neon core containing 10 electrons ($1s^2 2s^2 2p^6$). The elements in the third row, their atomic numbers Z, and their electron configurations beyond the neon core are: Na ($Z = 11$) $3s^1$; Mg ($Z = 12$) $3s^2$; Al ($Z = 13$) $3s^2 3p_x^1$; Si ($Z = 14$) $3s^2 3p_x^1 3p_y^1$; P ($Z = 15$) $3s^2 3p_x^1 3p_y^1 3p_z^1$; S ($Z = 16$) $3s^2 3p_x^2 3p_y^1 3p_z^1$; Cl ($Z = 17$) $3s^2 3p_x^2 3p_y^2 3p_z^1$; Ar ($Z = 18$) $3s^2 3p_x^2 3p_y^2 3p_z^2$.

1.3 H:F̈:

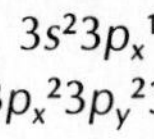

1.4 (*b*) :N̈:H (*c*) :N̈:F̈: (*d*) :P̈:Cl̈: (*e*) H:C̈:Cl̈: (*f*) H:C̈:C̈:H

with associated H atoms and lone pairs as drawn

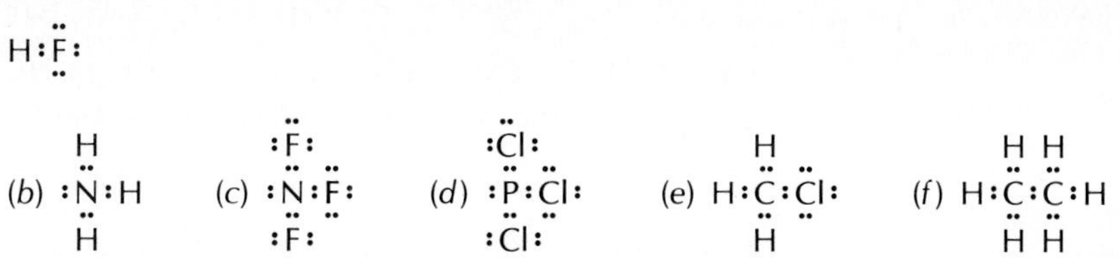

1.5 The formal charge of hydrogen in each species is 0. The formal charge of carbon is: (*b*) −1; (*c*) 0; (*d*) +1; (*e*) 0. In each species the net charge is equal to the formal charge of carbon.

1.6 $C_8H_{10}N_4O_2$.

1.7 (*b*) H—C̈=N̈—Ö—H (*c*) H—Ö—C̈=N̈—H (*d*) :Ö=C—N̈—H

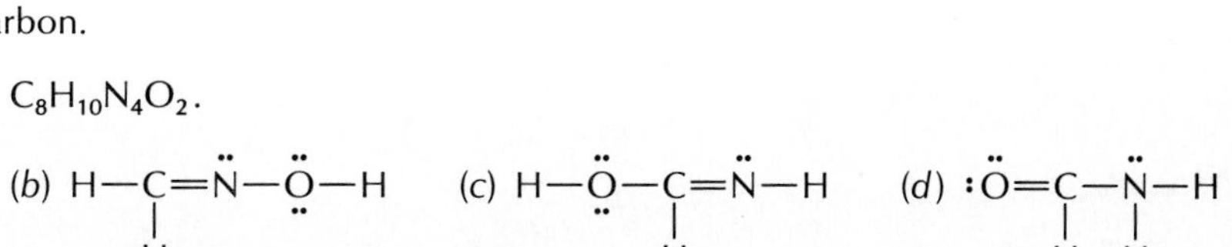

1.8 (b)

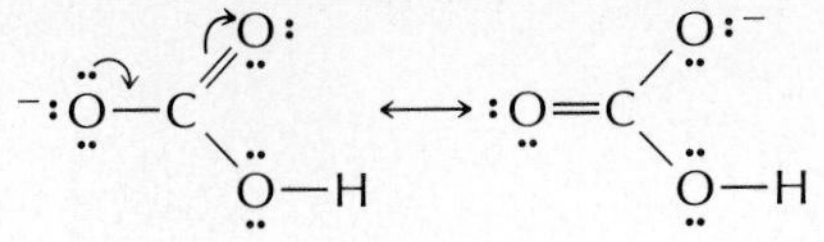

(c)

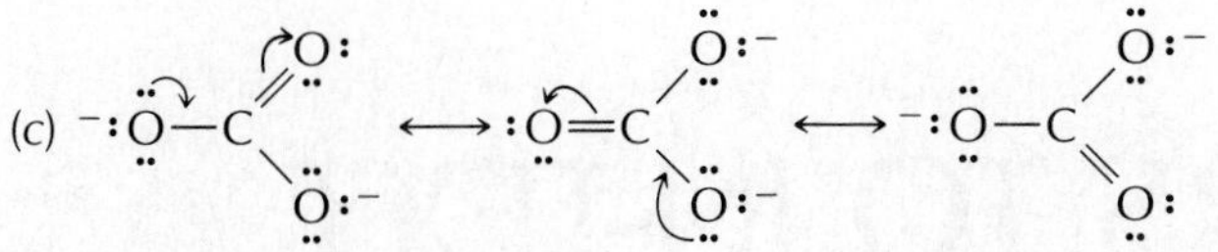

1.9 Tetrahedral (109.5°).

1.10 (b) Tetrahedral; (c) linear; (d) trigonal planar.

CHAPTER 2

2.1 All three carbons; 10 σ bonds; each C-H σ bond is C sp^3-H $1s$; each C-C bond is C sp^3-C sp^3.

2.2

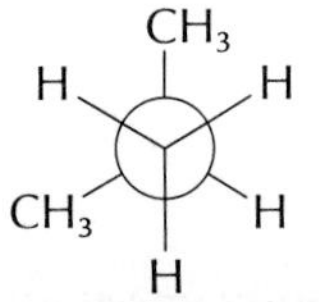

2.3 $CH_3(CH_2)_{26}CH_3$

2.4 Molecular formula $= C_{11}H_{24}$; condensed formula $= CH_3(CH_2)_9CH_3$.

2.5 $CH_3CH(CH_2CH_3)_2$ or

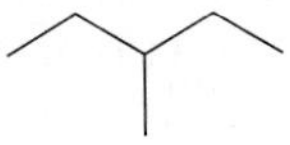

$(CH_3)_2CHCH(CH_3)_2$ or

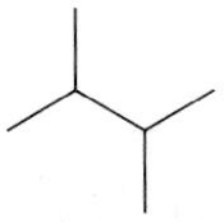

$(CH_3)_3CCH_2CH_3$ or

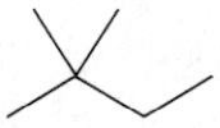

2.6 Undecane

2.7

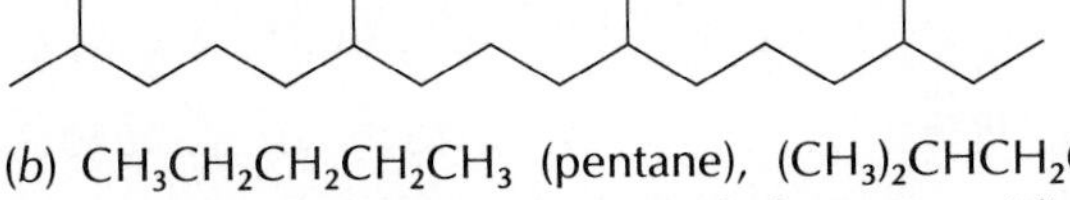

2.8 (b) $CH_3CH_2CH_2CH_2CH_3$ (pentane), $(CH_3)_2CHCH_2CH_3$ (2-methylbutane), $(CH_3)_4C$ (2,2-dimethylpropane); (c) 2,2,4-trimethylpentane; (d) 2,2,3,3-tetramethylbutane.

2.9 (b) 4-Ethyl-2-methylhexane; (c) 8-ethyl-4-isopropyl-2,6-dimethyldecane.

2.10 (b) cyclodecane; (c) 1,2-dicyclopropylbutane; (d) cyclohexylcyclohexane.

2.11 (b)

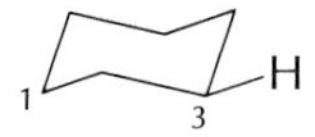

(c)

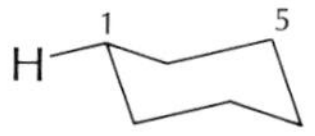

(d)

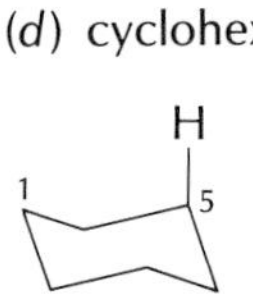

2.12

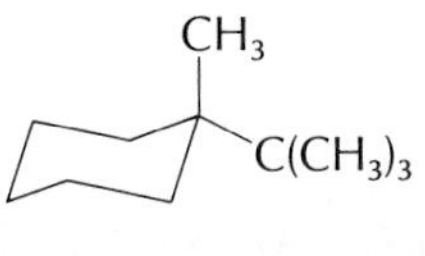

2.13 Methyl groups (CH_3) on same side of ring crowd one another.

2.14 (a) [structure: cyclohexane ring with $C(CH_3)_3$ and CH_3 substituents] (b) [structure: cyclohexane ring with $C(CH_3)_3$ and CH_3 substituents]

2.15 Prismane (C_6H_6); cubane (C_8H_8); adamantane ($C_{10}H_{16}$).

2.16 Octane (126°C); 2-methylheptane (116°C); 2,2,3,3-tetramethylbutane (106°C); nonane (151°C).

2.17 $CH_3CH_2CH_2CH_2CH_3 + 8 O_2 \longrightarrow 5 CO_2 + 6 H_2O$

CHAPTER 3

3.1 (b) [structure] $CH_3-\underset{\underset{CH_3}{|}}{\overset{\overset{CH_3}{|}}{C}}-\underset{\underset{CH_3}{|}}{\overset{\overset{CH_3}{|}}{C}}-CH_3 + Cl_2 \longrightarrow CH_3-\underset{\underset{CH_3}{|}}{\overset{\overset{CH_3}{|}}{C}}-\underset{\underset{CH_3}{|}}{\overset{\overset{CH_3}{|}}{C}}-CH_2Cl + HCl$

(c) [cyclopentane structure] $+ Cl_2 \longrightarrow$ [chlorocyclopentane structure] $-Cl + HCl$

3.2 $CH_3CH_2CH_2CH_2OH$ (1-butanol); $(CH_3)_2CHCH_2OH$ (2-methyl-1-propanol); $CH_3\underset{\underset{OH}{|}}{C}HCH_2CH_3$ (2-butanol); $(CH_3)_3COH$ (2-methyl-2-propanol).

3.3 $CH_3CH_2CH_2CH_2Cl$ (1-chlorobutane); $(CH_3)_2CHCH_2Cl$ (1-chloro-2-methylpropane): $CH_3\underset{\underset{Cl}{|}}{C}HCH_2CH_3$ (2-chlorobutane); $(CH_3)_3CCl$ (2-chloro-2-methylpropane).

3.4 $(CH_3)_4C$

3.5 Primary

3.6 $CH_3CH_2\underset{\underset{HO}{|}}{C}H\underset{\underset{CH_3}{|}}{C}HCH_2CH_2CH_3$ (secondary alcohol).

3.7 $H_3N:\overset{\frown}{+}H\overset{\frown}{-}\overset{..}{\underset{..}{Cl}}: \longrightarrow H_3\overset{+}{N}-H + :\overset{..}{\underset{..}{Cl}}:^-$

Base   Acid   Conjugate acid   Conjugate base

3.8 (b) 6.7×10^{-5}; (c) 1.8×10^{-4}; (d) 6.5×10^{-2}. Order of decreasing acidity is oxalic acid > aspirin > formic acid > vitamin C.

3.9 $H_2\overset{..}{N}:^- \overset{\frown}{+}H\overset{\frown}{-}\overset{..}{\underset{..}{O}}CH_3 \rightleftharpoons H_3\overset{..}{N}: + {}^-:\overset{..}{\underset{..}{O}}CH_3$

Stronger base   Stronger acid   Weaker acid   Weaker base

The position of equilibrium lies to the right.

3.10 $CH_3-\overset{..}{\underset{\underset{H}{|}}{O}}:\overset{\frown}{+}H\overset{\frown}{-}\overset{\overset{O}{\|}}{\underset{\underset{O}{\|}}{\overset{..}{O}SOH}} \rightleftharpoons CH_3-\overset{..}{\underset{\underset{H}{|}}{O}}{}^+ + {}^-:\overset{\overset{O}{\|}}{\underset{\underset{O}{\|}}{\overset{..}{O}SOH}}$

Methyl alcohol   Sulfuric acid   Methyloxonium ion   Hydrogen sulfate ion
(base)   (acid)   (conjugate acid)   (conjugate base)

3.11 (b) $(CH_3CH_2)_3COH + HCl \longrightarrow (CH_3CH_2)_3CCl + H_2O$
(c) $CH_3(CH_2)_{12}CH_2OH + HBr \longrightarrow CH_3(CH_2)_{12}CH_2Br + H_2O$

3.12 $CH_3CH_2CH_2\overset{+}{C}H_2$ $CH_3CH_2\overset{+}{C}HCH_3$ $CH_3\overset{+}{C}H\overset{}{C}H_2$ $CH_3\overset{+}{C}CH_3$

                                                                CH_3 CH_3

        (primary)   (secondary)   (primary)   (tertiary)

3.13 $CH_3\overset{+}{C}CH_2CH_3$
           |
          CH_3

3.14 $CH_3\overset{OH}{\underset{CH_3}{\overset{|}{C}}}CH_2CH_3$

3.15 $CH_3\overset{\cdot}{C}CH_2CH_3$
           |
          CH_3

3.16 The primary radical $CH_3CH_2CH_2\overset{\cdot}{C}H_2$ gives 1-chlorobutane; the secondary radical $CH_3\overset{\cdot}{C}HCH_2CH_3$ gives 2-chlorobutane.

3.17 (b) 3; (c) 5; (d) 3.

CHAPTER 4

4.1 (b) 3,3-Dimethyl-1-butene; (c) 2-methyl-2-hexene; (d) 2,4,4-trimethyl-2-pentene; (e) 3-isopropyl-2,6-dimethyl-3-heptene.

4.2

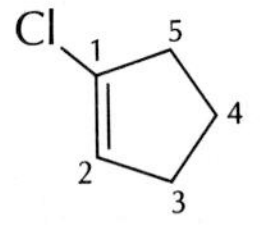

       1-Chlorocyclopentene 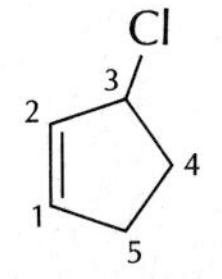   3-Chlorocyclopentene 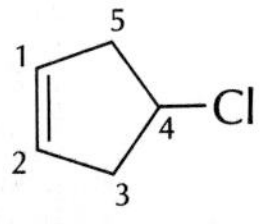   4-Chlorocyclopentene

4.3 (b) 3-Ethyl-3-hexene; (c) two sp^2-hybridized carbons, six sp^3-hybridized carbons; (d) three sp^2-sp^3 σ bonds, three sp^3-sp^3 σ bonds.

4.4 (b) Stereoisomers not possible; (c) stereoisomers possible.

4.5 (most stable) 2-methyl-2-butene > *trans*-2-pentene > *cis*-2-pentene > 1-pentene (least stable).

4.6 (b) $CH_3CH{=}CH_2$ (c) $CH_3CH{=}CH_2$ (d) $(CH_3)_3CC{=}CH_2$
                                                              CH_3

4.7 (b)

Major   and   Minor

(c)

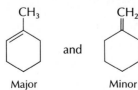

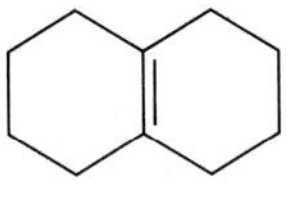

Major   and   Minor

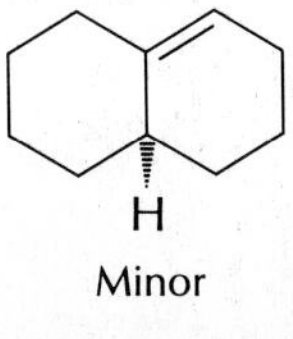

4.8 (*b*)

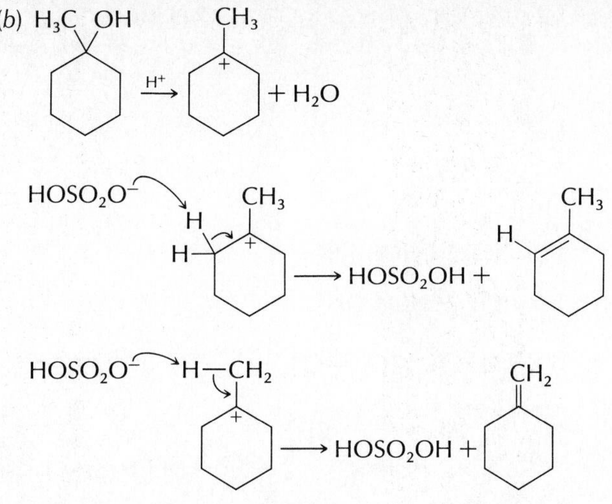

(*c*)

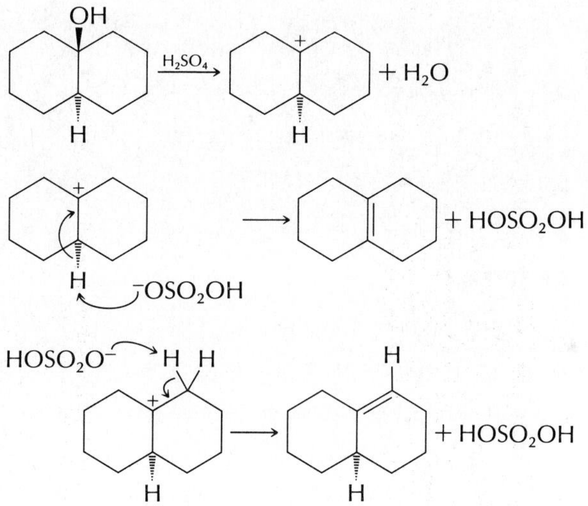

4.9 (*b*) $CH_2=CHCH_2CH_2CH_2CH_3$, *cis*-$CH_3CH=CHCH_2CH_2CH_3$, and *trans*-$CH_3CH=CHCH_2CH_2CH_3$; (*c*) $CH_2=CHCH_2CH(CH_3)_2$, *cis*-$CH_3CH=CHCH(CH_3)_2$, and *trans*-$CH_3CH=CHCH(CH_3)_2$; (*d*) *cis*-$(CH_3)_3CCH=CHCH_3$ and *trans*-$(CH_3)_3CCH=CHCH_3$.

4.10 (*b*) 1,3-Cyclooctadiene (conjugated); (*c*) 1,4-cyclooctadiene (isolated).

4.11 $CH_3CH_2CH_2C\equiv CH$ $CH_3CH_2C\equiv CCH_3$ $CH_3\underset{\underset{CH_3}{|}}{C}HC\equiv CH$

1-Pentyne	2-Pentyne	2-Methyl-1-butyne

4.12 $(CH_3)_3CCH_2CHBr_2$; $(CH_3)_3C\underset{\underset{Br}{|}}{C}HCH_2Br$; $(CH_3)_3C\overset{\overset{Br}{|}}{\underset{\underset{Br}{|}}{C}}CH_3$

CHAPTER 5

5.1

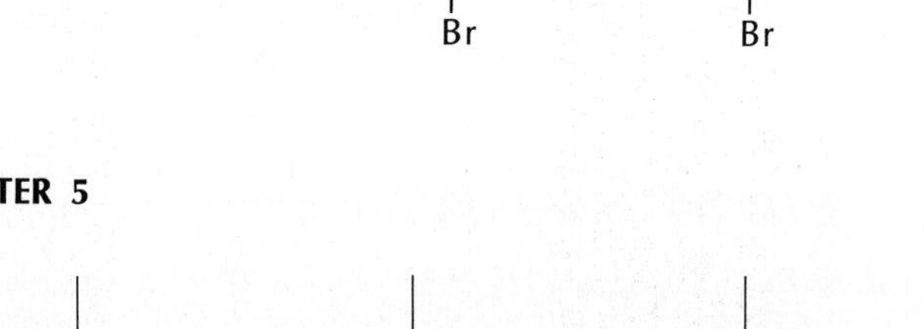

2-Methyl-1-butene	2-Methyl-2-butene	3-Methyl-1-butene

5.2 (b) $(CH_3)_2CCH_2CH_3$
       |
       Cl

(c) $CH_3CHCH_2CH_3$
       |
       Cl

(d) CH_3CH_2 —[cyclohexane ring with Cl]

5.3 (b) $(CH_3)_2\overset{+}{C}CH_2CH_3$

(c) $CH_3\overset{+}{C}HCH_2CH_3$

(d) CH_3CH_2—[cyclohexane ring with +]

5.4 $[(CH_3)_2CHCH_2]_3B$

5.5 $(CH_3)_2CHCH_2OH$

5.6 (b) $(CH_3)_3CCH{=}CH_2 \xrightarrow[\text{2. H}_2\text{O}_2,\text{ HO}^-]{\text{1. B}_2\text{H}_6} (CH_3)_3CCH_2CH_2OH$

(c)

$$CH_3{\diagdown}{}{\diagup}CH_3$$
$$C{=}C$$
$$H{\diagup}{}{\diagdown}H \quad \xrightarrow[\text{2. H}_2\text{O}_2,\text{ HO}^-]{\text{1. B}_2\text{H}_6} CH_3CH_2CHCH_3$$
$$\underset{\displaystyle OH}{|}$$

5.7
$$\overset{\displaystyle \underset{|}{CH_3}}{CH_3CCH_3} \quad \underset{|}{}$$
$CH_3\overset{\overset{\textstyle CH_3}{|}}{\underset{\underset{\textstyle Br}{|}}{C}}CH_3 \xrightarrow{\text{NaOCH}_2\text{CH}_3} (CH_3)_2C{=}CH_2 \xrightarrow{Br_2} (CH_3)_2\overset{\underset{\underset{\textstyle Br}{|}}{}}{C}CH_2Br$

5.8 $CH_3\overset{\overset{\textstyle CH_3}{|}}{\underset{\underset{\textstyle Br}{|}}{C}}CH_3 \xrightarrow{\text{NaOCH}_2\text{CH}_3} (CH_3)_2C{=}CH_2 \xrightarrow[\text{peroxides}]{\text{HBr}} (CH_3)_2CHCH_2Br$

5.9 2,4,4-Trimethyl-1-pentene

5.10 (b) $HC{\equiv}CH \xrightarrow[\text{2. CH}_3\text{Br}]{\text{1. NaNH}_2,\text{ NH}_3} CH_3C{\equiv}CH \xrightarrow[\text{2. CH}_3\text{CH}_2\text{CH}_2\text{Br}]{\text{1. NaNH}_2,\text{ NH}_3}$
$CH_3C{\equiv}CCH_2CH_2CH_2CH_3$ (or reverse order of steps).

(c) $HC{\equiv}CH \xrightarrow[\text{2. CH}_3\text{CH}_2\text{CH}_2\text{Br}]{\text{1. NaNH}_2,\text{ NH}_3} CH_3CH_2CH_2C{\equiv}CH \xrightarrow[\text{2. CH}_3\text{CH}_2\text{Br}]{\text{1. NaNH}_2,\text{ NH}_3}$
$CH_3CH_2CH_2C{\equiv}CCH_2CH_3$ (or reverse order of steps).

5.11 $HC{\equiv}CCH_2CH_3 \xrightarrow[\text{2. CH}_3\text{Br}]{\text{1. NaNH}_2,\text{ NH}_3} CH_3C{\equiv}CCH_2CH_3 \xrightarrow[\text{Lindlar Pd}]{\text{H}_2}$

$$CH_3{\diagdown}{}{\diagup}CH_2CH_3$$
$$C{=}C$$
$$H{\diagup}{}{\diagdown}H$$

5.12 $HC{\equiv}CCH_2CH_3 \xrightarrow[\text{2. CH}_3\text{Br}]{\text{1. NaNH}_2,\text{ NH}_3} CH_3C{\equiv}CCH_2CH_3 \xrightarrow{\text{Na, NH}_3}$

$$CH_3{\diagdown}{}{\diagup}H$$
$$C{=}C$$
$$H{\diagup}{}{\diagdown}CH_2CH_3$$

5.13 (b) $CH_2{=}CHCl \xrightarrow{\text{HCl}} CH_3CHCl_2$

(c) $CH_2{=}CHBr \xrightarrow[\text{2. H}_2\text{O}]{\text{1. NaNH}_2,\text{ NH}_3} HC{\equiv}CH \xrightarrow{\text{2HCl}} CH_3CHCl_2$

(d) $CH_3CHBr_2 \xrightarrow[\text{2. H}_2\text{O}]{\text{1. NaNH}_2,\text{ NH}_3} HC{\equiv}CH \xrightarrow{\text{2HCl}} CH_3CHCl_2$

5.14 $CH_3\overset{\overset{\textstyle O}{\|}}{C}CH_2CH_3$ (ketone); $CH_3\overset{\overset{\textstyle OH}{|}}{C}{=}CHCH_3$ (enol).

5.15 $CH_3(CH_2)_4C{\equiv}CCH_2CH_2C{\equiv}C(CH_2)_4CH_3$

5.16 (b) $(CH_3)_2\overset{\underset{\underset{\textstyle Br}{|}}{}}{C}{-}\overset{\underset{\underset{\textstyle CH_3}{|}}{}}{C}{=}CH_2$ (1,2-addition); $(CH_3)_2C{=}\overset{\underset{\underset{\textstyle CH_3}{|}}{}}{C}CH_2Br$ (1,4-addition)

(c)

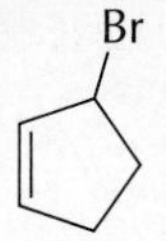

(same product from 1,2-addition and 1,4-addition)

5.17 Three isomers; their structures are

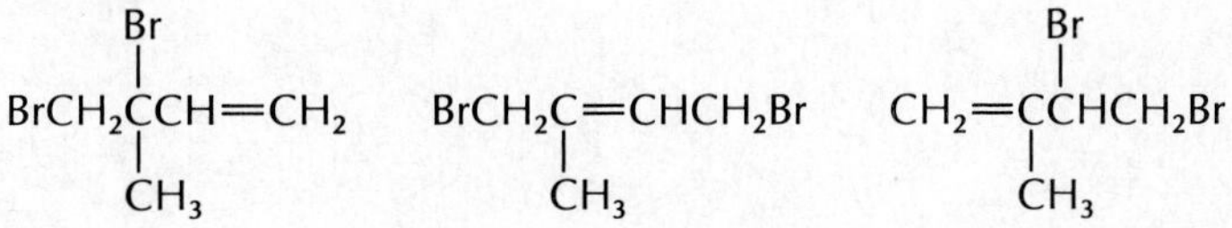

$$BrCH_2\overset{\overset{\displaystyle Br}{|}}{\underset{\underset{\displaystyle CH_3}{|}}{C}}CH=CH_2 \qquad BrCH_2\overset{}{\underset{\underset{\displaystyle CH_3}{|}}{C}}=CHCH_2Br \qquad CH_2=\overset{\overset{\displaystyle Br}{|}}{\underset{\underset{\displaystyle CH_3}{|}}{C}}CHCH_2Br$$

CHAPTER 6

6.1

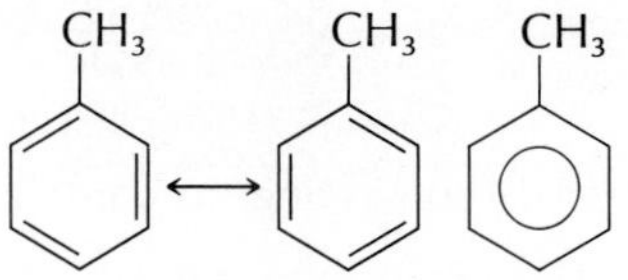

6.2 (b)

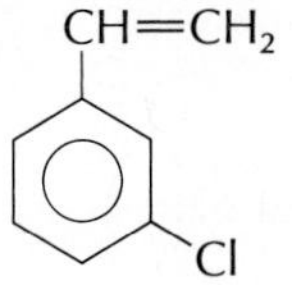

(c)

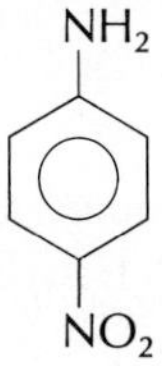

6.3

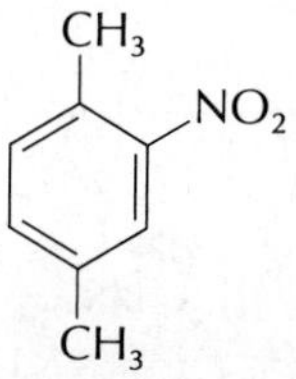

6.4

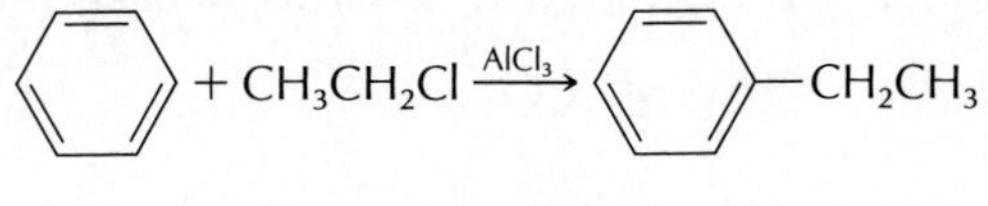

6.5

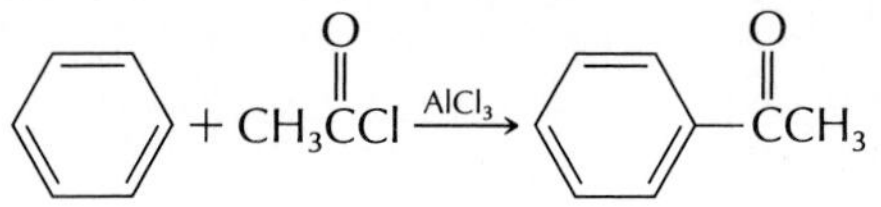

6.6 (b) Phenol; (c) aniline.

6.7 (b) meta; (c) ortho-para.

6.8

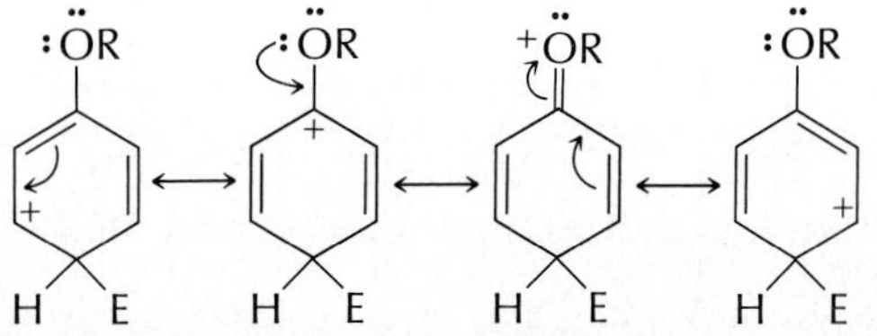

6.9 (b)

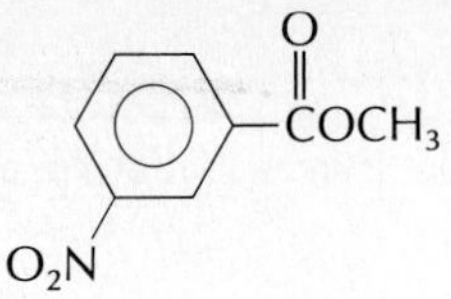

(c)

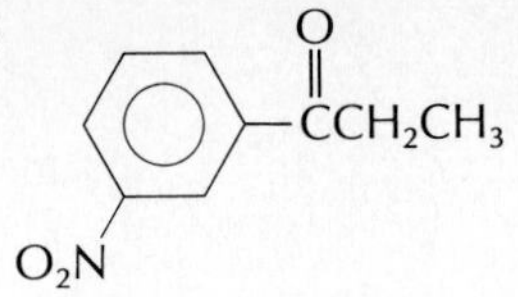

6.10 (b)

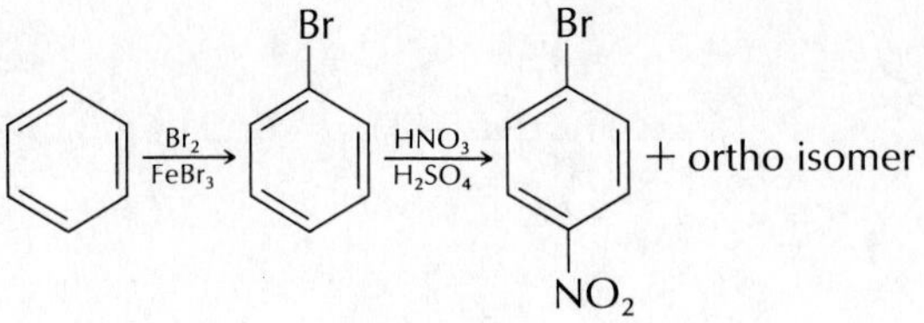

+ ortho isomer

(c)

+ ortho isomer

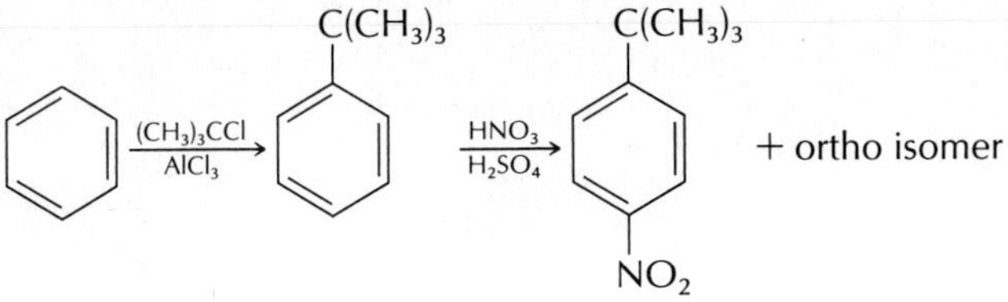

6.11 (b) Not aromatic (sp^3 carbon in ring); (c) aromatic (6 π electrons in ring).

CHAPTER 7

7.1 (a) Chiral; (b) chiral; (c) achiral.

7.2 (b) No chiral center; (c) C-4 is a chiral center; (d) C-2 is a chiral center.

7.3 (c)

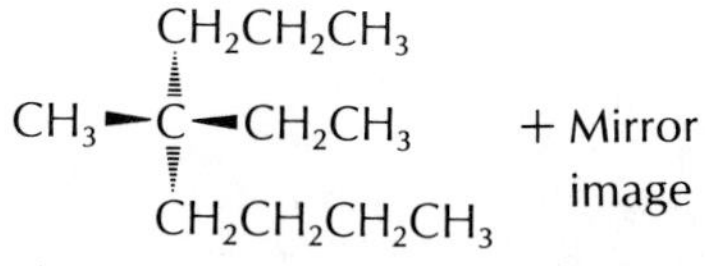

+ Mirror image

(d)

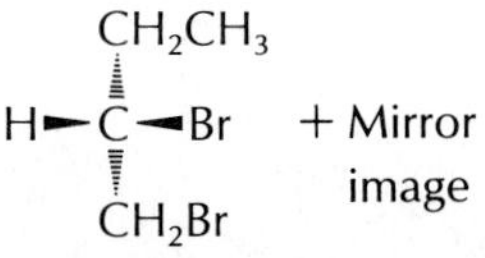

+ Mirror image

7.4 (c)

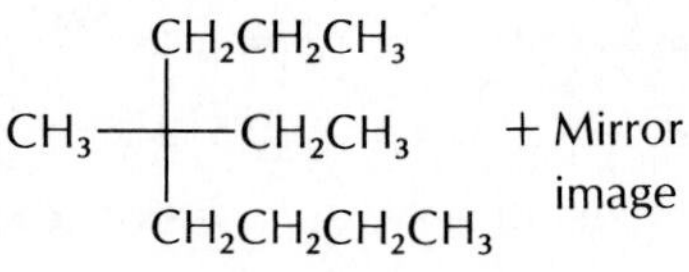

+ Mirror image

(d)

+ Mirror image

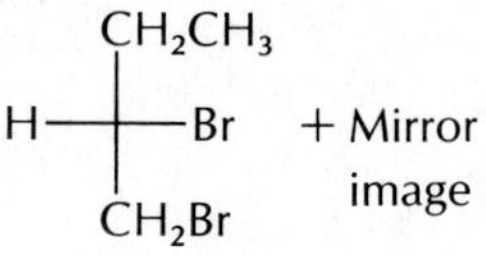

7.5 $[\alpha]_D$-39°.

7.6 (b) —CH_2OH (c) —$CH(CH_3)_2$ (d) —CH_2F

7.7 (b) R; (c) S; (d) S.

7.8 (b) Z; (c) E; (d) E.

7.9

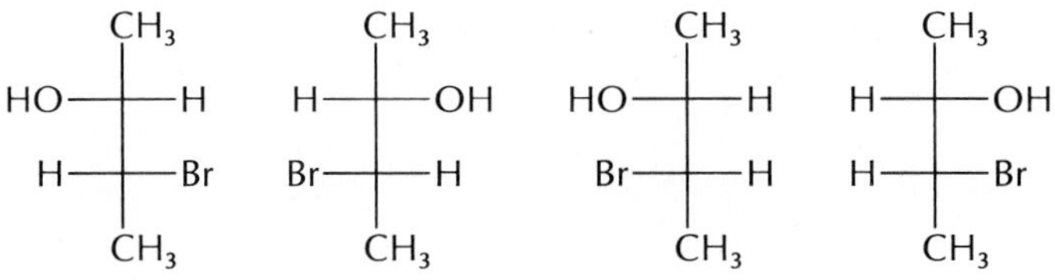

I   II   III   IV

7.10 2S,3R.

7.11

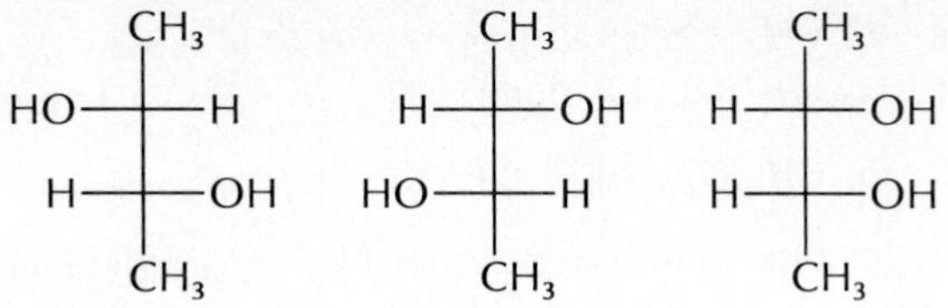

$$
\begin{array}{ccc}
\text{CH}_3 & \text{CH}_3 & \text{CH}_3 \\
\text{HO}\!-\!\!\!-\!\text{H} & \text{H}\!-\!\!\!-\!\text{OH} & \text{H}\!-\!\!\!-\!\text{OH} \\
\text{H}\!-\!\!\!-\!\text{OH} & \text{HO}\!-\!\!\!-\!\text{H} & \text{H}\!-\!\!\!-\!\text{OH} \\
\text{CH}_3 & \text{CH}_3 & \text{CH}_3
\end{array}
$$

7.12 Eight.

7.13 (S)-1-Phenylethylammonium (S)-malate.

CHAPTER 8

8.1

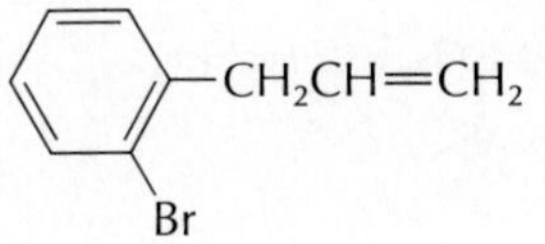

$$\text{C}_6\text{H}_4(\text{Br})\!-\!\text{CH}_2\text{CH}\!=\!\text{CH}_2$$

Aryl halide

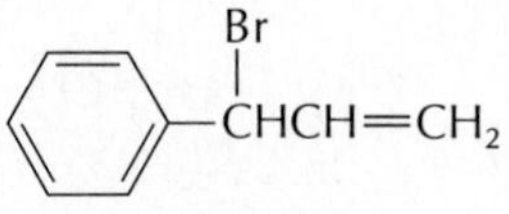

$$\text{C}_6\text{H}_5\!-\!\underset{\underset{\text{Br}}{|}}{\text{CH}}\text{CH}\!=\!\text{CH}_2$$

Alkyl halide

$$\text{C}_6\text{H}_5\!-\!\text{CH}_2\text{CH}\!=\!\text{CHBr}$$

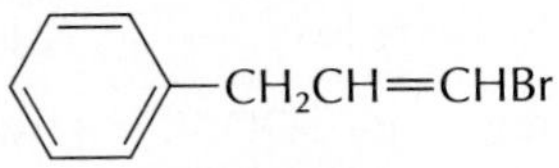

Vinyl halide

8.2 $\text{CH}_3\text{Br} + \text{HO}^- \longrightarrow \text{CH}_3\text{OH} + \text{Br}^-$

Substrate   Nucleophile   Product   Leaving group

8.3 (b) $\text{CH}_3\text{CH}_2\text{OC}(\text{CH}_3)_3$ (c) $\text{CH}_3\text{CH}_2\text{CN}$ (d) $\text{CH}_3\text{CH}_2\text{SH}$

8.4 $\text{ClCH}_2\text{CH}_2\text{CH}_2\text{CN}$.

8.5 Hydrolysis of (R)-(−)-2-bromooctane by the S_N2 mechanism yields optically active (S)-(+)-2-octanol. The 2-octanol obtained by hydrolysis of racemic 2-bromooctane is not optically active.

8.6 (b) 1-Bromopentane; (c) 2-chloropentane; (d) 2-bromo-5-methylhexane.

8.7 (b) 1-Methylcyclopentyl iodide; (c) cyclopentyl bromide; (d) *tert*-butyl iodide.

8.8 $(\text{CH}_3)_2\text{C}\!=\!\text{CHCH}_2\text{Cl}$

8.9 (b) $\text{C}_6\text{H}_{11}\!-\!\text{OCH}_2\text{CH}_3$ (c) $\text{CH}_3\underset{\underset{\text{OCH}_3}{|}}{\text{CH}}\text{CH}_2\text{CH}_3$

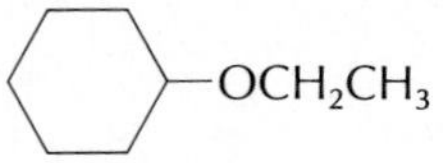

(d) *cis*- and *trans*-$\text{CH}_3\text{CH}\!=\!\text{CHCH}_3$ and $\text{CH}_2\!=\!\text{CHCH}_2\text{CH}_3$

CHAPTER 9

9.1 The frequency of red light ($\nu = 3.8 \times 10^{14}$ s^{-1}) is less than the frequency of violet light ($\nu = 7.5 \times 10^{14}$ s^{-1}).

9.2 A photon of violet light.

9.3 Benzyl alcohol.

9.4 (b) 5.37 ppm; (c) 3.07 ppm.

9.5 (b) Two; (c) one; (d) two.

9.6 (*b*) CH_3 ($\delta = 0.9-1.8$ ppm), CH_2O ($\delta = 3.3-3.7$ ppm); (c) CH_3 ($\delta = 0.9-1.8$ ppm); (*d*) CH_3 ($\delta = 0.9-1.8$ ppm), CHCl ($\delta = 3.1-4.1$ ppm).

9.7 (*b*) CH_3 (3), CH_2O (2); (c) CH_3 (only signal); (*d*) CH_3 (6), CHCl (1).

9.8 (*b*) One signal (singlet); (c) two signals (doublet and triplet); (*d*) two signals (both singlets); (e) two signals (doublet and quartet).

9.9 (*b*) CH_3 (triplet), CH_2O (quartet); (c) CH_3 (singlet); (*d*) CH_3 (doublet), CHCl (heptet).

9.10 $Cl_2CHCH(OCH_2CH_3)_2$

9.11 (*b*) Six; (c) three.

9.12

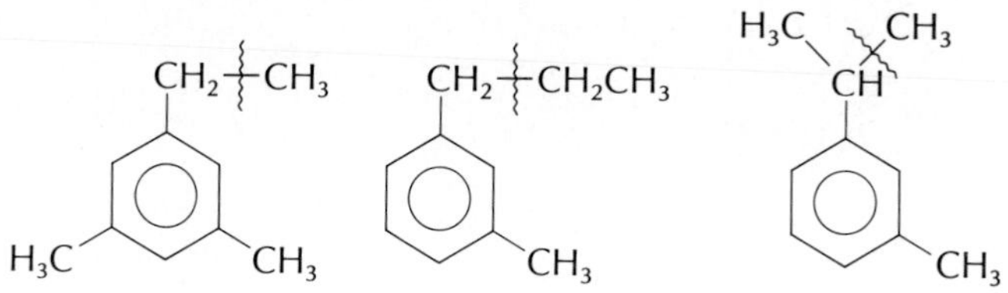

Base peak $C_9H_{11}^+$ (*m/z* 119) | Base peak $C_8H_9^+$ (*m/z* 105) | Base peak $C_9H_{11}^+$ (*m/z* 119)

9.13 (*b*) SODAR = 5, three rings; (c) SODAR = 4, two rings; (*d*) SODAR = 5, three rings; (e) SODAR = 4, two rings; (*f*) SODAR = 4, two rings.

CHAPTER 10

10.1 3,7-Dimethyl-3-octanol (tertiary alcohol).

10.2 Ethanol is oxidized, the chromium reagent is reduced.

10.3 The primary alcohols $CH_3CH_2CH_2CH_2OH$ and $(CH_3)_2CHCH_2OH$ can each be prepared by hydrogenation of an aldehyde. The secondary alcohol $CH_3\underset{\text{OH}}{\underset{|}{C}}HCH_2CH_3$ can be prepared by hydrogenation of a ketone. The tertiary alcohol $(CH_3)_3COH$ cannot be prepared by hydrogenation.

10.4 (*b*) $(CH_3)_2CHCH_2CH_2OH$

(c) 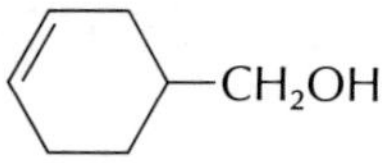 (*d*) 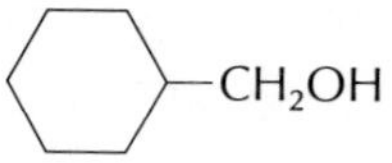

10.5 $2\ CH_3OH + 2Na \longrightarrow 2\ NaOCH_3 + H_2$

$CH_3OH + NaH \longrightarrow NaOCH_3 + H_2$

10.6 $2\ CH_3CH_2OH \xrightarrow{H_2SO_4} CH_3CH_2OCH_2CH_3 + H_2O$

10.7 (*b*) $CH_3\overset{\overset{\text{O}}{\|}}{C}(CH_2)_5CH_3$ (c) $CH_3(CH_2)_6\overset{\overset{\text{O}}{\|}}{C}H$ (*d*) 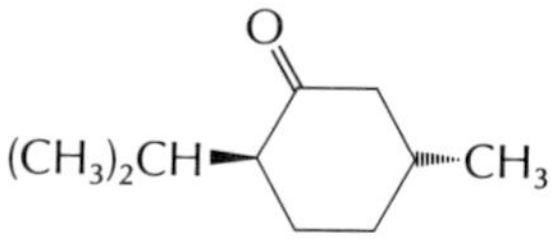

10.8 (*b*) 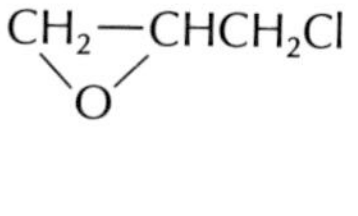 (c) 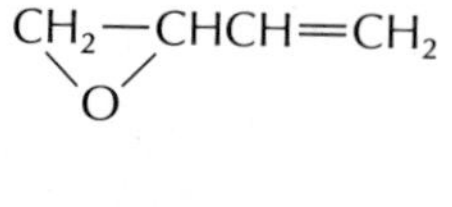

10.9 (*b*) $CH_3ONa + CH_3CH_2CH_2Br \longrightarrow CH_3OCH_2CH_2CH_3 + NaBr$
and $CH_3CH_2CH_2ONa + CH_3Br \longrightarrow CH_3OCH_2CH_2CH_3 + NaBr$
(*c*) $C_6H_5CH_2ONa + CH_3CH_2Br \longrightarrow C_6H_5CH_2OCH_2CH_3 + NaBr$
and $CH_3CH_2ONa + C_6H_5CH_2Br \longrightarrow C_6H_5CH_2OCH_2CH_3 + NaBr$

10.10 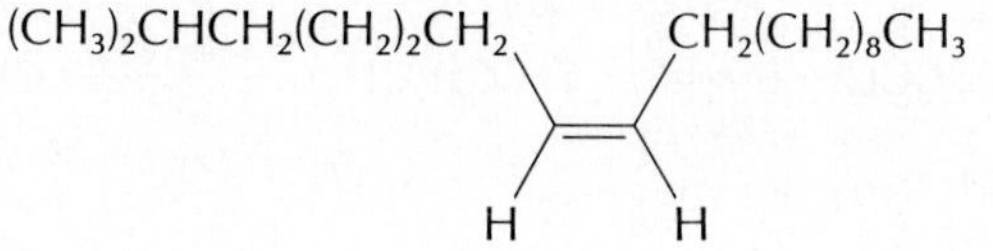

(*Z*-configuration)

10.11 $CH_3CH_2CH_2CH_2OCH_2CH_2OH$

10.12 $(CH_3)_2CHCH_2CH_2SH$

10.13 (*b*) 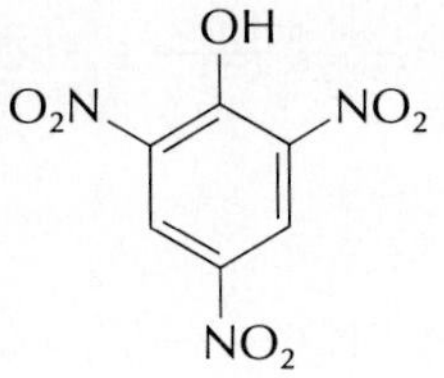

(*c*) 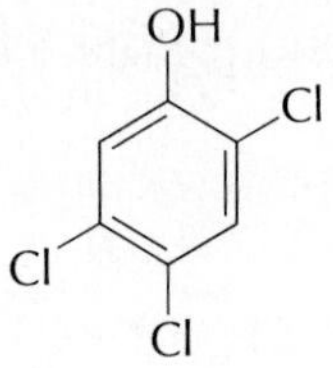

CHAPTER 11

11.1 (*b*) Trichloroethanal (or trichloroacetaldehyde); (*c*) 3-phenylpropenal; (*d*) 4-hydroxy-3-methoxybenzaldehyde (or 4-hydroxy-3-methoxybenzenecarbaldehyde).

11.2 (*b*) 1-Phenylethanone (methyl phenyl ketone); (*c*) 3,3-dimethyl-2-butanone (*tert*-butyl methyl ketone).

11.3 Benzyl alcohol can hydrogen bond to water better than benzaldehyde can. Both can bond to the hydrogens of water through their oxygens, but benzyl alcohol can also use its hydroxyl group proton to hydrogen bond to the oxygen of water.

11.4 $Cl_3C\overset{\overset{\displaystyle O}{\|}}{C}H + H_2O \longrightarrow Cl_3C\overset{\overset{\displaystyle OH}{|}}{C}HOH$

11.5 $CH_3\overset{\overset{\displaystyle OCH_3}{|}}{C}HOH$ (hemiacetal); $CH_3CH{=}\overset{+}{\ddot{O}}CH_3 \longleftrightarrow CH_3\overset{+}{C}H{-}\ddot{O}CH_3$ (carbocation); $CH_3CH(OCH_3)_2$ (acetal)

11.6 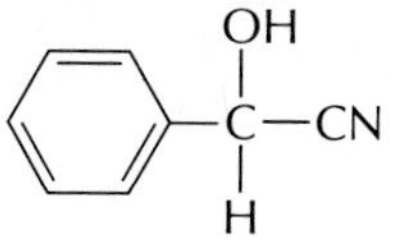

11.7 (*b*) $CH_2{=}CHCH_2MgCl$ (*c*) 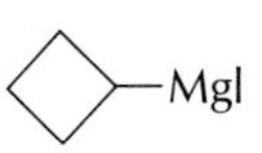

(*d*) 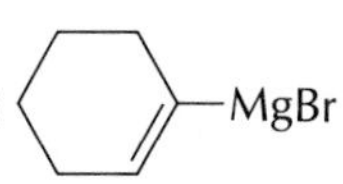

11.8 (*b*) $C_6H_5\underset{\underset{\displaystyle OH}{|}}{C}HCH_2CH_2CH_3$ (*c*) 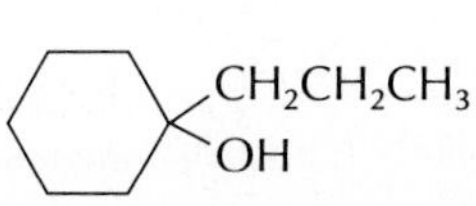

(*d*) $CH_3CH_2CH_2\overset{\overset{\displaystyle CH_3}{|}}{\underset{\underset{\displaystyle CH_3CH_2}{|}}{C}}OH$

11.9 (b) $CH_3MgBr + HCCH_2CH_2CH_2CH_3 \xrightarrow[\text{2. H}_3\text{O}^+]{\text{1. diethyl ether}} CH_3CHCH_2CH_2CH_2CH_3$

(with O double bonded to the H-bearing carbon of the aldehyde; product has OH on the CH)

and $CH_3CH + BrMgCH_2CH_2CH_2CH_3 \xrightarrow[\text{2. H}_3\text{O}^+]{\text{1. diethyl ether}} CH_3CHCH_2CH_2CH_2CH_3$

(aldehyde CH_3CH with O double bond; product OH)

(c) $C_6H_5MgBr + CH_3CH_2CH \xrightarrow[\text{2. H}_3\text{O}^+]{\text{1. diethyl ether}} C_6H_5CHCH_2CH_3$

(aldehyde with C=O; product OH)

and $CH_3CH_2MgBr + C_6H_5CH \xrightarrow[\text{2. H}_3\text{O}^+]{\text{1. diethyl ether}} C_6H_5CHCH_2CH_3$

(aldehyde with C=O; product OH)

11.10 $CH_3MgI + $ (phenyl)$CCH_2CH_3 \xrightarrow[\text{2. H}_3\text{O}^+]{\text{1. diethyl ether}}$ (phenyl)CCH_2CH_3 with OH and CH$_3$

$CH_3CH_2MgBr + $ (phenyl)$CCH_3 \xrightarrow[\text{2. H}_3\text{O}^+]{\text{1. diethyl ether}}$ (phenyl)CCH_2CH_3 with OH and CH$_3$

(phenyl)$-MgBr + CH_3CCH_2CH_3 \xrightarrow[\text{2. H}_3\text{O}^+]{\text{1. diethyl ether}}$ (phenyl)CCH_2CH_3 with OH and CH$_3$

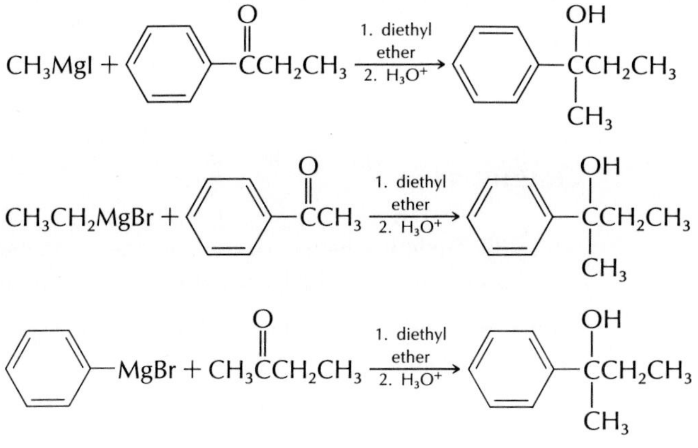

11.11 (b) Zero; (c) five; (d) four

11.12 (b) $C_6H_5\underset{\overset{|}{OH}}{C}=CH_2$ (c) cyclohexene with OH   (d) cyclohexene with OH and CH$_3$

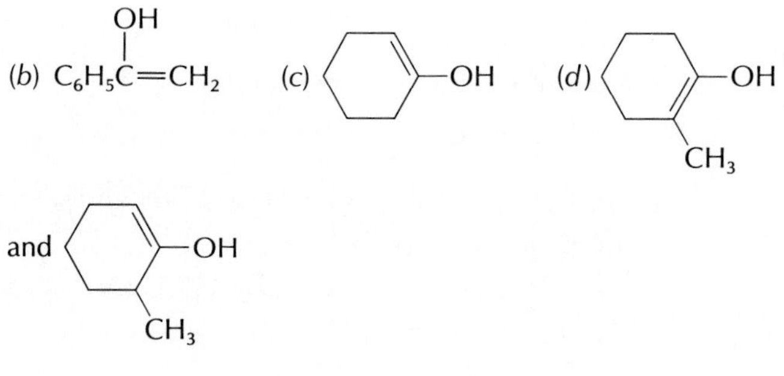

and cyclohexene with OH and CH$_3$

11.13 (b) $CH_3CH_2\underset{\overset{|}{CH_3}}{CH}\overset{\overset{HO}{|}}{CH}-\underset{\overset{|}{HC=O}}{\overset{\overset{CH_3}{|}}{C}}CH_2CH_3$ (c) $(CH_3)_2CHCH_2\overset{\overset{OH}{|}}{CH}-\underset{\overset{|}{HC=O}}{CH}CH(CH_3)_2$

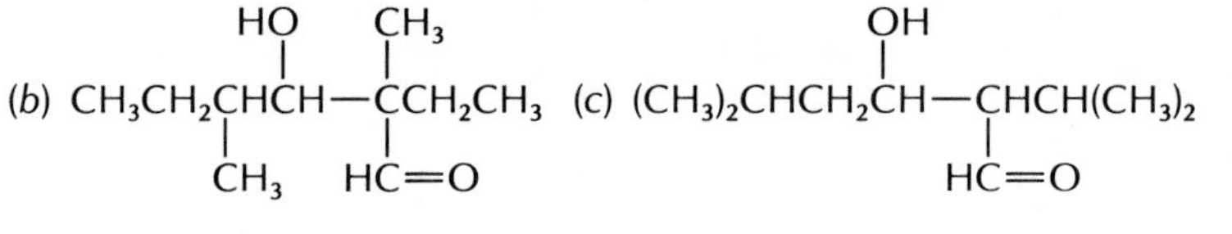

11.14 (b) $CH_3CH_2\underset{\overset{|}{CH_3}}{CH}\overset{\overset{HO}{|}}{CH}\xrightarrow{\alpha}\underset{\overset{|}{HC=O}}{\overset{\overset{CH_3}{|}}{C}}CH_2CH_3$ (c) $(CH_3)_2CHCH_2CH=\underset{\overset{|}{HC=O}}{C}CH(CH_3)_2$

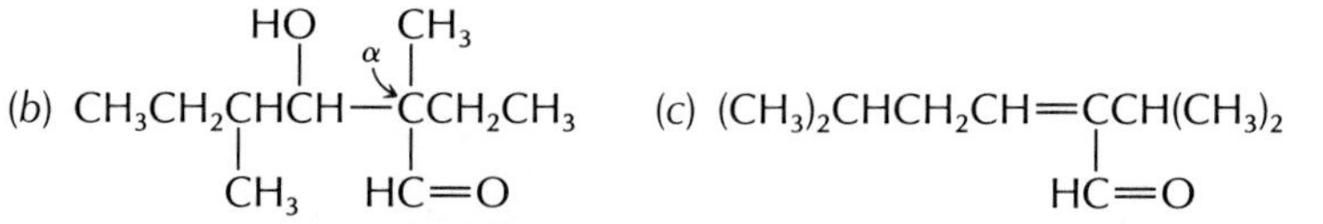

Cannot dehydrate; no protons
on α carbon atom

CHAPTER 12

12.1 (b) 3-Methylbutanoic acid; (c) 2,2-dimethylpropanoic acid; (d) 2-methylpropenoic acid.

12.2

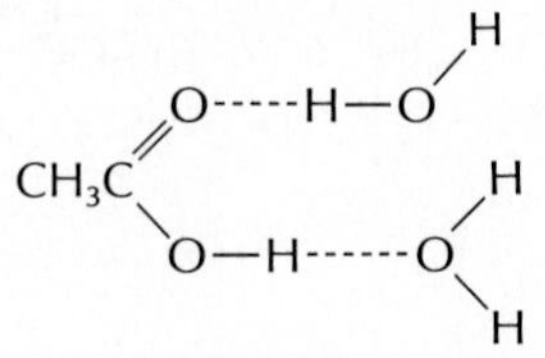

$$CH_3C \begin{matrix} O \text{-----} H-O \\ \\ O-H \text{------} O \end{matrix} \begin{matrix} H \\ \\ H \\ \\ H \end{matrix}$$

12.3 $pK_a = 3.48$

12.4 (b) CH_3CHCO_2H (c) CH_3CCO_2H
         $\quad\;\; OH$

12.5 (b) $CH_3CO_2H + (CH_3)_3CO^- \rightleftharpoons CH_3CO_2^- + (CH_3)_3COH$
(The position of equilibrium lies to the right.)
  (c) $CH_3CO_2H + Br^- \rightleftharpoons CH_3CO_2^- + HBr$
    (The position of equilibrium lies to the left.)
  (d) $CH_3CO_2H + HC\equiv C:^- \rightleftharpoons CH_3CO_2^- + HC\equiv CH$
    (The position of equilibrium lies to the right.)
  (e) $CH_3CO_2H + H_2N^- \rightleftharpoons CH_3CO_2^- + NH_3$
    (The position of equilibrium lies to the right.)

12.6 (b) $(CH_3)_2CHBr \xrightarrow[\substack{2.\ CO_2 \\ 3.\ H_3O^+}]{1.\ Mg,\ ether} (CH_3)_2CHCO_2H$

  (c) ⟨benzene⟩—Br $\xrightarrow[\substack{2.\ CO_2 \\ 3.\ H_3O^+}]{1.\ Mg,\ ether}$ ⟨benzene⟩—CO_2H

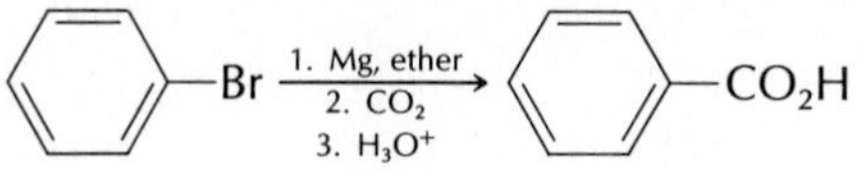

12.7 (b) $(CH_3)_2CHBr \xrightarrow{NaCN} (CH_3)_2CHCN \xrightarrow[heat]{H_2O,\ H_2SO_4} (CH_3)_2CHCO_2H$

  (c) Sequence not feasible because first step (nucleophilic substitution by cyanide on bromobenzene) fails.

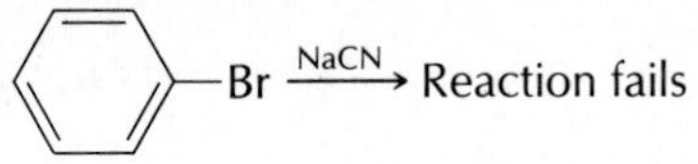

⟨benzene⟩—Br $\xrightarrow{NaCN}$ Reaction fails

12.8 (b) $C_6H_5CH_2CCl$ (c) $C_6H_5CH_2CH_2OH$ (d) $C_6H_5CH_2CO^-\ Na^+$

CHAPTER 13

13.1 (b) $CH_3CH_2CHCOCCHCH_2CH_3$ (c) $CH_3CH_2CHCOCH_2CH_2CH_2CH_3$
         $\qquad\quad C_6H_5 \quad C_6H_5 \qquad\qquad\qquad\quad C_6H_5$

  (d) $CH_3CH_2CH_2COCH_2CHCH_2CH_3$ (e) $CH_3CH_2CHCNH_2$
         $\qquad\qquad\qquad\quad C_6H_5 \qquad\qquad\qquad C_6H_5$

13.2 $\underset{\underset{\displaystyle O}{\|}}{C_6H_5C}Cl + H_2O \longrightarrow \underset{\underset{\displaystyle O}{\|}}{C_6H_5C}OH + HCl$

13.3 (b) $(CH_3)_2CHCH_2\underset{\underset{\displaystyle O}{\|}}{C}OCH_2C_6H_5$ (c) $CH_3O\underset{\underset{\displaystyle O}{\|}}{C}\!-\!\!\!\!\!\!\!\!-\!\!\underset{\underset{\displaystyle O}{\|}}{C}OCH_3$

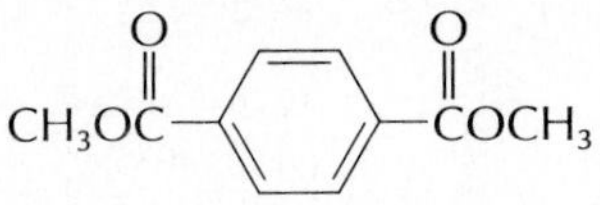

13.4 (b) $(CH_3)_2CHCH_2\underset{\underset{\displaystyle OCH_2C_6H_5}{|}}{\overset{\overset{\displaystyle OH}{|}}{C}}OH$

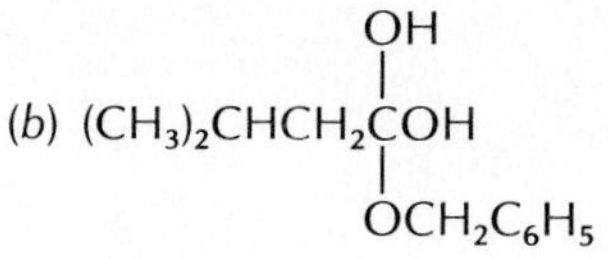

(c) $HO\underset{\underset{\displaystyle CH_3O}{|}}{\overset{\overset{\displaystyle HO}{|}}{C}}\!-\!\!\!\!\!\!\!\!-\!\!\underset{\underset{\displaystyle O}{\|}}{C}OH$ then $CH_3O\underset{\underset{\displaystyle O}{\|}}{C}\!-\!\!\!\!\!\!\!\!-\!\!\underset{\underset{\displaystyle OCH_3}{|}}{\overset{\overset{\displaystyle OH}{|}}{C}}OH$

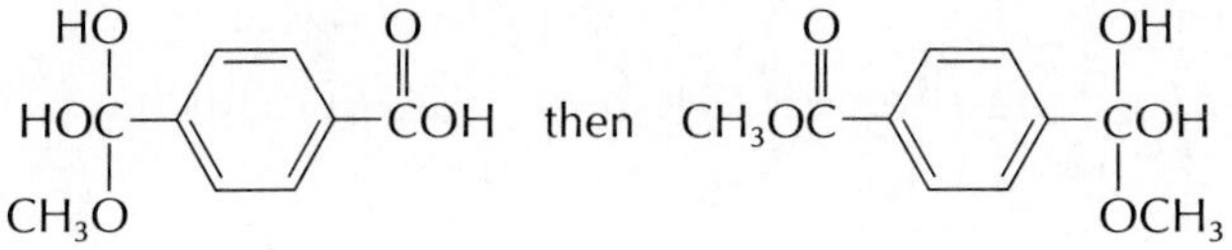

13.5 (b) $CH_3\underset{\underset{\displaystyle O}{\|}}{C}OH \xrightarrow{SOCl_2} CH_3\underset{\underset{\displaystyle O}{\|}}{C}Cl \xrightarrow[\text{pyridine}]{C_6H_5CH_2OH} CH_3\underset{\underset{\displaystyle O}{\|}}{C}OCH_2C_6H_5$

(c) $(CH_3)_2CH\underset{\underset{\displaystyle O}{\|}}{C}OH \xrightarrow{SOCl_2} (CH_3)_2CH\underset{\underset{\displaystyle O}{\|}}{C}Cl \xrightarrow[\text{pyridine}]{(CH_3)_2CHOH} (CH_3)_2CH\underset{\underset{\displaystyle O}{\|}}{C}OCH(CH_3)_2$

13.6

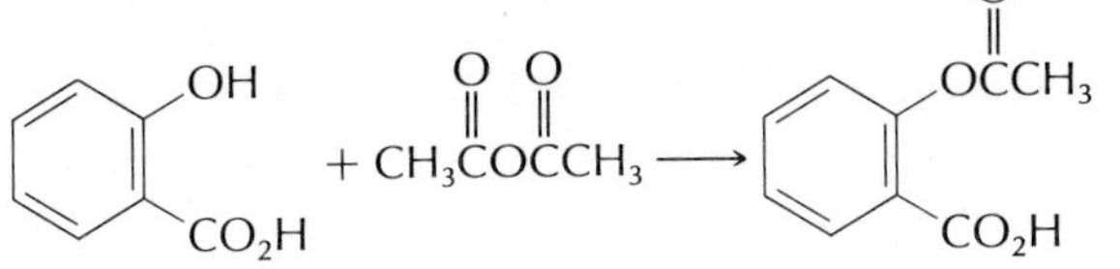

$+ CH_3\underset{\underset{\displaystyle O}{\|}}{C}O\underset{\underset{\displaystyle O}{\|}}{C}CH_3 \longrightarrow$

13.7 $CH_3(CH_2)_{14}\underset{\underset{\displaystyle O}{\|}}{C}OCH_2(CH_2)_{28}CH_3 + H_2O \xrightarrow{H^+} CH_3(CH_2)_{14}\underset{\underset{\displaystyle O}{\|}}{C}OH + HOCH_2(CH_2)_{28}CH_3$

13.8

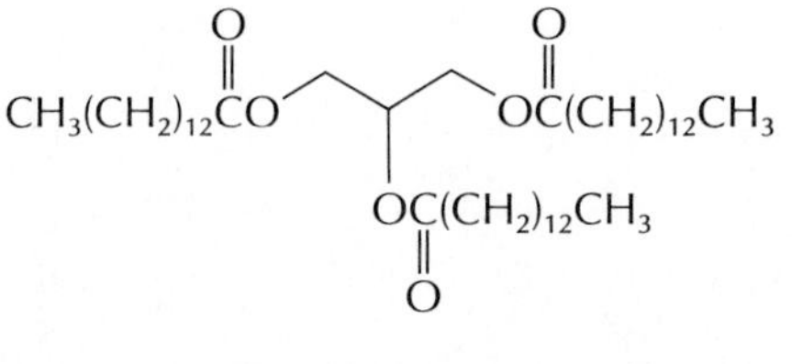

$CH_3(CH_2)_{12}\underset{\underset{\displaystyle O}{\|}}{C}O\underset{\underset{\displaystyle \underset{\displaystyle O}{\|}}{\underset{\displaystyle OC(CH_2)_{12}CH_3}{}}}{CH_2\!-\!CH\!-\!CH_2}O\underset{\underset{\displaystyle O}{\|}}{C}(CH_2)_{12}CH_3$

13.9 **Step 1:** Nucleophilic addition of hydroxide ion to the carbonyl group.

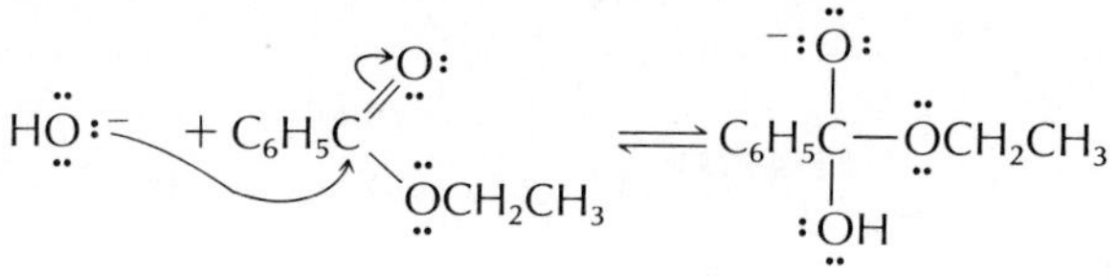

Step 2: Proton transfer from water to give neutral form of tetrahedral intermediate.

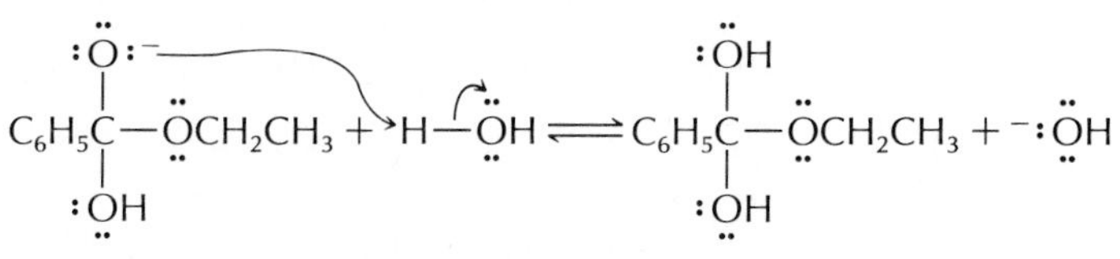

Step 3: Hydroxide ion–promoted dissociation of tetrahedral intermediate.

$$HO:^- + C_6H_5\overset{\overset{\displaystyle H-O:}{|}}{\underset{\underset{\displaystyle :OH}{|}}{C}}-OCH_2CH_3 \rightleftharpoons H\overset{..}{O}H + C_6H_5\overset{\overset{\displaystyle O:}{\|}}{C}\underset{\displaystyle OH}{\diagdown} + ^-:\overset{..}{O}CH_2CH_3$$

Step 4: Proton abstraction from benzoic acid and protonation of ethoxide.

$$C_6H_5\overset{\overset{\displaystyle O:}{\|}}{\underset{\underset{\displaystyle H}{|}}{C}}\underset{\displaystyle O}{\diagdown} + \quad ^-:\overset{..}{O}H \longrightarrow C_6H_5\overset{\overset{\displaystyle O:}{\|}}{C}\underset{\displaystyle O:^-}{\diagdown} + H\overset{..}{O}H$$

$$CH_3CH_2\overset{..}{O}:^- + H-\overset{..}{O}H \longrightarrow CH_3CH_2\overset{..}{O}H + ^-:\overset{..}{O}H$$

13.10 (b) $2CH_3CH_2CH_2CH_2CH_2MgBr + CH_3\overset{\overset{\displaystyle O}{\|}}{C}OCH_2CH_3$

(c) $2C_6H_5MgBr + H\overset{\overset{\displaystyle O}{\|}}{C}OCH_2CH_3$ (d) $2C_6H_5MgBr + C_6H_5\overset{\overset{\displaystyle O}{\|}}{C}OCH_2CH_3$

13.11

$$\left[\overset{\overset{\displaystyle O}{\|}}{C}O-\!\!\left\langle\;\;\right\rangle\!\!-\overset{\overset{\displaystyle CH_3}{|}}{\underset{\underset{\displaystyle CH_3}{|}}{C}}-\!\!\left\langle\;\;\right\rangle\!\!-O\overset{\overset{\displaystyle O}{\|}}{C}O\right]_n$$

13.12 (b) $CH_3\overset{\overset{\displaystyle O}{\|}}{C}Cl + 2\,CH_3NH_2 \longrightarrow CH_3\overset{\overset{\displaystyle O}{\|}}{C}NHCH_3 + CH_3\overset{+}{N}H_3\;Cl^-$

(c) $C_6H_5\overset{\overset{\displaystyle O}{\|}}{C}Cl + 2(CH_3)_2NH \longrightarrow C_6H_5\overset{\overset{\displaystyle O}{\|}}{C}N(CH_3)_2 + (CH_3)_2\overset{+}{N}H_2\;Cl^-$

13.13 $(CH_3)_2CH\overset{\overset{\displaystyle O}{\|}}{C}NHCH_3 + CH_3CH_2OH$

13.14 (b) $CH_3\overset{\overset{\displaystyle O}{\|}}{C}NHCH_3 + NaOH \longrightarrow CH_3\overset{\overset{\displaystyle O}{\|}}{C}ONa + CH_3NH_2$

(c) $C_6H_5\overset{\overset{\displaystyle O}{\|}}{C}NHC_6H_5 + H_2O + HCl \longrightarrow C_6H_5\overset{\overset{\displaystyle O}{\|}}{C}OH + C_6H_5NH_3^+\;Cl^-$

CHAPTER 14

14.1 (b) 1-Phenylethylamine   (1-phenylethanamine);   (c) N,N-dimethylpropylamine (N,N-dimethylpropanamine); (d) trimethylammonium bromide.

14.2 (b) N-Ethyl-4-isopropyl-N-methylaniline; (c) 3,4-difluoroaniline.

14.3 Primary amines (RNH_2) having two protons on nitrogen can participate in more hydrogen bonds than secondary amines (R_2NH) which have only one proton on nitrogen. Tertiary amine molecules lack protons on nitrogen and cannot form hydrogen bonds with one another.

14.4 Tetrahydroisoquinoline is a stronger base than tetrahydroquinoline. The unshared electron pair of tetrahydroquinoline is delocalized into the aromatic ring, and this substance resembles aniline in its basicity, whereas tetrahydroisoquinoline resembles an alkylamine.

14.5 $2[CH_3(CH_2)_6CH_2]_2NH + CH_3(CH_2)_6CH_2Br \longrightarrow$

$$[CH_3(CH_2)_6CH_2]_3N + [CH_3(CH_2)_6CH_2]_2NH_2{}^+ Br^-$$

$[CH_3(CH_2)_6CH_2]_3N + CH_3(CH_2)_6CH_2Br \longrightarrow [CH_3(CH_2)_6CH_2]_4N^+ Br^-$

14.6 (e) $CH_3CH_2CH_2CH_2CO_2H \xrightarrow{SOCl_2} CH_3CH_2CH_2CH_2\overset{\displaystyle O}{\overset{\|}{C}}Cl \xrightarrow{CH_3CH_2CH_2CH_2NH_2}$

$CH_3CH_2CH_2CH_2\overset{\displaystyle O}{\overset{\|}{C}}NHCH_2CH_2CH_2CH_3 \xrightarrow[2.\ H_2O]{1.\ LiAlH_4} CH_3CH_2CH_2CH_2CH_2NHCH_2CH_2CH_2CH_3$

14.7 First step: nitrate ring with HNO_3, H_2SO_4; last step: reduce with H_2, Ni or Fe, HCl or Sn, HCl.

14.8 $(CH_3)_3\overset{+}{N}CH_2CH_2OH\ HO^-$

14.9 (b)

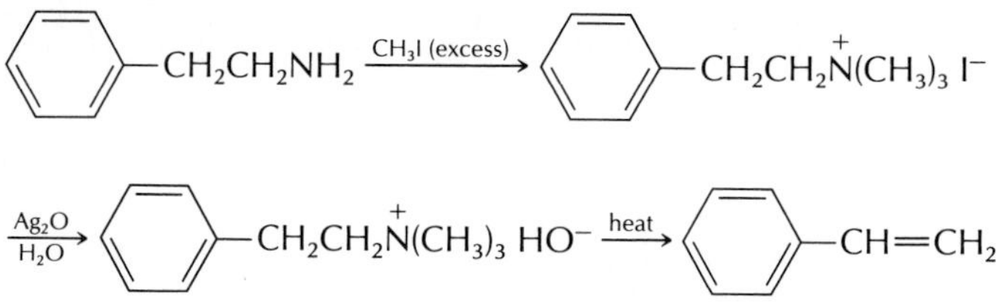

(c)

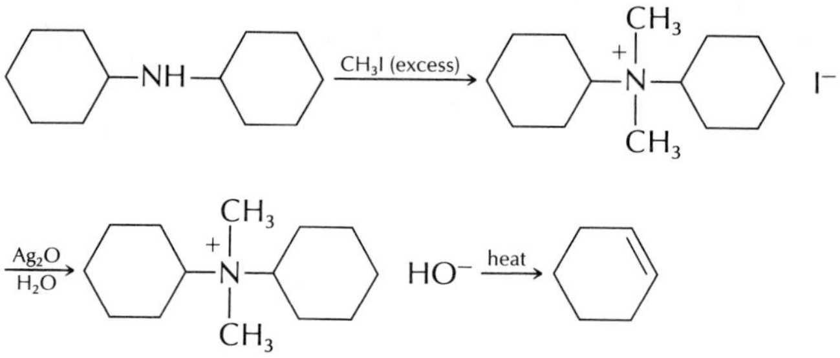

14.10 (b) $(CH_3)_3CCH_2\underset{\underset{\displaystyle CH_3}{|}}{C}=CH_2$ (c) $CH_2=CH_2$ (d) $CH_3CH=CH_2$

14.11 Intermediates: benzene to nitrobenzene to m-bromonitrobenzene to m-bromoaniline to m-bromophenol. Reagents: HNO_3, H_2SO_4; Br_2, $FeBr_3$; Fe, HCl then HO^-; $NaNO_2$, H_2SO_4, H_2O then heat in H_2O.

14.12

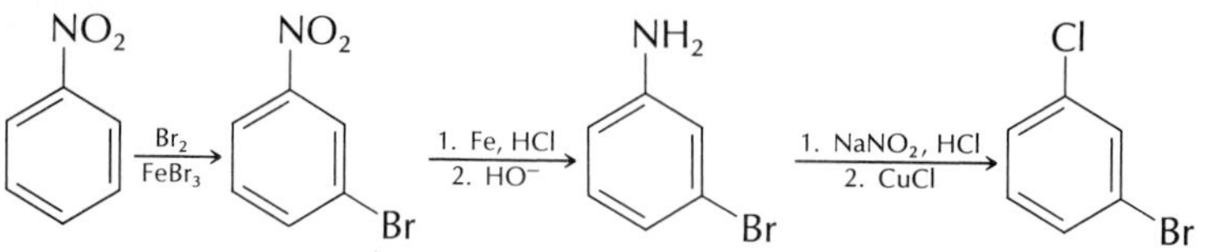

CHAPTER 15

15.1 (b) L-Glyceraldehyde; (c) D-glyceraldehyde.

15.2 L-Erythrose.

15.3

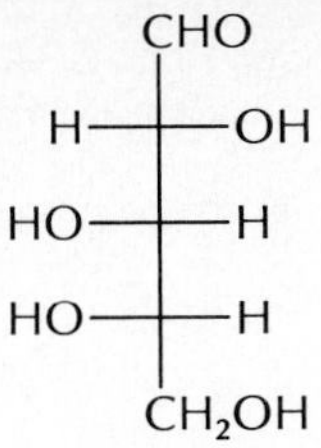

CHO
H——OH
HO——H
HO——H
CH$_2$OH

15.4   L-Talose.

15.5   (*b*)

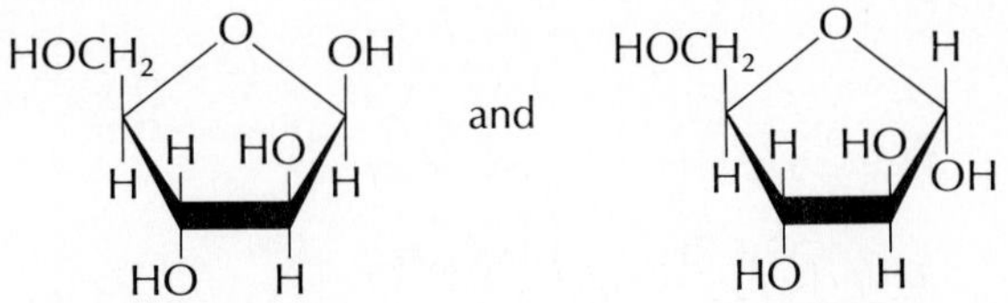

and

(*c*)

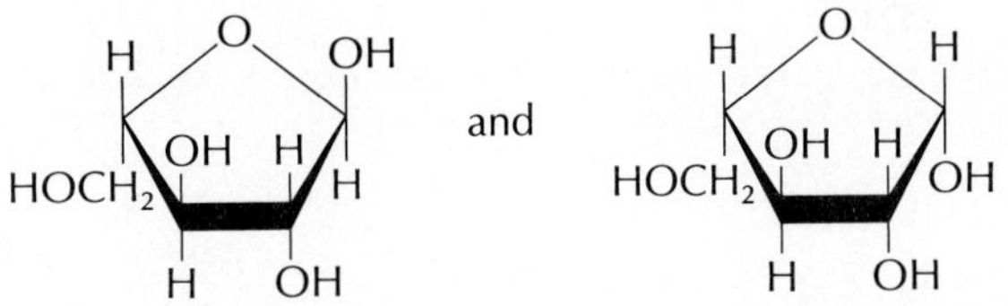

and

15.6

and

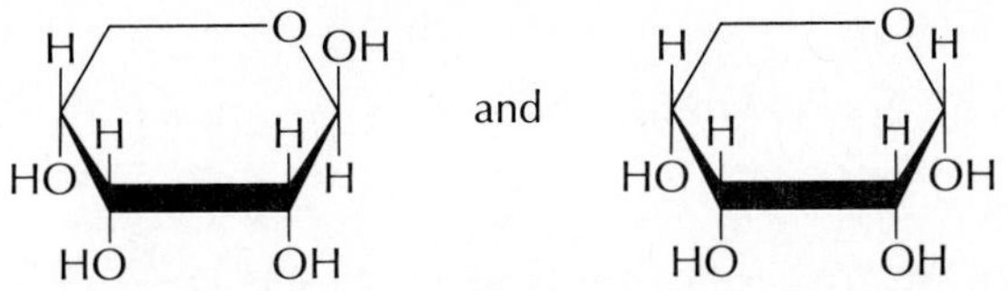

15.7   Adenosine: β-D-Ribofuranose;   Sinigrin: β-D-glucopyranose.

15.8   (*b*) Lactose is a reducing sugar;   (*c*) sucrose is not a reducing sugar;   (*d*) cellobiose is a reducing sugar.

15.9   (*b*) L-Erythrose;   (*c*) D-Altrose

CHAPTER 16

16.1 $\overset{+}{H_3}NCH_2CH_2CH_2\overset{\overset{O}{\|}}{C}O^-$

16.2   Isoleucine and threonine

16.3   (*b*) [phenyl]—CH$_2$CHCO$_2^-$ with $\overset{|}{{}_+NH_3}$ (*c*) [phenyl]—CH$_2$CHCO$_2^-$ with $\overset{|}{NH_2}$

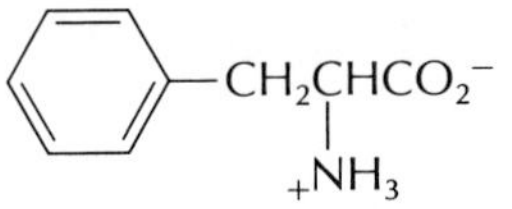

16.4 $(CH_3)_2CHCH\overset{\overset{O}{\|}}{} \xrightarrow[NaCN]{NH_4Cl} (CH_3)_2CHCHCN \xrightarrow[2.\ HO^-]{1.\ H_2O,\ HCl,\ heat} (CH_3)_2CHCHCO_2^-$

with NH$_2$ and $^+$NH$_3$ substituents

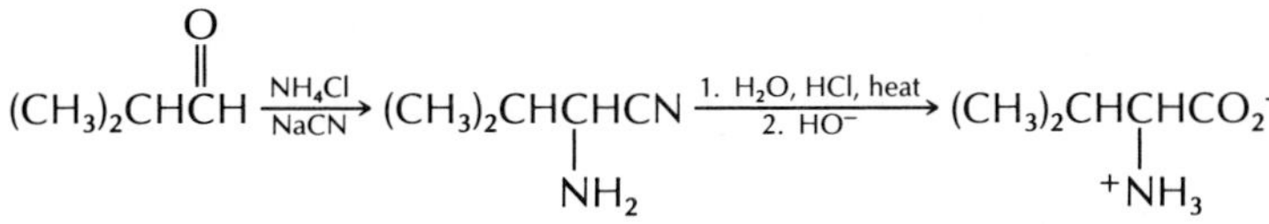

16.5   (*b*) $\overset{+}{H_3}NCHC\overset{\overset{O}{\|}}{}NHCHCO_2^-$ with CH$_3$ and CH$_2$C$_6$H$_5$ (*c*) $\overset{+}{H_3}NCHC\overset{\overset{O}{\|}}{}NHCHCO_2^-$ with C$_6$H$_5$CH$_2$ and CH$_3$

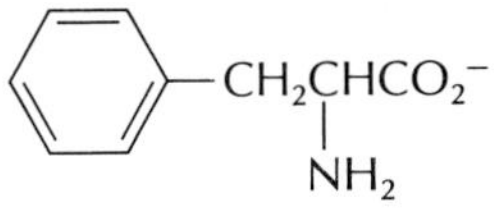

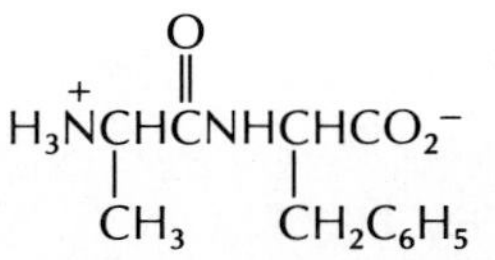

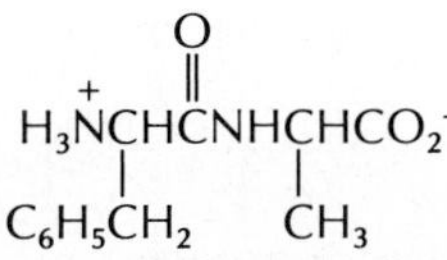

(d) $\overset{+}{H_3NCH_2}\overset{O}{\overset{\|}{C}}NHCHCO_2^-$
$\qquad\qquad\qquad\qquad | $
$\qquad\qquad\qquad\quad CH_2CH_2CO_2^-$

(e) $\qquad\qquad\qquad\qquad \overset{+}{H_3N}CH\overset{O}{\overset{\|}{C}}NHCH_2CO_2^-$
$\qquad\qquad\qquad\qquad\qquad | $
$\qquad\quad \overset{+}{H_3N}CH_2CH_2CH_2CH_2$

16.6 $\quad \overset{+}{H_3N}CH\overset{O}{\overset{\|}{C}}NHCH\overset{O}{\overset{\|}{C}}OCH_3$
$\qquad\quad | \qquad\qquad\quad | $
$\qquad -O_2CCH_2 \qquad CH_2C_6H_5$

16.7 After tyrosine, the next two amino acids are both glycine and give an unsubstituted PTH.

16.8 Arg-Pro-Pro-Gly-Phe-Ser-Pro-Phe-Arg + H_2O $\xrightarrow{\text{trypsin}}$
$\qquad\qquad\qquad\qquad\qquad\qquad$ Arg + Pro-Pro-Gly-Phe-Ser-Pro-Phe-Arg

Arg-Pro-Pro-Gly-Phe-Ser-Pro-Phe-Arg + H_2O $\xrightarrow{\text{chymotrypsin}}$
$\qquad\qquad\qquad\qquad\qquad\qquad$ Arg-Pro-Pro-Gly-Phe + Ser-Pro-Phe + Arg

16.9 Pro-Val-Leu-Phe-Pro-Ala

16.10 $\overset{+}{H_3N}CHCO_2^- + C_6H_5CH_2O\overset{O}{\overset{\|}{C}}Cl \longrightarrow C_6H_5CH_2O\overset{O}{\overset{\|}{C}}NHCHCO_2H$
$\qquad\quad | \qquad\qquad\qquad\qquad\qquad\qquad\qquad\qquad\qquad\qquad | $
$\qquad\quad CH_3 \qquad\qquad\qquad\qquad\qquad\qquad\qquad\qquad\qquad\quad CH_3$

$\overset{+}{H_3N}CHCO_2^- + C_6H_5CH_2OH \xrightarrow[\text{2. HO}^-]{\text{1. H}^+,\text{ heat}} H_2NCHCO_2CH_2C_6H_5$
$\qquad | \qquad\qquad\qquad\qquad\qquad\qquad\qquad\qquad\qquad\qquad\qquad | $
$(CH_3)_2CHCH_2 \qquad\qquad\qquad\qquad\qquad\qquad (CH_3)_2CHCH_2$

$C_6H_5CH_2O\overset{O}{\overset{\|}{C}}NHCHCO_2H + \qquad H_2NCH\overset{O}{\overset{\|}{C}}OCH_2C_6H_5 \xrightarrow{\text{DCCl}}$
$\qquad\qquad\qquad\qquad | \qquad\qquad\qquad\qquad | $
$\qquad\qquad\qquad\quad CH_3 \qquad\quad (CH_3)_2CHCH_2$

$\qquad\qquad\qquad\qquad\qquad\qquad C_6H_5CH_2O\overset{O}{\overset{\|}{C}}NHCH\overset{O}{\overset{\|}{C}}NHCH\overset{O}{\overset{\|}{C}}OCH_2C_6H_5$
$\qquad\qquad\qquad\qquad\qquad\qquad\qquad\qquad\qquad\qquad | \qquad\qquad | $
$\qquad\qquad\qquad\qquad\qquad\qquad\qquad\qquad\quad CH_3 \qquad CH_2CH(CH_3)_2$

$C_6H_5CH_2O\overset{O}{\overset{\|}{C}}NHCH\overset{O}{\overset{\|}{C}}NHCH\overset{O}{\overset{\|}{C}}OCH_2C_6H_5 \xrightarrow{\text{H}_2 \atop \text{Pd}}$ Ala-Leu
$\qquad\qquad\qquad | \qquad\qquad | $
$\qquad\qquad\quad CH_3 \qquad CH_2CH(CH_3)_2$

17.1 (b)

$$\begin{array}{c} CH_2O_2C(CH_2)_{16}CH_3 \\ | \\ CH_2O_2C(CH_2)_7CH{=}CH(CH_2)_7CH_3 + 3HO^- \longrightarrow \\ | \\ CH_2O_2C(CH_2)_{16}CH_3 \end{array}$$

$$\begin{array}{c} CH_2OH \\ | \\ CHOH + 2CH_3(CH_2)_{16}CO_2^- + CH_3(CH_2)_7CH{=}CH(CH_2)_7CO_2^- \\ | \\ CH_2OH \end{array}$$

(c)

$$\begin{array}{c} CH_2O_2C(CH_2)_7CH{=}CH(CH_2)_7CH_3 \\ | \\ CH_2O_2C(CH_2)_{16}CH_3 \qquad\qquad + 3HO^- \longrightarrow \\ | \\ CH_2O_2C(CH_2)_{16}CH_3 \end{array}$$

$$\begin{array}{c} CH_2OH \\ | \\ CHOH + 2CH_3(CH_2)_{16}CO_2^- + CH_3(CH_2)_7CH{=}CH(CH_2)_7CO_2^- \\ | \\ CH_2OH \end{array}$$

17.2   Lecithin contains both hydrophilic ("water-loving") and lipophilic ("fat-loving") structural units. Fat-type molecules bind to the lipophilic portion of lecithin and the fat + lecithin complexes are dispersed in the water.

17.3   (Z)-9-Octadecenyl (Z)-9-octadecenoate

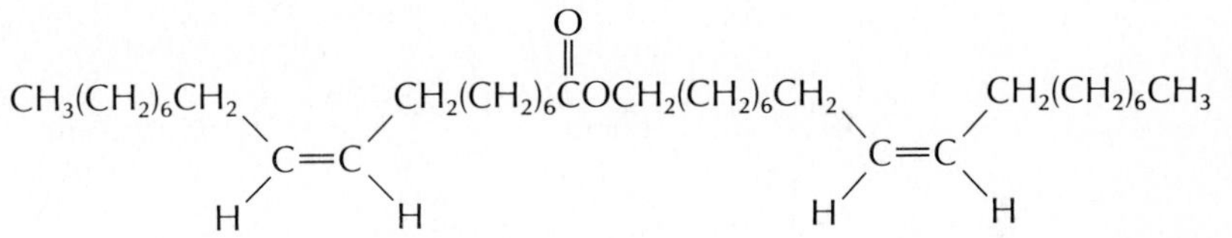

17.4   The structure of vitamin D_2 is the same as that of vitamin D_3 except that vitamin D_2 has a double bond between C-22 and C-23 and a methyl substituent at C-24.

17.5

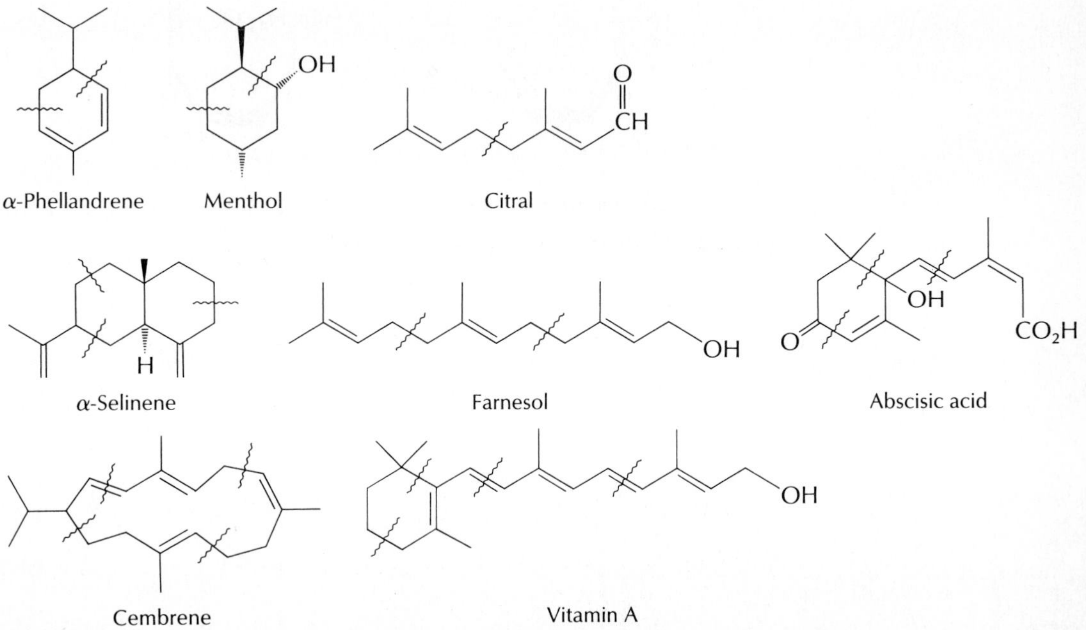

α-Phellandrene   Menthol   Citral

α-Selinene   Farnesol   Abscisic acid

Cembrene   Vitamin A

17.6

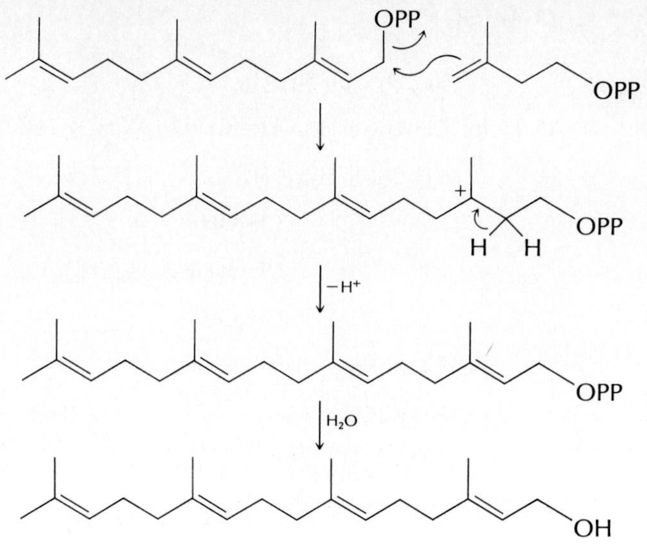

CHAPTER 18

18.1

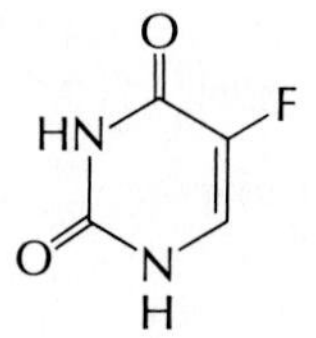

18.2 (*b*) Cytidine   (*c*) Guanosine

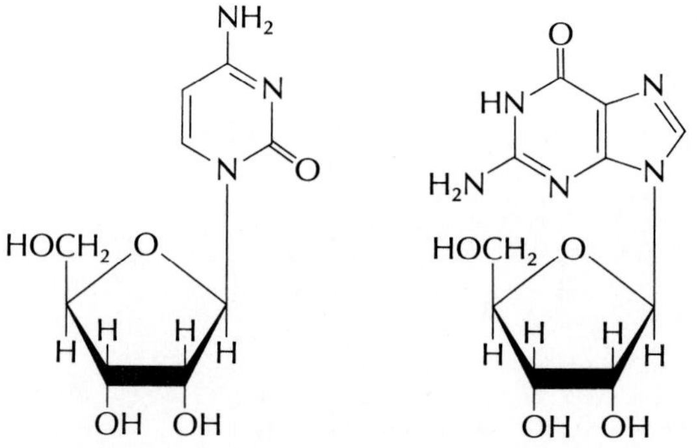

18.3 (*b*) Complementary strand is -T-A-C-C-T-G-A-; (*c*) -C-C-C-G-G-T-C-A-A-.

18.4 (*b*) Anticodon is CUG (aspartic acid); (*c*) AAA (phenylalanine); (*d*) UCC (arginine).

INDEX